T0350649

The Harvest
Ants collecting grain of *Cenchrus ciliaris* Linn.
Senegal, December, 1973.

THE USEFUL PLANTS
OF
WEST TROPICAL AFRICA

VOL. 2

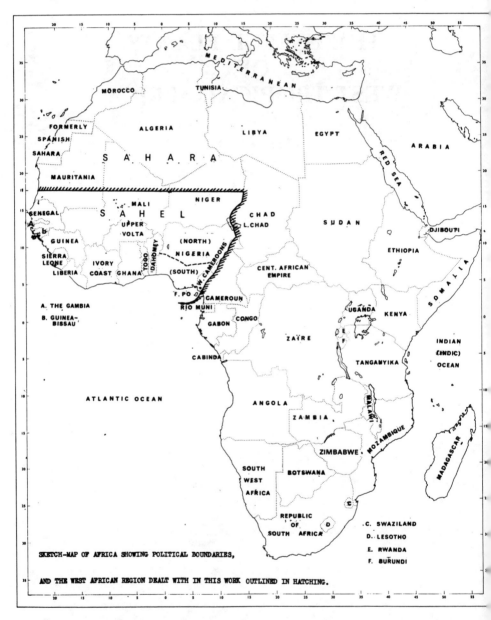

Sketch-map of Africa showing political boundaries, and the West African region dealt with in this work outlined in hatching. Regardless of recent changes in political state names, the same names as were used in Volume 1 have been retained here.

THE USEFUL PLANTS
OF
WEST TROPICAL AFRICA

Edition 2

H. M. BURKILL

VOL. 2
Families E–I

ROYAL BOTANIC GARDENS
KEW
1994

First Edition (1937) by J. M. Dalziel

The present volume is part of a revision of the First
Edition, and is published as a Supplement to
The Flora of West Tropical Africa (ed. 2, 1954–1972)
edited by R. W. J. Keay and F. N. Hepper
It has been compiled by
H. M. Burkill, O.B.E., M.A., F.L.S., C.Biol., M.I.Biol.
Honorary Research Fellow, Royal Botanic Gardens, Kew
lately of H.M. Overseas Civil Service
Director, Botanic Gardens, Singapore

Cover design by Eleanor Catherine

ISBN 0 947643 56 7

Database typeset by
BPC Whitefriars Ltd; printed and bound by BPC Wheatons
Ltd
Both members of the British Printing Company Ltd
for the Publishers
Royal Botanic Gardens, Kew,
Richmond, Surrey TW9 3AB

'And he gave it for his opinion, that whoever could make two ears of corn or two blades of grass grow upon a spot of ground where only one grew before, would deserve better of mankind, and do more essential service to his country than the whole race of politicians put together.'

Dean Jonathan Swift, 1667–1745: Gulliver's Travels; Voyage to Brobdingnag, Ch. 7 (1726).

TABLE OF CONTENTS

Introduction

Volume 1 of this work, *A Revision of Dalziel's Useful Plants of West Tropical Africa*, was published in 1985. It gave reference to some 1,500 species of plants in Families A to D occurring in the West African region. During the preparation of Volume 1 materials were concurrently being assembled for all parts of the flora to appear in subsequent volumes. It was anticipated then that Volume 2 would cover Families E to L, but such has been the increased accumulation of information since then, that, for logistic purposes, these families have been divided into two parts. Volume 2, as now published, deals with some 900 species of plants of Families E to I, including, of course, the economically important Gramineae (grasses) and Euphorbiaceae. Families J to L will follow in Volume 3.

The treatment of plants is in the same format as in Volume 1, that is alphabetically by families, by genera within families and by species within genera. Relevant citation of *The Flora of West Tropical Africa*, ed. 2, and of Dalziel: *The Useful Plants of West Tropical Africa*, ed. 1, is made as previously. The author's interpretation of 'usefulness' remains as explained in Volume 1, i.e., briefly that if a plant is used in any way, effectively or in fantasy, it is useful. The geographical area, as shown in the accompanying map, is unaltered. It has not been practical to keep abreast in the text with political name-changes: Burkina Faso remains Upper Volta, Benin remains Dahomey and Bioho is Fernando Po.

Vernacular names of plants occupy a large part of this work and problems encountered have been outlined in the introduction to Volume 1. The orthography used here is the same as previously, and for the convenience of the reader the alphabet is repeated below. A small number of extra accents has unfortunately been required, and the following list shows all the diacritical accentation used.

A	a	
B	b	
Ɓ	ɓ	a glottalised *b*
Ɓ	ƅ	an implosive *b*—for Niger Delta languages
C	c	
D	d	
Đ	ɖ	a post-alveolar *d*
Ɗ	ɗ	a glottalised *d*
Ɗ	ɖ	an implosive *d*—for Niger Delta languages
E	e	
Ɛ	ɛ	an open *e*
Ə	ə	a central vowel—as at the end of English sod*a*
F	f	
Ƒ	ƒ	a labial *f*
G	g	
Ɣ	ɣ	a voiced velar fricative—*gh*
H	h	
I	i	
J	j	
K	k	
Ƙ	ƙ	an ejective *k*

L	l	
M	m	
N	n	
Ŋ	ŋ	*ng* as in English si*ng*
Ɲ	ɲ	*ny*
O	o	
Ɔ	ɔ	a short vowel as in English n*o*t
Ọ	ọ	as the preceding in Nigerian languages
P	p	
Q	q	
R	r	
S	s	
Σ	ʃ	*sh*
T	t	
U	u	
V	v	
Ʊ	ʋ	a labial *v*
W	w	
X	x	
Y	y	or ʼy—a glottalised *y*
Z	z	
Ʒ	ʒ	a sibilant *zh*

Diacritical accents:

´ over a vowel: the standard acute accent—a rising tone in many languages

ʹ over d, k and t in Senegalese languages, hardens and shortens the letter

ˋ over a vowel: the standard grave accent—a falling tone in many languages

˵ over a vowel: a double grave accent making a deeper falling tone, e.g., when following a syllable already under a falling tone

˘ over a vowel shortens it

‾ over a vowel lengthens it, but now commonly superseded by a doubling of the vowel to show length

_ under a vowel: a low level pitch

˙ over a vowel: a mid-tone mark

^ over a vowel: a standard circumflex accent—a rising-falling tone in many languages

ˇ over a vowel: a falling-rising tone mark in Nigerian languages

ˇ over a consonant: in general producing a hardened aspiration, a fricative or sibilant

˙ over n: = ng as in si*ng*

. under a vowel: makes it more open

. under a consonant in Isekiri and Yoruba indicates a sibilant sound, e.g., ṣ = *sh*

ʼ apostrophe: in Hausa and Fula, a glottalisation of the following consonant, or when between two vowels, a glottal stop

¨ the German umlaut: ä = *ae*; ö = *ue*

¨ diaeresis over one of two adjacent vowels—vowels sounded separately

~ Spanish tilde—a standard nasalisation. In Yoruba a tone mark or contraction of two syllables, but now outdated by the use of two vowels each with its own tone mark

Ç below a letter, the Continental cedilla

Acknowledgements. This work was initiated in November, 1969, under Grant No. R2689 from the Ministry of Overseas Development of the United Kingdom Government. The grant was continued till December, 1977. Then it ceased. The work achieved during this period laid the foundation for the whole project, and when funding ended the preparations were such that abandonment was deemed out of the question. Thus the writer has continued the task honorarily. Some of the work of the period 1969–77 is reflected in the present volume. Relief assistance from the Bentham-Moxon Trust between April to December, 1978, and from January, 1981, to January 1984, and from the Leverhulme Trust Fund for the years 1979 and 1980, both for clerical services, is gratefully acknowledged. At a much later date (1992) the British Council, Lagos, Nigeria, very kindly made an honorarium available to have the Nigerian plant names checked by linguists.

The amenities of the Royal Botanic Gardens, Kew, have been made freely available to me, and I am grateful to Professor Ghillean Prance, the present Director, and to Professor G. Ll. Lucas, O.B.E., Deputy Director and Keeper of the Herbarium, and their predecessors in office for this; to Dr R. Polhill, Assistant Keeper, for handling the general executive and financial controls in connexion with printing, and to numerous colleagues of the Kew Herbarium for their frequent kindly help; and also to Dr S. Droop, of Edinburgh, for reading and editing the manuscript. I must also express my thanks to, and acknowledgement for help from, a number of persons outside of Kew: Dr Roger Blench, private research worker; Professor Kay Williamson, Department of Linguistics and African Languages, University of Port Harcourt, Nigeria; Dr. Joyce Lowe, Department of Botany and Microbiology, University of Ibadan, Nigeria; Mr M. Mann, School of Oriental and African Studies, University of London, U.K.; all for very many various assistances and patient encouragement, through whom much of the plant vernacular names lists have been derived and/or checked. The help of the following in scanning language lists is also gratefully acknowledged: THE GAMBIA: Mr. Foday Bojang, Department of Forestry (Mandinka); Dr Michael Jones, The Gambia College, and U.K. High Commission, and Mr R. J. McEwan, of IG Barshall, Ltd (Gambian languages); UPPER VOLTA: Dr David Baldry, Onchoceriasis Control Programme, W.H.O., and later of W.H.O., Geneva, Switzerland (Upper Voltan languages); MALI: Dr Pierre Garnier (Soudanian languages); GHANA: Mr A. A. Enti, Forest Enterprises, Legon, and Mr D. K. Abbiw, Department of Botany, University of Ghana (Akan languages); NIGERIA: Professor Ben Elugbe, Department of Linguistics and African Languages, University of Ibadan (Nigerian languages); Dr. Graham Furniss, School of Oriental and African Studies, University of London (Hausa); Mr I. D. Chapman, lately Department of Agriculture, Nigeria (Mambilla languages); Professor E. Vermeer, The George Washington University, U.S.A. (Tiv); Dr Pierre Verger, Fundação Pierre Verger, Bahia, Brasil (Yoruba); Dr Bruce Connell, Phonetics Laboratory, University of Oxford (Cross River languages); Miss Neneh Ilonah (Igbo); Dr O.-M. Ndimele and Mr O. G. Harry (Delta languages). To them, my thanks. The authorities for each plant name are shown as a name or initials at every instance unless three or more authorities quote the same word then they are lumped as *auctt.*

In connexion with the printing, I must especially mention Mr Allan Wyatt, of Messrs BPCC Whitefriars Press Ltd, for his interest, constant help and supervision in passing the manuscript through the press, and in particular for devising the computer processes for handling the plant vernacular names and exotic orthography. To him, I add my thanks to his wife, Mrs Mary Wyatt, for invaluable editorial assistance on the language lists to make them amenable to the computer regimes.

Funds for publication of this volume have come from roll-over revenue on the sales of Volume 1 and other Kew publications. To this the Overseas Development Administration of the U.K. Government added a subvention enabling the cost to the public to be kept within very reasonable bounds, and in particular to assist dissemination within the West African region for which the work has primarily been undertaken.

The Herbarium, Royal Botanic Gardens, Kew.
17 March, 1994.
H. M. Burkill

EBENACEAE

Diospyros abyssinica (Hiern) F White

FWTA, ed. 2, 2: 10.
West African: GHANA ADANGME ɡblɛtʃo (FRI) ADANGME (Abokobi) blonyatʃo (FRI)
ADANGME-KROBO blitcho (BD&H; DF) ɡblitʃo (FRI) GA blidzo (FRI)

A tree to about 17 m in the Region but attaining 35 m to the east in E
Cameroun and Gabon (5), with slender straight bole, of the dry deciduous
forest of Guinea to S Nigeria, and widely dispersed in Africa to Angola, Eritrea,
E Africa and Mozambique.
The wood is yellow with black streaks in the heart which may become entirely
black at the centre. It is medium hard, easily worked, but is reported in Kenya
to be not durable (1). It is used in Ghana to make pestles and mortars, and
Krobo prepare medicines from the bark and roots (3, 4). The wood is used in
Kenya as heavy loom shuttles in weaving sisal cloth (2).

References:

1. Battiscombe 1151, K. 2. Dale & Greenway, 1961: 172–3. 3. Irvine 1700, K. 4. Irvine, 1961: 576.
5. Letouzey & White, 1970, a, b: 29–31.

Diospyros barteri Hiern

FWTA, ed. 2, 2: 12. UPWTA, ed. 1, 346.
West African: GHANA AKAN-TWI aheneba nsatea = *prince's fingers* (FRI) **NIGERIA** EDO
ivín-ohá (Elugbe)

A tree to 7 m high, occasionally scrambling, of the evergreen forest of Ghana
and S Nigeria, and also in E Cameroun.
The timber is hard and finds local unspecified use (1). The stems are cut into
chew-sticks in Ghana (2, 3, 4). The thin fruit-pulp is edible, and is a minor item
of diet (1).

References:

1. Busson, 1965: 365. 2. Irvine 1153, K. 3. Irvine, 1961: 576. 4. Portères, 1974: 119.

Diospyros bipindensis Gürke

FWTA, ed. 2, 2: 11.

A tree reaching 20 m high with fluted trunk, and very hard fissured bark, of
the closed evergreen forest in W Cameroons, and extending to Zaïre and
Angola.
The sap-wood is white, heart-wood yellowish, sometimes black. The wood
has some elasticity and is used in E Cameroun and Gabon to make the bow of
crossbows (2).

1

The powdered root is taken in Congo (Brazzaville) as a poison-antidote, and expectorant in bronchial affections. A bark-macerate is used topically on localised pains (1).

References:

1. Bouquet, 1969: 107. 2. Letouzey & White, 1970 a, b: 35–39.

Diospyros canaliculata De Wild.

FWTA, ed. 2, 2: 14. UPWTA, ed. 1, 349, as *D. xanthochlamys* sensu Dalziel.
English: flint-bark (Ghana, Irvine).
West African: **LIBERIA** ·KRU· kéri (K&B) **IVORY COAST** AKYE baimbro (Aub., K&B) DAN wodri (K&B) ·KRU· garo (K&B) KRU-BETE bodri (K&B) bodrui (Aub., K&B) GUERE wanié (K&B) GUERE (Chiehn) kokaw (B&D) **GHANA** VULGAR twaberi (DF) AKAN-ASANTE asenkyiya (FRI) twabere = *cut turns yellow; refers to the slash turning yellow after exposure* (E&A) FANTE okusibiri (CV; FRI) WASA ankyeyi (BD&H; FRI) ɛsono afẽ =*eleplant's comb* (FRI) ɛsono ankenyi (FRI) kokodae (BD&H; FRI) twabere (E&A) ANYI-SEHWI ɛsono afẽ =*elephant's comb; ? from Akan-Twi* (FRI) NZEMA sanzamuki (CV; FRI) sanzamulike (CV; FRI) **NIGERIA** EDO olubaza (JMD) oribaza (JMD) YEKHEE oruboje (KO&S) YORUBA oriloje (KO&S) owe (KO&S)

A tree to 20 m high, with bole to 60 cm girth, straight (Nigeria, 7) or twisted (E Cameroun, 9), of the lowland evergreen rain-forest and in relic forest in savanna areas of Liberia to W Cameroons, and extending to the Congo basin and Angola.

The wood is hard and heavy, light yellow or slightly pink when first cut, turning bright yellow on exposure. It is used in Ghana for building and pit-props (6). Its resilience leads to its use in Ubangi as spear-shafts (12).

The outer layers of bark are jet black, very hard and brittle. The inner bark is yellowish and when freshly cut emits a smell of egg-yolk, turning in the course of time to violet (9). The bark-extract is strongly vesicant. It is a standard ingredient of arrow-poison mixtures in eastern Ivory Coast, and produces gangrene in the flesh around the wound facilitating penetration of other poisons, especially of *Mansonia* (Sterculiaceae), in the mixture (4, 8). Freshly pulped bark is commonly used in Ivory Coast in treatment of leprosy, application causing a sharp blistering and the disappearance of the leprous maculae (4, 8). Experimental work has not confirmed the efficacy of this (6). The bark is torrified in Congo (Brazzaville) and pounded with rock-salt and palm-oil to a paste which is applied to points of pain over the ribs after scarification (3). Bark infusion from the twigs has been shown to inhibit the growth of *Staphylococcus*, and has an anti-biotic activity comparable with that of penicillin. Bark is also active against *Streptococcus* and the diphtheria bacillus at dilution of 1–2%. The active substance is *plumbagine* (hydroxy-methyl-naphthoquinone) present at 2.25% in the trunk-bark and 0.12% in the leaves. A tannin, a saponin and a substance identical with *scopoletol* are also reported present. Bark-infusions have been shown to kill *Paramecium*, cause excitation and paralysis in fish, death in mice and hypertension in dogs. The action of the leaf-extract is similar but less marked (8). A trace of alkaloid has been detected in the bark of Nigerian material.

A root-decoction is taken in Ubangi for constipation (12).

The flowers are pleasantly fragrant.

The yellow fruit-pulp is caustic, oily and blackening in air. When rubbed on the skin it causes a burning sensation and turns the skin yellow (2), but nevertheless is used in Ubangi for tracing patterns on it which last for several months (12). In Central African Republic it is put into arrow-poison compounds (11), and a mixture of fruit and leaves with the root of *Parquetina nigrescens* (Asclepiadaceae) is cooked to a syrup to put on arrow and spear heads for elephant hunting. Death is said to be rapid (12). The antidote to this poison is to apply ground-nuts crushed with young shoots of tamarind in a poultice (12).

In S Nigeria the fruit and sap are used as a fish-poison (5), and in Central African Republic bark and seeds are similarly used (10).

References:

1. Adegoke & al., 1968. 2. Aubréville, 1959: 3: 159, as *D. xanthochlamys* sensu Aubréville. 3. Bouquet, 1969: 107–8, as *D. physocalycina* sensu Bouquet (probably *D. canaliculata* De Wild.). 4. Bouquet & Debray, 1974: 81, as *D. physocalycina* sensu Bouquet & Debray. 5. Dalziel, 1937: 349. 6. Irvine, 1961: 576, 578. 7. Keay & al., 1964: 336. 8. Kerharo & Bouquet, 1950: 177–8, as *D. xanthochlamys* sensu Kerharo & Bouquet; with references. 9. Letouzey & White, 1970, a: 43–48. 10. (?) 201, K. 11. (?) 2019, K. 12. Vergiat, 1970,a: 66, as *D. xanthochlamys* sensu Vergiat.

Diospyros chevalieri De Wild.

FWTA, ed. 2, 2: 12. UPWTA, ed. 1, 347.
West African: LIBERIA KRU-BASA pahn (C&R)

A shrub to 3 m high, in undergrowth of the closed-forest in Sierra Leone to Ghana.

The wood is bright yellow. No use appears to be recorded for it. The leaves are said to be used to decorate bows (1) and arrows (2) during country sports in Liberia. Is this a good luck charm? See *D. mespiliformis*.

References:

1. Cooper 200, K. 2. Cooper & Record, 1931: 99.

Diospyros cooperi (Hutch. & Dalz.) F White

FWTA, ed. 2, 2: 10. UPWTA, ed. 1, 349, as *Maba cooperi* Hutch. & Dalz.
West African: SIERRA LEONE LOKO pahai (NWT) MENDE tɛli (SKS) SUSU kuremoke (NWT) TEMNE ɛ-kbɔɔ (NWT ɛ-lin (NWT) LIBERIA KRU-BASA bluchu (auctt.) drebah (C; H&D) IVORY COAST ABE *n*gair (Aub.)

A tree to about 12–13 m high of the forest zone of Sierra Leone and Ivory Coast.

The long slender poles are used for hut-posts and rafters in Liberia, and the leaves are boiled to produce a black dye (1, 2).

References:

1. Cooper 92, 316, K. 2. Cooper & Record, 1931: 100, as *Maba cooperi* Hutch. & Dalz.

Diospyros crassiflora Hiern

FWTA, ed. 2, 2: 12. UPWTA, ed. 1, 347.
English: Benin ebony (trade, S Nigeria: Gray, Dalziel)
French: (trade, Gabon) ebénier véritable du Gabon (Walker); evila (Letouzey & White); (trade, Cameroun) mevini (Letouzey & White).
Portuguese: ébano
West African: NIGERIA BOKYI osibin (JMD; KO&S) EDO abokpo (auctt.) igiedudu (JMD) uhu-abokpo =*comb* (JMD; KO&S) EFIK nyareti *general for ebony* (JMD) EJAGHAM nyareti *general for ebony* (JMD) EJAGHAM-EKIN numnyareti *male of nyareti* (Gray) nyareh *female trees* (Gray) nyareti =*ebony, and applied to female trees* (Gray) EKPO nyareh (KO&S) nyareti (KO&S) YORUBA kanran (JMD: KO&S)

A tree to 20 m high with cylindrical bole to 2 m or more in girth, of the lowland rain-forest of S Nigeria, and extending to Zaïre.

Sap-wood is creamy to reddish yellow and very thick, producing bands of black. Heart-wood is black, very hard, but appears only in older trees over a certain girth. In Cameroun where the tree attains a larger stature (25 m high by 3.0–3.5 m in girth) this is said to be 95 cm girth (5). The heart-wood is the true ebony of commerce and there has long been an export trade from Benin, E Cameroun and Gabon which in the two last-named countries has now probably reached a state of over-exploitation (5). The exported ebony is valued for black-wood cabinetry, furniture manufacture and high-class carpentry. In Benin the Ejagham recognise four sorts of tree: *nyareti* is a 'female' tree yielding top quality ebony; *num nyareti* is the 'male' of this with second grade ebony; *nyareh* is 'female' with poor ebony but providing excellent fuel; and *num nyareh* is its 'male' yielding neither good ebony nor good fuel (4). Not all *Diospyros* species produce ebony, but in some a blackening appears on immersion in water for shorter or longer duration according to the species. The water also blackens. Such immersed timber loses its blackening on reexposure to air and drying. The true blackening of ebony is the result of necrosis in very small spots (2).

The wood of young trees is flexible and is used in Congo (Brazzaville) to make crossbows (3).

The bark is used in Gabon with the red heart-wood of *Pterocarpus soyauxii* Taub. (Leguminosae: Papilionoideae) to treat yaws (6, 7). In Congo (Brazzaville) it is applied to sores, perhaps in similar context, and a bark-decoction is taken in draught and by enema for ovarian troubles (3). Leaf-sap is instilled into the eyes for purulent ophthalmia (3).

Examination of the roots on Nigerian material has shown a strong presence of *alkaloids* (1).

References:

1. Adegoke & al., 1968: 13: 13–33. 2. Aubréville, 1959: 3: 157. 3. Bouquet, 1969: 107. 4. Gray 4/10, K. 5. Letouzey & White, 1970, a, b: 57–63. 6. Walker, 1953, a: 33. 7. Walker & Sillans, 1961: 155.

Diospyros dendo Welw.

FWTA, ed. 2, 2: 10. UPWTA, ed. 1, 346, as *D. atropurpurea* Gürke.

West African: **NIGERIA** EDO igiedudu *from Yoruba* (Dunnett; JMD) isán-ahiamewę =*faeces of bird(s)* (Elugbe) IGBO isi-jī *the head of the yam* (KO&S; KW) IJO-IZON ongblo (auctt.) YORUBA eshunshun (Symington; KO&S) igi dúdu =*black wood* (JMD; KO&S) **WEST CAMEROONS** KPE efindefinde (JMD) epindepinde *probably applies to several spp.* (JMD) finde-finde (JMD)

A tree to 17 m high, trunk often twisted and slightly fluted to 1 m girth, usually less, of lowland evergreen forest of S Nigeria and W Cameroons, and extending to Zaïre and Angola.

The wood is a 'brown ebony' with black streaks in the heart-wood which is very hard and amounts to about one third of the diameter of the main stem. It is potentially usable for timber, and it makes a good firewood (2).

The bark-slash is bright yellow. In W Cameroons the bark yields a dye and a tanning material (2).

The fruit is edible and is taken as a supplementary food (1, 2).

References:

1. Busson, 1965: 365. 2. Unwin, 1920: 439, as *D. atropurpurea* Gürke.

Diospyros discolor Willd.

FWTA, ed. 2, 2: 15.

English: butter fruit; mabolo; mabola persimmon; camogan ebony.

A tree to 30 m high of the lowlands and moderate altitudes throughout the forests of the Philippines, and introduced to many tropical countries for its edible fruits. It is present in the Region.

The tree is planted in the Philippines as a roadside tree where it provides excellent shade. The wood is cut and marketed. Sap-wood is reddish or pinkish, sometimes stained with grey. Heart-wood is streaked, mottled, or sometimes nearly black. It has been used in the Philippines to make best quality combs (1). The fruits have a large amount of edible pulp, which is rather dry and mealy. In SE Asia some selection has been carried out and superior races are known. Selection in West Africa should be rewarding to improve the taste.

The bark of stems and roots, and the leaves are cyanaphoric in the Philippines (2). An unnamed *alkaloid* is reported in the leaves and stems (3).

References:

1. Burkill, IH, 1935: 828–9. 2. Quisumbing, 1951: 1044. 3. Willaman & Li, 1970.

Diospyros elliotii (Hiern) F White

FWTA, ed. 2, 2: 15. UPWTA, ed. 1, 349, as *D. mannii* sensu Dalziel.
West African: SENEGAL DIOLA bu parnab (JB) bu sãnsãndi (JB) é bunétu (JB) DIOLA (Bayot) é tiru (JB) SIERRA LEONE MENDE si (*def.* -i) (auctt.) SUSU fibriabogontinyi (NWT) TEMNE an-gbɔka (FCD; S&F) ma-kboli-ma-wokər (NWT) ka-ruki (FCD) NIGERIA EBIRA-ETUNO akpa (KO&S) apa (KO&S) IGBO ulele (KO&S)

A tree to 10 m high with short bole to 1.3 m in girth, of swamp-forest from Senegal to S Nigeria and also in Central African Republic.

Sap-wood is white, heart-wood more or less black and in small amount. It is used for firewood and is probably suitable for implement handles (1). In Sierra Leone the flexible branches are used for spring-traps, and the twigs serve as chew-sticks (Pelly fide 2, 3).

The orange-yellow fruit has an edible pulp which is eaten in Sierra Leone (2).

References:

1. Dalziel, 1937: 349. 2. Irvine, 1961: 578–9. 3. Portères, 1974: 119.

Diospyros ferrea (Willd.) Bak.

FWTA, ed. 2, 2: 11.
English: ironwood (Burkill).
French: ngavi du fourré littoral (*'ngavi' of littoral thickets*, Aubréville).
West African: SENEGAL DIOLA (Bayot) é tikuñi (JB) MANDING-MANINKA ko gélin koː tree (JB) WOLOF sélah (JB) GHANA ADANGME ɡbletʃo (FRI) NIGERIA HAUSA kas kawami (KO&S) YORUBA paroko (KO&S)

A tree to 10–15 m high with a slender bole to about 60 cm in girth, of drier sites in closed-forest, forest margins, rocky savanna and coastal thickets, throughout the Region from Senegal to W Cameroons. It is an extremely widely dispersed species occurring in India and SE Asia, but curiously not in the intervening tropics of eastern Africa, and but rarely in E Cameroun and Gabon (2). There is a great diversity in form.

The wood in Asia is described as grey white with darker streaks in the heart, close-grained, hard, heavy and durable. It is liable to split. Where size permits it is used for boat-anchors, tool-handles, sheaths of weapons and for rafters. The fruit-pulp is edible when ripe. It is said to be a famine food in Madras, S India (1–3).

References:

1. Burkill, IH, 1935: 1380, as *Maba buxifolia* Pers. 2. Letouzey & White, 1970, a, b: 69–72. 3. Sastri, 1952: 78.

5

EBENACEAE

Diospyros gabunensis Gürke

FWTA, ed. 2, 2: 12. UPWTA, ed. 1, 347.

English: flint-bark tree (Ghana: Chipp, Irvine).

French: sanza minika à grandes feuilles (*'sanza minika' with the large leaves*, Aubréville, Ivory Coast); faux ebène (false ebony, Leeuwenberg).

West African: SIERRA LEONE MENDE mampa (DS) ŋwaŋyei (S&F) tɛli (FCD; Cole) LIBERIA KRU-BASA karhn (C&R) rookra (C&R) IVORY COAST ABE sanza minika (Aub.) GHANA AKAN-WASA ankyeryie (DF)

A tree to 20 m high with long, straight, slender, unbuttressed bole to 1 m girth, of lowland rain-forest, sometimes in very wet situations.

Sap-wood is light yellow, with heart-wood veined in black. The wood is strong. In Sierra Leone it is considered a secondary structural timber (2). In Liberia the tree is used for large house-poles, and is cut for wooden spoons, combs and small household articles generally for which its texture and hardness suit it (3–5).

The bark is smooth or slightly fissured, and is hard and brittle like glass, hence the name 'flint-bark tree' (1). In Liberia the inner bark is prepared as a decoction for use as an antiseptic wash for sores, wounds, etc., to which a decoction of leaves and/or bark is then applied as a poultice (4, 5).

References:

1. Chipp 247, K. 2. Cole, 1968,a. 3. Cooper 87, K. 4. Cooper 275, K. 5. Cooper & Record, 1931: 98–99.

Diospyros gracilens Gürke

FWTA, ed. 2, 2: 12, 14, as *D. nigerica* F White.

West African: NIGERIA EDO isahianmi (Kennedy) YORUBA igi dúdu = black wood; applies to most D.spp. (KO&S)

Tree to 25 m high, bole straight, fluted, slightly buttressed, to 1.7 m in girth, of the dense evergreen lowland rain-forest of S Nigeria, and also in E Cameroun and Gabon.

Sap-wood is cream coloured, and the heart-wood, as reflected by the Yoruba name, is black, which may perhaps be of some commercial value (1).

Reference:

1. Letouzey & White, 1970, a, b: 86–88.

Diospyros heterotricha (BL Burtt) F White

FWTA, ed. 2, 2: 15.

A shrub to 3 m high of the Congo basin. The fruits are edible for which it has been introduced and cultivated in the Region.

Diospyros heudelotii Hiern

FWTA, ed. 2, 2: 14. UPWTA, ed. 1, 347.

French: ngavi à petites feuilles (*'ngavi' with the little leaves*, after Abe, Ivory Coast, Aubréville).

West African: SENEGAL BADYARA mãntok (JB) KONYAGI a-ngunk (JB) a-nkunk (JB) MANDING-MANINKA yatété (JB) GUINEA-BISSAU FULA-PULAAR (Guinea-Bissau) malifu (D'O) SUSU iatêtê (JDES) láfa (D'O) SIERRA LEONE KISSI pesiwe (FCD; S&F) KONO difi-mɛsɛ (FCD) difinɛ (FCD) tɔnyamba (FCD) KRIO bush-baŋga (S&F) LIMBA bukbibili (NWT) LOKO tei-guro (S&F) têl-guru (FCD) MENDE ndɔkowulo (*def.* -wuli) *from ndɔko: to break wind;* wulo

6

(*def.* wuli, wui): *tree* (auctt.) *nd*ɔku-hinei = *male* (hinei) *break wind* [*tree*]; *cf. D. thomasii, female* (FCD) SUSU *g*bogontiŋyi (NWT; FCD) TEMNE ɛ-bup-kasi *from* bup: *smell;* kasi: *to break wind* (FCD) *an*-gboth (auctt.) *ma*-taŋk (NWT) TEMNE (Port Loko) ɛ-gbɔth (Pyne) **IVORY COAST** ABE *ŋ*gavi (Aub.) ABURE sanza brika (B&D) AKYE guè (A&AA) KYAMA blem'blain (Aub.) **GHANA** VULGAR oson-akenkye (DF) AKAN-BRONG duabiri (CV; E&A)

A tree to 20 m high of the lower storey of the evergreen forest or in secondary forest from Guinea-Bissau to Ghana. It is the commonest *D. sp.* in Sierra Leone (7).

The tree has a straight bole. The freshly cut wood is pink and sinks in water (6). The timber is hard (2). The twigs are chewed as a tooth-cleaner (3, 8), and the flexible stems are used for spring-traps (3).

The leaves and flowers emit a coarse smell which is likened in some Sierra Leonean names to that of rank flatulence (3).

The roots are used as a purgative in Guinea-Bissau (5).

The thin pulp surrounding the seed is edible and is sometimes eaten (2, 4).

Unspecified parts of the plant are used in Ivory Coast as a remedy for kidney-troubles, constipation and food-poisoning (1).

References:

1. Bouquet & Debray, 1974: 80. 2. Busson, 1965: 365. 3. Dalziel, 1937: 347. 4. Deighton 1475, K, 5. D'Orey 289, K. 6. Irvine, 1961: 579. 7. Savill & Fox, 1967: 106. 8. Portères, 1974: 119.

Diospyros hoyleana F White

FWTA, ed. 2, 2: 15.

A shrub or tree to 15 m high and vertical bole to 1 m girth, in understorey of the dense evergreen rain-forest of S Nigeria and W Cameroons and extending to Angola and Zambia.

In Congo (Brazzaville) the dry leaves powdered, or the leaf-sap, are sniffed up the nostrils for persistent headache. The pulped leaves are used for treating wounds and sores, and a leaf-decoction is given to women during pregnancy who have a history of miscarriage. Very small traces of *alkaloid* have been detected in the leaves. *Saponins* and *tannins* are present in the leaves, bark and roots. *Quinones* are present in small amount in the bark, in appreciable amount in the bark and absent from the leaves. *Steroids* and *terpenes* are present in the bark and roots, and not in the leaves (1, 2).

References:

1. Bouquet, 1969: 107. 2. Bouquet, 1972: 22, 89.

Diospyros iturensis (Gürke) Letouzey & F White

FWTA, ed. 2, 2: 14, as *D. alboflavescens* sensu F White. UPWTA, ed. 1, 347, as *D. insculpta* Hutch. & Dalz.

English: Benin ebony (loosely applied, S Nigeria, Dalziel).

West African: **NIGERIA** EDO uhu (JMD; KO&S) EJAGHAM-ETUNG nyangha (Catterall) IBIBIO nyàñá = *spread out* (KO&S) IGBO ákwìrìkwárísí = *ringworm* (KO&S; Ilonah) YORUBA esu (KO&S) oshu (KO&S) osun (KO&S)

A tree to 20 m high, with an irregular fluted bole to 1.3 m in girth, or branching from ground-level, on river-banks or lower storey of forest on firm soil in S Nigeria and W Cameroons, and extending to Zaïre and Angola.

Sap-wood is white. Heart-wood is black and is sometimes cracked at the centre (3).

The stems are used in Congo (Brazzaville) to make paddles and spear-shafts, pestles and mortars, carvings for masks and fetishes, and tourist bric-à-brac (2).

The roots are lightly torrified and the bark is pounded with rock-salt and palm-oil to produce a paste which is applied to points of costal pain and for bronchial affections after scarification (1, 2).

Tannin, steroids and *terpenes* are present in the leaves, bark and roots. *Quinones*, absent from the leaves, are present in the bark and in greater amount in the roots (3).

References:

1. Bouquet, 1969: 107, as *D. alboflavescens* sensu F White. 2. Bouquet, 1972: 22, 89, as *D. alboflavescens* sensu F White. 3. Letouzey & White, 1970, a, b: 93–97.

Diospyros kamerunensis Gürke

FWTA, ed. 2, 2: 11. UPWTA, ed. 1, 347.

English: Cameroon ebony (Unwin); African ebony (H-Hansen).

Trade: awran; otutu (Nigeria, H-Hansen).

West African: LIBERIA KRU-BASA dibah *general for Diospyres* (C&R) drebah (C&R) sefflay *general for Diospyros* (C&R) IVORY COAST ABE *ŋ*gavi (Aub.) 'KRU' guhé (Aub.) KRU-GUERE (Wobe) kaké (Aub.) GHANA VULGAR kokode (DF) omena (DF) AKAN-ASANTE *o*menowa (auctt.) *ɔ*menowa (FRI; E&A) WASA kokodae (aucct.) marawa (auctt.) *ɔ*mena (FRI) *ɔ*menawa (FRI; E&A)

A tree to about 16–17 m high with a straight bole to 1 m in girth, of the understorey of the closed-forest of Liberia, Ivory Coast and Ghana, and also in E Cameroun and Gabon.

The wood when fresh is pinkish with black heart which has at some time been exported from the Cameroons as 'Cameroon ebony' (10). It is heavy, hard, tough, resilient and does not float (6, 8, 9). The sap-wood can be impregnated. It tends to split during seasoning, but works well and takes a smooth finish (9). The wood is used in Ghana (7, 9, 11) for axe-handles, spear-shafts, telegraph-poles and hut-posts, and in Liberia (2, 3, 5) for rice-pestles, implement-handles and hut-posts. Drums are hollowed out of the wood in Liberia and drum-heads of hide are fastened by hoops made of the young saplings of this and other *D.* species which are particularly strong and pliable (4, 5).

The fruit pulp is edible (1, 9).

References:

1. Busson, 1965: 365. 2. Cooper 69, K. 3. Cooper 116, K. 4. Cooper 304, K. 5. Cooper & Record, 1931: 99. 6. Dalziel, 1937: 347 7. Greene 852, K. 8. Greene 928, K. 9. Irvine, 1961: 580. 10. Unwin, 1920: 415. 11. Vigne, 1363, K.

Diospyros mannii Hiern

FWTA, ed. 2, 2: 11.

French: ngavi à gros fruits (*'ngavi' with big fruits*, from Abe, Aubréville); faux ebénier (*false ebony*, Gabon, Walker); demi-deuil (*half mourning*, Gabon, Walker).

West African: SIERRA LEONE MENDE *n*dɔkowuli (Cole) sii (*def.*) (SKS) IVORY COAST ABE *ŋ*gavi (Aub.) ABURE essérériéni (B&D)

A tree to 20 m high with a straight long bole of the closed evergreen forest, sometimes in wet situations, of Sierra Leone to S Nigeria, and extending to the Congo area.

Sap-wood is pale lemon colour, heart-wood black. Heart-wood is used in Sierra Leone for making black carvings: it is also a furniture timber (2). In Gabon handles for knives are made of it. Before cutting up the wood it is left to soak in a stream for some time. This treatment gives it a stronger black colour (4).

In Ivory Coast the bark is pulped up and applied topically to fractured limbs after splinting (1). Bark-scrapings mixed with powdered red wood of *Pterocarpus soyauxii* Taub. (Leguminosae: Papilionoideae) are considered in Gabon to be good for chest ailments (3, 4).

References:
1. Bouquet & Debray, 1974: 80. 2. Cole, 1968, a. 3. Walker, 1953, a: 33, as *D. aggregata* Gürke. 4. Walker & Sillans, 1961: 157.

Diospyros melocarpa F White

FWTA, ed. 2, 2: 14.
West African: NIGERIA BOKYI asibeni (KO&S)

A tree to 30 m high with a straight bole, slightly fluted basally, to 1 m girth, of the dense evergreen rain-forest of S Nigeria, W Cameroons and Fernando Po, and also Congo (Brazzaville) and Zaïre and in a limited area of Gabon, not in E Cameroun.

The wood is pink and hard, becoming browner towards the centre and with cracks bordered with black.

Though the tree is recognised by the Bokyi of S Nigeria, no usage appears to be recorded.

Diospyros mespiliformis Hochst.

FWTA, ed. 2, 2: 12. UPWTA, ed. 1, 347–8.

English: ebony; West African ebony; swamp ebony; monkey guava; persimmon (The Gambia, Tattersall).

French: ebénier de l'Ouest Africain; plaqueminier (Bouquiaux); kaki de brousse (Bailleul).

Note: while the English and French names of West African Ebony are botanically correct, foresters and the timber trade apply these names to *Dalbergia melanoxylon* Guill. & Perr. (Leguminosae: Papilionoideae).

West African: SENEGAL BADYARA mãnkãnk (JB) BASARI a-gâka (K&A) a-ngãnga (JB) a-nkãnka (JB) BEDIK ma-ganga (after K&A, JB) ga-nganga (FG&G) FULA-PULAAR (Senegal) kuku (JB) kukudAé (K&A; JB) kukui (K&A; JB) kukuo (K&A; JB) kuuku (K&A) nèlbé, nèlbi (K&A; JB) pomponi (JB) tolé (JB) TUKULOR kukui (JB; K&A) kukuńé (K&A) kukuo (K&A) kuuku (K&A) nèlbé, nèlbi (K&A; JB) vèndu diènga (JB) KONYAGI ngang (JB) nkank (JB) MANDING-BAMBARA kukuo (JB) sunsun (K&A; JB) sunzun (JB) susu (JB) MANDINKA kukuo (K&A) MANINKA dabakala sunsun (K&A; JB) diombo (JB) sana sumsum (K&A) sana sunsu (JB) 'SOCE' kukuo (JB) NON gaholom (auctt.) SERER ken (K&A) ńâčiké (K&A after Aub.) nem (K&A) nen (K&A) nèn (JB) SONINKE-SARAKOLE dôba (K&A) WOLOF alom, alôm, alum (auctt.) doki (auctt.) kalum (K&A) THE GAMBIA FULA-PULAAR (The Gambia) nelberi (DAP) MANDING-MANDINKA kukuo (Hopkinson; ST) kukuwo (Hopkinson; DAP) WOLOF alom (DAP) halom (ST) GUINEA BASARI a-ngãngà (FG&G) KONYAGI a-ngank (FG&G) MANDING-MANINKA dabakala-sunsu (CHOP; FB) SUSU sunsu (JMD) MALI DOGON alugíle (CG) alukile (Aub.; FB) FULA-PULAAR (Mali) pupui (Aub.) MANDING-BAMBARA sunsu (A.Chev.) sunsun (auctt.) MANINKA dabakala-sunsu (CHOP) SONGHAI dué (A.Chev.; Aub.) tokoye (Aub.) TAMACHEK kania (Aub.) UPPER VOLTA BOBO kònon (Le Bris & Prost) DAGAARI gâ (K&B) gaa (GM) FULA-FULFULDE (Upper Volta) gagaahi (K&T) ganaahi (*pl.* -aaje) (K&T) nelbe (K&T) nelbi (K&T) GRUSI gẽá (*pl.* géesè) (GM) GURMA gaabu (GM) GYORE gãaká (*pl.* gãahi) (GM) HAUSA kagnia (K&B) kania (K&B) KIRMA onfra (K&B) KURUMBA akiria (Prost) MANDING-BAMBARA sunsu(-n) (K&B) DYULA sunsu(-n) (K&B) MOORE gãase (*pl.* gãase) (auctt.) SAMO (Sembla) tyïmī (Prost) SENUFO sunsu(-n) (Aub.; K&B) SONGHAI-ZARMA tokoye (A&B) IVORY COAST AKAN-BRONG komo (Aub.; E&A) BAULE bablé gualé (auctt.) blaguigole (K&B) kimi (B&D) DAGAARI (Wule) ga (Sidibe) KULANGO hion (K&B) 'MAHO' hamon sunsu (K&B) MANDING-MANINKA dabakala sunsu(-n) (Aub.; K&B) sana sunsun (K&B) sunzu(-n) (A&AA; B&D) SENUFO sunsu(-n) (Aub.; K&B) SENUFO-DYIMINI katio (K&B) TAGWANA siamboy (K&B) GHANA ADANGME nɔkɔtʃo (FRI) AKAN kusibiri (DF; E&A) AKAN-BRONG

EBENACEAE

komo (FRI; E&A) TWI okisibiri (aucct.) BAULE babliguale (FRI) DAGAARI ga (Gaisser) gaa
(GM) kirirema (FRI) DAGBANI kirirema (FRI) GA kirirema (FRI) okúshibli (KD) GBE-VHE
dɔkɔ (FRI) keyi (FRI) VHE (Kpando) keke (FRI) VHE (Pecí) keyi-keyi (FRI) GUANG-NCHUM-
BULU kirirema (Coull; FRI) KONKOMBA legabɔl (Gaisser; FRI) KUSAL gãaka (pl.
gãase) (GM) MAMPRULI gaa (Arana & Swodesh) MOORE taka (AEK; FRI) NANKANNI ginga (Lynn fide FRI)
gunya (Enti) POLCI-BULI gaabo (pl. gaasa) (GM) SISAALA kaliŋ (Blass) kaliŋ-nɛniŋ the fruit
(Blass) TOGO BASSARI bugau (Gaisser) gawelle (Gaisser) GANGAM gãa, gaam (pl. gande) (GM)
GBE-VHE doko (Aub. FB) jeti (Metzger) yeti (Metzger) KABRE nangalo (Gaisser; FB) tangala the
fruit (Gaisser) MOBA gaak (pl. gaati) (GM) NAWDM angalo (Gaisser) gaaga (pl. gaai) (GM)
tingalo (Gaisser; FB) SOMBA kâbu (Aub.) TEM (Tschaudjo) ningalo (Metzger) tigbata the fruit
(Metzer) tingalo (Metzer) YORUBA-IFE (TOGO) donko (Metzer) DAHOMEY BAATONUN nuibu
(Aub.; FB) uibu (Aub.) BIERI yesga (pl. yesi) (GM) GBE-FON kain (Aub.) kainui (Aub.) GEN
djéti (Aub.) GURMA gaabu (GM) POLCI-BULBA ñisbo (pl. ñisto) (GM) SOMBA kâbu (Aub.)
TAMARI pĩi (pl. mu) (GM) TAYARI kwĩhbu (pl. kwĩhna) (GM) WAMA kabu (pl. kana) (GM) YOM
gaayo (pl. gaai) (GM) garo (Aub.) NIGER ARABIC (Niger) diokane (Aub.) djiokan (Aub.)
djohan (Aub.) GURMA ogabu (Aub.) HAUSA kaéua (Aub.) kagnia (Aub.) kania (Leroux) kanyia
(Aub.) KANURI burgum (Aub.) SONGHAI dué (Aub.) tòkéy (pl. -à) (D&C) SONGHAI-ZARMA
tokoye (Aub.) TAMACHEK kania (Aub.) TEDA burkum (Aub.) NIGERIA ARABIC-SHUWA gughan
(JMD) jukhan (JMD) jukhan (JMD; KO&S) BADE ferdamu (FWHM) BIROM jɔ̀rɔ́p the fruit
(LB) tin ǹjɔ̀rɔ́p the tree (LB) FULA-FULFULDE (Nigeria) ɓalchi (pl. ɓaleje) = ebony wood; from
ɓale: black (JMD) jalambani (JMD) jalambori (JMD) kaiwahi (JMD) kaiwua (pl. kaiwaje)
(JMD) nelɓi (auctt.) GALAMBI iilá (Schuh) GERA dĩwlá (Schuh) GWARI kuci (JMD) HAUSA
kaiwa(-a) (auctt.) kánya (LB) kanyaà (auctt.) kanyan (auctt.) lubiya the fruit (JMD) madĩ the
dried fruit (JMD) moówàr bírii = the monkey's favourite - the fruit (JMD; A&S) IGALA obiudu
(JMD; KO&S) obiudu'adú = black kola; seems doubtfully correctly applied (H-Hansen; RB)
IGBO akawayi probably a general term for several spp. (JMD) onye-koyi (BNO) onye-ōji
= blackfellow, from oji: black (auctt.) KAMBARI kàrừndâ tree and fruit (RB) mừu moorừndâ the
fruit (RB) uurừndâ (RB) KANURI bergem (auctt.) bɔrgɔm (C&H; A&S) burgum (JMD) NGIZIM
(aB)vɔ́rdâmú (pl. -màmín) (Schuh) NUPE buswachi (auctt.) YORUBA igi dúdu = black wood
(auctt.)kanran (auctt.)

A tree to 30 m high with a straight bole to over 2 m in girth, of drier northern
borders of the humid rain-forest zone especially in wet situations, and in moist
places of the guinean and soudanian woodlands throughout the West African
Region, and generally widespread in such localities across Africa except in the
Congo Basin.

The tree is often kept unfelled when land is being cleared for farming for its
shade and fruit are valued. It is thus to be found in a state of tending if not
actually cultivated near villages. It is said to be a suitable species for reafforestation (7). Natural regeneration is good. Growth from seed is slow, and it
does not transplant easily (10). Ebina people in Adamawa, Nigeria, claim that a
tree planted in the compound prevents borers from eating away at cereal stalks
(25) [? in granaries].

The leaves are eaten in Niger (19). They are much relished by cattle, sheep
and goats in Senegal (1) and are taken sometimes by domestic stock in N
Nigeria (7). In Nigeria a leaf-infusion is taken as a mild laxative, and as a
vermifuge (2, 26), for fever and dysentery, and is applied to wounds as a
haemostatic (18, 29). Hausa chew the leaf, and fruit, or apply an infusion for
gingivitis and toothache (28). The liquid from leaves boiled with 1–2 guinea-
grains is drunk in Ghana as a cure for whooping-cough (Twi: nkoŋkoŋ) which
will be effected in 1–2 days; the guinea-grains are held to irritate the membranes
while the Diospyros makes the cure (Martinson fide 10, 11). The leaves are well-
known in Ivory Coast–Upper Volta to be haemostatic and cicatrisant for which
they are prepared as a dressing for cuts and wounds and to prevent infection;
the sap is instilled into the ear for otitis; and a leaf-decoction is taken as a
febrifuge and stimulant. A decoction of leafy twigs is taken in draught as a
poison-antidote (5, 15). In Senegal the leaves are prescribed for serious illnesses
such as pneumonia, infectious fevers, syphilis, leprosy and yaws (13, 14). Such
reliance is placed on this drug-plant that it is usually prescribed alone. Leaves
and fruit are used internally for menorrhoea and dysenteriform diarrhoea, and
externally for headaches, arthritis and dermal troubles (14). The use of the
leaves, with the bark, husks of tamarind and guinea-corn gruel, as a leprosy

remedy is recorded in N Nigeria (7, 26). Leaves in cold infusions are commonly taken in the soudanian region for dysentery, and a leaf-decoction by draught or as a wash for fevers (7).

The freshly cut wood is light pinkish-brown, slightly darker to the centre. Blackening of the heart-wood develops only after felling and appears to depend on edaphic characters, trees from savanna situations blackening while those from more thickly forested areas do not. Blackening is possibly a pathological process and burying is said to accelerate it. In trees cut down and left to season, the sap-wood is quick to decay. The heart-wood is durable though attack by large boring beetles is reported. Colouration varies somewhat with brown and green streaks. It is one of the ebonies of commerce, but it does not have the uniform black colour of classical ebony (7, 10).

It is hard, heavy and has a fine grain; dries rapidly but is inclined to distortion; strength is good, turning well but splits on nailing; it is moderately difficult to work and dulls tools quickly (17). It is used locally for tool-handles, gun-stocks, ploughs, hut-posts and rafters, combs, stools, walking-sticks, cudgels, carving and fancy goods, and for charcoal for cooking and for smoking fish (3, 4, 7, 10, 17, 20, 24). Tenda races of Senegal use the wood for chew-sticks (24), as well as the Bambara of Mali (Bazin fide 27). *Saponin* has been detected present (27).

Sawdust of this species, as also of most *Diospyros spp.*, causes dematitis after continuous contact. In Nigeria it is sometimes added to dog's food instead of sulphur to cure mange. Shavings of the wood with pods of *Acacia nilotica* (Linn.) Willd. (Leguminosae: Mimosoideae) and roots of *Borassus* (Palmae) are pounded in water and boiled for about two hours, after which the liquid is used in Nigeria to rinse the mouth for toothache. Sap from freshly felled trees, as also water from holes in the tree, or an infusion of the black heart-wood, are similarly used (2).

The bark contains a dark-coloured gum which is used in Ghana to mend broken pottery (10). The bark of both trunk and roots have medicinal use in Senegal, as have the leaves, for serious illnesses (13, 14). In Ivory Coast–Upper Volta trunk and branch-bark is taken by draught and added to baths in leprosy treatment (15), and in Nigeria a bark-infusion is used to wash sores, ulcers, etc. (2) and root-bark for skin-eruptions, such as itch (7, 29). The whole root may also be used (18). In the soudanian region root-bark is an ingredient of abortifacient prescriptions (7). The bark is considered a veterinary medicine (29), especially for horses, given pulverised as an effective vermifuge, and also in fumigation burnt along with old rags, etc. as a cough remedy (7).

The root, heated and powdered, is prepared in Ivory Coast–Upper Volta into tablets with salt and palm-oil which are administered for serious jaundice – the treatment causing vomiting and profuse diarrhoea (15). A root-decoction is taken as an anthelmintic and in cases of difficult childbirth (5, 15). In Tanganyika a root-decoction, with leaf-sap, is taken by draught for malaria and for scrofula (8).

The fruit-pulp is sweet and edible. Material from Senegal is reported containing 3.1% proteins and total carbohydrates 33.9% fresh weight (14). The fruits are eaten in all parts of Africa and are made into fruit-juice drinks. They have a flavour of the persimmon, *Diospyros kaki* Linn. In some parts it is dried before use, and in N Nigeria the dried fruit is known as *baro* (Hausa). In SW Africa it is stored in this form as a reserve against food shortage. The Hausa prepare a kind of soft toffee (*maɗi*) from the fruits, which along with some other similar edible fruits such as figs, dates, etc., are known as *lubiya*. In N Ghana the Grusi ferment the fruits to an alcoholic drink like mild palm-wine. In SW Africa a sort of 'brandy' is prepared from the fruit. (3, 4, 6, 7, 9, 10, 23, 30).

The flowers are fragrant and visited by bees for nectar. Hives are often placed in the branches (7).

The fruit-pulp is applied in Ivory Coast–Upper Volta to pottery to glaze and varnish it (15).

11

The seeds are much relished in Uganda by baboons (22).

Pharmacologically, *plumbagin* appears to be the principal drug present. This substance possesses anti-biotic, antihaemorrhagic and fungistatic properties, and is found in the root-bark to a concentration of 0.9% and but a trace in the leaves. *Tannin, saponin* and a substance probably identical to *scopolamine* are also present. Leaf-infusions have been shown to be pharmacologically active on *Paramecium*, fish and mice (14, 15, 18, 29). There is a high fluoride content (28).

The plant is credited with magical attributes. In Niger at Maradi it is held to be the sacred tree of the village (16). In the Diébougou district of Ivory Coast the Earth-Goddess is said to reside in a stone deposited at the foot of the tree near the village, and the stone and tree are reverenced as the residence of the protective spirit of the community (21). The Fula and the Wolof of Senegal prescribe the roots in ritual invocations to chase away evil spirits, effect cures and to obtain good fortune (14), and also use the plant in a medico-magical treatment of psychoses (13). The Pulaar of N Nigeria seek good luck in hunting by making a paste of the bark with a ram's belly-fat which is rubbed on their hunting bows (12) – cf. *D. chevalieri* De Wild.

References:

1. Adam, 1966, a. 2. Ainslie, 1937: sp. no. 137. 3. Aubréville, 1950: 442. 4. Aubréville, 1959: 3: 164. 5. Bouquet & Debray, 1974: 80. 6. Busson, 1965: 365. 7. Dalziel, 1937: 347–8. 8. Haerdi, 1964: 115. 9. Irvine, 1948: 265, 267. 10. Irvine, 1961: 580–3. 11. Irvine, s.d,a. 12. Jackson, 1973. 13. Kerharo & Adam, 1964, b: 438–9. 14. Kerharo & Adam, 1974: 399–401, with phytochemistry and pharmacology. 15. Kerharo & Bouquet, 1950: 175–6. 16. Leroux, 1948: 662. 17. Lucas, 1967. 18. Oliver, B, 1960: 25, 61. 19. Robin, 1947: 58. 20. Schnell. 1950, b: 261. 21. Sidibé, 1939. 22. Styles 279, K. 23. Watt & Breyer-Brandwijk, 1962: 389. 24. Ferry & al., 1974: No. 88. 25. Babuwa Tubra fide Blench, 1985. 26. Akinniyi & Sultanbawa, 1983. 27. Portères, 1974: 119. 28. Etkin, 1981: 81. 29. Oliver-Bever, 1982: 53. 30. Tattersall, 1978: 11.

Diospyros monbuttensis Gürke

FWTA, ed. 2, 2: 14. UPWTA, ed. 1, 348.

English: Yoruba ebony; walking-stick ebony (Dalziel).
West African: **IVORY COAST** AKYE békugbè (A&AA) niamiébaka = *the wood of God* (K&B; Aub.) nienébaka (K&B) BAULE gnamien-baka (A&AA) niamia baka (B&D) niamié baka = *the wood of God* (K&B) nienébaka (K&B) **GHANA** AKAN-ASANTE ada piri (auctt.) akyirinian (CV; FRI) FANTE atwirɛ-nantin (E&A) **TOGO** KPOSO etjannaka (Volkens) TEM (Tschaudjo) lia-nuwasaure = *francolin's spur; in allusion to the hard thickened peduncle remaining after the fruit has fallen* (Kersting; Volkens) **NIGERIA** IGBO okpu ọcha *from* ọcha: *white* (auctt.) YORUBA egungunekun (KO&S) erikesi = *pig's teeth,* (F White; KO&S) ògàn (auctt.) ògàn-ègbò (auctt.) ògàn-pa (Ross; Verger) ògàn-pupa = *red ogan* (auctt.)

A tree to 10 m high with short slender bole, often fluted, rarely straight, to 60 cm girth, of the drier semi-deciduous rain-forest of Ivory Coast to S Nigeria and galleried forest of Adamawa, and extending to northern Zaïre.

The sap-wood is white, turning slightly yellow on exposure, and heart-wood is pale yellow. The wood is hard, dense and durable. It is used in Nigeria to make clubs, walking-sticks and tool-handles (7) and scantlings (4). Similar uses are reported in Togo (4). The branches are flexible and are used to make traps (4).

A decoction of the bark and twigs, with the leaves of *Senna occidentalis* (Leguminosae: Caesalpinioideae) and *Lippia adoensis* (Verbenaceae), is taken in draught and put into baths by the Baule and Anyi of Ivory Coast as a leprosy treatment (3, 5). They also consider this to be good for fever-pains, stomach-ache and oedemas (3). The leaves are used by the Akye to treat chicken-pox (2).

The branches, particularly on the younger plants, are armed with short thick spines, to which perhaps the Yoruba name *pig's teeth* refers. These are made into an infusion in S Nigeria which is given, probably on the premise of sympathetic magic, to alleviate teething pains in children (7).

The buds (? flower-buds) are put into soup in S Nigeria (7), and the fruit has unspecified medicinal application (6).
Examination of the roots of Nigerian material failed to show the presence of any alkaloid (1).

References:

1. Adegoke & al. 1968. 2. Adjanohoun & Aké Assi, 1972: 117. 3. Bouquet & Debray, 1974: 80. 4. Dalziel, 1937: 348. 5. Kerharo & Bouquet, 1950: 176. 6. Millen 163, K. 7. White, F., 1957.

Diospyros physocalycina Gürke

FWTA, ed. 2, 2: 11.

A tree to 10 m high and trunk to 60 cm in girth, of the closed lowland rainforest in S Nigeria, and also in E Cameroun and Gabon.
This species is very close to *D. canaliculata* De Wild., and there has been confusion in distinguishing the two. References to usages of *D. physocalycina*, which all come from outside the recognised area of its limited distribution, have been cited under *D. canaliculata* De Wild.

Diospyros piscatoria Gürke

FWTA, ed. 2, 2: 10. UPWTA, ed. 1, 349.
West African: **SIERRA LEONE** LIMBA wotoi (Richards) LOKO wɔtɔi (NWT) MENDE lokumi (NWT) TEMNE ε-bomp-ε-robat (NWT) **GHANA** AKAN-ASANTE benkyi (auctt.) ɔtwetokesa (E&A) **NIGERIA** EDO igiedudu = *black wood; from Yoruba* (JMD) isánhiánmwẹ (Lowe; Elugbe) IJO-IZON ongbalo (auctt.) YORUBA igi dúdu = *black wood* (JMD; KO&S) isodudu (Richards; KO&S) keso (Richards) ogwagwa (Richards)

A tree reaching 30 m high with a long straight bole to 23 m long and low buttresses at the base, of lowland rain-forest, from Guinea to W Cameroons, and extending to Zaïre.
Sap-wood is white or off-white, pale pink or greyish, and heart-wood black sometimes with veins of greenish-brown. It is reported to be often attacked by a large boring beetle (1).
The fruit is used as a fish-poison in Nigeria (3) and Gabon (4). In E Cameroun and Gabon both bark and the fruit with persistent calyx are used (2).

References:

1. Irvine, 1961: 583. 2. Letouzey & White, 1970, a, b: 126–31. 3. Richards 3331, BM. 4. Walker & Sillans, 1961: 157.

Diospyros preussii Gürke

FWTA, ed. 2, 2: 14.

A tree to 10 m high, usually less, with bole to 60 cm in girth, of the evergreen rain-forest, often in swampy sites of SE Nigeria and in E Cameroun and Gabon.
Diosquinone, a crystalline naphthoquinone, is reported present in the bark. This substance has antibotic action against *Staphylococcus*. The bark possibly has some antiseptic value (1).

Reference:

1. Oliver, B, 1960: 25.

13

EBENACEAE

Diospyros sanza-minika A Chev.

FWTA, ed. 2, 2: 12. UPWTA, ed. 1, 349.

English: ebony; African ebony (trade); flint bark (Ghana).
West African: SIERRA LEONE MENDE ŋwaŋyei (*def.*) (DS; S&F) tɛli (FCD; Cole) LIBERIA DAN gboh (GK) rnjoko (GK) KRU-BASA kohr (C&R) GUERE (Krahn) de-tue (GK) IVORY COAST ABE erikissé (A.Chev.) sanza minika (Aub.) AKAN-ASANTE bakablé (B&D) AKYE *m*buobi (A.Chev.) *ŋ*guobi (A.Chev.) kusibiru (A.Chev.) ANYI asun séka (Aub.) esum seka (Aub.) esun seka (A.Chev.) sanza-minika (A.Chev.) KWENI damliziri (A.Chev.) KYAMA bomelé (Aub.) ntebrahia (A.Chev.) NZEMA sanza-minika (A.Chev.) GHANA VULGAR kusibiri (DF) AKAN-ASANTE bakablé (E&A) ɛsono-afẽe = *elephant's comb* (E&A) okusibiri (auctt.) FANTE okusibiri (Greene) TWI esono afee =*elphant's comb* (CJT; E&A) WASA ankyeri (auctt.) ankyeryie (BD&H; DF) dubima (Foggie) ɛsono-afẽ = *elephant's comb* (FRI; E&A) ɛsono-akyene (FRI) ANYI ɛsun seka (FRI) ANYI-SEHWI enkyenwune =*elephant's back* (CJT) NZEMA sanzaminike (Enti) xanzomuki (BD&H) zanzaminike (A&E)

A tree to 25(–40) m high, with straight, unbuttressed trunk 2(–3) m in girth, of the humid, evergreen forest of Sierra Leone to Ghana, and in E Cameroun and Gabon. It has not been reported in Nigeria.

The wood is not sharply demarcated into sap and heart-wood. The outer younger wood is greyish white, becoming pink with small streaks of black towards the centre, and sometimes under certain conditions becoming all black. It is very hard, heavy and difficult to work, fine-textured and durable (9, 11–14). In Sierra Leone it is considered a construction-timber of secondary importance (5). It is used in Ghana for heavy construction-work, telegraph-poles, turnery and rollers for textile manufacture (1), and has been valued for pit-props in the Tarkwa mines (4, 10, 11). It has some resilience for which it is good for implement-handles, e.g. pickaxes, hammers (Ghana, 11, 13), axes, shafts of harpoons, fish-spears, spears and stocks of crossbows (Liberia, 8; Gabon, 14). Small poles are somewhat flexible and are used as a spring in game-traps in Sierra Leone and Liberia, and the slender stems as wattles in mud-houses and for lacing into monkey-cages (7, 8, 12). The wood is used as fire-sticks in Liberia (6).

The bark is conspicuously black, and regularly and deeply fissured, characters by which the tree is readily identified, and which are recognised in the Sehwi name meaning 'elephant's back.' It is extremely hard and is difficult to cut with an axe, evoking the name 'flint bark.' It is highly resistant to decay and is said to survive when a tree dies in the forest as a hollow cylinder long after the wood within has rotted away (A Chevalier fide 2, 9, 11).

In Liberia the leaves are pulped and made into a poultice for application to areas of pains on the sides of the body (7). In Ivory Coast the plant (part unstated) is used to treat giddiness and attacks of epilepsy (3).

The fruit contains a gelatinous pulp of doubtful edibility (9).

References:

1. Anon. s.d., a. 2. Aubréville, 1959: 3: 168. 3. Bouquet & Debray, 1974: 81. 4. Burtt Davy & Hoyle, 1937: 40, as *D. nsambensis* Gürke. 5. Cole, 1968, a. 6. Cooper 71, K. 7. Cooper 271, K. 8. Cooper & Record, 1931: 99–100. 9. Dalziel, 1937: 349. 10. Greene 929, K. 11. Irvine, 1961: 584. 12. Savill & Fox, 1967, 106–7. 13. Taylor, 1960: 160. 14. Walker & Sillans, 1961: 158.

Diospyros soubreana F White

FWTA, ed. 2, 2: 14–15.
West African: IVORY COAST ABE *ŋ*gavi (F White) AKYE piakambo (auctt.) plakamo (auctt.) ʻKRUʼ guirégné (K&B) KRU-GUERE bihina (K&B) buina (K&B) GHANA AKAN-ASANTE akasaa (Foggie; E&A) tweto (E&A) ANYI-SEHWI bakabri (Foggie)

A shrub or tree to 5 m high, in undergrowth of high closed-forest, from Ivory Coast to S Nigeria.

The leaves are commonly used in Ivory Coast as a haemostatic dressing over cuts or even serious wounds, pulped or chopped up into a plaster (1, 2).
The fruit contains a gelatinous translucent pulp which is considered edible in Ghana (3).

References:

1. Bouquet & Debray, 1974: 8. 2. Kerharo & Bouquet, 1950: 178, as *Maba soubreana* A Chev. 3. Cansdale 9, BM; Foggie 66, K; Vigne 1925, K.

Diospyros suaveolens Gürke

FWTA, ed. 2, 2: 11.
West African: NIGERIA EDO isenue (KO&S) owokẹn (KO&S) IGBO akpupaja (KO&S) YORUBA erukoya (KO&S) èṣùṣu àpọ́n (Verger)

A tree to 30 m high, with a straight bole to over 1.5 m girth, fluted at the base, of the lowland rain-forest of S Nigeria and W Cameroons, and into E Cameroun and Gabon.
The sap-wood is white to pale yellow, heart-wood pinkish with black streaks. This is the largest of the Nigerian ebonies, but appears to have no particular usage.
The fruits have an edible pulp which forms part of the diet of the chimpanzee in W Cameroons (1).

Reference:

1. Gartlan 30, K.

Diospyros thomasii Hutch. & Dalz.

FWTA, ed. 2, 2: 14. UPWTA, ed. 1, 349.
West African: SIERRA LEONE KISSI *k*pililio (FCD; S&F) KONO difi-*m*ba (FCD) difinɛ (FCD) LIMBA (Tonko) gbilili (FCD) MENDE *n*-dɔko-wulo (FCD) *n*-dɔko-wu-wai (FCD) *n*-dɔku-hei = *female* (hei) *break wind [tree]; cf. D. heudelotii, male* (FCD) *n*-dɔku-wuli (-i) = *break wind* (*n*-dɔku) *tree* (FCD) *n*-dɔku-wulo-lɛli (FCD) tɛli (auctt.) SUSU bogonliŋxumbe (NWT) bɔguntinyi (NWT) TEMNE *ka*-bup-*ka*-si *from* bup: *smell;* si: *flatulence* (FCD; S&F) LIBERIA KRU-BASA dib-bah (C&R) gboe-kpay (C&R) guay-ve-ney (C&R) sefflay (C&R)

A tree to 10 m high of the rain-forest of Guinea to Liberia.
The wood is pink when first cut. The smaller branches are flexible and are used in Sierra Leone (6, 10), and in Liberia (1) for making spring traps. Saplings are used in Liberia to bind drum-heads to the hollowed log-base (2), and larger timbers are cut into oars (3).
Sap from the inner bark stains the wood reddish (4). Bark liquid (? sap or infusion, 5, 8) is used in Liberia for treating diarrhoea (3).
The newly-opened flush of leaves has a rank smell whereby this species, as also *D. heudelotii*, is given vernacular names in Sierra Leone referring to flatulence (6).
The fruit pulp is edible and is sometimes eaten (4, 5, 7, 8). It is said to be very sweet (9).

References:

1. Cooper 129, K. 2. Cooper 261, K. 3. Cooper 378, K. 4. Cooper & Record, 1931: 100. 5. Dalziel, 1937: 349. 6. Deighton 2245, K. 7. Deighton 2357, K. 8. Hutchinson & Dalziel, 1937: 54–55. 9. Pyne 97, K. 10. Savill & Fox, 1967: 106.

EBENACEAE

Diospyros tricolor (Schum. & Thonn.) Hiern

FWTA, ed. 2, 2: 12. UPWTA, ed. 1, 349.
West African: GHANA NZEMA akɔ (FRI)

A very much-branched bush to about 2 m high, of coastal thickets from Ivory Coast to S Nigeria, and also in Gabon.

The wood is white and very hard and seldom thicker than about 2 cm. It has been made into useful walking-sticks (Volkens fide 1). It is cut in Ghana to make chew-sticks (2, 3, 5).

The bark contains a crystalline naphthoquinone called *diosquinone* which has anti-biotic action on *Staphylococcus* (4).

The fruit pulp is edible (2, 3).

References:

1. Dalziel, 1937: 349. 2. Irvine 2157, K. 3. Irvine, 1961: 584. 4. Oliver, B, 1960: 25, 61. 5. Portères, 1974: 119.

Diospyros vignei F White

FWTA, ed. 2, 2: 15.
West African: GHANA AKAN-WASA ɛsonoayɛm kyene (FRI) NZEMA sanzamulike (FRI)

A shrub or tree to 3.5 m high, of the undergrowth of the evergreen forest of Liberia, Ivory Coast and Ghana.

The wood is recorded as yellowish (1).

Reference:

1. Irvine, 1961: 579.

Diospyros viridicans Hiern

FWTA, ed. 2, 2: 10.
English: Gaboon ebony (Irvine).
West African: IVORY COAST AKYE kékémi (Aub.; K&B) DAN bridié (Aub.; K&B) KRU-GUERE (Wobe) pitué (K&B; Aub.) KYAMA abuprô (Aub.; K&B) SENUFO-TAGWANA wen wenya bené (K&B) GHANA AKAN-ASANTE atwea (auctt.) wuse-ama (FRI)

A tree to 20 m high with a long straight bole attaining 1.3 m girth, of lowland rain-forest from Sierra Leone to W Cameroons, and extending to Zaïre.

A decoction of leafy twigs is taken in draught in Ivory Coast by the Tagwana for leprosy. Sometimes leaves of *Raphiostylis beninensis* (Icacinaceae) and bark of *Anogeissus leiocarpus* (Combretaceae) are added (2).

The fruits are edible. In W Cameroons they are eaten by chimpanzees (1).

References:

1. Gartlan 4, K. 2. Kerharo & Bouquet, 1950: 175, as *D. kekemi* Aubrév. & Pellegr.

Diospyros zenkeri (Gürke) F White

FWTA, ed. 2, 2: 14.
West African: NIGERIA BOKYI kambiri (KO&S)

A shrub or tree to 20 m high, bole to 60 cm in girth, of the lowland evergreen rain-forest, sometimes on stream-banks, in S Nigeria and W Cameroons, and extending to Zaïre.

Sap-wood white, heart-wood sometimes black (2). The inner layer of the bark is yellow, and like the inner layer of the pericarp, turns blue on exposure to air (1).

The fruit-pulp, and perhaps also the seeds, are an item of diet for the chimpanzee in W Cameroons (1).

References:

1. Gartlan 31, K. 2. Letouzey & White, 1970, a, b: 165–8.

ELAEOCARPACEAE

Elaeocarpus angustifolius Blume

FWTA, ed. 2, 1: 301.

English: utrasum bean tree (India, Sastri).

A large handsome tree of the evergreen rain-forest, to 20 m high, bole 10 m by 1.6 m diameter; of northeastern India, Assam, Burma, Malaya, Java and Celibes. It has been introduced into W Africa (as *E. sphaericus* Schum, 8; and as *E. ganitrus* Roxb., Sierra Leone, 4, but *FWTA* ed. 2 does not name any species, nor locality, 6).

The tree when grown standing on its own has a nicely shaped dense crown. It is sometimes cultivated in India as an ornamental (9). Its dispersal in Malaya is thought to be under man's agency, and it has been rated there as a good shade-tree (3).

The wood is particularly white, often shining, straight-grained, strong, tough but not durable (3, 5), though it is suitable for light carpentry.

The fruit is bright blue, a drupe 1–2½ cm across, the flesh greenish yellow, sour and edible (1, 3, 9). It has use in Indian medicine for mental disorders, epilepsy, hypertension, asthma and liver diseases (8, 9). Extract of the flesh is reported to have a depressing effect on the central nervous system, and a potentiation of hypnosis and analgesia. It is cardio-stimulant and smooth muscle-relaxant (8).

The fruit contains a stone which is elegantly tuberculed, marked with five vertical grooves (2). When cleaned of the flesh and polished, the stones are used in India as necklaces, bracelets and other ornaments, sometimes stained, sometimes set in gold (9). They are particularly important in rosaries worn by Hindu mendicants, followers of Siva, as they afford assistance to the attainment of Heaven and Siva's company. Such rosaries contain 32 or 64 stones (7) or 101 representing the number of eyes of Siva (3). A century ago there was a brisk trade in the stones from Singapore and Java to India, and Chinese in Indonesia devised a system of ring-barking their cultivated trees to obtain fruit of reduced size containing seed the size the market favoured (3).

References:

1. Bor, 1953: 160, as *E. ganitrus* Roxb. 2. Brandis, 1911: 102, as *E. ganitrus* Roxb. 3. Burkill, IH, 1935: 901–2, as *E. ganitrus* Roxb. 4. Deighton 2631, K. 5. Gamble, 1902: 113–4, as *E. ganitrus* Roxb. 6. Keay, 1958: 301. 7. Macmunn, 1933: 99, as *E. ganitri.* 8. Oliver-Bever, 1983: 23, as *E. sphaericus* Schum. 9. Sastri, 1952: 140, as *E. ganitrus* Roxb.

ELATINACEAE

Bergia capensis Linn.

FWTA, ed. 2, 1: 128.

An aquatic herb recorded only from The Gambia and Ghana in the Region, but probably it is more widespread: also from E Africa and Asia.

It has been found in Shai, Ghana, in a fetish pot and appears to be used as a fetish plant (1).

Reference:

1. Irvine 1643a, K.

ELATINACEAE

Bergia suffruticosa (Del.) Fenzl

FWTA, ed. 2, 1: 128. UPWTA, ed. 1, 29, as *B. guineensis* Hutch. & Dalz.
West African: SENEGAL ARABIC (Senegal) jerk (JB) jirk (JB) zerk (JB) FULA-PULAAR (Senegal) dalligéré (JB) dialli (JB) ñipé (JB) ñipéré (K&A; JB) TUKULOR ñoméré (JB) ñomré (JB) sagoro (K&A; JB) samtardé (JB) MANDING-MANDINKA dévé (JB) NIGERIA HAUSA baábán giíwaá = *elephant's indigo* (JMD; ZOG) bushi (JMD; ZOG) ɗaiɗoóyàr-ƙásà (JMD; ZOG) ɗandoyar kasa (ZOG) dushiya (auctt.) jishiya (JMD; ZOG) kasa (ZOG)

A much-branched shrub to 1 m high, recorded over the northern part of the Region from Mauritania to N Nigeria, and across Africa to Sudan, E Africa and on into western India.

The flowers attract bees (5), and may have some potential as a bee-plant. Cattle in Senegal are said to relish the plant greatly for fodder, and that it is also taken by donkeys, sheep and goats but not by horses. Cattle at Chad also graze it (1), and sheep and goats in Kordofan (3).

The plant is used in N Nigeria crushed with oil and applied for itch, craw-craw, etc. It is also taken internally for venereal and gastro-intestinal complaints. The root is used by the Hausa as an ingrediant in *dauri* prescriptions at weaning time to strengthen infants and to prevent sickness (4).

Fula fishermen of Fouta Toro on the Senegal River use the roots in a fish-lure to fix a movable stem of wood that oscillates in the flow of water making a characteristic noise that attracts fish to it. Hooks and line are fixed and the device is known as *koto koto*, an onomatopoeic name derived from the noise so made (6, 7): see also *Sporobolus sp. indet.* (Gramineae). Sudan material tested for molluscicidal activity has shown slight toxicity towards fresh-water snails, *Bulinus* and *Biomphalaria*(2).

References:

1. Adam, 1966,a. 2. Ahmed & al., 1984: 74–77. 3. Baumer, 1975: 88. 4. Dalziel, 1937: 29, as *B. guineensis* Hutch. & Dalz. 5. Keay, 1954: 128. 6. Kerharo & Adam, 1964,b: 412. 7. Kerharo & Adam, 1974: 401.

ERICACEAE

Agauria salicifolia (Comm.) Hook. f.

FWTA, ed. 2, 2: 2. UPWTA, ed. 1, 346.
West African: WEST CAMEROONS KPE bweli (JMD)

A shrub or tree to about 13 m high, of montane forest margins and grass-land at 1,300–3,300 m altitude in W Cameroons and Fernando Po, and widely dispersed in E, Central and southern Africa, and Madagascar.

The leaves are lethally toxic to both man and stock, even dead leaves causing vomiting, convulsions, respiratory difficulty and coma. These effects are similar to those caused by *andromedotoxin*, the poisonous principle in rhododendron. The roots also are toxic (5).

The leaves have anodynal properties. The sap from crushed leaves is rubbed on to scarified areas on the ribs as a remedy for a pain there in W Cameroons (2). The Chagga of E Africa rub the powder from dried and slowly carbonised leaves into scarifications to relieve rheumatism. The initial reaction is to cause severe pain and swelling. A piece of root covered with butter is applied to relieve lesser pains. Analgesic effect is thought to be due to the presence of *methyl-salicylic acid esters* (5). The Chagga also use the leaves as an insecticide and an arrow-poison antidote (1, 5). In Madagascar the grilled plant is reduced to a powder which is applied to ulcerous sores. It is also used to treat syphilis and for neuralgia (3).

The bark contains 0.03–0.04% *alkaloids*. The leaves contain twice as much, and both contain an abundance of *tannin* and a waxy substance (5). Pounded

18

bark is mixed with cold water which is then drunk in Kenya by the Masai as an aid to digestion after eating a surfeit of meat (4).

References:

1. Bally, 1937. 2. Dalziel, 1937: 346. 3. Debray & al., 1971: 69. 4. Glover & al. 1136, K. 5. Watt & Breyer-Brandwijk, 1962: 393, with references.

Philippia mannii (Hook. f.) Alm. & Fries

FWTA, ed. 2, 2: 2. UPWTA, ed. 1, 346.
West African: **WEST CAMEROONS** BAMILEKE para (Johnstone)

A heath-like shrub of open montane situations from 1500 m to 3700 m in W Cameroons and Fernando Po, and also in southern Africa.

ERIOCAULACEAE

Eriocaulon latifolium Sm.

FWTA, ed. 2, 3: 62.
West African: **GUINEA-BISSAU** FULA-PULAAR (Guinea-Bissau) orô *E.sp.indet.; perhaps this sp.,* (JDES) **GUINEA** FULA-PULAAR (Guinea) orro (Langdale-Brown) **SIERRA LEONE** MENDE negbe (*def.* -i) (LP; Dawe)

A robust aquatic perennial recorded from Guinea-Bissau, Mali, Guinea, Sierra Leone and Liberia, also in the Congo basin and Angola.

Eriocaulon pulchellum Koern.

FWTA, ed. 2, 3: 64.
West African: **SIERRA LEONE** MENDE foni *after* foni: *a wet meadow with running water; applied to all the vegetation therein* (FCD) SUSU tigɛrinyi (NWT)

A slender aquatic herb to about 6 cm high occurring from Guinea and Mali to Liberia.

The Mende word *foni* is strictly a grassy meadow where water trickles over rocks. All grasses occurring in such habitats are more or less also called *foni* by association, and with which this *Eriocaulon* is also included (1).

Reference:

1. Deighton 2054, K.

Mesanthemum radicans (Benth.) Koern.

FWTA, ed. 2, 3: 64–65.
West African: **SIERRA LEONE** LIMBA bumandi (NWT) LOKO fahauŋ (NWT) jɔbɔndi (NWT) MENDE bendara (NWT) g-bɔmi (NWT) TEMNE ə-bulik (NWT)

A robust perennial often forming loose mats in swamps and streams, from Senegal to S Nigeria and Fernando Po, and across the Congo basin to Uganda and Tanganyika.

ERYTHROXYLACEAE

Erythroxylum coca Lam.

FWTA, ed. 2, 1: 356.

English: coca; cocaine plant; coca leaf.

A shrub to about 2 m high, native of Peru, Bolivia and Colombia, now widely introduced to tropical countries including African territories, principally as *E. novo-granatense* (Morris) Hieron., now recognised as a variety of *E. cola*. This is a sea-level to 1,600 m altitude plant and more tolerant of hot conditions than *E. cola* var. *cola* which is montane.

The leaves contain 0.5–1.5% total alkaloid dry weight of which the most important of several present is *cocaine*. All however appear to be readily converted by chemical means to cocaine (4, 6, 7). This drug was amongst the earliest and more effective of local anaesthetics but is now superseded by derivatives and others with better characteristics.

The plant in Peru is 'The Divine Plant of the Incas' (5), used for royal and religious occasions and chewed by all to impart prolonged endurance. Cocaine has little or no action on the unbroken skin, but acts on mucous membranes of the mouth, eye, nose and stomach, depressing hunger and permitting feats of sustained effort. It is an addictive drug, 'snow' to modern addicts, acting on the central nervous system which is initially stimulated then later depressed, causing an increase in mental power, followed by a sense of calmness, happiness and abolition of fatigue.

The plant has been grown in Ivory Coast (2), Ghana (3) and W Cameroons (1) as a source of cocaine. In W Cameroons (1) the leaves are often chewed as a stimulant instead of kola nut. In SE Asia the leaf is chewed to get a plug of leaf in a carious tooth to quell toothache.

The plant can be grown and clipped into a neat hedge made the more decorative by its white flowers and red berries.

References:

1. Ainslie 1937: sp. no. 150. 2. Aubréville, 1959: 1, 364. 3. Burtt Davy & Hoyle, 1937: 41. 4. Henry, T.A., 1939: 97–113. 5. Mortimer, 1901. 6. Oliver, 1960: 7, 63. 7. Willaman & Li, 1970. Also see: Rivier (Ed.), 1981: 3 (2/3). Oliver-Bever, 1983: 7: 42–23.

Erythroxylum emarginatum Thonn.

FWTA, ed. 2, 1: 356.

A shrub or small tree to 6 m high, of coastal grassy savanna plains (Ghana, 4) and rocky hills (Nigeria, 5), in thickets and on stream-banks, occurring from Guinea to S Nigeria, and widespread in tropical Africa.

No use is recorded for the plant in West Africa. It has sweet-scented white flowers and red berries with glabrous foliage and has potential as an ornamental shrub to which use it has been put in Mozambique (3). In Zimbabwe where the plant grows to a big forest tree, the wood is used for interior work in houses, and fence-posts made of it are said to last four years (1).

In Kenya children are reported to eat the red sweet berries (2).

References:

1. Anon. 1179, K. 2. Bally B5882, K. 3. Faulkner 221, K. 4. Irvine 1961: 206. 5. Keay & al. 1960: 241.

Erythroxylum mannii Oliv.

FWTA, ed. 2, 1: 356.

Trade: landa (S. Leone, Savill & Fox; Liberia, Kunkel; Ghana, Irvine) a name from Yaoundé, Cameroun; dabé (Ivory Coast, Aubréville) from Akye.
West African: SIERRA LEONE MENDE *g*-bimini(-i) (auctt.) IVORY COAST ABE dahain (Aub., & ex K&B) AKYE ndabé (Aub.) ANYI pépécia (Aub.; K&B) DAN gantu (Aub.) kolégai (Aub.) GHANA VULGAR pepesia (DF) ANYI pεpεsia (FRI)

A tree to about 26 m high by 2 m girth, occasionally to 3 m in Sierra Leone (5), bole straight, slightly buttressed or fluted, of the closed high-forest, or more frequently of semi-deciduous forest, recorded from Sierra Leone, Ivory Coast, Ghana and W Cameroons and also in E Cameroun.

The tree is said to be confused by tree-searchers in Ivory Coast with *Margaritaria discoidea* (Baill.) Webster (syn. *Phyllanthus discoideus* (Baill.) M.-A., Euphorbiaceae) (1) where both species receive the same vernacular names. In Liberia there is the chance of confusion with *Combretodendron macrocarpum* (Beauv.) Keay (Combretaceae) (6).

The timber resembles that of *Guarea* (Meliaceae), is semi-hard to hard, heart-wood pale brown to reddish, even textured with straight to interlocking grain, durable and easily worked. It is a general construction-timber, and is considered suitable in Cameroun (4) and in Gabon (7) for furniture-manufacture and cabinetry, though in Ghana (2) and in Sierra Leone (5) it carries pith-flecks which make it unsightly. It is however considered usable in Ghana for veneers and plywood (2). The heart-wood may sometimes be brittle. In Sierra Leone it is a popular firewood (5).

In Ivory Coast the bark, taken from the east and west sides of the stem in a cryptic relic of sun-worship, is pulped with citron and maleguetta pepper for use in frictions for intercostal pains and pleurisy. A decoction of leafy twigs is reputed to be febrifugal. Traces of *alkaloid* have been recorded in the plant (3).

References:

1. Aubréville, 1959: 1: 366. 2. Irvine, 1961: 207. 3. Kerharo & Bouquet, 1950: 66. 4. Normand, 1937: no. 196. 5. Savill & Fox, 1967: 108–9. 6. Voorhoeve, 1965: 127. 7. Walker & Sillans, 1961: 159.

EUPHORBIACEAE

Euphorbiaceae

FWTA, ed. 2, 1: 364.
West African: SIERRA LEONE MENDE *g*-boni, *g*-buni *general term for E.spp. with viscid latex* (FCD) taŋgei (*def.*) *general term for miscellaneous spp. with white latex* (S&F)

A large family, trees, shrubs and herbs, of rain-forest, guinean, soudanian, and xerophytic habitats, most poisonous, some economically important, many lactiferous and are discerned by races in Sierra Leone, as indicated above, on their sort of gum, as is accepted probably generally elsewhere.

Acalypha brachystachya Hornem.

FWTA, ed. 2, 1: 409.

A slender, much-branched annual to 30 cm high occurring only in Nigeria and W Cameroons in the Region, but widely across tropical Africa and into Asia.

Leaf-sap is dripped into the eyes for headache in Gabon. This causes a burning sensation which is apparently deemed to be important. The burning may be intensified by the addition of sap from young shoots of *Cissus aralioides* (Vitaceae) (1, 2).

References:

1. Walker, 1953, a: 34. 2. Walker & Sillans, 1961: 159.

Acalypha ciliata Forssk.

FWTA, ed. 2, 1: 410. UPWTA, ed. 1, 134–5.

West African: **SENEGAL MANDING-MANDINKA** bañânkura (JB) nanaburo (JB) **GUINEA-BISSAU MANDING-MANDINKA** banhancúra (JDES; EPdS) **PEPEL** butchebetche (JDES) **GUINEA BASARI** a-nyèlɛgúŋgúŋ FG&G yèlɛgúŋgúŋ (FG&G) **MALI DOGON** mòno (CG) **GHANA AKAN-ASANTE** mfofoa (FRI; E&A) **GBE-VHE** (Kpando) ʋudɔ ʋudɔe (FRI) **NIGERIA ESAN** ifoki (Giwa) **IGBO** (Agukwu) abaleba-ji (NWT; JMD) **YORUBA** ẹfiri (SOA) jiwinni (auctt.) owu (IFE)

An annual herb to about 85 cm high, commonly on cultivated land and in forest shade, throughout the Region from Senegal to W Cameroons, and widespread across tropical Africa and into India.

Tenda people eat the leaves (8). The plant provides grazing for donkeys, cattle, sheep and goats in Senegal, but horses are said not to take it (1). It is a common weed of cultivation (3), and is reported as a weed of rice-fields in Bendel State, Nigeria (9). Its occurrence under planted cotton may not be altogether vicarious: for perhaps magical ends it is 'used to adopt good results' in S Nigeria (Dawodu in 6). In Ghana the mashed leaves are applied as a dressing to sores (4, 5). The plant might have expectorant and emetic properties as a succedaneum for *A. indica* Linn., which has numerous medicinal uses in India (7), and for the well-known ipecacuanha, *Psychotria ipecacuanha* (Brot.) A. Stokes. (Rubiaceae).

A trace of *alkaloid* has been detected in the leaves of Nigerian material (2).

References:

1. Adam, 1966, a. 2. Adegoke & al., 1968. 3. Egunjobi, 1969. 4. Irvine 525, K. 5. Irvine, 1930: 4. 6. MacGregor 255, K. 7. Oliver, 1960: 17, 42. 8. Ferry & al., 1974: no. 89. 9. Gill & Ene, 1978.

Acalypha crenata Hochst.

FWTA, ed. 2, 1: 410.

An annual herb to 50 cm high, recorded only from N Nigeria in the Region, but occurring across Africa to Sudan, Ethiopia, Somalia and Uganda.

It is everywhere a common weed of cultivated land in eastern Africa. In Sudan it provides some browsing for goats and sheep (1).

Reference:

1. Carr 820, K.

Acalypha hispida Burm. f.

FWTA, ed. 2, 1: 408.

A shrub, native of Malesia, and widely introduced to other tropical countries. It is grown in the Region as an ornamental for its decorative red catkins, and as a hedge.

It has medicinal uses in SE Asia. In Indonesia a root and flower-decoction is taken for blood-spitting, and the leaves are applied in poultices for leprosy. In Malaya a decoction of leaves and flowers is taken internally as a laxative and diuretic for gonorrhoea. The bark appears to act as an expectorant, and so to relieve asthma (1).

Reference:

1. Burkill, IH, 1935: 24.

Acalypha indica Linn.

English: Indian acalypha (Oliver).

An annual herb to about 80 cm high, of waste places and fields, recorded from S Nigeria, and probably occurring elsewhere in West Africa since it has anthropogenic tendencies towards settled areas. It is widespread throughout other parts of tropical Africa and across Asia to Polynesia.

In NE Africa the plant is grazed by sheep and goats (Sudan: 3; Ethiopia: 1).

The plant was at one time official in the *British Pharmacopoeia*. It has numerous medicinal uses in India and is official in the *Pharmacopoeia of India* as an expectorant. It has expectorant, emetic and laxative uses in the Philippines, the expressed leaf-sap being held to be a safe, certain and speedy emetic for children. In cases of constipation an anal suppository of the bruised leaves will at once relax the constricted *sphincter ani* muscle to give relief. Leaf-sap with lime-juice is applied to painful rheumatic affections (4).

Indian material contains the alkaloid *acalyphine*, an expectorant and emetic substance used to substitute ipecacuanha, which is also vermifugal. A resin and an essential oil are also present (2).

References:

1. Carr 898, K. 2. Oliver, 1960: 4, 43. 3. Peers KM3, K. 4. Quisumbing, 1951: 490–1.

Acalypha lanceolata Willd.

An herbaceous annual, widely dispersed in eastern Africa, now recorded in Senegal.

The pounded and powdered leaves are made into an ointment with castor-oil in Tanganyika for rubbing onto scabies (1).

Reference:

1. Haerdi, 1964: 91, as *A. glomerata* Hutch.

Acalypha ornata Hochst.

FWTA, ed. 2, 1: 409, incl. *A. nigritiana* Müll.-Arg.
West African: **NIGERIA** YORUBA ẹrà (AJC)

A shrub to 3 m high by streams and in open places of forest zone of N and S Nigeria and W Cameroons, and widespread across tropical Africa.

The stems are woven into baskets and fish-traps in Tanganyika (6) and Ubangi (10). Leaves and roots of Nigerian material have been found to show slight molluscicidal activity against the fresh-water snail, *Bulinus globulus* (11). The leaves are covered by soft sticky hairs and the bracts similarly covered together forming a cup perhaps act as insect-traps (2). The foliage is browsed by elephants in Kenya (5) and by stock in Tanganyika where the plant is thought to be non-toxic (9). It does, however, have medicinal applications in Central Africa. The cooked leaf is taken to relieve post-partum pains, and a root-decoction as a laxative (9). The powdered leaf together with the powdered flower-stem of *Psorospermum febrifugum* Spach, var. *ferrugineum* Keay & Milne-Redhead (Guttiferae) is used in Tanganyika as a healing application to circumcision wounds (3). Water in which leaves have been soaked is also used in Tanganyika to wash scabies on children (7), the root for leprosy (9), and the plant (part unspecified) in a medicine for infections of the umbilicus of new-born babies (8). In Ubangi a leaf-decoction is used in a hip-bath for piles, and a root-decoction is also drunk (10). The leaves in S Nigeria are compounded with the leaves of other drug-plants into a draught for children with rabies (1). This is perhaps an extension of a widespread belief found in Tanganyika of magical powers residing in the plant's roots (4, 9).

Plant-ash rubbed onto the chest for pain is an Ubangi treatment (10).

References:

1. Carpenter 1173, UCI. 2. Graham C.589, K. 3. Haerdi, 1964: 91. 4. Koritschoner 932, 1174, 1175, K. 5. Magogo & Glover 517, K. 6. Tanner 512, 1122, K. 7. Tanner 3491, K. 8. Tanner 5723, K. 9. Watt & Breyer-Brandwijk, 1962: 395. 10. Vergiat, 1970: 67. 11. Adewunmi & Sofowora, 1980.

Acalypha racemosa Wall.

FWTA, ed. 2, 1: 409. UPWTA, ed. 1, 135, as *A. paniculata* Miq.
West African: NIGERIA IGBO (Agukwu) odukwe (NWT; JMD) YORUBA ìlẹ̀wù (Dawodu; Verger)

A perennial herbaceous shrub to 3 m high occurring in forested country from Ivory Coast to W Cameroons and Fernando Po, and widespread across tropical Africa and Asia.

The plant with terminal red female inflorescence is possibly of some ornamental value.

No use of the plant is recorded in the Region. In E Cameroun, Sanguélima Prefecture, the plant has undefined medicinal use (1).

Reference:

1. Ndoumou 7, K.

Acalypha segetalis Müll.-Arg.

FWTA, ed. 2, 1: 410.
West African: GUINEA BASARI ε-syóɓò (FG&G)

An annual herb to 50 cm high, recorded widely from Senegal to W Cameroons, and occurring widespread across tropical Africa.

It is a weed of cultivation on the Vogel Peak massif of N Nigeria (2). Tenda people eat the leaves in sauces (1).

References:

1. Ferry & al., 1974: No. 90. 2. Hepper, 1965.

Acalypha villicaulis Hochst.

FWTA, ed. 2, 1: 409, as *A. senegalensis* Pax & K Hoffm., and *A. senensis* Klotzsch.
West African: SENEGAL BEDIK ga-ndɔrɔ (FG&G) GUINEA BASARI yámbɛrusyùsy (FG&G)

An herb to 1 m high from perennial woody base recorded from Senegal, Mali, Guinea-Bissau and N Nigeria, and occurring widespread across tropical and southern Africa.

The Tenda of Senegal lay chopped up leaves over a clean wound to hasten healing (1).
They soak the roots in water which is then used to bathe a suckling child that is constipated; the water may also be drunk (1). In Tanganyika a root-decoction is drunk as a cough-medicine (2).

References as *A. senensis*:

1. Ferry & al., 1974: No. 91. 2. Haerdi, 1964: 91.

Acalypha wilkesiana Müll.-Arg.

FWTA, ed. 2, 1: 408.

West African: **NIGERIA** YORUBA (Ijebu) aworoso (IFE)

A lax shrub to 5 m high, native of tropical Asia, and introduced to many tropical countries. It is grown in the Region, usually as a variegated-leaf cultivar, as an ornamental for its red catkins and leaf form. It makes a lax, tallish hedge or screen.

In Trinidad a leaf-poultice is deemed good for headache, swellings and colds. The leaf-extract is active against Gram + ve bacteria (1).

Reference:

1. Wong, 1976: 127.

Acalypha sp. indet.

FWTA, ed. 2, 1: 409, as *A.'sp.A'*.

West African: **SIERRA LEONE** MENDE gɛndui (NWT) TEMNE ɛ-kiŋkiŋ (NWT)

A plant recorded only rarely from Guinea and Sierra Leone.

Alchornea cordifolia (Schum. & Thonn.) Müll.-Arg.

FWTA, ed. 2, 1: 403. UPWTA, ed. 1, 135.

English: Christmas bush (Sierra Leone: Scott Elliot, Deighton; Ghana: Irvine; Nigeria: Oliver); Xmas tree (Sierra Leone: Kirk).

West African: **SENEGAL** BALANTA bulor (JB) bulora (JB) fiili (JB) pos (JB) BANYUN birénen (K&A) birérèn (JB) diakint (K&A; JB) si hanar (JB) si kãntã (JB) BASARI a-min (K&A; JB) ·CASAMANCE· bugôn (A.Chev.; K&A) bulora (auctt.) CRIOULO arcu (JB) brusus (JB) po de arco (JB) FULA-PULAAR bu dioy (JB) fu layab (JB) fu lé (JB) fu sub (JB) DIOLA (Bayot) o tépéré (JB) DIOLA (Diatok) busub (K&A) DIOLA (Diembéreng) nusuba (K&A) DIOLA (Pointe) budoy (K&A) DIOLA (Tentouck) fusubö (K&A) DIOLA-FLUP bu dioy (JB) bu sobam (JB) bu subam (JB) FULA-PULAAR (Senegal) bulora (Aub. fide K&A) diébanèdi (JB) garkasak (JB) ngimii (JB; K&A) holâta (Aub. fide K&A) lahédii (JB) lahéhi (JB) TUKULOR gargasaki (JB) n-gimii (K&A; JB) MANDING-BAMBARA ko-gira (JB) MANDINKA hira (auctt.) ioro (K&A) ira (K&A) yira (JB) yoro (JB) MANINKA budié (JB) diimo (JB) ko gira ko: tree (auctt.) ko ira ko: tree (Aub. fide K&A) ·SOCE· hira (K&A) inã (JB) ioro (K&A) ira (K&A; JB) MANDYAK bu gonga (JB) bu luru (JB) bugu (JB) MANKANYA be lora (JB) ·SAFEN· ndãgndãg (JB) SERER buda (JB) yira (JB) WOLOF lah (JB; K&A) **THE GAMBIA** FULA-PULAAR (The Gambia) garagasaki (DRR) MANDING-MANDINKA gulekemo (DRR) hira (DAP) jambokano (Fox) kabanunkano (DAP; DRR) surugba (DRR) WOLOF hira (DAP) **GUINEA-BISSAU** BALANTA bulóre (JDES) fiili (JDES) BIDYOGO ensúmbè (JDES) CRIOULO arco (pau d'-, pó d'-) (auctt.) DIOLA-FLUP bussobâme (DF; JDES) bussubâme (auctt.) FULA-PULAAR (Guinea-Bissau) charque (DF; JDES) djebanedje (pl.) (EPdS; JDES) garcassaque (JDES) gargassáqui (D'O) MANDING-MANDINKA irá (JDES) MANDYAK bugou (JDES) MANKANYA bugou (JDES) PEPEL ugonga (JDES) **GUINEA** BASARI a-min (FG&G) KISSI cholei (RS) tieulo (RS) KONO plêna (RS) KPELLE plêna (RS) SUSU bolonta (CHOP; RS) **SIERRA LEONE** BULOM (Sherbro) chok-lɛ (FCD) cɔkɔ (Pichl) GOLA njigi (FCD) njiri (FCD) KISSI cholê (FCD) KONO dui (FCD) lui (FCD) KORANKO yisamfire (NWT) KRIO krismas-tik i.e. *Christmas stick* (FCD) LIMBA budonɛbuati (NWT) ndonia (NWT) LOKO n-jeko (NWT; FCD) MANDING-MANDINKA yisaŋ (FCD) MENDE n-jekɔ (*indef.*) (NWT; FCD) sugui

25

EUPHORBIACEAE

(NWT) tekei (DS) susu bolontai (NWT; FCD) boloŋwuri (NWT) TEMNE an-thoma (NWT: FCD) VAI jamba-jamba (FCD) YALUNKA gɛrarasɛkhɛ-na (FCD) girakhasɛkhi-na (FCD) LIBERIA KRU-BASA blukoh (C&R) buka (Huntting) MANO flandê (RS) flanlah (Har.; RS) MALI MANDING-MANINKA ko guira ko: tree (A.Chev.; RS) hira (A.Chev.; RS) UPPER VOLTA MANDING-DYULA kotiâ (K&B) IVORY COAST ABE vidjo (A&AA) ABURE adièké (B&D) AKAN-ASANTE adiamba (K&B) diéka (B&D) yama (K&B) BRONG adiama (K&B) AKYE vigo (Aub.) 'ANDO' djèka Visser ngata koffi bodua Visser tango ahirè = cough medicine (Visser) ANYI diéca (K&B) BAULE diéca (K&B) DAN fon (K&B) FULA-FULFULDE (Ivory Coast) bulora (Aub.) holanta (Aub.) KRU-BETE burunéï (K&B) GREBO (Oubi) poro (A&AA) GUERE po (K&B; RS) poho (A&AA) polo (K&B) pro (K&B) GUERE (Chiehn) buru (K&B; B&D) buru kuè (B&D) KULANGO adiama (K&B) KWENI félémé (K&B) flinflin (B&D) KYAMA akuebégo (B&D) MANDING-DYULA kotiâ (K&B) MANINKA diangba (B&D) gramba (B&D) kodjiram (A&AA) koguira (Aub.) koira (Aub.) koya (B&D) yangba (B&D) 'SOCE' bugong (Aub.) 'NÉKÉDIÉ' bolu (B&D) bru (B&D) SENUFO-TAGWANA fémé (K&B) GHANA ADANGME gbloo (ASThomas fide FRI) gboo (FRI) ADANGME-KROBO bɔblɔ (JMD) gbɔgblɔ (JMD) AKAN-ASANTE agyama from agya: leaves; mma: children (auctt.) BRONG gyama (E&A) FANTE ogyamma (FRI) TWI a,ogyama (auctt.) ANYI-ANUFO gyaka (FRI) SEHWI gyaka (FRI) GBE-VHE ahame (OA) ayraba (JMD) VHE (Awlan) alas (JMD) VHE (Pecí) agyama (Bunting fide FRI) ahame (FRI) avovlɔ (FRI) ayraba from Twi: a-gyama (FRI) NZEMA gyeka (FRI; OA) TOGO NAWDM holentschess (Gaisser) TEM (Tschaudjo) tschifu (Volkens) tschufu (Volkens) YORUBA-IFE (TOGO) avovlɔ (Volkens) NIGERIA EBIRA-ETUNO oje (JMD) ose (JMD) EDO uwɔnmwɛ general for 'sauce' or 'stew' (auctt.) EFIK mbom (JMD) GWARI tahi (JMD) HAUSA bambami (auctt.) bambani (ZOG) bombana (ZOG) IGBO ubebe (Singha) ububo (NWT; JMD) IJO-IZON ipáín (KW) IZON (Kolokuma) ipáín (KW; KW&T) izɔ́n ípáín = Ijo's ípáín (KW&T) IZON (Oporoma) ɛpáín (KW) YEKHEE ukpaoromi (JMD) YORUBA èpa (auctt.) esin (JMD; IFE) ipa (auctt.) ipa-esin (JRA) WEST CAMEROONS DUALA dìbobongì (AHU; Ithmann) KPE bondji (JMD) dibobunjì (Waldau)

A sprawling, much-branched scandent shrub or tree to about 8 m high, of forest margins and secondary growth, often near the sea and fresh-water, widely dispersed throughout all countries of the Region, and widespread across tropical Africa.

The plant is endowed with many superstitious attributes. It is in flower and in fruit in Sierra Leone in December and is thus called 'Christmas Bush'; it is used in decorations. In forest regions the full ripening of the fruit is a sign that the rainy season is over (10). The wood is light, soft and perishable and is not normally used (9), but in The Gambia if large sizes are available benches are made of it (21). It is also used for sundry construction-work by Andos in Ivory Coast (28).

The leaves are not eaten, but they are freely used in medicine. They are taken in decoctions and in baths in Ivory Coast–Upper Volta as a sedative and anti-spasmodic, for epilepsy, headaches, cough, sore-throat and bronchial infections, and in eye-washes for conjunctivitis. The decoction has an action on the digestive tract and is prescribed for abdominal complaints such as dysentery, female sterility and as an emmenagogue and for venereal diseases. It is sometimes used as a febrifuge (2, 8, 19, 28). The leaves are administered against tachycardia in Senegal (18) with a related effect of dilation of the pupils (29). The Ijo of S Nigeria chew the leaves as an appetiser (27). The leaves are deemed to be cicatrisant and anodynal. Young leaves ground up with kaolin and mixed and strained are given in Ghana for infantile diarrhoea; the preparation must be made daily (30). In cases of dysentery the patient may squat in a bath of a diluted decoction of dried leaves of A. cordifolia and Ficus sur (Moraceae) for a quarter of an hour (30). A decoction of leafy twigs is applied as a wash for feverish chills, and rheumatic pains, also for sores and as an application to sore feet as a lotion or poultice (9–11, 22). A leaf-decoction is used as a mouth-wash for toothache in Nigeria (23) and in Gabon (26). In Congo (Brazzaville) it is used for dental caries (6) and the sap is applied topically to sores, ulcers and skin-infections. The pulverised dry leaf is also applied to ulcers and yaws, and, after previous treatment by baths or fumigation, to chancre. Powdered leaf and bark are similarly used in Sierra Leone, and the sap of leaves and fruit-juice is rubbed on the skin for ringworm (10). Powdered leaves are used on skin-infections generally in Ivory Coast and healing is said to be rapid (19). The pulped root is taken for blennorrhoea in Ivory Coast–Upper Volta (19) and for gonorrhoea in Nigeria (3) or in decoction as a wash for urethral discharges, and in a cold infusion of twigs and leaves with bark and root of *Mitragyna inermis* (Rubiaceae), natron and often with fresh lime-juice, by both men and women for the same purpose for vaginitis and metritis (6). A simple cold infusion of dried and crushed leaves is also used and is believed to be diuretic (10). The leaves are also held to be purgative. In S Nigeria they are boiled with castor-oil seeds for this effect (24). An infusion or decoction of leafy shoots, with lime-juice, is supposed to act similarly. In Ghana such a preparation is given by enema (10). In Gabon a leaf-decoction is an emetic (25, 26). In Ivory Coast the leaves are used as a haemostatic to stem prolonged menstruation, and a decoction of roots or leaves is given by vaginal application for post-partum haemorrhage (2). Young leaves are made into an haemorrhoidal suppositary in Congo (Brazzaville) (6), and powdered leaf and bark are applied to piles in Sierra Leone and are made into a wound dressing in Akwapim, Ghana, known as *dudo* (12).

Twigs are used in Ivory Coast as chew-sticks against toothache (8) and dental caries (28). The stem-pith is bitter and astringent. Young stems from which the leaves and cortex have been removed are commonly chewed, often with salt or natron, as a remedy for cough and sometimes for diarrhoea. The pith may also be rubbed on the chest (10). In Senegal it is eaten for chest-complaints and bronchitis, and if a cough is accompanied by bloody sputum, a leaf-macerate is drunk in addition, and occasionally a leaf-decoction is taken for one week as a drink and in a bath as a tonic (13, 16–18). The pith is chewed in Liberia to relieve sore-throat (9). The pith is also recognised as beneficial in Congo (Brazzaville) for various bronchial affections, such as whooping cough, cough, flu, bronchitis, etc. (6).

In Ghana root-bark with a little ginger in water and roasted corn powder is cooked with food for dysentery, or a decoction of the root with *Piper guineense* (Piperaceae), *Xylopia sp.* (Annonaceae), *Aframomum sp.* (Zingiberaceae) and natron (*kawu*) is taken (30).

The root-pith is made into a lotion, or is chewed, for thrush and mouth ulceration, or is placed hot on a carious tooth. Along with young leaves, peppers and white clay, the root-pith is compounded into an enema said to check abortion (10). Though, by contrast, the leaves are eaten in Gabon as an abortifacient (25, 26), and in Congo (Brazzaville) they are given as an emmena-gogue and to facilitate delivery (6). The roots are put into various Ivorean remedies for jaundice, leprosy and snake-bite (19), and leaves and root-bark are applied externally as a venom-antidote (15, 18) in Senegal.

The fruit is acidulous and is supposed to have laxative properties (10). A decoction of bruised fruit is taken in Nigeria to prevent miscarriage (3). The fruit is considered edible in Gabon (26). It is much favoured by birds (Camer-oun: 4; Gabon: 26), and is used as bait for trapping small birds in Gabon (26). In Nigeria the berry is used for eye and skin-diseases, and in USA it is a component of marine anti-foulings for application to metal surfaces (10, 20).

Various parts of the plant yield a black dye. Commonly the article to be dyed, e.g., matting and fabrics, is boiled with the fruit to which are added (in Nigeria) natron and the fermented extract either of *Parkia* pods (Leguminosae: Mimo-soideae) or of bark of *Bridelia ferruginea* (Euphorbiaceae). This dye is also used on pottery, bowls, calabashes and leather. In Cameroun and southwards the dye is prepared along with a decoction of *Mucuna flagellipes* (Leguminosae: Papilio-noideae), and in some cases the plant is steeped in the black mud of the forest or of streams (10). In S Nigeria only the leaves and petioles are used for dyeing and preserving fishing nets. The material is used either fresh or dry and it is said that dry leaves produce a darker colour (5). In Gabon bark and leaves are used to blacken cloth and pottery (26). The Ijo of the Niger Delta obtain an indelible dye by boiling leaves and fruit (27). The wood-ash is used in Guinea as an adjuvant in indigo dyeing (Portères fide 12).

Nigerian material screened for alkaloids showed none in the bark and but a trace in the roots (1, 20). Ivorean, Guinean and Congo (Brazzaville) materials are recorded as having traces in the root and stems (8, 18). Alkaloid of the *yohimbine* type (18) or *indole* type (29) seems to be present. *Saponins* and *tannin* are present in leaves and bark (7). A bitter principle, *alchorin*, is recorded as present in the leaf of Nigerian material (20), and leaves with petioles attached have about 10% tannin, while bark has about 11% (5). The tannin content is considered too low to raise any commercial interest, though in treating fishing-nets it probably contributes to their protection.

The leaf is used in West Africa for packing kola nuts and pipe-stems are made from the branches with the pith removed (10).

References:

1. Adegoke & al., 1968. 2. Adjanohoun & Aké Assi, 1972: 118–9. 3. Ainslie, 1937: sp. no. 18. 4. Bates 69, K. 5. Bennett, 1950: 132–4. 6. Bouquet, 1969: 108. 7. Bouquet, 1972: 22. 8. Bouquet & Debray, 1974: 82. 9. Cooper & Record, 1931: 50. 10. Dalziel, 1937: 135. 11. Deighton 2115, K. 12. Irvine, 1961: 211–2. 13. Kerharo & Adam, 1962. 14. Kerharo & Adam, 1964,b: 404–5. 15. Kerharo & Adam, 1963,a. 16. Kerharo & Adam, 1964,a: 434. 17. Kerharo & Adam, 1964,c: 291– 2. 18. Kerharo & Adam, 1974: 402–4, with phytochemistry and pharmacology. 19. Kerharo & Bouquet, 1950: 67–68. 20. Oliver, 1960: 18, 45. 21. Rosevear, 1961. 22. Scott Elliot 4197, K. 23. Thomas NWT.1878 (Nig. Ser.), K. 24. Thomas NWT.2305 (Nig. Ser.), K. 25. Walker, 1953,a: 34. 26. Walker & Sillans, 1961: 160. 27. Williamson, K. 45, UCI. 28. Visser, 1975: 47. 29. Sandberg & Cronlund, 1982: 190. 30. Ampofo, 1983: 22, 24. See also: Oliver-Bever, 1983: 56.

Alchornea floribunda Müll.-Arg.

FWTA, ed. 2, 1: 403. UPWTA, ed. 1, 136.

West African: **LIBERIA** MANO kai (Har.; JMD) **MALI** MANDING-MANINKA sumara fida (A.Chev.) **NIGERIA** EDO aramamila (?) (Dennett)

A leaning semi-scandent shrub or tree to 10 m high, of the undergrowth of the closed-forest, from Mali to W Cameroons and Fernando Po, and widespread across Africa to Sudan and Uganda.

Sap from the leaves and roots is applied in Congo (Brazzaville) to skin-affections and to circumcision wounds, and the leaves are eaten as a vegetable with meat or fish as an antidote to poison (natural or superstitious) (1).

The root has properties similar to those of Indian hemp *Cannabis sativa*, Cannabaceae, q.v.). There appears to be no use of the plant recorded for the West African region in this manner. In Zaïre the plant is well-known under the name *niando* to cause a sort of intoxication, sometimes mild, sometimes violent, and is used to induce a temporary excitement or vigour, physical or verbal, and for an aphrodisiac effect (3). It is also used in Gabon as an aphrodisiac and stimulant before fetish rites, sometimes admixed with *Tabernanthe iboga* (Apocynaceae) (8, 9). The plant thus has magical and superstitious uses. In Congo (Brazzaville) leaves are crushed on the fetish in the initiation of novices, and on setting out on a journey or entering a strange house leaves are crushed to ensure favourable circumstances (1). It is sometimes treated as a hunter's fetish (3). To increase physical activity, the roots are powdered and added to food, or are macerated for several days in palm-wine or banana-wine, and the liquid is drunk. Energy and excitement are greatly stimulated for festive dancing or acts of war, often leading to abandoned excesses. Root-parings dried in the sun, are mixed with salt and chewed for similar effect. The physiological response is first excitation, then depression according to dosage, individual temperament, habit, etc., and it may be fatal. The effect is considered to be as harmful as that of cannabis. The plant extract has a sympathicostenic action and increases the sensitivity of the sympathetic nervous system towards adrenalin. In test dogs small dosage produces hypotension followed by hypertension; in large dosage an increase of the blood pressure is followed by a strong decrease with only gradual recovery. Alkaloids are present in the roots and seeds. There are conflicting reports of the presence and absence of *yohimbine*, but chemotaxonomically its presence is not expected (1–7).

The only W African uses are in Ivory Coast where the roots are occasionally used as tooth-picks and an aphrodisiac (2) and in Nigeria for treatment of ophthalmia (6). The fruit of Nigerian plants tested for molluscicidal action on the fresh-water snail, *Bulinus globulus*, was found quite inert (10).

References:

1. Bouquet, 1969: 109. 2. Bouquet & Debray, 1974: 82. 3. Dalziel, 1937: 403. 4. Kerharo & Adam, 1974: 404. 5. Kerharo & Bouquet, 1950: 68. 6. Oliver, 1960: 45, 68. 7. Tyler, 1966. 8. Walker, 1953,a: 34. 9. Walker & Sillans, 1961: 160. 10. Adewunmi & Sofowora, 1980. See also: Oliver-Bever 1983: 56.

Alchornea hirtella Benth.

FWTA, ed. 2, 1: 403. UPWTA, ed. 1, 136.

West African: **SENEGAL** DIOLA-FLUP ba diënk (JB) MANDYAK be tira (JB) **SIERRA LEONE** KISSI koli-*g*boinda (FCD) LIMBA bwujanka (NWT) tuɛgbobo (NWT) LOKO ɛnyahɔ (NWT) suwɔrɔgɔ (NWT) tibi (NWT) tivi (NWT) MENDE tokeŋge *abbreviation of* tolokeŋge (FCD) tolo-geŋge (FCD) tolo-keŋge (FCD) tɔɔgingi (NWT) tukiŋgi (NWT) SUSU fibiaiɛmbe (NWT) kuliwuri (NWT) TEMNE tola-tamis = *spider's web kola* (JMD) ɛ-yɔla-ɛ-a-nes (FCD) **MALI** MANDING-MANINKA *ko*-fidaba *ko*: tree (A.Chev.)

A tree to 10 m high, under-shrub or lower storey of the closed-forest from Senegal to Ivory Coast, and in E Cameroun to Tanganyika and Angola. Two forms are present in W Africa: f. *hirtella*, and f. *glabrata* (Müll.-Arg.) Pax & Hoffm., distinguishable on the depth of pubescence. No distinction is drawn in respect of usage.

The plant has some analgesic uses. The root in decoction is taken in Ivory Coast as a sedative for pain, and the sap is applied topically (2). In Sierra Leone the plant (part not stated) is used for toothache (4), and the bark is scraped and ground with white 'lime' for application to craw-craw (3). In Congo (Brazzaville) leaf-sap is inhaled for headache or is applied over scarifications on the temples (1).

The roots in decoction are taken by draught for stomach-ache and as a purgative in Ivory Coast (2).

A small quantity of alkaloid has been reported present in the stem and root barks, including *yohimbine* (2) – see *A. floribunda* Müll.-Arg.

References:

1. Bouquet, 1969: 109. 2. Bouquet & Debray, 1974: 82. 3. Deighton 2131, K. 4. Thomas NWT.30, K.

Alchornea laxiflora (Benth.) Pax & K Hoffm.

FWTA, ed. 2, 1: 403. UPWTA, ed. 1, 148.

West African: **NIGERIA** EDO uwenuwen (AHU) IGBO ububo (BNO) URHOBO urievwu (Isawumi) YORUBA ijàn (JMD; Verger) ijàn funfun (Millson; Verger) ijàndú (Verger) pèpè (Isawumi; Verger) **WEST CAMEROONS** KPE longoso (AHU) KUNDU longoso (AHU)

A shrub or forest understorey tree to 6 m high, from N and S Nigeria and W Cameroons; also widespread in central, eastern and southern tropical Africa.

The leaves are recorded as amongst those used to preserve the moisture of kola nuts in packing (1).

The stems, and especially the branchlets, are used in Nigeria as chew-sticks (2).

29

The plant enters into a Yoruba incantation to make 'bad medicine' rebound on the sender (3).

References:

1. Dalziel, 1937: 148 2. Isawumi, 1978,a: 52/54. 3. Verger, 1967: no. 112; 1986: pers. comm. 9/12/86.

Aleurites moluccana (Linn.) Willd.

FWTA, ed. 2, 1: 368.

English: candle nut tree; Indian walnut.

A large tree of Malesia, now widely cultivated throughout the tropics.

The wood is light, not durable and of little value. Firewood and matchsticks seem to be the main uses in SE Asia. The seeds are used a little as food in Asia. The oil is extracted and the kernel residue is also eaten but only after roasting to destroy poisonous substances. The oil is a drying oil with an iodine value 140–146, greatly inferior to linseed oil. Component substances are *glycerides* of *linolenic, linolic* and *oleic acids*. The oil is usable for soap, paints, illuminations, etc.

Amanoa bracteosa Planch.

FWTA, ed. 2, 1: 371. UPWTA, ed. 1, 136.

West African: **SIERRA LEONE** MENDE *nj*agbone(-i) (auctt.) **LIBERIA** DAN hoto (JGA) **IVORY COAST** DAN hauto (Aub.) ·KRU· héré (Aub.)

A tree to near 20 m tall, and trunk to 1.70 m girth, but occasionally to 5 m (2), of the understorey of rain-forest of Sierra Leone and Liberia.

The wood is a uniform dark brown, hard and heavy, and without recorded use (2).

The seeds are eaten by antelope in the Nimba mountain area (1).

References:

1. Adam, 1971: 373. 2. Savill & Fox, 1967: 111.

Anthostema aubryanum Baill.

FWTA, ed. 2, 1: 416.

West African: **IVORY COAST** ABE mauli (A.Chev.) meuli (Aub.) ANYI sessé (Aub.) KYAMA anaya (B&D) **GHANA** VULGAR kyirikusa (DF) AHANTA kyirikasa = *hates talking* (E&A) kyrikesah (Port Develop. Synd.) ANYI sese (FRI) ANYI-AOWIN kyirikesa (DF) NZEMA kyirikesa (FRI; JMD) **NIGERIA** EDO uhruareze (auctt.) YORUBA òdògbo (auctt.)

A tree to 26 m high with a straight bole to 16 m long by 1.60 m girth, in freshwater swamp-forest and on river-banks, from Ivory Coast to S Nigeria, and in Gabon, Cabinda and Príncipe.

The wood is white and soft (2). It is a good firewood (9).

Latex is present in all parts of the tree, and is most abundant in the bark. It is very caustic and may cause blindness (7, 8, 9), though this may be temporary (1). Ophthalmia caused by latex is said to be treatable by instillation of human milk (6, 8). The latex is a drastic purgative, two or three drips being a dose for a large person in Gabon (8, 9) and to 3–9 drips according to weight of the patient and his resistance in Congo (Brazzaville) (3). It is not prudent to give to pregnant women, nor to children (3). Toxicity appears to be confined to the fresh state (5), but it is taken in a coagulated state in Congo (Brazzaville) at a 'thumbnail piece' with tapioca or ripe banana for a dose (3). The fresh latex is used to purge and to obtain diuresis in generalised oedemas in Ivory Coast (5).

In Benin (Nigeria) the latex is applied to sores (7).
The seeds are also purgative (8, 9).
The bark is used in Gabon as a fish-poison (9).
Tannin has been reputed to be present in the leaves but no other active principle has been recorded (4).

References:

1. Ainslie, 1937: sp. no. 35. 2. Aubréville, 1959: 2: 30. 3. Bouquet, 1969: 110. 4. Bouquet, 1972: 23. 5. Bouquet & Debray, 1974: 83. 6. Irvine, 1961: 212–3. 7. Kennedy 1737, K. 8. Walker, 1953,a: 34. 9. Walker & Sillans, 1961: 161.

Anthostema senegalense A. Juss.

FWTA, ed. 2, 1: 416. UPWTA, ed. 1, 136.
West African: SENEGAL BADYARA mambéré (JB) papodé (JB) BALANTA brasi (K&A) brozi (JB) kurozi (JB) suntz (K&A) BANYUN kiffiń (K&A) si fing (JB) si fis (JB) si laka fe tiir (JB) BASSARI a pè (JB) CRIOULO lete (po de l.) (JB) liti (po de l.) (JB) DIOLA bu mpétèñ (JB) bu pemba (JB) bubu-kunap (Aub. fide K&A) bufétèñ (JB) ka furfurak (JB) DIOLA (Bayot) du le (JB) DIOLA (Fogny) bu fenbân (K&A) bu pembó (K&A) funôn (K&A) DIOLA-FLUP bu fétèñ (JB) FULA-PULAAR (Senegal) bu fena (K&A) bu féna (JB) TUKULOR buro (JB) KONYAGI a-pè (K&A; FB) MANDING-BAMBARA fama (K&A; FB) fama dô (JMD fide K&A) fufurlâ (K&A) kofama (K&A) mano (K&A) MANDINKA fufurlâ (K&A) fufurlan (JB) fufurondom (JB) MANINKA ko fama ko: tree (auctt.) fama diẽ (JB) fama dô (JMD fide K&A) ko honna ko: tree (JB) mano (auctt.) MANKANYA be tafé (JB) be tasi (JB) 'SAFEN' godan (JB) WOLOF kindin (K&A; JB) kindir (JB) roh (JB) THE GAMBIA FULA-PULAAR (The Gambia) kukuyi DRR MANDING-MANDINKA fama DAP manayiro (DRR) GUINEA-BISSAU BIDYOGO cabete (JDES) CRIOULO pó de lete (JDES) FULA-PULAAR (Guinea-Bissau) bufena (EPdS; JDES) PEPEL minhále (JDES) tagi (DF fide JDES) GUINEA FULA-PULAAR (Guinea) mburo kiangole (Langdale-Brown) MANDING-BAMBARA fama (CHOP fide JMD) fama dion (CHOP fide JMD) ko fama (CHOP fide JMD) MANDINKA fama (CHOP fide JMD) fama dion (CHOP fide JMD) ko fama (CHOP fide JMD) SIERRA LEONE BULOM (Kim) sɛmgoɛ (FCD) BULOM (Sherbro) sɛm-dɛ (FCD; S&F) KISSI susianchindo (FCD; S&F) KORANKO famɛ (S&F) LOKO ndɛmbi (NWT) MENDE g-boni, g-buni general for Euphorbiaceae with viscid latex (FCD) n-debui (Macdonald) nyɛbu(-i) from nyɛ: fish; bu: under (auctt.) sɛma (def. sɛmi) (FCD; S&F) SUSU waniŋbeli (NWT) waniŋyi (FCD) TEMNE an-, ka-wan (auctt.) UPPER VOLTA MANDING-BAMBARA fama (JMD fide K&B) fama dion (JMD fide K&B) ko fama (JMD fide K&B) IVORY COAST DIOLA bubukunap (Aub.)

A shrub or tree of damp, swampy or flooded sites on river-banks, of the soudanian wooded savanna, coastal savanna or forest zone, occurring from Senegal to Ivory Coast.

In Sierra Leone and eastwards the plant grows into a large tree reaching 25 m high with a straight bole to 2 m girth (13). In Senegal and The Gambia its stature is appreciably less, and often is not taller than a shrub (9, 12). The wood is white in colour and light in weight. It sometimes is used for local building purposes. It works easily and is suitable for boxes and light carpentry (4). Poles are used in Sierra Leone for temporary fencing (1).

Latex is present in the bark, young shoots, leaves, flowers and fruit. It is very toxic, acrid and vesicant. It is harmful to mucous membranes and it is known to all races as being very dangerous to the eyes and capable of causing blindness (2, 4, 6, 8–10, 13, 14). It is used in Sierra Leone (3, 11, 13), as a bird-lime on account of its great tackiness. It is said to be similarly used in Senegal with the sap of Hibiscus cannabinus (Malvaceae) or allied species (4). In spite of its toxicity, it is held in high repute in Senegal, and with suitable precautions, is taken for expurgation and especially in those affections requiring treatment by strong action on the intestines, e.g., menstrual troubles, displacement of the afterbirth, leprosy, etc. Treatment may be effected by adding very small amounts of the latex to food, or of the pulped root. In Casamance (Senegal) diluted latex is given as an emetic in cases of poisoning, and a bark-macerate or infusion is used for all acute illnesses and to expel intestinal parasites. The young stems along other ingredients are used for neuralgia (6–9).

31

Leafy twigs and inflorescences are chopped up in the Casamance as a fish-poison (6, 9), and in rice-fields branches of leafy twigs are placed at water intake sluices to kill fish that eat the young rice plants (10). Conflicting results on insecticidal activity are reported (5).

Young leaves are ground up with flour in Sierra Leone; the paste is dried and this is taken as a laxative (3).

References:

1. Boboh, 1974. 2. Coll. omn, K. 3. Deighton 1342, K. 4. Dalziel, 1937: 136. 5. Heal & al., 1950: 118. 6. Kerharo & Adam, 1962. 7. Kerharo & Adam, 1963,a. 8. Kerharo & Adam, 1964,b: 408. 9. Kerharo & Adam, 1974: 404–5. 10. Kerharo & Bouquet, 1950: 68. 11. Macdonald 6/1922, K. 12. Rosevear, 1961. 13. Savill & Fox, 1967: 111–2. 14. Voorhoeve, 1965: 94.

Antidesma laciniatum Müll.-Arg.

FWTA, ed. 2, 1: 374–5. UPWTA, ed. 1, 136.
West African: **SIERRA LEONE** KISSI yɔm-kefo (FCD) KONO konya-kɔnɛ (FCD) LIMBA bufas (? bupas) (NWT) MENDE dɛmi (NWT) joijane (Aylmer ex JMD) kafa-guli (Fox) katale (NWT) nyɛga (FCD) nyɛgala (FCD) nyɛŋgala (FCD) SUSU kundexanfi (NWT) TEMNE ɛ-sɛgba (NWT) **LIBERIA** MANO bu yidi (Har.) **IVORY COAST** ABE ehetti (Aub.) etti (Aub.) **GHANA** GBE-VHE kpɔplɔti (FRI) **NIGERIA** EDO ogbomaton (AHU; JMD) IGBO (Umuahia) mmo = devil (AJC)

Small trees to about 15 m high, of two varieties distinguishable on the pubescence beneath the leaf: var. *laciniatum* shortly pubescent; var. *membranaceum* densely pubescent with spreading hairs; the former of moist deciduous and secondary forest from Ivory Coast to Fernando Po, and the latter of the rain-forest from Guinea to Togo. Both varieties occur in E Cameroun and the Congo basin.

The wood of both is yellowish-white and hard. Sap-wood of var. *laciniatum* is a light creamy-brown (4). The timber is used in S Nigeria for burning (2).

The bark is used in decoction in Congo (Brazzaville) for internal upsets (1). A leaf-decoction is prepared as a bath in Liberia to prevent miscarriage (Harley fide 4), and the leaves of var. *membranaceum* are used in equatorial Africa for scenting plates on which meat is served (Sillans fide 4). The stem and root of Nigerian plants were examined for molluscicidal action on the fresh-water snail, *Bulinus globulus*, and were reported inert (5).

The fruits are said to be not edible (3).

References:

1. Bouquet, 1969: 109. 2. Carpenter 545, UCI. 3. Dalziel, 1937: 136. 4. Irvine, 1961: 213. 5. Adewunmi & Sofowora, 1980.

Antidesma membranaceum Müll.-Arg.

FWTA, ed. 2, 1: 375. UPWTA, ed. 1, 136.
West African: **SENEGAL** DIOLA-FLUP bu bun a vana (JB) **IVORY COAST** 'KRU' watu (K&B) **GHANA** AKAN-ASANTE numanuma-gyama (E&A) **DAHOMEY** YORUBA-NAGO yanya holo (A.Chev.)

A tree, occasionally to nearly 20 m high, of the savanna from Senegal to W Cameroons, and widely distributed across Africa to Angola and Mozambique.

The bark has a great reputation in Ivory Coast amongst the 'Kru' as an aphrodisiac (1, 4). Powdered bark is used as a dusting on wounds in Tanganyika (3).

In Liberia the Mano use a leaf-decoction in a bath to prevent abortion (Harley fide 2), and a superstitious use is to tie a fruit to a stick inserted in the ground when a plot is cleared for farming to ensure that the rice will grow well

(2). It is possible that these practices, however, refer to *A. oblongum* (Hutch.) Keay rather than to this species.

References:

1. Bouquet & Debray, 1974: 83. 2. Dalziel, 1937: 136. 3. Haerdi, 1964: 91. 4. Kerharo & Bouquet, 1950: 68.

Antidesma oblongum (Hutch.) Keay

FWTA, ed. 2, 1: 375.
West African: LIBERIA KRU-BASA du-ah-dor (C&R)

A shrub or small tree to 3 m tall, of the rain-forest of Liberia and Ivory Coast.

The wood finds no use in Liberia, but milky juice from the bark is mixed with clay and used as a cosmetic (1).

References:

1. Cooper & Record, 1931: 50–51, as *A. membranaceum* Müll.-Arg.

Antidesma venosum E Mey. ex Tul.

FWTA, ed. 2, 1: 375. UPWTA, ed. 1, 137.
West African: SENEGAL DIOLA bo sõnt (JB) DIOLA-FLUP kéri (JB) THE GAMBIA FULA-PULAAR (The Gambia) keri (DAP) MANDING-MANDINKA dafingo (auctt.) WOLOF dafingo (DAP) GUINEA FULA-PULAAR (Guinea) kéri (auctt.) keridiaulé (Aub.) SIERRA LEONE LIMBA bu'kɔi (NWT) LOKO yau uru (? yanuru) (NWT) MANDING-MANDINKA bala (NWT) MENDE wɔjɛ (NWT) yawi (NWT) MALI DOGON bausiri dama (Aub.) UPPER VOLTA SENUFO sangoloma (K&B) IVORY COAST MANDING-MANINKA iokwa (B&D) koagnon (B&D) SENUFO sangoloma (K&B) GHANA AKAN-ASANTE ɱpepea (FRI; E&A) TWI mpepea (FRI; JMD) DAHOMEY SOMBA tetanté (Aub.) NIGERIA HAUSA kirni (ZOG) IGBO ọkọ̀lọ̀tọ̀ = chew-stick (Clusters) YORUBA arorò (Kennedy; Verger)

A tree to 10 m high, of the savanna, widespread from Senegal to W Cameroons, and elsewhere in tropical and southern Africa.

The poles are used in hut-construction in Malawi (8), and the wood as fuel in Gabon (16) and Tanganyika (23) where also tool-handles are made of it (18). The twigs are cut up for chew-sticks in N Nigeria (8, 22).

The bark in Kenya (2) and in Malawi (19) is used to prepare a cough-medicine, and in Tanganyika powdered bark is a wound-dressing and macerate of root and bark are used to wash syphylitic and gonorrhoeal eruptions (7). The roots are used in Shari to treat toothache (15) and in Tanganyika for stomach-pains (12), schistosomiasis (13) and for venereal disease (7, 14). The root is cooked in porridge and eaten for scabies (7).

Leafy twigs are prepared as a decoction for use on itch in Ivory Coast–Upper Volta by the Senufo (9). The leaves, or leafy twigs, commonly enter into prescriptions for stomach troubles (West Africa: 6, 8; E Africa: 11, 20, 22; S Africa: 17). The leaf in poultice is applied to the head for headache in Nigeria (6, 8, 21). The Manika of Ivory Coast mix honey with a leaf-decoction to take for constipation (1, 6). Leaf-sap is taken with other drug plants for diarrhoea and amoebic dysentery (7). In E Africa the leaf is used for toothache (11). Unspecified parts of the plant are used in Ivory Coast for mange, furuncles and costal pains (5).

Leafy twigs in Tanganyika are tied to the mesh of fishing nets as a fish-poison (3).

The leaf may contain an *alkaloid* (17, 22), and analysis of material from Congo (Brazzaville) has shown the presence of *saponin*, and of *steroids* and *terpenes* in the roots, and no other active principle (4).

The fruit is sweet when fully ripe (6) and is widely eaten throughout Africa, but there are suggestions that it is not always innocuous. In S Africa (17), and in the Usambaras of Tanganyika it has been recorded as toxic (10) – points which seem to merit further examination. The ripe fruit has been used as fish-bait (17, 19).

References:

1. Adjanohoun & Aké Assi, 1972: 119. 2. Bally 4672, K. 3. Bally 7518, K. 4. Bouquet, 1972: 23. 5. Bouquet & Debray, 1974: 83. 6. Dalziel, 1937: 137. 7. Haerdi, 1964: 92. 8. Irvine, 1961: 214. 9. Kerharo & Bouquet, 1950: 68. 10. Koritschoner 1596, K. 11. Tanner 3419, K. 12. Tanner 3887, K. 13. Tanner 3918, K. 14. Tanner 3958, K. 15. Turner 107, K. 16. Walker & Sillans, 1961: 161–2. 17. Watt & Breyer-Brandwijk, 1962: 397. 18. Wigg 261, K. 19. Williamson, J., 1955 (? 1956): 19. 20. Bally, 1937: 10–26. 21. Thomas NWT.48 (Nig. Ser.) K. 22. Portères, 1974: 114. 23. Watkins 153, K.

Antidesma vogelianum Müll.-Arg.

FWTA, ed. 2, 1: 375.
West African: **NIGERIA** IJO-IZON (Kolokuma) ịngɔlɔngɔlọ (KW; KW&T)

A tree to 10 m high, of the rain-forest of S Nigeria and W Cameroons, and in E Cameroun, Gabon, Congo basin and E Africa.

The Ijo of S Nigeria cook up the root with or without maleguetta pepper as an aphrodisiac, and the stem when crushed is used as a lamp (1).

Reference:

1. Williamson, K. 18A, UCI.

Argomuellera macrophylla Pax

FWTA, ed. 2, 1: 405.
West African: **SIERRA LEONE** KISSI peingyɔ = *spiny leaf* (FCD) **IVORY COAST** BAULE blakassi blakassa (B&D) ˙NÉKÉDIÉ˙ béiro titi (B&D) **GHANA** AKAN-BRONG mprepre (CV; FRI)

A shrub to 4 m high, of the forest zone from Sierra Leone to W Cameroons, and widespread across Africa to E Africa and Mozambique.

The leaf-sap is taken in Ivory Coast as a purgative and emetic in the treatment of poisonings and for ascites. Powdered dried leaves are sometimes taken as an aphrodisiac (1).

Reference:

1. Bouquet & Debray, 1974: 83.

Breynia disticha JR & G Forst. **var. disticha f. nivosa** (Bull.) Croizat

FWTA, ed. 2, 1: 368, as *B. nivosa* (Bull.) Small.
English: snow bush; ice plant.

A shrub reaching 2 m high with green, pink and white mottled foliage. The white leaves suggest ice, hence the English name. Native of Polynesia, but widely dispersed throughout the tropics and common in towns and gardens in the Region.

It makes a decorative shrub, and low hedge, standing well to clipping.

Bridelia atroviridis Müll.-Arg.

FWTA, ed. 2, 1: 370. UPWTA, ed. 1, 137.
West African: SIERRA LEONE KORANKO fira-bembε (S&F) MENDE bɔbwε (NWT) kpamanda (NWT) TEMNE ε-ruŋa (NWT) a-traka (NWT) IVORY COAST AKAN-ASANTE akwakorobo (B&D) ekpaï (B&D) AKYE tchikué (Aub. ex K&B) GAGU goolékréla (B&D) krinkrin (K&B) KRU-GUERE kuégé (K&B) GUERE (Chiehn) fekwo B&D) KWENI possoné iri (K&B) possonon iri (K&B) zonunélahua (B&D) KYAMA diangbrèmeya (B&D) ˈNÉKÉDIÉˈ kpoyo (B&D) GHANA ADANGME-KROBO asaraba = fever leaf (ASThomas; FRI) AKAN-ASANTE ɔpam-kotokrodu (BD&H; E&A) TWI ɔ-pamkotokrodu (BD&H; E&A) ɔ-pampotoporopoo (FRI) NIGERIA EDO ogangan (auctt.) oriaruzo (AHU) oviaruzo (EWF) IGBO aga (auctt.) YORUBA ààráṣá (auctt.) àṣá (auctt.) àṣáràgbà (auctt.)

A shrub or small tree to about 8 m high of evergreen, deciduous and secondary forests and extending into the savanna from Sierra Leone to W Cameroons and dispersed across Africa to Angola, Tanganyika and Kenya.

The heart-wood is dark brown, very hard and durable. It is used in house-building and for fuel.

A bark-decoction is used in Ivory Coast as a purgative, diuretic, aphrodisiac, for urethral discharges, fevers, dysenteric diarrhoeas, and for fever and rheumatic pains, its activity being ascribed to the presence of tannins and saponosides (1, 5). A bark-macerate is used in Zaïre for cough (6), and in Ubangi children and adults take a bark-decoction for difficulty in breathing (12). An extract of the stem of Nigerian plants has been tested for molluscicidal action on the freshwater snail, Bulinus globulus, and has been reported to be very effective, 100 ppm extract giving 100% mortality (13).

The leaves are purgative. An infusion is given to children in S Nigeria (11). In Ghana the leaves, known as fever-leaf, may be wilted for 1–2 days and then boiled up in water with some lime-juice added and the whole drunk. This is purgative, sudorific and febrifugal. Inhalations of the vapour from the boiling also induce sweating. The infusion is also used for bathing (8). A leaf-infusion known as ɔpampotoproko, Twi, is also used in Ghana as a purgative (2, 4).

The seeds are reported eaten in S Nigeria (10), and the wood used as a pillar for a loom (9).

The leaves are food of the African silk-worm larva, Anaphe spp., in Uganda (3, 4, 7) and in Zaïre (6).

References:

1. Bouquet & Debray, 1974: 83. 2. Burtt Davy & Hoyle, 1937: 42. 3. Dalziel, 1937: 137. 4. Irvine, 1961: 216. 5. Kerharo & Bouquet, 1950: 69. 6. Léonard, 1962, b: 35–36. 7. Maitland 478, K. 8. Thomas, AS D.193, K. 9. Thomas NWT.1796 (Nig.Ser.), K. 10. Thomas NWT.2031 (Nig.Ser.), K. 11. Thomas NWT.2163 (Nig.Ser.), K. 12. Vergiat, 1970: 68. 13. Adewunmi & Sofowora, 1980.

Bridelia ferruginea Benth.

FWTA, ed. 2, 1: 370–1. UPWTA, ed. 1, 137.
West African: GUINEA FULA-PULAAR (Guinea) dafi (Aub.) MANDING-MANINKA baboni (Aub.) da-fing saba = black mouth 'saba" [Landolphia ssp.] (RP) saba (Aub.) saga (Aub.) sagua (Aub.) sagua lé (Aub.) sagué (Aub.) MANINKA (Konya) sagba (RS) SIERRA LEONE KORANKO bembε (S&F) MENDE kui (def.) (Cole) MALI MANDING-BAMBARA sabua (JMD) sagba (JMD) sagua (Aub.) saguin (Aub.) MANINKA baboni (Aub.) saba (Aub.) saga (Aub.) sagua (Aub.) sagua lé (Aub.) sagué (Aub.) UPPER VOLTA FULA-FULFULDE (Upper Volta) babooni (K&T) cellepuri (K&T) kojuteki (K&T) seŋseyohi (K&T) HAUSA kirni (Aub.) kisni (Aub.) KIRMA g'bété (Aub.) tiblen felé (Aub.) MANDING-BAMBARA sagua (Aub.) DYULA babuni saba (Aub.) MOORE sagha (Aub.) IVORY COAST AKAN-BRONG barié (Aub.) BAULE séa (auctt.) tukwé (Aub.) DAN gôn (Aub.) KRU-BETE féféhi (RS) gli (Aub.) KULANGO irigo (Aub.) KWENI sagba (Aub.) MANDING-MANINKA saba (RS) sagba (A&AA; B&D) SENUFO-DYIMINI nakurugo (Aub.) NIAGHAFOLO tiakoroko (Aub.) TAGWANA nakru (Aub.) GHANA ? nyu (ASThomas) AKAN-ASANTE badiε (FRI; E&A) badiε-kɔkɔɔ (E&A) BRONG badiε-kɔkɔɔ (E&A) KWAWU bardiε-kɔkɔɔ (E&A) TWI badeε (FRI; E&A) badiε (E&A) badiε-kɔkɔɔ (E&A) ɔpam fufuo (FRI) BAULE sea (FRI) GA flatʃo (FRI) SISAALA doho (AEK) wallinjang (AEK; FRI)

EUPHORBIACEAE

warrinjung (AEK; FRI) TOGO ANYI-ANUFO yumpo (Volkens) GBE-VHE akamati (Aub.) MOBA hionmonli (Aub.) TEM (Tschaudjo) kolo (Volkens) YORUBA-IFE (TOGO) choluhae (Volkens) DAHOMEY BAATONUN bemebenku (Aub.) kpépéla (Aub.) pekpéla (Aub.) GBE-FON honsuk-okué (Aub.) uomo (Aub.) YORUBA-NAGO hira (Aub.) NIGER GURMA hedionbiga (Aub.) HAUSA kirni (Aub.) kisni (Aub.) NIGERIA BOKYI kensange abia (JMD; KO&S) FULA-FULFULDE (Nigeria) burburumhi (Taylor ex JMD: MM) gulumbi (MM) gulummehi (MM) lammulam-muki (J&D) marehi (auctt.) HAUSA dorowan birni = locust-bean tree of the town (MM) kaddafi = destroy poison (JMD) kas(-he) dafi = destroy poison (JMD) kirni the arrow poison antidote made from the bark; hence the tree itself (auctt.) kismi (JRA) kisni (auctt.) kizni (auctt.) kurni (ZOG) IGBO aga (Akubue & Mittal) olá (auctt.) KANURI zəndi (A&S) TIV kpine (Vermeer) YORUBA asaragba (IFE) ìrà (auctt.) ìrà ọdàn = 'ira' of open country (auctt.)

A shrub or straggly tree to 15 m high with crooked bole up to 1.80 m in girth, the commonest Bridelia species of the savanna woodland occurring from Guinea and Mali to S Nigeria, and throughout the wooded savanna regions of Africa.

It is fire-resistant (15). In Sierra Leone the tree is reported to drip water (? honey-dew, cf. B. micrantha/stenocarpa) throughout the dry season (Lane-Poole in 7, 17). The tree is the host-plant of several species of Anaphe, the African silkworm genus (12), and A. ambrizia (Butler) has been reared on it in Nigeria (9).

The wood is brown (10). It is said to be termite-proof, and is used for this reason in the soudano-guinean region to make granaries (2). In Sierra Leone it is used as a primary structural timber (6). The liquid obtained from an aqueous macerate of the wood is applied to locally-made pottery in Zaïre for glazing (12). In the Central African Republic it is recognised as a good firewood, indeed as a 'woman's firewood', that is one good for the hearth and cooking place, long-lasting while the housewife is away on other chores, and picking up quickly from sleeping embers with a hot flame and minimal amount of smoke (21).

The bark, leaves and roots are all ingredients of Yoruba agbo infusions, chiefly administered to children. The bark and the bright red infusion from it are commonly sold in Nigerian markets and shops for use as a mouth-wash and remedy for thrush in children (7, 8, 9). Similar use is made of a root-decoction in Ivory Coast (1). In Congo (Brazzaville) a bark-decoction is used for toothache (3), and in Ivory Coast for dysenteries and diarrhoea (11) or as a laxative (1). The bark has a great reputation in Ghana (5) as an antidote against poisons. In N Nigeria it is used against arrow-poison – chewed and applied to the wound which is then sucked. A bark-preparation called kirni is held to confer immunity against arrow-poisons and against syphilis. This and other Hausa names cited above meaning 'destroy poison' are given by transference to the tree itself (7). Igbo medicine-men use a cold-water extract of the bark with the stem of Costus afer (Costaceae) in the treatment of ose-nkenu, identified as minor epilepsy (23).

Root-bark is used in Togo on skin-diseases and eruptions (7), and stem-bark on skin-infections in Nigeria. The roots are used by the Yoruba as chew-sticks, while the Maninka of the Upper Niger (Guinea) grind the wood to a fine powder for use as a dentifrice (20). The stem and root are rich in tannin and are put into mouthwashes (2). Examination of boiled water extracts of Nigerian material has given positive action on Gram +ve bacteria, Sarcina lutea and Staphylococcus aureus, but no fungistatic action, nor action on Gram −ve bacteria (13). A bark-infusion at 1:100 has been found to have no effect on Daphnia nor fish, but a root-infusion at 1:200 was lethal to Paramecium (1). The Tiv of the Benue Plateau on Nigeria prepare a fish-poison from root-sap (18).

A leaf-extract in saline solution is reported to produce a marked reduction of blood-sugar in laboratory rats, and clinical trials have given a drop from 250 mg % to the normal less than 120 mg % which has been held for eight weeks of daily treatment (22). Decoctions of leaves, leafy twigs and bark are commonly used as purgatives, diuretics, aphrodisiacs, febrifuges and for urethral dis-charges, dysenteric diarrhoea and fever and rheumatic pains (4, 11). Activity is due to the presence of tannin and saponosides and the preparations may be taken by draught, in baths, fumigations, or topical applications. In Togo the root-bark is used for intestinal and bladder troubles (7) and in Nigeria bark is used to treat roundworm and cystitis (14). The grated bark may be taken mixed with tapioca flour to treat dysentery (2).

Tannin concentration in the bark is recorded as 3.0–3.5%, low in comparison with other sources, but of good quality and the bark is used in Togo and Ghana to dye cloth a fast reddish-brown if iron extract is used as a mordant, or purplish with alum (7, 9). The bark is used in Upper Guinea (2) and in Gabon (19) to blacken pottery. Leaves boiled up with rusty iron and shea-butter are used in Ghana to dye *adinkera* cloths a black colour (9). At Siguiri (Guinea) the teeth are blackened by a preparation of the plant which is called *dafing saba*: black mouth (Manding) (20).

A substance in the bark confers an adhesive quality to a decoction. Such an extract (Hausa: *makubar mahalba*: hunters' 'makuba') is used in N Nigeria, Togo and elsewhere to harden beaten laterite mud-floors, sometimes in a mixture with fruit-husks of *Parkia* (Leguminosae: Mimosoideae), and also to glaze pots after firing (7). In Upper Dahomey a mixture of this with clay is prepared which sets to a sort of cement of strength adequate to build terraces and houses (2). It is rain-proof and can be used for flat roofs of huts (9).

The bark is sometimes added to palm-wine to strengthen it (16).

References:

1. Adjanohoun & Aké Assi, 1972: 20. 2. Aubréville, 1950: 181. 3. Bouquet, 1969: 110–1. 4. Bouquet & Debray, 1974: 83. 5. Burtt-Davy & Hoyle, 1937: 42. 6. Cole, 1968, a. 7. Dalziel, 1937: 137. 8. Irvine, 1952, a: 30. 9. Irvine, 1961: 216–7. 10. Keay & al. 1960: 270. 11. Kerharo & Bouquet, 1950: 69. 12. Léonard, 1962, b: 38–39. 13. Malcolm & Sofowora, 1969. 14. Oliver, 1960: 20. 15. Savill & Fox, 1967: 112–3. 16. Schnell, 1950, b: 223. 17. Unwin, 1920: 51. 18. Vermeer 6, UCI, in litt. 19. Walker & Sillans, 1961: 162. 20. Portères, 1974: 114–5. 21. Roulon, 1980: 228, 230. 22. Iwu, 1980: 247. 23. Akubue & Mittal, 1982: 358.

Bridelia grandis Pierre ssp. grandis

FWTA, ed. 2, 1: 370. UPWTA, ed. 1, 137, incl. *B. aubrevillei* Pellegrin.
West African: SIERRA LEONE KONO bembe (S&F) KORANKO fira-bembɛ (S&F) MENDE ku (*def.* -i) (auctt.) LIBERIA DAN doaandoh (GK) doandoh (GK) duandoh (AGV) KRU-GUERE (Krahn) gbai (GK) IVORY COAST ? tchilpébi (A&AA) ABE chikuébi (JMD) chiukuébi (JMD) AKYE tchikuébi (auctt.) tugbibi *from* bi: *black* (auctt.) ANYI pankoko (K&B) KRU-GUERE kula béla (K&B) GUERE (Chiehn) kpakwé (K&B) KWENI zuguéné iahua (Aub.) GHANA VULGAR pamkotokurobe (DF) AKAN-ASANTE pamkotokrodu (FRI; E&A) NIGERIA EDO ogangan nugun (auctt.)

A tree of up to 33 m tall, bole buttressed or stilt-rooted, to 2.70 m in girth, straight and clear to about 17 m in large trees, of the closed high-forest from Sierra Leone to S Nigeria, and on into Gabon and Zaïre.

The wood is reddish-brown (Sierra Leone: 6), greyish-brown (Liberia: 7) to brown (2, 4). It is fairly hard, moderately heavy, durable and possibly termite-proof. It is used in Sierra Leone for construction-work (6) and furniture-making (7), and in Ivory Coast for canoes in preference over *B. micrantha* and *B. stenocarpa* (2). The stems are used in Gabon, where the timber is traded under the name of *asas* (cf. *B. micrantha*), for roof-joists (9). The timber makes a good firewood.

The bark is roughly fissured giving the appearance of blackness, evoking the word *bi*: black, in the Akye names (2). In Ivory Coast the bark is used either in decoctions or powdered in water as a powerful purgative, diuretic, aphrodisiac, febrifuge, and to treat urethral discharges, oedemas, dysenteric diarrhoea and as an anodyne for fever and rheumatic pains. Its activity is due to the presence of *tannins* and *saponosides* (3, 5). The bark in fumigations has superstitious use in Gabon to drive away evil influences (9).

The bark is sometimes added to palm-wine as a women's medicine in Liberia, perhaps to promote lactation (7), and an infusion of bark and leaves is given to newly-delivered women 'to cleanse' their milk (8, 9). In Ivory Coast the leaves

EUPHORBIACEAE

mixed with those of *Dichapetalum pallidum* (Dichapetalaceae) and *Cola chlamy-dantha* (Sterculiaceae) are used in vapour-baths and in enemas to relieve rheumatism (1).

References:

1. Adjanohoun & Aké Assi, 1972: 121. 2. Aubréville, 1959: 2, 47, as *B. aubrevillei* Pellegr. 3. Bouquet & Debray, 1974: 83. 4. Keay & al., 1960: 272–3. 5. Kerharo & Bouquet, 1950: 69, as *B. aubrevillei* Pellegr. 6. Savill & Fox, 1967, 113–4. 7. Voorhoeve, 1965: 94. 8. Walker, 1953, a: 35. 9. Walker & Sillans, 1961: 162.

Bridelia micrantha (Hochst.) Baill. **B. stenocarpa** Müll.-Arg.

FWTA, ed. 2, 1: 370, 762, as *B. micrantha* sensu Keay. UPWTA, ed. 1, 137, as *B. micrantha* sensu Dalziel.

Trade: asas; assas (francophone territories, from a Gabonese name, incl. *B. grandis* Pierre and *B. speciosa* Müll.-Arg. and *Dacryodes* Vahl, Dalziel).

West African: SENEGAL BALANTA brujé (K&A) BANYUN kélégen (K&A; JB) si légen (JB) BASARI o-mĕv (JB) BEDIK ga-mɛmɛ (FG&G) ga-méné (K&A; JB) DIOLA bukirun (auctt.) fu lékir (JB) kélikirèk (JB) u kiréy (JB) u likãnd (JB) DIOLA (Bayot) bulin (K&A) DIOLA (Fogny) bulikip (K&A) fuléké (K&A) fulékiir (K&A) fulékir éfu (K&A) DIOLA (Khombole) bulêgié (K&A) DIOLA (Tentouck) bulakir (K&A) DIOLA-FLUP bu lékiñ (JB) bu tuyolum (JB) bulikendi (K&A; JB) hu lakindi (JB) FULA-PULAAR (Senegal) dafi (K&A; JB) dafi saba (auctt.) gugri (JB) gurkiki (K&A) KONYAGI a-tya (JB) MANDING-BAMBARA dolé (K&A) sabua (JMD fide K&A) sagba (K&A) sagua (JB) saguã (JB) MANDINKA bisago (K&A) bisako (K&A; JB) bisayo (JB) wuli-kémo (K&A) MANINKA bako (JB) saguã (JB) vulikémo (JB) 'SOCE' bisago (K&A) bisako (K&A; JB) MANDYAK bu sak be tib (JB) MANKANYA be sak (JB) bo bamané é dunk (JB) SERER sim salakodi (K&A; JB) WOLOF sakin (K&A; JB) THE GAMBIA DIOLA wulakir (Hallam) FULA-PULAAR (The Gambia) marehi (DAP) MANDING-MANDINKA bisako (auctt.) busake (Hallam) busakio (DRR) sagba (DAP) WOLOF sagba (DAP) GUINEA-BISSAU BIDYOGO n-tongue (JDES) untongue (JDES) FULA-PULAAR (Guinea-Bissau) gúgri (EPdS; JDES) MANDING-MANDINKA bissaiô (JDES) PEPEL bissaque (JDES) GUINEA BASARI a-mèw (FG&G) SIERRA LEONE BULOM (Sherbro) yɛki-lɛ (FCD; S&F) KISSI sindio (FCD; S&F) KONO bembe (FCD; S&F) KORANKO bembe (NWT; S&F) bempe (JMD) pembe (NWT) LOKO kɔgo (NWT) meɛnge (NWT) MANDING-MANDINKA funi yomba (NWT) MENDE n-dɛvɛ-golo(-gowo) *incorrectly applied* (FCD) fawi (NWT) foni-ku (FCD) funi-wuli (NWT) igila (L-P) kpowui (Pyne) ku(-i) (auctt.) SUSU tolinyi, tulinyi, tulinje (auctt.) TEMNE a-, an-, ka-ta (auctt.) VAI gbeneh (DS) kuwie (FCD) YALUNKA tolil-na (DS) toli-na (KM) MALI MANDING-BAMBARA sabua JMD; FB) sagba (JMD; FB) UPPER VOLTA FULA-FULFULDE (Upper Volta) dafi (RS) dafi saba (RS) IVORY COAST ABE chikué (A.Chev.) chiukué (A.Chev.) AKAN-ASANTE opam (K&B) AKYE tchikue (auctt.) ANYI bacié (K&B) bianzua (K&B) egpa (K&B) epakotrubo (auctt.) FULA-FULFULDE (Ivory Coast) dafi (Aub.; RS) dafi saba (Aub.; RS) KRU-BETE apoï (K&B) KULANGO irigo sanga (K&B) KYAMA amangbréhia (auctt.) diembrémihia (auctt.) gringuinmia (B&D) MANDING-MANINKA kosagba (B&D) NZEMA ataba (A.Chev.; K&B) ekuané (A.Chev.; K&B) GHANA AKAN-ASANTE apakyisie (FRI) bare(ɛ) (TFC; FRI) ɔpam (FRI) BRONG ɔbadie (BD&H) KWAWU apakyisie (FRI) badie-fufuo (E&A) ɔ-bare(ɛ) (AEK; FRI) ɔ-pam (auctt.) TWI akese (OA) asunsane (OA) ɔ-pam (FRI) ɔ-pampotoporopoo (Twi Dict. fide FRI) pampotoprofo (OA) ANYI epakotrubo (FRI) GBE-VHE awɔdze (FRI) miblɛ (FRI) VHE (Awlan) akati (FRI) GUANG kàyásé (Rytz) SISAALA wallingjang (AEK; FRI) warrinjung (AEK; FRI) NIGERIA BOKYI kensange (JMD: KO&S) EDO ogangan (JMD; KO&S) ogangan neran (JMD) ogangan nugu (JMD) EFIK àfíá àkpáp = *white 'akpáp'* [Macaranga barteri, Euphorbiaceae] (RFGA) IGBO egede (Singha) ogaofja *from* ofja: *bush* (KO&S; KW) olá ozalu (NWT; JMD) IGBO (Awka) aga ofĭa = *'aga' of the bush* (JMD) aga-ojĭ = *black 'aga'* (JMD) IJO-IZON (Kolokuma) igbárágbára (KW; KW&T) IZON (Oporoma) igbárábárá (KW) YORUBA arasa (auctt.) aṣa (auctt.) àṣà *name used in actions for protection against one's enemies* (Verger) aṣa gidi (Thompson; JMD) fonufonu *name used in a remedy against worms* (Verger) ida ɔdàn (KO&S) ìrà *name for the bark in [treating?] cases of prolonged pregnancy* (Verger) irà (JMD; Singha) irà-ɔdàn (JRA) WEST CAMEROONS KPE bungo (AJC) mosenge (AHU fide JMD) KUNDU esenge (AHU fide JMD)

What has for long been treated as one species, *B. micrantha*, has more recently been recognised as consisting of two: *B. micrantha* as a savanna species widespread in the soudanian region of West Africa and into E Africa, and *B. stenocarpa* as a species of forest clearings in the rain-forest zone from Sierra

38

Leone eastwards as far as Angola (23). Morphologically they are very similar and it is probable that in areas where they occur together no distinction is drawn between them for usage. They are treated together here.

Trees to about 15 m high with an open spreading crown. *B. micrantha* is said to make a good shade-tree for coffee in Tankanyika (10), and is used as a shade-tree for cocoa in Sierra Leone (26). The wood is greyish to yellowish-white with a dark brown heart, hard, durable, heavy and termite-proof. When worked it takes a good polish, and larger timbers are good for indoor carpentry and furniture (5, 6, 11, 16, 24, 30, 32, 35). The wood is very durable in contact with the ground and in water. It is used for hut and fence-posts, and in Sierra Leone it is said to 'last 100 years' (7). It is also used in boat-building: in Urundi, boat-keels (*B. micrantha*, 24), in Zanzibar ribs for boats (34) and the Ijo river-folk of S Nigeria use the smaller sticks for canoe-seats (36, 39). In Nigeria the wood has been known as Yoruba and Benin Ironwoods (6, 30). It makes good firewood and charcoal, both species giving out intense heat. As a fuel for boiling oil-palm nuts to obtain the oil, it is preferred by the Ijo (36). The wood when boiled to a pulp has medicial use in Sierra Leone under the vernacular name *ejira* (28, 29), and in Nigeria (1) to treat open sores, or to foment them.

Leaves, bark and roots are purgative and are widely so used. The bark, after the removal of the corky outer layer is administered in various preparations in Senegal for stomach and intestinal complaints, and for conditions in which the stomach is considered the cause – sterility, amenorrhoea, dysmenorrhoea, beri-beri, oedemas, etc, and, in combination with other drug-plants for shock when rapid and spectacular success has been recorded (3, 17–20). Activity of the bark is said to be due to the presence of *tannins* and *saponosides* (3). Yoruba medicine-men use the bark for bringing to full term 'prolonged pregnancy', said to run sometimes up to three years (42), doubtless false pregnancy or obesity. The root may also be given for stomach-troubles as a laxative by eating food cooked in water in which the root has been boiled (6, 22). Similar uses are recorded in The Gambia (38). Root-sap instilled into the anus is an Ubangi treatment for oxyuris worm (40). Yoruba medicine-men recognise the plant as vermifugal (42). In Ghana a leaf-decoction is given for guinea-worm (43).

The fresh bark is slightly aromatic and is used in Gabon for stomach-aches (31, 32), and in Ivory Coast as a powerful purge in cases of obstinate constipation and poisoning (21), and, in curious contrast, to prevent abortion (16). Powdered bark is used in decoctions with palm-oil in Sierra Leone as a cough-medicine (16), and a hot decoction of bark and leaves may be held in the mouth for sores (8) and to relieve sore gums, toothache and cough (28, 41). The bark makes an Yoruba mouth-wash (15). In Malawi, a bark-decoction is applied to treat scabies (16).

In the inner bark and outer sap-wood there is a sticky substance which is adhesive. The inner bark with that of the baobab (*Adansonia digitata*, Bombacaceae) produces a glue used for boots and shoes in The Gambia (38). In northern Sierra Leone the stripped stems are rubbed onto newly-made rattan grain winnows (25), and in like manner, also in Sierra Leone, the sticky substance from the inner bark is applied to fresh wounds to form an effective binding of the tissues (26). The Ijo of S Nigeria use bark boiled in water to heal the circumcision wound of both girls and boys (36). In E Africa the bark is pounded to a paste which is used to stop cracks in doors, baskets, etc. (6), while such is used in The Gambia to caulk canoes (38).

The aromatic bark is used in Zaïre to flavour local tobacco (*B. micrantha*, 24).

The young foliage is taken as fodder in Zanzibar (34) though the older leaves are laxative (14). In Senegal, the Tenda use the leaves as a circumcision dressing, and the leaves are stuffed into cushions on which the new initiates sit to have their wound washed (37). Also in Senegal, the leaves and other parts of the plant are held to be enteralgic (19), while in Sierra Leone (9) the young leaves are eaten for headache. In eastern and southern Africa the root finds use as an analgesic (2, 33). The plant enters into a number of West African treatments for conjunctivitis, and in S Africa leaf-sap is used for sore eyes (33). Bark, leaves and roots are pounded in Nigeria for external use on bruises, boils, ulcers, dislocations and burns (1). The roots chewed with maleguetta pepper are held by the Ijo to be a cure for hernia (36).

The twigs and young leaves furnish a black dye used in Guinea (Pobéguin fide 6) and Ghana (4) to blacken pottery. The pounded bark of *B. micrantha* is used by the Tenda people of Senegal to redden pottery and baskets (37). In Zaïre the bark of *B. micrantha* gives a black dye and of *B. stenocarpa* a red dye for pottery (24), and in E Africa the former gives a brownish dye (12).

The fruits have a small amount of pulp which is edible. They yield a black dye.

In common with other *Bridelia spp.*, these trees are usually infested in Sierra Leone with red ants which 'milk' aphids living on the leaves and green shoots causing them to drip water (? honeydew) all through the dry season (28). The trees are also the food plant of African silk-worms, *Anaphe spp.*, recorded in The Gambia (27), Ghana (14, 16), Nigeria and Uganda (6) and Zaïre (*B. micrantha*, 24). Farming these silk-worms has been attempted in various parts of tropical Africa but without apparent success. The silk yarn is brownish in colour and can be woven mixed with cotton. The Yoruba name for the silk is *sãnyán* or *òwú sãnyán* (*òwú*: cotton; *sãnyán*: raw silk, coarse woven silk, silk cloth (Yoruba Dictionary, fide 6). Propagation of the food-plant is relatively simple from seed or cuttings, and people have been encouraged to establish the plant in hedges or as a boundary shrub, but since the silk-worm pupae are an item of diet the raw material has become scarce in the natural state (6). With considerable knowledge now gained on farming the commercial silk-worm, it may be of interest to look again at the possibility of cultivating the indigenous species.

The plant has magical attributes in Yorubaland as a protection against one's enemies (42), and mistletoe found growing on *B. micrantha* if placed in a new field is held by the Tenda people of Senegal/Guinea to have protective powers (37).

References:

1. Ainslie, 1937: sp. no. 58. 2. Bally, 1937. 3. Bouquet & Debray, 1974: 83. 4. Chipp, 1922, a: 56. 5. Dale & Greenway, 1961: 187. 6. Dalziel, 1937: 137. 7. Deighton 1670, K. 8. Deighton 2213, K. 9. Deighton 2353, K. 10. Doughty 19, K. 11. Eggeling & Dale, 1952: 117–8. 12. Greenway, 1941. 13. Haerdi, 1964: 92. 14. Irvine 303, K. 15. Irvine, 1952, a: 30. 16. Irvine, 1961: 217–8. 17. Kerharo & Adam, 1962. 18. Kerharo & Adam 1964, a: 408–10. 19. Kerharo & Adam, 1964, c: 296. 20. Kerharo & Adam, 1974: 405–6. 21. Kerharo & Bouquet, 1950: 70. 22. Lane-Poole 260, K. 23. Léonard, 1955: 373–4. 24. Léonard, 1962, b: 48–50. 25. Miszewski 4, K. 26. Pyne 125, K. 27. Rosevear, 1961. 28. Savill & Fox, 1967, 114–5. 29. Scott Elliot 4018, K. 30. Unwin, 1920: 335. 31. Walker, 1953, a: 35. 32. Walker & Sillans, 1961: 162. 33. Watt & Breyer-Brandwijk, 1962: 397. 34. Williams, RO, 1949: 154–5. 35. Williamson, J, 1955 (? 1956): 27. 36. Williamson, K A21, UCI. 37. Ferry & al., 1974: no. 93. 38. Hallam, 1979: 55. 39. Williamson, K. & Timitimi, 1983. 40. Vergiat, 1970: 68. 41. Portères, 1974: 115. 42. Verger, 1972: 38. 43. Ampofo, 1983: 55.

Bridelia ndellensis Beille

FWTA, ed. 2, 1: 371.
West African: GHANA ? huhǫe (Edy) GRUSI yelimadu (AEK)

A shrub or tree to 15 m tall, stilt-rooted, in understorey of dry forest, or fringing forest of savanna regions from northern Ghana to S Nigeria and across Africa to Sudan and Uganda.

The wood is white, hard and coarse-grained (3, 4). It is used in Zaïre for construction (5). The tree is resistant to annual burning (2). In Sudan the fruit is eaten (1).

References:

1. Andrews 1722, K. 2. Aubréville, 1950: 179, as *B. ferruginea* Benth. var. *orientalis* Hutch. 3. Eggeling & Dale, 1952: 117, as *B. ferruginea* Benth. var. *orientalis* Hutch. 4. Keay & al., 1960: 272. 5. Léonard, 1962, b: 42–44.

Bridelia scleroneura Müll.-Arg.

FWTA, ed. 2, 1: 370. UPWTA, ed. 1, 138.

West African: **UPPER VOLTA** GURMA hiedondiga (K&B) KIRMA tiblinfélé (K&B) MOORE tansaloga (Aub.; K&B) **GHANA** AKAN-ASANTE badiɛ-nua (E&A) BRONG badeɛ-nua (FRI; E&A) NANKANNI ba-udiga (Lynn; FRI) SISAALA warrinjung (AEK) **NIGERIA** FULA-FULFULDE (Nigeria) burburumhi (MM) gulumbi (MM) gulummehi (MM) marehi (MM) HAUSA dorowan birni = *locust-bean tree of the town* (RES; MM) kirni *an arrow-poison antidote made of the bark; hence also the tree itself* (auctt.) kismi (JRA) kisni (auctt.) kizni (JMD; ZOG) KANURI zə̀ndi (A&S)

A shrub or tree to 6 m high (13 m in Ghana, 4), of the guinean savanna from Guinea to S Nigeria and extending to Zaïre and eastern Africa.

The sap-wood is hard and yellowish white, and the wood makes an excellent fuel (4). The Gbaya of the Central African Republic recognise it as yielding good firewood, and dub it 'woman's firewood' because of its burning qualities in the hearth – see *B. ferruginea* (16). It is termite-proof (7). The bark is corky, blackening with age, and with a red slash (3, 5).

The plant (? the bark) is used in Ivory Coast as a purgative, diuretic, aphrodisiac, for urethral discharges, fevers, oedemas, dysenteric diarrhoea, and fever and rheumatic pains, taken as a draught, or in baths or by topical application. Activity is ascribed to the presence of *tannins* and *saponosides* (2, 6). The plant is used in Zaïre to treat dysentery and sores (7), and a root-decoction in Ubangi for diarrhoea (14). In N Nigeria there is unspecified medicinal use of the roots (9). In Tanganyika the roots are commonly used to treat sores (11, 12), one method prescribing that the powdered root be steeped in cold water for 15 minutes which is rinsed through the mouth before use. The powdered root is also cooked with sorghum gruel which is then drunk slowly for epilepsy (10). A root-infusion is taken in E Africa (1) and in southern Africa (13) for relief of abdominal pains and indigestion. Anodynal action on painful teeth is sought in Ubangi in the application of hot sap which has been expressed from wood-raspings (14), or of the root-scrapings themselves to the gums (15).

The leaves are used in Tanganyika for headache (12), and in Ubangi to make a poultice for dislocated joints (14).

The fruit has a layer of slightly succulent pulp which is eaten in various parts of Africa (West Africa: 3; Ghana: 3; Sudan: 8; eastern Africa: 13).

The bark is boiled in water in Zaïre to furnish a black dye for pottery (7).

References:

1. Bally, 1937. 2. Bouquet & Debray, 1974: 83. 3. Dalziel, 1937: 138. 4. Irvine, 1961: 218. 5. Keay & al., 1960: 272. 6. Kerharo & Bouquet, 1950: 70. 7. Léonard, 1962, b: 33–34. 8. Patel & El Kheir 57, K. 9. Shaw 61, K. 10. Tanner 451, K. 11. Tanner 1209, 4426, K. 12. Tanner 4160, K. 13. Watt & Breyer-Brandwijk, 1962: 397. 14. Vergiat, 1970: 68. 15. Portères, 1974: 115. 16. Roulon, 1980: 228, 230.

Bridelia speciosa Müll.-Arg.

FWTA, ed. 2, 1: 370.

West African: **NIGERIA** FULA-FULFULDE (Nigeria) burumburum (Chapman)

A tree about 10 m tall; in riparian situations of montane savanna; in N and S Nigeria, E Cameroun and Ubangi.

Sap-wood is white to orange, thin or to 5 cm thick. Heart-wood is greyish-brown and very hard (1). No usage is reported.

Reference:

1. Letouzey, 5514, 12008, K.

Bridelia stenocarpa Müll.-Arg.

See under *B. micrantha* (Hochst.) Baill.

Caperonia fistulosa Beille

FWTA, ed. 2, 1: 398, as *C. palustris* sensu Keay. UPWTA, ed. 1, 138, as *C. palustris* sensu Dalziel.
West African: MALI MANDING-BAMBARA foro (A.Chev.) furu (A.Chev.) NIGERIA IṢẸKIRI iyọrọ (Giwa)

Herb, erect to 1 m, or semi-procumbent, of damp sites, from Mali to S Nigeria, and widely dispersed to Zaïre, Zimbabwe and Malawi.
The stems contain a fibre used by some of the river-folk of the Niger in Mali as a textile, mainly for fishing-lines (2, 3). In Sudan they are used as ties in cattle-sheds (4).
The plant is a weed of damper cultivated land and in Zimbabwe is recorded as colonising the banks of dykes (1).

References:

1. Angus 1339, K. 2. Chevalier, 1920: 572, as *C. palustris* sensu A. Chevalier. 3. Dalziel, 1937: 138. 4. Evans Pritchard 81, K.

Caperonia serrata Presl

FWTA, ed. 2, 1: 398, as *C. senegalensis* Müll.-Arg.
West African: THE GAMBIA MANDING-MANDINKA balasafuno (Macluskie) kumare turo (Macluskie) SIERRA LEONE TEMNE a-tum-ɛ-runka (NWT) NIGERIA YORUBA ifo (Rowland)

A slender, semi-erect to prostrate herb, stems to 1.5 m long, of damp ground throughout the Region from Senegal to S Nigeria, and in E Cameroun to Zaïre and Ethiopia to Tanganyika and Uganda.
In The Gambia it is a weed of rice-padis and grass-swamps (1).

Reference:

1. Macluskie 11, 17, K.

Chrozophora brocchiana (Vis.) Schweinf.

FWTA, ed. 2, 1: 398.
West African: MAURITANIA ARABIC (Hassaniya) ta'amiya (AN) tahmie (AN) tahmiya (AN) SENEGAL ARABIC (Senegal) ta'amiya (JB) BADYARA yalélé (JB) BALANTA dèngènduna (JB) BANYUN buko on diakan (JB) BASARI o-ngal o-dépekel (JB) DIOLA ka kô (JB) ka vamburã (JB) DIOLA (Bayot) bu bon a tsimen (JB) FULA-PULAAR (Senegal) dialéré ruéré (JB) makal (JB) maké (JB) makì (JB) sotírkoñidié (JB) TUKULOR burodiõ (JB) dialavava (JB) falfal (JB) KONYAGI i-raka (JB) MANDING-BAMBARA sama ndéku (JB) MANDINKA taba tambög (JB) SERER dahafula (JB) mbafal a ñav (JB) musur (JB) ndadar lãg (JB) ndusur (JB) WOLOF n-diamat (K&A; JB) n-diamet (K&A) ndusur (JB) ratah (JB) MALI ARABIC (Mali) afaraq (RM) UPPER VOLTA FULA-FULFULDE (Upper Volta) duuru (K&T) NIGER FULA-FULFULDE (Niger) damay-gihi (ABM) duusur (ABM) HAUSA damagi (AE&L) dàmagì (AE&L) damagui (Bartha) damaigi (AE&L) dàmaigì (AE&L)

A low woody under-shrub, on sandy soil on the Sahel and savanna from Mauritania to Niger and N Nigeria, and extending to the Red Sea.

In Senegal the plant is recorded as causing vomiting and diarrhoea in stock and is thus not grazed except occasionally by sheep and goats (1), though in Niger it is sought after by goats and is taken at certain times of the year by cattle (3). In Mauritania it is used as a medicine for camels and sheep (6), and in the Hoggar region of central Sahara the plant ashes are applied to human and camels' sores (2, 5). Analysis of the chemical content shows no particular reason for a beneficial action as a wound-dressing; silica content (72.2%) is unusually high (4).

References:

1. Adam, 1966, a: 518. 2. Lapperrine, 1919. 3. Maliki, 1981: 48. 4. Maire, R., 1925. 5. Maire, H., 1933: 145. 6. Naegelé, 1958, b: 890.

Chrozophora plicata (Vahl) A Juss.

FWTA, ed. 2, 1: 398. UPWTA, ed. 1, 138.
West African: SENEGAL MANDING-BAMBARA ka vamburã (JB) sama ndéku (JB) SERER ndadar lãg (JB) ndahafula (JB) ndusur (JB) WOLOF n-diamat (K&A) n-diamet (K&A)

A semi-woody shrub recorded from Senegal and N Nigeria, and occurring in E Africa and into Asia.

The plant occurs in grassy savanna and as a post-cultivation weed. It is not grazed by stock in Senegal and is said to cause vomiting and diarrhoea (1). Nor is it grazed in Sudan (2), but camels and Grant's gazelle are recorded eating it in Kenya (4). It is known to be acrid and poisonous in India (5).

A purplish-blue dye is obtained from the fruits for use in E Africa (3).

References:

1. Adam, 1966, a: 518. 2. Andrews A.650, K. 3. Greenway, 1941. 4. Mwangangi 1381, K. 5. Sastri, 1950: 142–2, as C. rottleri Klotzsch.

Chrozophora senegalensis (Lam.) A Juss.

FWTA, ed. 2, 1: 398. UPWTA, ed. 1, 139.
West African: SENEGAL ARABIC (Senegal) aramache (JB) BANYUN bukoôdiakan (K&A) DIOLA bukao (K&A) DIOLA-FLUP bu diunka diunka (JB) bu kao (K&A) bu kav (JB) FULA-TUKULOR puré (K&A) MANDING-BAMBARA dabada (K&A) dabrada (K&A) ka vâbura (JB; K&A) sama ndeku (JB; K&A) MANDINKA dabada (JB) dabrada (JB) 'SOCE' mbolo (K&A; JB) MANDYAK lokotan (JB) NON baul (A.Chev.) SERER bol (K&A) ndadar lãg (lang) (JB; K&A) ndahafula (JB; K&A) ndiamat (K&A) ndusur (JB; K&A) WOLOF n-diamat (K&A) n-diamet (K&A) n-dusur (K&A) katakã (JB) THE GAMBIA MANDING-MANDINKA mborno (Wallace) GUINEA-BISSAU MANDING-MANDINKA tabatambom-ô (EPdS; JDES) MANDYAK lócótane (EPdS; JDES) MALI MANDING-BAMBARA dabada (A.Chev.) UPPER VOLTA BISA baselé (Prost fide K&B) MANDING-BAMBARA dabada (Laffitte fide K&B) MOORE oabgho-bim-pendo (K&B) NIGERIA HAUSA ɓauren kiyeshi *from* ɓaure: *Ficus;* kiyashi: *a species of small ant* (JMD; ZOG) dàmágìì (auctt.) damaigi (auctt.) walkin maciijii = *loin-cloth of the snake* (auctt.)

An undershrub, sometimes prostrate, in seasonally flooded flats and on river-banks of the savanna region from Senegal to N Nigeria, and in Shari.

The plant is not grazed by stock in Senegal except occasionally by goats and sheep. It causes vomiting and diarrhoea (1), but it is appreciated by camels (3). A leaf-macerate is used in the savanna region for taenia and ascaris and it is said to be very efficacious (2, 4, 8, 9). This is not bitter, but a preparation of the whole plant however is astringent, and is taken in N Nigeria boiled with cereal foods for diarrhoea (5). It is also used topically for rheumatism (5, 9). A root-decoction is a successful medicine in Senegal for suckling babies suffering from green diarrhoea (7). In Ivory Coast an enema is made of the plant for administration to rickety children and the plant has a general application in stomach-aches and for venereal diseases(8). In Nigeria it is used to treat syphilis and it is an ingredient of *rigakafi* (Hausa), a reputed preventive and cure. Some of the other ingredients are the roots and leaves of *Crotalaria spp.* (Leguminosae: Papilionoideae), *Portulaca oleracea* (Portulacaceae), *Feretia apodanthera* (Rubiaceae), etc. This preparation is also regarded as a remedy for mental derangement (5). A decoction is used as a body-wash by pregnant women (5).

The fruit-juice has a high reputation in Casamance in the treatment of blindness. The juice is dripped into the eyes. Some irritation is caused (6).

In some parts it provides a black dye for matting, etc.; the plant is mixed in a pot with the pods of *Parkia* and local natron along with strips of palm-leaf or other material to be dyed (5).

Plants of the genus *Chrozophora* have extra-floral nectaries, and one of the Hausa names suggests an association with ants (5).

References:

1. Adam, 1966, a: 518. 2. Bouquet & Debray, 1974: 83. 3. Chevalier, 1920: 573. 4. Chevalier, 1937, b: 171. 5. Dalziel, 1937: 139. 6. Kerharo & Adam, 1962. 7. Kerharo & Adam, 1964, c: 299. 8. Kerharo & Bouquet, 1950: 70. 9. Oliver, 1960: 22, 55.

Claoxylon hexandrum Müll.-Arg.

FWTA, ed. 2, 1: 401. UPWTA, ed. 1, 139.
West African: IVORY COAST AKYE lonkati (Aub.) GHANA AKAN-AKYEM sansammuro (FRI; E&A) ASANTE tetrete (Foggie fide FRI) turoturo (auctt.) TWI dubrafo-nin *cf.* dubrafo: *Mareya* (auctt.) NIGERIA EDO idemi (auctt.) IGBO odō = *wooden pestle* (BNO; KW)

A tree to 20 m high, of secondary regrowth in the high-forest zone from Sierra Leone to W Cameroons, and in E Cameroun.

The timber is white and soft.

When the fruits are ripe, it is considered by the Akyem to be near to *Omanheme*'s festival (*ohene–afahyɛ*; Christmas–New Year) (1, 2).

The seeds are used in play by children (1, 2).

The plant is used as a purgative in Ghana (2).

References:

1. Irvine 2688, K. 2. Irvine, 1961: 219.

Cleidion gabonicum Baill.

FWTA, ed. 2, 1: 406. UPWTA, ed. 1, 139.
West African: GHANA AKAN-ASANTE adafa (auctt.) *m*bawi (auctt.) notonoma (BD&H; FRI) *m*pawi (auctt.) *m*pawuo (auctt.)

A shrub or understorey tree to 16 m high of the closed-forest, and recorded only in Ghana in the Region, but occurring also in E Cameroun and Gabon.

The twigs are used as chew-sticks in Ghana (1–4).

References:

1. Chipp 101, K. 2. Irvine, 1930: 107–8. 3. Irvine, 1961: 219. 4. Portères, 1974: 116.

Cleistanthus spp.

FWTA, ed. 2, 1: 371, as *C. polystachyus* Hook. f.
West African: SIERRA LEONE MENDE fagbajos (SKS) *n*ja-gbɔhũ (FCD) *n*ja-gbɔhui (S&F) *n*ja-hahɛi (S&F) *n*ja-lahã (FCD) IVORY COAST AKYE uoré-uoré (Aub.)

Three species are involved under *C. polystachyus* sensu FWTA, ed. 2.: (a) *C. polystachyus* Hook. f., (b) *C. libericus* NE Br., and (c) *C. ripicola* J Léonard (3). All are small trees of damp, or riparian forest sites, the first named from Sierra

Leone to Nigeria and widespread to Angola, Uganda and Tanganyika, and the other two more locally restricted from Ivory Coast to Central Africa.
Their fruits are eaten in Ivory Coast. In places where the trees overhang riverbanks the falling fruit attract fish (1).
In *C. ripicola* from Congo (Brazzaville) *flavones* and *tannins* have been reported in the leaves, bark and roots, and *steroids* and *terpenes* in the bark and roots (2).

References:

1. Aubréville, 1950: 64, as *C. polystachyus* Hook. f. 2. Bouquet, 1972: 23, as *C. ripicola* J. Léonard.
3. Léonard, 1960: 430, 434, 438.

Cnidoscolus acontifolius (Mill.) Johnson

A much-branched shrub, 3–4 m high, stems thick and fleshy, leaf to 15 cm wide, 3–5 palmately lobed on peduncle to 12 cm long, native of C America, and now introduced and cultivated in Ghana and Nigeria.
The leaf can be cooked and eaten as a spinach. It is rich in protein and vitamin C. Propagation in Nigeria is being sponsored under a rural development scheme. The leaves bear some irritant hairs so picking requires a degree of care. They also contain cyanide which must be dispelled by boiling for at least half an hour (1). It is also under cultivation in Ghana for its edible leaves, and to grow as a hedge (3). It is propagated easily from cuttings and is ever-growing with a perpetual crop of leaves. It is without predatory pests in Africa (1).
The plant is closely related to *C. chayamansa* McVaugh, the *chaya* from ancient times (2).

References:

1. Lowe, 1989. 2. McVaugh, 1944: 467–8. 3. Newton 2918, K.

Codiaeum variegatum (Linn.) Blume

FWTA, ed. 2, 1: 368.
English: croton (of horticulture, not to be confused with the genus, *Croton* Linn.)

A shrub of eastern Malesia and the Pacific now distributed throughout the tropics as an ornamental for the brightly coloured and many-shaped leaves. Plants undergo somatic mutation and fascinating colour patterning can be obtained and propagated. Cuttings are normally easy to strike.

Croton aubrevillei J Léonard

FWTA, ed. 2, 1: 394, as *C.* sp. nr. *mubango* Müll.-Arg.
West African: IVORY COAST AKYE apétè (Aub.) BAULE fafu (B&D) KRU-BETE monocurihi (Aub.)

A small forest tree recorded only from Ivory Coast.
The wood is a clear yellow colour and hard (1).
The plant is reported taken internally as a remedy for constipation, stomachache and female sterility, and externally for pains in the sides and guinea-worm infection (2).

References:

1. Aubréville, 1959: 2: 90. 2. Bouquet & Debray, 1974: 83, as *C. mubango* sensu Bouq. & Deb., non Müll.-Arg.

45

EUPHORBIACEAE

Croton dispar NE Br.

FWTA, ed. 2, 1: 396.
West African: SIERRA LEONE LOKO bɔhɔnaga (NWT) MENDE gɛbot (NWT) sulɛ-mɔyɔ (FCD)

A straggling bush to 3 m high, or tree to 8 m, of the forestal area of Guinea, Sierra Leone and Liberia.

Croton eluteria (Linn.) Sw.

FWTA, ed. 2, 1: 396.
English: cascarilla; cascarilla bark.
French: cascarille.

A bush to about 4 m high, native of Bahamas, and now introduced into Nigeria, and perhaps elsewhere in the Region. The dried bark is the commercial source of cascarilla bark of pharmaceutical use as a tonic and aromatic bitter. It is however an inferior substitute for quinine bark. The world's main supply is from the Bahamas.

The bitter principle is *cascarillin*. The bark also contains two resins, an acid, and a volatile oil containing *eugenol, limonene, p-cymol* and a *terpene*.

Cascarilla bark used to be official in the *British Pharmaceutical Codex*, but is no longer.

References:

1. Farnsworth & al., 1969: 4, with references. 2. Oliver, 1960: 6, 58.

Croton hirtus L'Hérit.

FWTA, ed. 2, 1: 394.

An annual herb to about 60 cm tall, native of tropical America, becoming naturalised pan-tropically, and recently recorded in Sierra Leone where it has become established in pure stands in open waste places. It is able to crowd out weeds, and is apparently successful even in suppressing *lalang* grass (*Imperata cylindrica* (Linn.) Rauschel) (1). Its presence in other West African territories must be expected.

Reference:

1. Deighton 5551, K.

Croton lobatus Linn.

FWTA, ed. 2, 1: 394. UPWTA, ed. 1, 139.
West African: SENEGAL ARABIC (Senegal) hab el khechba (JB) hab el l'hesbal (JB) FULA-PULAAR (Senegal) ñakali (JB) vorudièh (JB) MANDING-MANDINKA kurdiãngdiãng (JB) SERER mbèt i tiãngay (JB; K&A) WOLOF ñakalé (JB) narhalé (K&A) ñaxali (K&A) **THE GAMBIA** MANDING-MANDINKA jambalulu *in allusion to the 5-pointed leaves* (Fox) nyri jejongho (Fox) **GUINEA-BISSAU** MANDING-MANDINKA curedjandjam-õ (JDES) **SIERRA LEONE** LOKO njɛwɔmbi (NWT) MENDE gɔmakfuri (NWT) **IVORY COAST** AKYE hitzasan (K&B) tototo (K&B) ARI tahatinta (Ivanoff fide K&B) BAULE bruété (K&B) KYAMA agbonsé (K&B) **GHANA** AKAN-TWI akɔnansa (FRI; E&A) gka (FRI) mfansu (FRI) DAGBANI aneta (Mellin fide FRI) GA wuɔnane = *fowl's foot; from the shape of the leaf* (FRI) GBE-VHE (Awlan) voligbe (FRI) HAUSA balasa natudi (Ll.W) **NIGERIA** HAUSA gaásàyaá (JMD) gaásàyár bareewa = *gasaya of the antelope* (JMD; ZOG) kiban kadangari (BM) námíjìn gaásàyaá = *male gasaya* (ZOG) námíjìn záakií-bánzaá (ZOG) IGBO òkwè-one *a kind of bean* (NWT; KW) YORUBA àjéofòlé (IFE; Verger) ẹ̀ru (auctt.)

46

An annual herbaceous plant becoming woody at the base, to nearly 1 m high, of open sandy places, often on river-banks, recorded throughout the West African region, and widespread across tropical Africa and Arabia.

The plant has a strong tap-root and laterals. It is not easy to root out in cultivated land (9). On sandy river-banks it may have some value against scour and erosion.

In Ivory Coast arrow-poison made from the plant is reputed to be the most poisonous used. The plant is simply crushed between stones with a little water and the arrows dipped in the paste. The leaves boiled in a little water are an enema for gynaecological affections, and mixed with palm-oil a leaf-paste is rubbed on to guinea-worm sores. Heated leaves are rubbed on to areas of costal and rheumatic pain, and a leaf-decoction by mouth, or a bark-decoction by enema is given as a purgative. The principal ingredient is *crotonic acid*, probably in combination with toxic albuminoids, and this is found throughout the plant (1, 6).

The plant has use in topical application to ulcers, sores, etc., and for headache. A decoction is reported as a children's medicine (2), but the purpose is not indicated. In Senegal the leaves are used with another undetermined plant for whooping-cough and all spasmodic coughing, and for stomatitis (5). In Nigeria it is used to assuage the pain of scorpion-sting (8). In The Gambia there is unspecified medicinal use (4).

Besides the presence of phytotoxins noted above, Brazilian samples are reported to contain tertiary and quaternary *alkaloids* and haemolytic *saponins* (3, 10). In N Ghana, however, the plant is reported to be a favourite food for fowls (7) and this anomaly may tie up edaphically with a non-toxic sample from Togo reported to be only slightly bitter and to contain only some tannin and sugar (Thoms fide 2).

In N Nigeria the plant has superstitious use against witchcraft: a person washing with a lotion prepared from it keeps evil at bay (2).

References:

1. Bouquet & Debray, 1974: 83. 2. Dalziel, 1937: 139. 3. Farnsworth & al., 1969: 7, 17. 4. Fox 134, K. 5. Kerharo & Adam, 1974: 408. 6. Kerharo & Bouquet, 1950: 71. 7. Lloyd Williams 833, K. 8. Musa Daggash FHI.24957, K. 9. Onwunyi 20, UCI. 10. Willaman & Li, 1970.

Croton longiracemosus Hutch.

FWTA, ed. 2, 1: 396. UPWTA, ed. 1, 139.
West African: NIGERIA YORUBA (Ondo) jesebe (Sankey; JMD)

A tree up to 25 m tall by 1.70 m girth, of the rain-forest in S Nigeria and W Cameroons, and also in Congo (Brazzaville). It is probably conspecific with *C. sylvaticus* Hochst., in which case the latter name has priority.

The wood is white, and the tree has a good bole, but the wood does not appear to be used (3) in the Region. In Zaïre it is a construction-timber (2).

In Congo (Brazzaville) the leaves are eaten as a tonic and restorative in the case of heavy effort and prolonged marching. The leaves are used to maturate furuncles (1).

References:

1. Bouquet, 1969: 112. 2. Léonard, 1962, b: 68–69. 3. Sankey 10, K.

Croton macrostachyus Hochst.

FWTA, ed. 2, 1: 394.
West African: NIGERIA MAMBILA kuh (Chapman)

A tree to about 16 m tall of the forested savanna and secondary bush from Guinea to W Cameroons, and widespread in tropical Africa.

The tree is considered useful as a shade-bearer over huts in Kenya (18). The wood is cream-coloured, very perishable and featureless (7, 8). Though it lacks hardness and toughness, it is used for tool-handles (11, 18) and as a building material and a fuel (5) in Kenya. It is used in Zaïre in carpentry though it is said to be difficult to saw (13). Pulping tests by the sulphate process gave a pulp of moderate strength which might be incorporated in writing paper or newsprint after bleaching. It was considered unsuitable for wrapping paper (6).

The bark-slash emits a peppery smell (15) and the bark of young branches exudes a colourless sticky slime (4). The bark is an ingredient of an effective purgative and vermifuge in Ethiopia (10). It is a purgative in E Africa and the presence of *crotin* is reported (17).

In Tanganyika leaf-sap is taken as an anthelmintic (2). An extract of leaves is used against itchy scalp in Ethiopia, where also a decoction of young leafy shoots is an ingredient of a prescription for jaundice (10), and for an eruptive disease resembling small-pox (19). In Uganda sheep and goats will not browse the young leaves (14). The old foliage apparently is not shunned. In Sudan a vegetable salt is prepared from the leaves (1).

The seeds contain a *resin* said to be more toxic than rotenone (10), and also a vesicant oil (19). The seed-oil is, of course, an extremely powerful purgative. The seed is recognised as poisonous in Ethiopia and is used as a fish-poison, and a preparation of the seed is instilled into the ear for ear-troubles (19). In Tigre the crushed leaves and seed admixed are drunk in water for tapeworm, the fruit is eaten and a root-decoction drunk for venereal disease, and the seed eaten to induce abortion (20).

The root is used in Kenya (12) and Tanganyika (16) for intestinal parasites.

The only recorded usage of the plant in West Africa is internally as a remedy for constipation, stomach-ache and female sterility, and externally for pains in the sides and guinea-worm sores (3). The part(s) of the plant are not specified.

Screening for anti-tumour activity of the plant has given negative results in one report, and in another some inhibiting effects on lung carcinoma in mice. The extraction of *crotepoxide*, a cyclohexane diepoxide, with anti-carcinoma properties is also reported (9).

The flowers are heavily scented and are unusually attractive to bees.

References:

1. Andrews A.1945, K. 2. Bally, 1937. 3. Bouquet & Debray, 1974: 83. 4. Breteler 485, K. 5. Chemkongo 1, K. 6. Chittenden & al., 1954. 7. Dale & Greenway, 1961: 191. 8. Eggeling & Dale, 1952: 120–2. 9. Farnsworth & al., 1969: with references. 10. Getahun, 1975. 11. Glover & al. 643, K. 12. Glover & al. 2596, K. 13. Léonard, 1962, b: 62–64. 14. Maitland 1230, K. 15. Styles 161, K. 16. Tanner 5157, K. 17. Watt & Breyer-Brandwijk, 1962: 400. 18. Wigg 1001, K. 19. Lemordant, 1971: 160–1. 20. Wilson, RT & Mariam, 1979: 30.

Croton nigritanus Sc. Elliot

FWTA, ed. 2, 1: 396.
West African: **SIERRA LEONE** LIMBA ndabia (NWT) LOKO čuaga (NWT) MENDE jingaomi (NWT) SUSU busui (NWT) tɔlɔŋwuri (NWT) xubɛswie (NWT) TEMNE ɛ-loŋ (NWT) kə-toru (NWT)

A shrub to 3 m high of the fringing forest in damp sites by rivers, from Guinea to N Nigeria.

In the Scarcies River area of Sierra Leone the plant is used as a compress on sores (1).

Reference:

1. Scott Elliot 4429, K.

Croton penduliflorus Hutch.

FWTA, ed. 2, 1: 396, in part (Sierra Leone) (Léonard).
West African: SIERRA LEONE TEMNE a-fɛt (NWT)

A tree to over 20 m high of the rain-forest area of Sierra Leone only.

Croton pseudopulchellus Pax

FWTA, ed. 2, 1: 394.

A shrub to about 4 m high, recorded only from Mali and N Nigeria, and occurring fairly widely distributed in East and South-central Africa.
No use is recorded for the Region.
In the sand-dune area of Inhambane, Mozambique, the plant is said to be an important fixer of dunes (2). The poles are durable and are fairly termite-proof. They are used in Tanganyika for hut-building (4).
The leaves and root are reported to contain the toxalbumin, *crotin* (6). Leaves are used in Tanganyika on syphilitic ulcers (1), and are applied to the chest for chest-affections (*kifua*, Swahili) (3). A root-decoction is also used for asthma (1). The root ground to a powder is taken as snuff for headcolds (5).
The flowers are sweet-smelling and attract many insects.

References:

1. Bally, 1937: 14. 2. Grandvaux Barbosa & de Lemos 8504, K. 3. Moggridge 45, K. 4. Semsei 552, K. 5. Tanner 3650, K. 6. Watt & Breyer-Brandwijk, 1962: 400.

Croton pyrifolius Müll.-Arg.

FWTA, ed. 2, 1: 396, as *C. penduliflorus* sensu Keay; excl. Sierra Leone.
UPWTA, ed. 1, 140, as *C. penduliflorus* sensu Dalziel.
West African: GHANA ADANGME-KROBO dudwatʃo (auctt.) AKAN-TWI nyamrem (auctt.)
NIGERIA YORUBA àjẹofòlé (IFE) amoroso (KO&S)

A tree to 20 m high by 1 m girth, of the closed and dry deciduous forest and secondary regrowth from Ghana to N and S Nigeria.
The wood is used for rafters in Ghana and an infusion of leaves is used externally for fever (2). The new leaf-flush is red and the young leaves are used in the Ondo province of Nigeria as a purge (1).

References:

1. Daramola FHI.32754, K. 2. Moor 66/306, K.

Croton scarciesii Sc. Elliot

FWTA, ed. 2, 1: 396.
West African: SIERRA LEONE MENDE tadueme (NWT) TEMNE àsá (NWT)

A shrub attaining 3 m high, on rocks beside and in stream beds, withstanding flooding, recorded only from Senegal to Ghana.
This common shrub affords harbour for the tsetse fly in N Ghana (1).

Reference:

1. Vigne 3864, K.

EUPHORBIACEAE

Croton sylvaticus Hochst.

FWTA, ed. 2, 1: 396, as *C. spp. A* and *B.*

Tree to 20 m or more high, bole to 1.5 m girth, of semi-deciduous savanna woodland and secondary regeneration, recorded only rarely from Guinea to S Nigeria, and more widely from Gabon to Angola and in E Africa.

The tree is sometimes used as a shade-bearer on coffee plantations in Zaïre, and wood shavings are used in a medicine for elephantiasis (1). Seeds and the seed-oil are used as purgatives in Gabon (2, 3).

References:

1. Léonard, 1962, b: 72–74. 2. Walker, 1953, a: 35, as *C. oxypetalus* Müll.-Arg. 3. Walker & Sillans, 1961: 164.

Croton tiglium Linn.

English: croton oil plant.

An erect or more or less spreading shrub or small tree, native of SE Asia, and introduced to the W African region. Its dispersal in Asia appears to be primarily anthropogenic for it is found usually around villages. It will grow on very poor soil and in Malaya it has been reported to have some potential in suppressing *lalang* grass, *Imperata cylindrica* (Linn.) Rauschel.

The plant is of very ancient cultivation being known in Sanskritic and early Chinese *materia medica* and the seeds and seed-oil have entered numerous official pharmacopoeias as powerful cathartics. As the oil is not stable, it has proved unreliable as a medication and has therefore been dropped from normal medical use.

All parts of the plant contain toxic substances. The leaves and the bark furnish arrow-poison to certain races in SE Asia. The leaves are reported used for poulticing snake-bite, and the roots to induce abortion. The oil has use in SE Asia as a fish-poison and for criminal adulteration of wells. It contains an extremely vesicant resin, *crotonol. Crotin*, a delayed-action poison which causes blood-clotting, is also present.

Symptoms of croton oil poisoning are firstly pain at the back of the throat, then in the anus. A dose of bismuth is an immediate antidote.

The oil can be used for soap-making and illumination, though in the latter instance noxious fumes are released causing, in a confined space, severe illness. The seed-cake retains the crotin and is thus not suitable for cattle-feed.

Reference:

Burkill, IH, 1935: 690–2, with references. Quisumbing, 1951: 498–501, with references.

Croton zambesicus Müll.-Arg.

FWTA, ed. 2, 1: 394. UPWTA, ed. 1, 139.

West African: **SIERRA LEONE** KRIO ajɛ-ofoŋla (FCD; S&F) MANDING-MANDINKA kurutinjengo (JMD) YORUBA-AKU jefulwal (JMD) **MALI** SONGHAI tonedibonhaïni (Aub.) **UPPER VOLTA** HAUSA koriba (Aub.) **IVORY COAST** ? fafo (Aub.) **DAHOMEY** GBE-FON djélélé (Aub.) YORUBA-NAGO adiékofolé (Aub.) **NIGER** HAUSA koriba (Aub.) SONGHAI tóndì-bòŋ héenì (*pl.* -ò) = *millet of the summit of the mountain* (D&C) **NIGERIA** EFIK étó òbùmà (Lowe) EJAGHAM mfam *properly the name of a juju cult* (JMD; KO&S) HAUSA ícen másàr = *Egyptian wood/tree* (ZOG) koriba (auctt.) KANURI mòròmórò (A&S; C&H) YORUBA àjé kò bàlé, àjé kòfòlé = *witches do not dare to perch on it; alluding to deterrent to sorcery* (auctt.)

A shrub or small tree to 16 m high, of fringing forest and savanna, from The Gambia to S Nigeria, and widely distributed elsewhere in tropical Africa.

The tree has a scaly bark and silvery leaves, rusty-scaly below, and has an attractive appearance. It is often planted in towns and villages. In times past it was planted as a fetish tree (Dahomey, 4). It has a reputation of conferring protection, and is often planted near the entrance of houses to ward off evil influences (Sierra Leone: 9; N Nigeria: outside the house of the Emir at Kontagora, 7). It is particularly protective against witches (Sierra Leone: 9; S Nigeria: 11). Its Yoruba name, àjẹ́ kò bàlé, means 'witches do not dare to perch on it', and the plant enters an incantation for the placation of witches (14). To the Ejagham of S Nigeria it is a symbolic tree, known, at least in its religious significance, as mfam (properly the name of the Juju cult). It is powerful to restore health to an important person, or the leaves drawn gently over the face of the dying to cause the spirit to pass painlessly (Talbot fide 8).

The wood is pale yellow, fine-grained, hard and gives a good polish (5, 8). The stems are used in parts of W Africa for hut-posts (10) and in Yoruba houses for beams in default of other timbers (8).

The bark-slash emits an aromatic smell (5) and an infusion of bark is used in Nigeria in cases of malaria (2).

The leaves are considered strengthening. A soup made of them is given to dysentery cases in S Nigeria (13), a leaf-decoction is used as a wash in both Nigeria and Sierra Leone and is taken internally for dysentery, fever, convulsions, etc. (8, 16), and in Nigeria for headache and as a vermifuge (2, 16). The shoots and roots are used in Sudan as a tonic, febrifuge and for menstrual pain (17). In Ghana the root passes as an aperient (6, 10).

The fruits, like the bark, are aromatic. They are used in the Adamawa region of Nigeria to spice food and to prepare a sort of scent (12). The seeds are said to have medicinal use in Togo (Beveridge fide 10). In Kordofan they are used to flavour tea (18)

An unspecified part of the plant is used in Zambia for nose-troubles (15)

Examination of Nigerian material has shown the presence of a trace of alkaloid in the stem and leaf (1).

References:

1. Adegoke & al., 1968. 2. Ainslie, 1937: IFI, sp. no. 119, as C. amabilis. 3. Anon. s.d., 10/6/1855, Herb. Hook., K. 4. Aubréville, 1950: 195. 5. Aubréville, 1959: 2: 88. 6. Burtt Davy & Hoyle, 1937: 43. 7. Dalziel 281, K. 8. Dalziel, 1937: 139. 9. Deighton 2675, K. 10. Irvine, 1961: 221. 11. Millen 14, K. 12. PFO FHI.48446, K, FBC. 13. Punch 20, K. 14. Verger, 1967: no. 27. 15. Wild 3102, K. 16. Akinniyi & Sultanbawa, 1983. 17. El-Hamidi, 1970: 279. 18. Hunting Tech. Services, 1964.

Croton sp.

West African: GHANA ADANGME-KROBO buko OA

A plant of the foregoing Krobo name, cited as C. membranaceus, which is not so far recorded in Ghana, is said to be used in treating retention of urine accompanied by haematuria: root cut up small and boiled, the liquor is given to the patient to drink (1).

Reference:

1. Ampofo, 1983: 39, as 'C. membraneceous.'

EUPHORBIACEAE

Crotonogyne chevalieri (Beille) Keay

FWTA, ed. 2, 1: 400.
West African: IVORY COAST KYAMA aboapombi (B&D)

A forest shrub to 4 m high recorded only from Ivory Coast and Ghana.
Examination for active principles in Ivorean material has found the presence only of a trace of *tannin* in the root-bark (1).

Reference:

1. Bouquet & Debray, 1974: 87.

Crotonogyne craterviflora NE Br.

FWTA, ed. 2, 1: 400.
West African: SIERRA LEONE LOKO ibowa (NWT) ndɔbɔrɔnga (NWT) MENDE *n*da-lolo *from* nda: *immature windfall;* tolo: *kola nut; inferring the fruit is inedible* (FCD) damainjɛnge (NWT)

An understorey shrub to 6 m high of forest in wet localities from Sierra Leone to Ivory Coast.
The inference of the Mende name, *nda-lolei*, is that the fruit is inedible (1).

Reference:

1. Deighton 3667, K.

Crotonogyne preusii Pax

FWTA, ed. 2, 1: 400.

A shrub or tree to 8 m high of the rain-forest in S Nigeria and W Cameroons, also in E Cameroun.
The plant is used in E Cameroun against diarrhoea (1).

Reference:

1. Leeuwenberg 7194, K.

Crotonogyne strigosa Prain

FWTA, ed. 2, 1: 400.

A shrub to 2 m high of the forest zone from Ivory Coast to W Cameroons and in E Cameroun.
It is known to the Kyama of Ivory Coast as highly poisonous for which there is no antidote (1).

Reference:

1. Bouquet & Debray, 1974: 83.

Cyrtogonone argentea (Pax) Prain

FWTA, ed. 2, 1: 399.
West African: WEST CAMEROONS LONG polopote (Ejiofor)

A forest tree to over 30 m tall, of the high-forest of S Nigeria and W Cameroons, and also E Cameroun and Equatorial Guinea.
The wood is used for furniture-making and general carpentry in Gabon. The seeds are edible. The grated bark mixed with sugar-cane juice and cooked with

palm-cabbage or ground-nuts is taken for stomach-ache. It has an acrid taste and is strongly purgative (2, 3). The root of Nigerian material has been found very effective as a molluscicide giving 100% mortality at 100 ppm concentration of extract when tested on the fresh-water snail, *Bulinus globatus* (1). Leaf-extract showed no activity.

References:

1. Adewunmi & Sofowora, 1980, as *Cryrtogonone* and *Cryptogonone* (both orthographic errors) *argentea* (Pax) Prain. 2. Walker, 1953, a: 35. 3. Walker & Sillans, 1961: 164.

Dalechampia ipomoefolia Benth.

FWTA, ed. 2, 1: 412–3.
West African: **SIERRA LEONE** LIMBA butɛgɛ (NWT) LOKO hĕhugiri (NWT) ndubu (NWT) MENDE dongonda (NWT) mɔbe (NWT) mɔli (NWT; FCD) tupwiya (NWT) TEMNE ɛ-kali (NWT) **IVORY COAST** BAULE atutuma kolbré (B&D) KRU-GUERE (Chiehn) wuzanizani (B&D) **NIGERIA** IGBO ngumiili (NWT)

A slender twiner of the grassy savanna occurring from Sierra Leone to S Nigeria, and extending to Zaïre.

The stems are covered with pubescent hairs but the hairs on the flowers and fruit are urticating (2, 3). The plant has use in Ivory Coast in topical applications to sooth costal and rheumatic pains (1).

References:

1. Bouquet & Debray, 1974: 83. 2. Morton SL.1530, FBC, and collectors. 3. Thomas, NWT.318, K.

Dalechampia scandens Linn. var. cordofana (Hochst.) Müll.-Arg.

FWTA, ed. 2, 1: 413.
West African: **SENEGAL** ARABIC (Senegal) tatrarèt (JB) FULA-PULAAR (Senegal) gité gelodi (JB)

A slender twiner of dry grassy savanna, recorded only from NW Senegal, and occurring in the Cape Verde Islands, and widely across Africa from Lake Chad to the NE, central and South Tropical Africa and into Arabia.

The plant has unpleasantly stinging hairs on the leaves, stems and fruits. No usage is recorded within the Region. In Tanganyika it has unspecified use as a children's medicine (3). The leaves of variety *hildebrandtii* (Pax) Pax are eaten in Kenya as a vegetable (1) and in Tanganyika the roots enter into magic (2).

References:

1. Graham 2053, K. 2. Koritschoner 1154, K. 3. Moggridge 142, K.

Dichostemma glaucescens Pierre

FWTA, ed. 2, 1: 416. UPWTA, ed. 1, 140.

A tree to about 13 m high of the forest zone in S Nigeria and W Cameroons, and into Zaïre.

The wood burns with a hot flame (4).

Though bark preparations are reported to be responsible for troubles connected with pregnancy (2), a bark-macerate is considered in Gabon to be a good tonic for suckling women (3, 4). It is also taken as an emetic (4).

A tisane of young leaves, or eaten as a salad, are taken in Congo (Brazzaville) for gastro-intestinal and liver complaints, and leaves with those of *Piper*

EUPHORBIACEAE

guineense (Piperaceae) and *Tetrochidium didymostemon* (Euphorbiaceae) for stomach-pain and spitting blood (1).

References:

1. Bouquet, 1969: 112. 2. Dalziel, 1937: 140. 3. Walker, 1953, a: 35. 4. Walker & Sillans, 1961: 165.

Discoglypremna caloneura (Pax) Prain

FWTA, ed. 2, 1: 403. UPWTA, ed. 1, 140.
West African: SIERRA LEONE MENDE gbɔvui-hini (*def.* -hinei) *from* bovui: *soft;* hini: *male* (S&F) LIBERIA KRU-BASA leh-bohn (C&R) IVORY COAST ABE akoret (Aub.; K&B) ABURE diéké (B&D) KRU-GUERE botué (K&B) GUERE (Chiehn) dikpalassu (B&D) KYAMA akoni (B&D) apohia (Aub.; K&B) didiréwabaka (B&D) mobohia (Aub.; ex K&B) GHANA VULGAR noma-noma gyama (DF) AKAN-ASANTE fetefrɛ (E&A) *m*omanamagyama (FRI; CJT) ogwam-dua (E&A) TWI fetɛfrɛ (E&A) ofretere (FRI) noma-noma gyama (DF) ogwam-dua (E&A) WASA dubrafo (BD&H; FRI) fetefre (DF; E&A) ANYI-SEHWI dein (auctt.) NIGERIA EDO okpághégūī (auctt.) uguomafia (JMD; KO&S) IGBO obinwaảnyị *from* nwaảnyị; *woman* (KO&S; KW) YORUBA akikagbá (KO&S; Verger) ṣòkùn ṣógwùo (Kennedy; Verger)

Tree to 30 m high by 2 m girth, bole straight to 15 m long, of the evergreen forest and a weed-species of regenerating forest, from Guinea to W Cameroons, and extending to Zaïre and into Uganda.

The wood is white or yellowish, moderately light, and too soft and perishable for building purposes, canoes, etc. (4, 7, 8), but it is easy to carve and is so worked in Liberia for devil-masks and food-bowls (4).

Wood-ashes are reported to produce sores on the skin (Moor fide 3, 5).

A decoction of crushed leaves is taken by draught in Ivory Coast as an expectorant in bronchial troubles (6) and the plant (part not stated) is used as an emeto-purgative in dysenteric diarrhoea and to help in difficult childbirth (2). In Congo (Brazzaville) a bark-decoction is given in draught to relieve convulsive coughing and intestinal pains in food-poisoning, and as an emetic; the bark-powder is used to treat sores (1).

Traces of *alkaloids* have been detected in the bark of the stems and roots (2).

The fruit is fleshy and pubescent, and reddish when ripe with a bitter taste (4). It is however sought after by birds which distribute the seeds (7), and in Ghana the seed (? fruit) is used as a bait in bird-traps (5).

References:

1. Bouquet, 1969: 112. 2. Bouquet & Debray, 1974: 83. 3. Burtt Davy & Hoyle, 1937: 44. 4. Cooper & Record, 1931: 51. 5. Irvine, 1961: 221–2. 6. Kerharo & Bouquet, 1950: 71. 7. Savill & Fox, 1967: 115. 8. Taylor, 1960: 162.

Drypetes aframensis Hutch.

FWTA, ed. 2, 1: 381.
West African: NIGERIA YORUBA tafia (KO&S) tanfia (Macgregor)

An understorey tree of the forested savanna zone, to 13 m high, recorded from Ghana and S Nigeria.

The wood is yellowish brown and very hard (1).

Reference:

1. Keay & al., 1960: 281–2.

Drypetes afzelii (Pax) Hutch.

FWTA, ed. 2, 1: 382. UPWTA, ed. 1, 140.

English: white kotowuli (from Mende, Sierra Leone, Small).

West African: **SIERRA LEONE** LOKO nde (NWT) ndɔkbai (NWT) yuŋge (NWT) MENDE boboi (?) (FCD) gokai (Aylmer) SUSU kantinyi (NWT) kiri (NWT) TEMNE an-gbongbo *the gum* (FCD) **LIBERIA** KRU-BASA duay-gre (C) **IVORY COAST** ABE mottikoro (Aub.) KRU-GUERE fua (Aub.) **GHANA** VULGAR betemaa (DF)

A tree to over 25 m tall, of the high closed-forest from Sierra Leone to Ghana and in E Cameroun. This species is confused with *D. aylmeri* Hutch. & Dalz. Sap-wood is white, heart-wood light brown. The wood is said to be hard, durable and resistant to termites (1).

Gum from the trunk is aromatic. Temne of Sierra Leone mix it with white clay and palm-oil to rub on the body as a scent (2).

The flowers are scented attracting many insects. Male flowers bear nectar (3, 4).

References:

1. Dalziel, 1937: 382. 2. Deighton 2904, K. 3. De Wilde 2614, K. 4. Jaeger, 1965: 109–10.

Drypetes aubrevillei Léandri

FWTA, ed. 2, 1: 381. UPWTA, ed. 1, 141, as *Drypetes sp.*

English: pepperbark (Liberia, Cooper); pepper stick (Liberia, Voorhoeve).

West African: **SIERRA LEONE** KONO putu-kɔne = *pepper tree* (S&F) MENDE pujɛ-wuli = *pepper tree* (S&F) **LIBERIA** KRU-BASA swam-beh = *sting pepper; alluding to the spicy bark* (C&R) **IVORY COAST** ABURE meniéua (B&D) DAN chleubu (Aub.) ·KRU· piatu (Aub.)

A forest understorey tree to 15 m high, rarely to 25 m, recorded from Sierra Leone, Liberia and Ivory Coast.

The wood is pale yellow, hard and heavy, with a rather fine texture, finishing well and taking a good polish. In Liberia it is used for timbers and house-poles (3) but it is probably not durable and is subject to fungus stain (4).

The bark has a peppery taste, hence the local English names. A decoction of macerated bark and fruit is used in Liberia as a liniment for application in fever, rheumatism and general fatigue (3). The powdered bark is used to treat a skin-condition called *dishcloth*, the symptoms of which are a blotchiness (2, 4). In Ivory Coast a pap made from the bark is taken as an expectorant and bronchial decongestant (1).

References:

1. Bouquet & Debray, 1974: 83. 2. Cooper 327, K. 3. Cooper 317, K. 4. Cooper & Record, 1931: 51–52 as *D. ovata* Hutch., with timber characters.

Drypetes aylmeri Hutch. & Dalz.

FWTA, ed. 2, 1: 381. UPWTA, ed. 1, 140 as *D. mottikoro* Léandri.

French: bois allumettes (i.e., match-wood, Ivory Coast, Aubréville).

West African: **GUINEA** AKAN-FANTE dua-samina (FRI; DF) WASA betimaa (Andoh; E&A) **SIERRA LEONE** LOKO pɛrehun (NWT) MENDE gokai (Aylmer) SUSU kolaxone (NWT) TEMNE ɛ-run (ɛ-ruŋ) (NWT) **LIBERIA** KRU-BASA duay-gray (C&R) MENDE to-ye (C&R) **IVORY COAST** ABE mottikoro (Aub.)

A tree to 13 m high of the closed-forest from Sierra Leone to Ghana.

Wood is white and splits easily, but it is said to be hard, durable and resistant to termites, hence its use in Liberia for hut-poles and planking (3–6). The ease of its splitting has suggested that it might be usable for match-splints (1, 2) as implied in the French name.

References:

1. Aubréville 533, K. 2. Aubréville 1959: 2: 56. 3. Cooper 315, K. 4. Cooper & Record 1931: 51, as *D. afzelii* Hutch. (?) 5. Dalziel, 1937: 381. 6. Irvine, 1961: 223.

Drypetes capillipes (Pax) Pax & K Hoffm.

FWTA. ed. 2, 1: 382.

A shrub or small tree to 4 m high of the forest in S Nigeria, and in E Cameroun to Zaïre.

No use is recorded in the West African region. In Congo (Brazzaville) a bark-decoction is used as a mouth-wash for severe toothache, and in enema for kidney-pains, and the leaves are used in topical massage for stiffness of the neck (1).

Reference:

1. Bouquet, 1969: 112.

Drypetes chevalieri Beille

FWTA, ed. 2, 1: 382. UPWTA, ed. 1, 140.
West African: **LIBERIA** MANO kpũ kpũ la (Har.) **IVORY COAST** ABE krahain (Aub.) KRU-GUERE kran (K&B) **GHANA** AKAN-ASANTE katrika (auctt.) **NIGERIA** EDO aghedan (KO&S) owe (KO&S) YORUBA aya (APDJ; KO&S) ità òyibo igbó (Kennedy) òsúnsún ìrò (JMD; KO&S), *Ondo dialect* (Verger)

A shrub or small tree to 10 m high, of the understorey of the closed-forest and secondary forest of the rain-forest zone, often by streams.

The wood is hard. Twiggy branch ends are used in Nigeria to make brooms (Kennedy fide 2).

Sap expressed from leaves and twigs is taken in draught by the 'Kru' and the Guere of Liberia for dysentery. The powdered leaves are sometimes used in Ivory Coast as a snuff for head-colds and sinusitis (3). The plant is also used for bronchial and intestinal troubles (1).

References:

1. Bouquet & Debray, 1974: 83. 2. Irvine, 1961: 223-4. 3. Kerharo & Bouquet, 1950: 71.

Drypetes floribunda (Müll.-Arg.) Hutch.

FWTA, ed. 2, 1: 381. UPWTA, ed. 1, 140, incl. *D. ovata* Hutch.
West African: **SENEGAL** DIOLA bu solèn (JB) fu kinora (JB) DIOLA (Bayot) é tiru (JB) DIOLA-FLUP suï-suï (JB) **GHANA** ADANGME-KROBO tɛtʃo = *stone tree* (BD&H; FRI) AKAN-ASANTE katrika (auctt.) TWI bedibɛsa = *if you eat it, it will soon be finished; i.e., it is so sweet* (FRI) GA tɛtʃo = *stone tree* (FRI) GBE-VHE (Awlan) klẽ (FRI) klẽti (FRI) VHE (Pecí) akpa (FRI) **NIGERIA** OGORI ako tagbesso (Taiwo) YORUBA ako tagbesso (Taiwo; KO&S) asokara (?) (JMD)

A tree to 10 m high of drier parts of the closed-forest and in savanna from Senegal to S Nigeria, and on to Zaïre.

The wood is white, very hard and strong, evoking the Krobo name *tɛtʃo* meaning 'stone-tree'. It is used in hut-construction, and by the Vhe of Ghana for yam-poles. It is a good firewood and is used for carving into spoons, etc. The wood has a sweetish taste and it is commonly used as a chew-stick, the Twi name being *bedibɛsã* meaning 'if you eat it, it will soon finish' (1-3). The tree has thorny evergreen shining leaves somewhat like the temperate holly. It is used by the Keta people of eastern Ghana as a 'Christmas tree' (1, 2). The fruits are about 1.25 cm in diameter and orange-red when ripe. The pulp is edible.

References:

1. Dalziel, 1937: 381. 2. Irvine 1961: 224. 3. Portères, 1974: 120.

Drypetes gilgiana (Pax) Pax & K Hoffm.

FWTA, ed. 2, 1: 382. UPWTA, ed. 1, 148.
West African: SENEGAL DIOLA-FLUP ka hota (JB) SIERRA LEONE LIMBA tɛŋɛnɔgɔ (NWT) MENDE gɛngɛmage (NWT) pondoɛmi (NWT) SUSU kɛkɔrisiri (NWT) kuluwuruanie (NWT) TEMNE ɛ-kala (NWT) IVORY COAST KRU-GUERE sampu (Aub.) GHANA ADANGME-KROBO tɛtʃo (FRI) AKAN-ASANTE katrika-nini (auctt.) FANTE adweaba (auctt.) harjuaba (Brent) TWI katerika (JMD; E&A) WASA adwea (FRI) adweaba (FRI) NZEMA adweaba (auctt.) NIGERIA YORUBA eke (KO&S) osunsunro (Macgregor; KO&S)

Tree to 13 m high, understorey of the high evergreen and deciduous forest from Guinea-Bissau to W Cameroons, and in E Cameroun. This species is easily confused with *D. chevalieri* Beille.
The wood is light brown and very hard (2).
The fruits, 1.2–2.0 cm diameter, are bright scarlet or orange. The pulp surrounding the seeds is sweet and is eaten. They are sometimes sold in markets in Ghana (1).

References:

1. Irvine, 1961: 224–5. 2. Keay & al, 1960: 282–3.

Drypetes gossweileri S Moore

FWTA, ed. 2, 1: 382. UPWTA, ed. 1, 140, as *D. amoracia* Pax & Hoffm.
English: horse-radish tree; as also *Moringa* (Dalziel).
West African: NIGERIA EDO okhuaba (auctt.) YORUBA agawo (Kennedy; KO&S)

A tree to 40 m high, with a clear straight hole to 30 m long, of the high forest of S Nigeria, and extending to Zaïre.
Any part of the tree when cut or bruised emits a pungent smell resembling that of horse-radish and methyl salicylate. The bark has a pungent taste. It is used in W Cameroons for gonorrhoea and toothache (4). In Gabon a bark-macerate or decoction with pimento is taken as an anthelmintic, and in Congo (Brazzaville) as a vermifugal enema (2). The bark has general use as a revulsive and treatment against pain: for rheumatism in Gabon (5); for headaches and all manner of body-pains in Congo (Brazzaville) (2); where also the bark-powder is taken as an aphrodisiac cooked with a banana and to relieve urethral discharges, and in decoction to make a bath against fever in young children.
In Congo (Brazzaville) the bark is reported to repel snakes. It is sufficient to keep a piece of bark in the hut-roof, or to pour about the hut some water in which bark has been boiled (2).
Bark and fruits are used as fish-poisons in Gabon (5).
In Nigeria the roots powdered with those of kola and *Celtis integrifolia* (Ulmaceae) are used to treat very deep wounds (1). *Saponin* has been detected in the roots and *alkaloids* up to 1% concentration in the bark and roots (3).

References:

1. Ainslie, 1937; sp. no. 140. 2. Bouquet, 1969: 113. 3. Bouquet, 1972: 24. 4. Dalziel, 1937: 140. 5. Walker & Sillans, 1961: 165.

EUPHORBIACEAE

Drypetes inaequalis Hutch.

FWTA, ed. 2, 1: 382.
West African: LIBERIA MANO kpŭ kpŭ la (Har.)

A shrub to 5 m high of the forest, only recorded in Sierra Leone and Liberia.

Drypetes ivorensis Hutch. & Dalz.

FWTA, ed. 2, 1: 381. UPWTA, ed. 1, 140.
West African: LIBERIA KRU-BASA kpahn-wi (C&R) IVORY COAST AKYE kokwanié (K&B)

A lower-storey tree of the closed-forest, to 5 m high, from Liberia to Ghana. In Liberia the wood reaches sufficient size for the making of small rice-mortars to about 30 cm diameter (5). Bark and fruits are used to prepare dressings to maturate boils and carbuncles (5, 6). The bark is considered toxic in Ivory Coast and is used by the Akye to prepare poison-bait for rats, mice and other noxious animals (3, 8).

The fruit-pulp is edible (2, 4, 7), and eaten by the Basa on the Liberian Coast and the Mano of the Nimba mountain region (1).

References:

1. Adam, 1971: 374. 2. Aubréville, 1959: 2: 57. 3. Bouquet & Debray, 1974: 83. 4. Busson, 1965: 163. 5. Cooper 418, K. 6. Cooper & Record, 1931: 51. 7. Dalziel, 1937: 140. 8. Kerharo & Bouquet, 1950: 72.

Drypetes klainei Pierre

FWTA, ed. 2, 1: 381.
West African: IVORY COAST KRU-GUERE zogré (Aub.)

A forest tree to 25 m high, bole straight to over 1 m girth, occurring in Ivory Coast and Gabon.

No use is recorded of the tree in Ivory Coast. In Gabon the bark and fruits serve as fish-poisons, and a macerate or decoction of fresh bark is used topically for rheumatism, and with a pimento is taken internally as an anthelmintic (1).

Reference:

1. Walker & Sillans, 1961: 165.

Drypetes leonensis Pax

FWTA, ed. 2, 1: 381.
West African: GHANA AKAN-WASA ɛsononankoroma = *elephant's knee* (auctt.) NIGERIA YORUBA ajoko (KO&S) pepe (Ross) tafia (KO&S)

An understorey tree of the high-forest, to 23 m high from Guinea to W Cameroons, and in E Cameroun.

The flowers are heavily scented.

Drypetes molunduana Pax & Hoffm.

FWTA, ed. 2, 1: 381.

A shrub or small tree to 10 m tall, of forest in S Nigeria and W Cameroons, and into E Cameroun.
The wood is hard (2) and is used in S Nigeria for house-building (1).

References:

1. Akpabla 1100, K. 2. Keay & al., 1960: 281.

Drypetes parvifolia (Müll.-Arg.) Pax & K Hoffm.

FWTA, ed. 2, 1: 382.
West African: IVORY COAST KRU-GUERE sampu (Aub.)

A shrub or small tree of the savanna forest from Sierra Leone to N Nigeria.

Drypetes paxii Hutch.

FWTA, ed. 2, 1: 381.
West African: NIGERIA YORUBA ata igbo (KO&S)

An understorey tree of the closed-forest, to 13 m high, occurring in S Nigeria, W Cameroons, and on to Gabon.
The bark has a peppery taste (1).

Reference:

1. Keay & al., 1960: 279–80.

Drypetes pellegrinii Léandri

FWTA, ed. 2, 1: 381. UPWTA, ed. 1, 141, incl. *D. vignei* Hoyle.
West African: IVORY COAST 'SOUBRE' kahibéhi (Aub.) GHANA VULGAR opaha (DF) AKAN-ASANTE ɔpaha-kɔkɔɔ (auctt.) TWI ɔpaha (auctt.)

A understorey tree of the high-forest, to 13 m tall by 1 m girth, locally abundant in Ivory Coast and Ghana.
The bark has a spicy taste and is used in Ivorean medicine (1).

Reference:

1. Aubréville, 1959, 2: 54.

Drypetes principum (Müll.-Arg.) Hutch.

FWTA, ed. 2, 1: 381.
West African: IVORY COAST ABE mottikoro (Aub.) GHANA AKAN-ASANTE ɔpaha (BD&H; FRI) NIGERIA YORUBA ude (KO&S)

An understorey tree of the closed-forest, to 15 m high, from Guinea to W Cameroons and Fernando Po, and in E Cameroun and Príncipe.

Drypetes roxburghii (Wall.) Hurusawa

FWTA, ed. 2, 1: 368, as *Putranjiva roxburghii* Wall.

A tree of India and Thailand that has been introduced into cultivation in the Region.

The timber is used for turnery in India, and the leaves serve as cattle-fodder. A decoction of leaves and fruit is taken for liver-complaints and fevers in India, and in Thailand for rheumatism. The hard stones of the fruit are threaded into necklaces.

Drypetes sp. indet.

West African: **IVORY COAST** ABURE atabinini (B&D)

Duvigneaudia inopinata (Prain) J Léonard

FWTA, ed. 2, 1: 415, as *Sebastiana inopinata* Prain.

A shrub or understorey forest tree of stream-banks and swampy localities, attaining 11 m height and recorded only from W Cameroons in the Region but also occurring in E Cameroun, Gabon and Zaïre.

The plant has creamy yellow flowers that are pleasantly scented. It is grown as an ornamental in parks and gardens in Zaïre (1).

The wood is white and soft. The bark contains an abundant white latex.

Reference:

1. Léonard, 1962, b: 140–2.

Elaephorbia drupifera (Thonn.) Stapf

FWTA, ed. 2, 1: 423. UPWTA, ed. 1, 141, as *E. drupifera* Stapf, in part, and 144, *Euphorbia leonensis* NE Br., in part.

West African: **GHANA** VULGAR adidi (DF) akan (DF) ADANGME-KROBO toro (FRI) tulo (FRI) AKAN-ASANTE akãn(-e) (FRI) FANTE kakan (FRI; RS) kanɛ (E&A) kankan (FRI) TWI akãn *a contraction of* akã aniwa: *to spoil, or to touch the eyes* (FRI; E&A) WASA adidi (auctt.) GA tùnyo (FRI; KD) GBE-VHE tredzo (auctt.) VHE (Awlan) tredzo (BD&H; FRI) VHE (Pecí) agyurlo (CV; FRI) tredzo (BD&H; FRI) NZEMA bɛnyibɔlɛ-baka = *a tree that spoils the eyes* (FRI) dodo (BD&H; FRI) **TOGO** ? tenjotjo (Pax fide JMD) GBE-VHE zoku (Aub.) NAWDM ssaldaza (Volkens fide JMD; RS) **DAHOMEY** GBE-FON sozo (Aub.) GEN adi (Aub.) VHE zoku (Aub.) **NIGERIA** IGBO ɔtamazi (NWT) YORUBA oro adete (Macgregor; KO&S)

A tree to 25 m tall with straight cylindrical bole in forest, or low-down branching in open situations (6). When young, cactiform and succulent. Of the closed-forest and open savanna. There has been confusion between this species and *E. grandifolia* as to their respective geographic distributions. This species occurs from Ghana to S Nigeria and extends to Zaïre. Information from references to this species from outside this area is referred to *E. grandifolia*.

In parts of Ghana the tree is considered sacred and is planted as a fetish tree (5). In Gabon, as with fleshy euphorbias generally, the plant is ascribed with fetish and magic properties on account of the bizarre form. It is planted in streets and before houses as a fetish-protector to keep away evil influences and thunder-bolts, and twisted around arrows, spears and pointed stakes. Above all it is necessary for the burial of human bones.

The wood is soft and is used in Ghana to smoke fish (5). It is said to be able to keep away termites (4), a reference more likely to imply a degree of resistance to termite attack.

The bark contains a copious white latex which is poisonous and dangerous to the eyes as is recognised in the Twi and Nzema names. Vernacular names meaning 'goggle-eye' are sometimes applied to cactiform euphorbias (4). The juice has many uses. Applied to the skin it causes swelling. There is a belief in its analgesic property (? revulsive). In Nigeria it is applied to scorpion-stings to relieve the pain (1, 4, 7). In Congo (Brazzaville) a bark-decoction is used as a mouth-wash for violent toothache (2). When applied to warts the latex is said to burn them away. It is used on ring-worm and taken in small doses with milk, egg or soup as a purgative (1, 4, 7). In S Nigeria a preparation known as *àgúnmu* (Yoruba) is partly composed of a dried extract of the juice (4). Sap expressed from raspings of the bark is rubbed over areas of craw-craw in Ubangi, after which the raspings are put on as a poultice (11). In Congo (Brazzaville) the latex is considered too strong a purge to give to children (2).

The latex is used in Ghana as a fish-poison. It is either put directly into the water or filled into shells and calabashes which are dropped into deeper holes in rivers where fish congregate (5). Action is fairly quick. Use as a fish-poison is also practiced in Congo (Brazzaville) (2).

The latex is used, along with others, to drip into the eyes of novitiates in secret society initiations in Gabon (8). Fang mediums use it in trances: if the *Alchornia floribunda* concoction which is taken to induce the trance is slow to act, some latex mixed with oil is wiped over the eyeball with a feather. This affects the optic nerve to produce bizarre visual effects and a general daziness (9, 10). While it is generally known to be dangerous to the eyes, even to the extent of causing blindness (7), these practices must recognise that instillation under certain conditions can be done without unacceptable permanent damage resulting.

Shavings of the stem used to be given to captives in Gabon during slave-raids to subdue them so that they lost any idea of escaping (8). Various euphorbiaceous plants were used, but principally this one. The leaves, pounded up with salt and onions are commonly used on guinea-worm sores to assist in the extraction of the parasite (4, 7).

The fruit is used in Ghana as a fish-poison (3).

Examination of the latex has indicated the presence of *diterpenes* of the *phorbol* and *ingenol* types. This could explain the plant's pharmacological activity. Recent work has shown that the esters of the two groups of diterpene alcohols commonly present in the Euphorbiaceae are responsible for the irritant, co-carcinogenic, piscicidal and toxicological properties (12).

References:

1. Ainslie, 1937: sp. no. 143. 2. Bouquet, 1969: 114. 3. Chipp 279, K. 4. Dalziel, 1937: 141, 144. 5. Irvine, 1961: 226–7, in part. 6. Keay & al., 1960: 253–4. 7. Oliver, 1960: 25, 61. 8. Walker & Sillans, 1961: 166. 9. Fernandez, 1972: 242–3. 10. Hargreaves, 1978: 21–30. 11. Vergiat, 1970: 69. 12. Kinghorn & Evans, 1974: 150–1.

Elaephorbia grandifolia (Haw.) Croizat

FWTA, ed. 2, 1: 423. UPWTA, ed. 1, 141, as *E. drupifera* Stapf, in part, and 144, *Euphorbia leonensis* NE Br., in part.

West African: SENEGAL BALANTA bañuô (JB) berendanda (JB) DIOLA bu koti (JB) DIOLA-FLUP bo koti (JB) kukoč (K&A) FULA-PULAAR (Senegal) mbura (JB) MANDING-BAMBARA mguana (JB; K&A) muana (JB) GUINEA-BISSAU BALANTA bagane-uône (JDES) berendenda (JDES) BIAFADA bidjaquedjaque (JDES) GUINEA FULA-PULAAR (Guinea) mbura (JMD; RS) SIERRA LEONE VULGAR ɔrɔ *from Yoruba-Aku* (JMD) BULOM (Kim) gbɛg (FCD) gbɛk (FCD) GOLA bondo (FCD) gbun (FCD) KISSI sondio (FCD; S&F) KONO sense (FCD; S&F) MENDE gboni (-i) (FCD; S&F) TEMNE an-paraŋ (FCD; S&F) ɛ-sot-koru (FCD; S&F) VAI boni (FCD) LIBERIA ˈKRUˈ boto (JMD; RS) MALI DOGON tényu (CG) UPPER VOLTA MANDING-BAMBARA faman (RP fide K&B) IVORY COAST ABE hié (RS; Aub.) yè (RP) yè hié (K&B) ABURE eya (B&D) AKAN-ASANTE doɗu (B&D) AKYE dan (K&B) dodo (auctt.) du (auctt.) ANYI dodo (Laffitte fide K&B) BAULE lôlô (A&AA) DAN dô, duô = *poison* (auctt.) KRU-GREBO

EUPHORBIACEAE

(Oubi) ulotu (RS) GUERE dohé = *poison* (RP; K&B) gbô (RP; K&B) gopo (A.Chev. fide K&B) gpo (K&B klatu = *forest tree; applied to toxic trees* (RP; K&B) GUERE (Wobe) g-borotu (A&AA) téwé (RP; K&B) NEYO gopo (auctt.) KWENI vuin (RP; JMD) KYAMA allé (Ivanoff fide K&B) MANDING-BAMBARA faman (RP) DYULA baga (RP; K&B) MANINKA fafa (B&D) faman (RP; K&B) 'PATOKLA' ulatu (RS) **GHANA** AKAN-TWI ɔkane (E&A)

A tree to about 13 m tall, of the forest and secondary formations from Senegal to Ghana.

The tree is widely planted as a fetish tree (Senegal: 8; Sierra Leone: 5; Mali: 4; Ivory Coast: 1, 8, 10) for protection against thunder-bolts and to ward off ghosts and evil spirits. It is grown as a hedge in Sierra Leone (4).

The leaves are used for sore-throat in Ivory Coast: softened over a fire and beaten, the sap is squeezed out. A leaf-extract is drunk or the dried powdered leaves are eaten for chest-complaints. The liquid from a bark-decoction is used to wash guinea-worm sores and the pulped bark is applied as a dressing (1).

The latex present in all parts of the tree is caustic. It is an ingredient of arrow-poison in Ivory Coast (10). As a fish-poison in Ivory Coast it is filled into a corked calabash which is thrown into water to be fished and the cork is pulled out by an attached string when the calabash has reached the appropriate position (8). The latex is recognised as being dangerous to the eyes. Tree-fellers in Ivory Coast will ring-girdle a tree about to be felled to a height above the cut for their own protection (2). The latex is a common ordeal-poison in Ivory Coast applied to the eye. Anyone admitting their crimes quickly receives an antidote – a decoction of dried leaves of this plant made in a young virgin's urine, or a decoction of a (unspecified) mucilaginous plant. Nevertheless latex in the eye can cause irreparable damage and government hospitals in Ivory Coast have recorded many cases of ophthalmia ascribed to such ordeal trials (8, 9).

Latex is particularly toxic to the mucosae. It causes blistering in the mouth. A few drops are placed on carious teeth in Casamance (Senegal) (6, 7) and in Sierra Leone (5). This facilitates extraction, perhaps by causing the gum to recede so that a broken stump is the more easily got at.

The latex is a drastic purgative, in general too violent for use and then only in grave situations (3, 6). Malingerers are said to use it to cause sickness and diarrhoea, but frequent taking of it has cumulative effect, and fatal cases of pericarditis and dropsy have been attributed to it (4). The latex is however taken in Ivory Coast in a prescription with food for poisoning (1). It is also applied to maturate furuncles (1).

Examination of the latex has indicated the presence of *diterpenes* of the *phorbol* and *ingenol* types. This could explain the plant's pharmacological activity. Recent work has shown that the esters of the two groups of diterpene alcohols commonly present in the Euphorbiaceae are responsible for the irritant, co-carcinogenic, piscicidal and toxicological properties (11).

References:

1. Adjanohoun & Aké Assi, 1972: 122. 2. Aubréville, 1959: 2: 28. 3. Bouquet & Debray, 1974: 83, as *E. drupifera*. 4. Dalziel, 1937: 141, 144. 5. Deighton 5047, K. 6. Kerharo & Adam, 1963,a: 7. Kerharo & Adam, 1974: 408–9. 8. Kerharo & Bouquet, 1950: 72, as *E. drupifera* Stapf. 9. Portères, 1935: 136, as *E. drupifera* Stapf. 10. Schnell, 1950,b: 222, under *E. drupifera* Stapf. 11. Kinghorn & Evans, 1974: 150–1.

Erythrococca africana (Baill.) Prain

FWTA, ed. 2, 1: 401. UPWTA, ed. 1, 142.

West African: **SENEGAL** BALANTA mbuhur mbéré (JB) DIOLA é férékéta (JB) MANDYAK be lasi (JB) **SIERRA LEONE** LIMBA budelemi (NWT) kunɔndi (NWT) LOKO bɛmba (NWT) kaiɛna (NWT ponjo (NWT) MENDE bɛgui (NWT) TEMNE ɛ-kan (NWT) **LIBERIA** MANO bu yiddi pulu (JMD) **GHANA** AKAN-TWI gyigyam forowa (BD&H; FRI)

A shrub or small tree to 3 m high or more, of deciduous or secondary forests from Senegal to S Nigeria.

Dried powdered leaves are added to food by the Mano of Liberia as a mild laxative (Harley fide 1). In Benin, S Nigeria, hunters use the leaves in cooking fresh meat to make it tender (Kennedy fide 2). Birds eat the fruit (Kennedy fide 2).

References:

1. Dalziel, 1937: 142 2. Irvine, 1961: 227

Erythrococca anomala (Juss.) Prain

FWTA, ed. 2, 1: 401. UPWTA, ed. 1, 142.
West African: GUINEA ? sakadhoëlly (Heudelot) SIERRA LEONE KISSI lumbuma-pɔmboma-yondoã (FCD) KRIO bush-lɛm *i.e., bush lime* (FCD) LIMBA budikiɛ (NWT) businesaŋ (NWT) butir (NWT) LOKO bandaga (NWT) supana (NWT) tenbuɛ (NWT) yandi (NWT) MENDE *n*dogbɔdumbi *corruption of* ndogbɔdumbele: *bush lime* (JMD) *n*dɔgbɔ-lumbe(-i) *corruption of* ndogbɔdumbele: *bush lime* (FCD) dɔgwɔdumbe (NWT) dɔkbɔlimbi (NWT) *n*doybodumbi (L-P) SUSU digiɛnwuri (NWT) siriwuri (NWT) tundewuri (NWT) TEMNE ɛ-banka (NWT) ɛ-baŋk-baŋk (FCD) ɛ-roks-ɛ-rokant NWT) IVORY COAST ABURE aprokondu (B&D) AKYE pa-*n*dédé (A&AA) BAULE bolidé yassua (B&D) kétiboaomi (B&D) DAN bliassé (K&B) GAGU koké (K&B) kokwè (B&D) 'KRU' daroagnéba (K&B) KRU-GUERE glédion (K&B) glévéréyé (K&B) gléyé (K&B) gléyoï (K&B) gliwé (K&B) GUERE (Chiehn) kélokwe (K&B) koré kwè (B&D) kroak kwé (B&D) kruékrué (B&D) KWENI goréiri (K&B) KYAMA apitiboé (B&D) NIGERIA EDO ehín-ógo *a kind of pepper* (Elugbe) YORUBA iyèré-igbò = *bush pepper* (EWF, ex JMD; Verger)

A spiny shrub to 3 m tall, in secondary growths of the forested zone from Guinea to W Cameroons and Fernando Po, and into E Cameroun.

This is the 'bush-lime' of the relic-forest savanna area of Sierra Leone. In the closed-forest zone 'bush-lime' is *Dovyalis zenkeri* Gilg (Flacourtiaceae).

The plant has analgesic and antiseptic properties. In Ivory Coast, leaf-sap is instilled into the nose, eyes or ears for sinusitis, colds, ophthalmias and inflammations of the outer ear; fruit-pulp is rubbed topically for localised pain; a decoction of leafy twigs is used to cleanse sores, ulcers, craw-craw and yaws ulcers; leaves powdered with maleguetta pepper furnish a snuff for chronic headaches (4, 7). The Igbo of Nigeria use the bark for arthritic conditions (8). Leaf-sap is dripped into the eyes for eye-troubles in Sierra Leone (5). The Akye of Ivory Coast reduce the leaves to a powder with those of *Cephaelis peduncularis* (Rubiaceae) and some kaolin and rub the mixture with a little water onto the body of infants suffering from malaria. The same preparation is rubbed on the neck for meningitis (2). A decoction of the plant may also be used on feverish children, and is used as a purgative (4).

In Guinea the plant has a high reputation as a specific against tapeworm (6).

Twigs and leaves are reported to contain 0.1% total *alkaloids* (4). Over 1% concentration is recorded for the roots and up to 1% in the bark (3), and a very strong presence has been found in the seeds of the Nigerian material (1).

In the Danané area of Ivory Coast the plant has magical power to protect the followers of the Snake Sect from snake-bite (7).

References:

1. Adegoke & al., 1968. 2. Adjanohoun & Aké Assi, 1972: 123. 3. Bouquet, 1972: 24. 4. Bouquet & Debray, 1974: 84. 5. Deighton 2652, K. 6. Heudelot 856, K. 7. Kerharo & Bouquet, 1950: 73. 8. Iwu & Anyanwu, 1982: 263–74.

Erythrococca chevalieri (Beille) Prain

FWTA, ed. 2, 1: 401.
West African: WEST CAMEROONS BAMILEKE mboù-è-ssob (anon.)

A forest shrub recorded from Guinea to W Cameroons, and on to the Congo basin.

In Congo (Brazzaville) the plant is eaten as a vegetable. It is held to be aphrodisiac, effective against blennorrhoea, and cicatrisant on sores. Bronchial troubles are treated by a draught of leaf-sap. Itch is also treated with leaf-sap. A tisane of the roots is taken to relieve stomach-pains (1). *Alkaloids* have been recorded present: leaves, a trace; bark, 0.1–0.3%; roots, 0.3–1.0% (2).

References:

1. Bouquet, 1969: 114. 2. Bouquet, 1972: 24.

Erythrococca welwitschiana (Müll.-Arg.) Prain

FWTA, ed. 2, 1: 401.

A forest shrub to 5 m high, recorded only from S Nigeria, and onwards to Zaïre and Angola.

No use is recorded in the Region. In Congo (Brazzaville) it is used in the same manner as is *E. chevalieri* (Beille) Prain (1): q.v.

Reference:

1. Bouquet 1969: 114.

Euphorbia baga A Chev.

FWTA, ed. 2, 1: 422. UPWTA, ed. 1, 145–6.
West African: MALI MANDING-BAMBARA baga = *poison* (A.Chev.)

A perennial herb with short erect underground or partly exposed stems from a stout woody root-stock, recorded only from Ivory Coast and N Nigeria.

The plant is poisonous, as is implied by the vernacular name. The fleshy stem may be mistaken for the tuber of certain edible asclepiads and it is recorded as causing headache and drowsiness if eaten (1).

Reference:

1. Dalziel, 1937: 145–6.

Euphorbia balsamifera Ait.

FWTA, ed. 2, 1: 422. UPWTA, ed. 1, 142–3.

English: balsam spurge.

French: euphorbe de Cayor (Dalziel, Schnell); euphorbe candélabre (Kerharo & Adam).

West African: MAURITANIA ARABIC (Hassaniya) afernane (Aub.; GR) afzrinam (Aub.) ifernane (Aub.; GR) SENEGAL ARABIC (Senegal) afdir (JB) afernan (auctt.) afzrinan (Aub.; JB) ifernan (auctt.) FULA-PULAAR (Senegal) badakaradié (JB) badékarèy (*pl.* budekardé) (K&A; JB) batukari (JB) budakaréï (JB) TUKULOR badakarey (K&A) badekarey (*pl.* budekardié) (K&A; JB) budakaréï (JB) MANDING-MANDINKA salan (JB) NON damoi (JB) SERER ndamô (K&A) ndamol (auctt.) SONINKE-SARAKOLE salan (K&A) WOLOF salam (K&A) salan (auctt.) MALI ARABIC (Mali) afernane (A.Chev.) DOGON gòmudu (Aub.; CG) SONGHAI bèrdé (*pl.* -a) (D&C) berré (A.Chev.; Aub.) wàa-ñá´ (*pl.* -ñóŋó) (D&C) TAMACHEK taghelt (Aub.) tigatt (Aub.) tighal (A.Chev.; Aub.) tighar (Aub.) UPPER VOLTA FULA-FULFULDE (Upper Volta) badagereehi (*pl.* -eeji) (K&T) barnaahi (*pl.* -aaji) (K&T) HAUSA agua (Aub.) aguau (Aub.) GHANA DAGBANI jiriyung (Coull) HAUSA aguwa (FRI) NIGER HAUSA agua (Aub.) aguau (Aub.) aguwa (AE&L) ayyara (AE&L) kwar-kar (Virgo) KANURI magara (Aub.) yaro (Aub.) SONGHAI berré (Aub.) TAMACHEK taghelt (Aub.) tigatt (Aub.) tighal (Aub.) tighar (Aub.) NIGERIA FULA-FULFULDE (Nigeria) agwaje (JMD) HAUSA aguwa (JMD; ZOG) alyara (JMD; ZOG) ayyara (auctt.) káágúwáá (JMD; ZOG) kwakka (JMD) úwár-yáàraá (JMD; ZOG) yaro (JMD; ZOG) SURA gùu *and other E.spp used for living fences* (anon. fide KW)

An erect shrub, much branched from base up, to 4 m high, of dry sandy sites from Mauritania to N Nigeria, its northern limit being the Sahel zone marking the southern boundary of the Sahara; and also in the Canary Islands and in Rio de Oro. It has been recorded in Somalia and is perhaps present in E Africa. The plant is commonly grown as a hedge and field-boundary marker. It is easily raised by cuttings and is said to be one of the best hedge plants for low rainfall areas, i.e. under 900 mm. It requires a deep sandy soil. It is not eaten by the stock, nor by termites (5, 13). In some cemeteries in the Accra plains of Ghana a plant is grown to mark each grave (7). It is grown in villages for superstitious as well as medicinal purposes (4). In the soundanian zone it is commonly grown along the sides of roads and tracks and on wind-blown dunes as a sand-binder (1, 2). Its ease of propagation, however, should not be allowed to lead to uncontrolled spreading and its restriction in the Sahel zone is recommended in order to improve pasturage (1).

The succulent branches carry a copious amount of latex. The plant, and more specifically the latex, are generally reported to be toxic. The latex is known to be dangerous to the eyes and the plant has been used for criminal purposes, in ordeal-poison and in arrow-poison (8, 9). Yet boys are said commonly to suck the young shoots, and in Senegal the shoots are cooked for food (4, 6). During the famine of 1972–74 in N Nigeria, there is record of the plant shoots being eaten in Kano Province, but on questioning after the famine was over villagers claimed it to be inedible (15). Though cattle do not browse the plant, it is taken in Senegal by camels, sheep and goats (1) and sheep will eat the fallen leaves (4). In a superstitious sense it is fed to camels in the Sahara as the lichen commonly covering the bark is said to enable them to see by night (10). The Fula of N Nigeria sometimes feed the latex to their cattle and other stock to promote fertility and to increase their milk (4). Superstition or no, this can hardly be practiced without a sense of responsible husbandry. It must appear that the toxicity of the plant needs examination in its edaphic and phenological context.

An aqueous macerate of the bark and roots is taken in draught in Senegal as a drastic purge (8, 9) and is administered as such in treatment of leprosy, syphilis, gonorrhoea, etc. requiring cleansing of the abdominal region. A decoction of the leafy twigs is used to wash the private parts in leucorrhoea and menorrhoea, and to treat ringworm. The roots in decoction are taken for intestinal parasites. Stems and roots are normal items of market trade in Dakar (9). Water in which leaves have been steeped is used in NW Senegal to wash fever patients, and a concoction of flowering branch ends is taken to expel worms (3).

The latex is used externally as an antitoxin on snake and insect-bites (8, 9) and to apply to guinea-worm sores (4, 12). It relieves toothache and gum troubles, and contains a gum-resin and a revulsive identified as *euphorbon* (14). Placed on a carious tooth it not only curbs aching but serves to loosen the tooth to facilitate extraction (4). It is placed on tsetse fly bites (11). It is possible that there is some analgesic action. The latex is viscously tacky and is used in N Nigeria to make a bird-lime (*madoro*: Hausa) to catch crickets, etc. It might have some potential as a 'fly-paper' as the latices of some other *Euphorbia spp.* which retain an adhesive quality have been successfully used to catch tsetse flies (4).

In N Nigeria the plant has veterinary use for horses suffering from *nanduhu* and *saminya* (Hausa). The horse's head is held in the smoke over burning dried twigs of the plant. Latex is compounded into a poultice with leaves of *Annona senegalensis* (Annonaceae) and latex of *Calotropis procera* (Asclepiadaceae) which is then applied to the indurations of 'yaws' to maturate them (4).

Superstitious connotations lie in the expression *alyara ka ba ni nono* (Hausa: *nono*, breast) used by Gobir girls at puberty, and in the term *taya-ni-goyo* (Hausa: help me carry the child) which is used for the plant, and is also the name of a skin-affection on the back of a woman due to carrying a child (Bargery fïde 4). In Senegal young women in parturition use a hot hip-bath containing fresh leaves or the boiled water of fresh leaves is used topically (Trochain fide 4, 7).

References:

1. Adam, 1966, a. 2. Aubréville, 1950: 176. 3. Boury 1962: 15. 4. Dalziel, 1937: 142–3. 5. Howes, 1946. 6. Irvine, 1952,a: 6: 32. 7. Irvine, 1961: 228. 8. Kerharo & Adam, 1964,b: 441–2. 9. Kerharo & Adam, 1974: 409–10. 10. Monteil, 1953: 95. 11. Oliver, 1960: 26. 12. Schnell, 1953,b: no. 28. 13. Trochain, 1940: 266. 14. Etkin, 1981: 82. 15. Mortimore, 1987.

Euphorbia calyptrata Coss. & Dur.

West African: **MAURITANIA** ARABIC (Hassaniya) ramad = *that which gives ophthalmia* (AN) rammade (AN) tinura (AN)

An annual, sometimes longer lived, of sandy rocky sites of the southern Sahara to the northern limit of the Region in Mauritania.

The plant is known to be highly poisonous and is not used in medicines. The Arabic (Hassaniya) name recognises its danger to the eyes (1).

Reference:

1. Naegelé, 1958, b: 890.

Euphorbia cervicornu Baill.

FWTA, ed. 2, 1: 421.

An erect semi-woody perennial to 1 m high, recorded only from S Nigeria, and in São Tomé and in Zaïre.

The plant occurs as a cultivate in Congo (Brazzaville) where it is often planted in villages. The latex is taken in palm-wine as an emetic and purgative: 1–2 drops is a dose; more is dangerous (1).

Reference:

1. Bouquet, 1969: 114.

Euphorbia convolvuloides Hochst.

FWTA, ed. 2, 1: 421. UPWTA, ed. 1, 143.

West African: **SENEGAL** FULA-PULAAR (Senegal) énèndé (JB) MANDING-BAMBARA daba da (JB) **MALI** DOGON áma íru dìi (CG) gèũ séŋe (CG) **UPPER VOLTA** FULA-FULFULDE (Upper Volta) murmurgehi (K&T) **IVORY COAST** MANDING-MANINKA suasingué (B&D) suaulingé (B&D) **NIGERIA** EDO asẹn-ulóko *from* ulóko: *iroko* (Elugbe) EFIK ẹtinkẹni-ẹkpo (FRI ex JMD) FULA-FULFULDE (Nigeria) endende (JMD) yindamhi (JMD) GWARI lukwe bebe = *dove's breast* (JMD) HAUSA nóoónòn kúrcíyáá = *dove's milk* (auctt.) IGBO ụdanị cf. *Uvaria chamae, Annonaceae* (JMD; KW) TIV mbasombol mingem (JMD) YORUBA amúyinú (Verger) ẹgélẹ, ègé-ilè *from* ègé: *cassava;* ilè: *ground* (JMD) ẹgélẹ (Verger) ẹmi-ilè *from* emi: *shea butter tree:* ilè: *ground* (JMD)

An herb, erect to 60 cm, of savanna and waste places from Senegal to N and S Nigeria and across Africa to Sudan and E Africa.

The plant is grazed by all stock in Senegal, though in the Lake Chad area it is not taken (1). It is readily eaten by duiker in Ghana (7).

An infusion of the dried leaves is used in Nigeria for dysentery. The plant latex has an astringent effect and is used internally for diarrhoea. It is said in a number of reports to be narcotic (3, 8). The plant has use in N Nigeria to prevent snake-bite or scorpion-sting: a portion is swallowed after chewing and the rest with the saliva is mixed with the juice of *Calotropis procera* (Asclepiadaceae) and rubbed on the hands. Salt, unless disguised by admixture with soup, etc. is said to have the opposite effect and to attract snakes (6). The plant is also said to be a remedy for scorpion-stings (4). Some Nigerian herbalists use the plant pulverised and mixed with palm-oil to smear on the rash of small-pox and chicken-pox which is said to dry up quickly under this treatment. An infusion of the plant is taken in Ghana orally and by enema as an aperient and for urethral discharges (6). A collyrium is used in Ivory Coast to relieve eye-troubles (5). Hausa herbalists near Kano use a plant-extract for treating asthma and bronchial inflammation (9). Perhaps on the Theory of Signatures, the latex applied to women's breasts is believed to increase lactation; it is also used as a love-charm.

The plant has anti-biotic activity and contains *tannin* (9). In a survey of Nigerian medicinal plants, no alkaloid was found in the leaves (2).

Yoruba medicine-men put the plant into a prescription designed to prevent witches beating anyone (10).

References:

1. Adam, 1966, a. 2. Adegoke & al., 1968. 3. Ainslie, 1937: sp. no. 153. 4. Anon 1487, Herb. Hook, K. 5. Bouquet & Debray, 1974: 84. 6. Dalziel, 1937: 143. 7. Ll. Williams 843, K. 8. Watt, 1967, with references. 9. Etkin, 1981: 84. 10. Verger, 1965: 243.

Euphorbia deightonii Croizat

FWTA, ed. 2, 1: 422.

West African: **GUINEA** FULA-PULAAR (Guinea) buro (CHOP) SUSU ganganhi (CHOP; Aub.) **SIERRA LEONE** KISSI nyɛŋgeô (FCD) KRIO ɔrɔ (FCD) MENDE *g̱*boni (FCD) *ngele-gonu* (FCD) **UPPER VOLTA** FULA-FULFULDE (Upper Volta) bura (Curasson) **IVORY COAST** BAULE dolo (auctt.) MANDING-DYULA baga-ni-fing *from* baga: *poison;* fing: *black* (A.Chev. ex JMD & ex K&B) **GHANA** ADANGME atroku (FRI ex JMD) BAULE dolo (FRI) GA blɔfo-aklaati (FRI) GBE-VHE esrɛ̃ (FRI) sra (FRI) **DAHOMEY** GBE-FON solo (A.Chev.; Aub.) **NIGERIA** EJAGHAM egakk (JMD) YORUBA ọrọ (JMD) ọró adete = *leper's 'ọrɔ́'* (Burton).

NOTE: *general for cacti-form and candelabra-form species*

A shrub, branching from the base without distinct trunk, to 6 m high, occurring in Sierra Leone, Ghana and S Nigeria.

The plant appears to be always cultivated. It is commonly grown as a hedge-plant, the lowest branches becoming buried in the ground thus providing a thick barrier to the base (2). In the Oyo district of S Nigeria it is used to mark the site of graves in Christian cemeteries (1).

The plant contains a copious white latex which is said to be poisonous. It is irritant in wounds. The Kissi of Sierra Leone used to smear it on to swords and spears for this end (3). It is also used on warts. The Yoruba name *oro adete*; leper's oro, suggests that it may be used for leprosy.

References:

1. Burton, s.n., Oct. 1949, K. 2. Croizat, 1938: 53–59. 3. Deighton 2609, K.

Euphorbia depauperata Hochst.

FWTA, ed. 2, 1: 421.

Herbaceous biennial, or subwoody perennial, stems to 1 m high from a woody perennial root-stock, of montane vegetation in Sierra Leone, Ivory Coast and W Cameroons, and occurring widespread in tropical Africa.

The plant contains an abundnant white latex. It is grazed by all domestic stock in Kenya (1).

Reference:

1. Glovei & al. 1005, K

EUPHORBIACEAE

Euphorbia desmondii Keay & Milne-Redhead

FWTA, ed. 2, 1: 422.
West African: NIGERIA ? (Jos) 'male' gu *cf. E. kamerunica:* 'female' gu (FNH)

A shrub or tree to 5½ m high, known only from N Nigeria to W Cameroons. On the Jos Plateau it is said to be the *male gu,* in distinction from the *female gu, E. kamerunica,* with which it may be planted to create a village pallisade (1). Its branches, unlike those of *E. kamerunica,* are not constricted into segments. The plant contains a copious latex.

Reference:

1. Hepper 1083, K.

Euphorbia forskaolii J Gay

FWTA, ed. 2, 1: 421, as *E. aegyptiaca* Boiss. UPWTA, ed. 1, 142, as *E. aegyptiaca* Boiss.
West African: MAURITANIA ARABIC (Hassaniya) um mulibeyna (AN) SENEGAL ARABIC (Senegal) amochar blum (JB) blum am vasar (JB) molibéné (JB) mu l'beina (JB) FULA-PULAAR (Senegal) énéndé (JB) SIERRA LEONE LIMBA bukbona (JMD) ukpona (NWT) MENDE kotuagbele (*def.* -i) (FCD) tolge (*def.* -i) (JMD) SUSU boxefɔrotai (JMD) UPPER VOLTA FULA-FULFULDE (Upper Volta) kosahi (*pl.* kosaaje) = *tree of milk* (K&T) NIGER HAUSA ƙurar shanu = *cattle dust* (Baumer) SONGHAI kólŋéy-hèeni (*pl.* -o) (D&C) kolŋey-wà (*pl.* -wàà) (D&C) NIGERIA ARABIC-SHUWA um lebeina (JMD) FULA-FULFULDE (Nigeria) ɓire baɗi (JMD) rafasa (JMD; ZOG)

A prostrate herb with stems to 30 cm long, of river-banks and sandy places in Senegal, Mali, Ghana and N Nigeria, and in Cape Verde Island and the drier parts of Africa to India.

In N Nigeria the plant is used as a purge and is regarded as a good vermifuge for tapeworm (2). Pounded and mixed with water it is sometimes applied to the head to relieve headache, or as a poultice for a fractured limb or traumatic swelling (2). Similarly in Kordofan a macerate in water is put on the forehead or the nape of the neck for headache, or topically in decoction for cracked ribs or bones in the hands or feet (4). It is also used as a galactogogue, inspired perhaps on the Theory of Signatures for its abundance of white milky sap. This idea is further established in the Arabic name *umm lebeina:* maker of milk (4). The sap is applied in Nigeria to warts and sores of guinea-worm and jiggers (3), and is also used in veterinary practice. The Hausa name *kùùráár sháánuu:* dust of cattle, suggests distribution of the plant through the agency of cattle for the plant does tend to invade pasturage where cattle have been. Cattle, however, are recorded as not grazing the plant in Senegal, though it is taken by sheep and goats (1). On the other hand, cattle in Kordofan will browse it a little even when the growth is at an advanced stage, and it seems that the grain will pass undamaged through the alimentary tract to colonise wherever the cattle roam (4).

Petrol-ether extracts of the whole plant of Sudanese material gave good molluscicidal action on the fresh-water snails, *Bulinus* and *Biomphalaria* (species not indicated), producing 100% mortality at 50 ppm extract. Alcoholic extracts showed only modest action on the former and none on the latter (5). The whole plant has a strong presence of *steroids.*

References (as *E. aegyptiaca* Boiss.):

1. Adam 1966, a. 2. Dalziel, 1937: 142. 3. Oliver, 1960: 26. 4. Baumer, 1975: 98. 5. Ahmed & al., 1984: sp. no. 17.

Euphorbia glaucophylla Poir.

FWTA, ed. 2, 1: 419.
West African: SIERRA LEONE TEMNE borikbori (NWT) yankaraibit (NWT)

An herbaceous perennial with prostrate branches occurring on sandy soil behind sea-beaches and recorded on the Atlantic seaboard from Senegal to Nigeria, and on to Angola.

The plant colonises sand-dunes as a pioneer down to high-tide level, often forming dense matted cover which doubtless acts to stabilise the sand (1).

Reference:

1. Rose Innes GC.Herb.30056, K; and many others.

Euphorbia glomerifera (Millsp.) Wheeler **E. hyssopifolia** Linn.

FWTA, ed. 2, 1: 419.
French: petit-lait (*E. glomerifera*).

Annual herbs, slender, erect to about 45 cm. Natives of S America and introduced, the first-named occurring in Senegal to Ghana, the second in Sierra Leone to W Cameroons.

Synonymy in literature is confused. Both are considered together here as uses appear to be alike. No usage is recorded within the Region. The following are from without. Plant sap is applied in America to warts, corns and indurations of the cornea (1). As *E. hypericifolia* the latex is used as purgative and as a caustic on skin-lesions (2).

References:

1. Hartwell, 1969,b: 159, 162, as *E. brasiliensis* Lam., *E. hyssopifolia* Linn. and *E. hypericifolia* Linn. 2. Watt & Breyer-Brandwijk, 1962: 611, as *E. hypericifolia* Linn.

Euphorbia heterophylla Linn.

FWTA, ed. 2, 1: 421.
English: annual poinsettia (Deighton).
French: euphorbe bicolore (Berhaut).
West African: GHANA GA kpá'kpla (KD) NIGERIA YORUBA ẹgẹ́lẹ (Egunjobi)

An erect semi-woody annual herb, to about 1 m high, native of tropical and sub-tropical America, now widespread in the tropics, and occurring in the Region throughout.

It is a weed of cultivation and waste land and is often troublesome. It is sometimes cultivated as an ornamental with forms bearing coloured upper leaves and bracts.

A decoction of the roots and bark has been used in northern Malaya for ague (2). The toxicity of the plant, especially of the root and latex, is recognised in E Africa. The latex is acrid and the toxic principle, neither alkaloid nor glucoside, is probably a *resin*, which can prove fatal (5).

The red colouring matter of the coloured leaves and bracts is *porcetin* (4).

Rubber content is recorded as: leaf 0.42%, stem 0.11%, root 0.06% and whole plant 0.77% (5). The plant, growing freely on the heavy alkaline clay soils (pH less than 9) of the Gezira Research Farm, Khartoum, Sudan, gave rise to the possibility of the plant being a source of rubber during the 1939–45 war, extraction being done by maceration. The extract was assayed at 5% rubber content (1, 3). The proposal does not appear to have been followed up.

References:

1. Andrews K.8, K. 2. Burkill, IH, 1935: 978. 3. Milne-Redhead, 1949: as *E. geniculata* Orteg. 4. Quisumbing, 1951: 502–3. 5. Watt & Breyer-Brandwijk, 1962: 408.

Euphorbia hirta Linn.

FWTA, ed. 2, 1: 419. UPWTA, ed. 1, 143.

English: Australian (or Queensland) asthma herb.

French: malnommée; Jean-Robert (Berhaut).

West African: SENEGAL BADYARA makoré sélu (JB) DIOLA-FLUP ku tim (JB) FULA-PULAAR (Senegal) èn èngil (JB) takèl pul (JB) TUKULOR takãpole (K&A; JB) KONYAGI tef e bidar (JB) MANDING-BAMBARA dabã dã (JB) daba da bélé (JB) daba da ble (JB; K&A) MANDINKA fara diato (JB) MANKANYA buko (JB) NDUT misdiombor (JB) mismuné (JB) mistélop (JB) SERER dad ondiis (JB; K&A) mbèl fo oy (JB; K&A) ndiis (JB) WOLOF mbal (K&A) mbalmbal (JB; K&A) homgölem (A.Chev. fide K&A) GUINEA-BISSAU FULA-PULAAR (Guinea-Bissau) taquelpôlhe (JDES; EPdS) SIERRA LEONE LIMBA fuŋkele (NWT) LOKO bumbuŋgo (NWT) buŋbuŋga (NWT) kpandi (NWT) ngainyi (NWT) ngeahal (NWT) njima (NWT) MENDE bɛlɛji (NWT) dowu (NWT) hɔndi (NWT) honti (NWT) kongolo (NWT) ŋaiɛndia (NWT) ndemi (NWT) nɛlima (NWT) nɛnɛwi (NWT) ŋɛtɛ (NWT) njaokwe (NWT) SUSU boxifɔrotai (NWT; JMD) konsui (NWT) mɔnɛmituri (NWT) mungunyi (NWT) njaiyafɔke (NWT; JMD) nyalɛfɔxɛ (NWT; JMD) tɔxɔfuxe (NWT) yarainyi (NWT) TEMNE ɛ-bit (NWT; JMD) ɛ-munakere (NWT) LIBERIA MANO tɔ a gbondo (Har.; JMD) tuagbono (JGA) MALI DOGON áma iru diì purugu (CG) UPPER VOLTA BISA gazingéré binné (Prost) FULA-FULFULDE (Upper Volta) d'abbirteeki (*pl.* -teed'e) = *tree for massage (shampoo)* (K&T) kosahi (*pl.* kosaaji) = *tree of milk* (K&T) HAUSA nonone kuchia = *milk of the pigeons* (K&B) MANDING-DYULA tuansingé (K&B) tuazingié (K&B) MOORE wallé bissum (K&B) SYEMOU karnguo-nĩ = *scorpion's breast water; i.e., refers to the latex* (PG) IVORY COAST ? déni-ba-singui (A&AA) ABURE orofa (B&D) AKAN-ASANTE akololo (B&D) akuododu (B&D) awaha (K&B) awin zinwin (K&B) AKYE atodu (A&AA) 'ANDO' akododo (Visser) ANYI aboko dodo (K&B) ako dodo (K&B) karagnawa (K&B) BAULE adododo (K&B) ako lulu (K&B) akodudu (K&B) ako-lôlo (A&AA) kumaguéssi (K&B) GAGU tao moa (B&D) tuawon (K&B) 'KRU' groboro (K&B) guabolo (K&B) KRU-BETE blableg-waré (K&B) GUERE gorobuo (K&B) gronohon (K&B) suonbohu (K&B) suongbou (K&B) GUERE (Chiehn) gbaneyanburu (K&B) sabrè uyé (B&D) NEYO dibue (K&B) KULANGO mama irifi (K&B) nofo nofo (K&B) KWENI sablé uyé (B&D) ziané vuin vuin (K&B) KYAMA duantapuin (B&D) MANDING-DYULA tuansingé (K&B) tuazingié (K&B) MANINKA suassingue (B&D) suaulingé (B&D) suaurézi (B&D) 'NÉKÉDIÉ' dobelu (B&D) SENUFO-DYIMINI niassingué (K&B) GHANA AKAN-ASANTE kakaweadwe (E&A) TWI ahinkogye (E&A) anim-akoa = *to know how to fight* (FRI; OA) nimakoa (OA) nufu nsu (OA) GBE-VHE notsigb̃e = *milk plant* (FRI; E&A) VHE (Awlan) ahinkodze (FRI) notsigbe = *milk plant* (FRI) NZEMA akubaa (FRI) SISAALA sikele (Blass) NIGERIA EDO ọzígban = *that which grows thorns* (Kennedy; Elugbe) FULA-FULFULDE (Nigeria) endamyel (J&D) HAUSA nóónòn kúrcíyáá = *dove's milk* (auctt.) IGBO (Asaba) ọ̀bụ ànị̀ = *ground fig tree* (NWT; KW) IGBO (Okpanam) ọ̀bụ ànị̀ = *ground fig* (NWT) IGBO (Owerri) ọba ala (AJC) IJO-IZON (Kolokuma) bou obíríma *the fruit* (KW&T) indóú béní díri *breast milk medicine* (KW&T) obírímá (KW; KW&T) YORUBA akun esan (IFE) buje (Egunjobi) ẹgẹ́le, ẹ̀gẹ́-ilẹ̀ *from* ẹ̀gẹ́: *cassava;* ilẹ̀: *ground* (Macgregor; JMD) ẹmile, ẹmi-ilẹ̀ *from* ẹmi: *shea butter tree;* ilẹ̀: *ground* (auctt.)

An anthropogenic ruderal herb, decumbent or erect to 40 cm tall, occupying open waste spaces, a weed of cultivation, road-sides, path-sides and a diversity of situations, occurring widespread throughout the Region, and dispersed pantropically and sub-tropically around the world.

The plant contains a relatively abundant white latex, and in common with other lactiferous species, a decoction of the whole plant, or of the leaves, or the latex itself is used by women as a galactogogue (Senegal: 14, 15; Ivory Coast: 6, 16: Ghana: 11, 33; Nigeria: 3, 12; Gabon: 25, 26; Congo (Brazzaville): 5; Tanganyika: 9; etc.). The plant has been an item of market trade in N Nigeria for this purpose. Preparations are massaged on to the breasts, or may be taken in draught. Laboratory experiments on immature female guinea-pigs showed that the plant induced mammary development and secretion of milk (6). At

Enugu Hospital, Ghana, the plant chewed with oil-palm kernels has been reported to act within 24 hours to produce a copious flow of milk (12).

The plant has a diuretic and purgative action. It is widely used in treatment of venereal discharges (Ubangi, Guinea: 20; Tanganyika: 4, 9, 27; Senegal: 14, 15: Ivory Coast: 6, 16). It is considered good for treating diarrhoea and dysentery (Senegal: 1, 14, 15; Guinea: 19; Ivory Coast: 2, 30; Ghana: 12, 33; Congo (Brazzaville): 5; Tanganyika: 22). Tests carried out in Conakry Hospital, Guinea, on patients with amoebic dysentery gave clear stool examinations and a cessation of bleeding after 48 hours treatment with a decoction of the fresh plant (16). Subsequent work has shown amoebicidal action of plant-extracts which had practically no toxicity towards man or guinea-pig (15).

The plant is known as a remedy for inflammation of the respiratory tract, and for asthma it has a special reputation causing bronchial relaxation. Extract or decoction of the flowering or fruiting plant was entered in the *British Pharmaceutical Codex* for this but has long since been surpassed. The plant has also sometimes been combined with bronchial sedatives in preparations for inhalation (8, 15, 19, 31). Hausa medicine-men near Kano prescribe the leaf for chewing for coughs (32). The plant shows anti-biotic activity. A number of substances has been detected in the plant: *tannins, gallic acid, quercetin, phenols, phyto-sterols, alcohols, alkaloid*, etc. (8, 15, 19, 20, 21) without obvious active principles. The pharmacology requires further study.

The plants used in E Africa as a treatment for *Oxyuris* threadworm (27) and leaf-sap as a gargle for angina (9). Sap is squeezed from the leaves into the eyes for styes and conjunctivitis, a practice that is known in E Africa (9, 23) and in Malaya (7) as well as in Ivory Coast (30), Nigeria (3, 8, 12) and Liberia (8). The latex is said to be capable of causing dermatitis. It is, however, thought to be cicatrisant and is applied to sores and wounds. It is used on ringworm in Indonesia (7). Tests have shown, some positive, some negative, anti-biotic action (15, 27). Ghanaian children trace a pattern on the skin with the juice and rub in charcoal. The design can be removed by washing (11). The Ijo of S Nigeria similarly create body-patterns which are said to be hard to wash off (28). They use the fruit for dyeing (29).

The plant is used in Congo (Brazzaville) to relieve costal pains, and in decoctions with other plants to make a soothing wash for infants with fever (5). It has analgesic application for headache in Ivory Coast by nasal installation (16), and for rheumatism as a liniment (3) and for pains in pregnancy (24) in Nigeria.

Latex applied to a thorn or other foreign body will assist in extraction, and applied to a poison-arrow wound will act as an antidote to the poison and facilitate removal of the arrow-head (8). The use of the latex on warts, whitlows and the like is worldwide (10). The leaves are used by the Lobi and Kulango of Ivory Coast as an antidote to snake and scorpion poisons (18).

The tender shoots are said to be edible and to serve in India as a famine-food. The plant is however considered to be slightly stimulant and narcotic (7, 21). Goats are recorded browsing it in Kenya but it is shunned by cattle and camels (17). The plant is considered aphrodisiac in Tanganyika (9), and a Fula man of N Nigeria will take an infusion of the pounded leaves before going to a woman to ensure she will love him (13).

References:

1. Adam, 1960,a: 369. 2. Adjanohoun & Aké Assi, 1972: 124. 3. Ainslie, 1937: sp. no. 154. 4. Bally, 1937. 5. Bouquet, 1969: 114. 6. Bouquet & Debray, 1974: 84. 7. Burkill, IH, 1935: 978–9. 8. Dalziel, 1937: 143. 9. Haerdi, 1964: 94. 10. Hartwell, 1969,b: 162. 11. Irvine, 1930: 191. 12. Irvine, s.d.,a. 13. Jackson, 1973. 14. Kerharo & Adam, 1964,b: 442. 15. Kerharo & Adam, 1974: 410–4, with phytochemistry and pharmacology. 16. Kerharo & Bouquet, 1950: 73–74. 17. Mwangangi & Gwynne 1055, K. 18. Oldeman 278, K. 19. Oliver, 1960: 7, 63–64. 20. Portères, s.d. 21. Sastri, 1952: 225. 22. Tanner 1330, K. 23. Tanner 4049, K. 24. Thomas NWT.1620 (Nig. Ser.), K. 25. Walker, 1953,a: 35. 26. Walker & Sillans, 1961: 167. 27. Watt & Breyer-Brandwijk, 1962: 408–11. 28. Williamson, K 53, UCI. 29. Williamson, K & Timitimi, 1983. 30. Visser, 1975: 75.15: 47 31. Wong, 1976: 127. 32. Etkin, 1981: 84. 33. Ampofo. 1983: 22, 53. See also: Oliver-Bever, 1983: 25–26.

EUPHORBIACEAE

Euphorbia hyssopifolia Linn.

West African: **SIERRA LEONE** MENDE kui (NWT)

See under *E. glomerifera* (Millsp.) Wheeler.

Euphorbia kamerunica Pax

FWTA, ed. 2, 1: 422. UPWTA, ed. 1, 143–4.
West African: **MALI** DOGON dulú (CG) **UPPER VOLTA** FULA-FULFULDE (Upper Volta) buro (K&B) GRUSI suru (K&B) MANDING-DYULA baga ni ting (Aub.) MOORE taxendo (K&B) **NIGERIA** ? (Jos) 'female' gu *cf. E. desmondii:* 'male' gu (FNH) BIROM gòndòŋ (LB) EFIK àkpà ṁbìèt (JMD; KO&S) FULA-FULFULDE (Nigeria) kwidehi (JMD; KO&S) widihi (JMD) HAUSA keraana (auctt.) keránáá (LB) kyànáráá (JMD; ZOG) kyaránáá (auctt.) IBIBIO ŋ̀wàt uwàt *from* wàt: *to drive, propel or scatter* (Lowe; BAC) IGBO abananya (BNO) KAMBARI màringû (RB) KANURI garuru (JMD) gururu (JMD) kimbilimbili (?) (JMD) TIV agondo (JMD; KO&S) karikassa (JMD) kariki (JMD) YORUBA ọrọ́ (KO&S; Verger) YUNGUR kukra (*pl.* kukras) *to this and other E.spp. generally* (RB)

A cactiform tree to 10 m high with a distinct trunk and spiny 3 or 4-angled branches constricted at intervals, of dry rocky country, widely distributed in the northern savanna region from Guinea to N Nigeria and W Cameroons: also in E Cameroun.

The wood is hard (1). The plant is one of the commonest hedge-pallisade plants of northern region where rainfall is limited to 1270 mm per annum. When properly tended it grows into an impenetrable stockade (1, 3, 6) and serves as a cattle-pound or to enclose villages, houses, fields, etc. On the Jos Plateau it is planted as a village pallisade mixed with *E. desmondii*, as female and male *gu* respectively (4 – see *E. desmondii*). The Lobi of Upper Volta plant it near their fetish huts (5).

The latex is highly purgative, a dose of only one or two drops being effective. It is sometimes used to purge cattle and horses (3, 5), but taken in excess death may swiftly follow as has been recorded by the use, apparently of this species, of the latex taken in trial by ordeal on the Benue River (3). It has been recorded in fact as one of the commonest ordeal-poisons amongst the Pabir and Busa tribes of Bornu, N Nigeria. The latex is smeared on a straw which is licked, saying the proverbial oath. The priest then administers a series of three draughts of the juice with water (to which he may have added an emetic as he thinks fit) which the accused then drinks (7).

There is widespread use of the latex to compound into arrow-poison, and into bait to kill noxious animals and to capture birds (1, 2, 3, 5). A milky juice known as *murtsunguwa* (Hausa) which is probably of euphorbiaceous origin is so used in N Nigeria (3).

On the Benue River, Nigeria, the latex mixed with fire-ash is rubbed on steel before heating and tempering for use with flint (Meek fide 3).

In Ivory Coast the plant is used medicinally for its vesicant action in treatment of leprosy. This action is said to arise from the presence of *triterpene* derivatives (2).

This species and closely related euphorbias have a common name to the Ebina of Adamawa, Nigeria, and a clan has adopted it as their own. The plant may be parasitised, perhaps by a loranth, which they use to cure deafness (8), probably ascribing magical attributes.

References:

1. Aubréville, 1950: 176. 2. Bouquet & Debray, 1974: 84. 3. Dalziel, 1937: 143–4. 4. Hepper 1084, K. 5. Kerharo & Bouquet, 1950: 75. 6. Howes, 1946. 7. Meek, 1931: 1, 173, as *E. barteri*. 8. Babuwa Tubra fide Blench, 1985.

Euphorbia lactea Haw.

French: euphorbe lactée (Berhaut).

A cactus-like shrub with 3- or 4-angled branches bearing a white marked area running through the middle of each face; native of India and Ceylon, and brought into cultivation in many tropical and subtropical countries. It is extremely rapid growing and is used for hedging in Florida and the West Indies. It is reported under cultivation in Senegal. In India it is used to make a hot poultice for rheumatism.

Euphorbia laro Drake

A shrubby species of Madagascan origin, now introduced to the Region. It is planted as hedges in the area of Cap Vert, Senegal (1).

Reference:

1. Aubréville, 1950: 177.

Euphorbia lateriflora Schum. & Thonn.

FWTA, ed. 2, 1: 422. UPWTA, ed. 1, 144.
West African: **SIERRA LEONE** KRIO blɛnyai *i.e., blind-eye* (FCD) MENDE *g*boni (L-P; FCD) **MALI** ·MUNIO· mágara (Barth) **GHANA** AKAN-TWI *n*kamfo-barima (FRI) GA kãŋ'fo (KD) kpĺɛmì (KD) ŋkãfo (KD) nkaŋfo (FRI) **DAHOMEY** ? rahié (?) (A.Chev.) **NIGERIA** ARABIC-SHUWA umm libeine (JMD) FULA-FULFULDE (Nigeria) cangalabali (JMD) cangalabanni (JMD) fidasaruɗehi (JMD) HAUSA bi sartse (JMD; Schuh) fid da saruɗɗa (JMD) fid da saruttsa (JMD) fid-dà-sártsè *from* fidda, *or* fitad da; *to take out;* sartse: *a splinter* (JMD; ZOG) KANURI tjaina (JMD; C&H) NGIZIM ágùrátlíima (*pl.* -àmín) (Schuh) YORUBA ẹnu ekure (JMD) ẹnu kò pa ire (JMD ẹnu-òpirè *a contraction of* ẹnu kò pa ire (JMD) ọrọ ẹnu kò piyè (JMD) ọrọ wẹ̀rẹ (JMD)

A shrub reaching 1.70 m height with near-vertical succulent branches, leaves shed, with abundant white latex, occurring in grassy savanna from Sierra Leone to W Cameroons.

The Yoruba name is derived from a proverbial expression *bí ẹnu ko ba è pa ire ko lè pa olóko*: 'since words or a form of speech does not hurt the crop it cannot harm the farmer.' The plant is commonly grown as a hedge and boundary-marker especially in the northern areas, in part because of a belief that snakes dislike it (1). It is planted in Ghana for other superstitious purposes also, and as a charm to protect farm crops (3). The Hausa name indicates the use of the latex to extract thorns and splinters, the part affected being bandaged with an application of the latex. The foreign body comes away the following day (1). The latex is powerfully caustic and it is on record to have been used by malingering soldiers to raise white blotches on the skin to simulate leprosy (1). In parts a piece of stem is added to arrow-poison prescriptions, and the latex to bait for catching guinea-fowl (1).

The plant, while not specifically planted for medicinal purposes, is used in local medicines. In Sierra Leone it is known as a 'cure-all' (2) under the Krio name *blainiyai*: blainy eye, of unexplained significance. The latex is used in N Nigeria as a remedy for head-lice, and also for ringworm on the scalp. A foot affected by guinea-worm or jiggers is soaked in a pot of infusion of the stems (1). The latex is a drastic purge, and is used especially as such in the treatment of syphilis. In Sierra Leone a plant decoction under the name of 'lesser sarsparilla' is taken as a blood-purifier. In N Nigeria the latex is taken with milk or cereals or liver causing purging and sometimes vomiting.

References:

1. Dalziel, 1937: 144. 2. Deighton, 1958, K. 3. Irvine, 1961: 229.

Euphorbia leucocephala Lotsy

A bush, laxly branched, to about 3 m high, native of C America, is reported under cultivation on the Jos Plateau, Nigeria, for its ornamental value. The flowering branch-ends bear a mass of small white bracts (leaflets) giving a decorative white blanket effect.

Euphorbia macrophylla Pax

FWTA, ed. 2, 1: 421.
West African: SENEGAL MANDING-MANDINKA kutumbo (JB)

A low shrubby or erect perennial to 50 cm tall or taller, of the dry northern savanna of Senegal and Ghana in the Region, and in Sudan and Tanganyika.
The latex has unspecified medicinal use in Ghana (1).

Reference:

1. Norman 1062, K.

Euphorbia milii Des Moul.

FWTA, ed. 2, 1: 419.
English: Christ's thorn.
French: épine du Christ; euphorbe écarlate (Berhaut).

A thorny much branched shrub, native of Madagascar, and now widely grown in the tropics as an ornamental for the bright red bracts subtending the flowers.

Euphorbia paganorum A Chev.

FWTA, ed. 2, 1: 422. UPWTA, ed. 1, 146.
West African: SENEGAL BASARI a-ñidin (K&A) MANDING-MANINKA omô omô (K&A) GUINEA-BISSAU BALANTA bague-uône (JDES) berendenda (JDES) BIAFADA bidjaque-dja-que (JDES)

A low fleshy shrub to 1.75 m high of dry rocky savanna from Senegal to N Nigeria.
It is planted in Mali as a fetish-plant (1). Its latex is abundant and very caustic. In Senegal the plant is reduced to ash, and decocted with the leaves of *Sarcocephalus latifolius* (Rubiaceae) and the liquid is used to wash the body and face for leprosy (4). In Ivory Coast the latex is a component of arrow-poison, and of a bait for trapping noxious animals. It is also used for leprosy (2). Vesicant action is due to the presence of *triterpene* derivatives (2, 4).
The flowers are melliferous and attract bees and ants (1, 3).

References:

1. Aubréville, 1950: 178. 2. Bouquet & Debray, 1974: 84. 3. Dalziel, 1937: 146. 4. Kerharo & Adam, 1974: 414–5, as *Nauclea latifolia*.

Euphorbia poissonii Pax

FWTA, ed. 2, 1: 422. UPWTA, ed. 1, 145.
West African: MALI FULA-PULAAR (Mali) pendiré (A.Chev.; Aub.) SONGHAI abári e sébuwa (Barth) táboru (Barth) UPPER VOLTA GURMA péni (A.Chev.) MOORE takésindo (A.Chev.; Aub.) GHANA ADANGME atroku (FRI) SISAALA salaw (AEK) TOGO TEM férémon (A.Chev.; Aub.) TEM (Tschaudjo) dididire (Volkens) DAHOMEY BAATONUN séséru (Aub.; JMD) DENDI lokoto (A.Chev. fide JMD; Aub.) YOM kelempendé (JMD; Aub.) NIGER DENDI lokoto (A.Chev. fide JMD; Aub.) FULA-FULFULDE (Niger) pendire (Aub.) GURMA péni (A.Chev.; Aub.) NIGERIA DERA gáarò (Newman) FULA-FULFULDE (Nigeria) buurohi (*pl.* buuroje) (JMD) gi'e rawandu = *dog's thorns* (JMD) hetjere (*pl.* ketje) (Westermann) GWARI magaba (JMD) HAUSA tínyaà (JMD; ZOG) tumniya (JMD; ZOG) túnyàà (JMD; ZOG) KANURI garuru (JMD; C&H) gururu (JMD) TIV koròkò (JMD) YORUBA òró elewé = *leafy 'oro'* (Burton) òró-adétè = *leper's 'oro'* (JMD)

An erect much-branched shrub to 2 m tall, branches thick, succulent, with leaves persisting at the top (cf. *E. lateriflora*), occasionally sub-spiny, of dry savanna from Guinea to N Nigeria.

The plant is sometimes cultivated, and often found in fields and villages in Dahomey (1).

A piece of the stem is commonly mixed with other ingredients in preparing arrow-poison from the seeds of *Strophanthus* (Apocynaceae). The latex is a powerful caustic, and its action seems to be solely as such for it is irritant to mucous membranes and in direct contact with tissue, but has no action on the heart, respiration or nervous system. However, cactiform euphorbias used for arrow-poison in Mali are said to have the name *kunkummia*, and at Timbuktu the Songhai name *abári e sebúwa* is derived from the belief that it causes the death of a hunted lion (Barth fide 2).

The Jukun of N Nigeria have used the plant-sap as a potent poison when added to food, drinking-water, kola nuts, etc., the antidote being cow's milk (Meek fide 2). In Sokoto some tribes use the plant as a fish-poison (2), an use also practiced in Dahomey (4). On the other hand, root-extract of Nigerian material, screened for molluscicidal activity on the fresh-water snail, *Bulinus globulus*, is reported as being inert (6).

The latex is said to be sometimes added to tobacco-snuff to increase its pungency. It is placed in a carious tooth to relieve toothache or to help to loosen a stump to render extraction easier. This species may be amongst those used for dehairing hides (2). The latex is purgative and is taken in Gabon in small doses on cane-sugar, or in palm-wine or soup (5).

A preparation in N Nigeria called *gunguma* (Hausa) or *lówu* (Nupe) containing the latex is used as a poison-bait for catching guinea-fowl, etc. (2).

The flowers are reported to be much visited by a large blue-black species of bee in the Katsina area of N Nigeria (3).

References:

1. Aubréville, 1950: 177. 2. Dalziel, 1937: 145. 3. Meikle 1370, K. 4. Portères, s.d. 5. Walker, 1953, a: 36. 6. Adewunmi & Sofowora, 1980: 57–65.

Euphorbia polycnemoides Hochst.

FWTA, ed. 2, 1: 421. UPWTA, ed. 1, 145.
West African: MALI DOGON gèũ seŋe ána (CG) NIGERIA HAUSA geron tsuntsaye = *bird's millet* (BM; JMD) nóónòn kúrcíyáá = *dove's milk* (JMD; ZOG)

An erect slender herb to 45 cm, in savanna and as a weed of cultivation in the drier parts of Senegal to N Nigeria and across the continent to the Red Sea and E Africa.

Its uses in N Nigeria are as for *E. convolvuloides*, and especially as a medicine for women after childbirth (2), probably to stimulate lactation (1). The latex is similarly used in Kordofan, but seemingly before confinement rather than after (3).

Sheep in Kordofan will graze a little (3); also other stock (4).

References:

1. Dalziel, 1937: 145. 2. Moiser s.n., 8/3/1922, K. 3. Baumer, 1975: 98. 4. Hunting Tech. Services, 1964.

Euphorbia prostrata Ait.

FWTA, ed. 2, 1: 421. UPWTA, ed. 1, 145.

French: petite teigne noir (Berhaut).
West African: **SIERRA LEONE** LIMBA bukbona (NWT) ukpona (NWT) MENDE kotuagbɛlei (def.) (FCD) tolgei (JMD) SUSU boxefɔrotai (JMD) **NIGERIA** ARABIC-SHUWA umlebeina (JMD) FULA-FULFULDE (Nigeria) ɓire baɗi (JMD) HAUSA kuùrar shaánuu = *dust of cattle* (JMD) rafasa (JMD) tafar biri = *monkey's armlet* (JMD) IGBO (Ogwashi) ọ̀kazī (NWT) YORUBA ewé bíyẹmi = *family fits me* (Verger)

A slender prostrate herb with stems to 20 cm long arising from a tough woody root-stock, of open places, of the American tropics, now pan-tropical and occurring throughout the West African region.

The plant is used in Nigeria for its astringent, vulnerary and anthelmintic properties (3). It is reported used by Igbo as a poultice for a broken arm (4). The stem and leaves are chewed by Masai in Kenya for gonorrhoea (1). It is used by Indians in S Africa for diabetes, and the latex in Australia for sores, and in N America as a snake-bite remedy (6). The latex is caustic. It is applied in Mexico to growths in the eye and in Venezuela to tumours and corns (2).

The whole plant is lactiferous and the yield of rubber has been recorded as 0.12% (6).

Under the Yoruba name *ewé ɓíyẹmi*: family fits me, it enters an incantation of hope to obtain several women at home (5).

References:

1. Glover & Samuel 2993, K. 2. Hartwell, 1969,b: 165. 3. Oliver, 1960: 26, 64. 4. Thomas NWT.2062 (Nig. Ser.), K. 5. Verger, 1967: no. 76. 6. Watt & Breyer-Brandwijk, 1962: 413.

Euphorbia pulcherrima Willd.

FWTA, ed. 2, 1: 419.
English: poinsettia.
French: poinsettia; étoile de Noel (Berhaut).

A shrub, native of tropical America, now pantropical, and widely distributed in temperate and subtemperate countries as a house-plant.

The plant is grown as an ornamental for the conspicuous crimson leaves surrounding the flowers. White and yellow-leaved forms exist. The latex is irritant. Leaf, bark and roots are markedly toxic, and the plant is used as fish-poison in China. Nevertheless the leaves are said to be eaten as a seasoning in Java (1). In screening trials of Nigerian material the root gave practically no activity on the fresh-water snail, *Bulinus globulus* (2). Some antibacterial activity has been found in the stem, leaf and flower (3).

References:

1. Burkill, IH, 1935: 980. 2. Adewunmi & Sofowora, 1980: 57–65. 3. Roia & Smith, 1977: 28–37.

Euphorbia schimperiana Scheele

FWTA, ed. 2, 1: 421.
West African: GUINEA-BISSAU FULA-PULAAR (Guinea-Bissau) dafeunina (JDES) MAND-ING-MANDINKA mantchmhô (JDES)

An herb (? perennial) to over 1 m high, of montane vegetation in W Cameroons only in the Region, but widely distributed in E Africa from Ethiopia to Zimbabwe, and in Arabia.

No usage is recorded in the Region. While it is recognised as a poisonous plant in Kenya (4), it is reportedly grazed by cattle and other domestic stock (3). A decoction of the leaf and root is taken in Tanganyika as a purge (2, 5). The plant has been used as a snake-bite remedy, and in Ethiopia against syphilis (5). The abundant white latex is applied in Kenya to wounds to make them fester (1).

References:

1. Bally Cory. Mus. 7706, K. 2. Bally, 1937. 3. Glover & al., 1030, K. 4. Glover & al., 2153, K. 5. Watt & Breyer-Brandwijk, 1962: 414.

Euphorbia scordifolia Jacq.

FWTA, ed. 2, 1: 421.
West African: MAURITANIA ARABIC (Hassaniya) tanut (AN) SENEGAL ARABIC (Senegal) l'embetêha (JB) mul beïne (JB) tanut (JB) MANDING-BAMBARA daba da (JB) WOLOF mbal (K&A) mburmbul (JB; K&A)

An erect or prostrate sub-woody herb recorded from the drier part of the Region from Mauritania to N Nigeria, and occurring in Cape Verde Islands and across Africa to the Red Sea.

The plant is grazed by sheep and goats in Mauritania (3), and by cattle, sheep and goats in Senegal (1), but on the other hand the plant has been reported dangerous for stock (Roberty fide 5) – perhaps differences of phenological and edaphic significance which should be examined.

In N Nigeria a plant-decoction is applied to women's breasts as a galactogogue (2). It is possible that the plant has some analgesic property for it is used in Nigeria on aching teeth, tsetse fly bites and for dysentery (4).

References:

1. Adam, 1966, a. 2. Elliott 195, K. 3. Naegelé, 1958, b: 890. 4. Oliver, 1960: 26. 5. Baumer, 1975: 98.

Euphorbia serpens H.B. & K.

FWTA, ed. 2, 1: 421.

A spreading prostrate annual to about 17 cm long, native of S America, and recorded from Sierra Leone.

The plant sap is applied to corns and warts in the Argentine (1).

Reference:

1. Hartwell, 1969,b: 165.

EUPHORBIACEAE

Euphorbia sudanica A Chev.

FWTA, ed. 2, 1: 422. UPWTA, ed. 1, 145, incl. 146, *E. trapaeifolia* A Chev. and *E. tellieri* A Chev.

French: euphorbe cactiforme (Senegal, Kerharo & Bouquet); euphorbe du Soudan (Berhaut).

West African: SENEGAL ? *m*bun (Rogeon) BASARI a ñidin (JB) FULA-PULAAR (Senegal) guõ no (JB) MANDING-MANINKA bamo (JB) hamo (JB) homõ-homõ (JB) pamo (JB) SONINKE-SARAKOLE fama (JB) MALI DOGON tendié (Rogeon; Aub.) FULA-PULAAR (Mali) guon no (Rogeon) MANDING-MANINKA homon-homon (Aub.) NIGER FULA-FULFULDE (Niger) guon no (Aub.) SONGHAI wàa-ñá` (*pl.* -ñóŋó) = *mother of milk* (D&C) NIGERIA BURA hira *perhaps this sp.* (anon.)

A fleshy shrub to about 2 m high of rocky situations in the dry savanna from Senegal to Niger.

In Niamey, Niger, it is planted along roadsides (1). The latex is caustic and is regarded as poisonous. A decoction of the plant reduced to ash with *Sarcocephalus latifolius* (Rubiaceae) is used in Senegal to wash the body and face in treatment for leprosy (2 – cf. *E. paganorum*).

References:

1. Dalziel, 1937: 146, as *E. tellieri* A Chev. 2. Kerharo & Adam, 1964, a: 417, as *Sarcocephalus esculentus*.

Euphorbia thymifolia Linn.

FWTA, ed. 2, 1: 421.
West African: SIERRA LEONE LIMBA bukbuŋa (NWT) ukpona (NWT) MENDE kotuagbɛlɛ (FCD) tolgei (FCD) toluga (?) (FCD) SUSU boxeforotai (NWT) bɔxɛforotai (NWT) TEMNE ɛ-kaiya (NWT) ɛ-puŋ (NWT)

An annual, spreading, prostrate herb, of open waste places, a pan-tropical, recorded throughout the West African region.

The plant contains an essential oil with a pungent odour and irritating taste, and containing, amongst others, *cymol, carvacrol, limonene, sesquiterpenes* and *salicylic acid* (7). The oil is put into medicinal soaps for treatment of erysipelas, into sprays to keep off flies and mosquitoes, and as a vermifuge for dogs and farm foxes (7). The leaves are pulped up with water into a paste for application to the head for headache in Sierra Leone (3). The plant is expectorant. The dried leaves and seeds are slightly aromatic. A decoction of the whole plant is used in Congo (Brazzaville) to relieve congestion in chest-affections (1), and it is used for coughs in Malaya (6). The dried leaves and seeds are also considered stimulant, astringent, anthelmintic and laxative, and are used in India for children with bowel complaints (7). There is a similar usage in SE Asia (2). In the Philippines, the leaves are commonly used in poultices on snake-bites (5), and in India juice expressed from the plant is taken internally and is applied topically for the same purpose (5). The plant-juice is a south Indian cure for ringworm and scurf (5, 7). A poultice is considered vulnerary, and is a dressing for broken bones (2, 5). The latex is vesicant. It is applied to warts in S America (4), and in India to the eyes to clear corneal opacity (5). In Trinidad a decoction of the plant is administered in flu and fever, for hypertension and venereal disease, as a galactogogue, post-partum depurant and abortifacient (8).

References:

1. Bouquet, 1969: 115. 2. Burkill, IH, 1935: 980–1. 3. Deighton 419, K. 4. Hartwell, 1969: 160. 5. Quisumbing, 1951: 508–9. 6. Ridley, 1906: 249. 7. Sastri, 1952: 227. 8. Wong, 1976: 130.

Euphorbia tirucalli Linn.

FWTA, ed. 2, 1: 419.

English: milk bush.

French: arbre de Saint Sébastien (Berhaut).

West African: SENEGAL SERER ndamol durubab (JB; K&A)

An unarmed shrub or small tree, native of Africa but not West Tropical Africa whither it is introduced and widely cultivated. In places it is sub-spontaneous.

It is usually seen as a hedge in all parts of tropical Africa. It is easy to propagate and appears to grow on almost any soil (7). It lacks thorns and this might be a limitation to its capacity to produce a good barrier, but its caustic irritant sap is an effective deterrant to small animals and human marauders from forcing a passage. This property must also protect it from browsing animals though sheep and the omnivorous goat are said to take it in Kordofan (17). The latex causes temporary blindness lasting several days (8, 13). It is planted in Gabon villages as a magic tree (12), and in Congo (Brazzaville) as a protection against lightning (4). In Malawi it is commonly planted on graves (15). In Tanganyika the plant is thought to keep away mosquitoes (2), and it is used as an insecticide in India (13). It is used as a fish-poison in Congo (Brazzaville) (4), E Africa (15, 18), Zanzibar (14) and in India and SE Asia (13). In the Philippines (3) and in Tanganyika (2) the latex is so used, but evidently a large quantity is necessary to stupefy fish.

The caustic latex is purgative in small doses: 2–3 drops at a time are given to adults only in Gabon (11, 12) with food. A dose of 3–4 drops is used in Congo (Brazzaville) as a purge for ascites and generalised oedemas (4). A root-decoction with other drug-plants is taken in Tanganyika for schistosomiasis and gonorrhoea (5). The plant also provides a purge for snake-bite there (13).

On the Theory of Signatures the jointed nature of the plant-stems lead to the use of the plant in SE Asia in a belief that it will assist healing of broken bones (3, 9). Similarly the white latex is considered in Tanganyika to be a remedy for sexual impotence (13). Young girls in Malawi smear their breasts with it to cause greater development, boys their penis as an added stimulant in masturbation, sometimes with dire results (16). The latex is used for skin-complaints in Indonesia (3), to raise blisters on syphilitic nodes in the Philippines and in many countries on warts, tumours, cancers, etc. (6).

Euphorbia latex contains a mixture of light hydrocarbons of molecular weight around 20,000, and after removal of the water the residue is a liquid oil (19). *Triterpenic alcohols* have also been identified in the latex (8). There have been plantings of *E. tirucalli* in Okinawa which on a spacing of 4 ft between bushes are expected to yield 5–10 barrels of oil p. acre p. year. Other *E.* species are under investigation in the USA and from *E. lathryis* a yield of 10 barrels of oil p. year in 9 months from planting is expected (19). West African material merits examination.

A form of rubber is obtained from the latex which contains a high proportion of resin: 14.3–15.7% *caoutchouc*, 75.8–82.1% *resin*, dry weight, in material from S Africa. The resin can be used in varnishes for certain purposes in the absence of better materials, but only at a high cost (1). Steam-distillation results in a better product which has found some favour in India in linseed-oil varnishes giving a tack-free glossy finish (10). The sap evidently has strong fixative power: mixed with animal hair it is used in Zanzibar (14) and on the coast of East Africa (18) for fastening knife-blades to wood handles and spear-heads to shafts.

The wood is white, close-grained and fairly hard. It finds use in India for rafters, toys and veneers. It gives a charcoal suitable for gun-powder (10) and for the manufacture of fireworks.

EUPHORBIACEAE

References:

1. Anon, 1944. 2. Bally, 1937. 3. Burkill, IH, 1935: 981–2. 4. Bouquet, 1969: 115. 5. Haerdi, 1964: 94. 6. Hartwell, 1969: 166. 7. Howes, 1946. 8. Kerharo & Adam, 1974: 416–7, with phytochemistry and pharmacology. 9. Quisumbing, 1951: 509–10. 10. Sastri, 1952: 227–8. 11. Walker, 1953, a: 36. 12. Walker & Sillans, 1961: 167. 13. Watt & Breyer-Brandwijk, 1962: 415. 14. Williams, RO, 1949: 256. 15. Williamson, J, 1955 (? 1956): 57. 16. Hargreaves, 1978: 21–30. 17. Baumer, 1975: 98. 18. Weiss, 1973: 190. 19. Wang & Huffman, 1981: 369–82.

Euphorbia unispina NE Br.

FWTA, ed. 2, 1: 422.
West African: SENEGAL BEDIK gi-mì (FG&G) GUINEA BASARI a-nyɛndín (FG&G)
MALI FULA-PULAAR (Mali) pendiré (A.Chev.) UPPER VOLTA GURMA péni (A.Chev.; K&B)
HAUSA tinya (JMD fide K&B) MANDING-DYULA wanda (K&B) MOORE takesindo (A.Chev.;
K&B) IVORY COAST MANDING-DYULA wanda (K&B) MANINKA fanfan (B&D) TOGO TEM
feremon (A.Chev.) TEM (Tschaudjo) dididire (JMD) DAHOMEY BAATONUN seseru (JMD)
DENDI lokoto (A.Chev.) YOM kelempendé (JMD) NIGER DENDI lokoto (A.Chev.) GURMA péni
(A.Chev.) NIGERIA BIROM yɛ́ɛ́p (LB) DERA gáarò (Newman) FULA-FULFULDE (Nigeria)
buurohi (pl. buuroje) (auctt.) gi'e rawandu = dog's thorns (JMD) hetjere (pl. ketje) (Westermann) GWARI magaba (JMD) HAUSA bida serti (RES) tínyàà (auctt.) tumniya (JMD; ZOG)
tunya (JMD; ZOG) KANURI garuru (JMD) gururu (JMD) SAMBA-DAKA nu-gullam (FNH) TIV
koṛoko (JMD) YORUBA oṛó adɛ̣tɛ̣ = lepers' 'oro' (JMD) WEST CAMEROONS FULA-
FULFULDE (Adamawa) borooje (FNH)

An erect shrub to 3 m high, on rocky hills in savanna from Western Senegal to W Cameroons.

The latex is caustic and vesicant. It is incorporated into arrow-poisons in Ivory Coast (1), Upper Volta (4), by the Fula in N Nigeria (3) and in Kordofan (5). It is added to bait in traps and Tenda medicine-men place it on scarifications to thicken them (2).

In Guinea, Mali and Ivory Coast medicine-men hold that the site of sleeping sickness is in the ganglia of the neck and ablation of them will effect a cure. They apply repeated dressings of the latex to achieve this (4). The latex is also used in Ivory Coast as a vesicant on areas of leprosy (1) and the Yoruba name, leper's oro, suggests a similar use in Nigeria. Vesicant action is due to the presence of triterpene derivatives (1).

The plant is used by Tenda singers at rituals in the belief that it improves the voice, and mistletoe from it is held to repel and even to frustrate the actions of sorcerers (2).

References:

1. Bouquet & Debray, 1974: 84. 2. Ferry & al., 1974: no. 94. 3. Jackson, 1973. 4. Kerharo & Bouquet, 1950: 75. 5. Baumer, 1975: 98, as E. venefica Tremaux.

Euphorbia spp. indet., 'cactiform'

West African: NIGERIA ANGAS gu -

The use of wax in making Ife and Benin bronzes in S Nigeria is said to be a 'lost' art. Sap from certain, not specified, Euphorbia spp. is reported as being substituted in Ekiti.

Reference:

Allison, 1969: 109.

Euphorbia sp. indet.

West African: NIGERIA FULA-FULFULDE (Nigeria) aguaje (Brotherton)

A plant of the above Fula name is fed by cattle herdsmen to their stock to promote fertility and to increase milk-yield. The name properly refers to *E. balsamifera* (q.v.) which is poisonous, so the present plant is unlikely to be the same.
Its presence in the bush is said to reduce the number of tsetse fly.

Reference:

Brotherton, 1969: 135.

Excoecaria grahamii Stapf

FWTA, ed. 2, 1: 416, as *Sapium grahamii* (Stapf) Prain. UPWTA, ed. 1, 163–4, as *S. grahamii* Prain.
West African: UPPER VOLTA BISA mahera (K&B) mara (Prost) BOBO dàganàga (Lebris & Prost) DAGAARI ignèl (K&B) FULA-FULFULDE (Upper Volta) nyan (K&B) pawdo (K&T) yarngal (K&B) GRUSI teul (K&B) HAUSA yazawa (K&B) KIRMA konyoro (K&B) pimpla (K&B) MANDING-DYULA lefora (K&B) MOORE sienhonlon (K&B) sukulu (K&B) tianon (K&B) SENUFO-KARABORO nakoronekon (K&B) TURUKA samplanan (K&B) IVORY COAST BAULE anufla-noku (B&D) MANDING-DYULA lefora (K&B) tara (Oldeman) GHANA DAGBANI pampiga (Graham; FRI) GRUSI tullu (Graham; FRI) HAUSA zaga rafi = *go round the stream* (auctt.) MOORE pulle (auctt.) TOGO ? fragugune (Pax) DAHOMEY BAATONUN dobogo (A.Chev.) NIGERIA FULA-FULFULDE (Nigeria) gurohi (JMD) inagoɗo (?) (JMD) nyanyarngal (Brotherton; Taylor) HAUSA yààzááwáá (JMD; ZOG) yààzááyáá (auctt.) NUPE nabbiri (JMD)

A semi-woody herb to about 1 m high from a creeping woody root-stock, of savanna and fringing forest, occurring from Guinea and Mali to Nigeria.
All parts of the plant contain a viscid latex which can be drawn out into threads. Root-latex is vesicant and powerfully caustic. Ground up with a little water it is applied on a stick to produce red or black marks on the face, causing swelling and ultimately tattoo marks (2, 6). Sometimes it is used for ritual scarifications (2, 6). In some districts the leaf is crushed with a little water which is applied to women's faces to make red or orange marks. In other places the more powerful root-latex is used. Boys raise blisters on the skin, Hausa boys, the while, reciting an imprecation *a rege zunufin uwa* to eliminate pre-natal evil influences (2). The extract of the pounded leaves is applied to guinea-worm sores or some drops of the latex to the blister to assist extraction of the parasite. A decoction of the whole plant is used in Ivory Coast–Upper Volta in baths for treatment of skin-affections and is taken internally for those diseases, e.g., leprosy, ascites, etc., requiring drastic purging (1, 4). Leaves and roots enter into a prescription for leprosy in N Nigeria along with other drug-plants and superstitious ceremonial (2, 6). The root is sometimes an ingredient of an arrow-poison intended for hunting, but probably not for warlike activity. The Karaboro of Ivory Coast prepare an arrow-poison called *kangouna* containing it (4), and also in Ivory Coast the plant is added as bait to kill pestilential wild animals (4). It is said to be immediately fatal to pigs (5). It has been used by the Hausa with criminal intent in food. Plant extract is very dangerous to the eyes.
The plant is so toxic that even the precaution of washing the hands after touching it is advocated (4).

References:

1. Bouquet & Debray, 1974: 86, as *Sapium grahamii*. 2. Dalziel, 1937: 163–4, as *S. grahamii* Prain. 3. Irvine, 1961: 251–3, as *S. grahamii* Prain. 4. Kerharo & Bouquet, 1950: 87–88, as *S. grahamii* Prain. 5. Oldeman 273, K. 6. Brotherton, 1969: 135, as *Euphorbia sp.*

Excoecaria guineense (Benth). Müll.-Arg.

FWTA, ed. 2, 1: 415–6, as *Sapium guineense* (Benth.) O. Ktze.
West African: NIGERIA YORUBA suru-fufu (JRA)

A shrub to under 2 m high of the dense forest occurring in Sierra Leone to W Cameroons, and extending to Zaïre.

Root-bark is used in Nigeria as a purge in constipation and also in kidney diseases. Stem-bark is also powdered and taken in small doses as a purgative. The bark-decoction is an emetic. The sap is acrid and poisonous (1). In Zaïre the leaves are used to treat sores (2).

References:

1. Ainslie, 1937: sp. no. 309. 2. Léonard, 1962, b: 150–1.

Grossera macrantha Pax

FWTA, ed. 2, 1: 762.

A shrub or tree to 6 m high, understorey of the high-forest, occurring only in W Cameroons and also southward to Zaïre.

In Congo (Brazzaville) powdered bark is dusted onto sores after they have been washed with the bark-decoction. This treatment causes severe smarting (1).

Reference:

1. Bouquet, 1969: 116.

Grossera vignei Hoyle

FWTA, ed. 2, 1: 398.
French: ohoué à grandes feuilles (Ivory Coast, Aubréville).
West African: GHANA AKAN-ASANTE *m*peduro (FRI; E&A) *m*peoro (BD&H) TWI dubrafo (BD&H)

A lower storey tree to 16 m high of the deciduous forest from Ivory Coast to W Cameroons.

The leaves have unspecified medicinal use in Ghana (Moor fide 1).

Reference:

1. Irvine, 1961: 230.

Hamilcoa zenkeri (Pax) Prain

FWTA, ed. 2, 1: 413.

A scandent forest shrub to 10 m high or tree to 15 m, recorded in W Cameroons and also in E Cameroun.

The bark contains a white latex.

Hevea brasiliensis (Willd.) Müll.-Arg.

English: rubber; Para rubber.
French: caoutchouc de Para; hevea.
Portuguese: seringueira.
West African: IVORY COAST KRU-AIZI amanį ke = *rubber tree* (Herault) GHANA AKAN amane-dua = *rubber tree* (FRI; E&A) NIGERIA IGBO rǫbà (BNO; KW)

A tree, 10–20 m tall, bole up to 1 m diameter, very variable and usually (especially grafted trees) very much less, native of the Amazon valley from

Eastern Peru to the Brazilian Atlantic seaboard. This is the source of the world's requirements of natural rubber ousting almost entirely all other sources except for very special uses. From a Brazilian forest product in the Nineteenth Century the world is now supplied by plantations principally in Malaya, Indonesia, Thailand, Indo-China, and Ceylon. From trees raised in Ceylon and Singapore the West African territories were sent seed in the closing years of the last century and in the beginning of this. Plantations and small-holdings are scattered along the seaboard from Guinea-Bissau to West Cameroons wherever rainfall is adequate and soil conditions are right. A minimum of 1500 mm rain per annum is required and this preferably well distributed throughout the year. A soil of pH 5.5–6.5 is desirable. A plantation industry of some importance has grown up in Liberia and considerable interest is being shown in other territories, e.g. Zaïre. *H. braziliensis* is extremely polymorphic and work of breeding and selection has resulted in clones and seeding families of outstanding yielding capacity to as much as ten times that of the unselected material. The product is made under industrial conditions into sheet, crêpe, concentrate, crumb etc., and can be classified according to clonal and soil origins, or compounded to various specifications for manufacturing requirements. As a small-holder's produce without the necessary finesse for this the essence is cleanliness and uniformity of production of sheet.

The rubber fraction of latex is a polymer of *isoprene*, C_5H_8, in extremely long chains with molecular weight from 500,000 to over 2,000,000. There is recurrent interest in cracking the chain to produce a substitute for petroleum (3). The Japanese in Malaya during World War II produced a combustable fuel by crude distillation of rubber coagulum, but engines burning it were soon rendered unserviceable by gumming up. It is certain the process can be improved.

Hevea wood is soft and light. It makes an excellent firewood. It can be sawn for making rough planks, and has a limited potential for pulping.

The earliest recorded selection of hevea by man was of the seeds for food when seeds from trees bearing the largest were taken in establishing new settlements along the rivers of the Amazon basin. Seed production under plantation conditions must be quite astonishingly large. Though the seed is not normally eaten, in SE Asia it was taken as an emergency food during World War II, 1942–45. Jungle pig relish them, and sheep in Gabon are said to eat them (2). Seed-cake can be fed to stock (1).

A semi-drying oil to as much as a 25% concentration is in the seed. By careful collection and quick processing oil with a low free fatty acid content can be expressed. Like other oil-rich seed, viability is short-lived and in germination for planting purposes the seed should be set immediately on falling from the tree.

References:

1. Anon 1930. 2. Walker & Sillans, 1961: 168. 3. Wang & Huffman, 1981: 369–82.

Hippomane mancinella Linn.

FWTA, ed. 2, 1; 368.
English: manchineel.
French: mancinillier.

A foreshore tree of tropical America, from 3 to 7 m high, now widely distributed in the tropics, and present under cultivation in W Africa.

The tree is frequently referred to in the chronicles of early explorers of the Americas as one of the most toxic trees of the New World tropics. The wood is yellowish-brown, variegated brown and black, medium hard and durable. Though the living plant is so highly poisonous, the dried timber loses this danger and is greatly valued for cabinetry. It takes a good polish with a fine-textured grain resembling walnut. The toxic substances lie in the sap and working the tree is hazardous for fear of contacting it. The result of so doing is a reaction like a burn on the skin, or severe conjunctivitis in the eyes which may lead to prolonged temporary or even permanent blindness (2, 3, 4). Wood-cutters in the Caribbean are said to fire the bole of the tree to be felled to char the bark. It is claimed that even resting or sleeping under the tree's shade, particularly if there is rain or dew, causes blistering and burning of the skin, headache and swelling of the eyes. Breathing of the wood-smoke also can cause headache. In spite of these very real risks the tree has been planted in the West Indies as a wind-break (1).

The leaves contain toxic elements and have been used for criminal poisoning. Admiral Horatio Nelson, R.N., is recorded as having suffered poisoning of water drawn from a spring saturated with machineel leaves by hostile Indians during the British expedition against Spanish Nicaragua (3) in 1780 from which he was invalided home at death's door.

The fruit, like Eve's apple, has a tempting appearance of edibility. Indeed, it resembles a red apple. Reports of its poisonousness vary, some that it is very dangerous causing death within 24 hours, and others of humans eating many fruit at one time and surviving. Goats are said to eat them avidly without ill-effect (3). Fruit falling into streams or rivers are readily eaten by fish, but those fish, if eaten by people, may cause food-poisoning (3). Like the akee apple (*Blighia sapida*, Sapindaceae) the degree of ripeness may be critical, and it is an aspect that should be investigated. In the event of poisoning a recommended antidote is to take some oil.

The most potent part of the tree is its sap. This is compounded into some S American arrow-poisons, and for its causticity it is used on warts, corns and cancers (2), and for dropsy and venereal diseases (3).

References:

1. Burkill, IH, 1935: 1177. 2. Hartwell, 1969, b: 167. 3. Lauter & al., 1952: 199–201. 4. Rao, 1974: 166–71, with phytochemistry.

Hura crepitans Linn.

FWTA, ed. 2, 1: 368.

English: sandbox tree; monkey's dinner bell (West Indies, Burkill).

French: bombardier (Kerharo & Bouquet; Berhaut); bois du Diable (Berhaut); crepiton (Anon, K.); sablier (Berhaut); sablier des Antilles (Kerharo & Bouquet).

West African: **IVORY COAST** BAULE dimbo (K&B) **GHANA** AKAN-FANTE abru koyin (BD&H; FRI)

A tree to 25 m tall with very spiny trunk and branches, native of tropical America and now widely dispersed throughout the tropics.

The seed-pod is explosively dehiscent when ripe by which it received the West Indian name and the French *bombardier*. If the unripe seed pod is pierced and boiled in oil, its explosiveness is lost. It then constitutes the "sandbox" for dusting sand in writing (a precursor of blotting paper) from which the other names derive (2, 4).

The tree is commonly planted along roads and in villages for shade in West Africa (Senegal: 1, 7; Ivory Coast: 6; Ghana: 3, 5; Dahomey, Nigeria, Cameroons: 4). It is similarly planted in many other countries. It has been used in SE Asia as a shade for cacao, and a prop for vanilla (2).

The wood is yellowish with a silky lustre, light in weight and of good strength. It has been much used in northern South America (2).

The bark has been used in S America for leprosy though with doubtful benefit. The latex is caustic and purgative due to the presence of two very toxic principles, *hurin* and *crepitin* (6). The latex is said to cause blindness. It has been used as a fish-poison and sometimes is added to S American arrow-poisons (2, 8).

The inflorescence has been used in the West Indies to make a jam (7). The seed contains *tannins, gallic acid* and *toxalbumin*. The kernel-oil is a golden yellow, semi-drying and is about 53% by weight of the kernel. The oil is used by the Baule of Ivory Coast as a purgative (6).

References:

1. Berhaut, 1967: 233. 2. Burkill, IH, 1935: 1203–4. 3. Burtt Davy & Hoyle, 1937: 46. 4. Holland, 1908–22: 612. 5. Irvine, 1961: 231. 6. Kerharo & Bouquet, 1950: 76,77. 7. Walker & Sillans, 1961: 168. 8. Moretti & Grenand, 1982: 154.

Hymenocardia acida Tul.

FWTA, ed. 2, 1: 377. UPWTA, ed. 1, 146.

French: digbé (Bouquiaux); coeurs-volants (Bailleul).

West African: SENEGAL BALANTA batu mendé (K&A; JB) BANYUN tipéo (K&A; JB) BASARI a-ndiimbémun (JB) a-nginbemur (after K&A; JB) BEDIK fufé (K&A; JB) gi-kemɛnyɔr (FG&G) gi-komoñir (JB) gui-komonyir (K&A) DIOLA bo sånd (JB) bu liu (K&A; JB) bu liuliu (K&A; JB) FULA-PULAAR (Senegal) kérèn kôdéi (JB) kérên kôdey (K&A) koren kôdé (K&A) korèñ kondé (JB) korô kôdey (K&A) péleti (K&A) péléti (JB) pélìtoro (K&A; JB) pénitoro (JB) pépitoro (JB) psellitoro (JB) TUKULOR korôkondi (JB) KONYAGI kot (JB) MANDING-BAMBARA grêgeni (K&A; JB) gringrin néréni (JB) kalakari (K&A; JB) komoní (JB) ya gualani logo (A.Chev.) ya guanani logo (JB) MANDINKA kôgutuménô (K&A) korôkôdo (K&A; JB) kukutumâdên (K&A) kunkutu-mâdu (K&A) MANINKA kankutuma ndimo (JB) tanóro (K&A; JB) tanóro irini (K&A; JB) ·SOCE· korôkôdo (K&A) kunku tumãndiô (K&A; JB) ·SOCE· (Niombato) kôgutuménô (K&A) kunkutu-mâdu (K&A) SERER ènkélèñ (JB) ngènkélèñ' (K&A; JB) ngèrkélèn' (JB) ·SUSU· barambara (JB) mérémériñi (JB) WOLOF enkéleñ (K&A; JB) THE GAMBIA FULA-PULAAR (The Gambia) dafi (DRR; DAP) MANDING-MANDINKA kabanumbo (DAP) kabunumbo (DRR) kali-kali (?) (DAP) kurukondo (DAP) kurunkondo (DRR) WOLOF kabanumbo (DAP) kabunumbo (DRR) GUINEA-BISSAU BIDYOGO oábi (JDES) FULA-PULAAR (Guinea-Bissau) caranconde (JDES) coronconde (EPdS) peliteró (JDES) MANDING-MANDINKA curencóndô (JDES) SUSU curencúnde (JDES) GUINEA BASARI a-ndém-ɓémun (FG&G) FULA-PULAAR (Guinea) pelitoro (Aub.) penitoro (Aub.) MANDING-MANDINKA tanóro (Aub.) tanóro irií (Aub.) MANINKA (Konya) timéni (Aub.) SUSU barambara (Aub.) karécugné (Farmar) mérémérigniyi (auctt.) SIERRA LEONE KORANKO kiŋkiliɛle (NWT) LOKO nyabɛre (NWT) MANDING-MANDINKA yɛgbɛ (FCD) MENDE fabanjuwi (JMD) fagbanjoi (def.) (JMD; S&F) fagbanjui (def.) (JMD; S&F) TEMNE ka-gbalkəntha (FCD) YALUNKA m-barambara-na (FCD) MALI FULA-PULAAR (Mali) pelitoro = a redman; i.e. an Arab, etc. (JMD) MANDING-BAMBARA grengréni (Aub.) gringrin néréni (Aub.) kali kali (FB) komoni (Aub.) ye gualani (FB) MANINKA tanóro (Aub.) tanóro irini (Aub.) UPPER VOLTA HAUSA ginyaro (K&B) jan itateha (Aub.) jan yaro (Aub.) MANDING-BAMBARA kalakari (K&B) DYULA kogno norubo, konion nunugbo = tree which kills the wife (K&B) SENUFO tifolo mongolo = tree which kills the wife (K&B) IVORY COAST ? timin-nu (A&AA) tunmin-nu (A&AA) AKAN-BRONG katéréka (K&B) BAULE brindé (K&B) kandié kwékwé (K&B) kpangui-kokorè (A&AA) painguokokolé (B&D) painkokolé (B&D) uoré uoré (auctt.) KULANGO paradio (K&B) LIGBI tumemi (K&B) LOBI angala (K&B) MANDING-DYULA kogno norubo, konion nunugbo = tree which kills the wife (K&B) MANINKA tanóro (Aub.; K&B) tanóro irini (Aub.; K&B) tiun (B&D) SENUFO tifolo mongolo = tree which kills the wife (K&B) SENUFO-TAGWANA kandiolé (K&B) kandioli (K&B) tumbala (K&B) YAORE guandeklé klé (K&B) GHANA AKAN-BRONG duakɔkowa (FRI) sabrakyie (auctt.) BAULE kalakari (FRI) worewore (FRI) DAGBANI dankunga (FRI) GBE-VHE adudze (auctt.) ziwuɛ tsi (BD&H; FRI) VHE (Pecí) dzêwuzi (FRI) freki (FRI) GUANG-NCHUMBULU kapiluwi (Coull) KONKOMBA kulangmau (JMD) NANKANNI prewia (Enti) SISAALA gbegelege (Blass) hadafiang (AEK; FRI) TOGO BASSARI kunang-kalange (JMD) GBE-VHE adudze (Volkens; FB) atidjain (Aub.) KABRE katschintschinga (Gaisser) ·KPEDSI· adekplatye (Volkens) TEM (Tschaudjo) tschenyenga, tschintschenga (JMD) YORUBA-IFE (TOGO) atidze (JMD) DAHOMEY BAATONUN séma (Aub.) senegu (Aub.) GBE-FON sotive (Aub.) GEN sobétigue (Aub.) YORUBA-NAGO oropa (Aub.) NIGER GURMA okuamonuana (Aub.) HAUSA jan itateha (Aub.) jan yaro (Aub.) NIGERIA BIROM lyàám syìnàŋ (LB) BOLE madiya (AST; JMD) FULA-FULFULDE (Nigeria) boɗehi = red (JMD) pattoyi = tree whose wood or fruits blister; e.g. when put into a fire (MM) samathi (Taylor) yawa sotoje meaning the wood is brittle and breaks sharply (JMD; KO&S) GWARI karagwehi (JMD) HAUSA ján ítaacéé = red tree (auctt.) ján yáárò (yáárò = red boy, or red fellow (auctt.) IGBO ikalaga (NWT; KO&S) IGBO (Uburubu) ag'ozala (NWT; JMD) wuzonye enyi amu (NWT; JMD) KAMBARI ùurùkpâ (RB) SAMBA-DAKA burrki (Chapman) YORUBA òrùpa (auctt.)

85

A tree to about 6 m high, gnarled and twisted with characteristic rough rusty-red bark, of the wooded savanna thoughout the Region from Senegal to W Cameroons, and widespread in tropical Africa.

The wood is light brown or pink, darkening to orange, close-grained, with conspicuous annual rings, and hard (12, 20, 21). It has been said to be brittle and good only for firewood (12, 39). This may be reflected in certain Ivorean names meaning 'The tree which kills the wife', i.e. when an unfaithful wife goes to collect some firewood, the tree shatters at her touch and a branch pierces her abdomen (27). The Gbaya of the Central African Republic recognise the tree as producing good firewood; indeed they classify it as 'woman's firewood' being good for the hearth and cooking place, long-lasting while the housewife is about other chores, yet reviving quickly from sleeping embers, with a hot flame and little smoke (47). The tree is used for house-posts in S Nigeria (38), and in Gabon where the wood is made into charcoal for blacksmith's work (40). In Kenya (11) and in Uganda (14) the wood is known for its hardness, denseness, durability and good resistance to termite-attack. It is used to make pestles and bark-cloth mallets. Charcoal made from the branches is powdered and rubbed on the head for headache in the soudanian region (5).

The foliage is browsed a little by cattle in Senegal (1). Young leafy shoots and the fruits may be eaten in Senegal as a supplementary food (19) and in Nigeria (18). Leaves and leafy shoots have a considerable medicinal use. When chewed they have an acid taste. Leaves are prepared into an infusion for use in Senegal for chest-complaints and small-pox and, with the roots, for deficiency diseases (22, 26). A macerate is given for gripe (24), and leaf-decoction used as an eye-wash (24, 26). A decoction with honey is taken in Guinea for biliousness (30). In Ivory Coast–Upper Volta a leaf-decoction is used in baths and draughts as a febrifuge, and leaf-powder is taken as snuff for headache or applied topically for rheumatic pains and toothache, or for the same purposes leaves may be pulped with an organic acidic substance such as citron juice or sap of *Piliostigma reticulatum* (Leguminosae: Caesalpinioideae) (27). Leaf-sap is used in Ivory Coast by installation in ears for otitis, eyes for ophthalmias and nose for headache, and by topical frictions for fever and rheumatic pains (8); or the dried and powdered young leaves may be sprinkled over sores after washing, and a leaf-decoction taken for stomach-ache or used as an eye-wash for conjunctivitis (3). In Tanganyika the leaves are chewed against cough and for 'swollen' tummy (36) and for serious stomach-ache (37), and the leaf-sap is squeezed into the infected eyes of children (34).

The bark is the source of a brownish red dye which is used in Gabon (40) on raphia work, and perhaps elsewhere in central Africa (15). The bark also contains *tannin* which is used in central Africa for tanning (12, 15). Concentration in Katanga-grown material has been reported to be 12% (27). Tannins are present also in the leaves and roots (7). Traces of *alkaloid* have been detected in stem-bark and leaves of Nigerian material but not in the roots (2, 26) and the alkaloid *hymenocardine* in the root-bark (43).

The bark has numerous medicinal uses. When chewed it is bitter and slightly astringent. It produces a copious salivation (22, 26, 31). It is chewed with kola in Sierra Leone for dysentery (13). An aqueous decoction is widely used for pulmonary affections in Senegal (22, 26) and in Guinea (30). In Kordofan a bark infusion, or of the fruit, is made into a boiling inhalation for breathing difficulty (47). It is also deemed good for colds (5). In Senegal powdered bark is added to food for an affection known as *pidal* (Fula-Fouladou) the symptoms of which are a piercing pain in the shoulders with spasmodic coughing, emaciation and sometimes dermatitis (24, 26). The Tenda compound the bark with that of *Zanha golungensis* Hiern (Sapindaceae) to make a poultice, covered with leaves of *Terminalia macroptera* Guill. & Perr. (Combretaceae), and changed daily for application to fractures (44). Powdered bark is taken in The Gambia for abdominal pains (33); for menstrual pains in Tanganyika (34); with palm-oil and rock salt for bloody vomit in Congo (Brazzaville) (6); for painful swellings and debility, and in baths and draughts for epileptic fits in Casamance (23); and against colic and in poultices on abscesses and tumours in Nigeria (4) and in Ubangi (31). The pulped bark is used in Congo (Brazzaville) to treat diarrhoea and dysentery, sterility and dysmenorrhoea of women, and cough; a bark-decoction is used to wash sores and to relieve ophthalmia and migraine, and pulped bark is applied to dermal infections such as parasites, itch, prickly-heat and leprosy (6). The inner bark is powdered and put on ulcers in Zambia (42), and in Tanganyika powdered bark with copper (or gold) filings is sprinkled on syphilitic chancres (16). Water in which the roots have been boiled is considered febrifugal in Nigeria (4) and in N Ashanti, Ghana (9). In NW Nigeria, Salka people use the plant (part not known) in medicine for children with fever (45). Root-bark is eaten with porridge in Tanganyika for malaria (16), and the plant (part not specified) is used in Mozambique for trypanosomiasis (28).

Bark and leaves are prescribed together in various ways. In N Nigeria they enter into a complex prescription to give strength on a journey (12), and in Ivory Coast–Upper Volta a preparation is used in baths and lotions to strengthen debilitated children, and in draught to relieve dysenteriform diarrhoea (27). A decoction is prescribed in Senegal for a treatment lasting 1–2 months for refractory cough and tuberculosis (24) and for deficiency oedemas (23).

The root is used in Senegal to make a mouth-wash for toothache which is said to give good results with treating stomatitis and pyorrhoea also (24). Sap from the roots is applied topically in Congo (Brazzaville) for earache and tooth-troubles (6). In southern Africa steam-inhalations from the root-powder is considered anti-enteralgic and cleansing. It is given in Senegal as a precursor of treatment for sterility, and can also be given safely during pregnancy (24, 26). In Tanganyika water in which roots have been soaked is considered a good medicine to give women during pregnancy (29). A root-decoction and leaf-sap is taken to prevent miscarriage (16), and powdered root is added to porridge for mothers whose breast-milk is 'too heavy' (35). In Ivory Coast the plant is used as a galactogogue, aphrodisiac and anti-dysenteric (8, 27). A root-decoction with leaf-sap is also an anti-dysenteric medicine in Tanganyika (16), and root and stem-barks are used in Ubangi as an emetic antidote to ordeal-poison (31).

Root-extracts have shown some insecticidal activity (17).

The fruit is edible and is sometimes eaten (10, 20). The ripe fruit is used in Zambia to cure deafness (32).

Tenda hunters consider it a hunter's charm (44).

References:

1. Adam, 1966, a. 2. Adegoke & al., 1968. 3. Adjanohoun & Aké Assi, 1972: 125. 4. Ainslie, 1937: sp. no. 188. 5. Aubréville, 1950: 186. 6. Bouquet, 1969: 116. 7. Bouquet, 1972: 24. 8. Bouquet & Debray, 1974: 84. 9. Burtt Davy & Hoyle, 1937: 46. 10. Busson, 1965: 163. 11. Dale & Greenway, 1961: 204. 12. Dalziel, 1937: 146. 13. Deighton 4162, K. 14. Eggeling & Dale, 1952: 130. 15. Greenway, 1941. 16. Haerdi, 1964: 94. 17. Heal & al., 1950: 119. 18. Hepburn 11, K. 19. Irvine, 1952,a: 31. 20. Irvine, 1961: 231-3. 21. Keay & al., 1960: 308-10. 22. Kerharo & Adam, 1962. 23. Kerharo & Adam, 1963,a. 24. Kerharo & Adam, 1964, b: 550-1. 25. Kerharo & Adam, 1964, c: 309. 26. Kerharo & Adam, 1974: 417-8, with references. 27. Kerharo & Bouquet, 1950: 77, with reference. 28. Lemos & Marrime 332, K. 29. Pirozynski P.556, K. 30. Pobéguin, 1912: 34. 31. Portères, s.d. 32. Richards, 19256, K. 33. Rosevear, 1961. 34. Tanner 307, K. 35. Tanner 1373, K. 36. Tanner 5258, K. 37. Tanner 6016, K. 38. Thomas NWT.2290 (Nig. Ser.) K. 39. Unwin, 1920: 135. 40. Walker & Sillans, 1961: 168. 41. Watt & Breyer-Brandwijk, 1962: 420. 42. White 2059, K. 43. Willaman & Li, 1970. 44. Ferry & al., 1974: No. 95. 45. Blench, 1985: pers. comm. 46. Baumer, 1975: 102. 47. Roulon, 1980: 228-9.

Hymenocardia heudelotii Müll.-Arg.

FWTA, ed. 2, 1: 377.

West African: SENEGAL KONYAGI ménãgol (JB) MANDING-MANDINKA koféta karo (JB) SIERRA LEONE MANDING-MANDINKA kolomaniŋbɛlɛ̃ (FCD) sukorogbele (NWT) MENDE fagbanjoi (FCD; S&F) fagbanjui (FCD; S&F) ɲja-fagbanjo (FCD) ɲja-fagbanjui (S&F) TEMNE ka-gbalkɔntha (S&F) am-balkunta (FCD) YALUNKA baramba-na (NWT) fɔtɔ-na (?) (FCD) kude-barambara (FCD) kude-barambara-na (FCD) **IVORY COAST** DAN gondé (Aub.)

A shrub or tree to 10 m high of riverain, savanna and fringing forests, locally abundant throughout the Region from Senegal to S Nigeria, and into Zaïre. The bark-sap is reported to be squeezed into the eye for ophthalmia in Sierra Leone (1).

Reference:

1. Deighton 4438, K.

Hymenocardia lyrata Tul.

FWTA, ed. 2, 1: 377.
West African: GUINEA FULA-PULAAR (Guinea) pelitoro pété (Aub.) SIERRA LEONE BULOM (Sherbro) gbathoŋ-dɛ (FCD; S&F) GOLA fagbanjo (FCD) KISSI lɔlɔ-kumsa (FCD; S&F) KONO fagbanji (FCD; S&F) MENDE fagbajoi (L-P) fagbanjo(-i) (FCD; S&F) fagbanjui (FCD; S&F) TEMNE ɛ-balkənta (NWT) ka-gbalkəntha (FCD; S&F) VAI gbara (FCD) LIBERIA MANO kma kma yidi (Har.)

Shrub or tree to 16 m high, of closed riverain forest or transition forest from Senegal to Ghana.

The wood is hard and pinkish (1). It is a popular fuel in Sierra Leone where the tree regenerates fairly readily under managed forestry and is thus regarded as a weed species (2).

References:

1. Irvine, 1961: 233. 2. Savill & Fox, 1967: 116–7.

Jatropha chevalieri Beille

FWTA, ed. 2, 1: 397.
West African: SENEGAL ARABIC (Senegal) gendefer (JB) FULA-PULAAR (Senegal) kollé diéri (JB) TUKULOR otenuba (K&A) vatèn u bet (JB) vatèn u bey (JB) waten böt (K&A) witen u böt (K&A) wuten u böt (K&A) MANDING-BAMBARA tiriba (K&A) MANDINKA sarak (JB) tiriba (JB) SONINKE-SARAKOLE tiriba (K&A) WOLOF otenuba (K&A) vitèn u mbet (JB) watenuböt (K&A) witen u bot (K&A) wuten u bot (K&A) MALI DOGON anyu sámu bugurii (CG) NIGER HAUSA zurma (AE&L) zùrmaa (AE&L) zurman (AE&L)

A shrub reaching to 1.5 m high of the sandy Sahel region of Mauritania and Senegal.

The plant is toxic to herbivores (4).

The latex is used as a popular vulnary in Senegal, and the roots are used for treating syphilis complications and leprosy. Indeed, the plant features regularly in the formulations of the Senegalese medicine-men who specialise in treatment of leprosy (1–3).

References:

1. Chevalier, 1937, b: 171. 2. Kerharo, 1967. 3. Kerharo & Adam, 1974: 418–9. 4. Naegelé, 1958,b: 878.

Jatropha curcas Linn.

FWTA, ed. 2, 1: 397. UPWTA, ed. 1, 147–8.

English: physic nut; pig nut; fig nut; purging nut; Barbados nut; pinhoen oil (the seed-oil).

French: purghère; pourguère; pignon d'Inde; pourguère de Cayor; médicinier béni (Berhaut); feve d'enfer (Berhaut).

Portuguese: purgueira.

West African: SENEGAL BALANTA ngum (JB) BASARI a-kemul kétié (K&A; JB) BEDIK akemul kété (JB) gi-kidi = hedge; from Fula (FG&G) DIOLA délégu (JB; K&A) purgèr (JB) purgèr éfiti (JB) purgèr yéténé (JB) yakirid (JB) DIOLA (Bayot) bu bon kashla (JB) FULA-PULAAR (Senegal) diuladukdé (K&A; JB) diulakukaĵi (K&A) kidi (K&A; JB) TUKULOR kidi (K&A; JB) KONYAGI bagha (JB) MANDING-BAMBARA bagani = small poison (K&A; JB) bila ñaraba (JB) kultii ni fli (JB) MANDINKA tabanani (K&A; JB) tabanano (K&A) ·SOCE· tabanano (K&A; JB) MANKANYA bu kulpa (JB) NDUT tébernani (JB) NON tabanani (K&A; JB) tuva (JB) SERER litrog (K&A; JB) niom (K&A) tuba (JB) SONINKE-SARAKOLE damabâtaô (K&A) ·SUSU· barhane (JB) WOLOF niom (K&A) tabanani (auctt.) **THE GAMBIA** FULA-PULAAR (The Gambia) chidigeh (Hallam) kidi (DAP; Hallam) MANDING-MANDINKA buyiro (Hallam) tababu-tabu (DRR) tabanani (DAP; Hallam) tubab tabo (Hallam; Bojang) WOLOF tabanani (auctt.) tabu (DAP) **GUINEA** BALANTA ngu-me (JDES) ungume (JDES) BASARI a-kemɔ́l kètiè = [making a] tough enclosure (FG&G) FULA-PULAAR (Guinea) kidi (auctt.) KISSI domd'jio (FB) KPELLE ninguéguéhiklu (FB) LOMA banigui (FB) borobali (FB) MANDING-MANINKA baga, bagha, baha = poison (Langdale-Brown; JMD) bani (FB) SUSU barhané (Aub.; JMD) **SIERRA LEONE** BULOM (Sherbro) katawo-lɛ (FCD) FULA-PULAAR (Sierra Leone) kidi (FCD) GOLA kata-wuli (FCD) KISSI bɔtɛmbã (FCD) KONO kata-kɔnɛ (FCD) KRIO fig-not i.e., fignut (FCD) LIMBA (Tonko) kusigbɔro (FCD) LOKO hɛmba (NWT) MANDING-MANDINKA baga (NWT; JMD) bagauro (JMD) bagha (JMD) baju (FCD) MENDE kata = a hedge (FCD) kata-wi (def. -wulo, -wului) = hedge tree (auctt.) SUSU bakhanɛ (FCD) bakhɛnɛ (FCD) TEMNE ɛ-sigoro-ropet (NWT) TEMNE (Port Loko) an-kidi (FCD) TEMNE (Sanda) an-kidi (FCD) TEMNE (Yoni) ɛ-nɔŋkɔ (FCD) YALUNKA kidi-safu-na (FCD) **MALI** BOBO daga-naga (A.Chev.) manan-naga (A.Chev.; Aub.) MANDING-BAMBARA adé the seed (A.Chev.) bagani = small poison (A.Chev.; Aub.) iridigué (A.Chev.) **UPPER VOLTA** BOBO manan naga (Aub.) MANDING-BAMBARA lagani (Curasson fide K&B) SENUFO nakuo (K&B) **IVORY COAST** AKAN-ASANTE taluka (B&D) BRONG propro (K&B) AKYE mpopo (A&AA) ·ANDO· ploplo (Visser) ANYI propro (K&B) BAULE a ploplo (A&AA) ploplo (B&D) propro (auctt.) GAGU sakoli (K&B) KRU-BETE badaguigui (auctt.) bategné ni (K&B) GUERE (Chiehn) sacré (K&B) KULANGO adibalaga (K&B) KYAMA samanobo (B&D) SENUFO nakuo (K&B) SENUFO-TAGWANA belia (K&B) **GHANA** ADANGME kuadidi (-tʃo) from Twi: akua; slave; didi; ear, tʃo: tree (FRI) ADANGME-KROBO kitigblɛtʃo (FRI) kutugblɛtʃo (FRI) AKAN-ASANTE abrɔtɔtɔ (OA) akaneadua from Twi: kanea; lamp; alluding to the use of the seeds (FRI) nkrang-gyadua (E&A) FANTE aborɔkyiraba (FRI; OA) adaadze (FRI; OA) adadze (FRI) kaneadua (FRI) TWI aborɔtɔtɔ (FRI) abrɔtɔtɔ (OA) akaneadua (FRI; E&A) GA kitigblɛtʃo (FRI) kplukatʃo = bedbug tree (FRI) kutugblɛtʃo (FRI) kwaadidi-tʃo (FRI) kwadiditʃo (FRI) ŋkpúluka (KD) plukateo (OA) GBE-VHE babatsi = termite plant (FRI; OA) gbomagboti (FRI) kpɔtî = fence tree (FRI) VHE (Awlan) abrɔtɔto (Bunting fide FRI) kpɔtî = fence tree (FRI) VHE (Pecí) abrɔtɔto (FRI) NZEMA takploka (OA) tapleka (FRI; OA) **TOGO** GBE-VHE kpŏti (Volkens) ·MISAHÖHE· wabati (Volkens) **NIGERIA** ANAANG mbubok (JMD) ANGAS bilit (RES) ARABIC-SHUWA habb el meluk (JMD) BIROM gàrà the plant (LB) kparak gàrà the fruit (LB) EDO oru-ẹbo (JMD) EFIK étó m̀kpà = tree of death (auctt.) FULA-FULFULDE (Nigeria) kolakolaje (JMD) kolakolaje debbe 'female' of Ricinus communis Linn. (JMD) kolkolwaahi (J&D) kwolkwolaje (JMD) FULFULDE kokolaji (Chapman) GWARI kwotewi (JMD) HAUSA bíí ní dà zúgúú (auctt.) cii ní da zúgúú from zúgúú: a shroud; suggesting danger from its use (JMD; ZOG) halallamai for the leaves which are used to wrap the feet when staining with henna (auctt.) IBIBIO étó-m̀kpà (JMD) IDOMA ọcígbede (RGA) IGBO bulu olu (JMD; BNO) olulu-idu (Singha; BNO) òwulù idu from òwulù: cotton (JMD; KW) IGBO (Agulu) ugbolo (NWT; JMD) IGBO (Obu) okwata (NWT; JMD) ugbolu (NWT; JMD) IGBO (Umuakpo) okweni (DRR) JUKUN (Wukari) ho (Shimizu) TIV ígádàm (pl. ágádăm) (auctt.) URHOBO urieroh (Isawumi) YORUBA bòtújẹ̀ (auctt.) bòtújẹ̀ pupa pupa: red (JRA) làpá làpá (JRA; IFE) lóbòtújẹ̀ (JMD; IFE) lóbòtújẹ̀ pupa (JRA) olóbòntújẹ̀ (JMD) olobotujẹ (Verger) YORUBA (Ijebu) ṣẹngirùn (JMD) YORUBA (Ijesha) bòtújẹ̀-úbò (JMD) úbò (JMD)

A shrub or tree to 6 m high, native of the American tropics and now dispersed and naturalised throughout the Tropics.

It is easily propagated by cuttings and by seed, and it is one of the commonest plants used for making hedges in W Africa. It can be cut and trimmed to any height. It is often planted over graves, and sometimes to mark the limits of fields (30). These practices are observed also in other parts of Africa and as a hedge elsewhere in the tropics. It is held to be resistant to termite-attack, a feature noted in the Vhe name *babatsi*: termite plant (14). In Gabon it is grown as a support for vanilla and pepper vines (26). One should expect that it has a similar potential for yams. The stem and twigs are used in Nigeria (37) and in N

Cameroun (Malzy in 33) as chew-sticks. Foaming results. The watery sap is put onto fresh cuts and sores at the corner of the mouth (37).

The young leaves are eaten in Java. Some say that animals will not touch the leaves; others that hungry animals will, but this must be rare as the leaves are reported to contain *hydrocyanic acid* (7). An infusion, hot or cold, of the leaves is also taken internally for fever. They are mashed up by the Tenda people for poultices, and a macerate is taken as an emetic (31). The Baakpe of Cameroon Mountain drink the decoction with beer as a diuretic for rheumatism, and in Nigeria an infusion of young leafy shoots is used for most urinary complaints. In Gabon a leaf-decoction is considered good for cleansing the kidneys and releasing bile (3, 26). In Nigeria a decoction of leaves with natron is used by women as a wash for a month before childbirth. In The Gambia the leaves are used to make a mouthwash. In Ivory Coast heated leaves are applied to relieve pain (19, 30). A compress of leaves is applied for toothache in Kordofan (39). In Ghana the leaves are commonly an ingredient in enema preparations, and are prepared, along with oil-palm fruit, for an injection administered to weakly children. In S Nigeria they are a remedy for jaundice, applied by rectal injection (1). For immediate application after snake-bite, sap from the leaf is recommended in Ghana. It is said to dilute the snake-venom and to stop it from entering the blood-stream. For emergency use herbalists recommend that leaves be charred and powdered and held in store (40). The leaves are also widely used to treat guinea-worm sores as a lotion made with the crushed leaves in hot water, or by applying the ashes of burnt leaves (3). In Senegal an aqueous leaf-decoction is taken by mouth for bronchitis (15, 16), and a decoction of dried leaves in The Gambia to relieve coughing (32). The leaves and the green viscid sap are rubefacient. Pounded leaves are applied to sluggish ulcers, and the sap of dried leaves is applied direct to cuts and bleeding wounds as a styptic (5, 17, 22, 26, 30, 40). The pounded leaf has been used in India as a repellant of house-flies (27), and in Ghana the leaves are used to fumigate a house for bed-bugs (cf. the Ga name meaning *bed-bug tree* (10, 14). In Indonesia the leaf is applied to hard tumours (13). In Congo (Brazzaville) the sap is instilled into the outer ear for otitis (5), and in Ivory Coast it is given by enema for blennorrhoea and by draught for jaundice (19). The sap is given by the Tiv for the Benue Plateau, Nigeria, to a child with fever (24).

Leaf-sap with water is a plaything for children who blow bubbles with it (10, 11). It can be used to make soap. It is irritant and causes inflammation in the eyes. It is used to put in a hollow tooth and on to bee and wasp-stings (27). Mixed with salt it is rubbed on teeth to clean them (10). Sap from the leaf-petiole or the bark is applied in The Gambia to wounds as a drying agent and antiseptic (32). In India young twigs are used as a dentifrice (28). In Ivory Coast the sap is rubbed on young children's gums to help in teething (19), and the latex is given to newly-born babies affected by tetanus (2).

The sap yields a black dye, the bark dark blue and the leaves grey (12). Extracts are used as a marking-ink and dyes, and the stain is said to be indelible (10). But fastness was far from the case in Japanese soldiers' uniforms dyed in a concoction of the young foliage in Thailand during World War II, 1942–45, since the colour ran in monsoon rains with grotesque results. The latex, used in Ivory Coast for external treatment of sores (30), is said to contain 10% *tannin* and turns brown and brittle on drying (14).

Examination of powdered leaves of Sierra Leone material showed no alkaloid, nor saponin present (23), but a trace of *alkaloid* is reported in Indian material (18).

A root-decoction with natron or salt is used in W Africa for gonorrhoea (3, 10, 38), and with flour for dysentery (10). Root-bark, dried and pulverised, is applied as a dressing for sores, and mixed with guinea-grain is rubbed on gums to relieve the spasms of infantile tetanus (10, 14). The powdered root is taken internally for worms (3). In Gabon a root-macerate is taken for spermatorrhoea, or pieces of the stem may be chewed for the same purpose (25, 26).

The bark contains a wax which is a mixture of *melissyl alcohol* and its ester, *myricylmelissate* (19, 27). The bark is commonly used in the Philippine Islands as a fish-poison where the plant bears the same name as *Derris* (Leguminosae: Papilionoideae) (7, 21, 27).

The dried fruits are powdered and taken with food in The Gambia as a vermifuge and to void excessive fat by inducing diarrhoea and vomiting; they are also taken for stomach-ache (32). The fruit and the seed are reported to contain a contraceptive principle (35).

The seeds contain a palatable pleasant-tasting kernel like that of the sweet almond believing a latent toxicity. Poisoning is of the irritant type causing nausea, abdominal pain, vomiting, diarrhoea, depression and collapse (27). A purgative dose is 3–4 seeds (5, 10, 19, 25–27, 35). Any more than this is dangerous. For treatment of ascites they are crushed and boiled with cereal pap, or roasted in ashes and mixed with natron or extract of wood-ashes and taken with water or milk (10), and in Congo (Brazzaville) as many as 8 seeds per day may be given but buffered with peanuts or sugar-cane (5). Seeds enter into medication for serious illnesses such as syphilis and leprosy (30). For treatment of the former, seeds are crushed and mixed with cereal foods and left to ferment for two nights (10). They are used by Igbo in S Nigeria for arthritic troubles (36). The crushed seed or the seed-cake in palm-oil is used in Gabon as a raticide (25, 26). In Nigeria the seed is sometimes incorporated along with *Euphorbia* latex in a mixture (*gunguma*, Hausa; *lówu*, Nupe) to poison corn as a bait for guinea-fowl (10). Some tribes of the soudanian region add the nut to *Strophanthus* (Apocynaceae) seeds for arrow-poison (4, 10). In Sudan the seed of a *Jatropha sp.*, known locally as *habat el-mollok* and identified on the basis of this name as *J. curcas*, has been found to have high molluscicidal toxicity (29). The seed also has insecticidal potential (35).

The seed is about 35% husk, 65% kernel, and the oil content of the kernel is 50–58%. The oil is semi-drying. Tenda of Senegal use the seed-oil to put a shine on pottery (31). It has been used in the woollen industry in England and in the preparation of Turkey Red. The Portuguese established a plantation industry on the Cape Verde Islands to supply oil to SW Europe, but the main planting there is to assist in reafforestation and to maintain a plant cover. The oil is indifferent for lubrication of machinery owing to the speed with which it dries, but it is good for candle and soap manufacture. It can be burnt as an illuminant and at one time was used for street lighting around Rio de Janeiro (7). It burns without smoke, and even a seed spiked on a stick and dipped in palm-oil will burn and keep alight in a strong wind (14, 26). Commercial names for the oil are: seed oil; hell oil, pinhoen oil, pulza oil, pulguiera oil, purging-nut oil, curcas oil; and pharmaceutically: *oleum ricini majoris* and *oleum infernale.*

The oil, like the rest of the plant is purgative, and its clinical use is mainly external. It is the oily extract called *kufi* in Hausa, which is used as a rubefacient for rheumatic conditons, and for itch and other parasitic skin-diseases. In Yoruba the plant is called *làpálàpá*, meaning 'ringworm', because the oil causes an irritative rash on the skin of children (10). An oily preparation is applied to tumours in Mauritania (13), and in Ivory Coast to scarifications around filaria blisters (19).

The oil consists mainly of *stearic, palmitic, myristic, oleic, linoleic* and *curcanoleic acids*. A phytotoxin, *curcin*, is present in the seed and remains in the cake on expression rendering the cake unusable as cattle-food, but it is satisfactory as a fertiliser. This substance is related to *ricin* of *Ricinus* and to *crotin* of *Croton tiglium.* A resin is also present in the seed and this is carried over with the oil in extraction. This is a vesicant causing redness and pustular eruptions. (6, 7, 9, 10, 14, 17, 20, 27, 34.) There is record of the oil being used as an adulterant of cooking oils with drastic purgative effects (10, 27).

Ash from the burnt plant is used to produce a lye for soap-making. A vegetable salt is also extracted. (8, 10, 15.)

EUPHORBIACEAE

References:

1. Adams, RFG, 1943. 2. Adjanohoun & Aké Assi, 1972: 126. 3. Ainslie, 1937: spp. no. 195. 4. Aubréville, 1950: 174. 5. Bouquet, 1969: 116–7. 6. Bouquet & Debray, 1974: 84. 7. Burkill, IH, 1935: 1268–70. 8. Busson, 1965: 164. 9. Chevalier, 1935: 848. 10. Dalziel, 1937: 147–8. 11. Duckworth, 1947. 12. Greenway, 1941. 13. Hartwell, 1969,b: 167. 14. Irvine, 1961: 233–6. 15. Kerharo & Adam, 1963, a. 16. Kerharo & Adam, 1964, b: 124. 17. Kerharo & Adam, 1964, c: 311. 18. Kerharo & Adam, 1974: 419–21, with phytochemistry and pharmacology. 19. Kerharo & Bouquet, 1950: 77–78. 20. Oliver, 1960: 29, 68. 21. Quisumbing, 1951: 512–6, with many references. 22. Schnell, 1960: no. 110. 23. Taylor-Smith, 1966: 539. 24. Vermeer 27, UCI. 25. Walker, 1953, a: 36. 26. Walker & Sillans, 1961: 170. 27. Watt & Breyer-Brandwijk, 1962: 420–2. 28. Williams, RO, 1949: 316–7. 29. Amin & al., 1972. 30. Visser, 1975: 75–15: 47. 31. Ferry & al., 1974: no. 96. 32. Hallam, 1979: 57. 33. Portères, 1974: 127. 34. Wong, 1976: 130. 35. Oliver-Bever, 1982: 54. 36. Iwu & Anyanwu, 1982: 263–74. 37. Isawumi, 1978: 118. 38. Sharland, 1978. 39. Hunting Tech. Services, 1964. 40. Ampofo, 1983: 33, 43.

Jatropha gossypiifolia Linn.

FWTA, ed. 2, 1: 397. UPWTA, ed. 1, 148.

English: wild cassava; wild cassada; red fig-nut; red physic nut.

French: médicinier; médicinier batard; médicinier rouge; médicinier sauvage.

West African: SENEGAL DIOLA purgèr émé (JB) purgèr yémé (JB) FULA-PULAAR (Senegal) lumulum (JB) MANDING-BAMBARA baga, baghia, baha = poison (JMD) satana (JB) MANDINKA baga, bagha, baha = poison (JMD) MANINKA baga (JB) SERER ardiana (JB) fadilèp (JB) fop (JB) WOLOF lumolum (JB) GUINEA-BISSAU CRIOULO pau raio (EPdS) SIERRA LEONE BULOM (Sherbro) katawo-lɛ (FCD) FULA-PULAAR (Sierra Leone) kidi (FCD) GOLA kata-wuli (FCD) KISSI bɔtɛmbã (FCD) KONO kata-kɔnɛ (FCD) KRIO fig-not (FCD) fig-not-flawa (FCD) LIMBA (Tonko) kusigbɔro (FCD) MANDING-MANDINKA baju (FCD) MENDE kata = hedge (FCD) kata-wulo (def. -wuli, -wului) = hedge tree (auctt.) SUSU bakhanɛ (FCD) bakhɛnɛ (FCD) TEMNE an-kidi (FCD) TEMNE (Yoni) ɛ-nɔŋkɔ (FCD) YALUNKA kidi-safu-na (FCD) MALI MANDING-BAMBARA santanan (A.Chev.) UPPER VOLTA MANDING-BAMBARA santamân (K&B) DYULA baga = poison (K&B) IVORY COAST AKAN-ASANTE koagayi (K&B) MANDING-DYULA baga = poison (K&B) MANINKA baga = poison (K&B) GHANA ADANGME kitigblɛtʃo (FRI) AKAN-ASANTE kaagya (auctt.) FANTE aborɔkyiraba (FRI) akandedua (FRI; A&E) GA eŋmebii = small palm-nut (BD&H; FRI) kpitikpitʃo (FRI) GBE-VHE babatsi (FRI) gbomagboti (BD&H; FRI) ŋkrakpɔti = Accra fence tree (auctt.) NIGERIA EDO orúrúebo (JMD; Elugbe) IGBO olulu idu (BNO) IGBO (Umuahia) akimbogho (JMD) YORUBA bòtújè pupa (JMD) làpálàpá pupa from pupa: red; = cut and kill, cut and kill red (JMD; Verger) lóbòtújè (JMD) lóbòtújè pupa (Phillips; Dawodu) olóbòntújè (JMD) olóbòntújè-pupa (EWF)

A shrub to barely 2 m high, of tropical American origin, now dispersed to most tropical countries, and to West Africa perhaps dating from the time of the Slave Trade. It is established spontaneously along the coastal area and inland around villages.

The foliage is often tinted purplish or reddish. The flowers are dark red. The plant is grown as an ornamental and is frequently established as a village hedge for fire-protection. The plant also has other protective virtues against lightning, and in an Yoruba incantation to prevent someone from being beaten (7), and to scare off snakes (4). The viscid sap is said to be poisonous. The leaves are used in Ghana as a purgative, while the leaf-sap is applied to the tongues of babies for sores (? thrush) (4). In Mexico a leaf-decoction is taken to cleanse the blood and for venereal disease (6), and in Trinidad for diarrhoea, venereal disease and indigestion, and in poultice for sores and piles (8). The leaf contains *tannins* and *histamines*.

The yellowish-brown pith of old stems is sold in Ghana markets as a medicine to cure a headache. It is wrapped in a clean cloth and inserted into the nostrils of the patient causing sneezing (4, 6).

In India bark-decoction is taken as an emmenagogue (6).

The root contains *jatrophine*, a toxic alkaloid (8).

Oil from the seeds is a powerful purgative and emetic like that of *J. curcas* Linn. In the Abidjan area of Ivory Coast the seeds are taken as a purgative (1, 5). In Madura, Indonesia, 20 seeds taken after roasting is considered to be a single dose for an adult. The oil has been used to treat leprosy. It is also an illuminant (2, 6). *Curcin* and an emetic are present (8).

The whole plant has been popularly used in Costa Rica for cancers (3).

References:

1. Bouquet & Debray, 1974: 84. 2. Burkill, IH, 1935: 1270–1, with references. 3. Hartwell, 1969,b: 167. 4. Irvine, 1961: 236. 5. Kerharo & Bouquet, 1950: 78–79. 6. Quisumbing, 1951: 516–7, with references. 7. Verger, 1967: no. 133. 8. Wong, 1976: 130.

Jatropha multifida Linn.

FWTA, ed. 2, 1: 397. UPWTA, ed. 1, 148.

English: French physic nut; Spanish physic nut; coral plant.

French: médicinier (Aubréville); médicinier d'Espagne (Berhaut); arbre au corail (Berhaut).

West African: SIERRA LEONE BULOM (Sherbro) katawo-lɛ (FCD) FULA-PULAAR (Sierra Leone) kidi (FCD) GOLA kata-wuli (FCD) KISSI bɔtɛmbã (FCD) KONO kata-kɔnɛ (FCD) KRIO fig-not (FCD) LIMBA (Tonko) kusigbɔro (FCD) MANDING-MANDINKA baju (FCD) MENDE kata (FCD) kata-wulo (FCD) SUSU bakhanɛ (FCD) bakhɛnɛ (FCD) TEMNE an-kidi (FCD) TEMNE (Yoni) ɛ-nɔŋkɔ (FCD) YALUNKA kidi-safu-na (FCD) **NIGERIA** EDO ebósa (AHU) IGBO olulu idu (BNO) YORUBA bòtújè (JRA; IFE) bòtújè-pupa (JRA) ẹgẹ Phillips) làpálàpá (JRA; IFE) lóbòtújè (JRA; IFE)

A bush or tree to 6 m tall, native of tropical America, and now dispersed pan-tropically.

It is a village ornamental tree in the Region and is often planted in hedges, and as an object of superstition (5). It is said, like *J. gossypiifolia*, to keep off snakes (8).

The leaves and leaf-sap are purgative, but not so violently as of *J. curcas*. In the Antilles it is said that 10–12 leaves lightly cooked and eaten in a salad provide a non-griping purge, but the leaf contains a *saponin* and an active toxic principle, *jatrophin* (11). Leaves and fruit are boiled and taken internally or used externally in a bath as a febrifuge in Nigeria, and an infusion of young leaves is used for most urinary complaints (1).

Root-bark and the roots are ground up as a wound-dressing, and the root is taken internally for worms and, with salt added, for gonorrhoea in Nigeria (1).

The root produces long tubers which can be eaten like tapioca root (*manihot*) after roasting (3, 4).

The red inflorescences are attractive, and are an item of demand by florists in the Philippines (9).

The seeds are strongly purgative. They are sometimes used in Ivory Coast–Upper Volta (2, 8). They have a nutty flavour and there are records of fatalities from eating them. A single one is sufficient to produce violent illness (11). An antidote is said to be a glass of white wine or other stimulant with lime-juice (9, 10). Seed-oil amounts to about 30% and is know as Pinhoen Oil in Brazil. It has properties similar to the oil of *J. curcas*, and has been used as an illuminant.

There is latex throughout the plant and the rubber content of leaves has been reported as 0.52% (11). Latex has been used in the Caribbean on cancers (6).

References:

1. Ainslie, 1937: spp. no. 195. 2. Bouquet & Debray, 1974: 84. 3. Burkill, IH, 1935: 1271, with references. 4. Busson, 1965: 164. 5. Dalziel, 1937: 148. 6. Hartwell, 1969,b: 167. 7. Irvine, 1961: 236–7. 8. Kerharo & Bouquet, 1950: 79. 9. Quisumbing, 1951: 517–8, with references. 10. Walker & Sillans, 1966: 169. 11. Watt & Breyer-Brandwijk, 1962: 422.

EUPHORBIACEAE

Jatropha podagrica Hook.

FWTA, ed. 2, 1: 397.
English: gouty-stalked jatropha.

A fleshy sub-woody plant to about 70 cm high with a much-swollen and knobbly stem – hence the Latin specific name from *podagra*: gout; native of Central America, and dispersed to many tropical countries as an ornamental. It is commonly grown in W African gardens for its showy red flowers.

The plant is used in Ghana and Nigeria as an antipyretic, diuretic, choleretic and purgative. *Tetramethylpyrazine* has been isolated from the stem. This substance is known from fermented soya and cacao beans. It has antibacterial activity and appears to act as a nerve-blocking agent (1).

Reference:

1. Oliver-Bever, 1983: 40–41.

Jatropha sp. indet.

West African: NIGERIA HAUSA barbalu (RES)

This plant is probably *J. curcas* Linn.

Keayodendron bridelioides (Mildbr.) Léandri

FWTA, ed. 2, 1: 382, as *Casearia bridelioides* Mildbr. (*Drypetes sassandraensis* Aubrév. *nom. nud.*).
West African: IVORY COAST ? dakpandeu (Aub.) kohaingue (Aub.) NIGERIA YORUBA ebo (Sankey; KO&S)

A tree of the lowland rain-forest, reaching 35 m height, bole to 2.5 m girth, straight, cylindrical, to 22 m long to lowest branching. Recorded from Ivory Coast to S Nigeria, and in E Cameroun and Gabon.

The wood is brown to dark red, hard, heavy, easily worked and not too heavy for joinery (2–5).

Phytochemical tests show the presence of traces of *saponosides* (1).

References:

1. Bouquet & Debray, 1974: 159, as *Casearia bridelioides*. 2. Irvine, 1961: 82, as *Casearia bridelioides* Mildbr. 3. Keay & al., 1960: 287. 4. Léandri, 1958. 5. Sankey 17, K.

Klaineanthus gaboniae Pierre

FWTA, ed. 2, 1: 413.
West African: NIGERIA IJO-IZON (Kolokuma) ekpélá (KW&T) ekpélá (KW) IZON (Oporoma) okpólóta (KW)

An understorey forest tree to 30 m tall by 1.30 m girth, bole crooked, fluted. Recorded from S Nigeria and W Cameroons, and extending into E Cameroun, Gabon and São Tomé.

The wood is pale yellow, soft and light (1), and is used as a building material in SE Nigeria (2).

References:

1. Dechamps 59, K. 2. Williamson, K & Timitimi, 1983.

94

Macaranga Thouars

West African: IVORY COAST ABE tofé (Aub.) ANYI egba (Aub.) 'KRU' pépié (K&B) KRU-BETE feféï (K&B) GUERE bahué (K&B) GUERE (Chiehn) lulu (K&B) KWENI fonofa (Aub.) KYAMA abué (Aub.)

In Ivory Coast a clear distinction between species is not always drawn, and the foregoing names apply to the genus as a whole. Regeneration on abandoned land is rapid. Some species are thorny and they form impenetrable barriers in a matter of months (1).

Reference:

1. Aubréville, 1959, 2: 78.

Macaranga barteri Müll.-Arg.

FWTA, ed. 2, 1: 407–8. UPWTA, ed. 1, 149, incl. *M. rowlandii* Prain.
West African: SIERRA LEONE BULOM (Sherbro) hɔp-lɛ (S&F) hɔ̃p-lɛ (FCD) GOLA sei (FCD) KISSI falabo (FCD; S&F) KONO papu (FCD; S&F) KORANKO farabaŋ (S&F) LOKO ɲdɛvɛ (FCD; S&F) MENDE ɲdɛvɛ (-i, -guwi) (auctt.) ɲdɛvɛ-gbɔ (Macfoy & Sama) dewei-golei (SKS) ndewei (auctt.) ndɛwɛi (-guwi) (JMD; S&F) SUSU balambalamyi (NWT) fotete (NWT) futete (NWT) jagale (NWT) TEMNE *ka*-fɛp (auctt.) VAI hala (FCD) YALUNKA filege-na (FCD) **LIBERIA** MANO gu (Har.) **IVORY COAST** ABE tofé (Martineau) tofé dola (Aub.; K&B) ADYUKRU librébré (Aub.; K&B) 'KRU' péoma (K&B) KRU-GUERE banuan mokon = *good leaf for nanny-goats* (K&B) **GHANA** ADANGME-KROBO akafekafei (Farmar; FRI) AKAN opam-kokoo (DF; E&A) AKAN-AKYEM ampe noa (FRI; E&A) ASANTE gyapam (FRI) ɔpam (auctt.) ɔpam fufuo (FRI) ɔpam-kɔkɔɔ (FRI; JMD) ɔpesare = *inhabiting the grassland* (auctt.) TWI ɔpam (auctt.) ɔpam-kɔkɔɔ (auctt.) GBE-VHE owurudu (auctt.) **NIGERIA** EDO ohaha (JRA; KO&S) ohaha ẹze (JMD) ohaha nehun (JMD) EFIK àkpáp (auctt.) ESAN ohoha (KO&S) IBIBIO àkpáp (KO&S) IGBO ọwariwa (JMD; KO&S) ọwọlewa (JMD) IJO-IZON aran (JRA; KO&S) YORUBA ààràsá (auctt.)

A shrub or tree to 20 m high by 1.30 m girth, of forest, secondary jungle and grassy savanna, common throughout the Region from Guinea to S Nigeria, and Equatorial Guinea. The Ghanaian Akan name ɔpɛsare, meaning 'it loves the grassy plains', takes note of the plant's habit of occurring where the forest is dying out (5). It is a coloniser of farm bushland, and its span of life is, in Sierra Leone, 30–40 years (9). It is a common but relatively unimportant weed species of managed forestry.

The wood is light, white and soft (6). It is a popular firewood. The poles are used in Sierra Leone in the construction of hut-walls (9).

Bark and leaf, either powdered or in decoction, are used in Nigeria as a vermifuge (2). The plant is used as a vermifuge and febrifuge in Congo (Brazzaville), and in mixture with other *Macaranga spp.* to relieve cough and bronchitis (3). In Sierra Leone there is a use of the leaves for gonorrhoea (Edwardson fide 5), and in Ivory Coast the plant (part not stated) is taken in draught as an aperient and anti-anaemic tonic (4). Nigerian material has been examined for *alkaloids*: none was found in the bark and a trace in the leaves (1).

In S Nigerian markets the leaves have been used as wrappers for salt, etc. (8).

The Guere of Ivory Coast put the leaves into magical powders (7).

References:

1. Adegoke & al., 1968. 2. Ainslie, 1937: sp. no. 218. 3. Bouquet, 1969: 117. 4. Bouquet & Debray, 1974: 85. 5. Irvine, 1961: 237. 6. Keay & al., 1960: 265. 7. Kerharo & Bouquet, 1950: 79. 8. Mullen 28, K. 9. Savill & Fox, 1967: 117–8.

EUPHORBIACEAE

Macaranga beillei Prain

FWTA, ed. 2, 1: 407.
West African: IVORY COAST KRU-GREBO (Oubi) léona (A&AA)

A shrub or tree to 5 m high, of the forest zone of Ivory Coast.
The leaves, pounded with the aerial parts of *Scleria barteri* (Cyperaceae),
water and vegetable salt from the ashes of the bark of *Calpocalyx aubrevillei*
(Leguminosae: Mimosoideae), and wrapped in leaves of *Thaumatococcus daniel-lii* (Marantaceae), are cooked and the water is drunk as a cough medicine in
Ivory Coast (1).

Reference:

1. Adjanohoun & Aké Assi, 1972: 127.

Macaranga heterophylla (Müll.-Arg.) Müll.-Arg.

FWTA, ed. 2, 1: 407. UPWTA, ed. 1, 149.
West African: SENEGAL BANYUN kisâdor (K&A) kisandor (K&A) kisândor (JB) DIOLA-
FLUP a diilékoy (JB) diakõnt (JB) FULA-PULAAR (Senegal) bulañiña (JB) MANDYAK be mièl (JB)
GUINEA-BISSAU BIAFADA badje (EPdS; JDES) FULA-PULAAR (Guinea-Bissau) bulanhinha
(D'O; DF) GUINEA KISSI palafo (FB) KONO uagni-tugulé-lâ *from* uagni: *squirrel*; tugulé: *kola
tree*; lâ: *leaf*; i.e. = *leaf of the squirrel's kola tree* (RS) KPELLE olo (FB) LOMA dévèye (FB)
uasangbâlo (RS) MANDING-MANINKA bolounan (FB) SIERRA LEONE BULOM (Sherbro) hɔp-lɛ
(auctt.) GOLA sei (FCD) KISSI falabo (FCD; S&F) KONO papu (FCD; S&F) KORANKO farabaŋ
(S&F) kadila (NWT) LOKO ndɛvɛ (FCD; S&F) ngaisipa (NWT) MENDE ndɛvɛ (-i, -guwi) (FCD;
S&F) ndɛwɛ (-i, -guwi) (auctt.) fɔfɔ (-i) (L-P; SKS) TEMNE ka-fɛp (FCD; S&F) VAI hala (FCD)
YALUNKA filege-na (FCD) IVORY COAST AKYE biango (auctt.) DAN van (K&B) vangoné
(K&B) ˙KRU˙ tué (K&B) KRU-GUERE batué (K&B) grapuyé (RS) GUERE (Wobe) kbaï (RS)
MANO gbénago (RS) GHANA AKAN-ASANTE ɔpam (auctt.) TWI ɔpam-nua (auctt.) GBE-VHE kedi
(auctt.)

A shrub or tree to 10 m tall, of secondary forest in wet localities from
Casamance (Senegal) to Togo.
The wood is light and soft. The tree exudes a yellow translucent gum when
cut (3). Ash from the wood of the trunk and branches is used in Guinea as an
alimentary salt (2, Portères fide 4).
Various parts of the plant are purgative. A root-decoction is taken by draught
in Casamance for amenorrhoea and the effect is said to be rapid. It is also
prescribed as an emmenagogue and abortifacient (5, 6).
In Ivory Coast the Dan use the plant to treat snake-bite, and other races take
a bark-decoction by draught and in baths for cough (1, 7). In Sierra Leone a
decoction of young leaves is taken for gonorrhoea (Thomas fide 4).

References:

1. Bouquet & Debray, 1974: 85. 2. Busson, 1965: 164. 3. Dalziel, 1937: 149. 4. Irvine, 1961: 237–8.
5. Kerharo & Adam, 1963, b. 6. Kerharo & Adam, 1974: 421–2. 7. Kerharo & Bouquet, 1950: 79.

Macaranga heudelotii Baill.

FWTA, ed. 2, 1: 408. UPWTA, ed. 1, 149.
West African: SIERRA LEONE BULOM (Sherbro) hɔp-lɛ (FCD; S&F) GOLA sei (FCD)
KISSI falabo (auctt.) KONO papu (FCD; S&F) KORANKO farabaŋ (S&F) LIMBA bɔɔlugɔ (NWT)
bukotia (NWT) bupɛl (NWT) butompo (NWT) butumpu (NWT) kutɛgea (NWT) LOKO benyu
(NWT) bewunguru (NWT) ndɛvɛ (auctt.) feke (NWT) koaŋgi (NWT) teke (NWT) waɛlɛ
(NWT) yingewuru (NWT) yonjoɛŋ (NWT) LOKO (Magbile) tokbe (NWT) MENDE ndɛvɛ (-i,
-guwi) (auctt.) ndɛwɛ (-i, -guwi) (auctt.) SUSU hankai (NWT) kalambalaini (NWT) nɛrɛsaki
(NWT) salai (NWT) salansalainyi (NWT) selenyi (NWT) seri (NWT) sunɔne (NWT) taŋani
(NWT) wurɛkainyɛ (NWT) TEMNE an-; ka-fɛp (auctt.) VAI hala (FCD) YALUNKA filega-na
(FCD) MALI MANDING-MANINKA sunkuda na bala (A.Chev.) IVORY COAST ABURE etchoa

96

(A.Chev.) AKYE abo (A.Chev.) ANYI ekua (A.Chev.) AVIKAM eko (A.Chev.) KYAMA abué (A.Chev.) NZEMA ɛson (A.Chev.) **GHANA** NZEMA apazina (BD&H; FRI) mpuazinra (BD&H; FRI)

A scrambling shrub or tree to 8 m high in marshy sites and on river-banks of the closed-forest, recorded across the Region from Senegal to S Nigeria. The wood is soft. In Ghana the plant enters into a Nzema medicine for diarrhoea (1).

Reference:

1. Irvine, 1961: 238.

Macaranga hurifolia Beille

FWTA, ed. 2, 1: 407. UPWTA, ed. 1, 149, as *M. huraefolia* Beille.
West African: **GUINEA** KONO kû (RS) LOMA dâkollé (RS) MANO gû (RS) **SIERRA LEONE** BULOM (Sherbro) hɔp-lɛ (auctt.) GOLA sei (FCD) KISSI falabo (FCD; S&F) KONO papu (FCD; S&F) KORANKO farabaŋ (S&F) karalaŋ (NWT) LOKO *nd*ɛvɛ (FCD; S&F) MENDE *nd*ɛvɛ (-i, -guli, -guwi) (auctt.) ndɛwɛ (-i, -guwi) (auctt.) TEMNE *ka*-fɛp (FCD; S&F) VAI hala (FCD) YALUNKA filege-na (FCD) **IVORY COAST** ABE tofé (DF; K&B) ANYI egba (K&B) GAGU urapapa (B&D) 'KRU' pépié (K&B) KRU-BETE féféi (K&B) GREBO (Oubi) péo (RS) GUERE bahué (K&B) gaife (RS) glin (RS) GUERE (Chiehn) fafé (K&B) lulu (K&B) KWENI fonola (Aub. fide K&B) KYAMA abué (K&B) **GHANA** AKAN-ASANTE ɔpam (BD&H; FRI) ɔpam-fufuo (Martinson) TWI ɔpam (FRI) NZEMA kpazida (FRI) **NIGERIA** EDO ohaha (Ross; KO&S) IGBO ọwariwa (JMD; KO&S) ọwọlowa (JMD) ùbé (NWT) IGBO (Idumuje) owolewa (NWT)

A scrambling shrub or tree to 12 m high, of secondary jungle, from Sierra Leone to W Cameroons.

Sap-wood is white, heart-wood pale lemon, tinged purple. The wood is of medium texture and easily worked, usable for many of the same purposes as soft pine. It is commonly used for firewood, and occasionally for hut-poles though it is probably not resistant to insects and decay (2–4, 6).

The bark is sometimes used as a purgative (3). An aqueous macerate of the leafy twigs, often with *Baphia nitida* (Leguminosae: Papilionoideae), is used by the Guere of Ivory Coast as a purgative for all gastro-intestinal affections. The Shien consider it is a good medicine for cough especially tubercular haemophthysis for which the patient should chew bits of bark or take powdered twigs in palm-wine accompanied by friction of the back and chest with leaf-pulp. Kru claim the root in decoction to be extremely diuretic and give it in enema to relieve oedemas in pregnant women (1, 5).

References:

1. Bouquet & Debray, 1974: 85. 2. Cooper & Record, 1931: 52, with timber characters. 3. Dalziel, 1937: 407. 4. Irvine, 1961: 238. 5. Kerharo & Bouquet, 1950: 79–80. 6. Savill & Fox, 1967, 117.

Macaranga monandra Müll.-Arg.

FTWA, ed. 2, 1: 407, incl. 408, *M. sp. A* sensu Keay; which is probably this species. UPWTA, ed. 1, 149.
West African: **NIGERIA** BOKYI kensange abia (Catterall) IGBO ọwariwa (JMD) ọwọlowa (JMD)

A shrub or tree to 16 m high, of the forest of S Nigeria and W Cameroons and extending widely across Africa to Tanganyika and Angola. *M. sp. A* of *FWTA*, ed. 2, is recorded from Sierra Leone and Liberia.
The pith of the branches exudes when cut a sticky jelly (3).
In Congo (Brazzaville) a decoction of bark with that of *Pentaclethra ectveldeana* (Leguminosae: Mimosoideae – a species not recorded in W Africa) is

given by draught to women to treat for sterility (1). *Tannins, steroids* and *terpenes* are recorded present in leaves, bark and roots (2).

References:

1. Bouquet, 1969: 117. 2. Bouquet, 1972: 25. 3. Breteler 232, 1311, K.

Macaranga occidentalis (Müll.-Arg.) Müll.-Arg.

FWTA, ed. 2, 1: 407. UPWTA, ed. 1, 149.
West African: WEST CAMEROONS KPE awo-wo (Dunlap)

A tree to 20 m, occurring on the margin of montane forest in S Nigeria, W Cameroons and Fernando Po.
The wood is soft (1).

Reference:

1. Keay & al., 1960: 263–4.

Macaranga schweinfurthii Pax

FWTA, ed. 2, 1: 407. UPWTA, ed. 1, 149, as *M. rosea* Pax.
West African: NIGERIA IJO-IZON (Kolokuma) ọfọ́fọ̀ (KW) WEST CAMEROONS DUALA njon-bwele (AHU) KUNDU boka (AHU)

A shrub or straggling tree of the forest zone of S Nigeria and W Cameroons, and extending widely across Africa to NE and E Africa.

The young stems are reported to yield a copious white sticky mucilage in Kenya (4) and in Malawi (5). The exudate is further described as a pink jelly from the pith and a colourless slime from the bark in E Cameroun plants (2) and a dark red resin in Nigeria (3).

Flavones and *tannins* are reported present in Congo (Brazzaville) material, and *saponin, tannins, steroids* and *terpenes* in leaves, bark and roots of *M. rosea* (1).

References:

1. Bouquet, 1972: 25. 2. Breteler 872, K. 3. Keay & al., 1960: 264. 4. Proctor 712, K. 5. Richards 4524, K.

Macaranga spinosa Müll.-Arg.

FWTA, ed. 2, 1: 408.
West African: LIBERIA KRU-GUERE (Sapo) pobwe (Eggeling) IVORY COAST ABE tofé (Aub.) ABURE aulinvéni (B&D) KYAMA bazinga fufué (B&D)

A shrub or tree to 20 m high of the closed-forest and secondary jungle from Liberia to W Cameroons and Fernando Po, and extending to Angola.

In Ivory Coast the plant is used in treatment of dysentery and coughs (3). In Congo (Brazzaville) it is used in mixture with other species in draughts, frictions and vapour-baths for broncho-pneumonial affections, asthma, cough, headache, feverish aches and pains, rheumatism, and for liver and stomach complaints; taken with maleguetta pepper it is deemed to be aphrodisiac; a bark-decoction is used as a mouthwash and gargle for toothache, thrush and gum-troubles (1).

Saponins, tannins, steroids and *terpenes* have been detected in the bark and roots (2).

References:

1. Bouquet, 1969: 118. 2. Bouquet, 1972: 25. 3. Bouquet & Debray, 1974: 85.

EUPHORBIACEAE

Macaranga spp. indet. Two species recorded in Liberia appear to be as yet unnamed.

1. Macaranga sp. indet. quoad Cooper 102 and 393.

UPWTA, ed. 1, 149.
West African: **LIBERIA** KRU-BASA garfoe (Cooper 102, 393; C&R) zar-zreh-wehn-ye (Cooper 103, 395; C&R)

A tree to 16 m high by 30 cm diameter, common in secondary bush and abandoned farmland. The wood is similar to that of *M. hurifolia* Beille. The dried young leafy shoots are taken mixed with rice for bronchitis (1).

2. Macaranga sp. indet. quoad Cooper 103 and 395.

A rapidly growing tree on abandoned farmland, with wood similar to that of *M. hurifolia* Beille and of no recorded use. A decoction of macerated bark is taken as a poison-antidote (1).

Reference:

1. Cooper & Record, 1931: 53.

Maesobotrya barteri (Baill.) Hutch. **var. barteri**

FWTA, ed. 2, 1: 374. UPWTA, ed. 1, 149 as *M. barteri* Hutch. in part, and *M. edulis* Hutch. & Dalz.

var. sparsiflora (Sc. Elliot) Keay

FWTA, ed. 2, 1: 374. UPWTA, ed. 1, 149 as *M. barteri* Hutch. in part, and 150 as *M. sparsiflora* Hutch.
English: bush cherry (Liberia, Voorhoeve).
West African: **SIERRA LEONE** KISSI sowundio (JMD; Fisher) KORANKO kɛleniŋko (NWT) MENDE ɲjayai (DS) tɔyɛ(-i) (auctt.) SUSU forɔkwe (NWT) TEMNE an-titi (auctt.) **LIBERIA** MANO käi (JMD) **IVORY COAST** ABE wuniogpa (auctt.) wuniongpagba (A.Chev.) ADYUKRU kokobri (Aub.; K&B) AKAN-ASANTE patroa (B&D) AKYE abidjakué (A.Chev.; K&B) habizakué (A.Chev.; K&B) ANYI kautié-kualé (A.Chev.; K&B) kuatié-kualé (A.Chev.; K&B) DAN kanlé (K&B) ·KRU· bugbonotu (K&B) KRU-GUERE zahalo (A&AA) KYAMA kopié (Aub.; K&B) padréa (B&D) NZEMA kadé (Aub.; K&B) **GHANA** AKAN-FANTE apotorowa (auctt.) apɔtrewa (auctt.) kwadrega (FRI) podirire (FRI) apɔtrewa (E&A) asansambro (auctt.) asansammuro (FRI) WASA apɔtrewa (auctt.) kwadrega (FRI) podirire (FRI) apotorowa (auctt.) ANYI-AOWIN apɔtrewa (auctt.) NZEMA kwadrega (CV) podirire (auctt.) **NIGERIA** BOKYI ntum kache (KO&S) EDO orúrú (auctt.) IGBO miri ogu *from* miri; *water* (JMD; KO&S) IGBO (Umuahia) mmiri ogwu (AJC) YORUBA ɔdun (Kennedy) olóhun *the fruit* (JMD) olówùn *the fruit* (auctt.) orówo (auctt.) YORUBA (Ikale) obomodu (Kennedy)

A tree to 10 m high, of the understorey of the high rain-forest, var. *barteri* occurring in Sierra Leone, S Nigeria and W Cameroons, E Cameroun and the Congo basin, and var. *sparsiflora* dispersed only within the Region from Guinea to Ghana.

Var. *barteri* is low-branching and the timber is of little value. It is attacked by termites (4), but placed in a house it may be protected and is used in Benin as pegs to anchor roofs (9). The wood is light and burnt for fuel (5).

The bark of both varieties is sweetish and can be chewed as a stomachic (4). The bark of var. *sparsifolia* is pounded with white clay and is taken in Liberia for diarrhoea (4) and in Ivory Coast for stomach-ache (1). A bark-decoction of var. *barteri* is taken in Congo (Brazzaville) for dysentery, urethral discharge and as an aphrodisiac (2), while a bark-infusion is used in Liberia for gonorrhoea (4). A bark-decoction is an abortifacient in Ivory Coast (6), where the plant is also considered good for jaundice and respiratory troubles, and the sap to be haemostatic, healing and ecbolic (3). In Congo (Brazzaville) a bark-decoction serves as a wash for measles cases and the bark enters a prescription for

99

soothing coughs (2). The mushed leaves are applied as a wound-dressing in Ghana (Vigne fide 5), and a leaf-decoction is given in Ivory Coast to dispel giddiness (1).

The fruits of var. *barteri* are succulent black-purple berries about 1 cm long, edible and staining the tongue. They are sold in Yoruba markets in S Nigeria under the name *olówun*. The fruits of var. *sparsiflora* are bright red, succulent, and edible with a slightly acid and refreshing taste (4, 10). A jam can be made from them, and a filling for tarts (4, 5). The fruits are however not given to twins in Sierra Leone for whom they are considered unlucky (8). In the same country, perhaps also superstitious, but the significance is not explained, the plant is used to make a cage in dwellings for twins (7).

References:

1. Adjanohoun & Aké Assi, 1972: 128. 2. Bouquet, 1969: 118. 3. Bouquet & Debray, 1974: 85. 4. Dalziel, 1937: 149. 5. Irvine 1961: 239. 6. Kerharo & Bouquet, 1950: 80. 7. Thomas NWT.3099, K (Arch.). 8. Savill & Fox 1967, 110. 9. Unwin 1920: 414, as *Oruru*. 10. Voorhoeve, 1965: 94.

Maesobotrya dusenii (Pax) Hutch.

FWTA, ed. 2, 1: 374.

A tree to 15 m high with dense crown, of the rain-forest area of S Nigeria, W Cameroons and Fernando Po, and also E Cameroun and Equatorial Guinea.

The fruit contains an edible pulp of a sourish taste and is used in Gabon to make a jam (1).

Reference:

1. Walker & Sillans, 1961: 170.

Maesobotrya floribunda Benth. **var. vermueleni** (De Wild.) J Léonard

FWTA, ed. 2, 1: 374.

A small forest tree to 4 m high recorded only in S Nigeria and W Cameroons, and in E Cameroun to the Congo basin.

No use is recorded within the Region. In Congo (Brazzaville) a decoction of the bark is taken by draught and in baths for leprosy (1).

Reference:

1. Bouquet, 1969: 119.

Maesobotrya staudtii (Pax) Hutch.

FWTA, ed. 2, 1: 374.
West African: **NIGERIA** BOKYI ntum kache (Catterall; JMD)

An understorey tree to 8 m high of the rain-forest of S Nigeria and W Cameroons, and in E Cameroun to Zaïre.

The fruit is edible with an acid taste and is said in Gabon to make delicious jam (1).

Reference:

1. Walker & Sillans, 1961: 170.

Mallotus oppositifolius (Geisel.) Müll.-Arg.

FWTA, ed. 2, 1: 402. UPWTA, ed. 1, 150.

West African: SENEGAL KONYAGI ndămbuv (JB) MANDING-MANDINKA kohiri ndihō (JB) sama kati (JB) MANINKA bano (JB) THE GAMBIA MANDING-MANDINKA ko saffono = *washing soap* (Hayes; DAP) SIERRA LEONE MENDE tambaloi (NWT) TEMNE *ma*-tu-*ma*-ro-bat (NWT) YALUNKA hude-na (?) (FCD) IVORY COAST AKAN-ASANTE nia nia froé K&B BRONG tomida (K&B) 'ANDO' tominda (Visser) BAULE tomenda (B&D) tumenda (A.Chev.; K&B) tumina (B&D) tunda (B&D) tunsa (K&B) GAGU hon hon (B&D) KRU-BETE klawizi (K&B) kora izi (K&B) GUERE niaékola (K&B) niukra (K&B) GUERE (Chiehn) auraté (K&B; B&D) béléguie (B&D) blégué (B&D) bligié (K&B) brigué (B&D) KULANGO sigué (K&B) KWENI gatagbwé (B&D) 'NÉKÉDIÉ' uraté (B&D) GHANA ADANGME satwe (FRI) AKAN-ASANTE anyanforowa (FRI; OA) FANTE saka (auctt.) seratadua (auctt.) KWAWU anyanforowa (JMD; E&A) nyanyamforowa (FRI; E&A) nyanyanfrowa (JMD; E&A) TWI anyanforowa (auctt.) anyanyanforowa (FRI; E&A) sortchuetchu (ASThomas) GA kájà (KD) tátadua (KD) tɛtɛdwaatʃo (FRI) teteedua (OA) tɛtɛedua (FRI) GBE-VHE sroti (OA) ɲyati (FRI; JMD) VHE (Awlan) sroti (Eady; FRI) VHE (Kpando) akati (FRI) NZEMA fuampuaheleba (OA) puampuaheghra (FRI) NIGERIA EDO uhosa (JRA) HAUSA káfàr mútúwaà = *the foot of death* (auctt.) IGBO kp'akakwo (NWT) kpǫkǫkwa (Chester; JMD) mgboko (NWT) okpǫ-kirinyan (auctt.) okwe okpǫ-kirinyan *from* okwe: *a leaf; alluding to medicinal use* (JMD) ukpo (BNO) IGBO (Agukwu) oli ǫro (NWT) IGBO (Obu) akụkwǫ natoriegoto (NWT) IGBO (Owerri) nne okpokirinya = *female string leaf* (AJC) IJO-IZON (Kolokuma) fúrú ípáín (KW; KW&T) YORUBA ẹjá (auctt.) YORUBA (Ijebu) ipa (Okusi fide AJC)

A shrub or small tree to 13 m high, common in drier types of forest and in secondary regrowth throughout the Region and widespread across tropical Africa to Madagascar.

The plant is a common weed of rice-fields in Bendel State of Nigeria (21).

The leaves are taken by Hausa in cold infusion for tape-worm (5, 6). A decoction is a vermifuge in Ivory Coast–Upper Volta (10). Crushed or chewed leaves are used in Ivory Coast (18), Nigeria (1, 6, 12) and Ghana (9, 22) as a haemostatic on wounds and sores; also for burns (15). This quickly arrests bleeding. The Ijo of S Nigeria use the young foliage, and if the wound is large add 'pepper' (16, 17). In Ivory Coast–Upper Volta the leaves are compounded into an ointment with *karite* butter for application to sores and ulcers (3, 10). The plant is held to have anodynal properties. To treat headache the patient in Nigeria holds his head over a pot in which the leaves are boiling and inhales the steam (6). In Ivory Coast–Upper Volta leaf-sap is instilled up the nostrils for headache (3, 10), and the 'Ando' of Ivory Coast kneed the leaf into a quid for toothache (18). In Nigeria and Ghana the roots with guinea-grains are prepared as an enema for lumbago (1, 6).

All parts of the plant are used for dysentery (1, 6), and in Ghana and Nigeria the leaves are added to suitable food (6, 9). Crushed leaves are applied in Ghana to inflamed eyes during an attack of small-pox (13), and in S Nigeria the plant (part not specified) is used to treat sore eyes (7). Leaf-sap is an eye-instillation in Ubangi for eye-troubles (19) and in Congo (Brazzaville) it is administered in attacks of epilepsy (2). Crushed leaves are placed between the toes by the Ijo for *tinea* infection (17). Leaves along with those of other plants are boiled and the draught is given to women in Akropong, Ghana, with urinary troubles during pregnancy (9). A decoction of leaves and root is taken by mouth in Ivory Coast–Upper Volta for anaemia and general fatigue (10). With other plants it is considered aphrodisiac (3, 10). The root is used as an aphrodisiac in Tanganyika (11). Leaf-sap and root-decoction are drunk in Tanganyika for pneumonia (8).

Unspecified parts of the plant are used in Nigeria to treat small boys suffering from *amudzu* convulsions (14), and in Ivory Coast for venereal diseases, dysentery, leprosy, chicken-pox and female sterility (3).

The twigs are commonly used as chew-sticks (4, 6, 9, 18, 20). The stems are used in Nigeria as yam-stakes and for bows in Owerri Province (9). The thinner stems provide a binding material (6).

The seed is poisonous, an antidote being *Piliostigma thonningii* (Leguminosae: Caesalpinioideae) (19).

Rottlerin has been recorded present in the leaves and bark (12). This is the active principle in the drug known as *kalama*, official in pharmacopoeias as a taenicide, and obtained from the closely related *M. philippinensis* Müll.-Arg. – q.v.

References:

1. Ainslie, 1937: spp. no. 220. 2. Bouquet, 1969: 119. 3. Bouquet & Debray, 1974: 85. 4. Burtt Davy & Hoyle, 1937: 48. 5. Dalziel 725, K. 6. Dalziel, 1937: 150. 7. Dawodu 75, K. 8. Haerdi, 1964: 94. 9. Irvine, 1961: 239–41. 10. Kerharo & Bouquet, 1950: 80. 11. Koritschoner 1307, K. 12. Oliver, 1960: 8, 30, 42, 70. 13. Thomas, AS D.66, K. 14. Thomas, NWT.1649 (Nig. Ser.), K. 15. Thomas, NWT.1847 (Nig. Ser.), K. 16. Williamson KW.38, UCI. 17. Williamson 46, UCI. 18. Visser, 1975: 75–15: 49. 19. Vergiat, 1970: 76. 20. Portères, 1974: 129. 21. Gill & Ene, 1978. 22. Ampofo, 1983: 21.

Mallotus philippinensis Müll.-Arg.

English: kamala (from Hindi, the produce).

A small tree occurring widespread from NW Himalayas to eastern Australia, which has been introduced to West Africa for cultivation in some francophone territories.

Kamala is the powder consisting of glandular and stellate hairs rubbed off the ripe fruit. The stellate hairs contain a strong dye which can be mordanted to give fast orange and flame colours. *Rottlerin* and *iso-rottlerin* are present in the glands and these are taenifugal. Kamala has been official in numerous pharmacopoeias though nowadays it is not so much favoured. The drug remains in veterinary use. It has a long tradition in eastern medicine for the treatment of skin-troubles.

The seed kernel contains 35–36% oil which is quick drying, and might be of use in paint-manufacture.

References:

1. Burkill, IH, 1935: 1396–7, with references. 2. Oliver, 1960: 8. 3. Quisumbing, 1951: 521–3, with references. 4. Sastri, 1962: 229–33 with references.

Mallotus subulatus Müll.-Arg.

FWTA, ed. 2, 1: 402. UPWTA, ed. 1, 150.
West African: SIERRA LEONE LOKO haka (NWT) NIGERIA EDO ebuhosa (AHU; JMD) uhosa (JRA) EFIK nne okpokri inyang (AJC) IGBO (Asaba) kpǫkǫwa (NWT) IGBO (Uburubu) ifie ogume (NWT) odoji (NWT)

A shrub or tree to 5 m high of the forested zone from Sierra Leone to W Cameroons and Fernando Po, and on to Zaïre.

The roots, leaves and fruit are all ground up and prescribed for medication in Nigeria for dysentery (1). Seeds are ground and taken by draught for stomach-pains (6). The roots with grains of paradise (*Aframomum granum-paradisi*, Zingiberaceae) are used as an enema for lumbago (1). In Congo (Brazzaville) pains in the sides and lumbar region are treated by application of sap from the bark into scarifications in the area of the pain (2). Wet leaves are pulped and applied to wounds as a styptic in Nigeria (1), and the plant (part not stated) is given to young boys for *amudzu* convulsions (4).

The seeds are ground to a powder by the Igbo of S Nigeria which is used to mask the faces of young men and girls (5).

A trace of *alkaloid* has been recorded in the roots, and abundant *saponins* and a little *tannin* in the bark and roots (3).

References:

1. Ainslie, 1937: sp. no. 220. 2. Bouquet, 1969: 119. 3. Bouquet, 1972: 24. 4. Thomas, NWT.1649 (Nig. Ser.), K. 5. Thomas, NWT.2074 (Nig. Ser.), K. 6. Thomas, NWT.2075 (Nig. Ser.), K.

Manihot dichotoma Ule

FWTA, ed. 2, 1: 413.

English: jequie rubber tree; tequie manincoba.

A small tree to about 5 m high, native of Brazil. It was reported in 1906 to be capable of better yield of rubber than *M. glaziovii*, and seed was obtained and distributed in the Far East, and at some time which is not recorded it reached West Africa as a rubber plant. It has no place as a rubber-producer, nor does it appear to have any other recorded use.

Reference:

1. Burkill, IH, 1935: 1408–9, with references.

Manihot esculenta Crantz

FWTA, ed. 2, 1: 413. UPWTA, ed. 1, 150–4, as *M. utilissima* Pohl.

English: tapioca; cassava; cassada (U.S.A.).

French: manioc; mandioc; cassave.

Portuguese: mandioca.

West African: SENEGAL BALANTA dion-a (FB) mciila (Wilson fide Pasch tiila (JB) BANYUN dõko (K&A) dsónko (Koelle fide Pasch) BASARI ɔ-bâdara (K&A) ɔ-bândara (JB) BEDIK bantára (Tersis fide Pasch; FG&G) DIOLA diôko (JB) dôko (JB) é sana (auctt.) FULA-PULAAR (Senegal) bãntara (JB; FB) kapé (JB) GANJA njõngã (N'Diaye-Correard fide Pasch) KONYAGI bãntara (JB) MANDING-BAMBARA banani uku (JMD; JB) bani uku (JB) maninku (JB; K&A) MANDINKA bamã-nku (Delafosse fide Pasch) bãntam-ñanãmbo (JB) pelôfé (Correira fide Pasch) MANINKA bananku (JB) MANKANYA kãlofi (Koelle fide Pasch) tuk (Trifkovic fide Pasch) SERER kab (JB; K&A) mbuloh (JB) mbuloh gôr *bitter varieties* (JB) nâbi (K&A) ñambi (JB) puloh (JB) puloq (K&A) putok (*pl.* mutok) (Koelle fide Pasch) SONINKE-SARAKOLE ku (Delafosse fide Pasch) ·SUSU· yoka (JB) WOLOF gniambi *misapplied; this is properly, the wild yam, Dioscorea* (A.Chev.; FB) nâbi (K&A) ñàmbi (Diop. Fal fide Pasch) pullóóx (Diop. Fal fide Pasch) puloh (AST; JB) puloq (K&A) THE GAMBIA FULA-PULAAR (The Gambia) nyambe (Hopkinson fide JMD) MANDING-MANDINKA banta nyambo *a c.var name after* ban-tang: *Ceiba* (Hopkinson fide JMD) banta nyambo koyo *a white var.* (JMD) banta-nyambo fingo *a black var.* (JMD) bentang nyambo (Hopkinson fide JMD) mandioka *a Haitian name, and applied to a soft-fleshed variety good for fufu* (auctt.) nyambo (Hopkinson fide JMD) WOLOF nyambe (Hopkinson fide JMD) GUINEA-BISSAU BALANTA quiíla (JDES) BIDYOGO mondeogo (Koelle fide Pasch) mondeoko (Koelle fide Pasch) FULA-PULAAR (Guinea-Bissau) bántara (JDES) cápè (JDES) MANDING-MANDINKA banta nhanhambô (JDES) MANKANYA etuco (JDES) PEPEL kemand (Koelle fide Pasch) GUINEA BADYARA bantámyambi (Koelle fide Pasch) BAGA (Kalum) ayoka (Koelle fide Pasch) BASARI bandárá (Tersis fide Pasch; FG&G) ɔ-ŋáɓ ɔ-rín = *yam of the Ceiba tree* (FG&G) FULA-PULAAR (Guinea) bantãde (Labouret fide Pasch) bantara (CHOP; Labouret fide Pasch) KISSI yãmbàliŋ (FB; Heydorn fide Pasch) KONO tanga (Koelle fide Pasch) KONYAGI bantara (FG&G) u-nkwav u-kan = *yam of the kapok (Ceiba or Bombax)* (FG&G) KPELLE manã (FB; Castelhain fide Pasch) LANDOMA tandioro (Koelle fide Pasch) LIMBA batánka (Koelle fide Pasch) LOKO bátanga (Koelle fide Pasch) LOMA gulu manankui (FB) mánaku (Koelle fide Pasch) ui manankui (FB) MANDING-MANINKA banan gu (FB) banan ku (CHOP; FB) bananku gué *a sweet form* (CHOP) bara bananku *a bitter form* (CHOP) MANO beí (Becker-Donner fide Pasch) bõi (Koelle fide Pasch) SUSU dsõkãi (Koelle fide Pasch) kondárabīna (Koelle fide Pasch) yõkài *from Caribbean 'yuca'* (auctt.) YALUNKA bantara (Koelle fide Pasch) SIERRA LEONE BULOM yek (Dalby fide Pasch) yeke (Koelle fide Pasch) yoca (Dalby fide Pasch) BULOM (Kim) iɛge (FCD) iɛke (FCD) yeg (Dalby fide Pasch) BULOM (Sherbro) pu-langei *from Mende, in allusion to an 'European' origin; a bitter var.* (FCD) yɛkɛ-a-potho *a bitter var.* (FCD) yɛkɛ-lɛ (auctt.) FULA-PULAAR (Sierra Leone) bantara (FCD; JMD) GOLA gbada (FCD) gbára (FCD; Koelle fide Pasch) kégbàla (Westermann fide Pasch) KISSI yambai-ma purleŋ *a bitter var.* (FCD) yambalẽ (FCD; JMD) yambalẽ-fulalẽ *the roots* (FCD) KONO kono-lange *a c.var.* (Fisher) taŋga (FCD; JMD) taŋka (FCD) yokaa (Dalby fide Pasch) KORANKO banake (FCD) banake kowulen *a sweet, red-stemmed c.var.* (Glanville) bundo-borobo *a sweet c.var.* (Glanville) kawonko = *fingers, in allusion to the slender roots: a sweet var.* (Glanville) sa banake *from sa: snake; a sweet red-stemmed var. said to be efficacious when taken raw in snakebite* (Glanville) tankaso (JMD) KRIO kasada (FCD) LIMBA ba limba

103

EUPHORBIACEAE

(Glanville) babudo *a heavy yielding var.* (Glanville) ko fufe *a white var.* (Glanville) ko pintan (JMD) ku pintan ko (Clarke fide Pasch) LIMBA (Safroko) kupintaŋ (FCD) LIMBA (Tonko) bataŋga (FCD) LIMBA (Warawara) bapintaŋ (FCD) LOKO bataŋga (FCD) MANDING-MANDINKA banaŋgu (FCD) batere *c.var. with white stem and root* (Glanville) MENDE be tange *bitter and poisonous raw, but good for fufu* (Glanville) bele (Fisher) fande tange = *cotton-leaved cassava, a heavy yielding var.* (Glanville) kanda, kandei *bitter and poisonous raw causing giddiness, but good for 'fufu'* (auctt.) kono tange (-i) (auctt.) mayugbe *very sweet, even raw; most palatable* (Glanville; Fisher) mende tange (-i) (Glanville) nguwo tange (-i) (auctt.) pu tange (-i) (Glanville; Fisher) shenge *bitter and poisonous raw, but good for 'fufu'* (Glanville; Fisher) sowin *a.var.* (Fisher) tanga, tangeì (*or* lange, -i) (auctt.) tanga gbame (-i) *a sweet c.var.* (Fisher) tanga gboli *a sweet, brown c.var.* (Glanville) tanga gowe (-i) *a white c.var.* (Fisher) tanga leli *a black c.var.* (Glanville) tanga mende (Fisher) tanga wai (t. wawai) *a bitter var.* (FCD) tise *a.var.* (Glanville; FCD) yo bonge *very sweet, valued for its 'fufu'* (Glanville) SUSU yoka (FCD; Dalby fide Pasch) yoka fore (NWT) TEMNE ɛ-kanda *a bitter var.* (JMD) ɛ-rɔgbɔ (NWT) samouya *very sweet, even raw; most palatable* = *Mende:* mayugbe (Glanville) a-yokâ (*pl.* ɛ-yokâ *from Caribbean 'yuca'* (aucct.) *a*-yoka abi *a c.var.* (NWT; Glanville) a-yoka-afera *a white c.var.* (Glanville) a-yoka-a-kono *a heavy-yielding c.var. with rounded lobes to its leaves,* = *Mende:* fande tange (Glanville) a-yoka-a-ronketh *bitter and poisonous raw, causing giddiness, but good for 'fufu'* = *Mende:* kande (Glanville) a-yoka-a-yim *a brown c.var.* (Glanville) YALUNKA kondek-habi-na (FCD) kunda khabi-fore *a black var.,* *slightly bitter* (Glanville) kundakhabi-fikai *a white var., sweet, watery, not keeping well* (Glanville) kundakhabi-na (FCD) **LIBERIA** BANDI ndsurui (Koelle fide Pasch) DE lwiío (Koelle fide Pasch) KPELLE dozeéteī *a magic name* (Seidel fide Pasch) KRU-BASA bōe (Koelle fide Pasch)GREBO sodo (Innes fide Pasch) soôlo (Koelle fide Pasch) MANO bai, bei (Har.) VAI dumbai (JMD) gbasa (Heydorn) nyɔrɔ (Hair fide Pasch) **MALI** DOGON angá ra (CG fide Pasch) banaku (CG fide Pasch) vòso *from Bambara:* woso, *sweet potato* (CG) MANDING-BAMBARA bananku (auctt.) bandugu (auctt. fide Pasch) bâniku (auctt. fide Pasch) SENUFO gbörr (Knops fide Pasch) fu (Delafosse fide Pasch) SENUFO-BAMANA banankungo (Cheron fide Pasch) SONGHAI bananku (Prost fide Pasch) SONINKE bántàra (Koelle fide Pasch) SONINKE-BOZO mana-n-ku (Monteil fide Pasch) **UPPER VOLTA** BISA banki (Prost fide Pasch) BOBO sɔ-míngí = *tree yam* (Le Bris & Prost) FULA-FULFULDE (Upper Volta) leggal roogo *the plant* (K&T) roogo *the root* (K&T) GRUSI-LYELA bandeku (Prost fide Pasch) GURMA panyuma (Prost fide Pasch) tăngùbo (Surugue) KIRMA kpengbayo (Prost fide Pasch) KRU-GUERE (Chiehn) bu (K&B) MOORE banduku (Prost fide Pasch) SAMO bwananku (Prost fide Pasch) SAMO (Sembla) kpánkà (Prost) kpánkkà (Prost fide Pasch) SENUFO-TUSIA gbende (Prost fide Pasch) SONINKE-PANA bananku (Prost fide Pasch) **IVORY COAST** ABE *m*bɛdɛ (Dumestre fide Pasch) ADYUKRU *m*bosi (Dumestre fide Pasch) AKAN-ASANTE banki (Dolphyne fide Pasch) bédé (B&D) kóturuwááánka (Koelle fide Pasch) BRONG agba (Dolphyne fide Pasch) bankye (Dolphyne fide Pasch; E&A) AKYE bébé (Ivanoff fide K&B) *m*bèdè (A&AA) bɛndɛ (Dumestre fide Pasch) bindè (A&AA) ALADYA bɛdɛ (Dumestre fide Pasch) 'ANDO' agba (Visser) ANYI agba (Delafosse fide Pasch) bundrè (Delafosse fide Pasch) ANYI-AFEMA agba (Dolphyne fide Pasch) ARI fɛdɛ (Dumestre fide Pasch) AVIKAM vɛdɛ (Dumestre fide Pasch) BAULE agba (auctt.) DAN báa (Donneux fide Pasch) bē (Koelle fide Pasch) 'EOTILE' bɛdɛ (Dumestre fide Pasch) GWA ōmɛdɛ (Dumestre fide Pasch) 'KRU' sógūlo (Koelle fide Pasch) KRU-AIZI jeshi (Dumestre fide Pasch) jheshi (Herault) BETE boō (*pl.* bòòjā) (Marchese) gōnyì (Pageaud) GREBO (Plapo) solo (Thoiré fide Pasch) GUERE baha (A&AA) GUERE (Chiehn) bu (K&B; B&D) gbukwé (B&D) NEYO gebisokle *a white var.* (Thomann fide Pasch) gebye *a poisonous var.* (Thomann fide Pasch) sokole *a red var.* (Thomann fide Pasch) KWENI būū (Gregoire fide Pasch) KYAMA bébé (Ivanoff fide K&B) *m*bédé (JMD; Dumestre fide Pasch) bédé bro *c.var.* (Hédin fide JMD) bédé mango *c.var.* (Hédin fide JMD) bédé nana *c.var.* (Hédin fide JMD) bédé putu *c.var.* (Hédin fide JMD) MANDING-DYULA bànagu (Derive fide Pasch) bànåkừ (Derive fide Pasch) bēde (Kastenholz fide Pasch) gbéndé (Derive fide Pasch) MANINKA banaku (A&AA) gbendé (A&AA) NAFANA doo (Delafosse fide Pasch) NZEMA bɛdɛ (Dolphyne fide Pasch) SENUFO-DYIMINI gbende (Delafosse fide Pasch) TAFILE gbedi (Delafosse fide Pasch) TAGWANA gbél (Herault, Mlanhoro fide Pasch) **GHANA** ADANGME-KROBO agbeli (FRI; Lowe) agbeli kani (FRI) fɛdɛ (Dumestre fide Pasch) AKAN-AKYEM bankye (E&A) ASANTE bankye *lit.: it came not long ago* (auctt.) bedie kofie (FRI) FANTE bankye (auctt.) bankyi (Brew fide Pasch) TWI bankye (auctt.) AKPAFU adya-bōdyé (Seidel fide Pasch) *i*-gbèḍì (*pl.* à-) Kropp) AVATIME àgbèlì (Kropp) DAGBANI bankye (JMD) kambonjule *a white var.* (Gaisser) GA bantsí- *with varietal names suffixed* (KD) bɔi (FRI) duadé (FRI; KD) duadé atomo (FRI) duadé bulu = *foolish cassava; because it is slow in maturing* (FRI) duadé pamplo (FRI) gã-duade (FRI) kakâbā (FRI) klaba (FRI) nkãni (FRI) ntibli (FRI) obroni bediawuo = *if the whiteman eats it, he dies; from Twi* (FRI) odonti (FRI) odzoŋkoli (FRI) GBE-VHE agbeli (FRI) VHE (Peci) agbedidze-nkani (FRI) agbediloto kutumpa (FRI) agbedi-xe (FRI) agbelidze (FRI) agbeliloto (FRI) bede kofi (FRI) gbedi (FRI) gbeli (FRI) mampon-gbelî *c.vars.* (FRI) GRUSI-BUILE kpaherung (Prost fide Pasch nyungńsa (Koelle fide Pasch) nyūtísa (Koelle fide Pasch) GUANG agbedi (Westermann fide Pasch) bànchè (Rytz)

104

bank'e (Westermann fide Pasch) dùá (Rytz) GUANG-NKONYA agbēdí (Seidel fide Pasch) KON-KOMBA akombanul (JMD) LEFANA agbedi *from Gbe-Vhe,* (Höftmann fide Pasch) ɔ̃-bànci *(pl. lě-)* (Kropp) a-bandzi (Höftmann fide Pasch) obántye (Seidel fide Pasch) ɔ́lɔ́dzutu (Höftmann fide Pasch) LIKPE ābezí (Seidel fide Pasch) LOWILI abanyī (Seidel fide Pasch) MAMPRULI kukulajo (JMD) MOBA jangbwanguyuya (JMD) NZEMA bɛdɛ (FRI) SANTROKOFI odya-bodyé (Seidel fide Pasch) SISAALA gbanko (Blass) **TOGO** ANYI-ANUFO análô (Krass; & fide Pasch) bánci (Krass; & fide Pasch) AVATIME agbèḍi (Funke fide Pasch) BASSARI banking *a red var.* (JMD) kambohando *a white var.* (JMD) GBE-VHE agbeli (Kumah fide Pasch) bōki (Koelle fide Pasch) GURMA-KASELE agbana (Westermann fide Pasch) dogbono (Westermann fide Pasch) rogo *from Hausa* (Westermann fide Pasch) KABRE mbaumhae *a white var.* (JMD) KEBU gímu-zú-dí (Wolf fide Pasch) yumu zúdo (Seidel fide Pasch) KONKOMBA kam bon yulè (Froelich) KPOSO agbélī (Seidel fide Pasch) KUSAL zanbenyu-re (Lässig fide Pasch) NAWDM bondoronde *a white var.* (JMD) NYANGBO agbedi (Heine fide Pasch) TEM bónfe (Koelle fide Pasch) TEM (Tschaudjo) kungola *a red var.* (JMD) TOBOTE akamboasínó (Westermann fide Pasch) YORUBA-IFE (TOGO) ākobētē (Seidel fide Pasch) **DAHOMEY** GBE-FON agbede (Koelle fide Pasch)agberi (Koelle fide Pasch) felīe (Delafosse fide Pasch) FON (Gũ) akasa (Delafosse fide Pasch) faliña *from Portuguese: farinha: flour* (Delafosse fide Pasch) FON (Maxi) fária (Koelle fide Pasch) GURMA foho-nõri (Prost fide Pasch) foko-õbu (Prost fide Pasch) HWEDA ōkũte (Koelle fide Pasch) POLCI-BULBA baa-nyure (Prost fide Pasch) SOMBA difua nuã (Prost fide Pasch) YOM nu-gɔmya (Prost fide Pasch) **NIGER** DENDI rogo (Prost fide Pasch) róógó ài (Tersis fide Pasch) HAUSA rogo (Robin) SONGHAI òróggò *(pl. -à)* (D&C) SONGHAI-ZARMA lōga (Prost fide Pasch) rogo (FB) **NIGERIA** ? ibòbòdí (Elugbe) AGOI iyemi (Cook) AGWAGWUNE iwà ʔloan-word *from Efik* (Cook; BAC) AGWAGWUNE (Abini) usi-ekpet (Cook) AGWAGWUNE (Etuno) iburu (Cook) AKPA òyìrà (Armstrong; Oblete fide RB) AKPET-EHOM (Ubeteng) e-emi (Cook) EHOM (Ukpet) aiemi (Cook) ANGAS ɓwèr-yóm (Kraft) AOMA ibòbòdìí (Elugbe) ARABIC-SHUWA baghūt (JMD; Roth-Lely fide Pasch) garîsa (Letham fide Pasch) BADE dáwíyan (Kraft) BAKPINKA ayemi (Cook) BARAWA-GEJI dòyàmàsà (Kraft) BASSA (Kaduna) igbingbi (Koelle fide Pasch) BATA-BACAMA mbái (Kraft) mbàytó (Kraft) BETTE-BENDI ú-lógò *(pl. í-)* (Ugbe fide KW) ú-lógò fùng *a white-rooted c.var.* (Ugbe fide KW) ú-pànyá *(pl. í-) a c.var. with reddish inner bark* (Ugbe fide KW) BIROM kit còkọt = *tree yam* (LB) rókò (LB) BOGHOM mâm (Kraft) BOKYI ologo (Osang fide RB) panya (Osang fide RB) BOLE dóyà (Kraft) BURA mbai (anon) ngali *from Arabic* (Trenga fide Pasch) CHOMO mbaì (Shimizu) DENDI róógò *(indef. sing.)* (Tersis) DGHWEDE mbàyà (Kraft) DOKO-UYANGA iyemi (Cook) EBIRA (Igu) enapa (Koelle fide Pasch) EBIRA-ETUNO ẹcókà (Ladefoged fide Pasch) ōdsóka (Koelle fide Pasch) EDO ígaí (Elugbe) igari (JMD; Melzian fide Pasch) igari-ikeripia *a bitter var.* (JMD) igari-makole *a sweet var.* (JMD) EFIK àkparamkpa iwá *a bitter var.* (auctt.) ibo iwá = *Ibo (Igbo) yam* (JMD; FRI) iwá (Lowe; Adams fide Pasch) iwá únèñé = *yam of the Igbo* (JMD; Goldie fide Pasch) ọ̀fọp iwá *the tubers* (RFGA; Lowe) òkpọ ọfọp iwá *the plant* (RFGA; Lowe) ENGENNI ọbúrabụ (Elugbe) EPIE èḍìôbộ (Elugbe) ERUWA ùmũfié (Elugbe) ESAN ebọbọzi (Elugbe) FALI mbàyà, mbáyà, mbáyá *from Fulfulde* (Kraft) mbàyì (Kraft) FULA-FULFULDE (Nigeria) mba *(pl. mbaji)* (auctt.) bantarawal (Westermann fide JMD) mbawol *a single tuber* (Taylor) GA'ANDA bâi (Kraft) mbày (Kraft) GA'ANDA (Boka) mbày (Kraft) GHOTUO ibòbòdi (Elugbe) GLAVDA mbàyà (Kraft) GUDE mbàyá (Kraft) GUDU mbòy (Kraft) GUDUF gàlísá (Kraft) mbàyà (Kraft) GWANDARA-CANCARA rógo (Matsushita) GITATA àrógo (Matsushita) KARSHI rógo (Matsushita) KORO rógo (Matsushita) NIMBIA rógo' (Matsushita) TONI rógo (Matsushita) GWARI loogo (Banfield fide Pasch) HAUSA agingira, ganjigaga, gantamau, ƙaraza *vars. maturing in 2-3 years* (JMD) akul *a tuber long in the ground and probably unwholesome; eaten only in famine* (JMD) bara fiya, jagugo *vars. maturing in 6 months* (JMD) baushin kurege = *'baushe' of the squirrel* (JMD; ZOG) ba-zaƙa, ɗan-zaƙi *sweet vars. eaten raw* (JMD) bi naso *a damp land var.* (JMD) butsara, ɗanbana, ɗan jigawa, ɗan madi, ɗan walu, gama gari, ganji, róógon tuudu *common vars. maturing in 12 months* (JMD) dóóyàr kúdù = *yam of the South* (auctt.) kaaraajii (JMD; ZOG) kaaraaza (JMD; ZOG) kàndíirin béégúwáá = *staff of the porcupine* (JMD) róógon magariya *a var. grown in scrub; from* magariya: *Zizyphus, a typical scrub plant* (JMD) róógon sabara *a var. grown in scrub;* from sabara: *Guiera, a typical scrub plant* (JMD) róógòò, róógon rafi *apply generally* (auctt.) ta ɓe *a marshy land var.* (JMD) HUBA mbái (Kraft) HWANA mbòy (Kraft) IBIBIO okúmūdog (Koelle fide Pasch) ICHEVE kàsélà (Anape fide KW) pànyá = *real (Cassava); i.e. the preferred var. grown everwhere* (Anape fide KW) IDOMA oliya (Armstrong) oyila (Armstrong) IGALA àbáchà (RB) àbáchà lógò *a c.var.; cf. Hausa,* róÕgòò: *cassava* (RB) ābása (Koelle fide Pasch) IGBO akpu (auctt.) akpu (Armstrong fide Pasch) akpụ ǹkọ̀lọ̀ (KW) akpụ ǹkọ̀lọ̀ inū *a bitter var. after* inū: *bitterness* (DA fide JMD; KW) akpụ ǹkọ̀lọ̀ ūsọ *a sweet var. after* ūsọ: *sweetness* (DA fide JMD; KW) akpụ nwaelumelu *a bitter var.* (DA fide JMD) akpū (ụ)-jī *a second year tuber for planting* (KW) ji-akpu *a sweet var.* (JMD) ji-akwu (Iwu) jì-gbò (Green, Igwe fide Pasch) jì-goo (JMD) oi-akpu (JMD) oke akpụ = *big male akpụ; a bitter var.* (DA fide JMD; KW) okpolo *the root* (JMD) IGBO (Asaba) akpu awara nọku (DA fide JMD) IGBO (Awka) àbàchà (NWT; JMD) IGBO (Awo-Idemmiri) akphụǹkọ̀rọ̀ (Armstrong) jñakphũ (Armstrong) IGBO (Ezinehite) ìgbụrụ (Armstrong) ji aphū

EUPHORBIACEAE

(Armstrong) ọ̀dọ̀g̣hàrà (Armstrong) IGBO (Ohuhu) jìgbọ̀ (Armstrong) IGBO (Umuahia) iwa *loan-word from Efik* (DA fide JMD) IGBO-UKWUANI ìmàlakà ? *loan-word from Urhobo* (Armstrong fide Pasch; KW) IJO-IZON (Egbema) imidáka (Tieno fide KW) IZON (Kolokuma) abáburú (KW&T) abáburú umgbou *the stem* (KW&T) ịzọ́n ábábúrú = *Ijo cassava* (KW) NEMBE ébiabúru (Kaliai fide Pasch) íbiabúru (KW) IṢEKIRI mīdáka (Koelle fide Pasch) ISOKO ẹgú (Elugbe) IVBIE ávbiorẹ̀ (Elugbe) JAKU rógò (Kraft) JEN wụ̀nkà (RB)JIRU mbàì (RB) JUKUN (Wukari) irogò (Shimizu) KAKA kukum (RB) KAMBARI gbámgbárá (RB) KAMBARI-SALKA ààgbámggbàrà (RB) KAMWE mbàyá (Mohrang fide Pasch; Kraft) KAMWE (Kiria) mbáyà (Kraft) KANURI doya (Noel fide Pasch) gàlísà ? *from Arabic* (C&H) garisa (JMD) ngàdàlá *a sweet c.var.* (C&H) KAREKARE dóyà (Kraft) KATAB-KAGORO kòsèy *from Hausa:* cassava cake (Gerard fide Pasch) KORO rugu (RB) KYIBAKU mbôu (Kraft) LENYIMA jakpu (Cook) LEYIGHA gemonjeen (Cook) LOKE kè-báájên (*pl.* yè-) (Elugbe) ebajenti (Berry & Guy) kebajen (Berry & Guy) LUBILO erenge (Cook) MALA kalogozhi (*pl.* ạ-) (RB) MARGI mbúy (Kraft) MARGI (South) mbàyú (Kraft) mbùy (Kraft) MARGI-PUTAI dùwáyà (Kraft) dúwyà (Kraft) MATAKAM mbáyúk gàlìsà (Kraft) MBEMBE ọ́-rángkirí (Cook) MIYA gwíyìm (Kraft) NDE árōganti (Koelle fide Pasch) éluākárara (Koelle fide Pasch) NGAMO dóya (Kraft) NGGWAHYI mbái (Kraft) NGIZIM dáuyâ (*pl.* -áyin) *from Hausa:* dóoyaà: *yam* (Schuh) NUPE rógò (JMD; Banfield fide Pasch) rógò dzúrúgi *a red c.var.* (JMD; Banfield fide Pasch) NZANGI mbàyé (Kraft) OBULOM íjápù (Cook) ODUAL ẹleebo (Gardner) ọkpukoro (Gardner) OGONI-GOKANA gbèbãã̀ (Brosnahan fide Pasch) OKPAM-HERI (Emarle) òt-ālìbò (Elugbe) OLULUMO è-wá ? *loan-word from Efik* (Cook; BAC) OLULUMO (Ikom) ì-wá (Cook) PERO dóyà (Frajzyngier) gbólà (Frajzyngier) rógo (Kraft) POLCI-BULI dóyà (Kraft) ZUL màngkị́nì (Kraft) RESHE rigeshe (Harris fide Pasch) SAMBA-DAKA pii-gọọ (RB) SANGA marọgo (RB) SURA roogò (anon. fide KW) TANGALE ɗóyà (Kraft) TERA rógò (Newman) TERA-PIDLIMDI dóyà (Kraft) TIV dowase (Abraham fide Pasch) duàsə (KW) duaso (Terpstra fide Pasch) ivambèyon (JMD) lougou *from Hausa* (Abraham fide Pasch) rogo (Terpstra fide Pasch) rougou (JMD) vambèyon (JMD) UBAGHARA (Biakpan) iburu (Cook) UBAGHARA (Ikun) iburu (Cook) UHAMI-IYAYU ikpáki *a c.var.* (Elugbe) paki (Elugbe) UKUE kpáki *a c.var.* (Elugbe) UKUE-EHUEN kpáki *a c.var.* (Elugbe) UMON iwa *loan-word from Efik* (Cook; BAC) UNEME ibɔ̀rɔ̀jî (Elugbe) URHOBO igu mádáka *bitter form* (JMD) imidáka (Elugbe) mádáka (JMD; (Koelle fide Pasch) mádáka kole *sweet form* (JMD) UVBIE ìbi dákà (Elugbe) WANDALA mbàyà (Kraft) YALA (Ikom) ijikāpā (Ogo fide RB) YALA (Ogoja) ịkpāleke (Onoh fide RB) YATYE àkwọ́m (Armstrong) YEDINA kárahó (Lukas) maha *the fruit* (Lukas) YESKWA ádsira gbare (Koelle fide Pasch) YẸKHEE (Auchi) abọbọzíi (Elugbe) YẸKHEE (Avianwa) ẹko (Elugbe) YORUBA ẹ̀gẹ́ *the plant* (auctt.) gbāgbuda, gbaguda *cassava, the article* (auctt.) ōgege (Koelle fide Pasch) YORUBA (Bunu) áágugóaayó (Koelle fide Pasch) YORUBA (Egba) ogége (Koelle fide Pasch) YORUBA (Jumu) agugoyo (Koelle fide Pasch) YORUBA (Yagba) dereẹ́fe (Koelle fide Pasch) YUNGUR mbal *from Fulfulde* (RB) ZAAR gyèdî (Kraft) **WEST CAMEROONS** VULGAR ngabo (JMD) BAFOK ngawa (JMD) sanagal *sweet var.* (JMD) BAFUT kàsàvà (*pl. bi-*) (Mfanyam) BAMILEKE shùàsom (Tchamda) KPE mionde *bitter var.* (JMD) KUNDU kasara *bitter var.* (JMD) LONDO ewa (JMD) LONG sanagal *sweet var.* (JMD) MBONGE kasara *bitter var.* (JMD) NGEMBA kàsárə̀ (Chumbow) kasarə wumbaŋnə = *red cassava; cultivars* (Chumbow) TANGA ewa *bitter var.* (JMD) WOVEA makawama *bitter var.* (JMD)

Products: GUINEA MANDING-MANINKA sinkoro *the flour* (CHOP) **SIERRA LEONE** ? (N Province) kondo bala *cassava, peeled, parboiled, sun-dried* (Glanville) to *cassava made into balls and sun-dried ready for boiling and eating* (Glanville) to bagadalì *parched cassava made into dough, sometimes with addition of sugar or honey, then sun-dried* (Glanville) **GHANA** GA duade agwao *the boiled whole root* (JMD) dzi-dzi *boiled cassava meal* (JMD) kokonté *the dried flour* (KD), or *the peeled root split and dried* (JMD) **DAHOMEY** ? (Lower Dahomey) gari *the flour* (Adande) gari okû = *cassava flour of the sea; flour of inferior quality* (Adande) **NIGERIA** EDO ọbọbọzíi *the tuber sliced and left in water till the following day.* (Elugbe) EFIK ńtọ́rọ̀rọ̀ *soaked cassava, from Ibibio* (Adams fide Pasch) HAUSA fufu *cassava meal* (JMD) gari *a coarse dry-ground meal* (JMD) gurka, huhu, kaneku, kwaki *from Yoruba:* páki; *grades of cassava meal* (JMD) IDOMA ikwu líikwu *dried balls of flour* (Armstrong) IGBO akpu-ǹkọ̀lọ̀ = *shredded cassava, or fufu; a general term* (KW) garri *coarse dry meal* (Iwu) IJO-IZON (Kolokuma) ẹfébụá *the tubers* (KW&T) ofonía, ofoníyan *dried lumps of 'farina' prepared for chewing* (KW&T) ọkpụúnkọrọ *unpounded fufu* (KW&T) osún *the starch prepared in a ball to eat* (KW&T) YORUBA bẹjú *cassava bread* (JMD) eba *a pudding made with cassava flour* (Oke) fùfú *cassava meal* (JMD) gāri *a coarse dry-ground meal* (JMD) gari yiyan *powdered root with palm-oil* (IFE) gbáguda, gbágbúda *cassava flour* (auctt.) isu gbáguda *stem-pith* (IFE) kpokpo gari *processed cassava with particles larger than gari* (Oke) ògì-gbáguda *cassava starch* (JMD) páki *cassava meal* (JMD)

A shrub to 4 m high with a markedly articulated stem and large fleshy to woody tubers, native of Brazil and now dispersed throughout the world tropics.

The tuber is the source of the farinaceous meal variously called cassava, tapioca, manioc, etc. This plant had assumed such importance in the New World long before the Voyages of Discovery from Europe of the Fifteenth and Sixteenth Centuries that it had been dispersed in the American tropics from 25° south of the Equator to 25° north. European voyagers requiring victuals for ships' stores soon found that it was not possible to use the roots as with age they accumulated hydrocyanic acid and became toxic. The roots had to be prepared into a form of meal, and it is recorded that Columbus met native ships at sea provisioned with cassava bread. The Portuguese trafficking in slaves to Brazil from Angola stocked their ships with *farinha de pao*. It is probable that the Portuguese were the first foreigners to bring the plant to Africa (Angola), but its general dispersal in the Old World seems to have been held back by the use of non-viable products in ships' stores, and it was perhaps not till the Nineteenth Century with slaves returning from the Americas that the plant really became established in West Africa. Clearly the evidence of vernacular names indicates an exotic and a recent origin. The Mandinka of Senegambia use the Haitian name 'manioc' in their *mandioka*. The Susu and Temne of Guinea and Sierra Leone use the Caribbean word *yoka* and the Sherbro recognised its 'European' origin in *pu-langei*. The Ghanaian Akan name *bankye* means 'it came not long ago.' In Nigeria the Efik ascribe the coming of the plant to the neighbouring Igbo people, *ibo iwa*, Igbo's yam, and Hausa to a southern origin, *doyar kudu*, yam of the south. Just as in the Americas where the plant has spread till it has reached the limit of its climatic tolerance, checked, in the north at least, by winter frost, so in West Africa it has spread to a limit of minimal rainfall, reaching only as recently as 1935 its northernmost limit of 750 mm precipitation in the savanna belt across Senegal to Nigeria (6).

The plant is extremely variable, especially in the roots. The main distinction of races lies between 'bitter' and 'sweet.' In the bitter races a cyanogenetic glycoside, *linamarin*, occurs throughout the plant tissues, and this by enzymatic action is reduced to glucose, acetone and hydrocyanic acid, and in the roots the last-named may amount to upwards of 0.01% concentration. In the sweet races the glycoside occurs only in the roots in much lower amounts, and then principally in the root-skin, facilitating easy removal by paring in preparation for eating. Concentration of hydrocyanic acid varies also according to soil type, moisture, temperature and age, unfavourable conditions leading to higher amounts (6, 13, 19, 30). A high potash availability is desirable for improved starch-production and lower hydrocyanic acid level (14).

The sweet races are a short season crop, the roots being harvested in 6–9 months. The roots of some cultivars are edible raw, and the sucrose content may amount to 17%. The bitter races, which are grown to a very much greater extent, may be harvested only after 3 years but never usually before 2 years. The roots of some cultivars attain a very large size even up to 2 m long. They contain a latex in which toxic substances reside, and by careful preparation involving washing, drying and heating, the toxicity caused by the hydrocyanic acid present is eliminated. The roots are the principle source of a starchy meal that is the farina or tapioca of commerce. *Fufu, gari* and many other local preparations are made from the roots of bitter cultivars, and for these there are often special selections yielding superior products. Characteristically gari swells to 3–4 times its volume when water is added. This may contribute to its popularity as a food giving a feeling of repleteness, but other than the calorific value of the starch it has little nutritional value being low in protein, and lacking in vitamins. In areas where much is grown, protein deficiency diseases are prevalent. The carbohydrate is in a gelatinous form making it easily digestable and has for this reason been used as an arrowroot substitute. It is official in many pharmacopoeias (6, 20, 21, 22, 26). Production of starch from the roots is an industrial exercise, and when supplies from Java were denied to Britain during the war, 1942–45, considerable quantities of high grade starch were produced from

Nigeria. The starch is used in the manufacture of food-stuffs, textiles, paper-products, plywoods, veneers, adhesives, glucose, dextrin, etc., and in a smaller way for dyes, drugs, chemicals, carpets, linoleum, alcohol, etc. (8, 14). Dried roots and tapioca provide good stock feed (5, 14, 18, 21). The starch is the standard material in laundering linen and cloth to confer stiffness, and after ironing a smooth appearance (29).

The leaves of sweet races are eaten as a vegetable and are reported to be of good nutritional value, high in protein, minerals and vitamins, comparing very well with other leaf-vegetables (6, 13, 25, 28, 30, 31). These are important assets in territories suffering from poverty and malnutrition. Though serious poisoning of a population dependant on tapioca as a staple food is rare, there is a constant underlying risk of associated diseases such as goitre and cretinism, and possibly diabetes, arising from a regular mild toxification. There is here an excellent case for research into selection of cultivars and into husbandry methods both of which can contribute to elimination of the dangerous glycosides. In parts of the Cameroons plants are specially grown kept constantly cut-back in order to yield leafy shoots for consumption (11). Harvesting the foliage of course reduces tuber-production. If both leaf and root are required the farmer must exercise a fine judgement as to husbandry and priorities. Work has indicated that a leaf-harvest at 2 to 3 months interval gives a fair balance (30). The plant is not devoured by locusts and it is recorded that timely planting under threat of attack has staved off famine (10). The plant is also browsed by stock (12). Leaves and leafy twigs are used sometimes in W Africa to make a vegetable salt (6).

In West and Central Africa the powdered leaves are applied as a compress to the head for headache and in fever (7, 28). In Nigeria cooked and pulped leaves have been applied to tumours (1). The Ijo of the Niger Delta squeeze leaf-sap into wounds as a haemostatic (27). In Gabon the leaves are pounded and rubbed into small cuts for healing, and also on to chicken-pox pustules and other mild skin-affections (24, 25), and in Ubangi for yaws and small-pox (28). The leaf is applied to chicken-pox pustules in Senegal (15). Similarly leaf-sap is applied in Congo (Brazzaville) to small-pox pustules, and it is there also used as an eye-ointment for ophthalmia and to kill filaria (4), while in Ubangi a leaf-poultice is used to treat abcesses caused by jiggers (28). Latex from the stem is used in the Ivory Coast as an eye-instillation for conjunctivitis (16). Bruised leaves or pounded stems placed across an ant-run will, it is claimed in Gabon, disperse the ants immediately (25).

The seeds (of the bitter races ?) contain an oil which is a drastic purgative (26), and conversely are said to be edible in Gabon (25). This latter report must undoubtedly refer to the sweet races.

The root has a number of medicinal uses. The pulp makes a good poultice, and is used in Nigeria on burns, ulcers, jigger sores, etc. (1, 20). In Congo (Brazzaville) sap from root-scrapings is used on skin-affections (4). In Ubangi root-ash in anal suppositories is used for *Oxyuris* infection (28). The Ijo of the Niger Delta pound the peeled tuber to a paste which is applied twice daily to fresh wounds. It is said to arrest the pain and to promote rapid healing (27). A root-macerate of the sweet races is considered in Gabon to be a good galactogogue and with leaves of *Piper umbellatum* (Piperaceae) added to be good treatment for pains in pregnancy (24, 25).

The plant is used in Gabon as a temporary shade for young cacao and coffee. It is also planted as hedges (25). Strains resistant to mosiac virus disease with high root-yield have been raised in both Nigeria and Ghana (2). An interesting observation in Ivory Coast is recorded regarding the collecting of tapioca flour by honey bees, *Apis mellifera*, in the absence of pollen flowers (17), presumably as a source of sugar.

In the Loma country of Guinea there is a superstitious invocation not 'to beat a friend [sic!] with a stem of tapioca, for if he is a sorcerer, he will die' (3). The plant enters into an Yoruba incantation to get money quickly (23).

References:

1. Ainslie, 1937: sp. no. 222, as *M. utilissima* Pohl. 2. Anon, 1939,a. 3. Appia, 1940: 360. 4. Bouquet, 1969: 119. 5. Burkill, IH, 1935: 1411–20, with history and many references, as *M. utilissima* Pohl. 6. Busson, 1965: 164–74, with analyses of tubers, starch and leaves, as *M. utilissima* Pohl. 7. Dalziel, 1937: 150–4. 8. Furlong, 1942. 9. Holland, 1908–22: 600–6, as *M. utilissima* Pohl. 10. Irvine, 1952, a: 24. 11. Irvine, 1956: 36–37. 12. Irvine, 1961: 242. 13. Johnson & Raymond, 1965. 14. Kay, 1973: 24. 15. Kerharo & Adam, 1974: 422–4, with phytochemistry and pharmacology. 16. Kerharo & Bouquet, 1950: 81, as *M. utilissima* Pohl. 17. Kullenberg, 1955: 1129. 18. Modebe, 1963,b. 19. Oke, 1966, a. 20. Oliver, 1960: 8, 70. 21. Quisumbing, 1951: 523–4, with many references. 22. Spickett & al., 1955. 23. Verger, 1967: no. 88. 24. Walker, 1953, a: 36–37, as *M. utilissima* Pohl. 25. Walker & Sillans, 1961: 171–2, as *M. utilissima* Pohl. 26. Watt & Breyer-Brandwijk, 1962: 423, as *M. utilissima* Pohl. 27. Williamson, K. 59, UCI. 28. Vergiat, 1970: 76. 29. Diarra, 1977: 44. 30. Lancaster & Brooks, 1983. 31. Iwu, 1986: 142.

Manihot glaziovii Müll.-Arg.

FWTA, ed. 2, 1: 413. UPWTA, ed. 1, 150.

English: Ceara rubber; tree cassava.

French: maniçoba; céara; caoutchouc de Céara; manioc de Céara.

Portuguese: manicoba.

West African: SENEGAL DIOLA bônkos (JB) SERER gavsu (JB) ngavsu (JB) WOLOF ngavsu (JB) SIERRA LEONE KRIO rɔba *i.e.*, rubber (FCD) LOKO (Magbile) bororamga (NWT) MENDE nafaŋge (NWT) GHANA GBE-VHE agbeli ngwa (JMD) aŋeti = *rubber tree* (auctt.) NIGERIA FULA-FULFULDE (Nigeria) kolkolwaahi (J&D) YORUBA gbaguda (IFE)

A tree reaching 10 m high with short crooked stem and bushy branching, native of NE Brazil, and now widely dispersed under cultivation in the tropics. The original interest in the tree was as a rubber-bearer. Living seedlings were amongst the material of rubber plants collected by Cross, as the agent of the India Office in London, during his visit to Brazil in 1876, the same year as Wickham made his classical collection of *Hevea brasiliensis* (Euphorbiaceae) seed. Cross's material of Ceara rubber was established at Kew and later in Ceylon, Singapore and Malaya. Plants failed at the two latter places but seed was raised in Ceylon and planting material appears to have reached West Africa before 1888 where the drier climate and sandy soil of The Gambia was considered the most suitable. The plant comes from an area that is seasonally very dry and the soil is poor, in distinction from *Hevea brasiliensis* which requires a relatively high and regular rainfall and an unimpoverished soil. To some extent they are complementary and successful plantations of Ceara rubber have been established in the dry climate areas of E Africa and sandy savanna zone of W Africa. Ceara rubber is said to be of good appearance, but resin content at 3–12% is too high, and in quality, quantity and economics it could not compete with para rubber. As a plantation crop it ceased to be important during the second decade of the Twentieth century, but regained an ephemeral status in 1942–45 in World War II (1, 2).

The plant is still widely grown in the Region as an ornamental and as a hedging-barrier plant. The Fula of N Nigeria use the latex as a glue for paper (4).

The leaves yield a white plastic substance which is not rubber (1). Hydrocyanic acid is produced in them, but this is dissipated by heat and they are eaten cooked as a vegetable in Gabon (9) and in E Africa (3).

The root is rich in starch, but it is hard and woody. It also produces hydrocyanic acid. In time of dearth only is it eaten and in the same way as tapioca (Ghana, 6; Gabon, 9). The stem and root enter into a Nigerian remedy for skin-infections, but screening for bacteristatic and fungistatic action showed none (7), and examination of the stem for molluscicidal activity against freshwater snails gave very limited value (10).

The plant has been used with some success in E Africa in hybridising work with *M. esculenta* to raise resistance in the latter to mosaic virus disease (5, 8).

EUPHORBIACEAE

The flowers are rich in honey and are freely visited by bees (1), but the value of honey from this source is questioned, though beeswax may be a more useful product (2).

The seeds contain an oil which is about 35% of the kernel, or 10% of the whole seed. The oil is clear, yellow and slow-drying (1).

References:

1. Burkill, IH, 1935: 1409–10. 2. Holland, 1908-22: 598–600. 3. Irvine, 1961: 242–3. 4. Jackson, 1973. 5. Jennings, 1957. 6. Ll.-Williams 580, K. 7. Malcolm & Sofowora, 1969. 8. Nichols, 1947. 9. Walker & Sillans, 1961: 170. 10. Adewunmi & Sofowora, 1980: 57–65.

Manniophyton fulvum Müll.-Arg.

FWTA, ed. 2, 1: 400. UPWTA, ed. 1, 154.

West African: **SIERRA LEONE** BULOM (Sherbro) kanthinkɔ-lɛ (FCD) KISSI yoleŋ (FCD) KONO yoo (FCD) KORANKO kɔnrogoro (NWT) LIMBA bumumwi (NWT) bunawalu (NWT) LOKO njoroi (NWT) pɔŋ (NWT) MENDE njole(i) (auctt.) njolo (FCD) SUSU baminkame (NWT) kamwii (NWT) kubali (NWT) TEMNE an-fal (FCD) ɛ-fil (NWT) ra-mɛr-ra-sip (FCD) **LIBERIA** MANO fai (JMD) fei (JMD) **IVORY COAST** AKAN-ASANTE kolomodia (B&D) ANYI frafrabié (K&B) 'KRU' topué (K&B) KRU-BETE dobuï (K&B) GUERE zohé (K&B) zoobo (K&B) **GHANA** AKAN-ASANTE hunhun (Andoh) WASA etwi homa (auctt.) **NIGERIA** EDO ebómé = palm-frond, but cf. ebúmwẹhẹn; pumpkin leaves (Elugbe) IGBO ege (auctt.) IGBO (Umuahia) ageregre (AJC) egereghe (AJC) IGBO (Umudike) egreregre (Tuley)

A straggling bush or liane to 30 m long with stems to 10 cm thick, of the closed-forest or mixed deciduous evergreen forest from Sierra Leone to W Cameroons, and on to Zaïre and Angola.

The foliage is recorded as browsed by goats in S Nigeria (3).

The stem of the liane yields a red sap which rapidly goes tacky in air. The plant, on the Theory of Signatures, is widely used for affections in which blood is manifest. In Congo (Brazzaville) it is considered haemostatic and cicatrisant on wounds, and good for treating dysentery, piles, haemophthysis and dysmenorrhea (1). In Ivory Coast it is used for painful menses (2). The red stem-sap is used topically in Ivory Coast–Upper Volta on herpes and other dermal infections (7), while the leaf-sap is similarly applied, even to areas of leprosy, in Congo (Brazzaville) (1). In Liberia, the powdered dried leaves are sprinkled on sores (5). A stem-decoction is drunk in Congo (Brazzaville) (1) and in Sierra Leone (6) for blennorrhoea or gonorrhoea.

A decoction of young shoots and roots is considered a sovereign remedy for cough and bronchitis and a twig-decoction for all abdominal complaints in Ivory Coast (2). Leaf, stem and root-decoction is drunk in Sierra Leone for stomach-ache and the bark and stem are chewed for cough (6). The husk of the nut is also chewed in Sierra Leone for cough (4). The plant is used in Ivory Coast for whooping cough (2) and the sap is made into a cough linctus in Congo (Brazzaville) (1) where it is applied also to a carious tooth. Leaf-sap is an ear-instillation in Zaïre for ear-troubles (8).

Young leaves with kola nut and seeds of maleguetta pepper are taken in Congo (Brazzaville) for tachycardia, and sometimes leaf-sap with sugar-cane sap and other drug-plants for insanity (1).

The stems possess considerable strength and are used in Zaïre (8) and Ubangi (10) to make traps for hunting. The stem-bark yields a strong fibre which resists rotting. It is used by fishers and hunters in Gabon (9), Zaïre (8) and widely in central Africa (4). Leaves of Antiaris africana (Moraceae), which form the wrapping of a witch-doctor's magic baton in the Central African Republic, are stitched together with this fibre (11). Fibre-extraction is usually done by drying in the sun and beating. Retting is not normal (9).

The root-bark, dried and powdered, is poisonous. An antidote is Piliostigma thonningii (Leguminosae: Caesalpinioideae) (10).

110

The seeds are oil-bearing. Cooked nuts are an item of market trade in S Nigeria; the large white kernels, c. 2.5 cm diameter, are mealy and edible when well-boiled. Oil-content of Central African material has been recorded as about 50%. The oil is yellow and tasteless, thickening on exposure in thin films, suggesting a possible use in paint manufacture, but having an iodine value of only 101 (4). Good quality drying oil has an iodine value of 125 to over 200.

References:

1. Bouquet, 1969: 119. 2. Bouquet & Debray, 1974: 85. 3. Carpenter 347, UCI. 4. Dalziel, 1937: 154. 5. Daniel & Okeke 25, K. 6. Deighton 2336, K. 7. Kerharo & Bouquet, 1950: 81. 8. Léonard, 1962, b: 172–4. 9. Walker & Sillans, 1961: 172. 10. Vergiat, 1970: 77, as *M. africanum* Müll.-Arg. 11. Motte, 1980: 126.

Maprounea africana Müll.-Arg.

FWTA, ed. 2, 1: 416.

A shrub or tree to about 6 m high of the savanna, recorded only in N Nigeria and W Cameroons, but extending widespread eastwards into central and east Africa.

The plant is widely known for its purgative properties. A macerate of root-bark is so used in Tanganyika in limited doses, large amounts being lethal (4). Both stem and root-bark are used in Congo (Brazzaville) for treating constipation and for other affections deemed to require catharsis, e.g., ascites, generalised oedemas, female sterility and irregular menstrual cycle (2). The preparation is laced with honey or juice of sugar-cane, and care is necessary to control the size of the dose. It is also used in Congo (Brazzaville) as a vermifuge, and as a vaginal douch for uteritis and vaginitis, or the rolled-up leaves may be placed as a vaginal suppository for the same purpose (2). The leaves are chewed and the pulp, which is bitter, is swallowed in Zaïre for stomach-complaints (6), and a preparation of the young twigs is used in Tanganyika to relieve constipation (7).

The roots, bark and leaves are said in Gabon to be emetic and diuretic (8). Toothache is treated in Zaïre (1, 5) by chewed or chopped up leaves. An ointment made from powdered bark in palm-oil is applied in Congo (Brazzaville) to leprous maculae, small-pox pustules and to skin-infections generally (2). In Zaïre the roots are used against syphilis, blennorrhoea and leprous sores (5).

Tannins have been reported present in the leaves, bark and roots, a little *flavones* and *saponin* in leaves, and *saponin* in larger amount in the bark and roots (3).

References:

1. Aubréville, 1950: 198. 2. Bouquet, 1969: 120. 3. Bouquet, 1972: 25. 4. Haerdi, 1964: 94. 5. Léonard, 1962, b: 145–7. 6. Persson 80, K. 7. Pirozynski P.185, K. 8. Walker & Sillans, 1961: 172.

Maprounea membranacea Pax & K Hoffm.

FWTA, ed. 2, 1: 416.
West African: NIGERIA EDO isán-ahismwẹ = *faeces of birds*(Elugbe)

A graceful tree with conspicuous red young foliage, to 17 m high, of forest regions usually in secondary formations, and sometimes in savanna, only recorded in S Nigeria in the Region, but extending southwards to Zaïre and Angola.

The wood is used in Gabon to make mortars and the bruised young stems are laid in houses to drive away cockroaches (5).

The bark is a violent purgative and is used in Ubangi (4) and in Congo (Brazzaville) (1) against constipation and in the latter country also for ascites, generalised oedemas, female sterility and irregular menstrual cycle, affections deemed to require catharsis. Either root or stem-bark may be used laced with honey or juice of sugar-cane, and care is necessary to control the size of the dose. It is also used in Congo (Brazzaville) as a vermifuge, and as a vaginal douch for uteritis and vaginitis, or the rolled-up leaves may be placed as a vaginal suppository for the same purpose (1). In Ubangi a child's face may be washed with a leaf-decoction against mattery eyes (6). The dried powdered leaves are used in Gabon to cicatrise wounds, especially those of circumcision (5). In Congo (Brazzaville) an ointment of the powdered bark compounded in palm-oil is applied to leprous maculae, small-pox pustules and infections of the skin generally (1).

The root-decoction is poisonous and is used in Ubangi against syphilis (6).

Congo material is reported to contain a little *flavone* in the roots, and in the leaves, bark and roots *tannins, steroids, terpenes* and *saponin* (2).

The seed (? or the conspicuously red fruits) are reported eaten by birds in Zaïre (3).

This plant resembles *Psorospermum spp.* (Hypericaceae), and in Ubangi is held to be 'female' towards them (6).

References:

1. Bouquet, 1969: 120. 2. Bouquet, 1972: 25. 3. Léonard, 1962, b: 144–5. 4. Portères, s.d. 5. Walker & Sillans, 1961: 173. 6. Vergiat, 1970: 77.

Mareya micrantha (Benth.) Müll.-Arg. ssp. micrantha

FWTA, ed. 2, 1: 404. UPWTA, ed. 1, 154–5, as *M. spicata* Baill., and 155 as *Necepsia* sp. fide Harley 1143 (K).

English: 'number one' (Sierra Leone: Dalziel, Smythe; Fernando Po: Daniell).

West African: **SIERRA LEONE** BULOM (Sherbro) ŋwaŋwa-lɛ *from* ŋwa: *bitter* (FCD; S&F) KISSI lamba (FCD; S&F) KONO ŋɔnaŋgɔna (FCD; S&F) KORANKO warɛ-warɛ (S&F) KRIO nɔmba-wan *i.e., number one; a corruption of* ŋwaŋwa *(Mende)* (FCD; S&F) LIMBA ɛkbɔgɔbagawandi (NWT) kukbɔgo (NWT) kukpɔbo (NWT) LOKO giribɛ (NWT) mačɔnjɔ (NWT) ngalamum (NWT) ŋwaŋwa (FCD); ŋwa: *bitter* (S&F) walewale (NWT) wɔŋwa (NWT) MENDE maŋwa (NWT) ndewe (NWT) ŋwaŋwa (*def.* -i) = *bitter-bitter;* from ŋwa: *bitter* (auctt.) wakwe (JMD) wãnwã (Fisher) waŋwei (NWT) wongwai (Aylmer) wonwai (JMD) wua wua (DS) SUSU barambaramyi (NWT) buisigbuisigbwe (NWT) TEMNE *an*-gbeseŋ (auctt.) **LIBERIA** KRU-BASA chaw (C&R) chew (C&R) MANO wana (Har.; JMD) MENDE nguangua (C&R) **IVORY COAST** ? koékpagni (A&AA) ABE gba-gba (A. Chev.; K) AKAN-ASANTE du frafo = *itis poisonous* (K&B) nofé (B&D) AKYE oya (Aub. ex K&B) oyia (Aub.) ʼANDOʼ ado bla fo = *to do wrong to a woman* (Visser) ANYI bagba (K&B) essan (K&B) pété péré (K&B) BAULE bodiè (B&D) GAGU guala (K&B) ʼKRUʼ trahain (K&B) KRU-BETE tiamui (K&B) GREBO (Nyabo) tiému (K&B) GUERE tehin (K&B) GUERE (Chiehn) bogbo (K&B; B&D) KWENI sanbra dikra (B&D) siambralapa (B&D) simbla hakana iri (K&B) **GHANA** AKAN-ASANTE enanta (auctt.) odubrafo *from* odu(-ru): *medicine;* (ɔ)brafo: *executioner; alluding to its poisonous properties* (auctt.) FANTE odubrafo *from* odu(-ru): *medicine;* (ɔ)brafo: *executioner; alluding to its poisonous properties* (FRI; A&E) TWI odubrafo *from* odu(-ru): *medicine;* (ɔ)brafo: *executioner; alluding to its poisonous properties* (auctt.) WASA odubrafo *from* odu(-ru): *medicine;* (ɔ)brafo: *executioner; alluding to its poisonous properties* (FRI; E&A) **NIGERIA** EDO uhosa (Ross; KO&S) IGBO oke-owuma *from* oke: *male* (Tuley)

A common shrub or understorey tree to about 17 m high of the closed-forest and in secondary regrowth, occurring from Guinea to W Cameroons and Fernando Po, and extending to Zaïre.

The leaves and fruit are bitter, and are very poisonous causing drastic purging. The Mende name *ŋwaŋwui* or *nwangwai*, meaning 'bitter-bitter', has been adapted in Sierra Leone Krio to *nɔmba-wan*, i.e., 'number one', and this term has been carried to other parts of West Africa, in respectful acknowledgement of the plant's dangerous toxicity (7, 13). Even in Fernando Po it carries this name where it has been recorded that the plant is a powerful drastic purgative dreaded greatly by the natives for its poisonous qualities, having killed many people (8). In Ghana Akan peoples know it as 'medicine executioner' from *odu(ru)*: medicine and *(ɔ)brafo*: executioner (9). The Basa of Liberia use it as an ordeal-poison (2).

A decoction of leaves, or the leaf-sap, is known everywhere as a violent purgative and abortifacient. A dozen leaves boiled in a pint (c. half a litre) of water is said to be a safe dose in Ghana, but any stronger is dangerous, and in Ghana and Liberia it is never given to pregnant women, nor to children, nor old persons (7, 9). After an abortion a single leaf in food ensures removal of the afterbirth, and a small dose is repeated for some days (7). In Ivory Coast–Upper Volta medicine-men consider it a valuable drug-plant prescribing it for diseases the treatment of which requires violent action on the intestines, e.g., ascaris infection, blennorrhoea and especially leprosy (3, 10). The bark with other drug-plants is used topically for leprosy in Liberia (7). In Sierra Leone water in which dried leaves have been boiled is given to children as a worm-treatment (13).

An over-dose or criminal poisoning will cause complete exhaustion by purging (5, 16), and leaves produce hypotension in dogs (12), and root-extracts coma and paralysis of the respiratory centres in rats (15). Antidotal treatment has been reported to be red palm-oil (5) or raw cassava (7).

The plant has analgesic and anaesthetic properties. It is widely used internally as an anodyne for headache and stomach-pains, and externally topically on fractures, stiffness, sprains, lumbago and rheumatism; also on sores and ulcers. The powdered root is used in Ivory Coast–Upper Volta on snake-bite and the stings of venomous animals (10). Leaves are crushed up to a paste with a little soil and water in Ivory Coast for rubbing into areas of rheumatic pains and for kidney pains (1), and a paste of leaves, pepper and citron is applied, with frequent renewal, to a guinea-worm sores (10). The pounded bark, sometimes mixed to a paste with a sort of white clay, is applied topically and a leaf-decoction taken internally for craw-craw (7). In Sierra Leone burnt leaves are ground up with clay for application to the body for scabies (13). Also in Sierra Leone the leaf in decoction is taken for malaria, and the bark along with that of *Zanthoxylum gilletii* (syn. *Fagara macrophylla*, Rutaceae) and the leaf of *Microdesmis puberula* (Pandaceae) and new black soil from on top of a termite heap is ground to a paste for paralysis of the limbs in topical massage (11). Fermented leaves, with rum and coconut, are taken by draught in Ghana for cough and weakness and in small doses for blackwater fever (7, 9).

Though the plant is very plainly pharmacologically active, there remains some doubt regarding the identity of the active principles. Sierra Leone material has shown alkaloids absent but *saponins* present (14). Ivory Coast material has not shown the presence of any simple active principle, neither alkaloid, saponin, nor bitters, but other work suggests the presence of toxic *triterpenic* substances of the *cucurbitacin* group (3, 10).

The wood is white, soft and perishable. It sinks when fresh, but later floats (4). The stems are commonly used for yam stakes.

Birds avoid eating the fruit.

References:

1. Adjanohoun & Aké Assi, 1972: 130. 2. Barker 1273, K. 3. Bouquet & Debray, 1974: 85. 4. Burtt Davy & Hoyle, 1937: 48. 5. Cooper 404, K. 6. Cooper & Record, 1931: 53. 7. Dalziel, 1937: 154–5. 8. Danniell s.n., 28/12/1855, K. 9. Irvine, 1961: 243–4. 10. Kerharo & Bouquet, 1950: 81–82. 11. Massaquoi, 1973. 12. Oliver, 1960: 30, 70. 13. Savill & Fox, 1967, 118–9. 14. Taylor-Smith, 1966,a. 15. Taylor-Smith & Chaytor, 1966. 16. Visser, 1975: 75–15: 49, as *M. spicata* Baill.

Margaritaria discoidea (Baill.) Webster

FWTA, ed. 2, 1: 387, as *Phyllanthus discoideus* (Baill.) Müll.-Arg. UPWTA, ed. 1, 156–7, as *P. discoideus* Müll.-Arg.

Trade: adjansi (Ivory Coast, Dalziel).

West African: SENEGAL BALANTA kifiro (JB) BANYUN kifiro (K&A) DIOLA bu hilèn (JB) é bondiõng (JB; K&A) DIOLA (Brin/Seleki) bufembèlo (K&A) DIOLA-FLUP bu ilèn (JB) buhilen (K&A) hu linga (JB) MANDING-BAMBARA bakoko (JB) bakolo (K&A) ńéńé (JB, & ex K&A) suruku (JB; K&A) suruku ńéńé (JB) MANINKA kéri (auctt.) NDUT sahrèr (JB) THE GAMBIA

EUPHORBIACEAE

FULA-PULAAR (The Gambia) keri (DAP) MANDING-MANDINKA bakoro(?) (DAP) jonkongkong(?) (DAP) WOLOF bakko(?) (DAP) jonkongkong (DAP) GUINEA MANDING-MANINKA dion kon-kon (CHOP) **SIERRA LEONE** BULOM (Sherbro) nɛŋkon-dɛ (FCD; S&F) FULA-PULAAR (Sierra Leone) bori (Jordan) keri (FCD) GOLA tido (FCD) tilo (FCD) tiro (FCD) KISSI cholondo (FCD; S&F) solondo (FCD; S&F) KONO tisoɛ (FCD; S&E) tusuɛ (FCD; S&F) KORANKO yɛgerɛ (S&F) LIMBA (Tonko) kumɛtɛ (FCD) LOKO čuanji (NWT) tihu (FCD; S&F) MANDING-MANDINKA bakɔŋkɔŋ (FCD) MENDE kɔŋgo-lijo(-i) (FCE; S&F) kongolo-tijoi (FWHM) ngongo-lijoi (S&F) tijo(-i) (auctt.) SUSU nɛtɛ (? mɛtɛ) (FCD) TEMNE an-, ka-saka (auctt.) VAI sawa (FCD) YALUNKA silamɛsi-na (FCD) **LIBERIA** KRU-BASA blar-du-gbah (C) MANO wã-yiddi (Har.; JMD) **MALI** ? surunku gniégnié (A. Chev.) FULA-PULAAR (Mali) kéri (Caille) MANDING-BAMBARA bakoko (Aub.) **UPPER VOLTA** MANDING-DYULA barambara (K&B) **IVORY COAST** ABE diom'bi (A.Chev. & ex K&B) lié (auctt.) mussan-hué (A.Chev. & ex K&B) rié (auctt.) ABURE adjansi (A.Chev.; K&B) AKYE bon (auctt.) diom'bi (A.Chev., ex K&B) mussan-hué (A.Chev. & ex K&B) ANYI pépéschia (Aub. & ex K&B) BAULE kwékwé sia (B&D) pétéhésia (Aub.; K&B) FULA-FULFULDE (Ivory Coast) kéri (Aub.) KRU-GUERE (Chiehn) glé (B&D) wiliniangré (B&D) zagrogramaï (B&D) KYAMA brakassa (A.Chev. & ex K&B) MANDING-BAMBARA bakoko (Bégué ex K&B) DYULA barambara (K&B) MANINKA bakoho (A&AA) bakoko (A&AA) kéri (Aub.) NZEMA pepezia (A.Chev. & ex K&B) SENUFO-TAGWANA katien (K&B) **GHANA** VULGAR pepea (DF) AKAN-ASANTE benkyi (auctt.) ɔpepea (auctt.) BRONG pepea (auctt.) FANTE nkuku (FRI) KWAWU pɛpea (E&A) TWI ɔpepea (E&A) WASA pepea (E&A) pepe-esia (BD&H; FRI) ANYI pɛpɛsea (FRI) ANYI-ANUFO pɛpɛsea (FRI) AOWIN pɛpɛsea (auctt.) GBE-VHE adzadze (Howes; FRI) NZEMA pɛpɛsea (FRI) **TOGO** TEM (Tschaudjo) kamfua (Volkens) kongonga (Volkens) YORUBA-IFE (TOGO) dantivi (Volkens) GBE-FON bafla (Aub.) **NIGERIA** EDO asẹ́ivin =the kernel is cracked (auctt.) IGBO isi m̃kpị = head of billy-goat (Kennedy; KO&S) URHOBO ololo (KO&S) YORUBA àṣàṣà (auctt.) àṣàṣà-odan (Ross) awẹ (Dodd) **WEST CAMEROONS** KPE wovengaenga (Maitland)

A tree reaching to over 30 m high by 1 m girth, of damper sites of the savanna and closed secondary forests, common throughout from Senegal to W Cameroons, and widespread elsewhere in tropical Africa.

The tree with its spreading crown is grown in villages on the Cameroon Mountain as a shade tree (21). It is one of the known hosts of *Armillaria heimii* Pegler, the causitive fungus of collar-crack disease of cacao, so that its presence on cacao plantations as a shade-bearer may not be desirable.

The bole is clear and unbuttressed. It can be sawn into planks and used for ordinary building purposes (6, 8–10). Sap-wood is yellowish, heart-wood pinkish-white to brownish-red, hard, heavy, of medium texture, not difficult to work and is suitable for cabinetry, finishing smoothly and taking a fine polish. It is not resistant to insects and decay (9, 10, 15, 24). It is cut for roof-shingles in Ghana (5) and chew-sticks (6), and in the Congo region it is used for boat-building as the wood is said to bend ideally after steaming for shaping ribs (10). It is a good firewood (10–12), and charcoal is made of it in Sierra Leone (24). The wood-ash yields a vegetable salt (15).

The bark is stringy and fibrous. It is commonly used in West Africa as a purgative (Senegal: 19; Sierra Leone: 10, 24; Liberia: 8, 10; Nigeria: 22), and is held to be anthelmintic in central Africa (28). It has anodynal properties. Dried and powdered it is rubbed on the body in febrile conditions (10, 22, 24), and in central Africa a bark-preparation is rubbed into scarifications as a stimulant and tonic (28). The Fula of Sierra Leone use the bark for toothache (16). In central Africa a decoction is used for the relief of post-partum pains, and bark-ash with local salt and palm-oil, which at the moment of application produces a burning sensation, is a topical embrocation for lumbar pain (28). In Congo (Brazzaville) a bark-decoction is used to relieve stomach and kidney complaints and to facilitate parturition (3). Treatment with a bark-decoction is said to bring about a more-or-less prolonged arrest of the menstrual cycle (3), though conversely in central Africa it is said to be taken to relieve amenorrhoea (28). In Malawi powdered bark-extract is applied to swellings and inflammation (15).

The roots are used as a purgative in Senegal, and are deemed to have stimulatory properties (18, 19). In Ivory Coast they are used as chew-sticks, and in this manner also be aphrodisiac (20). They are used in Guinea for stomach-ache (7) and are commonly applied in E Africa for treatment of internal troubles (15, 25), stomach-ache (26), and heart-pains (27). Root-decoctions are used in Tanganyika for schistosomiasis, gonorrhoea and hard abscesses, and, with leaf-sap added, for malaria (14).

The young leaves are fed by Fula to their cattle in Sierra Leone (16), and the foliage, flowers and fruit are reported to be appreciated by antelope in Nigeria (12). In some parts of Tanganyika it is said that the tree is specially planted to provide fodder (13).

An eye-wash is prepared in Ivory Coast–Upper Volta from a leaf-decoction (20). The leaves are said in Malawi to cure earache, and in Zaïre to be able to stanch wounds (15). Leaf and twig poultices are applied to ulcers caused by jiggers in Tanganyika (2, 28), and in Congo (Brazzaville) to furuncles and abcesses to hasten maturation (3). A leaf-decoction is taken in Ivory Coast for blennorrhoea and for poisoning, while a wash of the decoction is a stimulant in case of general fatigue (1). In Ubangi a decoction of roots and leafy twigs is also used for blennorrhoea (23). The young shoots make chew-sticks in Ghana (15, 30).

The flowers are fragrant and in E Africa are much visited by bees and other insects. The fruit is edible, and is eaten in West and East Africa. It is a relished food of the guinea-fowl and the francolin. Seed in Kenya have attracted the bushbuck (12).

Many alkaloids have been isolated from this plant, and of them *phylochrysine* and *securinine* are the most important. The former is a central nervous system stimulant which may account in part for the plant's stimulatory properties (4, 19, 29). The bark yields about 10% *tannin* (28).

References:

1. Adjanohoun & Aké Assi, 1972: 134, as *Phyllanthus discoideus* (Baill.) Müll.-Arg. 2. Bally, 1937: as *P. discoideus* Müll.-Arg. 3. Bouquet, 1969: 122, as *P. discoideus* (Baill.) Müll.-Arg. 4. Bouquet & Debray, 1974: 86, as *P. discoideus* (Baill.) Müll.-Arg., with references. 5. Brand 408, K. 6. Brand 924, K. 7. Chevalier, 1920: 560, as *Fluggea obovata* Baill. 8. Cooper 436, K. 9. Cooper & Record, 1931: 55, as *P. discoideus* Müll.-Arg., with timber characters. 10. Dalziel, 1937: 156–7. 11. Deighton 1624, K. 12. Eggeling & Dale, 1952: 136. 13. Greenway & Polhill 11724, K. 14. Haerdi, 1964: 95, as *P. discoideus* (Baill.) Müll.-Arg. 15. Irvine, 1961: 245–6, as *P. discoideus* Müll.-Arg. 16. Jordan 209, FBC, K. 17. Keay & al., 1960: 290., as *P. discoideus* (Baill.) Müll.-Arg. 18. Kerharo & Adam, 1963, b: as *P. discoideus* Müll.-Arg. 19. Kerharo & Adam, 1974: 424–7, as *P. discoideus* Müll.-Arg., with phytochemistry and pharmacology. 20. Kerharo & Bouquet, 1950: 83–84, as *P. discoideus* Müll.-Arg. 21. Maitland 1138, K. 22. Oliver, 1960: 34, 77, as *P. discoideus*. 23. Portères, s.d.: as *P. discoideus* Müll.-Arg. 24. Savill & Fox, 1967, 121–2, as *P. discoideus*. 25. Sensei FH.2916, K. 26. Tanner 2788, K. 27. Tanner 3744, K. 28. Watt & Breyer-Brandwijk, 1962: 426, as *P. discoideus* Müll.-Arg. 29. Willaman & Li, 1970: as *P. discoidcus* Müll.-Arg. 30. Portères, 1974: 133, as *P. discoideus* Müll.-Arg. See also: Oliver-Bever, 1983: 51, as *P. discoideus* (Baill.) Müll.-Arg.

Martretia quadricornis Beille

FWTA, ed. 2, 1: 372–3.
West African: SIERRA LEONE TEMNE ɛ-soɛnt (NWT)

A small forest tree of swampy sites from Sierra Leone, Ivory Coast and S Nigeria, and in Ubangi.

Micrococca mercurialis (Linn.) Benth.

FWTA, ed. 2, 1: 402.
West African: SENEGAL NDUT bôn (JB) SIERRA LEONE MENDE uja (NWT) TEMNE bɛramokil (NWT)

An annual herb becoming woody below, of open waste places and a weed of cultivation occurring throughout the West African region in the soudanian, guinean and forest zones and across Africa, Asia to northern Australia.

The plant is eaten as a vegetable in Gabon (3). There is no record of consumption in the West African region. A study of chromosome numbers shows a polyploidal sequence of 2n = 20 in the dry soudan zone of Senegal, 2n = 40 in the guinean zone of Ivory Coast and Cameroun, and 2n = 60 in the dense forest area of Gabon (2). It is postulated that this may be related to climatic factors, i.e. length of growing time, and accounts for the Gabonese and not those elsewhere using the plant as food.

In Congo (Brazzaville) the plant is used to treat children with fever, and the plant-sap is instilled into nose, eyes or ears to treat headache, filariasis of the eye or otitis respectively (1).

References:

1. Bouquet, 1969: 120. 2. Champault, 1970. 3. Sillans, 1953, b: 84.

Mildbraedia paniculata Pax ssp. **occidentalis** J Léonard

FWTA, ed. 2, 1: 397.
West African: IVORY COAST KRU-GUERE klawhéré (K&B)

A shrub to 3 m high, of the undergrowth of the dense forest of Liberia and Ivory Coast, and also in Zaïre.

Leaf-decoction is used in draughts and baths for serious cases of jaundice in Ivory Coast (1).

Reference:

1. Bouquet & Debray, 1974: 85.

Necepsia afzelii Prain

FWTA, ed. 2, 1: 405. UPWTA, ed. 1, 155.
West African: SIERRA LEONE LIMBA butoge (NWT) butugama (NWT) LOKO popama (NWT) MENDE joki (NWT & ex JMD) tɛvɛ gofai (FCD) tɛvɛ-kofei (FCD) SUSU burɛgɔendi (NWT) nyɛlɛlenyi (NWT) TEMNE kə-lɔt (NWT) LIBERIA KRU-BASA zah (C&R) IVORY COAST KRU-GREBO (Tepo) meshirié (JMD)

A shrub or tree to 14 m tall and bole to 30 cm girth, of the understorey of the closed-forest from Sierra Leone to W Cameroons, and in E Cameroun.

The wood is lemon-yellow, moderately hard and heavy resembling the temperate boxwood. It is easily worked, takes a fine polish, but is liable to decay and fungus stain. It has good qualities for turnery, and is also found suitable for rice-pestles in Liberia (1, 2). It is used for spoons and small mortars and pestles for grinding tobacco in Ivory Coast (3).

References:

1. Cooper 76, K. 2. Cooper & Record, 1931: 54. 3. Dalziel, 1937: 155.

Neoboutonia mannii Benth.

FWTA, ed. 2, 1: 404, incl. as *N. diaguissensis* Beille = *N. mannii* Benth. var. *diaguissensis* (Beille) Pax & K Hoffm., and as *N. glabrescens* Prain = *N. mannii* Benth. var. *glabrescens* (Prain) Pax & K Hoffm. UPWTA, ed. 1, 156, as *N. glabrescens* Prain.
West African: LIBERIA KRU-BASA gbo-ah (C&R) NIGERIA BOKYI keperu-mbang (KO&S) kepuru-mbang (Catterall ex JMD) EJAGHAM ejuraip (KO&S) OLULUMO (Ikom) kenpuru mbang (Catterall)

A tree of open spaces and secondary jungle of the forest zone, to 23 m high; of three varieties, occurring in Guinea (var. *diaguissensis*), Liberia (var. *glabrescens*) and S Nigeria and W Cameroons (vars. *mannii* and *glabrescens*) and extending into E Cameroun and Equatorial Guinea and the Congo area.

The wood is soft and bends easily. For this quality it is used to make casks for palm-oil in Liberia. It is also sawn into planks (var. *glabrescens*, 3).

Root-bark is taken as a cathartic purge in Congo (Brazzaville), pulped up or in decoction and always with palm-kernel [? *Elaeis*] or macerated tapioca root, but never is it given to pregnant women unless abortion is intended (1). No active principle has been detected (2).

References:

1. Bouquet, 1969: 121, as *N. africana* Müll.-Arg. 2. Bouquet, 1972: 26, as *N. africana* Pax. 3. Cooper 438, K.

Neoboutonia melleri (Müll.-Arg.) Prain var. **velutina** (Prain) Pax & K Hoffm.

FWTA, ed. 2, 1: 404, as *N. velutina* Prain. UPWTA, ed. 1, 156, as *N. velutina* Prain.
West African: **WEST CAMEROONS** BAMILEKE chako (Johnstone)

A shrub or small tree to 6 m high, of montane forest on stream-sides in W Cameroons, and also E Cameroun.

Oldfieldia africana Benth. & Hook. f.

FWTA, ed. 2, 1: 368. UPWTA, ed. 1, 156.
English: African oak.
French: chêne d'Afrique.
West African: **GUINEA-BISSAU** CRIOULO pau bicho *from* 'beech' (GeS) **SIERRA LEONE** KISSI tanalia (S&F) KONO mbaimba (S&F) KRIO oak (JMD) MENDE kpacha (DF ex Imp.Inst.) kpaola(-i) *from* kpao: bamboo-wine, lai: *leaf* (auctt.) TEMNE tortorza (AHU) ka-tɔsa (FCD; S&F) **LIBERIA** DAN rla (AGV; GK) KRU-BASA bawpoh (C) gauh (C) pau-lai (C ex AGV) saye (C&R, ex JMD) slah (C) slaugh (C) GUERE (Krahn) sain-tue (AGV) MENDE pau-lai (C&R) **IVORY COAST** VULGAR dantué *from Guere* (RS) fu *from Abe, the timber* (JMD) ABE fu (A.Chev.; K&B) AKYE ɛson (A.Chev.) fu (A.Chev. ex K&B) sué-ngoran (A.Chev.) ANYI ésui (A.Chev.; K&B) étui (A.Chev.; K&B) BAULE ahien'gré (A.Chev.) ·KRU· blahon (K&B) habétu (A&AA) hirahiré (K&B) KRU-GREBO hiérahiré (RS) hirahiré (Aub.) GREBO (Oubi) takuétu (A&AA) GUERE dantué (auctt.) dentué (auctt.) KYAMA anguaran (A.Chev.) MANDING-DYULA sséhiri (A.Chev.)

A large tree to over 30 m tall with a straight, clear unbuttressed bole to 17 m long by up to 5 m girth, occurring in the forested zone from Guinea-Bissau to Ivory Coast, and attaining its greatest stature in Sierra Leone.

The wood is very heavy, hard, tough and strong. Heart-wood is dark brown to reddish brown, sap-wood lighter. It is an excellent construction-timber though difficult to work when dry. It takes a good polish. It is durable in contact with water and has been used for dock-gates and bridges, and in naval construction for key-parts of ocean-going boats such as keelsons. At one time it was exported from Sierra Leone to Britain as a teak-substitute under the name of 'African teak', a name more appropriately applied to *Chlorophora excelsa* (Moraceae). *African oak*, or in the French translation *chêne d'Afrique*, is the preferable name. Nor should these be applied to *Lophira alata* (Ochnaceae), ambiguities which make for confusion. (4, 9, 10, 15–18). Though the timber appears to be less abundantly available from Sierra Leone than formerly, it remains an item of export from G. Bissau (11), Liberia (9, 18) and Ivory Coast (13). Other African uses are for sawing into planks, beams, joists and boxes (7–10), and when small for house-posts (6).

117

In Ivory Coast the tree is considered a powerful fetish tree with efficaceous medicinal virtues. The bark is added by medicine-men to prescriptions for the increase of potency. The 'Kru' add a decoction of bark to baths and draughts for blennorrhoea as a pelvic decongestant and Guere bathe the circumcision wound in a bark-decoction as an antiseptic and haemostatic followed by a dressing of powdered bark to hasten healing, and to treat sores (1–3, 13).

The bark-slash emits a slightly watery sap which is bitter (14). The bark boiled up with palm-oil into an ointment is used in Liberia to treat lice infestation (7). The leaves are similarly used on the head and loins for lice and crabs (5). The leaves are also bitter and in Sierra Leone are crushed and inserted into incisions on wine-palms (*Raphia* sp.) to drive off bees which are after the palm-wine (12), an use which perhaps explains the Mende name *kpao laï* bamboo wine leaf.

Seeds and bark are pounded together for use in Liberia as a pesticide (8), though the seeds, which are oily, are said to be a relished food of monkeys and other animals (15).

References:

1. Adjanohoun & Aké Assi, 1972: 132. 2. Aubréville, 1959: 2: 30–32. 3. Bouquet & Debray, 1974: 85. 4. Cole, 1968,a. 5. Cooper 88, K. 6. Cooper 111, K. 7. Cooper 295, 439, K. 8. Cooper 319, K. 9. Cooper & Record, 1931: 54–55, with timber characters. 10. Dalziel, 1937: 156. 11. Gomes e Sousa, 1930: 82. 12. Irvine, s.d. 13. Kerharo & Bouquet, 1950: 83. 14. Kunkel, 1965: 152. 15. Savill & Fox, 1967: 119–21. 16. Schnell, 1950, b: 222, 258. 17. Unwin, 1920: 26–27, 51, 67, 75. 18. Voorhoeve, 1965: 98.

Pedilanthus tithymaloides (Linn.) Poit.

FWTA, ed. 2, 1: 368.

French: herb à cors; herbe à bordures; ipéca de S. Domingue (Berhaut).

A semi-succulent shrubby plant to over 1 m high, native of West Indies and dispersed by man throughout the tropics, and occurring as a garden ornamental in the West African region.

It is commonly grown as a hedge. The plant contains a milky sap which is used in various countries on warts and corns (2). The root is a powerful emetic, and is so used in the West Indies. The leaves arranged in rows like the legs of a centipede result in the use of the sap in Malaya for treatment of centipede bite and scorpion stings, and also for skin diseases (1).

References:

1. Burkill, IH, 1935: 1683. 2. Hartwell, 1969,b: 168.

Phyllanthus acidus (Linn.) Skeels

FWTA, ed. 2, 1: 388.

English: Otaheite gooseberry; Indian gooseberry (Keay).

French: surelle; cerisier de Tahiti; surette de la Martinique (Berhaut).
West African: SENEGAL CRIOULO azedina (JB) GUINEA-BISSAU CRIOULO azedinha (JDES) GHANA HAUSA dunyan (BD&H FRI)

A shrub or small tree, frequently erroneously ascribed to Indian, Madagascan or Malesian origin, but correctly of the coastal region of NE Brazil (3), whence it has been dispersed to the Caribbean and to countries of the Indic basin, and more lately to West Africa.

The fruits are fleshy, acid and astringent. In various countries they are pickled or made into preserves. The root is an active purgative and the seeds are cathartic (1). Fruit-juice is used in Ghana as an eye-instillation for eye-troubles (2).

References:

1. Bailey, 1901: 1318, as *P. distichus* Muell. 2. Irvine, 1961: 248, as *P. wildemannii* Beille. 3. Webster, 1957: 68.

Phyllanthus alpestris Beille

FWTA, ed. 2, 1: 387.
West African: SIERRA LEONE MENDE juki (NWT) TEMNE ɛ-tɔŋwaŋ (NWT)

A shrub to nearly 2 m high of upland areas from Guinea to Ivory Coast.

Phyllanthus amarus Schum. & Thonn.

FWTA, ed. 2, 1: 387. UPWTA, ed. 1, 156.
West African: IVORY COAST ABURE sambéfa (B&D) AKAN-ASANTE féfiama (B&D) AKYE *m*pégui baté (A&AA) MANDING-MANINKA kodrè (A&AA) NIGERIA EFIK óyómókésó ámànké èdèm (AJC) HAUSA geérón-tsúntsaàyeé = *bird's millet* (JMD) IGBO ub'akofe (NWT) IGBO (Asaba) buchi oro (NWT) IGBO (Umuahia) ngwu (AJC) YORUBA dobisowo (Dawodu, & ex JMD) èhin olobe (auctt.) eyín-olobe (Macgregor; JMD)

An herb, common from Sierra Leone to S Nigeria and Fernando Po, and widespread elsewhere in tropical Africa. A weed of cultivated land and in waste spaces. It is said to have sand-binding properties (7).
It is a plant of general medicinal application. In Yorubaland it features in an incantation 'against disease' (9). It is an ingredient of the *agbo* prescription in Lagos (6). An infusion of leaves is used in the Ibadan area for haemorrhoids (8). A similar use with the addition of citron juice and powdered fruits of *Xylopia aethiopica* (Annonaceae) is made by Maninka in Ivory Coast (1). The latter people also reduce the plant to a paste and prepare vaginal suppositories to induce menstruation, and the powdered plant administered by enema to infants is considered a poison antidote by the Akye (1). The plant is also used in Ivory Coast in difficult childbirth, for oedemas and to counter costal and fever pains and sore-throat (5). In Ubangi a decoction of the whole plant is taken for blennorrhoea (10), and in Trinidad for oliguria and venereal diseases (11). In Congo (Brazzaville) sap is taken in draught for costal pains, tachycardia, blennorrhoea and female sterility, by ear-instillation for otitis, and by topical application to maturate furuncles and abcesses (4). A cold leaf-infusion is taken in Kenya to relieve stomach-pains (3).
Leaves rubbed between the hands give a lather (2).
The ripe fruit and seeds are considered vermifugal in Congo (Brazzaville) (4).
The plant has a number of magical uses in Ivory Coast (5).

References:

1. Adjanohoun & Aké Assi, 1972: 133. 2. Bally 4828, K. 3. Bally 6056, K. 4. Bouquet, 1969: 121. 5. Bouquet & Debray, 1974: 85. 6. Dalziel, 1937: 156. 7. Irvine, 1930: 335. 8. Lowe 2212, UCI. 9. Verger, 1967: no. 91. 10. Vergiat, 1970: 77. 11. Wong, 1976: 130.

Phyllanthus beillei Hutch.

FWTA, ed. 2, 1: 388.
West African: SENEGAL DIOLA bu rèm a dita (JB) DIOLA-FLUP bu sēngsēng amata (JB)

A shrub to 3 m tall, of foothill savanna in a few localities from Senegal to S Nigeria, and the Cameroun Republic and E Africa.

No use is recorded within the Region. The root is reported used in Tanganyika as an aphrodisiac (1).

Reference:

1. Koritschoner 961, K.

Phyllanthus capillaris Schum. & Thonn.

FWTA, ed. 2, 1: 387.
West African: SIERRA LEONE KISSI bokpone (FCD) LOKO kbagɔ (NWT) GHANA AKAN-TWI bɔmagu wakyi (E&A) NIGERIA IGBO (Agukwu) kpukpu kwonwa nagu (NWT) YORUBA iranje (Phillips)

A shrub to 1.70 m high, of the grassy savanna from Guinea to W Cameroons and Fernando Po, and widespread elsewhere in tropical Africa.

No usage is recorded for the West African region. In Kenya a decoction of the whole plant is taken as a remedy for vomiting (2), and crushed roots are eaten for stomach-ache (1).

References:

1. Broadhurst-Hill 560, K. 2. Magogo & Glover 1032/A, K.

Phyllanthus fraternus Webster

FWTA, ed. 2, 1: 388, as *P. niruri* sensu Keay. UPWTA, ed. 2, 157, as *P. niruri* sensu Dalziel.

French: herbe au chagrin (Berhaut).
West African: SENEGAL FULA-PULAAR (Senegal) lébèl (K&A; JB) WOLOF ngégian (JB) ngètsal (JB) IVORY COAST AKAN-ASANTE bomaguenti (B&D) BAULE sumaguéssi (B&D) GHANA ADANGME-KROBO ofɔbiokpab (JMD) AKAN-ASANTE bo mma gu w'akyi = *to carry the children behind, or on the back* (FRI; E&A) TWI bɔ mma gu w'akyi = *to carry the children behind, or on the back* (FRI) GA ombatoatshi (JMD) NIGERIA EFIK óyómó-kẹ-isó-ámàn-kẹ-èdèm (JMD) NUPE ebó zunmaggi (JMD) ebógi (JMD) YORUBA ẹ̀hìn olobe (JRA; Verger) yoloba *probably an abbreviation of* eyín olobe (Easmon)

An herbaceous weed of roadsides, cultivated land, waste places of the forest and savanna, generally rare in Senegal, Ivory Coast, Ghana and Nigeria. It is widely distributed in Asia and in the West Indies. It is probably native to Western India and Pakistan.

The curious way in which the flowers and fruit are borne on the lower side of the leaf-stalk results in Ghanaian vernacular names likening the plant to a child carried pick-a-back (5, 7), a feature noted also in Malay names (3).

The whole plant is very bitter due to the presence of the alkaloid *phyllanthin*. Alkaloids *hypophyllanthin* and *quercetin* are also present amongst a number of active substances (2, 8, 9, 12). Plant-extracts are poisonous to fish and frogs (3, 8, 10, 13). *Saponin* is present but, however, no record has been seen of the plant's use as a fish-poison, though very strong molluscicidal activity is reported from the root of Sudanian material – a petrol-ether extract giving 100% mortality of fresh-water snails, *Bulinus* and *Biomphalaris*, at a concentration of 25 ppm (12). Notwithstanding the bitter taste all stock are said to browse it in Senegal, at Lake Chad (1), in Kordofan (11) and elsewhere (5). It has been cited as a fodder for horses in Java (5).

At one time a plant-extract was officially used in the Government hospital at Accra to allay griping in dysentery, and other painful spasmodic affections of the intestines (5, 7). It found use in cases where the administration of opium and morphia was not advisable. The leaves are rich in potassium by which they possess a strongly diuretic property. It is also given for dysentery in Kordofan (11). The pounded leaves are taken in Ghana for gonorrhoea and (unspecified) parts of the plant are used for constipation (7). In Ivory Coast it is used in cases

EUPHORBIACEAE

of difficult childbirth, for costal and fever pains, sore-throat and oedemas (2). In India, Malesia and the Philippines it has a miscellaneous usage for stomach-troubles, dropsy and urino-genital diseases. Fresh roots are said to be beneficial in jaundice and are taken with milk as a galactogogue. Poultices are applied to caterpillar urticarias, skin-complaints, oedematous swellings and ulcers; the latex to sores and ulcers, and mixed with oil, for ophthalmias; young leaves for cough in children, hiccups, etc. (3, 9, 10). Young leaves have been used in India for mild forms of intermittant fever. In parts of the West Indies the plant has a reputation as a remedy for malarial fever (5). Examination of plant extracts on experimental malaria showed no activity (4, 8). Antidiabetic substances have been recorded in Indian material (8).

A black dye is obtained from the stems and leaves which is used for cotton cloth and to make ink in E Africa (6), India (9) and in Indonesia (3).
References as *P. niruri* Linn.:

1. Adam, 1966, a. 2. Bouquet & Debray, 1974: 85. 3. Burkill, IH, 1935: 1718–9. 4. Claude & al., 1947. 5. Dalziel, 1937: 157. 6. Greenway, 1941. 7. Irvine, 1930: 337. 8. Kerharo & Adam, 1974: 427–8. 9. Krishnamurthi, 1969: 34–35. 10. Quisumbing, 1951: 525–7. 11. Baumer, 1975: 110. 12. Ahmed & al., 1984: 74–77. 13. Wong, 1976: 130. See also: Oliver-Bever, 1983: 52.

Phyllanthus maderaspatensis Linn.

FWTA, ed. 2, 1: 388.

A herb or woody undershrub to over 1 m high, of grassy savanna, and a weed of cultivated places. Recorded from limited localities in Senegal, Ivory Coast, Ghana, Togo and N Nigeria; occurring in E Africa and widely dispersed in tropics and subtropics of the Old World.

No use is recorded within the Region. In Somalia it is recognised as poisonous and if taken causes pain in the stomach (4). Cattle in Kordofan will browse it a little but only while still green (7). The whole plant is pounded and boiled for washing areas of the body affected by scabies in Tanganyika (5). Roots are boiled and the liquor is taken for constipation (6), and they are used for snake-bite treatment (1). Smoke from the burning plant is used in Kenya to kill caterpillars boring into maize cobs (3). In India the leaves in infusion are used for headache (2).

The seeds are said to have laxative, carminative and diuretic properties. A clear deep yellow oil can be extracted from them, which contains *myristic, palmitic, stearic, oleic, linoleic* and *linolenic* acids and *a-sitosterol.* The defatted seed-cake contains a fibrous mucilage which can be hydrolysed to *galactose, arabinose* and *rhamnose* and *aldobionic acid.* A reddish brown colouring matter, *maderin,* and an essential oil are reported present (2).

References:

1. Koritschoner 1030, K. 2. Krishnamurthi, 1969: 35, with references. 3. Magogo & Glover 1057, K. 4. Peck 71, K. 5. Tanner 1431, K. 6. Tanner 3416, K. 7. Baumer, 1975: 110.

Phyllanthus muellerianus (O Ktze) Exell

FWTA, ed. 2, 1: 385. UPWTA, ed. 1, 157, as *P. floribundus* Müll.-Arg.; and 158, as *P. wildemannii* sensu Dalziel.
West African: SENEGAL FULA-PULAAR (Senegal) mama momoti (JB) GUINEA-BISSAU FULA-PULAAR (Guinea-Bissau) mámámómóti (DF) GUINEA MANDING-MANINKA tri (CHOP) SIERRA LEONE KISSI hɛlichuinda (FCD) hɛlifinda (FRI) KONO kundunɔmɔɛ (FCD) LIMBA wiɛtɛlɛ (NWT) LOKO lapoka (FCD) mapɔka (NWT) sibera (NWT) woniabunka (NWT) MANDING-MANDINKA furuka (NWT) MENDE naojakwe (NWT; JMD) ngwoyɔ-jɔkwi (JMD) nɔni-wulo-jalakɔ (FCD; JMD) nyoiwojakoi (SKS) woɛjakoe (NWT) wonowulojalakoi (FCD) SUSU bɛlakhɔ-suri-suri (FCD) surdsuri (NWT) TEMNE ɛ-gbɛlɛ (auctt.) UPPER VOLTA MAND-ING-DYULA suca koma (K&B) sunum burossona (K&B) IVORY COAST ABURE ayé essi (B&D) tavéti (B&D) ADYUKRU nifi atier (Aub. & ex K&B) AKAN-ASANTE aobé (B&D) awabé

(K&B) boidien (B&D) dinigo (B&D) guédié (B&D) AKYE awotchin (A&AA) ANYI agiépra (K&B) agigié (K&B) BAULE abakadié (B&D) akuakodié (K&B) kréguiéguié (K&B) GAGU bubé gbo (K&B) 'KRU' didie (K&B) KRU-BETE bélilikmé (K&B) bluméï (K&B) GREBO (Oubi) ulidilé (A&AA) GUERE badua (K&B) gbaruho (A&AA) GUERE (Chiehn) uriagré (B&D) urignagre (K&B) wuliniagié (K&B) KWENI boduchié (K&B) diare soguesu (K&B) vonivraurè (B&D) KYAMA dodo (B&D) kotien fienfie (B&D) 'MAHO' dokuani juan (K&B) MANDING-DYULA suca koma (K&B) sunum burossona (K&B) MANINKA diokomanzu (B&D) duokonin nidsu (B&D) gnoro dagba (A&AA) kaagbwé (B&D) SENUFO-NIAGHAFOLO tobakadavi (K&B) TAGWANA kogo kadiéré (K&B) **GHANA** ADANGME-KROBO potopole-boblo *after* potopole: *a bird which eats the fruits* (auctt.) AKAN-ASANTE awobɛ (auctt.) TWI awobɛ (auct.) **TOGO** TEM (Tschaudjo) irubre-imbre (Gaisser) **NIGERIA** EDO igbehẹn = *thorns of a fish* (Ross; Elugbe) EFIK ñkañá (JMD) HAUSA kumci = *a thicket; from the habit* (JMD; ZOG) májíríyár kúrumii = 'majiriya' *[Erythrina] of the forest* (JMD; ZOG) IGBO (Idumuje) anya nnụnụ = *bird's eye* (NWT) ẹgu eza (NWT) IGBO (Uburubu) anya nnụnụ = *bird's eye* (NWT) egu eza (NWT) ogo azu (NWT) TIV yargum (JMD) YORUBA arunjeran (auctt.) arunyeran (JRA)

A shrub or climber, occasionally arborescent, of deciduous and secondary forests from Guinea-Bissau and Mali to West Cameroons and Fernando Po, and widespread in other areas of tropical Africa.

The stem seldom becomes large. In Bendel State of Nigeria it is reported as a weed of rice-fields (22), plainly by lack of timely cultivation. In Kenya it is said to yield excellent firewood if not over 15 cm diameter: any stems over that become hollow (9). A clear potable water may be obtained from the cut stem (Migoed fide 6) and this sap is used in Sierra Leone (7, 8) to relieve ophthalmia, and in Nigeria (2, 14) for pain in the eyes, or to remove a foreign body (6). A piece of twig is sucked in Ivory Coast–Upper Volta to prevent toothache (13), and in Nigeria twigs are used as chew-sticks after removal of the surface spines (21).

The bark yields a brown dye which is used in E Africa for dyeing mats (10) and fishing-lines (9). From the whole plant a black dye is obtained. The bark is sometimes added to palm-wine to render it strongly intoxicating, which amongst some tribes of the Cameroons may give rise to a sort of frenzy (Mildbraed fide 6). In Congo (Brazzaville) dried bark-powder is taken for colds and sinusitis (4). In Nigeria the root-bark in decoction is given as an alterative (2), and is used in Ubangi on swellings (20).

The roots are widely used for intestinal troubles. The root is cooked with maize meal for severe dysentery in Ghana (6, 12), or with other drug plants (23). Powdered root-charcoal with palm-oil is taken in Congo (Brazzaville) for stomach-upsets and as an antemetic (4). In Tanganyika water in which the roots have been pounded is drunk for diarrhoea (17), or have been boiled is given as an enema for stomach-pains (16). In Nigeria a root-decoction is used as a febrifuge (2), and an infusion of roots and leaves is given to children in Togo suffering from eruptive fevers (Kersting fide 6). A decoction is taken in draught and made into a vapour-bath on the exposed genitals as a pelvic decongestant, and by draught for anaemia in Ivory Coast–Upper Volta (13). The powdered roots are used in Ivory Coast–Upper Volta as a snuff together with a bark-decoction in draught and enema for throat-troubles with glandular fevers (13). In Nigeria the young root, with young leafy twigs, is given for jaundice (2) and as a mild purgative and to treat dysentery and urethral discharges (6, 15). In Tanganyika a root-decoction is taken by draught for hard abscesses, and the powered dried root and bark is sprinkled on wounds as a dressing (11).

Leaves are an occasional supplement cooked with food or in soup in Sierra Leone and S Nigeria (6, 18). In Ghana (12) and in Nigeria (2) leaves boiled with the palm fruit are given as soup to women after delivery. This preparation is also a general tonic. In Ivory Coast the leaves are eaten in the belief that they promote male fertility (1), and a leaf-decoction is taken in Congo (Brazzaville) for anaemia (4). An infusion of young shoots is said to be a remedy for chronic dysentery (2) and for chest-complaints (6), and in Sierra Leone a leaf-decoction for constipation (7). Freshly pounded leaves are a wound-dressing which is commonly used in the Region (2, 6), and leaf-sap is used as a wash for fevers and skin-eruptions (6). Leaf-sap is a widely used instillation for eye-troubles (1, 5, 6, 12, 13, 15, 20), with in Ivory Coast–Upper Volta the leaves made into an eye-pad on the lids (13). Leafy twigs prepared with a pulp are rubbed topically on the body in Ivory Coast–Upper Volta to cure paralysis (13). A leaf-decoction serves in Ubangi as a mouth-wash for toothache after which the cooked leaves are applied to the gums (20).

For cough, the root in Ghana is cut into small pieces with those of *Psychotria calva* (Rubiaceae) and *Harrisonia abyssinica* (Simaroubaceae) and decocted and the liquid drunk (23). This prescription is also given for whooping-cough.

An infusion of the flowers is cooling and gently aperient (2).

The fruits are edible and are eaten by some people. The pulp provides a hair-fixative used in Ubangi (3, 20).

Analyses have shown no alkaloids, but the presence of non-toxic *tannins* (13, 18).

The plant has superstitious attributes. In Sierra Leone it is held to be taboo to the Ture clan of the Temne (NW Thomas fide 6). In parts of Sierra Leone the leaves, mixed with cow-dung, are rubbed on the floor of huts (6). In Gabon it is believed to have the power to lift taboos and ritual interdictions (19).

References:

1. Adjanohoun & Aké Assi, 1972: 135. 2. Ainslie, 1937: sp. no. 272, as *P. floribundus*. 3. Aubréville, 1950: 189, as *P. floribundus* Müll.-Arg. 4. Bouquet, 1969: 122. 5. Bouquet & Debray, 1974: 86. 6. Dalziel, 1937: 157–8. 7. Deighton 529, K. 8. Deighton 1097, K. 9. Graham 1666, K. 10. Greenway, 1941: as *P. floribundus* Müll.-Arg. 11. Haerdi, 1964: 96. 12. Irvine, 1961: 246–7. 13. Kerharo & Bouquet, 1950: 84, as *P. floribundus* Müll.-Arg. 14. Macgregor 159, K. 15. Oliver, 1960: 76–77. 16. Pirozynski P.18, K. 17. Tanner 3022, K. 18. Thomas NWT.2167 (Nig. Ser.), K. 19. Walker & Sillans, 1961: 174, as *P. floribundus* Müll.-Arg. 20. Vergiat, 1970: 78, as *P. floribundus* Müll.-Arg. 21. Isawumi, 1978,c: 168. 22. Gill & Ene, 1978. 23. Ampofo, 1983: 17, 24.

Phyllanthus niruroides Müll.-Arg.

FWTA, ed. 2, 1: 387.
West African: **GUINEA-BISSAU** FULA-PULAAR (Guinea-Bissau) bubúnguel (EPdS) **SIERRA LEONE** MENDE ɛreboe (NWT) **IVORY COAST** AKAN-ASANTE bomaguaki (K&B) pamagwèké (B&D) BRONG bomaguaki (K&B; E&A) ANYI bomagua kéné (K&B) BAULE sugniassi (K&B) sumaguéssi (B&D) sumasi = *seeds in the back* (K&B) DAN ti (K&B) ·KRU· siaho (K&B) KRU-GUERE tienwé (K&B) GUERE (Chiehn) lolo (K&B) KULANGO lumbodiataka (K&B) KWENI alaman kridjian (K&B) m'bli (K&B)

A semi-woody herb to 60 cm tall occurring from Guinea to S Nigeria, and in E Cameroun to Zimbabwe, often a weed of cultivated land and in waste places.

In Ivory Coast it has a special place in the pharmacopoeia and as a fetish plant. A decoction of the whole plant in draught is given as a diuretic and purgative in jaundice, and with pimento added it is used in enema for slightly uncomfortable menses. Sap from the pulped plant is taken in draught with citron juice and by enema to facilitate child-birth. Pulp is rubbed on and the sap is taken by mouth in cases of bronchitis in infants. Sap is instilled in the ears for otitis, and it may have expectorant soothing properties in treating cough (3). It is used also for costal pain, fever pain, oedemas and sore-throat (1). In Tanganyika leaf-sap is rubbed on specially for stomatitis, and is an ingredient of a medicine for hookworm (2).

The plant is put into many superstitious formulae used by snake-charmers in Ivory Coast. It is an ingredient of a medicinal nostrum sprinkled about houses during small-pox epidemics. Medicine-men use a 7-days macerate of the leaves to wash their faces to improve business (3).

References:

1. Bouquet & Debray, 1974: 85. 2. Haerdi, 1964: 96. 3. Kerharo & Bouquet, 1950: 84–85.

EUPHORBIACEAE

Phyllanthus odontadenius Müll.-Arg.

FWTA, ed. 2, 1: 388.

West African: GUINEA-BISSAU FULA-PULAAR (Guinea-Bissau) bubúnguel (JDES) SIERRA LEONE LIMBA koyiŋk (NWT) GHANA ADANGME-KROBO oforbiokpabi = *you carry on your own back* (ASThomas) AKAN-TWI bɔ mma gu w'akwi = *carry children behind; alluding to the flowers on the back of the leaves as children are carried on their mother's back* (FRI; E&A) NIGERIA IGBO ŋwileʲi (NWT) IJO-IZON (Kolokuma) eré ófóníbúɔ = *female fowls legs* (KW&T)

A sub-woody herb to 1 m high, a common weed of the forested area from Guinea-Bissau to W Cameroons and Fernando Po, and to Sudan and Angola.

The way in which the flowers and fruits are borne on the underside of the leaves gives rise to the Ghanaian names likening the plant to a child carried pick-a-back.

The leaves are chewed with guinea-grains in Ghana to cure hiccough (1).

The Ijo of the Niger Delta have a superstitious use of the plant to drive away bad spirits: a person who suffers fever every evening which is attributed to the spirit of a dead person should urinate on the plant, pick the leaves for addition to a bath with local soap, and be relieved of the spirits (2).

References:

1. Thomas, AS D.196, K. 2. Williamson, K 65, UCI.

Phyllanthus pentandrus Schum. & Thonn.

FWTA, ed. 2, 1: 387. UPWTA, ed. 1, 158.

West African: SENEGAL ARABIC (Senegal) tenèchmarit (JB) FULA-PULAAR (Senegal) lébèl (JB) lékon féro (JB) tamalel (K&A) tiamalèl (JB) varbubé (JB) MANDING-BAMBARA bô-diara (JB; K&A) kokonin dolo (K&A) NDUT garga hélèl (JB) SERER mbof rôg (JB) médin (JB) ndimatèl (JB; K&A) ngodiil nô basil (JB; K&A) omboy (K&A) sombar (K&A) sumbānd o mboy (JB) sumbar o mboy (JB) WOLOF garab si tiar (JB) garap si tiaw (K&A) neutélé (K&A) THE GAMBIA MANDING-MANDINKA dakongo (Fox) wulakonakaro (Fox) NIGERIA HAUSA farin chap (Grove) geérón ítàceé (JMD ZOG) géérón tsúntsaàyéé = *bird's millet* (JMD; Z0G) geéroó (JMD; ZOG) hátsín tsúntsààyéé = *bird's corn* (JMD) IGBO (Agukwu) igweligwe (NWT; JMD) IGBO (Awka) nsi-ngwesu (NWT; JMD) YORUBA èhin olobe (JRA)

A woody herb to about 50 cm tall, of dry sandy savanna of the Sahel and soudanian regions, often in cultivated land.

A decoction of the whole plant is taken in Senegal internally for syphilis and is used externally on syphilitic chancres (5, 6). In N Nigeria a decoction taken internally is believed to relieve earache, and a watery extract of the ashes obtained from burning it is also instilled in the ear (4). Root, leaves and tender shoots are used in Nigeria as a deobstruent and diuretic; the roots and leaves, generally in powder or decoction, are given for over-secretion of bile and jaundice, and an infusion of young shoots for chronic dysentery (2). Root and leaf-decoctions are used in Nigeria for fever; they are said to have a good tonic effect when taken repeatedly (2). Test of extracts of the whole plant for effective action on experimental malaria have proved negative (3, 6, 7). An infusion made from bruised leaves is given in Nigeria for gonorrhoea and a leaf-decoction as a stomachic for colic and all abdominal pains, and for dysentery and dropsy (2). A root-macerate is taken in draught in Senegal for sterility (6). In southern Africa sap squeezed from the leaf is put into a new-born baby's mouth in the belief that it aids suckling, and to act as a mild purgative (7).

All stock is reported to graze the plant in Senegal. It is not taken in the Chad area (1). The seeds are eaten by birds (4).

References:

1. Adam, 1966, a. 2. Ainslie, 1937: sp. no. 273. 3. Claude & al., 1947. 4. Dalziel, 1937: 158. 5. Kerharo & Adam, 1962. 6. Kerharo & Adam, 1974: 428. 7. Watt & Breyer-Brandwijk, 1962: 427.

Phyllanthus petraeus A Chev.

FWTA, ed. 2, 1: 388.
West African: SIERRA LEONE LOKO purindi (NWT) MENDE yambava (NWT)

A shrub to over 2 m high of damp sites by streams from Guinea to Liberia.

Phyllanthus physocarpus Müll.-Arg.

West African: NIGERIA EDO nowate (Dennett)

A rare straggling tree to 8 m high recorded from Príncipe and once near Benin City.

Phyllanthus reticulatus Poir.

FWTA, ed. 2, 1: 387. UPWTA, ed. 1, 158.
English: 'sour grapes' (the fruit, Freetown market, Sierra Leone, Morton & Anderson).
West African: SENEGAL DIOLA dibiribi (JB) dibribi (JB) FULA-PULAAR (Senegal) dibribi (JB) MANDING-BAMBARA balã balã (JB) MANKANYA be gof (JB) NDUT bolopan (JB) tiabèl (JB) 'SAFEN' susan (JB) SERER diankus (JB) ñamb'in (JB) ñamd'in (JB) ngubor fa mbé (JB) sikidi (JB) WOLOF firfiron (JB) gérhay (JB) susal (JB) SIERRA LEONE KORANKO filakoke (NWT) MENDE nwɔniwulo-jaakoei (FCD) tacumbui (NWT) TEMNE ɛ-gbɛlɛ (FCD; JMD) a-timpoes (NWT) MALI ? turière gô (Aub.) GHANA AKAN-FANTE nkokobro (auctt.) TWI awobɛ (auctt.) NIGERIA HAUSA alambu (JMD; ZOG) alambu na tudu *i.e., of the dry ground, as opposed to the swamp* (JMD; ZOG) báƙin alambu = *black 'alambu'* (JMD; ZOG) kalambu, ƙalambu (JMD; ZOG) YORUBA iranje (JMD)

A shrub to 3 m high, of the savanna forest, often on river-banks, throughout the Region from Senegal to N and S Nigeria, and widespread elsewhere in tropical Africa. The two varieties recognised in West Africa, var. *reticulatus* and var. *glober* Müll.-Arg., are treated together here.
The wood is said to be very hard and is used by some peoples in E Africa for the threshing flail (6). Stems are used in N Nigeria as roof-binders in conical huts (6). Twigs are used as chew-sticks in Ghana (12, 13), as chew-sticks and toothbrushes in Kenya (16) and Tanganyika (Dragendorff fide 25), and as toothbrushes in southern Africa (24). The wood is suitable for tinder and in Tanganyika is fixed into a fire-drill (8). Sap from the stem is blown into the eyes to cure soreness in Ghana (13).
The foliage and young shoots are browsed by all stock in Kenya (18) and Tanganyika (24). In Ghana a soup made of leaves boiled with palm fruits is given to Asante women after child-birth (6). The leaves and bark are reputed to be diuretic and cooling in Sudan (3) and southern Africa (24). The powdered leaf is used in South Africa for topical application to sores, including venereal sores, burns, suppurations and skin-chafes (24). In Tanganyika mashed leaves are rubbed over the body of a malaria patient (11). The powdered leaves are compounded with cubebs and camphor in India into tablets for sucking for bleeding gums (25).
The fruits are edible. They are traded in Sierra Leone Freetown market as *sour grapes* (17) and may be occasionally eaten in E Africa (Kenya, 2), but perhaps serve only as an emergency food (6). The fruit and root have been recorded as used criminally in Tanganyika (24).
The plant can become an invasive weed of cultivated land. It is reported in Gabon to be toxic to poultry (22, 23), but the part(s) of the plant is not indicated. Birds in Tanganyika are reported to eat the ripe fruit (4).
A red or black dye is obtained from the bark and roots which is used in Sudan (3) and E Africa for dyeing fishing lines (1, 5, 7, 9, 10, 24, 26).

The root is purgative (14) and has a variety of uses in Tanganyika. A decoction enters into hookworm medicine (15), and water in which the root has been boiled is taken as a male aphrodisiac (21), to increase fertility (2), to treat headache (19), for dysmenorrhoea (11), for hard abscesses (11) and with the leaf-sap as an antispastic (11). In Zanzibar the plant is considered a remedy for anaemia and intestinal haemorrhage (24).

References:

1. Bally 4706, K. 2. Bally 5790, K. 3. Broun & Massey, 1929: 143–4. 4. Bullock 29, K. 5. Dale & Greenway, 1961: 215. 6. Dalziel, 1937: 158. 7. Donald 65, K. 8. Gillett 17938, K. 9. Greenway 5099, K. 10. Greenway, 1941: sp. no. 197. 11. Haerdi, 1964: 96. 12. Irvine 76, K. 13. Irvine, 1961: 247–8. 14. Koritschoner 1453, K. 15. Koritschoner 1642, K. 16. Mathew 6313, K. 17. Morton & Anderson SL.1163, FBC. 18. Mwangangi & Gwynne 1147, K. 19. Tanner 1478, K. 20. Tanner 3200, K. 21. Tanner 3720, K. 22. Walker, 1953, a: 37. 23. Walker & Sillans, 1961: 174. 24. Watt & Breyer-Brandwijk, 1962: 427. 25. Portères, 1974: 133. 26. Weiss, 1973: 184.

Phyllanthus rotundifolius Klein

FWTA, ed. 2, 1: 388.
West African: SENEGAL ARABIC (Senegal) bedrine (JB) FULA-PULAAR (Senegal) kodiolel Allah (JB) kodioloñ Allah (JB)

An annual herb to about 40 cm high, found across the West African Sahel, particularly in proximity to rivers, and extending on to India.
The plant seems to have no recorded use, though the Fula name suggests belief in Divine providence. Cattle in Senegal do not graze it (1).

Reference:

1. Adam, 1966,a.

Phyllanthus sublanatus Schum. & Thonn.

FWTA, ed. 2, 1: 388.
West African: GHANA GA áwò báňtuéntsi (KD) NIGERIA IGBO nwarigu (NWT)

A semi-woody herb to 50 cm high, widespread from Mali to S Nigeria. A weed of cultivation, and found in Sierra Leone invading cultivated swampland (1).

Reference:

1. Deighton 5509, K.

Phyllanthus urinaria Linn.

FWTA, ed. 2, 1: 388.

A small sub-woody herb occurring in restricted distribution as a weed of habitation in Sierra Leone, Ghana and S Nigeria, and rarely elsewhere in Africa but widely distributed in Asia to Australia.
No usage is recorded in the Region. The plant has unspecified medicinal use in Mozambique. In India it has been used as a fish-poison, and in India, Indo-China, Malaya and Indonesia it has a miscellaneous use for jaundice, diarrhoea, dysentery, dropsy, gonorrhoea, menorrhoea, children's cough and appetiser, kidney-trouble, syphilis, tumours, diaphoresis, abortion, etc. (1–5).
In Australia chewed or soaked leaves are rubbed into scarifications on the forehead to relieve 'congestion in the head' (5).

126

Insignificant biotic action on *Escherichia coli* by the plant stem has been recorded (5). A bitter principle and an *alkaloid* have been recorded present (4).

References:

1. Beille, 1927: 5, 586. 2. Burkill, IH, 1935: 1718–9. 3. Hartwell, 1969,b: 169. 4. Krishnamurthi 1969: 36. 5. Watt & Breyer-Brandwijk, 1962: 427.

Phyllanthus spp. indet.

West African: IVORY COAST GAGU bubéba (B&D) KYAMA penguinbi (B&D) MANDING-MANINKA dio fuinizon (B&D)

Plagiostyles africana (Müll.-Arg.) Prain

FWTA, ed. 2, 1: 414.

A shrub or tree reaching 16 m tall by 1.30 m in girth, of the lowland rain-forest in S Nigeria and W Cameroons, and extending to Zaïre.

No usage is recorded within the Region.

The wood is a light yellowish white. It is cut in Gabon to make spoons, combs and hair-pins, and a wood-decoction is taken in the belief that it promotes milk-production (4).

The bark contains a white, or sometimes yellowish and viscid (2), latex.

The bark is used in Gabon (4) and in Congo (Brazzaville) (1) for chest-affections, and in the latter country for fever pains and is considered revulsive in topical application in a palm-oil or clay vehicle for more or less all localised pains. A bark-decoction or leaf-sap is given in Congo (Brazzaville) by draught or enema – quantity according to age – to infants with worms or enlarged spleen. Suckling babies are dosed by application of the preparation to the mother's nipples. The bark is somewhat vesicant. It is taken to remove thread-worm and sap is administered as an eye-wash for filaria and conjunctivitis. Pulped bark is applied to snake-bite and is rubbed over areas of itch.

The plant is said to be capable of reducing the rate of heart-beat, and sap is given in Congo (Brazzaville) in draught as a tranquilliser in cases of fits of insanity and in agitation while an aqueous bark-decoction is used as a body-wash on the patient (1).

The leaves are pounded for use as a fish-poison in Gabon and the bark is made into talismans for trappers, or (by an interesting extension of sophistry) to attract suitors (4). In Congo (Brazzaville) the plant is also endowed with powerful fetish properties for secret societies and the bark is put into all manner of philtres and charms, benign, malignant, protective or offensive (1).

The plant's cambial tissue is used in Zaïre in a macerate for ear troubles and deafness (3).

References:

1. Bouquet, 1969: 122–3. 2. Keay & al., 1960: 255. 3. Léonard, 1962, b: 132–3. 4. Walker & Sillans, 1961: 174.

Plukenetia conophora Müll.-Arg.

FWTA, ed. 2, 1: 410, as *Tetracarpidium conophorum* (Müll.-Arg.) Hutch. & Dalz. UPWTA, ed. 1, 164, as *T. conophorum* Hutch. & Dalz.

English: African walnut (the fruit, Lowe).

Trade: conophor oil; awusa; n'gart (the oil, Bray).

West African: SIERRA LEONE ? musyabassa (GFSE ex JMD) KRIO awusa *from Yoruba* (FRI) NIGERIA EDO ókhuę = *walnut* (auctt.) IGBO ǫ̀kpa (NWT) ụkpà = *walnut* (BNO) IGBO (Owerri) oke okpokirinya = *male string leaf* (AJC) IGBO (Uburubu) ǫkụmù *from* ụmù;

children; = *babies call babies; alluding to its use in fertility cult* (NWT & ex JMD) ọmụmù (NWT) YORUBA asala (Lowe) awùsá (auctt.)

A woody liane to over 30 m long, of the bushy savanna in Sierra Leone and from Dahomey to W Cameroons and Fernando Po, and extending to Zaïre. Though well-recorded from Sierra Leone and cultivated there, it is apparently not indigenous (7): it is not recorded from Liberia, Ghana and Togo. Its presence in Sierra Leone may be due to returning slaves for it is known to the Krio by its Yoruba (Nigerian) name.

The liane is cultivated in Sierra Leone on newly cleared land principally for the nuts, but the leaves and young shoots are edible and are eaten, often with rice (4, 6, 8). Cultivation trials have been carried out in Nigeria (9). The leaves are considered a headache cure in S Nigeria (5, 6), and have magical use to wash children to cause their mothers to conceive, the Igbo name meaning *babies call babies* (11). In Gabon consumption of the seeds by husbands of wives already pregnant is believed to mitigate the risk of miscarriage (13).

Nigerian material has been screened for *alkaloids*, a trace of which is recorded in the bark (1).

The fruit is a capsule 6–10 cm long by 3–11 cm wide containing sub-globular seeds 2–2.5 cm long with a thin brown shell resembling the temperate walnut, hence the English name. The seed kernel is edible. Eaten raw they have a bitter flavour not unlike the kola nut and are considered to be tonic (3, 4, 9, 12, 13) and aphrodisiac (14). More usually the kernels are roasted and eaten in the general diet, or added to cakes (14). The kernels are oil-bearing yielding 48–60% of a light golden coloured oil with a taste resembling linseed oil. Composition is *linolenic acid* 64%, *palmitic* and *stearic acids* 15%, *oleic acid* 11% and *linoleic acid* 10% (4). This is conophor oil, or in the paint and varnish trade *awusa* or *n'gart.* It is edible and could be used in food preparations. It is unsuitable for soap-manufacture, and being quick drying it is certainly usable in the paint industry provided there is a certain supply and the kernels are free from excessive free fatty acids. Fresh oil has an iodine value of 190 (4) which is excellent for a drying oil, but the seeds do not store well and deterioration caused by enzymatic action needs to be prevented at the time of collection by heat-treatment (3). The oil has medicinal use in Nigeria in massages (2).

The cake left after expression of the oil contains 45% protein. It has local uses for food and is obviously a good source of protein. It can safely be fed to stock (3). The plant, presumably the kernel, is a good source of vitamins (10).

References:

1. Adegoke & al., 1968, as *Tetracarpidium conophorum*. 2. Ainslie, 1937: sp. no. 334, as *Tetracarpidium conophorum.* 3. B[-ray], G.T., 1947: with many references. 4. Busson, 1965: 176–8, as *Tetracarpidium conophorum* Hutch. & Dalz., with kernel and oil analyses. 5. Carpenter 285, UCI. 6. Dalziel, 1937: 164. 7. Deighton 5018, K. 8. Holland, 1908–22: 611. 9. Irvine, 1961: 255–6, as *Tetracarpidium conophorum* (Müll.-Arg.) Hutch. & Dalz. 10. Okiy, 1960: 121, as *Tetracarpidium conophorum.* 11. Thomas, NWT.2072 (Nig. Ser.), K. 12. Walker, 1953, a: 38, as *Tetracarpidium conophorum* Hutch. & Dalz. 13. Walker & Sillans, 1961: 177, as *Tetracarpidium conophorum* Hutch. & Dalz. 14. Iwu, 1986: 145.

Protomegabaria macrophylla (Pax) Hutch.

FWTA, ed. 2, 1: 373.

A forest tree to 17 m high occurring in Ghana, S Nigeria and W Cameroons and into E Cameroun, Gabon, Príncipe and São Tomé.

It is perhaps but a variety of *P. stapfiana* (q.v.) with the same properties.

Protomegabaria stapfiana (Beille) Hutch.

FWTA, ed. 2, 1: 373. UPWTA, ed. 1, 158.

Trade: senan (the timber, Ivory Coast).
West African: **SIERRA LEONE** KONO ɡbaŋgfoboi (S&F) bogboi (S&F) MENDE ɡbɔgbɔ(-i)
(FCD; S&F) **LIBERIA** DAN dó (GK) KRU-BASA wah (C&R) **IVORY COAST** ABE mbraua
(Aub.) djilika (A.Chev.) djirika (A.Chev.) ABURE sanié (A.Chev.) AKAN-FANTE emuain-kain
(A.Chev.) emuin-kwin (A.Chev.) AKYE sénan (A.Chev.) ANYI assa-boguié (A.Chev.) guahélé
(Aub.) AVIKAM bapi (A.Chev.) bopi (A.Chev.) ꞌKRUꞌ klaklé (Aub.) KRU-GREBO sulé (Aub.)
KYAMA siédzo (A.Chev.) NZEMA emuain-kain (A.Chev. fide JMD) emuin-kwin (A.Chev. fide
JMD) sulo-koba (A.Chev. fide JMD) **GHANA** AKAN-FANTE koain (JMD) WASA agyahere
(auctt.); *if the tree is stilt-rooted* (CV) kwintan *if the tree is not stilt-rooted* (CV) subète (FRI)
subèto (FRI) ANYI agyehyele (FRI) NZEMA agyehyele (FRI)

A rain-forest tree to 30 m high with bole up to 2 m girth, low-branching, not
buttressed in closed-forest or with prop-roots, locally abundant in swampy sites
from Sierra Leone to W Cameroons, Gabon, Príncipe and São Tomé.

Sap-wood is pale brown; heart-wood red or pinkish-brown, speckled or
veined, often spongy and brittle. It does not appear to resist decay, splits easily
and is classified as a semi-hardwood (3–5). It is used for firewood (1) and sawn
for planks (2, 3) in Liberia, and seems to be usable for carpentry and certain
structural work (4).

References:

1. Cooper 109, K. 2. Cooper 294, K. 3. Cooper & Record, 1931: 53–56, with timber characters. 4.
Dalziel, 1937: 158. 5. Savill & Fox, 1967, 122.

Pycnocoma angustifolia Prain

FWTA, ed. 2, 1: 405.
West African: **SIERRA LEONE** MENDE mbɔyɛ (*def.* -i) (FCD)

A shrub, usually of forest undergrowth, recorded only from Sierra Leone,
Liberia and Ivory Coast.
In Liberia the sap is known to arrest bleeding (1).

Reference:

1. Deighton 3656, K.

Pycnocoma cornuta Müll.-Arg.

FWTA, ed. 2, 1: 405. UPWTA, ed. 1, 158.
West African: **GHANA** ADANGME-KROBO kofia-kofia (FRI) AKAN-ASANTE kofie-kofie
(auctt.) TWI ahlɔefɔti (Eady) kafiɛ-kafiɛ (E&A) WASA nomnefie (BD&H; FRI) GUANG-GONJA
kpante-feta (FRI) **NIGERIA** IGBO omenkpume (NWT) IGBO (Idumuje) ewe otolo (NWT)
inyinta (NWT) IGBO (Uburubu) otolo (NWT) YORUBA suru-oke (JRA)

A stout shrub to 5 m high of undergrowth of deciduous forest and open dry
forest, occurring from Ghana to N and S Nigeria.
The stem, root, bark and leaves are strongly purgative (1, 6) and the root,
sometimes added to soup, is used in S Nigeria as a laxative (4, 5). Both stem and
root are used in Ghana to prepare a strong purgative which if taken in large
doses can be fatal (2, 3). The ground bark, with lime-juice and water, is used in
Ghana as an emetic, repeated doses being effective for 'swelling of the abdomen'
(Mackay fide 3). The bark is sometimes used criminally, and is put in palm-wine
to discourage theft. The bark, with others, is used as a fish-poison (3). The

Ghanaian Akan name, *kofie-kofie*: 'go home, go home', perhaps implies acknowledgement of the dangerous toxicity of the plant (cf., *P. macrophylla* Benth.).

References:

1. Ainslie, 1937: spp. no. 294. 2. Burtt Davy & Hoyle, 1937: 50. 3. Irvine, 1961: 249. 4. Thomas, NWT.2173 (Nig. Ser.), K. 5. Thomas, NWT.2298 (Nig. Ser.), K. 6. Vergiat, 1970: 88.

Pycnocoma macrophylla Benth. var. **macrophylla**

FWTA, ed. 2, 1: 405. UPWTA, ed. 1, 159.

West African: IVORY COAST ? kloapuo (A&AA) AKYE baffeu (Aub.; K&B) 'KRU' wuba (K&B) wulola (K&B) GHANA ADANGME-KROBO kofia-kofia (FRI) AKAN-ASANTE kofie kofie = go-home, go-home (auctt.) WASA nomnefie (FRI) GUANG-GONJA kpante-feta (FRI) NIGERIA YORUBA suru-oke (JRA)

A shrub to about 1.70 m high of undergrowth in high-forest, occurring from Ivory Coast to W Cameroons and Fernando Po, and extending to Zaïre.

That the plant is known to be very poisonous is perhaps implied in the Ghanaian Akan name, *kofie-kofie*: go home-go home (5). All parts are strongly purgative. Administration causes immediate and prolonged intestinal shock. It is however used as a drastic purge (4; Ivory Coast: 1, 3, 7; Ghana: 5, 6, 8; Nigeria: 2). Root-bark in Ivory Coast it is not given to children under 15 years old. To stop purging an adult patient to whom it has been administered is given raw [?oil-]palm nut to eat (1). The bark is used in combination with others as a fish-poison, and it is said to be used criminally put in palm-wine to discourage theft (4). The wood is used to make boys' tops in Ghana. *Tannin* is perhaps present, and the fruits are used in Natal to tan leather (4–6).

The presence of the plant is held to be an indication of poor soil (5, 6).

References:

1. Adjanohoun & Aké Assi, 1972: 136. 2. Ainslie, 1937: sp. no. 294. 3. Bouquet & Debray, 1974: 86. 4. Dalziel, 1937: 159. 5. Irvine, 1930: 360–2. 6. Irvine, 1961: 249. 7. Kerharo & Bouquet, 1950: 85. 8. Thomas, AS D.127, K.

Ricinodendron heudelotii (Baill.) Pierre

FWTA, ed. 2, 1: 391, 393. UPWTA, ed. 1, 159–60, as *R. africanum* Müll.-Arg.

English: African wood-oil nut tree; African nut tree; African wood (Liberia, Kunkel).

Trade: Anglophone territories – erimado (from Yoruba); also – okwen (but not preferred because of liability of confusion with the Edo name for *Brachystegia eurycoma* Harms, Leguminosae: Caesalpinioideae). Francophone territories – essang; essessang; eho (Ivory Coast).

West African: SENEGAL DIOLA bu kĕnkaré (JB) bu makurĕg (JB) bu makurèn (JB) MANDING-MANDINKA bôn kuôforo (JB) GUINEA KISSI bo-gboho (FB) KONO plo (RS) KPELLE gbolo (FB) MANDING-MANINKA gboloba-bulu (FB) MANO kô (RS) SUSU tonta (RS) TOMA boroï (FB) gboloye (FB) gporo (RS) SIERRA LEONE KISSI gbo (S&F) kpo (S&F) KONO gbɔɛ (FCD; S&F) gbwɔyɛ (FCD; S&F) KORANKO gbɔrɛ (S&F) MENDE gbolei (S&F) gbɔlɔ (FCD) gboloi (Cole) gbolei (auctt.) kpɔlei (S&F) kpɔlɔ (FCD) TEMNE ka-kino (FCD; S&F) ka-sigbɔrɔ (FCD; S&F) LIBERIA DAN ko (GK) KRU-BASA koor (C&R; RS) GUERE (Krahn) karro-tu (GK) MANO ko (JGA) koo (JGA) MENDE gbolei (C&R) IVORY COAST ABE akpi (A&AA) eho, ého (auctt.) hobo hapi (LC) ABURE poposi (auctt.) ADYUKRU ṃbob (auctt.) ṇbob (A.Chev.) essandaille (Aub.; K&B) esseng ṇdaye (A&AA) AKAN-ASANTE akpi (B&D) uama (B&D) wamba (K&B) BRONG api (K&B) AKYE akpi (A&AA) akwi (A.Chev.; RS) isain (LC) tsain (auctt.) ANYI akwi (auctt.) api (A.Chev. ex B&D) hacbiuagpi (auctt.) haipi (LC) BAULE akin (B&D) akpi (B&D; A&AA) akporo (B&D) DAN goodi (auctt.) GAGU kô (K&B) 'KRU' katotu (Aub.; RS) 'KRU' (Lower Cavally) nbob (LC) KRU-GREBO (Oubi) karatu (RS) kohué (A.Chev.; RS) kotué (auctt.) GUERE kotue (K&B) ko-ué (RS) GUERE (Chiehn) kô (K&B) kua (B&D) KULANGO api (K&B) KULANGO (Bondoukou) haipi (LC) KWENI kô (auctt.) KYAMA

popossi (auctt.) popossi ya (B&D) propossi (Aub.) NZEMA sosahu (A.Chev.; K&B) **GHANA** ?
alokpo (Eady) flekpo (Eady) okao koodo (LC) VULGAR wama (DF) ADANGME-KROBO ekpedi
(BD&H; FRI) AHANTA ɔwama (E&A) wama (E&A) AKAN-ASANTE awama (LC) awoma (LC)
epi (Taylor) owama (auctt.) FANTE ɔwama (auctt.) TWI a-, o, ɔ-wama (auctt.) ɔnwama (auctt.)
wamba (auctt.) WASA asoma (auctt.) wama (auctt.) wamba (BD&H) ɔwamma (auctt.) ANYI
asɔma (FRI) hakpiwaka (FRI) ANYI-AOWIN epui (auctt.) epuwi (auctt.) SEHWI ɛ-nwany(e) (FRI)
ANUFO epuwi (auctt.) ɛwama (FRI) ɛnwany(e) (FRI) nwuama (TFC; FRI) GBE-VHE e-kpedi
(JMD) VHE (Pecí) kpedi (auctt.) NZEMA asɔma (FRI; Taylor) awuma (TFC; LC) engwanle
(Taylor) ɛ-wama (JMD) ewan (auctt.) ngwama (auctt.) ngwani (JMD) ɛ-nwany(e) (FRI)
NIGERIA ABUA okpɔnum (JMD) BASSA (Kwomu) wawan kurmi (Chapman) EDO erinmadu
epo (Ross) ọkhuẹn (auctt.) ọkhuẹn-nebo (JMD; LC) ọkhuẹn-n'fua (JMD) ọkhuẹn-seva (JMD;
LC) EFIK ṅsásáŋá (auctt.) HAUSA wááwán-kúrmìi = *cloth of the forest* (auctt.) IGALA òdē (RB)
odede (H-Hansen; RB) IGBO òkwè *a kind of bean* (auctt.) IJO-IZON okengbo (auctt.) ovovo (LC;
Kennedy) IṢEKIRI okue (LC; KO&S) MBE okwar (Nicklin) URHOBO eke (auctt.) YORUBA ekku
(LC) erín madò (auctt.) olóbò igbó (Kennedy) ọmọdọn (LC; JMD) ọrọmọdọn (JMD) póto
pòto (auctt.) putu putu (JRA; JMD) putu putu funfun *from* funfun: *white* (JMD) **WEST
CAMEROONS** DUALA bònjaɔsaɔ (Ithmann) KOOSI isange (JMD) KPE esangasanga (JMD)
wongasanga (JMD) wonjangasanga (JMD) KUNDU bonjasanga (JMD) njangsang (JMD)
wonjasanga (JMD) LUNDU bosisang (JMD)

NOTE: *Names from Senegambia to Ghana refer to ssp.* heudelotii. *Nigerian names are of ssp.*
africanum *(Müll.-Arg.) J. Léonard*

A fast-growing tree to 50 m high by 2.70 m girth, bole straight with short
buttress, of the fringing, deciduous and secondary forests, common throughout
the semi-dry wooded-savanna zone of the Region from Lower Casamance of
Senegal to West Cameroons and Fernando Po; and to Zaïre, Angola and
Tanganyika. Two varieties are recognised: var. *heudelotii* in Ghana and west-
wards, var. *africanum* (Müll.-Arg.) J Léonard, Nigeria and eastwards (23).
  The tree regenerates readily from the stump, and it comes up freely on old
farms. In open light spaces it will bear fruit in the seventh to tenth year, and its
rapidity of growth is proverbial. It grows spontaneously from seed and is often
preserved in the neighbourhood of forest villages. Hunters recognise the tree as
attractive to wild animals which eat the fallen fruits. There is a belief that
'collar-crack' disease will occur on a cacao farm if the tree is cut down (17). In
Gabon people of the interior relish a small white mushroom *dibindi*, Eshira)
which grows on the dead trunks (28).
  The wood is dull white, fibrous, soft, light and perishable. Density is 0.327
(26). It is used for rough planks and coffins. The wood is very buoyant and is
used for fishing-net floats and rafts for heavy timbers. Its sawdust is extra-
ordinarily light and is suitable for life-saving belts. Because of its ease of
working it is carved into fetish-masks, spoons, ladles, plates, platters, bowls,
dippers, stools, etc. (13, 16, 17, 25, 27). The wood is used in Zaïre for making
drums which are said to be very sonorous, and it is carved to make the whole or
the resonant parts of musical instruments in S Nigeria, Gabon and Angola (17,
19, 28). In the Lower Niger and Cross Rivers area in SE Nigeria a log is carved
out to make a giant xylophone called in Mbe: *ogbang*. Wood of other trees is
attached to give a variety of tones: *mankwaro* (*Mangifera indica*, Anacardiaceae)
for high notes, *ntweno* (sp.?) for middle notes and *nkomni* (sp.?) for bass notes
(31). The wood is currently recommended in Ghana for use in insulation (5) and
the sawdust is no doubt suitable for sun-helmets (17). The wood is indifferent as
fuel (17) since it burns with great rapidity (7), but the ash is used in Guinea for
the preparation of a vegetable-salt in soap-making and in indigo dyeing
(Portères fide 20), and in Sierra Leone Mendes use it for soap-making (18). The
wood is perhaps suitable for paper-pulp (14, 20, 26).
  Root-bark is used in Nigeria when ground up and mixed with pepper and salt
for constipation (6). Temne of Sierra Leone tie to the body bark which has been
beaten and warmed for elephantiasis (18). In Liberia a bark-liquor is taken by
pregnant women to relieve pains and to prevent miscarriage. It is also taken by
women 'to kill a worm which is in the bowels and which prevents them from

breeding' (15). In Ivory Coast a root-bark-decoction taken by mouth is considered a powerful anti-dysenteric, and stem-bark is taken by enema to prevent abortion (21). Stem-bark-decoction is also used to wash and cicatrise sores (4). In Gabon a bark-decoction is taken for blennorrhoea (28), and in Congo (Brazzaville) it is taken by draught for cough, blennorrhoea, painful menstruation and as a poison-antidote; in lotions and baths to strengthen rachitic children and premature babies, and to relieve rheumatism and oedemas; and pulped the bark (also the leaves) is applied to fungal infections, to maturate abscessess, furuncles and buboes, and expressed sap is instilled to the eye for filaria and ophthalmias (9). Examination of the bark of Nigerian material revealed no alkaloid present (3), and in Congo (Brazzaville) material no active principle was found (10).

A leaf-decoction is taken by draught and in baths in Ivory Coast as a febrifuge (21), and leaves are used to treat dysentery, female sterility, oedemas, and stomach-pains (11). Leaves and stems have been reported to contain an unnamed *alkaloid* (30).

The roots in Ivory Coast are considered aphrodisiac (11).

The fruit is 2–3-lobed with a thick hard shell. In quantity it has a smell of over-ripe apples. Bats, hornbills and rodents are said to assist in dispersal (27). The seeds are edible, but not everywhere valued as food. The kernels are eaten boiled in water (7) or in sauce like ground-nuts in Ivory Coast (4) or mixed with fish, meat and other vegetables (12). In Gabon they are roasted and made into a paste (28). They are recorded as being eaten by the Efik of Nigeria (2) and in Liberia (22). The kernel is oil-bearing and contains about 47% by weight of oil consisting of the following fatty acids: *eleostearic* 44%, *oleic* 16%, *palmitic, stearic, linoleic* and *linolenic* 10% each (Ivory Coast material, 12). The oil is light yellow, drying and has a sweet taste (13, 17, 21). It is usable in varnish and to make soft-soap (17, 18, 25) and it has industrial application in water-proofing materials (21). Decortication, however, is not easy, and as the shell amounts to 37% of the weight of the seed the total amount of oil may be as low as 14% of the whole seed.

The seed contains small amounts of toxic substances said to be a *resin* (29) which renders the residual cake unfit for use as a cattle-food though the cake should be a good nitrogenous agricultural fertiliser. The West African use of the seed, husk and latex as a remedy for gonorrhoea and diarrhoea may rest on the action of this substance (29), as also the use in the Nimba Mountain area for treating amoebic dysentery (1).

The seeds are used in rattles in *bundu* dances in Sierra Leone (25). They are also used in Nigeria in a game played by the Igbo called *òkwè* (24) and another in Cameroun called *songo* (8, 17), both forms of mancala. It will be noted that the Igbo word applies equally to the game as to the tree. Indeed, *izūòkwè* is to play any game using seeds or pieces as counters, as, for example, draughts might be so called.

Superstitious applications of the tree are practised on the Liberian/Ivory Coast border region where hunters cover their faces with bark ash. This, they hold, enables them to kill all elephants they see (21).

References:

1. Adam, 1971: 376, as *R. africanum* Müll.-Arg. 2. Adams, RFG, 1947: as *R. africanum*. 3. Adegoke & al., 1968. 4. Adjanohoun & Aké Assi, 1972: 137. 5. Anon., s.d., a. 6. Ainslie, 1937: sp. no. 302, as *R. africanum*. 7. Aubréville, 1959: 2: 76. 8. Bates, s.n., 21/3/1921, K. 9. Bouquet, 1969: 123. 10. Bouquet, 1972: 26. 11. Bouquet & Debray, 1974: 86. 12. Busson, 1965: 175–6, with seed analysis. 13. Chalk & al., 1933: 43, as *R. africanum* Müll.-Arg. 14. Cole, 1968,a. 15. Cooper 457, K. 16. Cooper & Record, 1931: 56, as *R. africanum* Müll.-Arg. 17. Dalziel, 1937: 159–160. 18. Deighton 1121, K. 19. Gossweiler 4898, K. 20. Irvine, 1961: 249–50. 21. Kerharo & Bouquet, 1950: 86, as *R. africanum* Müll.-Arg. 22. Kunkel, 1965: 176. 23. Léonard, 1961: 397–401. 24. Murray, HJR, 1952. 25. Savill & Fox, 1967: 123–4. 26. Schnell, 1950, b: 258, as *R. africanum*. 27. Taylor, 1960: 163. 28. Walker & Sillans, 1961: 175–6, as *R. africanum* Müll.-Arg. 29. Watt & Breyer-Brandwijk, 1962: 428, as *R. africanum* Müll.-Arg. 30. Willaman & Li, 1970. 31. Nicklin, 1975: 149, as *Ricinodendron sp.*

Ricinus communis Linn.

FWTA, ed. 2, 1: 410. UPWTA, ed. 1, 160–3.

English: castor plant; castor oil plant; palma Christi (an ornamental c.var., Dalziel).

French: ricin.

Portuguese: ricino.

West African: SENEGAL ARABIC (Senegal) auïrirt (JB) aveyrur (JB) avreivar (JB) avriur (JB) vavlgirt (JB) BALANTA entegay (JB) tora (JB) BANYUN bi gès (K&A; JB) CRIOULO diakdiak (JB) DIOLA bi gès (JB) bukot (JB; K&A) FULA-PULAAR (Senegal) diakula (JB) kekamédii (JB) kékemédii (JB) TUKULOR dimbé iligala (K&A) dimbèyligala (K&A; JB) MANDING-BAMBARA subaga baña (JB) subara bana (JB) *n*tomontigi (JB; K&A) MANDINKA tigi ñiré (JB) tumum suma (K&A) tumun suma (JB) MANDYAK bu purura (JB) SERER batarpolì (JB) damal (JB) lamapèti (JB) lampèti (JB) lapet (K&A) matarmboli (JB) mbatarmboli (JB; K&A) mbatrapoli (JB) SONINKE-SARAKOLE tigińéré (K&A) WOLOF héhèm (JB) kerhom (K&A) xerhem (K&A) xerhom (K&A) GUINEA-BISSAU BALANTA entôgai (JDES) torra (JDES) BIAFADA buorai (DF) CRIOULO djague-djague (JDES) FULA-PULAAR (Guinea-Bissau) djácula (JDES; EPdS) MANDYAK bupurura (DF) GUINEA FULA-PULAAR (Guinea) diakula (CHOP) SIERRA LEONE KISSI tanahala (FCD) tandahala (FCD) KONO kasawe (FCD) KORANKO marabaŋk-eyobayambe (NWT; JMD) KRIO kastarɔil (FCD) LOKO tawabawa (NWT; JMD) MENDE bɔnde (NWT; JMD) *n*gele-bɔndɔ (-i) = *sky okra* (FCD; JMD) SUSU fore (FCD) limbiaxule (NWT; JMD) mbiakula (NWT; JMD) TEMNE an-fental (FCD; JMD) LIBERIA MANO gbana yidi = *thunder tree* (Har.) MALI MANDING-BAMBARA subarabana *a small-seeded var.* (Vuillet) TAMA-CHEK fueni (JMD) UPPER VOLTA MANDING-DYULA tomotigui (K&B) IVORY COAST AKAN-ASANTE attindé (B&D) AKYE atondu (K&B) ·ANDO· atèndè (Visser) BAULE atendé (B&D) aterré (B&D) KRU-GUERE puzu (K&B) KYAMA n'téké (Ivanoff) ·MAHO· yuma (K&B) MANDING-DYULA tomotigui (K&B) MANINKA dioma dioma (B&D) GHANA ADANGME-KROBO kumɛ-lo (Bunting) AKAN-FANTE abrɔnkruma = *whiteman's okra* (Bunting; FRI) adede *n*kuruma *from* nkuruma: *Hibiscus esculentus, okra* (auctt.) esuso *n*kuruma (FRI; JMD) sunsum *n*kuruma (FRI; JMD) TWI abɔnkruma = *whiteman's okra* (FRI; JMD) adade-*n*kuruma *from* nkuruma: *Hibiscus esculentus, okra* (FRI; JMD) adede-*n*kuruma = *crow's okra* (auctt.) ANYI-AOWIN ateende (FRI) GA abrɔnnkruma = *whiteman's okra* adade nkuruma *from* ŋkuruma: Hibiscus esculentus, *okra* (auctt.) adede ŋkrú (KD) adede ŋkruma (auctt.) adedenkruma (OA) GBE-VHE dzɔŋbati (FRI) kasuwelti, kasuwɔlti *i.e., 'castor-oil tree'* (FRI) lɔŋgɔ (FRI) VHE (Awlan) dzɔŋdalẽ (FRI) VHE (Pecí) atɔŋgɔ (FRI) yevu-tɔŋgɔ (FRI) TOGO GBE-VHE dzegbele (Volkens) dzongbati (Volkens) lõngɔ (Volkens) KABRE assimballɔ *a large-leaved var.* (Gaisser) TEM (Tschaudjo) dendelle *a small-leaved var.* (Gaisser) sau *a large-leaved var.* (Gaisser) NIGER SONGHAI zìrm-ñá ` (*pl.-*ñóŋó) (D&C) NIGERIA ARABIC-SHUWA hurua (JMD) khiruwi (JMD) BIROM vyɔ̀ŋrɔ (LB) BURA ameru (anon.) whadlawhadla (anon.) EDO eraogi (JMD; Singha) EFIK étígí únèŋè = *Igbo's okra* (auctt.) étó árán ukẹbẹ (JMD) FULA-FULFULDE (Adamawa) kolakolaaji (RB) FULFULDE (Nigeria) derre (*pl.* derreje) (JMD) kola kolaaji gorɗe = *male 'kolakolaaji'*; cf. *Jatropha which is female* (JMD) kolkolwaahi (J&D) zurmaje (JMD) HAUSA cìkà-cídàà, cìkà-gídàà *fill the house - referring to the many seedings that appear wherever it is planted* (JMD; ZOG) dán kwásárè (JMD; ZOG) darman (RB) kulakula (JMD; ZOG) zirman (Schuh) zurma, zurmâŋ *to stretch out; e.g., the hand - in allusion to the shape of the leaf* (auctt.); ? = zurman mutane *indigenous vars.* (JMD; ZOG) zurman nasara = *'zurma' of the Europeans - exotic vars.* (JMD; ZOG) IDOMA ájɔŋggɔ (Armstrong) IGBO ɔgba (JMD) ògìlì, ògìrì *the seeds* (JMD; Singha) ògìlì úgba (KW) ogiri-ugbo (Iwu) ụgba (BNO) IGBO (Asaba) osisi-ògìlì *from* osisi: *tree* (JMD) IGBO (Owerri) ògìri ọ̄gba = *cow's coarse grass* (JMD) IGBO (Umuahia) ògìri-arọ (JMD) IJO-KAKIBA (Kalabari) alamba-ngbọle (JMD) KAMBARI-SALKA àlásà méryô̄'ō̄ (RB) NGIZIM dlábùwàk (Schuh) NUPE kpamfini gulu = *vulture's okra* (JMD) SURA puul (anon. fide KW) TIV harev *the plant itself* (JMD) ihurua dzengo (JMD) *i*-jija (JMD) masev *the plant itself* (JMD) sherai jongo (JMD) showara jongo (JMD) YORUBA eso lárà *the fruit or seeds* (IFE) ilárà (JMD) ilarum (JMD) láà (JMD) làpálàpá adẹ̀tẹ̀ (JMD) lárà (JMD; Singha) òróro-lárà *the oil* (JMD) WEST CAMEROONS BAFOK bokuri (DA) LONG bokuri (DA) MBONGE boku-balondo (DA)

An annual herb or perennial shrub to 3–4 m high, but capable of producing a stout trunk and becoming a small tree to 10 m tall, very polymorphic, seldom, if ever, wild, but subspontaneous or cultivated throughout the West African region. The plant is thought to be originally of African origin. Its seeds have been found in Egyptian tombs dated 4,000 B.C., and the plant is known to have been in general cultivation under the Pharaohs. From Egypt the plant has spread throughout tropical Asia and along the Mediterranean. The Romans, first meeting it, likened the seeds to the sheep-tick, giving it the same Latin name, *ricinus*, hence the present generic name. The plant is now pantropical (6, 10, 19, 24). The date of its arrival in West Africa seems likely to be relatively recent for it has acquired several borrowed names, the Vhe of Ghana using a corruption of English, and other races taking from the names for *okra*. An exotic origin is also manifest in the qualification of names by foreign epithets, e.g. 'whitemen' and to the Efik of 'Igbo' – see the vernacular name list above.

The leaves are not grazed by stock (1). They are eaten by women in Nigeria as an emmenagogue, and are applied as a poultice to the breasts to increase milk-flow (2). The leaves are recognised in Libya and Somalia as a powerful galactogogue (28), though conversely in Gabon they are applied to arrest a too copious flow (26). In Senegal the leaves are applied in massage to areas of varicose veins in pregnant women (15, 17), and are taken by draught of a decoction for schistosomiasis and ascites with, at the same time, fumigation of the lower stomach by dried inflorescences scattered onto hot but dead embers (16, 17). In Ivory Coast–Upper Volta the leaves are used in frictions (29) as a revulsive and vesicant for pneumonia and febrile conditions (5, 18). There is a suggestion that the leaves may have some anodynal property. In some parts they are pounded and applied as a poultice to swellings and a leaf is tied to the forehead for headache (10, 14, 30). They are a treatment for haemorrhoids in Ethiopia (32). In India they are used for rheumatism, lumbago, sciatica, etc, and in Somalia for rheumatism (28). They are also used for skin-diseases and on framboesia, and in Tanganyika on carbuncles and wounds (3). In Togo the leaves of a large-leafed variety, known as *sau*, are crushed to provide an eyewash (10), an use that is also practised by the Hausa (23). The leaf in decoction is laxative (10), and a preparation is used in enema in Gabon (25). In Adamawa, Nigeria, liquid in which leaves and potash have been boiled is administered for two days as a cure for jaundice (30). The leaves rendered soft and pulpy by heat are applied to guinea-worm sores to facilitate extraction of the parasite (2, 10). In Gabon, the leaves crushed and mixed with 'false shea butter' (? *Lophira alata* Banks, Ochnaceae), or with a slice of lime, are used to massage epileptic children. This preparation is also an emollient (26–28).

The Eri silkworm, *Attacus ricini*, is raised on the leaves of the castor plant in India. This moth has been completely domesticated and is the source of widespread cottage-industry in the Bengal region (6, 9).

The leaves are rich in potassium nitrate. *Ricinine* is also present, but this is not so markedly toxic as *ricin* present in the seeds. *Rutin* has also been reported present at a concentration of 0.2% (13, 17, 27, 28).

The twigs are used in Nigeria as an abortive irritant (2). In Tanganyika sap from the twigs and the fruit is dripped into new wounds as a dressing (11). The bark is somewhat fibrous and is used for stitching wounds in Zimbabwe and placing over sores (28). The stems have been examined for the making of paper-pulp. The bark is valueless for this but debarked and defibred stems gave a paper suitable for books, and though it was not so strong as pinewood paper might substitute it for certain purposes, and as a filler for longer fibred pulps. Indian material is said to rank next to bamboo in order of merit (7).

Root-extract is prescribed by Hausa medicine-men near Kano as a mouth-wash and for toothache (33). In Ghana, the root is well-cleaned and then ground to a paste in a little water which is instilled into the nose for headache (35). The root-bark is a powerful purgative. It is used in veterinary medicine in Nigeria, and a bolus, 3–5 cm in diameter, with chillies and tobacco leaves is said to be an excellent remedy for gripes in horses (2).

The fruit capsule is 3-seeded, usually spiny, occasionally smooth. The seeds are generally mottled-grey, with brown-purple streaks, and have a white swelling at the narrow end, but they are very variable in size and colouration. Variation has been used as a basis for classification below specific rank and it seems probable that the smaller-seeded forms are the African aboriginals. Size of seed, however, appears to be of lesser importance than soil and climate for oil production, and, indeed also, on an individual plant the yield between branches varies. In general, oil from India is considered superior industrially than West African oil. The plant seems more adapted to small-holding culture than to estate production, but it has been grown widely in all areas of the Region, and in Dahomey especially it has become in important crop. The smaller-seeded varieties are a 4–5 months crop, the larger-seeded requiring 7–10 months according to variety, soil and climate. As ripening is irregular and the capsules are explosively dehiscent, it is usual to harvest the whole fruiting spikes a week or so before ripening and leave them to sun-dry on a winnowing area where the seed and husk can be separated after dehiscence. In some parts of the world special cultivars have been raised for specialist purposes, especially the oil for pharmaceutical use. In N Nigeria (22) and in Malaya (6) the existence of races with edible seeds is claimed, though these are more likely to be famine foods and such reports require cautious acceptance – but see below.

The seed consists of approximately 20% or more of shell, and 80% or less of kernel, and the kernel contains 58–66% oil with a considerable variation. The oil is obtained by expression, with variations of grinding, heating, boiling and solvation. Top quality oil is taken from the first crushing of the cold kernels. This is colourless and is free of the toxic *ricin*. Residual cake is further pressed and/or heated to give oil variously coloured and of inferior quality. The special characters of castor oil are its high specific gravity, viscosity, optical rotation and acetyl value, and it differs from other vegetable oils of commerce in being soluble in alcohol, but insoluble in light petroleum and other mineral oils. These properties are due to its high percentage of *ricinolein*, the glyceride of ricinoleic acid. *Palmitin*, so common in other vegetable oils is absent, and *olein* is present in only small quantity (6, 10).

The oil is non-drying and viscous, almost odourless with an acrid roughish after-taste. It burns with a clear light and its primary use in antiquity must have been as an illuminant and as an unguent, before it came to be used medicinally as a purgative. It has, however, been used medicinally from an early time. In Sanskritic medicine in India it was used as a purge and anodyne for rheumatism. Dioscorides in the First Century A.D. knew of the oil as a vermifuge, emetic and purge. The oil has very many industrial uses (6, 8, 21, 28). It is stable and its non-drying property makes it useful as a plasticiser in the lacquer industry. It is a good lubricant for machinery, and as it does not solidify till very low temperatures are reached nor break down at high temperatures it is a reliable lubricant for aero engines. It serves as a brake-fluid, in nylon manufacture, for the production of turkey-red oil in dyeing, the production of 'rubber substitutes', etc. Under heat with a catalyst it can be transformed into a drying oil suitable for mixing with linseed oil, tung oil, etc. for paint and linoleum production. It is used in soap-manufacture though usually with other oils. It gives to soap a transparency. *Undecyclic acid* used in fungistatic preparations is obtained from the oil (20). It is emollient and antibacterial. It has been official in many national pharmacopoeias. It has insecticidal properties good for treating timbers against termites and other pests, as an antimalarial additive to kerosene-based spray, as an embrocation against body-vermin and parasitic skin-diseases (6, 10, 21) and chest-pains (30). Insecticidal action is ascribed to the *ricinine* present, not to the *ricin* (4).

On expression of the oil, the highly toxic *ricin* remains in the cake rendering it poisonous as a food-stuff. The cake can however be used as an agricultural fertiliser. It is rich in proteins and so has an use in the manufacture of plastics (28). In recent years it has been found that the meal submitted to heat loses its toxicity and can then be used as a feed in a limited way. Poultry in Brazil have been satisfactorily fed with it as 2.5–10.0% of the total diet. Pigs have been found unable to digest it perhaps because of their monogastric digestive system being unable to digest the high fibre content. Protein is 39%; fibre 19%. The report has no reference to other stock (22). The seed, or the whole fruit pod, is sometimes given in small measure with hay to various stock in Kordofan (31), but the report makes no mention of any pre-treatment. In Onitsha Province, Nigeria, detoxification of the seed is being undertaken by boiling the decorticated cotyledons without the embryo for an hour wrapped in banana leaves, fermenting for 2 days, and then grinding to a paste. Ash from burnt, threshed palm-oil bunches is added to discourage flies. This product is wrapped in leaves and sold in Igbo markets under the names of ogili-isi (Onitsha) and ogili-ugba (Awka) (22), or ogiri (36). This is a dark brown viscous substance rich in glutamic acid, methionine, cystine and tryptophan, all amino-acids of importance to human metabolism and good health.

Industrial workers exposed to the bean may suffer two types of poisoning: one caused by the ricin itself, the other an allergy engendered by a substance termed castor bean allergen. Ricin is a toxalbumen causing coagulation of the blood. In small amounts illness is caused, and in large doses death, but administered in small and increasing amounts over 4–6 weeks an immunity builds up giving a thousandfold tolerance. Pharmaceutically immunisation is obtained by administration of ricin detoxified by formaldehyde. The allergen works in the opposite direction, the sensitivity becoming greater with continued exposure. Symptoms are varied and can affect skin, nose, lungs, etc. The effects of both together to some extent cancel out, but quite obviously caution is indicated (21, 22).

A plant of such diverse and powerful action has naturally attracted superstitious attributes. It has entered witchcraft and is used by 'discerning' wizards in Ghana (14). It is held to prevent injury by lightning. The Mano of Liberia call it thunder tree (12). Similarly Gabonese vernaculars mean thunder tree or thunder trap, the plant supposedly conducting the lightning flash safely to the ground (26).

The ornamental cultivar known as 'Palma Christi' with brilliant red inflorescences is found in the drier parts of the Region, and this has been introduced elsewhere and is a house-plant in temperate regions. In S America the castor plant is often grown around houses in the belief that it dispels mosquitoes. It has also been grown on sand-dunes as a sand-binder (10).

Israeli chemists have devised a process to produce a plastic from the plant as a starting point (Calder fide 34).

References:

1. Adam, 1966, a. 2. Ainslie, 1937: sp. no. 303. 3. Bally, 1937. 4. Bezanger-Beauquesne, 1955. 5. Bouquet & Debray, 1974: 86. 6. Burkill, IH, 1935: 1907–12. 7. Chittenden & Coomber, 1948. 8. Chadha, 1972: 26–47. 9. Chadha, 1972: 327–53, under Silk and Silkworms. 10. Dalziel, 1937: 160–3. 11. Haerdi, 1964: 97. 12. Harley s.n., 12/5/1932, K. 13. Henry, 1939: 13–16. 14. Irvine, 1930: 370–1. 15. Kerharo & Adam, 1963, b. 16. Kerharo & Adam, 1964, b: 572. 17. Kerharo & Adam, 1974: 428–32, with phytochemistry and pharmacology, and many references. 18. Kerharo & Bouquet, 1950: 86–87. 19. Mauny, 1953: 717. 20. Oliver, 1960: 9, 49. 21. Quisumbing, 1951: 529–33, with many references. 22. Raymond, 1961. 23. Robinson, 1913. 24. Trochain, 1940: 266. 25. Walker, 1953, a: 37. 26. Walker & Sillans, 1961: 176. 27. Watt, 1967. 28. Watt & Breyer-Brandwijk, 1962: 428–30, with pharmacology and many references. 29. Visser, 1975: 75–15: 50. 30. Babuwa Tubra fide Blench, 1985. 31. Baumer, 1975: 111. 32. Lemordant, 1971: 166. 33. Etkin, 1981: 82. 34. Hunting Tech. Services, 1964. 35. Ampofo, 1983: 36. 36. Iwu, 1986: 135. See also: Scarpa & Guerci, 1982: 117–37.

Sapium aubrevillei Léandri

FWTA, ed. 2, 1: 415.
West African: **IVORY COAST** ·KRU· grégrée (Aub.) ·KRU· (Duékoué) cocoti (Aub.) KRU-BETE losiokos (Aub.)

A medium to large-sized forest tree of secondary formations, recorded only from Ivory Coast and Liberia.

The wood is light greyish-brown and soft. The bark exudes a little white latex (2, 3). A root-decoction drunk three times a day is considered in Ivory Coast to be aphrodisiac (1).

References:

1. Adjanohoun & Aké Assi, 1972: 138. 2. Aubréville, 1959: 2: 102. 3. Voorhoeve, 1965: 94.

Sapium carterianum J Léonard

FWTA, ed. 2, 1: 415, as *S. cornutum* sensu Keay, p.p.
West African: SIERRA LEONE KONO peteteyamba (?) (FCD)

A forest shrub, recorded only in Sierra Leone and Liberia, and confused with *S. cornutum* Pax which occurs only east of the Region.

Sapium ellipticum (Hochst.) Pax

FWTA, ed. 2, 1: 415. UPWTA, ed. 1, 163.
West African: SENEGAL MANDING-MANDINKA famadihõ (JB) GUINEA MANDING-MANINKA fama dion (Aub.) UPPER VOLTA MANDING-DYULA badulon (K&B) IVORY COAST ABURE bonyuromé (A.Chev.; K&B) ADYUKRU tatahiro (A.Chev.) tato-iri (Aub. ex K&B) BAULE tomi (Aub. & ex K&B) vlanuku (K&B) BOBO conguromé (Aub.) KULANGO endirem (K&B) KYAMA aguaya (K&B) MANDING-DYULA badulon (K&B) MANINKA korundi (B&D) tubake (B&D) SENUFO-TAGWANA kégné (K&B) GHANA BAULE tomi (FRI) NZEMA ketebontore (DF) NIGERIA YORUBA (Ilorin) aloko-agbọ (Clarke)

A tree reaching usually to about 15 m high in dry situations, but to 40 m by over 2 m girth in damp sites, usually on stream-banks, of the savanna and forest zones, widely distributed from Senegal to W Cameroons and Fernando Po, and across tropical Africa into Natal.

In spite of the size that the tree may attain, the wood is little used in West Africa. The bole is often fluted, and shape is poor. The wood is pale brown, somewhat coarse and fibrous. In East Africa it is said to be not durable in the ground and is liable to borer attack (5, 7), but in Zaïre the wood has been reported to contain silicaceous concretions in the medullary rays and it is possible that the presence of these may confer resistance to marine borers if the timber is used in sea-water (8). Nigerian timber is described as tough (14, Jones & Onochie fide 13). It is used in Uganda to make anvils on which bark-cloth is beaten out (7), and in Tanganyika for purposes where hardness is necessary (9); the strong stems are used in Malawi as the main parts in maize storage bins (Townsend fide 13) and to make mortars (Clements fide 13), and in Kenya for general building (20). Timber from the Kivu Province of Zaïre is exploited for its fine texture and elsewhere in Zaïre it is used to make tom-tom drums (15).

The bark is recognised as a purgative in Congo (Brazzaville) (3). In Ivory Coast a preparation (? from the bark) is taken as a drastic purge and is known for its toxicity. This is sometimes taken by draught for ascites and leprosy and externally for guinea-worm (4, 11). The bark is used in Zaïre on eczema (15) and in central Africa a decoction is used as a mouthwash in cases of scurvy and stomatitis (23). A white slightly gummy latex is present in the younger twigs and leaves. This is used to make a glue in Gabon (22). On skin it is very caustic. It is an ingredient of an arrow-poison in Kenya (1, 6), and has been suspected of criminal use (23). In Tanganyika a leaf-preparation is used to relieve pains in the head, chest, shoulders and back, and a preparation of dried leaves is applied to maggot-infested wounds (2, 23). A leaf-preparation is also used for sore-eyes and abdominal swelling (23).

A root-concoction is prepared as a fomentation in E Africa for enlarged spleen in babies (23) and is taken by draught for malaria (10). The root is laxative (21). In Tanganyika it is eaten pulped with leaves in water for worms

EUPHORBIACEAE

(19), and root-sap is instilled into painful eyes (17). It also provides a cough medicine (12, 18). In Zaïre pulped roots are held to be a cure for stammering (15).

The fruit are said to be eaten in Tanganyika (19). Birds seek them out (7, 15, 16). The fruit is, however, recorded as containing a white latex, as do the younger growths, and this is used in a Kenyan arrow-poison (6). Human consumption of the fruit merits a cautious approach.

References:

1. Aubréville, 1950: 196. 2. Bally, 1937. 3. Bouquet, 1969: 124. 4. Bouquet & Debray, 1974: 86. 5. Dale & Greenway, 1961: 218–9. 6. Dalziel, 1937: 163. 7. Eggeling & Dale, 1952: 141. 8. Frison, 1942. 9. Jefford & al. 198, K. 10. Haerdi, 1964: 98. 11. Kerharo & Bouquet, 1950: 87. 12. Koritschoner 1303, K. 13. Irvine, 1961: 251. 14. Keay & al., 1960: 299–300. 15. Léonard, 1962, b: 153–6. 16. Siwezi 43, K. 17. Tanner 3941, K. 18. Tanner 5041, K. 19. Tanner 6041, K. 20. Templer T.6, K. 21. Wallace 458, K. 22. Walker & Sillans, 1961: 176–7. 23. Watt & Breyer-Brandwijk, 1962: 435.

Sebastiana chamelaea (Linn.) Müll.-Arg.

FWTA, ed. 2, 1: 415.

West African: **NIGERIA** IGBO ile agwọ = *snake's tounge* (NWT; KW) iluloogwo obizi (NWT) ọnụngo (NWT)

A slightly woody herb, annual or perhaps lasting over one year, reaching to about 60 cm high, of open grassy localities, somewhat anthropogenic, from Ghana to N and S Nigeria, and extending into E Cameroun and Central African Republic, and widespread across India, SE Asia and N Australia.

No use is recorded for the plant in Africa. In India a plant-decoction taken with *ghee* (clarified butter) is considered tonic, and this preparation is applied to the head for vertigo. The plant-sap is astringent and the plant is used in India in treatment for syphilis and diarrhoea (1, 2). Leaves and stems of Nigerian material tested for molluscicidal activity showed none at all (3).

References:

1. Burkill, IH, 1935: 1988. 2. Chadha 1972: 263. 3. Adewunmi & Sofowora, 1980: 57–65.

Securinega virosa (Roxb.) Baill.

FWTA, ed. 2, 1: 389. UPWTA, ed. 1, 146, as *Fluggea virosa* Baill.

French: balan des savanes (Bailleul).

West African: **SENEGAL** ARABIC (Senegal) l'emleise (JB) BALANTA bi osì (JB) BANYUN sauda (K&A) savda (JB) BASARI a-nambarisitèn (JB) tok u nangal (JB) DIOLA ba hérèr (JB) é bukèr (JB) fu ñéñé (JB) fu sabèl (JB) DIOLA (Fogny) fusabel fuñéñé (K&A) FULA-PULAAR (Senegal) tem-belgoréy (K&A) tianbèlgorèl (JB) tièmbèlgorèl (JB; K&A) TUKULOR kéki (JB) tem-belgoréy (K&A) tièmbèlgorel (JB; K&A) MANDING-BAMBARA balamanantièn (JB) baram baram (K&A) diéné (JB; K&A) karam karam (JB; K&A) nkoloninge (K&A) nkoloningié (K&A) surku mañéñé (K&A) tiéné (JB) MANDINKA balam balam (K&A) mbiri baram (K&A) MANINKA barin barin (JB) mburum barãg (JB) mpalampala (JB) ·SOCE· burum bara (K&A) burum barã (JB) burum baran (K&A) NDUT bolapan (JB) bulapan (JB) SERER baram baram (K&A) farâg farâg (JB; K&A) maymayin (JB) mbaram mbaram (JB; K&A) mbaram param (JB) SONINKE-SARAKOLE talintia (K&A) talintié (K&A) WOLOF héên (K&A) kên (De la Croix ex K; K&A) kêng (JB) kiên (K&A) **THE GAMBIA** MANDING-MANDINKA brumbarongo (Fox) **GUINEA** BASARI a-nèmbrèsyét (FG&G) MANDING-MANINKA barinbarin (Aub.) kuindié (Aub.) suruku gué gué (Aub.) **SIERRA LEONE** MENDE tigwi (JMD) **MALI** DOGON sɛgɛlɛ (CG) segere(m) (Aub.) **UPPER VOLTA** FULA-FULFULDE (Upper Volta) cami (K&T) sirmuuhi (K&T) sugun lagaahi (*pl.*-aaji) (K&T) MANDING-DYULA bala-bala (K&B) **IVORY COAST** BAULE niassulé baka (K&B) KRU-BETE genakwo (K&B) KULANGO sokulénié (K&B) KWENI buregnemïé (K&B) MANDING-DYULA bala-bala (K&B) MANINKA mokokoama (B&D) mokro-doma (B&D) mpalampala (K&B) SENUFO garadiéma (K&B) SENUFO-TAGWANA katia (K&B) **GHANA** AKAN-ASANTE nkanaa (FRI; E&A) DAGBANI susuwulugu (AEK & ex FRI) GA gbekɛbii-able-tʃo (FRI) wɔlɔmɔtʃo (FRI) GBE-VHE hesre (FRI; JMD) hesse (BD&H) VHE (Pecí)

hlose (FRI) HAUSA tsa (FRI) NIGER ARABIC (Niger) kartié kartié (Aub.) kartjé-kartjé (Aub.)
HAUSA tsa (AE&S) KANURI dagkirto (Aub.) NIGERIA ARABIC-SHUWA dabalab (JMD) kartjik-
artij (JMD) BADE zandanu (FWHM) BIROM dìnáká (LB) hícen gádóó (LB) tsáá (LB) tswáá
(LB) FULA-FULFULDE (Nigeria) camal (JMD; J&D) cambe (JMD) came (JMD) HAUSA gussu
(JMD; ZOG) gwíiwàar kàréé = monkey's knee (JMD; ZOG) itachen-gado = bed-tree; from
itache: tree; gado: bed, alluding to the use of the stems for bed making (JMD) tsa (auctt.)
tsúwàawún kàréé = dog's testicles (ZOG) tswa (JMD; ZOG) IDOMA ǫkplá (Armstrong) IGBO
njisi ntà = small njisi(NWT) KANURI shimshim (JMD; C&H) YORUBA iranjé (JMD; Verger)

A shrub or small tree to 4 m high with numerous branches arising from the
base and spirally arranged upwards, somewhat angular, of wooded savanna and
transition-forest, throughout the Region from Senegal to S Nigeria, and widespread across tropical Africa to India and Australasia.

The plant has attractive foliage and white waxy berries. Its bushy nature
ideally lends itself to ornamental purposes, and it is commonly grown in N
Nigeria as a hedge (11, 23, 41).

The wood is reddish-yellow, close-grained, said to be durable and to yield
good fuel and charcoal in E Africa (9, 12). The timber, when large enough, is
used in Tanganyika to make chair-legs (Watkins fide 23), but the relatively
smaller stature of the tree in West Africa limits the use of the wood to fuel and
charcoal (23) and poles for huts in N Nigeria (10). In Gabon it is intentionally
grown for these purposes (42). The tough virgate stems are commonly used in N
Nigeria to make beds, hence the Hausa name, itachen-gado: tree of the bed, and
also fishing-stakes, wicker-traps, for part of roof-structures (paalol) of Fula
huts, to reinforce granaries, etc. (11, 24). Similar uses occur in India (8). They
are woven into 'shelves' in Ethiopean houses (7) and in Zanzibar split for
basketry (45, 46). Twigs in Kenya are cut for toothbrushes (15). Powdered
charcoal is used in the Philippines as a cicatrisant on wounds (35), and the
wood-ash by the Masai of Kenya to clean milk-containers (14). A gum is
obtained from the stems which has been used in Ghana for sealing envelopes
(22).

The leaves are considered laxative. They are used in decoction in Nigeria (34)
and Ivory Coast–Upper Volta (30). The Hausa in Ghana use them to treat
venereal disease, and the Vhe for constipation (20). A leaf-macerate is added to
baths and applied in massage in Senegal as a stimulant in fatigue and for
stiffness (28). In W Asante a decoction of the plant is used as a lotion to impart
strength (23), and in Nigeria a 2-hour hot water infusion of the whole plant is
said to be stimulant taken as a broth (2). A leaf-decoction is used in S Nigeria
by the Yoruba in draughts or baths for fever (Macgregor fide 23). Leaf-sap in
Tanganyika is administered in cases of epilepsy, and with other drug-plants as a
tranquilliser in insanity (17). In Zimbabwe the leaves (11) and in Gabon the
young shoots with salt and maleguetta pepper (42) are considered aphrodisiac.
Worms in sores are destroyed in India by application of leaf-sap, or the leaf
mashed into a paste with tobacco (8, 43). The leaves yield a dye used in
Zanzibar to dye palm-fibres black (16).

The alkaloids, viroallosecurinine, virosecurinine and virosine, have been
recorded from the leaves (29, 44). Notwithstanding the presence of these active
principles, the foliage is more or less readily browsed by stock in Tanganyika
(19), by elephants (9), goats and sheep (32) in Kenya, and by giraffe and
rhinoceros in Uganda (5).

The bark is astringent. In some parts of the Ivory Coast it is considered too
poisonous to use medicinally. In other parts it is used even in children's
medicine (30). It is sometimes used as a fish-poison in Nigeria (2) and in some
francophone West African territories (34), but toxicity to fish appears to be low
(30). Bark-tannin has been recorded as 8.9% and is used for tanning in India
(8). A black dye is also obtained from the bark in India (8) and in Malaya for
dyeing matting. Bark-extracts have shown some lethal toxicity to mice (8).
Besides tannin and a dye stuff, two alkaloids, one of them flueggeine, have been
detected at 0.4–0.6% concentration (8, 29, 34).

The root is the most commonly used and most pharmacologically active part of the plant. It is used in Senegal by itself or in conjunction with other drug-plants often to provide synergistic effect for liver, bile, kidney and urino-genital complaints, diuresis, renal stone, schistosomiasis, rheumatism, venereal diseases, orchitis, dysmenorrhoea, frigidity, sterility, arthritis, etc. (25–27, 29). With *Anacardium occidentale* (Anacardiaceae) it is an aphrodisiac and an elixir of longevity (28). The root-sap with fat is a recognised soothing ointment and the root-pulp an analgesic in Nigeria (34). In Ivory Coast the plant is used as a purgative and anti-dysenteric, analgesic for costal and fever pains, for ophthalmias and headache and as a sedative and soporific for children (4, 30). In Tanganyika water in which roots have been boiled is taken for schistosomiasis (36), and given to nursing mothers whose milk is unsuited for the child (38), or whose baby is sickly at birth or to mothers who have still-born babies (39); also for infestation of intestinal worms and for infected ears (40). A root-decoction is taken in Kenya (31) and in Tanganyika (37) for stomach-ache, and dysmenorrhoea (17).

Securine is the main alkaloid present, and as many as ten alkaloids closely related to it have been determined present in the plant (1, 4, 8, 24).

The flowers attract bees (21) and insects (13). Thge fruit is edible when mature enough to fall from the bush (11). It is eaten in West Africa (3, 26) and in Senegal is often prescribed in medicine without ill-effect (26). Mixed with pulses the fruit is eaten in NE India to relieve digestive disorders (47). It is eaten in E Africa (11; 31) and in Indonesia (18). In Zanzibar women are said to eat the fruit in order to promote fertility (46). In Kenya the fruit is fed to hens (33). A red dye is obtained in Zanzibar by pounding the fruit in a little hot water. The dye is used as red ink (16).

References:

1. Adegoke & al., 1968. 2. Ainslie, 1937: sp. no. 161, as *Fluggea microcarpa*. 3. Aubréville, 1950: 190, as *S. microcarpa* (Blume) Pax & Hoffm. 4. Bouquet & Debray, 1974: 86. 5. Brooks s.n., 12/6/1960, K. 6. Burkill, IH, 1935: 1027, as *Flueggia virosa* Baill. 7. Carr 884, K. 8. Chadha, 1972: 268–9. 9. Dale & Greenway, 1961: 220. 10. Dalziel 207, K. 11. Dalziel, 1937: 146. 12. Eggeling & Dale, 1952: 128–9, as *Flueggia virosa* Baill. 13. Gillett 14097, K. 14. Glover & al. 75, K. 15. Glover & al. 2572, K. 16. Greenway, 1941: as *Flueggia virosa* Baill. 17. Haerdi, 1964: 98. 18. Heyne, 1927: 902, as *Flueggea virosa* Baill. 19. Hornby 39, K. 20. Irvine 215, K. 21. Irvine 4577, K. 22. Irvine, 1930: 205, as *Fluggea virosa* Baill. 23. Irvine, 1961: 253–4, with leaf-analysis. 24. Jackson, 1973. 25. Kerharo, 1967. 26. Kerharo & Adam, 1962. 27. Kerharo & Adam, 1964, b: 575. 28. Kerharo & Adam, 1964, c: 320. 29. Kerharo & Adam, 1974: 432–4, with phytochemistry and pharmacology. 30. Kerharo & Bouquet, 1950: 75–76, as *Flueggia virosa* Baill. 31. Mwangangi 1289, K. 32. Mwangangi 1483, K. 33. Mwangangi & Napper 1344, K. 34. Oliver, 1960: 82. 35. Quisumbing, 1951: 533–4. 36. Tanner 1329, 3115, K. 37. Tanner 3394, K. 38. Tanner 4266, K. 39. Tanner 4420, K. 40. Tanner 5623A, K. 41. Thornewill 85, K. 42. Walker & Sillans, 1961: 177, as *S. microcarpa* (Blume) Pax & Hoffm. 43. Watt & Breyer-Brandwijk, 1962: 417, as *Fluggea virosa* Baill. 44. Willaman & Li, 1970. 45. Williams, RO 179, K. 46. Williams, RO, 1949: 268–9, as *Flueggia virosa*. 47. Jain & Dam, 1979: 55. See also: Oliver-Bever, 1983: 50–51.

Spondianthus preussi Engl.

FWTA, ed. 2, 1: 372. UPWTA, ed. 1, 164, incl. *S. ugandensis* Hutch.
West African: LIBERIA KRU-BASA bu-en-waye (C&R) IVORY COAST ABE djilika (auctt.) ABURE kianga (auctt.) kiangua (auctt.) AKAN-ASANTE tianga (B&D) tuanga (B&D) AKYE schiédzo (Aub.; K&B) KRU-GUERE buangbu = *rat-poison*; *from* buang: *rat*: bu: *devil, devilry*; *hence poison* (K&B) kootue (K&B) KYAMA agboboba (auctt.) GHANA VULGAR tweanka (DF) AKAN-TWI tweanka (E&A) WASA tweabowuo (DF) tweanka (auctt.) ANYI-ANUFO kyeanga (FRI) SEHWI kyeanga (FRI) NZEMA tweanga = *a dog does not touch it*; *i.e., the fruit, or it will die* (auctt.) NIGERIA EDO orho (KO&S) EFIK íbók-ékù = *rat-poison* (JMD; KO&S) IJO-IZON opolata (KO&S) YORUBA òbo èkúté = *mouse's vagina* (Verger) owe (JMD) WEST CAMEROONS DUALA ebai *from E Cameroun* (E; Ithmann) KPE bojunde (JMD) wujonde (JMD)

Trees to about 30 m tall; of two varieties: var. *preussii* in swampy areas within rain-forest, from Liberia to W Cameroons and Fernando Po, and extending to Zaïre, and var. *glaber* in fringing-forest from Guinea, Ivory Coast, Nigeria and

W Cameroons, and widespread across Africa to Sudan, Tanganyika, Uganda, Zaïre and Angola.

The wood is brownish, strongly speckled, hard, and heavy (2, 4, 11). In Liberia it is sometimes sawn into planks (6) and in Uganda var. *glaber* is commonly used to make dugout canoes (8).

All parts of the tree are poisonous. It is recorded as being one of the most toxic plants in Ivory Coast (12), rodents, cattle and man being particularly susceptible. It has been used for criminal purposes (3, 7, 12), and in many territories (Liberia: 5, 6; Ivory Coast: 12; Ghana: 7, 10; Nigeria: 13; W Cameroons: 7; Gabon: 18) the bark is cooked into a suitable bait as a rodent and verminicide. Bark-sap may also be used, which on fish is recorded to have an immediate effect (17). The seeds (2) and the powdered bark (1) are known in Ivory Coast to be a violent and sudden poison to dogs. It said in Ivory Coast that vermin will also die just by sniffing a poisoned corpse (12), and it is held that the toxicity is so easily communicated to the meat of poisoned animals that arrow-poisons used in hunting do not normally contain it. It has however been used in Ivory Coast in hunting elephants (3).

The foliage is widely known as being dangerous to cattle (Ivory Coast: 12; Nigeria: 13; Katanga: 14, 15; Uganda: 7). Trespassing stock have been poisoned without apparently showing any significant pathological post-mortem signs (7). It is known however that the poisonous stage is in the young growth, the toxicity being lost in dried leaves (14). Phytochemical assays have shown the presence of an active principle of a triterpenic substance of the *cucurbitacin* group (3).

Birds appear to be able to eat the fruit with impunity (9). This fact may have some significance in a curious Yoruba *Odu* incantation involving erotic rats, birds and this plant for the cure of cough (16).

References:

1. Adjanohoun & Aké Assi, 1972: 139. 2. Aubréville, 1959: 2: 62. 3. Bouquet & Debray, 1974: 87. 4. Burtt Davy & Hoyle, 1937: sp. no. 50. 5. Cooper 140, K. 6. Cooper & Record, 1931: 56. 7. Dalziel, 1937: 164. 8. Eggeling 590, K. 9. Eggeling & Dale, 1952: 140. 10. Irvine, 1961: 254–5. 11. Keay & al., 1960: 294. 12. Kerharo & Bouquet, 1950: 88. 13. Oliver, 1960: 37, 84. 14. Staner s.n., K. 15. Staner, P., 1932. 16. Verger, 1967: no. 140. 17. Walker, 1953, a: 37. 18. Walker & Sillans, 1961: 177.

Suregardia ivorense (Aubrév. & Pellegr.) J Léonard

West African: **IVORY COAST** AKYE tassa (Aub. & FP)

An understorey tree to 25 m high by 1.25 m girth recorded only in the dense forest of S Ivory Coast.

Several species are cultivated in India for their highly scented flowers. One other species, *S. occidentalis* (Hoyle) Croizat (syn. *Gelonium occidentale* Hoyle, *FWTA*, ed. 2, 1: 413) occurs in West Africa, Ghana to Nigeria, but no usage is recorded.

Tetrochidium didymostemon (Baill.) Pax & K Hoffm.

FWTA, ed. 2, 1: 414, incl. *T. minus* Pax & K Hoffm. UPWTA, ed. 1, 164.

French: arbre à savon du Gabon (Gabon, Walker).

West African: **GUINEA-BISSAU** CRIOULO pau branco (GeS; JMD) **SIERRA LEONE** KISSI chayiliŋ (FCD; S&F) nyɛlɔ (FCD; S&F) KONO fɛŋgɔnɛ (FCD; S&F) KORANKO wuliyaŋgɛ (S&F) MENDE sɔlɛ (auctt.) sɔlɛ-gbɔlo (*def.* s. gboli) *with red stems* (auctt.) sɔlɛ-golo (FCD) sɔlɛ-guli (-wuli) *with white stem* (FCD; JMD) **LIBERIA** KRU-BASA plor-plor (C&R) MANO tũ bu (Har.) **IVORY COAST** ABE uologpaue (Aub.; K&B) ABURE atuan (B&D) AKAN-ASANTE anénédua (B&D) AKYE nguépé (A.Chev., & ex K&B) ANYI aïrofu (K&B) améné (Aub.) anéré (K&B) echirua (auctt.) BAULE aoarafo (B&D) DAN tumbu (K&B; Aub.) KRU-BETE sagugra (K&B) KYAMA anéné (B&D) bledwé (B&D) kotiem brédué (Aub. & ex K&B) **GHANA** VULGAR anenedua (DF) fihankra (DF) kyekyerantina (DF) AKAN-ASANTE anenedua (auctt.) kyereantena (FRI) asansammuro (FRI) BRONG fiankra (auctt.) TWI abogyedua = *bearded tree;*

from its hanging inflorescences (FRI) kyekyerantena (FRI; E&A) kyikyilantena (BD&H; FRI) ANYI amene (FRI) ekyirua (FRI) **NIGERIA** EDO iheni (auctt.) YORUBA ọfún òkè (KO&S; Verger)

A shrub or tree to 25 m high, bole up to 2 m in girth, occasionally more, with caracteristic zigzag branchlets, common in secondary jungle throughout from Guinea to W Cameroons and Fernando Po, and extending to Zaïre, Uganda, Tanganyika, Angola and São Tomé. *T. minus* of *FWTA*, ed. 2, recorded from Sierra Leone and S Nigeria is a 'witches' broom' form of *T. didymostemon* (13).

Sap-wood yellowish white to colourless, heart-wood pinkish, and they are soft, perishable and of very limited use. In Ghana the poles are used in building (Vigne fide 10), and in Guinea-Bissau for carpentry (9). The wood in Gabon is rated semi-hard and is used for plywood (18a) and for making huts in Zaïre (14). In Liberia the wood is not used (5, 7). The tree is considered a weed species of managed forestry in Sierra Leone (16).

The bark is used in Gabon often beaten up as a soap for washing clothes, from which use it gets the French name *arbre à savon* (soap tree), and also to produce a mouthwash for toothache (17, 18a). Soaked in water or rum in Liberia the liquor is taken as a purgative (5). A bark-decoction is taken in Ivory Coast–Upper Volta as a febrifuge and purgative (12), and is given either by draught or enema to young children with distended stomach due to constipation who do not eat properly and who cry much: reaction is said to be immediate, soothing and purgative (3). Pounded bark is applied to swellings in Liberia (6), and sap or bark-pulp to poultice abscesses, furuncles, buboes etc., and to massage points of pain in the joints, or over the sides and kidneys in Congo (Brazzaville) (1). The bark in infusions is applied to rheumatic limbs in Zaïre, and to combat fleas; scrapings are also made into suppositories (14). A little sap (? bark) in palm-wine or on a banana, or a bark-decoction is considered in Congo (Brazzaville) to be an excellent medicine for enlarged spleen in babies, the liquid being administered via the mother's nipples. Larger dosage is violently purgative and is used with caution for stomach-complaints, food-poisoning, ascites, general oedemas, etc. The leaves may also be eaten cooked as a vegetable. This treatment is also prescribed for blennorrhoea, haematuria and as a vermifuge (1).

The bark contains an abundant lactiferous sap described as rusty coloured (8), reddish (Nigeria: 11) to colourless (E Cameroun: 4). The pharmacological action of the bark is probably mainly due to this sap. It is used in Zaïre after expression from the bark on leprous areas and on buboes, and is instilled into the eye to kill filaria. It is rubbed into small incisions on the tummy of infants to treat for constipation (14). The sap (14), as are the flowers (15), is strong-smelling.

The twigs are used as chew-sticks in West Africa, or are sucked for the sweet sap in the bark (8, 10, 19).

The leaves in Gabon are applied hot to limbs with pain caused by rheumatism or yaws (18b), and in Zaïre they are used as a dressing on the limbs becoming enlarged as symptoms of certain illnesses (14).

The plant (part unstated) enters into a treatment for whooping cough and convulsive coughing in Congo (Brazzaville) (1).

A trace of alkaloid has been detected in the leaves and bark of material from Congo (Brazzaville). In a closely-related species, *T. congolense*, J Léonard, as much as 1% total *alkaloids* has been found in the bark and roots, and a little *saponin* in the bark (2).

Seed-oil is used in massage in Nigeria (Ainslie fide 10). The purpose is not given.

References:

1. Bouquet, 1969: 124. 2. Bouquet, 1972: 26. 3. Bouquet & Debray, 1974: 87. 4. Breteler 1884, 2885, K. 5. Cooper 377, K. 6. Cooper 437, K. 7. Cooper & Record, 1931: 57. 8. Dalziel, 1937: 164.

9. Gomes e Sousa, 1930: 82. 10. Irvine, 1961: 256. 11. Keay & al., 1960: 287–9. 12. Kerharo & Bouquet, 1950: 89. 13. Léonard, 1962, a: 34. 14. Léonard, 1962, b: 134–6. 15. Purseglove P.2430, K. 16. Savill & Fox, 1967, 110. 17. Walker, 1953, a: 36, as *Hasskarlia didymostemon* Baill. 18a. Walker & Sillans, 1961: 178. 18b. ibid, 178, as *T. minus* Pax & Hoffm. 19. Portères, 1974: 139.

Tetrorchidium oppositifolium (Pax) Pax & K Hoffm.

FWTA, ed. 2, 1: 414.
West African: LIBERIA KRU-BASA plor-plor (C&R) IVORY COAST 'KRU' salo koatu (K&B)

A shrub to 3 m high of the closed-forest from Liberia to W Cameroons, and also in Gabon.
The bark has a disagreeable smell. Shavings soaked in palm-oil are used to rub on to areas of rheumatic pain in Gabon. The bark is also considered a good remedy for eye-troubles (3, 4). The plant is used on the Liberian border of Ivory Coast by the 'Kru' as a febrifuge (1, 2).

References:
1. Bouquet & Debray, 1974: 87. 2. Kerharo & Bouquet, 1950: 89. 3. Walker, 1953, a: 36, as *Hasskarlia oppositifolia* Pax. 4. Walker & Sillans, 1961: 178.

Thecacoris stenopetala (Müll.-Arg.) Müll.-Arg.

FWTA, ed. 2, 1: 372.
West African: SIERRA LEONE MENDE kbokile (NWT) kɔngɔli (NWT) TEMNE ɛ-luntewor (NWT) ɛ-til (NWT)

A forest shrub to 4 m high occurring from Sierra Leone to West Cameroon and Fernando Po.
The plant has been recorded used for toothache in Sierra Leone (1).

Reference:
1. Thomas, NWT.30, K (Arch.).

Tragia Linn.

FWTA, ed. 2, 1: 410–2. UPWTA, ed. 1, 165.
English: 'nettle'.
French: liane brûlante (*T. senegalensis* Müll.-Arg., Berhaut).
West African: SIERRA LEONE BULOM (Sherbro) tintis-tuntun-dɛ *for all stinging plants* (FCD) KONO fase-mɛsɛ *for all stinging plants generally* (FCD) KORANKO mɛlimuida (? mɛliminda) *T. volubilis* Linn. (NWT) LIMBA kusomi (NWT) LOKO maŋaŋge *T. volubilis* Linn. (NWT) mɔɛ *for all stinging plants* (FCD) nyaŋge *T. sp.A.* (NWT) LOKO (Magbile) bwe ɛnge *T. tenuifolia Benth.* (NWT) MENDE gɔndomɔɛ *T. volubilis Linn.* (NWT) kasodobui *T. sp.A.* (NWT) keyui *T. volubilis* Linn. (NWT) mɔli, mɔli-mumu, mɔli-nyɛnyɛ *for all stinging plants* (FCD) njaubui *T. volubilis* Linn. (NWT) TEMNE a-thenthres *for all stinging plants* (auctt.) GHANA AKAN-ASANTE batafo sasono = *bush-pig's nettle; for all T.spp.* (FRI) nsansono *general for all urticating T.spp.* (E&A) TWI nsansono *for all T.spp.* (E&A) NIGERIA IGBO àbàlagwọ = *cow-itch of the snake; T. spp.* (FRI & ex JMD) agbala ileñkịtà = *dog's tounge cow-itch; T. volubilis* Linn. (NWT) agbalọ ile nkịtā = *dog's tongue agbalo* (NWT ex JMD; KW) YORUBA èsin *T. spp.* (JMD) èsinsin *T. spp.* (JMD) fara *T. vogelii* Keay (Dawodu)

Herbaceous or sub-woody, mostly twining, of savanna, scrub or secondary vegetation. *T. wildemanii* Beille (incl. *T. akwapimensis* Prain) is sparingly bushy. A few species have wider distribution than the West African region (*T. volubilis* Linn., *T. benthamii* Bak., *T. tenuifolia* Benth., *T. preussii* Pax). The others are all rather limited occurrence within the Region. All except little-known *T. polygonoides* Prain of Ivory Coast are armed with stinging hairs on the leaves, stems, or flowers or fruit. Their stinging habit appears to engender superstitious ideas towards them, e.g., *T. tenuifolia* Benth. has been used by the Igbo to prepare a magical wash for use before going to war (2). See also under *T. benthamii* Bak.

Nigerian material has been examined for *alkaloids,* and these have been detected in the leaves and roots (1).

References:

1. Adegoke & al., 1968. 2. Thomas, NWT.1855 (Nig. Ser.), K.

Tragia benthamii Bak.

FWTA, ed. 2, 1: 412.

English: climbing nettle (Malawi, Cameron).
West African: IVORY COAST AKAN-BRONG agnansompo (K&B) agnansono (K&B) ANYI assuatotoma (K&B) BAULE atutuma (K&B; B&D) DAN yu (K&B) yulé (K&B) KRU-BETE wonfredidi (K&B) GUERE suinzo (K&B) GUERE (Chiehn) uinzeni (K&B) wenzani (K&B) KULANGO inlęyo (K&B) ngimaléyo (K&B) GHANA ADANGME-KROBO gbieha (ASThomas) AKAN-ASANTE nsasun (Andoh) FANTE nsasun (Andoh; A&E) TWI batafo sasõno = *bush pig's nettle* (FRI) NIGERIA EDO ibabidǫn (Hambler) IGBO abalagwǫ (NWT) abwala (NWT) YORUBA èsìsì funfun (Verger)

An herbaceous twining or trailing plant occurring from Ivory Coast to W Cameroon and Fernando Po, and extending widespread to Sudan, E Africa, Zaïre and Angola.

The plant is covered with stinging hairs. The sting can be very painful and the pain is said to last about ten minutes (2). A leaf-mash is used in Ghana on sores on the arms, which have resulted in swollen armpit glands, causing reduction of the swelling (7). At Okpanam, S Nigeria, the plant is used to wash wounds (8). It is known to Baule women in Ivory Coast as an abortifacient or to promote child-delivery. The property of hastening parturition has been shown under medical supervision, but without skilled attention its use can be dangerous as causing too vigorous response and damage to the uterus (1, 3). The plant is also used in enemas for blennorrhoea (6).

It enters into criminal use in Ivory Coast in a mixture of crushed glass and other drug-plants known as *gou-ga*: The Devil's packet, to the Guere, and pulped with the heart of a young dog and the skin of an electric eel, it is said to have magical power to induce madness according to the 'Kru' (6). In Ghana fetishmen mix leaves with eggs for washing themselves, and they also lay the plant on the ground on which to spread their sacred objects (4, 5). The plant enters into a Yoruba Odo incantation against death separating husband and wife (9).

References:

1. Bouquet & Debray, 1974: 87. 2. Collenette 261, K. 3. Fritz & Gazet du Chatelier, 1967. 4. Irvine 845, K. 5. Irvine, 1930: 416. 6. Kerharo & Bouquet, 1950: 89. 7. Thomas, AS D.191, K. 8. Thomas, NWT.1697 (Nig. Ser.), K. 9. Verger, 1967: no. 72; pers. comm. 9/12/86.

Tragia preussii Pax

FWTA, ed. 2, 1: 412.

A twining sub-shrub of the forest region of S Nigeria and W Cameroons, and extending into the Congo basin.

The leaves are said in *FWTA,* ed. 2 to be without stinging hairs, but this is contradicted by other authors. In Ubangi it is described as strongly urticant, and heated leaves are rubbed as a revulsant over areas of rheumatic pain and on the body in fever. If a new-born baby fails to cry, it is soon induced to do so by being rubbed with the stinging leaves! Also cooked leaves are laid over abscesses. There is a belief that leaves bundled into a package and dropped into a river attract fish.

Powder of the crushed seed capsule is dangerous to the eyes.

Reference:

Vergiat, 1970: 79.

Tragia spathulata Benth.

FWTA, ed. 2, 1: 412.

West African: **SIERRA LEONE** BULOM (Sherbro) tintis-tuntun-dɛ (FCD) MENDE mɔli-mumu (-i) (FCD) mɔli-nyɛnye (-i) (FCD) TEMNE a-thentres (FCD) **GHANA** AKAN-ASANTE kwaku nsanson (FRI) **NIGERIA** YORUBA ẹsisi (EWF)

A slender twiner recorded from Sierra Leone to S Nigeria.

The plant is armed with stinging hairs. Leaves mixed with those of *Senna occidentalis* (Linn.) Link (Leguminosae: Caesalpinioideae) are pulped and the sap squeezed out is instilled into the nose in Ghana as a headache cure (1).

Reference:

1. Irvine 634, K.

Uapaca acuminata (Hutch.) Pax & K Hoffm.

FWTA, ed. 2, 1: 390.

West African: **NIGERIA** OLULUMO (Ikom) odan (Catterall)

A tree to 25 m high by 1.30 m girth, with stilt-roots to as high as 5 m, of rain-forest of S Nigeria and W Cameroons, and to Cabinda.

This species has previously been recognised as a variety of *U. heudelotii* (*FTA* 6, 1: 639, 1912). Its characteristics and applications are probably closely similar (q.v.).

Uapaca chevalieri Benth.

FWTA, ed. 2, 1: 390.

French: rikio des montagnes (Ivory Coast, Aubréville).

West African: **SIERRA LEONE** KONO kondi (Fox) KORANKO dombɛ (S&F) dumbɛ (S&F) **IVORY COAST** FULA-FULFULDE (Ivory Coast) ialagué (Aub.)

A tree to 20 m high usually in damp sites by montane streams of Guinea, Sierra Leone and Ivory Coast.

Uapaca esculenta A. Chev.

FWTA, ed. 2, 1: 392. UPWTA, ed. 1, 165.

English: the fruit – sugar plum; (Liberia, Cooper); 'medlar' (Ghana, Irvine).

French: the tree – rikio noir.

West African: **GUINEA** KONO zong'o (RS) **SIERRA LEONE** KONO suanɛ (S&F) MENDE ɲja-kondi (-i) (auctt.) **LIBERIA** DAN soang-ti (GK) KRU-BASA beyor (C) GUERE (Krahn) karro (GK) **IVORY COAST** ABE borikio (auctt.) AKYE nanby = *black 'rikio'* (auctt.) ANYI alohua (auctt.) KRU-GREBO (Oubi) boué (RS) GUERE bué, kantu (RS) KYAMA admellébié (auctt.) MANDING-MANINKA somon (RS) MANO sana (RS) ·SOUBRE· kébi (auctt.) **GHANA** AKAN-WASA kontan-miri (DF; E&A) kuntammiri = *dark kuntan* (FRI) ANYI alohowa (FRI)

A tree to 20 m high by 2 m girth with stilt roots up to 4 m high, of damp rain-forest from Sierra Leone to S Nigeria.

EUPHORBIACEAE

Sap-wood is pale yellow, heart-wood light pink. It makes a good charcoal, favoured by blacksmiths in Ghana (3). The wood is hard and is used for carpentry (6), and in Liberia for canoes and sawing into planks (2).

The bark is black in appearance, resulting in the Akye name in Ivory Coast meaning 'black rikio' (1) and the Wasa name in Ghana 'dark kuntan' (3), but the slash is reddish-brown beneath (4) and the inner bark contains a red sap (2). The fruit is much larger than those of *U. guineensis* and *U. heudelotii.* It reaches 5 cm long by 2.25 cm in diameter. It is edible and resembles the temperate medlar in flavour. It is popular in Liberia where it is known as the 'Sugar Plum' (2, 5).

References:

1. Aubréville, 1959: 2: 40. 2. Cooper 147, K. 3. Irvine, 1961: 257. 4. Keay & al., 1960: 277. 5. Kunkel, 1965: 202. 6. Schnell, 1950, b: 222.

Uapaca guineensis Müll.-Arg.

FWTA, ed. 2, 1: 390. UPWTA, ed. 1, 165–6.

English: red cedar; sugar plum (the fruit, Liberia, Cooper & Record).

French: faux palétuvier; palétuvier d'eau douce (Gabon, Walker).

Trade: rikio (the timber: Liberia: Voorhoeve; Ivory Coast: Schnell).

West African: SENEGAL BALANTA kufa (JB) toro (JB) DIOLA bu begel (JB) é pot mag (JB) fu lalagat (JB) ka begel (JB) tikõk (JB) tikon (JB) DIOLA-FLUP voñi voñi (JB) yaga (JB) FULA-PULAAR (Senegal) mãntiãmpo (JB) tiãgol (JB) yalage (JB) MANDING-BAMBARA samo (JB) somo (JB) somõ (JB) MANDINKA yalagèy (JB) MANINKA samo (JB) MANDYAK bu pal (JB) MANKANYA be pal (JB) **GUINEA-BISSAU** MANDING-MANINKA iala-guéi (EPdS) **GUINEA** KONO zong'o (RS) KPELLE hong-o (RS) MANO sana (RS) TOMA kudi (RS) **SIERRA LEONE** BULOM (Sherbro) tuo-lɛ (auctt.) GOLA njoro (FCD) KISSI kaaŋgo (FCD; S&F) KONO suanɛ (FCD; S&F) KORANKO dombɛ (S&F) dumbɛ (S&F) nerɛ-kerɛ (S&F) sɔmɛ (NWT; JMD) LOKO gɔndi (JMD) kondi (kɔndi) (auctt.) MANDING-MANINKA somo (FCD) somɔŋ (FCD) MENDE nja-kundi = *water 'kundi'* (JMD) kondi (-i) (auctt.) kundi (-i) (auctt.) me-kundi = *edible 'kundi'* (JMD) SUSU jagala (JMD) jagale (NWT) TEMNE an-lil (auctt.) kɔ-sukbe (NWT) VAI lɔŋgbɔ (FCD) YALUNKA khɔkhɔŋyalagɛ-na (FCD) yalage-na (KM) **LIBERIA** DAN soang nasa (GK; AGV) swoang-nasa (AGV) KRU-BASA be-yor (auctt.) GUERE (Krahn) brue (AGV) MENDE kindi (C&R) **MALI** MANDING-BAMBARA sofiro (A.Chev.) MANINKA somɔ (A.Chev.) somon (A.Chev. RS) **UPPER VOLTA** MANDING-DYULA somo (A.Chev.) somon (A.Chev.) **IVORY COAST** ABE borikio (A.Chev.) rikio (auctt.) ABURE alaba (A.Chev.) alobo (A.Chev.) edan (A.Chev.) man (A.Chev.) olonbo (B&D) orobo (A.Chev.; B&D) tiom'bi (A.Chev.) AKYE nan (auctt.) ANYI alohua (A.Chev.) éléhoha (auctt.) elékhua (A.Chev.) kahio (A.Chev.) kayo (A.Chev.) AVIKAM edji-bari (A.Chev.) enebien (A.Chev.) 'KRU' uméné (Aub.; RS) uonmelon (JMD) KRU-GUERE bué (RS) kantu (RS) GUERE (Wobe) kahié (A.Chev.; FB) KULANGO (Bondoukou) kahio (A.Chev.) kayo (A.Chev.) KYAMA alébié (auctt.) NZEMA alokaba (A.Chev.) 'SOUBRE' kéhi (Aub.; RS) **GHANA** AHANTA alocoba (BD&H) AKAN-ASANTE kuntan (ayctt. TWI kontan (E&A) kuntan (CJT; E&A) WASA kuntan (auctt.) ANYI alohowa (FRI) ANYI-AOWIN ɛlɛhowa (FRI) SEHWI ɛlɛhowa (FRI) NZEMA alokoba (auctt.) **NIGERIA** ABUA agɔm (JMD) ɔghɔm (DRR) BAKPINKA nkpana (McLeod) BOKYI odáng =*charcoal* (Kennedy; BAC) odáng-kace *the tree of which* odáng *(charcoal) is made* (BAC) EDO oyɛn (auctt.) EFIK m̀kpènék (Adams; Lowe) ǹkpènék (JMD) EJAGHAM oriang (Kennedy; KO&S) EJAGHAM-ETUNG oren (Kennedy) EKPEYE achi (JMD) ENGENNI ile (JMD) FULA-FULFULDE (Nigeria) bakurehi (*pl.* bakureji) (JMD) HAUSA kafaffago (JMD) wawan kurmi = *cloth of the forest;* the leaves, in allusion to their use to wrap kola nuts (JMD) IBIBIO m̀kpènék (KW) óbúbít m̀kpènèk = *black m̀kpènèk* (KO&S; BAC) IDOMA apǒ (Armstrong) IGBO obia (auctt.) ubia (auctt.) IGBO-IKWERE (Isiokpo) nderite (JMD) IJO-IZON (Kolokuma) ilě (auctt.) IZON (Oporoma) elé (KW) KAKIBA (Ibani) ile (LAKC) NEMBE ile (LAKC) NUPE senchi (JMD) TIV ishase (JMD) shasun (JMD) YORUBA abo ɛmidó = *female 'emido'* (auctt.) àjɛgbé (JMD) ujobe (JMD) yèré (JMD; Verger) yèyé (JMD; Verger) **WEST CAMEROONS** DUALA bòsambi (Ittmann)

A tree to 30 m high, rarely more, bole to 4 m girth by 13 m long, usually much less, more or less fluted with large stilt-roots up to 3 m high, and dense low-

146

branching crown, in high-forest and riverain forest, widespread from Senegal to S Nigeria, and to Congo basin and Uganda.

The stilt-rooted habit of this riverside species has raised the suggestion that it is able to stabilise river-banks, to curb wash and to break the flood movement of water in dry-forest zone rivers (11).

Sap-wood is whitish tinged red, heart-wood is red to reddish-brown. It is hard, durable and moderately heavy, and when quarter-sawn is attractively figured with a silver grain (4, 6, 9, 12). The timber has at one time been exported in a small way from Liberia as 'false mahogany' (4), but present-day use appears to be entirely local. It has been felled in Sierra Leone for saw-milling but this too has ceased in recent years and the tree is now considered a weed-tree (9). The wood is easy to work and is used in local carpentry, in planks, beams, furniture, etc. It has been used for dugout canoes in several countries of the Region (Liberia: 3; Ghana: 6; Niger Delta, Nigeria: 7). The stilt-roots and branches are suitable for boat-ribs (5, 11) and the Munchi use them for the feet or props of beds (5). The wood produces a good firewood and charcoal. In Gabon the charcoal is used for heating flat-irons (14).

The leaves which may reach 30 cm long by 15 cm wide are used in Nigeria to wrap kola nuts, hence the Hausa name for the leaves, *wawan kurmi*: cloth of the forest (5); cf., the leaves of *U. togoensis*, similarly used in S Nigeria. The leaves pulped with palm-oil are used in Congo (Brazzaville) to maturate furuncles and to relieve migraine and rheumatism, and in massage onto children late in learning to walk (1).

The bark has a sickly odour when fresh, and on young twigs exudes a red sticky sap which dries as a gum (3, 5). This is used as a dye in Sierra Leone (9). A bark-preparation is used in Gabon to dye fishing-lines (14). A decoction of root-bark is taken by mouth or by enema in Ivory Coast (2) and in Congo (Brazzaville) by draught (1) for oedemas and gastro-intestinal troubles. In Congo (Brazzaville) bark-decoctions are also used for female sterility, tooth-troubles, rheumatism, piles and a strengthener to rachitic or premature babies (1), and in Gabon in enemas as an antemetic, in lotions with salt for skin-complaints (13, 14) and in powder form for nasal cancer (13). A bark-preparation is used for leprosy in Ivory Coast (2).

The roots are considered aphrodisiac in Gabon (13, 14,), to be good for male impotence in Nigeria (15), and in Ivory Coast to be aphrodisiac and anti-abortive (2). A preparation is recommended for young women in labour (2). In Sierra Leone inhalation of steam from water in which roots are boiled is said to relieve headache (9) and in Congo (Brazzaville) a tisane made from the roots is given for flu with headache and to relieve fever-pains while the affected parts are embrocated with the lees (1). A root-tisane is also considered expectorant and good for rhino-pharyngeal and pulmonary affections (1).

The flowers and bark are sometimes put into N Nigerian arrow-poisons as ingredients supposed to make the flesh putrify around the wound (5).

The fruit has a sweetish edible pulp, that from forest areas being more esteemed than from the savanna (5). It is eaten raw and the pulp is made into a refreshing drink in Nigeria (15).

In Liberia the fruit is known as the 'sugar plum' (3) and is said to be rich in vitamins (8). It is considered in Nigeria that places in rivers immediatedly near or overhung by these trees are the best sites for fishing (11). Presumably fish are attracted by the fallen fruit. Unripe fruit while the capsule is still green and spongy is used in Liberia in cough-medicine (3, 4).

References:

1. Bouquet, 1969: 125–6. 2. Bouquet & Debray, 1974: 87. 3. Cooper 298, K. 4. Cooper & Record, 1931: 57, with timber characters. 5. Dalziel, 1937: 165–6. 6. Irvine, 1961: 257. 7. King Church 46, 49, K. 8. Okiy, 1960: 121. 9. Savill & Fox, 1967, 125–7. 10. Taylor, 1960: 166. 11. Unwin, 1920: 51, 334–5. 12. Voorhoeve, 1965: 101. 13. Walker, 1953, a: 38. 14. Walker & Sillans, 1961: 179. 15. Iwu, 1986: 145.

EUPHORBIACEAE

Uapaca heudelotii Baill.

FWTA, ed. 2, 1: 390. UPWTA, ed. 1, 166.

French: rikio des rivières (Ivory Coast, Aubréville).
West African: **GUINEA-BISSAU** PEPEL bichine (JDES fide EPdS) SUSU iagale (JDES fide EPdS) **SIERRA LEONE** KISSI bondilo (FCD; S&F) KORANKO nerɛ-kerɛ (S&F) MANDING-MANDINKA kɔ-some (FCD) MENDE nja-kondi(-i) = the water 'kondi' (auctt.) koondi (SKS) TEMNE an-lil (NWT) YALUNKA bolokhoni-na (FCD) **LIBERIA** DAN soang-puh (GK) KRU-BASA jah-dah (C) **IVORY COAST** ABE rikio (A.Chev.) ABURE sannaba (A.Chev.) AKYE nobi (A.Chev.) ANYI alohua (A.Chev.) FULA-FULFULDE (Ivory Coast) ialagué tiangol (Aub.; RS) KRU-GUERE (Wobe) niguri-kahie (A.Chev.) KULANGO (Bondoukou) nion (A.Chev.) KYAMA dambrohia (Aub.; RS) MANDING-BAMBARA ko-somon (A.Chev.) MANINKA ko-somo (Aub.; RS) **GHANA** AKAN-ASANTE kuntan (FRI; E&A) kuntan-akoa (auctt.) FANTE kweidzabra (FRI) TWI kuntan (CJT) WASA kuntan (auctt.) GA kweidzabalã (FRI) NZEMA alokoba (JMD; CJT) dua baa (FRI) kpakyi-alokoba (CJT) **TOGO** YORUBA-IFE (TOGO) oli (Volkens) **NIGERIA** BOKYI odang (JMD; KO&S) EDO ǫyęn (auctt.) IDOMA apo (Odoh) IGALA ápǭ (Odoh; RB) IGBO ibia-ile (KO&S) obia-ile (JMD) IJO-IZON ileé (KW) KAKIBA (Ibani) ile (JMD) IṢEKIRI ighęn (JMD; KO&S) NUPE senchi (auctt.) URHOBO otehor (JMD; KO&S) YORUBA abo ęmidó (JMD) àkún (auctt.) yèré (JMD; Verger) yèyé from òyé: intelligence (Verger)

A tree to 30 m tall, bole to 3 m girth, stilt-rooted to 3 m high, dense crown, low-branching limiting bole length to 7–8 m, of riparian situations in the closed and fringing-forests from Guinea and Mali to W Cameroons, and extending to Zaïre. This species has very similar characteristics and uses as has *U. guineensis*.

The stilt-roots, and in addition adventitious aeration roots, of the tree on stream-banks encourage accumulation of silt and the formation of solid ground countering erosion (5, 8). Fish-traps may be placed against the stilt-roots in Ghana, and contribute to further deposition of silt (6).

The sap-wood is light-coloured tinged with red; the heart-wood is reddish-brown with grain resembling mahogany but a bit harder, more fibrous and less easily worked (6). It is durable and suitable for carpentry and general construction, and is used in buildings, cut for planks and timbers (3–5). It splits well and makes good firewood and charcoal. It is cut for building poles in N Asante (7). It has been suggested for barrel-staves (5). The stilt-roots and the branches are used for ribs of boats and parts of canoes (5, 8).

The inner bark contains a sticky red sap (3, 4). A bark-extract is used in Gabon to dye fishing-lines and is used on skin-affections and as an emetic (9). In Congo (Brazzaville) a stem or root-bark-decoction is given for female sterility, food-poisoning, mouthwash and gargle for tooth-troubles, for rheumatism and localised oedemas and rachitic and premature babies in baths, and for piles by enemas (1).

The leaves, or the bark, like those of *U. guineensis*, are pulped with palm-oil in Congo (Brazzaville) for application to furuncles, migraines and rheumatism, and for massage on to children late in learning to walk (1). A quantity of *saponin* is present in the leaves, but without other active principles (2). The fruit is edible.

The plant (part not specified) is used by Yoruba medicine-men in medication to promote intelligence: òyé, from which the Yoruba name is derived (10).

References:

1. Bouquet, 1969: 125. 2. Bouquet, 1972: 26. 3. Cooper 179, K. 4. Cooper & Record, 1931: 57–58. 5. Dalziel, 1937: 166. 6. Irvine, 1961: 258. 7. Taylor, 1960: 166. 8. Unwin, 1920: 334. 9. Walker & Sillans, 1961: 179–80. 10. Verger, 1972: 39.

Uapaca paludosa Aubrév. & Léandri

FWTA, ed. 2, 1: 390. UPWTA, ed. 1, 166.

French: rikio des marais (Ivory Coast, Aubréville).
West African: **IVORY COAST** ABE rikio (DF) KYAMA dambrohia (auctt.) **GHANA** AKAN-ASANTE kuntan-nua (CJT; E&A) TWI kuntan (CJT) WASA kuntan (auctt.) GUANG-GONJA bóyè this sp. prox. (FRI) NZEMA alokoba (CJT) **NIGERIA** YORUBA ujobe (Farquhar)

148

A tree to 20 m high, bole straight to 2 m girth with strongly developed stilt-roots up to over 3 m in height, of the swamp-forest from Ivory Coast to S Nigeria, and in E Cameroun to Congo (Brazzaville).

Sap-wood is pinkish-white, heart-wood darker with wood-vessels which in longitudinal section appear very shiny as if they contained a gum (3). The wood in Ghana is said to furnish good firewood (2).

No medinical use is recorded for the tree in West Africa. In Congo (Brazzaville) a tisane of the root is taken as an expectorant in rhino-pharyngeal and pulmonary affections, and for flu with headache, fever and fever-pains a tisane with massage applying the lees; root or stem-bark in decoction is drunk for female sterility, dysentery and food-poisoning, and is used as a mouthwash and gargle for tooth-troubles, as a vapour-bath for rheumatism and localised oedema, as an enema for piles, and in baths as a strengthener for rachitic children and premature babies; and leaves or bark pulped with palm-oil are applied to furuncles and for migraine and rheumatism, and in massage to children late in learning to walk (1).

The fruit is edible (2).

References:

1. Bouquet, 1969: 125. 2. Burtt Davy & Hoyle, 1937: 51. 3. Taylor, 1960: 167.

Uapaca staudtii Pax

FWTA, ed. 2, 1: 390. UPWTA, ed. 1, 166.

Trade: bosambi; rikio (West Cameroons, Dalziel).

West African: NIGERIA BOKYI oda (KO&S) EDO ọyẹn (Elugbe) EFIK ńkpènék (KO&S) IBIBIO obubit nkpenek (KO&S) YORUBA abo-àkún = *female 'akun'* (auctt.) àkún (auctt.) WEST CAMEROONS DUALA bòsambi (JMD) KUNDU bosambi (JMD) LUNDU bihambi (Mildbraed) disambi (JMD)

Tree to 30 m tall, bole often short and crooked, with prop or stilt-roots and a heavy crown, along streams and coastal forests, sometimes in almost pure stands in S Nigeria and W Cameroons, and in E Cameroon. Sometimes it is regarded as a fresh-water mangrove.

The heart-wood is red, darkening to red-brown with red rays in transverse section; sap-wood is paler, of medium coarse grain, fairly hard and heavy, sinking in water, polishing readily and easily worked. It is a good timber for general purposes, very durable and said to be termite-proof. It is useful for interior work in houses and for some sorts of furniture such as tables, chairs, etc. It has been proposed for railway-sleepers and barrel-staves. It is converted into charcoal (1).

The fruit is said to be edible (1).

Reference:

1. Dalziel, 1937: 166.

Uapaca togoensis Pax

FWTA, ed. 2, 1: 390 UPWTA, ed. 1, 166, incl. *U. somon* Aubréville & Léandri.

West African: SENEGAL DIOLA bu begel (JB) MANDING-BAMBARA samo (JB) MANINKA somo (Aub.) somon (Aub.; AS) GUINEA-BISSAU FULA-PULAAR (Guinea-Bissau) mant-champo (JDES) MANDING-MANDINKA iála-guei (JDES) PEPEL bichime (JDES) bissime (JDES) SUSU iágalê (JDES) GUINEA FULA-PULAAR (Guinea) yalagué (Bouronville) MANDING-MANINKA somo (Aub.) somon (Aub.; RS) SUSU yagalé (RS) TOMA kudi (RS) SIERRA LEONE BULOM (Sherbro) tuo-lɛ (FCD) GOLA ɲjoro (FCD) KISSI kaaŋgo (FCD) KONO suanɛ (FCD) KORANKO bambe (NWT) LOKO kondi (FCD) MANDING-MANDINKA somo (FCD) somɔŋ (FCD) MENDE kondi (FCD) TEMNE aɲ-lil (FCD) yalagɛ-na (FCD) VAI kain kolan (DS) lɔŋgbɔ (FCD) YALUNKA khɔkhɔŋyalagɛ-na (FCD) yalage-na (FCD; KM) MALI MANDING-BAMBARA ko-somon (Dubois) MANINKA somo (Aub.) somon (auctt.) UPPER VOLTA MANDING-DYULA

EUPHORBIACEAE

somon (A.Chev.) **IVORY COAST** ABE akoré (A.Chev.; RS) AKYE akoré (A.Chev.; RS) BAULE béla (Aub.) bila (Aub.; RS) FULA-FULFULDE (Ivory Coast) ialagué (Aub.; RS) KRU-GUERE (Wobe) pɔ-tihuɛ (A.Chev.; RS) MANDING-MANINKA somon (auctt.) sumonko (B&D) SENUFO somon (Aub.) **GHANA** AKAN-ASANTE kuntan ɛsirem *i.e.,uapaca of the savanna* (TFC; FRI) GA kweidzablã (FRI) GBE-VHE adziro-ku(-e) (BD&H; FRI) dziro (FRI) dzobe (FRI) dzogbedzro (FRI) jobejiro (BD&H) **TOGO** BASSARI tschelengɔ (Gaisser) GBE-VHE egba (Volkens) SOMBA muhotihuo, uabu (Aub.) TEM (Tschaudjo) kedscheling (Volkens) kidgeling (Volkens) YORUBA-IFE (TOGO) nagudi (Volkens) **DAHOMEY** BAATONUN sâru (Aub.) uadu (Aub.) SOMBA muhoti-huo (Aub.) uabu (Aub.) **NIGERIA** BOKYI odáng = *charcoal* (KO&S) EDO ọyẹn (Elugbe) FULA-FULFULDE (Nigeria) bakurehi (KO&S) HAUSA kafaffago (auctt.) IGALA àpõ (H-Hansen; RB) IGBO obia (KO&S) KANURI gòrámfî (auctt.) NUPE senchi (Yates; KO&S) SAMBA-DAKA bungya (Chapman) TIV shase (JMD; KO&S) YORUBA àjẹgbé (KO&S)

A tree to 13 m high, bole short, usually without stilt-roots on dry sites, but may be stilted in damp localities, of savanna regions from Senegal to W Cameroons, and in E Cameroon to Central African Republic. Sometimes in Sierra Leone it forms nearly pure stands (7).

The wood is reddish brown, hard, heavy and beautifully figured (4, 8), but it appears to be used only as fuel providing an excellent firewood and charcoal (3, 4, 6).

The leaves, which may be as much as 25 cm long by 15 cm wide, have been used in S Nigerian markets to wrap kola nuts (2). In Ivory Coast they are decocted with leaves of *Annona arenaria* (Annonaceae) to make a restorative wash used in general fatigue (1). The Maninka of Ivory Coast use the sap expressed from young twigs on furuncles, and also take it in the belief that this can result in having fine children (1).

The fruits are edible and are recorded as eaten in Sierra Leone (5) and in Gabon (Sillans fide 6).

References:

1. Adjanohoun & Aké Assi, 1972: 140. 2. Ajayi FHI.36098, K. 3. Aubréville, 1950: 192, as *U. somon* Aubrév. & Léandri. 4. Dalziel, 1937: 166. 5. Deighton 4257, K. 6. Irvine, 1961: 258. 7. Savill & Fox, 1967, 124. 8. Taylor, 1960: 168.

Vernicia cordata (Thunb.) Airy Shaw

FWTA, ed. 2, 1: 368, as *Aleurites cordata* (Thunb.) R.Br.

English: Japanese tung oil tree.

French: noix de Bancoil; noix chandelle; abrasin (Berhaut).

A tree of Japan, grown and cultivated there for its seed-oil used principally as an illuminant, for lubrication of machinery, and water-proofing fabrics. It is a drying oil with an iodine value of 148–160. It has been introduced to Senegal (1). It requires a temperate climate and is likely to succeed only at higher elevations in West Africa.

Reference:

1. Berhaut, 1967: 209, as *Aleurites cordata*.

Euphorbiacea indeterminata

West African: **NIGERIA** IJO-IZON (Kolokuma) dísin (KW&T)

An undetermined tree of the above Izon name is used in the Delta region of Nigeria for making canoes (1).

Reference:

1. Williamson, K & Timitimi, 1983.

150

FLACOURTIACEAE

Byrsanthus brownii Guill.

FWTA, ed. 2, 1: 197.
West African: **SIERRA LEONE** LOKO yubɛŋ (NWT) MENDE kwɛga (NWT) TEMNE anaŋkaragbɔnko (NWT) rosɔs (NWT)

A small tree or shrub of riverain forest recorded from The Gambia, Guinea and Sierra Leone.

Caloncoba brevipes (Stapf) Gilg

FWTA, ed. 2, 1: 188. UPWTA, ed. 1, 46.
West African: **SIERRA LEONE** MENDE gɛŋɛ (def. -i) (FCD; S&F) kenne (def. -i) (DS) kenye (def. -i) (SKS; S&F) **LIBERIA** KRU-BASA klehn (C&R) MANO kênê kênê (Har.) **IVORY COAST** KRU-GREBO dule (Aub.) GUERE butue (Aub.) GUERE (Wobe) dolié (Aub.)

A forest tree to 16 m high recorded only from Sierra Leone, Liberia and Ivory Coast.

The tree grows with a strong, durable, slender clean bole in Liberia and is used there for house-posts (2, 4, 5). In Ivory Coast it is a small tree with twisted trunk and much branched (1).

The inner bark and the leaves are used in Liberia for headache either in a poultice or by draught in a decoction (4, 5).

The seeds are used in Liberia for treating skin-diseases, craw-craw and scrofula, either by the expressed oil (4) or pulverising them to a powder and making a paste (3).

The mucilage surrounding the seeds is eaten in Sierra Leone (6), and the ripe seeds in Ivory Coast (1).

References:

1. Aubréville, 1959: 18. 2. Cooper 89, K. 3. Cooper 173, K. 4. Cooper 331, K. 5. Cooper & Record, 1931: 26, with timber characters. 6. Deighton 3830, K.

Caloncoba echinata (Oliv.) Gilg

FWTA, ed. 2, 1: 188. UPWTA, ed. 1, 46.
French: gorli (from Mende, a common term throughout francophone West Africa).
West African: **GUINEA** TOMA kuêuéui (RS) **SIERRA LEONE** VULGAR gorli *from Mende* KISSI komehumda (FCD; S&F) kulukɛnyɔ (FCD; S&F) nyanyandɔ (S&F) KONO fen-kɔne (S&F) kumiŋ-kaŋya (S&F) kumiŋ-kɔne (S&F) kumiŋ-kɔnɛ (FCD) LOKO nikawumbi (NWT) MENDE gɔli (def. -i) (auctt.) TEMNE kʌ-nanafira (NWT) **LIBERIA** KRU-BASA dooh (C&R) flan-chu (C&R) **IVORY COAST** ABURE ɡbétélibé (B&D) ·KRU· katupo (Bardin; RS) **GHANA** AKAN-KWAWU ɔkyerɛ (FRI) TWI ɛsono ankyi (FRI) WASA ɔbafufu (auctt.) wanka (FRI; E&A) GBE-VHE gbogble (FRI) NZEMA marhɛbomuane (FRI) **WEST CAMEROONS** VULGAR gorli *in francophone Africa*

A shrub, or small tree to 8 m high of the understorey of the closed-forest from Guinea to Western Nigeria.

The wood is very hard and is used for walking-sticks, tool-handles, and small articles like combs (2, 3, 5), which in Ghana evokes the Twi name meaning *elephant's comb* (5).

The roots are considered in Ivory Coast to be emmenagogic (1), but the principal application of the plant in all parts of its area of occurrence is in the treatment of leprosy and dermal infections. A decoction of leafy twigs is used in Ivory Coast to wash sores (6) and by enema and in baths for small-pox (1). The root, bark and seeds are used in local medicine in Liberia, mostly for treating skin-diseases (2), and a lotion of the plant is used in Guinea and Sierra Leone for pustular eruptions of the skin (3). The seeds contain about 46% of an oil. Crushed seeds yield a vegetable butter known as *gorli butter*, and the oil, *gorli oil*, can be extracted by pressure, ether extraction or by heating to 120°C (19), and consists of about 87.5% *chaulmoogric acid*, one of the classical substances used in leprosy therapy. Indian material contains also *hydnocarpic acid* which has been found to be the more active principle, and this unfortunately is lacking from gorli oil. These substances have, however, now been abandoned for sulphone drugs which are easier to handle and are better tolerated by the patient (1).

Gorli oil is not edible. The seeds when chewed taste sweetish leaving a peculiar after-taste and may result in nausea, vomiting and irritation of the mucous membrane of the stomach. The oil which is a hard white crystalline fat is suitable for soap and candle-making (3). In Sierra Leone the oil is used as a hair-dressing (4), and because of its pleasant smell as a perfume (7), and to apply to pustular eruptions (8). In Ivory Coast the pounded seeds are used sometimes with success against lice and mange (6).

The fruit-pulp is considered edible in Sierra Leone and is the source of a sweet drink (4).

References:

1. Bouquet & Debray, 1974: 90. 2. Cooper & Record, 1931: 26–27. 3. Dalziel, 1937: 46. 4. Deighton 2114, K. 5. Irvine, 1961: 75, with seed analysis, but lacking oil content. 6. Kerharo & Bouquet, 1950: 39. 7. Pyne 147, K. 8. Savill & Fox, 1967: 128. 9. Schnell, 1950,b: 236.

Caloncoba gilgiana (Sprague) Gilg

FWTA, ed. 2, 1: 189. UPWTA, ed. 1, 47.
West African: **SIERRA LEONE** MANDING-MANDINKA singo (NWT) MENDE gɛnɛ (*def.* -i) (S&F) kenye (*def.* -i) (S&F) SUSU kokwi (NWT) **IVORY COAST** AKYE kauanunguessé (Aub.) **GHANA** AKAN-ASANTE asratoa (E&A) aweawu (FRI; E&A) sissiru (Soward) FANTE nwoantia (CV; FRI) KWAWU ɔkyerɛ (FRI) WASA kotowiri (auctt.) GBE-VHE (Awlan) efiohle (Volkens fide FRI) gbogble (FRI) **TOGO** GBE-VHE efiohlɛ (Volkens) **NIGERIA** YORUBA iroko ojo (KO&S)

A straggling shrub or small thorny tree to 13 m high of the deciduous and fringing forest from Sierra Leone to S Nigeria.

The wood, hard and light brown, takes a good polish and is suitable for inlay and cabinetry. It is used for house-building and for firewood (1, 2). The large showy white flowers with conspicuous yellow stamens render it worthy of cultivation as a garden ornamental. The fruits contain numerous seeds surrounded by a thin pulp, and are considered edible (2).

References:

1. Dalziel, 1937: 47. 2. Irvine, 1961: 75, 77.

Caloncoba glauca (P. Beauv.) Gilg

FWTA, ed. 2, 1: 189. UPWTA, ed. 1, 47.
West African: **NIGERIA** EDO otienme *cf.* otien; *cherry* (KO&S; Elugbe) otiosa (Ross fide JMD) IGBO ụdalā ènwè = ụdalà (Chrysophyllum delevoyi, Sapotaceae) *of the mokeys* (KO&S; KW) YORUBA kánkán dìká (auctt.)

A small tree to 15 m tall of the fringing forest from Ivory Coast to W Cameroon, and extending to Zaïre.

The wood is red (reddish-white, 6), fine-grained and suitable for turnery, tool-handles, etc. (5).

The fruit is edible (6). The seeds contain an oil similar to that from *C. echinata* with concentration of 30% (5) to 36% (1), containing glycerides of *chaulmoogric acid*, but of *hydrocarpic acid* its presence is not clear (1). The oil has been used in Nigeria for pustular skin eruptions, and also for treating leprosy but efficacy has not been confirmed (7). The oil contains *cyanocardin*, a cyanogenetic substance which hydrolyses to *cyanocardase* (2). In Gabon the crushed seed cooked in water is used as poison-bait for rats in food storage-rooms and the oil is used as a syphilis remedy (9, 10).

In Congo (Brazzaville) leaf-sap is put on the temples to relieve migraine. The plant is held to be aphrodisiac, and leaf-sap diluted in water is sprinkled over tombs to keep the spirits of the dead away (3). Tannins are present in the bark and roots, steroids and terpenes in the leaves, bark and roots, and an appreciable quantity of hydrocyanic acid in the leaves, bark and roots (4).

The plant, part not stated, is recorded as being part of the diet of the chimpanzee and gorilla in Rio Muni (8).

References:

1. Adriaens, 1946. 2. Bardin, 1937, b: 356–65. 3. Bouquet, 1969: 126. 4. Bouquet, 1972: 26. 5. Dalziel, 1937: 47. 6. Irvine, 1961: 77. 7. Oliver, 1960: 21, 50. 8. Sabater & Sanford 6061, K. 9. Walker, 1953,a: 38. 10. Walker & Sillans, 1961: 180.

Caloncoba welwitschii (Oliv.) Gilg

FWTA, ed. 2, 1: 188.

A shrub or tree to 10 m high of the evergreen forest of Nigeria and W Cameroon, and of E Cameroun, Gabon, Zaïre and Angola.

No usage appears to be recorded for the W African region. In Central Africa it finds considerable medicinal use. In Congo (Brazzaville) the leaves and bark are made into poultices to maturate absesses and into plasters for bronchial affections and rheumatism; leaf-sap is instilled into the nose for headache; pounded leaves are applied to reduce swellings in bone-fractures before fixing splints; they are sometimes used to relieve trypanosomiasis, and powdered leaves are sprinkled on the severed umbilical cord of the new-born baby; pulped bark with palm-oil is used to treat itch, bark-juice in draught or by enema is taken for internal complaints and to expel parasites, and the plant is prescribed as a vermifuge to kill body-lice and as a snuff for head-colds (2). In Tanganyika a root-decoction with the roasted ground-up seeds of *Cucurbita pepo* Linn. is taken for epilepsy (4). In Gabon the fruit-pulp is considered edible and it seems that there, as in the W African Region, the plant has no medicinal uses (5, 6).

The fruit [presumably the seed for its contained oil] is used in Ubangi for leprosy (7). The seed-oil has been used in Zaïre for leprosy, and the presence of both *chaulmoogric* and *hydnocarpic acids* has been shown (1). In material from Congo *tannins, quinones, steroids* and *terpenes* have been found in the roots, and the two last-named also in the leaves and bark. *Hydrocyanic acid* is reported present in quantity in leaves, bark and roots (3). The seed, dried and reduced to powder, is poisonous, an antidote to which is given as *'Berlinia acuminata'* (7), probably *B. grandiflora* (Leguminosae: Caesalpinioideae).

References:

1. Adriaens, 1946: with references. 2. Bouquet, 1969: 126–7. 3. Bouquet, 1972: 27. 4. Haerdi, 1964: 70. 5. Walker, 1953,a: 38. 6. Walker & Sillans, 1961: 181. 7. Vergiat, 1970,a: 74–80.

Caloncoba sp. indet.

West African: GHANA NZEMA akonzikozokwa (FRI)

An unidentified plant of the above vernacular is recorded as having an unspecified medicinal use in Ghana (1).

Reference:

1. Irvine, 1961: 77.

FLACOURTIACEAE
Camptostylus mannii (Oliv.) Gilg

FWTA, ed. 2, 1: 187.

A forest tree to about 16 m high occurring in eastern Nigeria, W Cameroon and Fernando Po, and into E Cameroun to Angola.

The wood is red and hard (3).

A bark-decoction is used in Congo (Brazzaville) as a febrifuge, to dress sores and as an eye-instillation for filaria. The roots have a sharp piquant smell and powdered root is made into snuff for head-colds and headache, and is also used as a tranquilliser in mental cases (1). Traces of *flavones* and *tannin* have been reported present in the roots (2).

References:

1. Bouquet, 1969: 126. 2. Bouquet, 1972: 27. 3. Keay & al., 1960: 106.

Casearia barteri Mast.

FWTA, ed. 2, 1: 198, incl. *C. dinklagei* Gilg. UPWTA, ed. 1, 48, incl. *C. dinklagei* Gilg.
West African: **LIBERIA** KRU-BASA nu-eh-blay-chu (C&R) **IVORY COAST** AKYE kobété-trumon (Aub.) 'SOUBRE' guézu (Aub.) **GHANA** NZEMA punum (CV; FRI) **NIGERIA** EDO akpano-ẹzẹ *from* ẹze: *waterside*; akpano: *a common term for Rubiaceous shrubs* (auctt.) ukpakuzọn (auctt.)

A tree to about 20 m tall, of humid and swamp-forests – hence the Edo epithet *waterside* – occurring from Sierra Leone, where it is locally abundant, to W Cameroon, and into E Cameroun and Gabon.

The branches are used to make chew-sticks (4) in S Nigeria and Ghana (5). The leaves are given (? in decoction) in Liberia as a purge (2). The bark-slash exudes a little gum (3). Tests have shown (part not stated) the presence of *saponins* (1).

References:

1. Bouquet & Debray, 1974: 159. 2. Cooper 408, K. 3. Keay & al., 1960: 115, as *C. dinklagei* Gilg. 4. Macgregor R243, K. 5. Portères, 1974: 115.

Casearia calodendron Gilg

FWTA, ed. 2, 1: 198, *C. inaequalis* Hutch. & Dalz.
West African: **IVORY COAST** SENUFO-TAFILE kalakari (Aub.)

A tree of the closed-forest and old farmland, to about 20 m high with straight, slender, clear bole, recorded from Sierra Leone to S Nigeria.

Sap-wood is whitish, and bark-slash exudes a little gum. (1).

Reference:

1. Keay & al., 1960: 117.

Dasylepis brevipedicellata Chipp

FWTA, ed. 2, 1: 186, incl. *D. assinensis* A Chev.

A shrub or tree to 10 m high by 1 m in girth of the rain-forest of Ivory Coast and Ghana.

The wood is yellow and very hard (1, 2).

References:

1. Burtt Davy, & Hoyle, 1937, as *D. racemosa* sensu BD & H. 2. Irvine, 1961: 77.

154

Dasylepis racemosa Oliv.

FWTA, ed. 2, 1: 186. UPWTA, ed. 1, 47.
West African: WEST CAMEROONS BAMILEKE shiyatu (Johnstone fide JMD)

An understorey tree of montane rain-forest, to 15 m high, and known only from eastern Nigeria and W Cameroon in the Region, and from Zaïre and E Africa.

Dissomeria crenata Hook. f.

FWTA, ed. 2, 1: 194. UPWTA, ed. 1, 48.
West African: GHANA AKAN-ASANTE papa-yɛ (BD&H; FRI) BRONG abadua (auctt.) GUANG abotiya (Rytz) kùbòtìyà (Rytz) obotiye (Rytz) GUANG-GONJA abotiyɛ (FRI) obotiyie (FRI)

A small tree to 13 m high, of fringing-forest, and stream-sides in the savanna. Occurring in Guinea to N Nigeria. The tree has spikes of attractive white flowers.

Dovyalis zenkeri Gilg

FWTA, ed. 2, 1: 90, incl. *D. afzelii* Gilg and *D.* spp. A, B and C.
West African: SIERRA LEONE KONO gbɔ *the fruit* (FCD) gbɔ-wai *the bush* (FCD) gbɔ-wai-kɔnɛ *from* kɔnɛ: *tree; for the bush* (FCD) gbɔ-wali *from* wali: *thorn; for the bush, as opposed to the fruit* (FCD) KRIO bush-lime (Cole) LIMBA koyerete (FCD) LOKO (Magbile) ejombe (NWT) lukwe (NWT) MENDE gbɔ, kpɔ (*def.* -i) (FCD) kukwi (NWT) kpɔ-wulo (FCD) TEMNE a-sonko (FCD) IVORY COAST KWENI troobéni (B&D)

A shrub or small tree to 6 m tall of the closed-forest of Sierra Leone, Ivory Coast and S Nigeria, and into E Cameroun.
The fruits are pale orange-yellow when ripe, to 7.5 cm in diameter resembling a lime; hence the Krio name. In other areas of Sierra Leone (relic-forest savanna) 'bush-lime' refers to *Erythrococca anomala* (Juss.) Prain (Euphorbiaceae) (3). The fruit appears to be the important part of the tree for the root of the vernacular names in Sierra Leone refers to the fruit while for the plant itself the root is qualified. Nevertheless, uses of the fruit seem to be so far little recorded. The leaves are mashed in cold water which is drunk with 'lime-juice' for gonorrhoea. The root is scraped and cut up small and boiled in water to which 'lime-juice' is added and the liquid is drunk for any sickness (2). In these usages it is possible that 'lime-juice' may be not of *Citrus*, but of the *bush-lime* itself (3), an ambiguity that requires resolving. In Ivory Coast sap of the plant is taken in draught for nausea and oedema of the legs (1).

References:
1. Bouquet & Debray, 1974: 90. 2. Deighton 2562, K. 3. Cole, 1977.

Flacourtia flavescens Willd.

FWTA, ed. 2, 1: 189. UPWTA, ed. 1, 47.
English: Niger plum.
West African: UPPER VOLTA MOORE kitenga (Aub.) SENUFO diaramini (Aub.) sofara (Aub.) IVORY COAST SENUFO diaramini (Aub.) sofara (Aub.) GHANA AKAN-TWI piti piti (BD&H; FRI) DAGAARI voalu (AEK; FRI) GA amugui (auctt.) GBE-VHE (Awlan) adeliãgo (FRI) adenega (BD&H; FRI) nlewe (FRI) MOORE kwakadili (AEK; FRI) SISAALA botong (AEK; FRI) WALA jarem piala (AEK) saram piala (FRI) NIGERIA YORUBA kánkán dìká (JMD; Verger) osérè (JMD)

FLACOURTIACEAE

NOTE: the vernaculars of this and of Oncoba spinosa tend to be confused because of the similarity between the leaves and long spines of the two species, but they are easily distinguished by the absence of petals in the flower of this species, and the showy flowers of the other.

A shrub or small tree of dry coastal or northern savanna, often favouring rocky situations, occurring from Senegal to N Nigeria, and into equatorial Africa to Angola.

The stems are thorny and the plant if cut back makes a good hedge (1, 2). The fruit is red or purple when ripe and about 1.2 cm in diameter resembling a small plum. They are edible and are also added to medicinal prescriptions (1). The stem is chewed as an astringent to cure diarrhoeic conditions in Ghana (3, 4), though in northern Ghana the Sisaala (6) and in Upper Volta the Moore (5) use the plant as a purgative, and the latter also as a cholagogue.

References:

1. Dalziel, 1937: 47. 2. Howes, 1946: 51–87. 3. Irvine, 1930: 204. 4. Irvine, 1961: 77–79. 5. Kerharo & Bouquet, 1950: 39. 6. Kitson 514, K.

Flacourtia indica (Burm. f.) Merr.

FWTA, ed. 2, 1: 189.

English: Governor plum; Madagascar plum (Howes).

A shrub, common in India and in SE Asia and introduced into Africa, and is particularly widespread in E Africa.

The main interest in the plant lies in its edible fruit which can be eaten raw and is usually of a pleasant flavour. It is said to lack ascorbic acid (7). In its native India it is used to make preserves, and in Sierra Leone it is found to produce first-class jelly, and an excellent rosé wine (10). Trial planting at Newton, Sierra Leone, showed poor flowering and fruiting (3) – an obvious case for investigation by breeder/selectionists.

The leaves are eaten as a vegetable in Madagascar and leaves and bark are used (? flavouring) in the making of rum (2). The leaf is said to be carminative, expectorant, tonic and astringent and to be useful for asthma and in some gynaecological remedies (7). Sap from fresh leaves and tender stalks is useful in infantile fevers (6). In Tanganyika the leaf-sap is taken for diarrhoea, and with addition to a root-decoction for schistosomiasis and malaria; the root-decoction, and the cooked root are taken for hydrocoele (4). Bark triturated in oil is used in Madagascar as an antirheumatic liniment, and in India a bark-infusion is a gargle for hoarseness, etc. (7). A root-decoction is used in southern Africa to relieve body pains (7), and in Malawi for pneumonia (9).

Plant extracts have given negative anti-biotic tests (7).

The plant is used in many parts of the world to give a close impenetrable hedge, or a good wind-break. Frequent clipping is tolerated, and it will grow on poor rocky soil (1, 5, 8).

References:

1. Burkill, IH, 1935: 1022. 2. Decary, 1946: 57. 3. Deighton 4876, K. 4. Haerdi, 1964: 71. 5. Howes, 1946: 51–87. 6. Quisumbing, 1951: 626–7. 7. Watt & Breyer-Brandwijk, 1962: 444. 8. Williams, RO, 1949: 266–7. 9. Williamson, J, 1955 (1956?): 59. 10. Akiwumi, 1986.

Flacourtia vogelii Hook. f.

FWTA, ed. 2, 1: 189.

English: berry tree (Sierra Leone, Deighton).
West African: **SIERRA LEONE** KRIO blak-bɛri (S&F)

A shrub or tree reaching 10 m high of forested areas from Sierra Leone to W Cameroon and on into Central African Republic.

156

The fruits are red and fleshy, pleasant to eat. The plant is much cultivated for them by the Krio in Sierra Leone (1).
The plant can be cut and trained into a hedge (2).

References:

1. Deighton 2621, K. 2. Howes, 1946: 51–87.

Flacourtia sp. indet.

West African: SIERRA LEONE YALUNKA sungutindabala-na (FCD)

An unidentified gathering (Deighton 4170, K) from Musaia, Sierra Leone.

Homalium africanum (Hook. f.) Benth.

FWTA, ed. 2, 1: 196, incl. *H. molle* Stapf and *H. sarcopetalum* Pierre. UPWTA, ed. 1, 49, incl. *H. molle* Stapf.
West African: **SIERRA LEONE** LIMBA kamandi (NWT) LOKO nguma (NWT) MENDE kolo-gale (*def.* -i) (JMD) nye gale (*def.* -i) = *fish bone; applies also to other spp.* (JMD) SUSU kumɔdiri (NWT) sarayenyi (NWT) wuriboge (NWT) TEMNE ɛ-lopa (NWT) **IVORY COAST** ABE bleufu *? this sp.* (JMD) AKAN-FANTE akonibia (A.Chev. fide JMD) ANYI akohima (A.Chev. fide JMD) NZEMA akonibia (A.Chev. fide JMD) **GHANA** AHANTA ɛsono nankroma = *elephannt's legs* (E&A) wirokom (BD&H) NZEMA wirokom (FRI; CJT) **NIGERIA** EDO ekalado ęzę (auctt.) IGBO (Umuahia) oyuru uguru = *lives through the Harmattan* (AJC) IJO-IZON (Kolokuma) kịmị gbásá = *human-being's stick* (KW&T)

Tree to 33 m tall of humid rain-forest to secondary forest, usually in swampy or damp localities occurring from Guinea to W Cameroon, and widespread to Angola and across Africa to Mozambique.
The bole is short with low branches and is slightly fluted. The wood varies from brown (as *H. molle* Stapf, 3) to dull white (as *H. africanum* (Hook. f.) Benth., 3), and is hard and suitable for carpentry (2, 4). In SE Nigeria it is used for yam stakes (5).
A trace of alkaloid has been detected in the bark and roots of material from Congo (Brazzaville) (1).

References:

1. Bouquet, 1972: 44. 2. Dalziel, 1937: 49. 3. Keay & al., 1960: 113–4. 4. Sleumer, 1973: 281–9. 5. Williamson, K & Timitimi, 1983.

Homalium angustifolium Sm.

FWTA, ed. 2, 1: 195.
West African: SIERRA LEONE MENDE gai (Macdonald) yé (Macdonald) TEMNE ɛ-gbɛse (NWT)
NOTE: *The H. spp.* appear to be not separately distinguished in S. Leone and the vernacular names may apply to any of the others represented, *H. africanum, H. angustifolium, H. longistylum, H. smythei,* (Deighton)

Tree to 5–10 m high of the dense rain-forest to 2000 m elevation, occurring in Guinea, Sierra Leone and Liberia.
The wood is hard.

FLACOURTIACEAE

Homalium dewevrei De Wild. & Th. Dur.

FWTA, ed. 2, 1: 196, as *H. angustistipulatum* Keay.

A tree of the rain-forest, galleried or dry forest, of swamps or flooded sites occurring in Cameroun to Zaïre, and recorded but once in Western Province of Ghana.
The wood is very hard (1).

Reference:

1. Irvine, 1961: 83, as *H. angustistipulatum* Keay.

Homalium letestui Pellegr.

FWTA, ed. 2, 1: 196, incl *H. skirlii* Gilg, p. 1: 197. UPTWA, ed. 1, 49, as *H.dolichophyllum* Gilg.
West African: SENEGAL FULA-PULAAR (Senegal) tiubétad (JB) GUINEA-BISSAU FULA-PULAAR (Guinea-Bissau) tchubètáde (JDES) SIERRA LEONE KISSI teinpiando (?tienpiando) (S&F) MENDE kɔli-galɛ (*def.* -i) = *leopard's bone* (auctt.) kɔtu-wulo (FCD) SUSU nyaŋgi (NWT) YALUNKA wudi-na-akhɔ-na (FCD) LIBERIA KRU-BASA bro-kpar = *hard as bone* (C&R) IVORY COAST ABE *m*bléufu (Aub.) mélé fu fu (Aub.) BAULE di iroa (Aub.) KYAMA ahubé (Aub.) diunankahia (Aub.) GHANA AKAN-ASANTE duakowaboba (CJT) mmepɔmdua (FRI; CJT) ɛsononankoroma = *elephant's knee* (auctt.) TWI osononankroma (DF; E&A) WASA osonnankroma (DF) ɛsononankoroma = *elephant's knee* (auctt.) BAULE di iroa (FRI) NZEMA miarhire (auctt.) NIGERIA EDO akporo (auctt.) ekalado (JMD) kalado (JMD) kalado akporo (Lamb) IGBO akpurukwu (KO&S) NUPE rokò bachi (auctt.) YORUBA abo àkó = *female 'ako'* (auctt.) òtúrù (Verger) YORUBA (Ondo) akoledo (Lancaster)

A tree with a long straight slender bole attaining about 27 m height, occasionally up to 33 m, and to 1 m girth, of dense rain-forest, transition, semi-deciduous, galleried and secondary forests of lowlands and foothills in Senegal to Nigeria and Fernando Po, and also into central Africa to the Congo basin. It prefers proximity to running water.

The tree has magnificent clusters of rose-coloured flowers. The fruits are also showy and the young leaf-flush is red before turning green. The tree is thus attractive and worthy of cultivation (1, 4, 5).

The sap-wood is light brown and hard. Heart-wood is light brown, hard, heavy and with moderately fine texture (7). Because of its strength and hardness it is commonly used for house-posts and is cut for small timbers (3) and boards (6). Several vernaculars alluding to *bone* reflect the wood's hardness. The wood is also used in carpentry.

A bark-slash will produce a little clear sap which darkens on exposure (6). In Ivory Coast sap from the bark is used in enemas for the treatment of generalised oedemas while lees from the bark are rubbed over the area (2). In Gabon a bark-decoction with other drug-plants is taken by draught for orchitis, and bark-scrapings enter a prescription given to a newly-delivered woman (9).

The Yoruba of Nigeria call on the plant in an incantation against small-pox (8), while the bark, finely ground to a powder, is blown by Liberian witch-doctors into a dragon's lair to stupefy it before slaying it (3).

References:

1. Aubréville, 1959: 3: 26. 2. Bouquet & Debray, 1974: 159. 3. Cooper & Record, 1931: 27. 4. Dalziel, 1937: 49. 5. Irvine, 1961: 83. 6. Savill & Fox, 1967: 232. 7. Taylor, 1960: 300. 8. Verger, 1967: No. 8. 9. Walker & Sillans, 1961: 383.

Homalium longistylum Mast.

FWTA, ed. 2, 1: 196, as *H. aylmeri* Hutch. & Dalz.; *H. macropterum* Gilg and *H. sp. A* (Vigne 1006), p.l: 197. UPWTA, ed. 1, 49, as *H. alnifolium* Hutch. & Dalz. and *H. aylmeri* Hutch. & Dalz.
West African: SIERRA LEONE MENDE kɔli-galε (*def.* -i) = *leopard's bone* (FCD) IVORY COAST ABE akohissi (auctt.) KYAMA ahiapopo (Aub.) GHANA AKAN-ASANTE ɔfam (BD&H; FRI) NIGERIA EDO aba (JMD) ekalado (JMD) kalado (auctt.) okuaba (JMD; KO&S) EJAGHAM-ETUNG ekang (Catterall fide JMD) YORUBA akólédò (KO&S); *but cf.* Cordya platythyrsa, *Boraginaceae* (Verger)

A tree to 30 m tall with long straight slender bole to 1.30 m in girth slightly buttressed, of the dense rain-forest to savanna formations, sometimes in swampy sites at low elevation and up to 1500 m. Occurring from Guinea to Nigeria, and widespread aross Africa to Angola and Mozambique.
The wood is very hard, yellowish-white. The poles are used for hut-posts in Sierra Leone and are said to last in the ground for 5 years or more (1).

Reference:
1. Deighton 2826, K.

Homalium smythei Hutch. & Dalz.

FWTA, ed. 2, 1: 195, incl. *H. aubrevillei* Keay, p. 1: 196. UPWTA, ed. 1, 49.
West African: SIERRA LEONE MENDE kɔligalε (*def.* -i) (FCD; DS) kɔtu-wulo (FCD) TEMNE ε-lan-ε-ra-wuto (NWT) LIBERIA KRU-BASA bro-kpar (C&R) day-quehn (C&R) plehju-eh (C&R) IVORY COAST ·KRU· docla (Aub.)

A tree of the primary dense rain-forest, galleried forest or secondary growth, from Senegal to Ivory Coast, in lowlands and to 1850 m altitude.
The trunk is slender, straight and long. The tree seems to attain its tallest at 25 m high in Ivory Coast (1), but to about 10 m elsewhere. The stems are strong and used for hut-poles in Liberia (2) and Sierra Leone (5). The wood is very hard and good for pestles for rice-mortars (5). In Liberia it is burned for charcoal (2). The wood is yellow and is sawn up for small timbers (4).
Bark-ash mixed with palm-oil is rubbed on to the back in Liberia to relieve back-pains but 'in rubbing it in, however, the fingers must not be used: only by using the big toe to apply the salve will relief be soon effected'! (3, 4).

References:
1. Aubréville, 1959: 3: 28, as *H. aubrevillei* Keay. 2. Cooper 324, K. 3. Cooper 469, K. 4. Cooper & Record, 1931: 27–28. 5. Deighton 1923, K.

Homalium stipulaceum Welw.

FWTA, ed. 2, 1: 195–6, as *H. neurophyllum* Hoyle. UPWTA, ed. 1, 49, as *H. neurophyllum* Hoyle.
West African: GHANA AKAN-ASANTE ɛsononankoroma (CJT; E&A) TWI osononankroma (DF; E&A) WASA osononankroma (DF) ɛsononankoroma = *elephant's knee* (auctt.)

A tree of the lowlands and foothills of the humid rain-forest and transition forest, galleried forest and secondary growth, growing to about 30 m tall, occurring in Liberia, Ivory Coast and Ghana.
The bole is straight and slender and has a very hard wood varying between yellowish-white (2, 3) or light brown (5) to rosy (4). The bark-slash is slightly scented and exudes a brown gum (1, 5).

FLACOURTIACEAE

References:

1. Aubréville, 1959: 3: 26, as *H. neurophyllum* Hoyle. 2. Burtt Davy, & Hoyle, 1937: 124 as *H. neurophyllum* Hoyle. 3. Irvine, 1961: 84, as *H. neurophyllum* Hoyle. 4. Sleumer, 1973: 276–9. 5. Taylor, 1960: 301, as *H. neurophyllum* Hoyle.

Lindackeria dentata (Oliv.) Gilg

FWTA, ed. 2, 1: 189. UPWTA, ed. 1, 47.
West African: SIERRA LEONE KISSI bɛwo (FCD; S&F) LOKO heŋwa (NWT) maiyuŋbe (NWT) ngaingainge (NWT) MENDE bɔku (*def.* -i) (FCD; S&F) fulo (FCD) fulo-*kpɔkpo* (*def.* f. *kpɔkpɔi*) (FCD; S&F) toya-hina (*def.* t-hinei = *male 'toya' [Maesobotrya]* (auctt.) SUSU simɛ (NWT) xaienyi (NWT) IVORY COAST AKYE dédébroguissé (Aub.; A.Chev.) DAN boua (Aub.) GHANA AKAN-ASANTE asratoa (BD&H) NIGERIA IGBO oru kpakǫkwa (NWT)

A tree to 16 m tall of the forest lower storey occurring from Guinea to W Cameroon, and from E Cameroun to Sudan and Angola.

The fruits are spiny, orange-yellow or bronze coloured when ripe containing a few shining black seeds with a red aril. On opening they give off a smell of methyl-salycilate (7). Children in S Nigeria are reported to suck the seeds (5). The seed yields an oil used in Ubangi to treat yaws and leprosy (7).

A leaf-decoction is used in Ubangi to kill fleas (7). In Congo (Brazzaville) the plant enters a prescription to relieve mental affections, and a root-decoction is taken by sniffing for headache (1).

The fibrous bark emits a strong smell of *hydrocyanic acid* when slashed (3). Some has been detected in the leaves and a strong presence in the bark and roots (2). *Alkaloids* have been found in the leaves, stem, bark and roots (2, 6), and some *saponin* in the leaves, bark and roots (2).

A root-decoction is drunk and root-ash is rubbed on areas of oedema in Ubangi (7).

The plant yields a vegetable salt (7).

References:

1. Bouquet, 1969: 127. 2. Bouquet, 1972: 27. 3. Gilbert, 10544, K. 4. Portères, s.d. 5. Thomas, NWT.1669 (Nig. Ser.), K. 6. Willaman & Li, 1970. 7. Vergiat, 1970: 81.

Oncoba brachyanthera Oliv.

FWTA, ed. 2, 1: 188.
West African: LIBERIA MANO bai sa bls (Har.) IVORY COAST ? (Rasso) to unto (Aub.)

A shrub or small spreading tree to 10 m high with sweet-scented showy white flowers, occurring from Sierra Leone to S Nigeria and in Gabon.

The plant has decorative value. The dry fruits are used as snuff-boxes (2) and in Liberia as castanets (1). The wood is pale yellow and hard (3).

References:

1. Harley, 1170, K. 2. Irvine, 1961: 79. 3. Keay & al., 1960: 104.

Oncoba spinosa Forssk.

FWTA, ed. 2, 1: 187–8. UPWTA, ed. 1, 47–48.

English: snuff-box tree; wild white rose; (Ainslie); fried egg tree (USA, Meninger).

French: arbre à tabatières; (Aubréville) – from the African use of the fruit as snuff-boxes; oncoba tabatières (Bailleul).

West African: SENEGAL BANYUN si lind (JB) BASARI a-ñékedo, a-ñékôdo (JB) a-nékodo (K&A) BEDIK gi-tioven (K&A) gi-tyofèn ganak (auctt.) DIOLA ka tãnda (JB) ka tiãda (K&A after JB) FULA-PULAAR (Senegal) dègdèg (JB) sarnana (JB) KONYAGI a-ngubanan (JB) MANDING-

160

BAMBARA sara bara (auctt.) sira bara (JB) MANINKA ko bara ni (auctt.) sara bara (JB) MANDYAK be mével (JB) NDUT mud (JB) unguti (JB) NON ogèl (JB) ogun (auctt.) SERER lumbuĭ (auctt.) lumbuti (JB) mburkul (JB) nduyuf (JB) WOLOF mburkul (K&A after JB) mur (auctt.) ndumbuj (Aub.; K&A) nduyuf (K&A) palkiu (Aub.; JB) palkiu (K&A) **GUINEA** BASARI a-nyékɛdiɔ (FG&G) MANDING-MANINKA ko bara ni (Aub.) **SIERRA LEONE** MENDE gbuwe (JMD) ndogbɔ-dumbele = *bush lime* (JMD) **MALI** MANDING-BAMBARA sara bara (JMD) sira bara (JMD) MANINKA *ko* bara ni *ko*: tree (Aub.) **UPPER VOLTA** SENUFO toro sogo nani (Aub.) **IVORY COAST** MANDING-MANINKA toanegosoro (B&D) SENUFO toro sogo nani (Aub.) **GHANA** AKAN-TWI asratoa (OA) asratoa-dua = *snuff-box tree* (FRI; E&A) DAGAARI kpuri (FRI) kpuri-tia (FRI) GA asrato (OA) asratɔ = *snuff-box* (FRI) GBE-VHE asragui (OA) asrãgui (FRI) NABT monomorka (auctt.) **TOGO** BASSARI butjesu (Volkens & Gaisser) GBE-VHE kpoe (Volkens & Gaisser) NAWDM fendira (Volkens & Gaisser) TEM kongowura (Volkens & Gaisser) TEM (Tschaudjo) kongofira, kruta (Volkens & Gaisser) **NIGERIA** BIROM epúnuŋ *the fruit* (LB) tin púnúŋ *the plant* (LB) FULA-FULFULDE kokochiko (Chapman) HAUSA kóókóócikoo, kóókóó-cikóó = *a rattle* (auctt.) IDOMA icákiricá (Armstrong) IGBO okpoko (KO&S) JUKUN amurikpa (JMD; KO&S) KAMBARI-SALKA màkpàtalíkpɔ'ɔ̀ (RB) YORUBA kánkán diká (auctt.) pònṣé, pònṣéṛé (JMD; Verger) **WEST CAMEROONS** BAMILEKE takwa (Johnstone)

A thorny shrub, much branched to 3–4 m high, or small tree to 13 m high under conditions of higher rainfall, of deciduous, secondary and fringing forest from Senegal to W Cameroon, and generally widely distributed in tropical Africa and Arabia.

The large white flowers are sweetly scented and the plant is an attractive ornamental found occasionally planted in West Africa in gardens and villages. It has been tried as hedging (8) but with indifferent results (6). The plant has been taken into cultivation in the USA and becomes deciduous at places where the temperature falls to near 0°C (13). The wood is hard and light brown, takes a good polish and is suitable for inlay and cabinetry (6). Small chips of wood are boiled in water by the Tenda which they drink for a week for stomach-ache and loss of appetite; the chips are then cast away at a cross-roads (15).

The fruits, which reach a diameter of about 6 cm, mature with a hard shell of a rich red-brown colour and sometimes with surface sculpturings. All over Africa these fruits are hollowed out for use as snuff-boxes, or filled with pebbles or hard seeds as rattles – hence the English *snuff-box tree*, French *arbre tabatière* (snuff-box tree) and many West African vernaculars referring to these uses. The hollowed-out fruit are used in Ghana in the Twi game of *saara* (9). Children put the whole fruit into a fire and when the fruit fizzes, they throw them to burst with a loud report (9). In Kenya the dried fruit serve as spinning tops (2).

The fruit-pulp is edible with a rather sour acidulous taste and is eaten particularly in NE Africa. The Tenda take it with karite butter, *Vitellaria paradoxa* Gaertn.f. var. *parkii* (G Don) Hepper (Sapotaceae) for stomach-ache and loss of appetite (15). The seeds contain 35–37% of a brownish-yellow fixed oil (4). It has been used for treatment of leprosy but the oil contains no chaulmoogric acid and is considered useless against the leprosy bacillus (1, 6, 14). The oil is drying and might find some application in making paint and varnishes (14), but separation of seed from pulp appears to pose difficulties (6). The seed-oil has also been used in Nigeria for other skin-complaints and taken internally for fever (1).

In Ivory Coast the plant has a good reputation as an aphrodisiac (5). A decoction of leafy twigs is also used to wash sores (12). The use of a leaf and root-decoction is recorded for West Africa for urethral discharges and a root-decoction for dysenteries and bladder conditions (6, 14). In Tanganyika leaf-sap and root-decoction in conjunction with *Schrebera trichoclada* Welw. (Oleaceae, but not in W Africa) is taken for vertigo (7). In Senegal the root is considered anti-dysenteric, and the dried powdered root taken with meat is a strengthening tonic (10, 11).

The plant has a great reputation in Congo (Brazzaville) as a panacea for all sicknesses and as protector against evil influences and spirits. No part of the plant must be cut without making an offering and explaining to the plant the reason and expectations (3).

FLACOURTIACEAE

Traces of *flavones* in the leaves, of *tannins, steriods* and *terpenes* in the leaves, bark and roots, have been reported in material from Congo (Brazzaville) (4). The plant has given negative results in anti-biotic tests (11, 14).

References:

1. Ainslie, 1937: sp. no. 253. 2. Bally 6062, K. 3. Bouquet, 1969: 127. 4. Bouquet, 1972: 27. 5. Bouquet & Debray, 1974: 90. 6. Dalziel, 1937: 47–48. 7. Haerdi, 1964: 75. 8. Howes, 1946: 51–87. 9. Irvine, 1961: 79–80. 10. Kerharo & Adam, 1964,a: 423. 11. Kerharo & Adam, 1974: 483, with references. 12. Kerharo & Bouquet, 1950: 39. 13. Meninger, 1948: 249–51. 14. Watt & Breyer-Brandwijk, 1962: 446, with references. 15. Ferry & al., 1974: No. 100.

Ophiobotrys zenkeri Gilg

FWTA, ed. 2, 1: 189.
West African: **IVORY COAST** AKYE uolobo (Aub.) urogbo (Aub.) **GHANA** VULGAR abuana (DF) AKAN-ASANTE abuana (BD&H) TWI akwana (E&A) **NIGERIA** BOKYI bofan (auctt.)

A tree to 35 m high by 1.70 m in girth of the rain-forest and mixed deciduous forest and recorded from Ivory Coast to W Cameroons, and occurring also in E Cameroun and Gabon.
The base of the tree is buttressed with a long clean straight bole. The wood is hard, yellow, and sinks in water when fresh (1, 2). It is used for making mortars in Ghana (2).

References:

1. Burtt Davy & Hoyle, 1937: 53. 2. Irvine, 1961: 80–81.

Phyllobotryum soyauxianum Baill.

FWTA, ed. 2, 1: 190–1.

A shrub or small tree to 4 m high of the forest in S Nigeria and W Cameroons, and E Cameroun.
The wood is hard and has a marked mentholated smell (1, 2). The flowers are visited by bees (1).

References:

1. Brenan 9306, K. 2. Richards 5153, K.

Scottellia chevalieri Chipp

FWTA, ed. 2, 1: 186. UPWTA, ed. 1, 48.
French: akossika à grandes feuilles (Aubréville).
West African: **IVORY COAST** ABE akosika (A.Chev.; Aub.) ABURE eddé (A.Chev.) AKYE séké (A.Chev.) KYAMA n'droya (B&D) tim'bania (A.Chev.) **GHANA** VULGAR tiabutuo (DF) AKAN-ASANTE tiabutuo (auctt.) WASA kruku (CJT) ANYI-SEHWI dein (auctt.) GBE-VHE sakpa (FRI)

A tree of the transition forest, to 50 m tall, 1.80 m girth, clear bole, occurring in Ivory Coast and Ghana, and relatively common in places.
Sap-wood is white; heart-wood yellowish-white, moderately hard and heavy. It is suitable for carpentry and is easily polished, but splits readily. It is sometimes used in Ghana in building houses.

References:

1. Aubréville, 1959: 3: 12. 2. Burtt Davy & Hoyle, 1937: 53. 3. Dalziel, 1937: 48. 4. Irvine, 1961: 81. 5. Taylor, 1960: 169.

Scottellia coriacea A Chev.

FWTA, ed. 2, 1: 187. UPWTA, ed. 1, 48, incl. *S. kamerunensis* Gilg.
French: akossika à petites feuilles (Aubréville).
Trade: odoko (from Yoruba: Sierra Leone, Saville & Fox; Ghana, Dept.
For., Irvine).
West African: SIERRA LEONE MENDE *g*bui (S&F) *n*golo-gali (?) (S&F) LIBERIA KRU-
BASA mehr-chu (C&R) ne-mor-ba-de (C&R) IVORY COAST AKAN-FANTE aburuhi (A.Chev.)
aburuki (A.Chev.) AKYE bakaza (A.Chev.) NZEMA aburuhi (A.Chev.) aburuki (A.Chev.)
GHANA VULGAR koroko (DF; FRI) AKAN-ASANTE tiabutuo (CJT) WASA kruku (CJT) ANYI-
SEHWI dein (CJT) NIGERIA EDO emuẹfuohai (KO&S; Elugbe) emwenfuohai (auctt.) ENGENNI
uguokpa (JMD; KO&S) ESAN eranfohi (KO&S) IGBO akporo (KO&S) YORUBA òdòko (auctt.)
YORUBA (Ondo) eluro-orunghe (Kennedy fide JMD)

Tree of the lowland rain-forest to 30 m high, tall straight bole with slight
fluting at the base, occurring in the Region from Sierra Leone to W Cameroons,
locally abundant, and into E Cameroun and Gabon.
Wood is whitish to pale yellow, hard, fine-grained, splits easily (2, 5, 7, 9, 10,
12, 13). Stems are used for hut-posts (6, 8, 9) but the wood is not durable and is
liable to insect-attack (13) and decay (9). Preservation processes are possible
and it is currently recommended in Ghana for general joinery, turnery, wood-
work and general purposes (1). It peels well and can be made into plywood and
veneer. Logs are reported to float when green (11).
The wood is also used for making small domestic articles, such as spoons,
ladles, combs, water-bailers, etc. in Gabon (14). Yorubas of S Nigeria use the
ashes in soap-making (9).
The bark-slash is very wet and exudes a clear sap with a strong smell of bitter
almonds (Sierra Leone, 12) or an amber-coloured gum (Nigeria, 10). The fresh
wood also carries the marked smell. The bark is considered of great importance
in witchcraft in Liberia when ju-ju amulets are losing their power. It is also an
ingredient in legal oath-taking and an infallible panacea used by all Liberian
herbalists (7). The bark is used in draught in Congo (Brazzaville) to relieve
stomach-ache (3).
The root has strong purgative and diuretic properties used in Ivory Coast to
treat general oedemas (4).

References:

1. Anon. s.d., a. 2. Aubréville, 1959: 3: 14. 3. Bouquet, 1969: 128. 4. Bouquet & Debray, 1974: 90.
5. Burtt Davy & Hoyle, 1937: 53. 6. Cooper 292, K. 7. Cooper & Record, 1931: 28, with timber
characters. 8. Dalziel, 1937: 48. 9. Irvine, 1961: 81–82. 10. Keay & al., 1960: 97, 99. 11. Sankey 16,
K. 12. Savill & Fox, 1967: 129. 13. Taylor, 1960: 169. 14. Walker & Sillans, 1961: 182.

Scottellis leonensis Oliv.

FWTA, ed. 2, 1: 186.
West African: SENEGAL DIOLA-FLUP bu diékèk (JB) SIERRA LEONE KORANKO fira-
mana (S&F) MENDE *n*golo-gali (*def.* -i) (S&F) *n*golo-kala (*def.* -kalei) (FCD; S&F) *n*gookai
(*def.*) (S&F)

A small tree to 13 m high of swamp-forest with strongly scented white
flowers, recorded from the Casamance of Senegal to Liberia, and in the Central
African Republic.

Scottellia mimfiensis Gilg

FWTA, ed. 2, 1: 186.
West African: NIGERIA EDO emuefuohai (KO&S) IJO-IZON ewono (KO&S)

A waterside tree of the lowlands of S Nigeria attaining 10 m height, and in
Cameroun, where it reaches 30 m.

FLAGELLARIACEAE

Flagellaria guineensis Schum.

FWTA, ed. 2, 3: 51. UPWTA, ed. 1, 467.

West African: IVORY COAST AKAN-ASANTE gabafae (B&D) AKYE kujugbun-baté (A&AA) KYAMA kakaté (B&D) GHANA AKAN-ASANTE ahõmantiaantia (FRI; E&A) FANTE biribiɛ (FRI) TWI mɛrɛbia (FRI) GA bábabia (FRI; KD) NZEMA nghafalɛ (FRI)

A tall, tough, herbaceous climber from a rhizomatous root-stock, leaves ending in a hooked or twining extension of the midrib; of forest margins, and waste places scrambling over other vegetation, usually near water, from Ivory Coast to S Nigeria, and across Africa to Somalia, Central and Southern tropical Africa, and Mascarenes.

The plant occurs as a weed on tree crop plantations in E Africa and is said to be resistant to herbicides (5). It is a weed invader of rice-fields in the Bendel state of Nigeria (3). The strong supple stems are commonly used in Tanganyika (9) and Zanzibar (11) to make fish-traps. In Gabon the stems are used in hut-construction, and forest people prepare necklace cords from them (8).

In E Africa the whole plant is prepared into medicine for skin-diseases and veldt sores (1) and for refractory leg ulcers (9). Ash from the calcined plant mixed with castor-oil is rubbed into scarifications on the temples as an analgesic for headache in Tanganyika (6), and ash is placed in Kenya over a rupture to cure it (7). The stems enter into rain-making ceremonies in Tanganyika (9).

The pulped leaf is applied in Ivory Coast to carious teeth, while a leaf-decoction is used as a mouthwash (2). Forest people in Gabon hold the leaf to be aphrodisiac (8). The Kyama of Ivory Coast claim the leaf to be a good cure for chronic gonorrhoea (2).

The fleshy berry is widely used in E Africa as a cure for veneral disease (10; Kenya, 4; Tanganyika, used externally, 9; Zanzibar, 11). The fruit-pulp is highly irritant to the skin, and made up as a poultice it is applied to skin infections (9).

References:

1. Bally, 1937. 2. Bouquet & Debray, 1974: 91. 3. Gill & Ene, 1978: 182–3. 4. Graham 1531, K. 5. Kamiri EA.17002, K. 6. Haerdi, 1964: 193. 7. Magogo & Glover 1071, K. 8. Walker & Sillans, 1961: 183. 9. Watt & Breyer-Brandwijk, 1962: 447. 10. Weiss, 1979: 44. 11. Williams, RO, 1949: 267.

FRANKENIACEAE

Frankenia pulverulenta Linn.

FWTA, ed. 2, 1: 198.

West African: SENEGAL ARABIC (Senegal) ddaeifoe (JB) ddesmoe (JB) l'ehdeiboe (JB) l'emleffoe (JB)

A spreading, sparsely-branched herb to a few centimeters high, of halophylic, and dry sandy situations in coastal Senegal, and fairly widely dispersed in NE Africa and warmer coastal areas of the Old World.

In NE Africa where it occurs in salt pans and on sandy soil, it often forms almost pure stands, and provides browsing for sheep and goats (1, 2).

References:

1. Andrews 2702, K. 2. Jackson 3929, K.

GENTIANACEAE

Canscora decussata (Roxb.) Roem. & Schult.

FWTA, ed. 2, 2: 300.

A slender, much-branched, annual herb to 45 cm tall, of damp sites across the Region from Senegal to W Cameroons, and widely dispersed across Africa into tropical Asia and Australia.

The plant is bitter, pungent and oleaginous to the taste. In Indian medicine it is regarded as laxative, alterative and tonic. Freshly expressed sap is prescribed for insanity, epilepsy and nervous debility (3). In the Philippines a decoction of the whole plant is taken as a tonic and for stomach pain, and the leaves are used in infusion as a substitute for tea (2). The plant's bitter and tonic properties are recognised in Senegal (1).

References:

1. Berhaut, 1975, b: 54–56. 2. Quisumbing, 1951: 717–8. 3. Sastri, 1950: 65.

Canscora diffusa (Vahl) R.Br.

FWTA, ed. 2, 2: 300.

A much-branched annual herb, erect to about 60 cm high, of damp sites throughout the Region from Senegal to W Cameroons, widespread across tropical Africa, and into tropical Asia and Australia.

In India this plant is used as a substitute for *C. decussata* (q.v.) (1).

Reference:

1. Sastri, 1950: 65.

Centaurium pulchellum (Sw.) EHL Krause

FWTA, ed. 2, 2: 300.

Portuguese: fel-da-terra (Crioulo, Cape Verde).
West African: SENEGAL CRIOULO fel-da-terra (RdoF)

An erect annual herb 10–25 cm high in damp depressions of sand-dunes, not common in Mauritania, Senegal, Togo and Niger, but common in N and NE Africa and the Mediterranean region.

No usage is recorded, but plants of the closely related 'Centaurium aggregate' are known from olden times for their strong febrifugal action (1). The W African material merits examination.

Reference:

1. Schunck de Goldfeim, 1945: 152–3, as *Erythrea centaurium*.

Exacum oldenlandioides (S Moore) Klack.

FWTA, ed. 2, 2: 298, as *E. quinquenervium* Griseb. UPWTA, ed. 1, 424, as *E. quinquenervium* Griseb.
West African: SENEGAL DIOLA-FLUP èlguñay (JB) GUINEA-BISSAU DIOLA-FLUP ele-gúnhae (JDES; EPdS) GUINEA MANDING-MANINKA nkomi-mabolini (Brossart ex A.Chev.)

An annual herb up to 75 cm tall, of sandy grassland, rice-padis and wet places, throughout the Region from Senegal to W Cameroons, and widespread in tropical Africa to Zanzibar, Madagascar and Southern Africa.

In Tanganyika, sap is squeezed from the pulped up plant and drunk 'for gnawing pain' in the stomach (3). In Guinea the plant is used as a purgative (1, 2).

References:

1. Berhaut, 1975, b: 62–63. 2. Dalziel, 1937: 424. 3. Tanner 3664, K.

Hoppea dichotoma Hayne

A small tufted annual herb to 12 cm high, recorded in the Region only in Mauritania and Senegal, but occurring also in Ethiopia and into Asia.

In India the plant is used to treat piles and snake-bite, and the roots for epilepsy (1).

Reference:

1. Sastri, 1959: 117.

Neurotheca loeselioides (Spruce) Baill.

FWTA, ed. 2, 2: 298.
West African: SIERRA LEONE TEMNE ɛ-kbunkabaki (NWT) ɛ-kwɛkədi (NWT)

An annual herb, erect and branched to ½ m high, of grassy savanna often in damp sites, and a component of the 'foni' community in Sierra Leone; common throughout the Region and into central Africa, Madagascar and tropical S America.

Though known to Temne in Sierra Leone, no usage is reported.

Schultesia stenophylla Mart. **var. latifolia** Mart.

FWTA, ed. 2, 2: 299–300.
West African: SENEGAL CRIOULO fel de tera (JB) DIOLA-FLUP é gãng (JB) GUINEA-BISSAU CRIOULO fel de tera (JDES)

An annual herbaceous plant, erect to 50 cm high on damp waste sites, species native of Brazil, and represented by var. *latifolia* in Senegal, The Gambia, and Sierra Leone in the Region, and extending into Chad.

Sebaea oligantha (Gilg) Schinz

A herb, 5–6 cm high, saprophytic, white to reddish without chlorophyll, recorded in central tropical Africa, and now known to occur in Ivory Coast to Nigeria (3).

The plant appears to be a parasite of tapioca attacking the root (1) and probably other euphorbiaceous plants (2).

References:

1. Boutique, 1972: 51–52. 2. Dawkins 695, K. 3. Raynal, 1967: 207–19.

GERANIACEAE

Geranium ocellatum Cambess.

FWTA, ed. 2, 1: 157.

English: cranesbill (general for the genus).

An erect annual to about 1.30 m high, recorded from 2,000–3,000 m elevation in W Cameroons, and occurring widely in montane situations in Africa and to the western Himalayas.

The plant has astringent and diuretic properties which have medicinal application in India. The root is rich in tannins. Several American European and Indian species of this genus yield tannins and root dyes. This species and the other W African species, *G. arabicum* Forssk. (syn. *G. sinense* Hochst.) which also occurs in montane locations of S Nigeria, W Cameroons and Fernando Po, should be examined.

Monsonia nivea (Dec'ne) Dec'ne

FWTA, ed. 2, 1: 157. UPWTA, ed. 1, 39.
West African: NIGER TAMACHEK tazereut (B&T)

A hoary herb with long woody rhizome recorded from Mauritania and Mali, and across northern Africa to Arabia.

Monsonia senegalensis Guill. & Perr.

FWTA, ed. 2, 1: 157. UPWTA, ed. 1, 39.
English: pink-flowered crane's bill (Rhodesia, Ventner).
West African: SENEGAL ARABIC (Senegal) ragèm (JB) FULA-PULAAR (Senegal) ndusurno (JB) ndusuru (JB) WOLOF diidii (JB)

A semi-woody herb occurring across the dry northern part of the Region from Senegal to Niger and N Nigeria, and across Africa to Egypt and Ethiopia and into Western India.

The plant is grazed by all stock in Senegal (1), and in Kordofan, but here only at the start of grain-set (5, 6). It has medicinal use as an emmenagogue in Guinea (3). In N Nigeria the woody root has been noted as used by Fula as an occasional ingredient in strophanthus arrow-poison (2).

Several *Monsonia spp.* have been used in southern Africa, mainly for dysenteries and intestinal haemorrhage. Various substances have been reported present but *tannins* appear to be the most likely active principles (4).

References:

1. Adam, 1966, a. 2. Dalziel, 1937: 39. 3. Pobéguin, 1912: 48. 4. Watt & Breyer-Brandwijk, 1962: 451. 5. Baumer, 1975: 107. 6. Hunting Tech. Surveys, 1964.

GESNERIACEAE

Achimenes spp.

FWTA, ed. 2, 2: 382.
English: achimenes.

A group of about 50 spp. of rhizomatous herbaceous plants of the American tropics, and now cultivated worldwide as ornamentals, many being especially valued as house-plants and for trailing basket-plants. The most used are: *A. candida* Lindl., flowers creamy white marked red and yellow to 12 mm long; *A.*

erecta (Lam.) HP Fuchs, with trailing stem and tubular red flower; a tetraploid species and dominant parent of many hybrids; *A. longiflora* DC., with trailing stem, flower large, tubular to 7 cm across, very variable lavender to purple-blue with white, source of many cultivars; and *A. grandiflora* (Schieke) DC., the popular 'hot water plant', violet purple flower, source of several cultivars.

Saintpaulia ionantha H Wendland

FWTA, ed. 2, 2: 382.
English: saintpaulia; African violet.

A perennial stemless plant bearing a crown of hirsute cordate leaves and flowers in stoutly peduncled cymes, of the hilly region of Tanganyika, and now cultivated worldwide as an ornamental.

The flowers are variable, dark violet to pale blue to white, and innumerable variant strains have been raised. Propagation is usually by seed.

The genus has several other showy-flowered species, also from E Africa, which may be found in cultivation.

GNETACEAE

Gnetum africanum Welw.

FWTA, ed. 2, 1: 33.
West African: NIGERIA EFIK áfàng (Udofia fide Lowe) IBIBIO áfàng (Udofia fide Lowe) IGBO ọ̀kazị̄ (auctt.) ụkazi (Lowe) YALA (Ogoja) eruru (Lowe) YORUBA (Ondo) àjáàbalè (Lowe) ajakobale (Lowe)

NOTE: These names may also refer to C. buchholzlanum

A rain-forest liane of SE Nigeria and W Cameroon, and extending into tropical central Africa as far south as Angola.

Wherever it occurs the leaf is valued as a tasty vegetable, usually eaten finely shredded for addition to soup or made up into condiments, or even taken raw (1, 2, 3, 5, 7, 9). The plant is not cultivated and the leaf is collected as a forest-product. It is a common article of market-trade. The leaf has medicinal uses. In Nigeria it is taken for enlarged spleen, for sore throat and as a cathartic (9). In Ubangi it is eaten against nausea and is considered antidotal against arrow-poison based on *Parquetina nigrescens* (Asclepiadaceae) (6). In Congo (Brazzaville) the chopped-up leaf serves as a dressing on furuncles to hasten maturation (1). There are undefined medicinal uses in Mozambique (8).

The stem is supple and strong. In Zaïre it is made into traps and nooses for catching game, and into straps for porterage (4). In Congo (Brazzaville) the stem cut up into small pieces produces a tisane taken to ease childbirth and to reduce the pain (1). Lianous species in SE Asia yield a potable sap. The African species should be looked at in this respect.

The fruit-pulp is eaten in Ubangi (6) and the seed in Zaïre (4).

References:

1. Bouquet, 1969: 131. 2. Busson, 1965: 97–98. 3. Lowe, 1984: 99–104. 4. Robyns, W, 1948: 11–12. 5. Sillans, 1953,b: 85. 6. Vergiat, 1970,a: 83. 7. Walker & Sillans, 1961: 184. 8. Watt & Breyer-Brandwijk, 1962: 458.

Gnetum buchholzianum Engl.

FWTA, ed. 2, 1: 33. UPWTA, ed. 1, 1.
West African: NIGERIA IGBO ọ̀kazị̄ (BN) ụkasi (Swarbrick fide Lowe) **WEST CAMER-OONS** DUALA ikōko (Swarbrick fide Lowe) KPE mokaka ko (Maitland fide JMD) mokako (Maitland fide Lowe)
NOTE: names under *C. africanum* may also apply.

A forest liane of the Cameroun and Congo (Brazzaville) and into SE Nigeria where it was first recorded as recently as 1957 (4), but even so its Nigerian occurrence was not known at the Kew Herbarium till 1971 when leaf was observed on sale in Shepherd's Bush market, England, whence it had been air-freighted from Nigeria for West African residents in this London suburb! (3).
'Unde etiam vulgare ... dictum "Semper aliquid novi Africam adfere".' ('Whence it is commonly said ... that Africa always offers something new': Pliny, the Elder, AD. 23-79 in *Historia Naturalis*, II, VIII, 42.)
The leaf is eaten as a greatly appreciated pot-herb (Nigeria: 4; Congo (Brazzaville): 1; Cameroun: 2). As for *G. africanum*, the chopped-up leaf is used in Congo (Brazzaville) as a dressing on furuncles to hasten maturation, and a tisane of the cut-up stem is taken to ease childbirth and to reduce the pain (1).

References:

1. Bouquet, 1969: 131. 2. Busson, 1965: 98–99, with chemical analysis. 3. Forman, 1971. 4. Lowe, 1984: 99–104.

GOODENIACEAE

Scaevola plumeri (Linn.) Vahl

FWTA. ed. 2, 2: 315. UPWTA, ed. 1, 424.
West African: SENEGAL SERER ndiob diuam (JLT; JB) WOLOF hel u buki (JB) **GHANA** ADANGME wogbɔ (FRI) GBE-VHE ngɔli fɔyi = *blue fruit of the spirits; from fɔyi: a blue fruit,* ngɔli: *spirits* (FRI)

A woody shrub to 2 m high with thick succulent branches and prominent leaf-scars, of sandy soil and dunes behind the sea foreshore; common along the coastline from Senegal to Nigeria, and on to Angola.
The branches are ascending or prostrate. The underground roots may coppice. The plant is a good sand-binder helping to fix coastal sand-dunes (1, 2, 4, 5). On the coast of Zaïre, it is said to fix dunes sometimes to 5 m high (6). It often grows in association with *Ipomoea pes-caprae* (Convolvulaceae) (5) and *Sporobolus spicatus* (Gramineae) (6).
The leaf is used in Senegal in poultices, lotions and decoctions to maturate inflammations, as an emmenagogue and diuretic, and to treat thickening of the cornea and for conjunctivitis (1). The leaf and other parts of the plant are prepared as a syrup for treating venereal diseases. A weak dose promotes sweating, a stronger use is purgative, and more still emetic (1). Leaves put into water are used in Ghana for washing fever patients (3).
Stem-bark is bitter and tonic (1). Pith is regarded in Senegal as good against diarrhoea, and to be aphrodisiac (1).
The fruit is blue and fleshy. The Gbe-Vhe name (see above) associates it with the edible fruit of *Vitex grandifolia* (Verbenaceae): *fɔyi*, and is said to be edible also, but is acrid to the palate (7).

References:

1. Berhaut, 1975,b: 78–80. 2. Dale & Greenway, 1961: 230. 3. Irvine, 1961: 727. 4. Onochie FHI 33488, K. 5. Schnell, 1953,a: fasc. 16. 6. Vanden Berghen, 1979: 185–238. 7. Watt & Beyer-Brandwijk, 1962: 458.

GRAMINEAE

Gramineae

FWTA, ed. 2, 3: 349 *et seq.*
English: grass, in general.
French: herbe, in general.
Portuguese: gramíneas; capim.
West African: SENEGAL MANDING-MANINKA nyânga *ruderal grasses, esp. on. impoverished soil* (GR) **THE GAMBIA** MANDING-MANDINKA koos, kous *general for cereals, except maize* lansango *tall grasses* nyo *corn in general* wa *grass generally* **SIERRA LEONE** KISSI denoyeno *a general term for bush pathside grasses* (FCD) KONO mināsabine *small grasses* (FCD) LOKO waga *bush grasses* (JMD) MENDE gbonje *general for grasses in S Mende area* (FCD) mbowi *loosely for certain grasses* (FCD) foni, fɔni *an ecological term for flat outcrops of rock colonised by grass - hence the grasses themselves* (JMD; FCD) fovo *(def. -i: fovui) general for tall grasses* (auctt.) ngala, ngalɛ *(def. -i) general for tall grasses, e.g., Pennisetum purpureum, Chasmapodium caudatum* (JMD; FCD) ngala, ngale *(def. -ei) general for tall grasses* (FCD) lɛti *grass generally* (auctt.) pisui *creeping grasses* (FCD) puvɛ *(def. -i) for small grasses generally* (FCD) tomo *a general term* (JMD) tugbɛ, rugbɛlɛ *general for several small grasses and sedges* (FCD) yowo, yuwo *creeping grasses* (FCD) SUSU foni *rock outcrops and grasses thereon - see Mende* TEMNE a-gbel *for small-grained grasses: after gbel: the fundibird, or waxbill* (auctt.) kariŋ *creeping grasses* (NWT) ɛ-thaŋkɛ *a general term for stout grasses with matured flowering culms* (FCD) a-wo, an-wo *for tall grasses; applied before flowering* YALUNKA fuŋfure-khamɛ-na *general for grasses with large twisted awns* (FCD) tamedi-sara-na *for a number of fodder grasses* **MALI** SONGHAI hori *grass in general* (A.Chev.) suba, subu *grass in general* (A.Chev.) **UPPER VOLTA** BOBO mana *cereals* (Le Bris & Prost) FULA-FULFULDE (Upper Volta) burgu *water grasses* (K&T) ɓurundi *a loose term for certain annual spp.* (K&T) huɗo *grass in general* (K&T) kuɗol *(pl. kuɗi) blade of grass* (K&T) yantaare *for large perennial savanna spp.* (K&T) GRUSI-KASENA yara *cereals* (Prost) KURUMBA ayana *cereals* (Prost) MOBA nyali *cereals* (Prost) SONINKE-PANA nya *cereals* (Prost) **GHANA** AKPAFU ɔ̀-kúá *(pl. sì-.) grass* (Kropp) GUANG kuyu *(pl. ayu) cereals* (Rytz) LEFANA ālākpī *grass* (Kropp) SISAALA yaŋ *grass* (Blass) **NIGER** FULA-FULFULDE (Niger) ndiiriiri *grazing grasses* (ABM) fuɗaalo *newly sprouted grass* (ABM) geenal *grass completely dried up, or hay* (ABM) huɗo *any fresh grass* (ABM) huɗo nyaako *grass at full maturity* (ABM) huɗo woggo *short soft grazing grasses* (ABM) jokko *tall-culmed, noded grasses* (ABM) keccum *new grass after the rains begin* (ABM) rimo *a range of grazing grass* (ABM) sanaalo *growing grass after the first flush growth* (ABM) wiigaalo *mature grass at anthesis* (ABM) woggo *young grass, not yet long and without noded culms* (ABM) **NIGERIA** ARABIC-SHUWA koreib, kreb *grass with edible grain* (JMD) nyamaya *applies to several small grasses, cf., Eragrostis, and collectively equivalent to Hausa:* tsìntsíiyáá DERA gáatí *a sort of corn* (Newman) pépèl *coarse grass for matting* (Newman) pílí *thatching grass* (Newman) EFIK m̀bíét *grass* (Lowe) ńnyányáñá *any strong grass* (Lowe) EPIE ìkpòò (Kpolovie) GWANDARA-CANCARA cànwá *grass in general* (Matsushita) GITATA tsàŋ̀wā *grass in general* (Matsushita) KARSHI càŋwā *grass in general* (Matsushita) KORO càŋwā *grass in general* (Matsushita) NIMBIA ábo *grass in general* (Matsushita) TONI sàŋ̀wā *grass in general* (Matsushita) HAUSA alkamura *grass in general* (Abrahams) bunsurun daaji = *bush he-goat - applied to several grasses of objectionable qualities* (JMD) bunu *grass for thatching* (JMD) chiga, chika, tsíigàà tsííkàà *any spike of grass* (JMD) ciyaawàà *grass in general* (Matsushita) daatsi *in general for grass used chopped up with clay for building* (JMD) ɗamba *general for swamp grasses* (JMD) daura *loosely for swamp grass* (JMD) falfoli *pithy grasses* (JMD) gaji *grasses used for planting* (JMD) karankɓau *tall marsh grasses* (JMD) katsaura *grasses used to fill camel saddles* (JMD) laɓanda *weak thatching grass* (JMD) lambani *loosely for swamp grass* (JMD) machara *spp. with hollow reeds* (JMD) milla *loosely for swamp grass* (JMD) roba *tall marsh grasses* (JMD) roùbáá *group term for several tall marsh grasses* (JMD) rumayya *spp. used for weaving straw mats* (JMD) sabko búbbúúkùwáá *group term for many short grasses* (JMD) sangari *for thick-culmed grasses* (JMD) silka *any spike of grass* (JMD) tafa ɗuwa *spp. used for weaving straw mats* (JMD) tattakiya *loosely for swamp grass* (JMD) tsamba *general for swamp grass* (JMD) tsíntsíiyáá = *broom: general for any grass with brush-like inflorescence* yama *loosely for grass* (JMD) zaanaa *a matting for fencing enclosures: see Hyperthelia dissoluta, Gramineae* IDOMA ácí = *grass* (Armstrong fide RB) IGALA àlichā, áichā *a roofing-grass collected in bunches and placed directly on the roof* (RB) égbé *grass, in general, incl. sedges* (RB) igwó *a roofing-grass, bound in rolls on the ground and then laid on the roof* (RB) íkélé-àkpúkpà *grass producing canes of use for roof-struts* (RB) ɔtachɔ *any large bamboo* (RB) IGBO àchàlà, àchàrà *general for any tall or coarse grass; also for straw* (BNO) ata = *grass; applied principally to Loudetia spp.* (BNO) azara = *grass; applied principally to*

Hyparrhenia spp. (BNO) ikute ala *general for spp. with Paspalum-like inflorescence* (AJC) IJO-IZON (Egbema) tuké *grass generally* (Tiemo fide KW) IZON (Kolokuma) ẹtụnwẹ́ín *heap of floating grass* (KW&T) KANURI cídûk *grass generally* (C&H) garbazam *a group term for spp. similar to Monocymbium* (JMD) shèshē *spp. for horse fodder* (C&H) títī *thatching grass* (C&H) PERO púrè *spp. for roofing* (ZF) SURA cùk-kúr *sharp-edged leaf spp.* (anon. fide KW) TIV acho *a general term* (JMD) ícá nyăm *a type of grass (not specified)* (KW fide RB)icàn *a type of grass (not specified)* (KW fide RB) igbó tōhō *grass* (KW fide RB) igomoŭgh *a type of grass (not specified)* (KW fide RB) ikangel a ika *grasses with digitate inflorescences* (JMD) YORUBA bẹrẹ *of general application for several spp.* (JMD) eéran *general for fodder grasses* ọkà ẹ̀sin *fodder grasses* ọparun *general for canes or giant grasses*

Parts and products: **MALI** TAMACHEK ichiban *edible grain* (A.Chev.) **GHANA** AKPAFU ɔ̀-tátá-yùè *(pl. sì-.) chaff* (Kropp) AVATIME símùnùsí *chaff* (Kropp) **NIGERIA** HAUSA fura *pounded & cooked grain* (Schuh) KANURI kasha edible grain of Several spp. NGIZIM áagăw *(pl. -wàwín) pounded & cooked grain* (Schuh) aɗa-k âw *a head of grain* (Schuh) ámbàawà *(pl. -àwin)* a stalk of which the pulp is sweet, e.g. sugar cane and some sorghum (Schuh) áptâ fine-ground flour (Schuh) bákâ a covered heap of corn/grass, or a shelter with corn-straw or matting over a frame (Schuh) kàstà *(pl. -àtín)* a corn-stalk flute (Schuh) sáptó *(pl. -ótín)* a pile of grass or corn stooked for storage (Schuh) táɓsâk *(pl. -ásin)* a sheaf of long grass prepared for thatching (Schuh) PERO úrúbi chaff (ZF)

The grasses make up the most important economic family of the plant kingdom. It contains the cereals (wheat, rice, maize, barley, oats, etc.) which are the staple for almost all mankind, and the fodder grasses upon which virtually all stock and game subsist. Larger species provide building materials. The family as a whole is a source of diverse valuable by-products. To a remarkable extent grasses and mankind, and grasses with stock and game have each evolved a sort of commensalism. Grasses dominate the savanna, which, in an African context, is not only striking but is ecologically and economically important. In a general way, species are catagorised in the African concept, as, indeed, universally, into growth forms, ecological forms and communities, and by other conspicuous characteristics. Vernacular names for these are shown in the lists above.

There are some 640 species in the Region, of which over one-half has some economic attribute duly recorded in the following pages.

Acrachne racemosa (Heyne) Ohwi

FWTA, ed. 2, 3: 395.

A tufted annual grass to 60 cm tall, of disturbed soils in Senegal and Mali in the Region, and occurring in the Asian tropics to N Australia, and in the W Indies.

It is said to provide a little grazing at all stages of growth in the arid state of Rajasthan, India (1).

Reference:

1. Gupta & Sharma, 1971: 59.

Acritochaete volkensis Pilger

FWTA, ed. 2, 3: 448.

A weakly trailing annual grass with culms to 1 m long; of shaded sites in montane locations of N Nigeria and W Cameroons, and across to Ethiopia.

The grass lacks leaf and is of little value, but it is reported to be grazed by stock on Mt Elgon, Uganda (1).

Reference:

1. Snowden 1229, K.

GRAMINEAE

Acroceras amplectens Stapf

FWTA, ed. 2, 3: 435. UPWTA, ed. 1, 519.
West African: THE GAMBIA MANDING-MANDINKA jajeo (Pirie fide JMD) nyaro (Macluskie) GUINEA-BISSAU BALANTA lábar (EPdS; JDES) SIERRA LEONE KORANKO kɔbɔlɔ (JMD) LOKO babarawo (NWT) MENDE mbowi general for certain types of grass (JMD) pisui (JMD) yuwi (JMD) SUSU sunyugi (JMD) TEMNE ka-rin-ka-sui (JMD) MALI DOGON diì vòònu (CG) haratébé (A.Chev.) FULA-PULAAR (Mali) mbu niari (A.Chev.) MANDING-KHASONKE sogo (A.Chev.) soko (A.Chev.) SONINKE-SARAKOLE kitibué (A.Chev.) UPPER VOLTA GURMA bugau (A.Chev.) kiu pété (A.Chev.) GHANA DAGAARI bamɔ́dò (FRI) MOORE komúdo (FRI) NIGERIA HAUSA géérón tsúntssayéé = bird's millet (JMD; ZOG)

A weak-stemmed annual grass, decumbent scrambling to 90 cm high, of marsh or shallow water, common throughout the Region from Senegal to N and S Nigeria, and Sudan and the Congo basin.

The plant is a weed of rice-padis in The Gambia (4), and also of hill rice where it is considered a serious pest (5). In suitably wet areas it forms large meadows, sometimes floating on water from a prostrate root (3). It grows very good fodder and hay (2, 3) relished by cattle and horses (1).

References:

1. Adam, 1966, a. 2. Beal 28, K. 3. Dalziel, 1937: 519. 4. Macluskie 8, K. 5. Terry 3161, K.

Acroceras gabunense (Hack.) WD Clayton

FWTA, ed. 2, 3: 436, as Commelinidium gabunense (Hack.) Stapf.
West African: SIERRA LEONE MENDE ngale (def. -i) (Cole)

A scrambling perennial grass, rooting from a prostrate base to 60 cm high, of forest shade in Sierra Leone to W Cameroons, and across central Africa to Uganda and Angola.

It is considered useful as fodder in Sierra Leone (1).

Reference:

1. Cole, 1968.

Acroceras zizanioides (Kunth) Dandy

FWTA, ed. 2, 3: 435. UPWTA, ed. 1, 519.
West African: GUINEA-BISSAU FULA-PULAAR (Guinea-Bissau) québè-fárô (JDES) SIERRA LEONE KONO bakabinε (FCD) KORANKO kɔbɔlɔ (NWT) LOKO kbaraga (NWT) MANDING-MANDINKA kɔbɔrɔ́ (FCD) MENDE g-bati (FCD) m-bowi general for certain types of grasses (NWT) kotopoi (FCD) k-pagɔ (FCD) pisui (NWT; JMD) yuwi (NWT; JMD) SUSU barεkɔre (NWT) mεlkɔre (NWT) sunyugi (NWT; JMD) TEMNE ε-kbil (NWT) ka-rin-ka-sui (NWT; JMD) YALUNKA tamidisεra-na (FCD) MALI MANDING-BAMBARA ngon (Rogeon fide JMD) NIGERIA HAUSA géérón tsúntssayéé = bird's millet (JMD) YORUBA iyè-ẹtu (Dawodu; JMD)

A perennial grass, scrambling from prostrate base to over 1 m high, of moist and shady soils, or in shallow water, common through the Region, and widespread southwards to Angola and in Uganda and Tanganyika; also in India and tropical America.

The plant produces good forage and hay relished by cattle and horses (2, 3).

A trace of alkaloid has been detected in the leaf (1).

References:

1. Adegoke & al., 1968. 2. Adam, 1954: 87. 3. Adam, 1966,a.

Aeluropus lagopoides (Linn.) Trin.

FWTA, ed. 2, 3: 382.

An halophytic, stoloniferous, mat-forming grass 20–30 cm high, of salt pans, salt-marsh, foreshore dunes, and inland brackish sites, in Mauritania, and widely dispersed from the Mediterranean and NE Africa to the Middle East and into India.

It is salt-tolerant and because of its matting habit it is sand-binding on foreshore sand-dunes. In the Middle East it provides good grazing for all stock including camels. Stock are said in Iraq to obtain their salt requirement and to thrive and to grow fat on it (1). In Rajasthan, India, its value as fodder is recognised as it will grow on inhospitable saline soil where nothing else will (2).

References:

1. Abd.-ar-rizaq Barbuti, s.n., K. 2. Gupta & Sharma, 1971: 59.

Alloteropsis cimicina (Linn.) Stapf

FWTA, ed. 2, 3: 449.

English: carpet grass (India, Chadha).

A tufted annual grass, or longer-lived if under wet conditions; culms 60–90 cm high; of open dry or moist soils, recorded only in N Nigeria, but occurring widespread in Sudan, E and southern Africa; also in the Asian tropics.

Leaf bulk is not abundant, and though cattle graze it, it is of limited value (Sudan: 6, 8; Uganda: 3; India: 2, 5). In parts of India with heavier rainfall, it grows a good pasture, and converts into hay (2). In Malaya plants are reported to contain *coumarin* which puts off cattle from browsing it (1). Stock in Tanganyika do not much like it (4, 7).

Because of its low spreading habit, it is recommended in India as a soil-binder for anti-erosion purposes, and along watercourses to stop scour (2).

References:

1. Burkill, IH, 1935: 104–5. 2. Chadha, 1985: 187–8. 3. Dyson-Hudson 22, K. 4. Emson 255, K. 5. Gupta & Sharma, 1971: 59–60. 6. Harrison 900, K. 7. Marshall 41, 48, K. 8. Peers K.025, K.

Alloteropsis paniculata (Benth.) Stapf

FWTA, ed. 2, 3: 449. UPWTA, ed. 1, 519.

West African: SIERRA LEONE MENDE ndiwi (JMD) yani (JMD) yobɔvatui (NWT) SUSU yane (NWT) TEMNE ka-fon-ka-an-yari (Glanville) kɔ-sinkr NWT ka-sota (Adames); ? misapplied (FCD) MALI SONGHAI hori *general for grass* (A.Chev.) subu *general for grass* (A.Chev.)

An annual grass, culms often decumbent, rising to 1.2 m high; of moist soils and damp places in Senegal to S Nigeria and across central Africa to Zanzibar and Mauritius.

The plant is a weed of tidal rice-padis in Sierra Leone (2), said to be non-salty, still within the fresh-water limit (4). On drier sites it forms large tufts with tillers, and is a good fodder (1, 3).

The roots of the plant in Ghana are reported to be scented (5).

References:

1. Adam, 1966, a. 2. Adames 154, K. 3. Dalziel, 1937: 519. 4. Glanville 227, K. 5. Irvine 3006, K.

173

GRAMINEAE

Alloteropsis semialata (R Br.) Hitchc.

FWTA, ed. 2, 3: 448.

A tufted perennial grass, culms 60–90 cm high; of savanna territory in Senegal to W Cameroons, and widely dispersed elsewhere in the African and Asian tropics.

It is reported as not eaten by cattle in Malawi (3), but to be valued as fodder in Assam (1, 2).

References:

1. Bor, 1960: 277. 2. Chadha, 1985: 185. 3. Fenner 224, K.

Anadelphia afzeliana (Rendle) Stapf

FWTA, ed. 2, 3: 501. UPWTA, ed. 1, 519–20, as *A. arrecta* Stapf.

English: thatchgrass (Sierra Leone, Cole).
West African: SENEGAL DIOLA mu git (JB) SERER bati (JB) fati (JB) WOLOF ñañtãg (JB) SIERRA LEONE BULOM (Sherbro) pui-lɛ (FCD) KISSI ɡboŋgbonelɔ (FCD) KORANKO kulusabinyi (NWT) MANDING-MANDINKA tikolo-mɛsɛ (FCD) MENDE foni *a general term* (auctt.) foni mayambe (JMD) fovo (FCD) TEMNE a-nɛpəl (*pl.* ɛ-nɛpəl) (auctt.) ɛ-nɛpəl-ɛ-bana (JMD) ɛ-nɛpəl-ɛ-bira (NWT; JMD) ɛ-nɛpəl-ɛ-rək-rək (FCD) YALUNKA ɡbolesɛhrɛ-na (FCD) NIGERIA HAUSA bayan maraya = *back of the cob antelope* (Thatcher) YORUBA bɛrɛ *applied also to other spp.* (JMD)

A perennial tufted grass with slender wiry weak-stemmed, genuflexed and basally branched culms rising to 1–2 m tall; in the savanna favouring damp and marshy places; common across the Region from Senegal to N and S Nigeria, and extending southwards into the Congo basin.

The haulm provides good thatching (1, 4, 5). In the Northern Province of Sierra Leone it is described as the best (6–8) and in places in the Region it is cultivated for this purpose (5).

When young it is grazed (2, 4).

Chemical analysis of the root of Nigerian material has shown a trace of alkaloid (3).

The Mende name *foni* applied to it is but a generic epithet for the community in which it may be found.

References:

1. Adam, 1954: 97. 2. Adam, 1966, a. 3. Adegoke & al., 1968: 13–33. 4. Cole, 1968,a. 5. Dalziel, 1937: 519. 6. Deighton 4443, K. 7. Glanville 66, FBC. 8. Jordan 651, K.

Anadelphia bigeniculata WD Clayton

FWTA, ed. 2, 3: 502.
West African: GUINEA FULA-PULAAR (Guinea) fugolo (Adames)

An annual grass, 30–60 cm tall; of ironstone outcrops; known only in Guinea and Sierra Leone.

No usage is recorded, but the grass is known to Fula herdsmen.

Anadelphia leptocoma (Rendle) Stapf

FWTA, ed. 2, 3: 501.

English: thatch grass (Sierra Leone, Jordan, Cole).
West African: SIERRA LEONE KISSI kilaichiɛyo (FCD) MENDE foni *a general term* (auctt.) foni-mayambe (Fisher) TEMNE an-bunthi (Glanville) ɛ-nɛpəl-ɛ-bana (FCD) NIGERIA YORUBA bɛrɛ (MacGregor)

174

A perennial grass, culms 1–1½m tall; of low-lying savanna from Mali to S Nigeria.

It is a thatch grass but not as good as *A. afzelii* (1, 3–5).

It furnishes a little grazing (3).

The Mende name *foni* is incorrectly applied to it in mistaken allusion to *A. afzelii*; while the latter may be said to be a member of the *foni* community, *A. leptocoma* is not (2).

References:

1. Cole, 1968,a. 2. Deighton 2116, K. 3. Fisher 50, K. 4. Jordan 566, K. 5. Jordan 648, FBC, K.

Anadelphia trepidaria (Stapf) Stapf

FWTA, ed. 2, 3: 501.

West African: **GUINEA** FULA-PULAAR (Guinea) tchelbi Adames

An annual grass with culms reaching to 60cm tall; of ironstone outcrops; known only in Guinea.

No usage is recorded, but Fula of Fouta Djallon in Guinea lump it with the same name as is given to *Hyparrhenia subplumosa* (q.v.). It probably has the same usages in grazing and thatching.

Andropogon Linn.

FWTA, ed. 2, 3: 482-3.

West African: **UPPER VOLTA** FULA-FULFULDE (Upper Volta) coobol (K&T) soobol *general for A.spp.* (K&T) yantaare *general for the large perennial savannah grasses* (K&T)

Andropogon africanus Franch.

FWTA, ed. 2, 3: 488.

West African: **NIGERIA** FULA-FULFULDE (Nigeria) raneraneho (Saunders) yakawre (Saunders)

A caespitose perennial grass with culms to 2 m high; of swampy sites and seasonally flooded plains, occasionally amongst rock outcrops; at upland altitudes; throughout the Region from Mauritania to N and S Nigeria, and in E Africa and the Congo basin to Angola.

Fula cattle-men in Nigeria recognise the grass as giving good cattle-grazing pasture (1).

Reference:

1. Saunders 31, 44, K.

Andropogon amethystinus Steud.

FWTA, ed. 2, 3: 485, incl. *A. pratensis* Hochst.

A perennial tufted or straggling shortly rhizomatous grass with culms to about 30 cm high; of upland evergreen forest or montane grassland at 2,000 to 3,800 m elevation in W Cameroons and Fernando Po, and extending into NE, E and to southern Africa, and in S India.

It is an useful grazing grass in Uganda, much liked by all stock (2), yet in Kikuyu and Masailand it is said to be bad for cattle (1).

References:

1. Grant 1238, K. 2. Snowden 1465, K.

GRAMINEAE

Andropogon auriculatus Stapf

FWTA, ed. 2, 3: 488.
West African: SIERRA LEONE BULOM (Sherbro) pui-lɛ (FCD)

A caespitose perennial grass with culms to 1½ m high; on sandy beaches to upland situations between Senegal to S Nigeria, and in E Cameroun.
It is an important forage grass in Senegal (1). On Sherbro Island of Sierra Leone it is a co-dominant grass (2), and on the Obudu Plateau in Nigeria it is an important constituent of the Loudeto-Hyparrhenetum ranchland grass community (3); see *Hyparrhenia diplandra.*

References:

1. Adam, 1954: 88. 2. Deighton 2480, K. 3. Tulley, 1966: 899–910.

Andropogon canaliculatus Schumach.

FWTA, ed. 2, 3: 486.

A tufted perennial grass, culms erect to 2 m high; of damp and swampy sites; in the Region from Mali to N Nigeria, and in E Africa.
At Pong Tamale, Ghana (5), and at Entebbe, Uganda (8), it is recorded as being highly palatable to cattle which take it to the exclusion of all other grass species. Elsewhere it is generally grazed by domestic cattle (Kenya: 4, 7; Tanganyika: 3), becoming unpalatable when old and dry. Buffalo in Kenya take it (6). Commonly it is a short savanna grass of insufficient bulk to be of much importance (2, 9).
Analysis of Kenyan material shows that the grass is relatively rich in *protein* (13.36%, dry wt.) and is not unduly fibrous (31.96% dry wt.). Calcium and phosphorus are adequately present (2).
Larger-culmed plants are used in Kenya for thatching (7). In Bamako market, Mali, the culms are commonly sold for matting, screens and roofing (1).

References:

1. Diarra, 1977: 42. 2. Dougall & Bogdan, 1960: 241–4. 3. Emson 94, K. 4. Glover & al. 1870, K. 5. Ll. Williams 838, K. 6. Magogo & Glover 612, K. 7. Magogo & Glover 821, K. 8. Maitland 77, K. 9. Snowden 1361, K.

Andropogon chinensis Merrill

FWTA, ed. 2, 3: 486, as *A. ascinodis* CB Cl.
West African: SENEGAL BEDIK *nyi*-syil (FG&G) GUINEA BASARI ɔ-nírí (FG&G) UPPER VOLTA FULA-FULFULDE (Upper Volta) yantaare (*pl.* jantaaje) (K&T) MOORE pita (Scholz) NIGERIA BUSA (Bokobaru) senteni (Ward) FULA-FULFULDE (Nigeria) gamari boderi (Ward) HAUSA dargaza DA. Samaru kiara natudu (Ward) YORUBA (Ilorin) bere (Ward)

A coarse tufted perennial grass with culms to 1.5 m high; of poor shallow stony soils and old fallows; throughout the Region from Senegal to N and S Nigeria, and in E and southern Africa and across Asia to India and China.
The grass provides grazing while still young (1, 3, 6; Kenya: 7). It is taken by buffalo in Kenya (5). The haulm is used as thatch (1, 2, 4, 6).

References:

1. Beal 33, K. 2. DA. Samaru 28, K. 3. De Leeuw 1301A. 4. Ferry & al., 1974: sp. no. 101, as *A. ascinodis.* 5. Leuthold 60, K. 6. Ward 0025, 0038, K. 7. Wilson 30, K.

176

Andropogon fastigiatus Sw.

FWTA, ed. 2, 3: 485. UPWTA, ed. 1, 525, as *Diectomis fastigiata* Kunth.

West African: GUINEA FULA-PULAAR (Guinea) fugolo (Adames) NIGERIA HAUSA bayan maraya = *cob antelope's back* (JMD) jam bauje, yama *applied loosely* (JMD) KANURI garbazam (JMD)

A tufted annual grass with culms to 1 m high; of shallow or sandy soils and dry fallows; throughout the Region from Mauritania to N and S Nigeria, and in E Africa, and in the tropics generally.

The grass when young is readily taken by stock (Senegal: 1; Ghana: 3; Dahomey: 7; Nigeria: 5, 8; Kordofan: 4). When mature its value is indifferent and the awns are liable to injure the mouth (4). It dies out early in the dry season (6).

In N Nigeria huts may be thatched with it (9).

In the South Province of Zambia the grass is reported to form distinct belts around termite mounds (2), suggesting perhaps that termites harvest the grain.

References:

1. Adam, 1966, a, as *Diectomis fastigiata*. 2. Astle 2269, K. 3. Beal 25, K. 4. Baumer, 1975: 96, as *Diectomis fastigiata* (Sw.) Beauv. 5. Dalziel, 1937: 525, as *Diectomis fastigiata* Kunth. 6. Onycago- cha FHI.7776, K. 7. Risopoulos 1262, K. 8. Saunders 8, K. 9. Ward L.150, K.

Andropogon gabonensis Stapf

FWTA, ed. 2, 3: 488.

A robust perennial grass with culms to 3½ m high; known in the Region only in W Cameroons, but occurring also to the Congo basin and Angola.

The young growth is grazed by buffaloes (1).

Reference:

1. Walker & Sillans, 1961: 185.

Andropogon gayanus Kunth

FWTA, ed. 2, 3: 488. UPWTA, ed. 1, 520.

English: tambuki grass (Sierra Leone, Scott-Elliot).

West African: SENEGAL BEDIK mɔ-diidi (FG&G) ma-kas, ɔ-kas *var. pubescens* (FG&G) ga-ndány *var. glabra* (FG&G) gɨ-nyidi (FG&G) ɛ-tiùɓ *var. glabra* (FG&G) DIOLA é-buk (JB) FULA-PULAAR (Senegal) dagué *a c.var.* (A.Chev.) guélori (A.Chev.) kiene (A.Chev.) MANDING- BAMBARA badoba (JB) mussa waga (A.Chev.) uaba (A.Chev.) vaba (A.Chev.; GR) waga (A.Chev.) waga-gué *a c.var.* (A.Chev.) zara (A.Chev.) SERER o nduy (JB after A.Chev.) yèv (JB) WOLOF cicca (JLT fide JMD) khat, hat (A.Chev.; JB) soya (A.Chev.) **THE GAMBIA** MANDING-MANDINKA wa *general for tall grasses* (auctt.) **SIERRA LEONE** BULOM (Sherbro) pui-lɛ (JMD) LIMBA kabusa (GFSE) MALI FULA-PULAAR (Mali) dagué *a c.var.* (A.Chev.) guélori (A.Chev.) kiené (A.Chev.) MANDING-BAMBARA mussa waga (A.Chev.) uaba (A.Chev.) waba (A.Chev.) waga (A.Chev.) zara (A.Chev.) KHASONKE wako *a c.var.* (A.Chev.) MANINKA nguon (Lean) **UPPER VOLTA** BISA késsé (Scholz) FULA-FULFULDE (Upper Volta) danye (K&T) ɗayye (K&T) kagarire (Scholz) lanyere (K&T) ranyere (K&T) GRUSI soporé (A.Chev.) GURMA mokiri *a c.var.* (A.Chev.) MOORE mofogo (A.Chev.) mopaka (A.Chev.) mopoko (A.Chev.; Scholz) pita *the stems only* (A.Chev.) **GHANA** DAGBANI purim pielega (JMD) **NIGER** FULA-FULFULDE (Niger) sooɓre (ABM) yayyere (ABM) SONGHAI sùb-ñá (*pl.* -ñóŋó) (D&C) **NIGERIA** BIROM eré (LB) FULA-FULFULDE (Nigeria) ɗadeppure (J&D) palawal (JMD) welho (Saunders) FULFULDE kalawal (Chapman) HAUSA gambà, gámba, gámbàà (auctt.) gírmán dáréé ɗáyá = *growth of one night* (Bargery; ZOG) jimfi (JMD) wááwán rúwá = *cloth of the water* (AST; ZOG) IDOMA ekpo (Odoh) IGALA ìkpò (Odoh; RB) IGBO ikpò (BNO; KW) IGBO (Agukwu) ìkpò agū̄ = *leopard's ikpo* (auctt.) JUKUN (Wukari) séfun-kwe (Shimizu) KAMBARI-SALKA ìikubɔ̂ (RB) ɔ̀ɔkpɔ̀hɔ̀mɔ̀ (RB) KANURI sugu (JMD) sugu kal (JMD) súwù (C&H) súwù búl *a tall grass, sp. indet., but similar to this sp.* (C&H) súwù kâl *a large form*

GRAMINEAE

(C&H) súwúkàl (A&S) NGIZIM mâdlbák (*pl.* -àbín) (Schuh) TIV igomough (JMD) YORUBA èrùwà = *come and carry* (Verger) èrùwà funfun (auctt.)

A tufted perennial grass, polymorphic with W Africa being the main centre of variation; four varieties are recognised: var. *gayanus,* culms to 4 m high, glaucous appearance and fastigiate habit, of seasonal swamps and flood-plains; var. *tridentatus* Hack., culms 1.2 to 1.8 m high; var. *polyclados* (Hack.) WD Clayton (as var. *squamulatus* (Hochst.) Stapf in *FWTA,* ed. 2, 3: 488) common, culms to 3 m high, of the Soudanian savanna; and var. *bisquamulatus* (Hochst.) Hack., culms to 3 m high, of the Soudanian savanna. Classification is based on spikelet variation (9,a). All are represented throughout the Region, the first two also in Sudan, the third in NE to S Africa, and the fourth in Sudan and the Central African Republic. All varieties are treated together here.

They all give a good and palatable forage especially while still young (1, 2,a,b, 5, 6, 10, 15, 16, 18,a,b, 20, 22, 23, 25, 28, 31). The leaf-sheaths and pseudo-petioles bear nectaries producing sugar analysed as being a mixture of *sucrose, glucose, fructose* and small amounts of *maltose* and *raffinose* (8). This may perhaps account for the ready palatablity of the grass to stock. Horses do not always take it (2,a), or do so only when the growth is still quite tender (15). It is said to be slow to dry out after the rains in Niger (18,b). The grass is indigenous throughout northern Nigeria merging towards its southern limit with the closely related *A. tectorum.* (14). It is recognised as one of the best perennial pasture grasses in N Nigeria (14). In Guinea it appears to be a termitophile species being present in the Soudanian savanna as a principal constituent of a vigorous ant-hill flora and restricted solely to the zone of activity of the termites (*Bellico-termes spp.*) (2,b). It rapidly invades fallows and newly opened soil (1, 5, 21). It is a weed of arable land in The Gambia (26). By the rapidity with which it regenerates after a burn it establishes itself as the dominant grass of the annually burnt savanna (1). Because of its vigour it is well adapted to short-term ley pastures since it is quick to establish, and can be eradicated easily by cultivation for the following crop (19).

A high proportion of plants of var. *tridentatus* are diploid (2n = 20) whereas other varieties are tetraploid (9,b). Genetical breeding work may result in improved strains suitable for a wide climatic range (5, 14).

The grass is planted in N Senegal as a hedge around habitations and pounds (1). The robust culms reaching to 3 m or more high by 2 cm diameter are used to make palisades and enclosures (10, 13,a, 23). They serve as building material for walls (27), the circular bands of a conical thatched roof (10) and for thatching (23). Mats, called *zaanaa,* in Hausa in Nigeria (10, 15, 22, 30) and *sekko* in Upper Volta (17) and Mali (11) are commonly made for matting and screens. In northern Sierra Leone the culms are fashioned together into improvised bee-hives for taking swarms (7), whereas in Central African Republic bundled stems are lit as a torch to smoke out a hive for taking the honey (12).

An infusion of the grass after toasting is given as an aid to digestion, especially for children, in S Nigeria (4).

Tenda children use the culm as a toy, heating it over a fire, then breaking it across a stone when it bursts with the crack of a gun-shot (13,b).

In a cryptic way Yoruba invoke the spirit of the grass to come dancing as a cure against bleeding eyes (29).

References:

1. Adam, 1954: 88, as var. *bisquamulatus* Hack. 2a. Adam, 1966, a: as var. *bisquamulatus.* 2b. ibid., 1966,a: as var. *genuinus.* 3. Adam, 1968, a: 927, as var. *gayanus.* 4. Ainslie, 1937: sp. no. 30. 5. Baumer, 1975: 86. 6. Beal 34, K. 7. Boboh, 1974: in litt. 24/4/1974. 8. Bowden 1971: 77-80. 9a. Clayton, 1972: 3: 484. 9b. ibid., 1972: 488. 10. Dalziel, 1937: 520. 11. Diarra, 1977: 42. 12. Fay 5413, K. 13a. Ferry & al., 1974: sp. no. 102. 13b. ibid., 1974: sp. no. 103. 14. Foster & Munday, 1961: 313. 15. Golding 28, K. 16. Harrison 84, K. 17. Kintz & Toutain, 1981. 18a. Maliki, 1981: 52, sp. no. 58. 18b. ibid., 1981: 53, sp. no. 64. 19. Ogor & Hedrick, 1963: 146. 20. Peers M.010, K. 21. Roberty, 1953: 449. 22. Saunders 6, K. 23. Schnell, 1953,b: no. 48. 24. Scott-Elliot 4261, K. 25. Scott-Elliot 5932, K. 26. Terry 3026, K. 27. Thornewill 185, K. 28. Trapnell 821, 857, as var. *polyclados.* 29. Verger, 1967: sp. no. 71. 30. Ward L.134, K. 31. Ward 149, K.

Andropogon lima (Hack.) Stapf

FWTA, ed. 2, 3: 485.

A densely tussocked perennial grass with culms to 1 m high; of montane grassland in W Cameroons, and widely dispersed in Sudan, Ethiopia, E African highlands and Malawi.

It is recorded as seemingly eminently suitable for high altitude (3,000–4,000 m) herding of domestic stock in Uganda (2), but in Kenya in the Aberdares (2,800 m) it is said to be too coarse for grazing though it is taken by bush-buck while still young (1).

References:

1. Grant 1245, K. 2. Snowden 1477, K.

Andropogon macrophyllus Stapf

FWTA, ed. 2, 3: 488.
West African: SIERRA LEONE MANDING-MANDINKA tõfõ (FCD) tõfo-ke (FCD) YALUNKA kofu-na (FCD) GHANA AKAN-ASANTE ɛsire (CV; E&A)

A robust perennial grass with culms to 3½ m high; recorded from Guinea to Fernando Po.

In Sierra Leone it is recognised as a fodder grass, palatable to cattle, but less so than is *A. tectorum*, which is the more common (1).

Reference:

1. Deighton 4235, 4431, 4526, K.

Andropogon perligulatus Stapf

FWTA, ed. 2, 3: 486.

A tufted perennial grass, culms erect to 1½ m high; of swampy locations; occurring throughout the Region from Senegal to N Nigeria, and in Ethiopia, E and south central Africa. It is very similar to *A. canaliculatus* (q.v.) and may be but a variant (1).

No attribute is recorded for it in the Region. *Flavonic glycosides* have been identified in Zambian material (2).

References:

1. Clayton & Renvoize, 1982: 782–3. 2. Grassi 46.35 (Harborne G.124), K.

Andropogon pinguipes Stapf

FWTA, ed. 2, 3: 488. UPWTA, ed. 1, 520.
West African: SENEGAL FULA-PULAAR (Senegal) ngétiétiadi (K&A) MANDING-BAMBARA bohdô (auctt.) MANDINKA banôbo (K&A) SERER golôbâ (K&A) *n*-golumban (A.Chev.; JB) WOLOF gomgom (JB) gongon (A.Chev. after JMD, fide K&A) taf (A.Chev.; K&A)

An annual grass with culms reaching 1–3 m high; recorded only in Senegal.

It is a good forage plant when still young (2, 3), and is browsed even when in flower (1). The stems are used in Senegal as ties for thatch (3). They are items of merchandise in Bamako (Mali) markets sold for making mats [*sekko*], screens, doorway covers and thatching (4).

References:

1. Adam, 1954: 88. 2. Adam, 1966, a. 3. Dalziel, 1937: 520. 4. Diarra, 1977: 42.

GRAMINEAE

Andropogon pseudapricus Stapf

FWTA, ed. 2, 3: 486. UPWTA, ed. 1, 520.

West African: SENEGAL SERER mafar mbil (JB) ndianyué (JMD) WOLOF ndiangua (JMD) **THE GAMBIA** FULA-PULAAR (The Gambia) sheioko (Frith) MANDING-MANDINKA fula nyantango (def.) (Frith; JMD) kombo wa messengo (def.) (Hayes fide JMD) wulo konno nyamo (def.) (JMD) WOLOF njanga (JMD) **UPPER VOLTA** BISA monkatjin (Scholz) FULA-FULFULDE (Upper Volta) siiwuko (K&T) si'uko (K&T) yantaré (Scholz) GRUSI-KASENA kjiempuh (Scholz) tangolo (Scholz) MOORE kaloiéga (Scholz) njianparagha (Scholz) yandéparaga (Scholz) **GHANA** DAGBANI daziman (Ll.W) **NIGERIA** ARABIC-SHUWA geishsh an naar (JMD) BUSA (Bokobaru) funu kiango (Ward) FULA-FULFULDE (Nigeria) gamari boderi (Ward) tidingho (JMD) HAUSA ján baƙo (JMD; ZOG) ján baújeé (JMD) ján daatsii *from* daatsii: *a general name for grass used chopped up and mixed with clay for building* (JMD) ján rámnoá (JMD) ján raúnáó (JMD) laßanda *applied to weaker thatching grasses* (JMD; ZOG) mai gindin biri = *with a monkey's backside* (Ward) KANURI katjin (JMD; C&H) ngirmasan (JMD) NUPE dogó (JMD) TIV acho *a general term* (JMD)

A tufted annual grass with culms to 1.2 m high; of shallow or sandy soil, and common on dry fallow land; occurring throughout the Region from Senegal to N and S Nigeria, and into Chad and E Cameroun, and in Mexico and Brazil.

It is a common grass on irrigation bunds and river-banks in N Senegal (2). It is grazed by stock, especially when still young (1, 2, 7, 9). It is valued as a winter fodder (5), and in Senegal the aftermath regrowth is considered especially useful (2).

The haulm is widely used for thatching (3–6, 8, 9). In Nigeria it is chopped up for puddling with clay as a building material (5). In Mali it is one of several grasses used to make mats and screens, and is commonly on sale in markets in Bamako for this purpose (6).

References:

1. Adam, 1960,a: 365-6. 2. Adam, 1966, a. 3. Dalziel 490, K. 4. Dalziel 8418, K. 5. Dalziel, 1937: 520. 6. Diarra, 1977: 42. 7. Risopoulos 1252, K. 8. Ward L.148, K. 9. Ward 0024, K.

Andropogon schirensis Hochst.

FWTA, ed. 2, 3: 486, incl. *A. dummeri* Stapf. UPWTA, ed. 1, 520.

West African: NIGERIA FULA-FULFULDE (Adamawa) hahaendenoh (Pedder) FULFULDE (Nigeria) lawrehe (Saunders) HAUSA rumiya (Taylor) yambiu (Freeman)

An erect tufted perennial grass, culms to 2 m high; of shallow soils over rocks, usually well-drained, but into swamp margins on deciduous bushland and wooded grassland to montane grassland, 100 to 3,400 m altitude; throughout the Region from Senegal to W Cameroons, and over tropical Africa to S Africa.

The grass provides grazing for stock while young and tender (1, 2, 4; Uganda: 7; Kenya: 5). Buffalo readily eat it in Tanganyika (3).

The roots are aromatic (6, 8). Phenolic substances have been detected in Nigerian plants (9).

The grass has use as thatch in Kenya (5).

In the Eastern Province of Uganda thickets of the grass occur around termitaria (10) suggesting that termites collect the grain.

References:

1. Adam, 1966, a. 2. Beal 41, K. 3. Burtt 4605, K. 4. Dalziel, 1937: 520. 5. Glover & al. 576, 787, K. 6. Scholz 120, K. 7. Snowden 1167, K. 8. Ward 0034, K. 9. Wit & al. 450 (Harborne G.145), K. 10. Wood 573, 958, K.

Andropogon tectorum Schumach. & Thonn.

FWTA, ed. 2, 3: 488. UPWTA, ed. 1, 520.

English: horse grass (Sierra Leone, Cole).

West African: SENEGAL BEDIK diamél (FG&G) MANDING-BAMBARA dãz0 (JB) MANDINKA badoni (JMD) SERER humir (JMD) **GUINEA-BISSAU** FULA-PULAAR (Guinea-Bissau)

180

uaba (JDES) PEPEL djagalhe-quentche (JDES) **GUINEA** BASARI a-nàwén (FG&G) KONYAGI luri (FG&G) ɔ-sampan (FG&G) MANDING-MANINKA wara (Brossart) **SIERRA LEONE** KISSI bendaŋ (FCD) bendẽ (FCD) KRIO bobogia (FCD) hɔs-gras *i.e., horse grass* (FCD) LOKO (Magbile) bɔbɔ (NWT) MANDING-MANDINKA tõŋfɔ (FCD) MENDE fovo (*def.* -i) *general for this sort of grass* (FCD) ngongɔi (JMD) SUSU yobainyi (NWT) TEMNE am-bobo (auctt.) ɛ-tanke (NWT) YALUNKA nɔmi-na (FCD) **MALI** MANDING-MANINKA wara (Brossart) **UPPER VOLTA** FULA-FULFULDE (Upper Volta) dayye koobi (K&T) **NIGERIA** HAUSA gámbàà (JMD; ZOG) IGBO ìkpò (BNO; KW) IGBO (Umuahia) ikpuru otọ (AJC) YORUBA ẹ̀rùwà dúdú (auctt.)

A robust perennial grass with culms to 3 m high, often stilt-rooted; of shady sites under trees; common across the Region from Senegal to W Cameroons, and on into Central African Republic.

This grass has close affinity to *A. gayanus*, but with distribution rather to the southward where it is an important pasture grass (8). It is readily grazed before anthesis by all stock (1, 2, 5, 7, 11). It is deemed especially good fodder for horses (3, 4).

The culms are commonly cut in Sierra Leone for use in fencing (5). They are stout and pithy, and are used for matting and for roofing (4, 11). The culms are a common item of merchandise in Bamako markets in Mali sold for matting, screens and roofing (6).

In Ubangi, a root-decoction is given to women for stomach-ache (10).

The grass may have some potential as a source of honey: bees in Abeokuta Province of S Nigeria have been observed collecting pollen (9).

Tenda in SE Senegal have ritual use of the grass during funerals (7).

References:

1. Adam, 1966, a. 2. Carpenter AJC.820 (UIH 4186), UCI. 3. Cole, 1968,a. 4. Dalziel, 1937: 520. 5. Deighton 856, K. 6. Diarra, 1977: 42. 7. Ferry & al., 1974: sp. no. 104. 8. Foster & Munday, 1961: 313. 9. Onochie 8140, K. 10. Vergiat, 1970,a: 83–84. 11. Walker & Sillans, 1961: 185.

Andropogon tenuiberbis Hack.

FWTA, ed. 2, 3: 485.

A coarse tufted perennial grass with culms to 2.5 m high; of swamps and water-logged sites from Senegal to N and S Nigeria, and into Sudan and Zaïre.

The grass is used for thatching in Tanganyika (1).

Reference:

1. Davies D.967, K.

Anthephora ampullacea Stapf & CE Hubbard

FWTA, ed. 2, 3: 457.

A perennial grass, tussocking, culms to 1 m high; of sandy and waste places in Guinea and N and S Nigeria, and in the Congo and Angola.

The grass is hardy and remains alive throughout the seasons. By its robust habit it has some potential erosion prevention in sandy soils.

Anthephora effusa (Rendle) Stapf

FWTA, ed. 2, 3: 457.

A straggling perennial grass, culms to 90 cm high; of open hillsides in montane situations of Ghana, N Nigeria and W Cameroons, and in central and SW Africa.

GRAMINEAE

Anthephora nigritana Stapf & CE Hubbard

FWTA, ed. 2, 3: 457.
West African: NIGER SONGHAI díirí (*pl*. -o) (D&C) NIGERIA HAUSA kashin ɓera = *dung of the mouse* (Taylor)

A tufted perennial grass, culms to 1.5 m high; of dry rocky situations in Niger and N Nigeria and NE, E, S central and S Africa.
The grass is a weed of arable land, fallows and shifting cultivations in Sudan (1, 7). It has been under trial for fodder at agricultural stations in N Nigeria (Kano: 6; Mokwa: 5; Argungu: 4). It is leafy and has good dry season survival and thus is of some value for rough grazing, but it has a low seed production which must limit its use in pasture (2).
The grain is collected in Kordofan as a famine-food (3).

References:

1. Blair 5, K. 2. Foster & Munday, 1961: 314. 3. Hunting Tech. Surveys, 1968, as *Anthephora spp*.
4. Latilo 62751, K. 5. Scholz 134, K. 6. Taylor 6, K. 7. Wickens 1921, K.

Anthephora pubescens Nees

FWTA, ed. 2, 3: 457, as *A. hochstetteri* Nees.

A tufted perennial grass, culms to 1 m high; of rocky sites in deciduous bushland of Mali, and extending to NE and E Africa.
The grass is said to be liked by cattle in Uganda (2).
The grain is collected in Kordofan as a famine-food (1).

References:

1. Hunting Tech. Services, 1964: as *Anthephora spp*. 2. Liebenberg 1840, K.

Aristida Linn.

FWTA, ed. 2, 3: 378.
West African: NIGERIA FULA-FULFULDE (Nigeria) ulumboju *applies to all spp., but some have special terms* (J&D)

A large genus worldwide, of poor soils and low rainfall areas. Thirteen species are recorded for the Region. Utility varies, with climate and locality. They are of importance in the Sahel for pasturage, but of little use in the Soudan (1, 2).

References:

1. Adam, 1966: 522. 2. Roberty, 1953,a: 449.

Aristida adscensionis Linn.

FWTA, ed. 2, 3: 379. UPWTA, ed. 1, 520.
West African: MAURITANIA ARABIC (Hassaniya) lhaïyet lehmar (AN) SENEGAL SERER hétièb (JB) mbol tièb (JB) MALI DOGON dúŋu bɛɛ (CG) TAMACHEK allomoze (JMD) tezenat (JMD) UPPER VOLTA FULA-FULFULDE (Upper Volta) selɓo (*pl*. celɓi) (K&T) GHANA MOORE mɔtodo = *bitter grass* (FRI) NIGERIA FULA-FULFULDE (Nigeria) selbi (J&D) wicco tenemeje (Saunders; JMD) HAUSA bàzàyyánáó (auctt.) datsi a *general term* (JMD; ZOG) gatsaura (JMD) katsaura (JMD; ZOG) lale shamuwa (Golding) tsíntsíyár dutsee = *brush of the rocks* (auctt.) wútsíyàr kurege baki = *black squirrel tail* (Kennedy) YORUBA ɔkà olongo *corn of 'olongo' [a kind of bird]* (Dodd; JMD) YORUBA (Ilorin) iru ofe (Ward)

An annual grass forming erect or sprawling tufts to 1 m high, very variable, of dry soil and sub-desert waste places of the Sahel across the Region, and throughout the tropics.

182

The plant is a quick growing pioneer of waste places. In India it assists in stabilising sand-dunes (9). It is adventive on irrigation bunds and river-banks in the rice growing area of Senegal (2), and fixes the sand-dunes north of Dakar (4).

Reports on grazing value are varied, but on the whole it supplies a useful fodder of indifferent or of good quality wherever it grows, especially while green. It is available in both the rainy and the dry seasons, and in places is reckoned to be very important at the latter time for lack of better. An analysis of Ghanaian material has shown a high proportion (13%) of protein, and only 2% silica-free ash (1, 5). But sometimes the grass may be bitter when stock will not touch it – hence the Moore name in Ghana: *mɔtodv* meaning 'bitter grass' (10). As for other *Aristida spp.* (cf. *A. sieberana, A. mutabilis*), the awned spikelets may prove troublesome rendering the plant unpalatable when in seed, but after shedding, the dried straw may be taken. In Somalia attacks of coughing in sheep have been shown to be caused by awns sticking in the throat (12). In Tanganyika the awned seed cause irritation to cattle's eyes (13), and in Kordofan injury to cattle's mouth and damage to the hide (3).

The culms are used in Lesotho to make brooms, but of poor quality (8). In Nigeria (5) and in Kenya (6) they are used for thatching, and in the former territory for weaving into matting and making into sieves. The grass is one of the group known in Hausa as *katsaura* traditionally used in Nigeria (5) and Kordofan (3) for stuffing camel saddles.

Hausa in N Nigeria claim that the plant (part not stated) pounded with white onions is a lactogene for women (7).

Fula sorcerers in N Nigeria use the plant in offensive magic (*lekki baatal*) (11), but the circumstances and method are not recorded.

References:

1. Adam, 1954: 88. 2. Adam, 1960,a: 366. 3. Baumer, 1975: 86–87. 4. Broadhurst 27, K. 5. Dalziel, 1937: 520. 6. Glover & al. 1985, K. 7. Golding 32, K. 8. Guillarmod, 1971: 413. 9. Gupta & Sharma, 1971: 61. 10. Irvine 5007, K. 11. Jackson, 1973. 12. McKinnon S/24, S/86, K. 13. Marshall 28, K.

Aristida funiculata Trin. & Rupr.

FWTA, ed. 2, 3: 379. UPWTA, ed. 1, 520.

West African: SENEGAL SERER dohandok (JB) WOLOF galé kiam (A.Chev.) **MALI** DOGON holu (A.Chev.) FULA-PULAAR (Mali) kelbi (A.Chev.) selbéré (A.Chev.) MANDING-BAMBARA ngasan (A.Chev.) nkassa (A.Chev.) KHASONKE kasso (A.Chev.) **UPPER VOLTA** FULA-FULFULDE (Upper Volta) fitaako (K&T) huɗo raneeho = '*raneeho*' *grass* (K&T) selɓo (*pl.* celɓi) (K&T) MOORE sudumore (A.Chev.) SAMO bissi (A.Chev.) **NIGER** FULA-FULFULDE (Niger) korom (ABM) HAUSA fari n'tchaua (Bartha) **NIGERIA** HAUSA datsi *a general term* (Golding)

A tufted annual, culms wiry to 45 cm tall, of dry sandy places across the northern part of the Region from Mauritania to Niger and Nigeria, and extending through Sudan into India.

In Mauritania the plant forms extensive areas on the clay substrate between the dunes (2), and may thus contribute a little to fixing them. It is grazed a little but mainly whilst young (1, 7, 9). While in fruit it is usually shunned since the twisted awns are liable to cause soreness in the mouth (2). The ripe flower-spikelets are readily detached and the awns become intertwined so that small clumps of inflorescences are rolled about in the wind. The straw after the spikelets are shed may once again become palatable to stock. In Kordofan (4) and Sudan (3, 8) the dry grass contributes to the available browsing. Whilst herdsmen in Niger particularly seek out pasture where this plant occurs after the rainy season, the faeces of their cattle which have grazed it for an extended period have a characteristic brown colour (9).

The straw is used in Mali to plait into hats and wrist bands (5). In N Nigeria it is incorporated with mud for making hut-walls (6).

References:

1. Adam, 1966,a. 2. Adam, 1966,b: 338. 3. Andrews 3169, K. 4. Baumer, 1975: 87. 5. Dalziel, 1937: 520. 6. Golding 31, K. 7. Gupta & Sharma, 1971: 61–62. 8. Harrison 956, K. 9. Maliki, 1981: 51.

Aristida hordacea Kunth

FWTA, ed. 2, 3: 379.
West African: UPPER VOLTA BISA bonguburu (Scholz) FULA-FULFULDE (Upper Volta) butakureje (K&T) NIGERIA HAUSA wútsíyàr bera = *mouse tail* (RES)

An annual grass, culms to 45 cm high, on dry grass-land, deciduous bush savanna, old cultivations and waste places across the Region from Senegal to N Nigeria, and widespread over tropical Africa.

It provides a little browsing for cattle in Senegal (1). It is a wet season grass and available only in the rains.

Reference:

1. Adam, 1966,a.

Aristida kerstingii Pilger

FWTA, ed. 2, 3: 379.
West African: UPPER VOLTA FULA-FULFULDE (Upper Volta) selɓo (*pl.* celɓi) (K&T) NIGERIA HAUSA datsi *a general term* (Taylor)

An annual or short-lived perennial with culms to 90 cm high, of degraded and disturbed land, common throughout the Region from Senegal to N Nigeria.

The grass is reported to be a good source of fodder at Tamale in Ghana (1), and to be unappreciated by stock in N Dahomey (2). Such divergence may be phenological.

References:

1. Ll. Williams 601, K. 2. Risopoulos 1254, K.

Aristida mutabilis Trin. & Rupr.

FWTA, ed. 2, 3: 381. UPWTA, ed. 1, 521.
West African: MALI FULA-PULAAR (Mali) kelbi (A.Chev.) SONGHAI ibirsiagué (A.Chev.) pufo sufo (A.Chev.) TAMACHEK allomoze (A.Chev.) amadzarnɛ (A.Chev.) okras (A.Chev.) UPPER VOLTA FULA-FULFULDE (Upper Volta) selɓo (*pl.* celɓi) (K&T) NIGER HAUSA kasura (Virgo) SONGHAI sùb káaréy (*pl.* -à) = *white grass* (D&C) NIGERIA ARABIC-SHUWA gau (JMD) HAUSA datsi *a general term* (JMD; ZOG) gatsaua (Golding) kas maƙaru *in allusion to the sharp awns which cause injury to horses' mouths* (JMD) katsaura (JMD; ZOG) KANURI baya (JMD; C&H)

A loosely-tufted annual grass, culms to 60 cm tall, of open waste spaces and deciduous bush-savanna, common from Mauritania to N Nigeria, and extending eastwards across Africa to Somalia, the Middle East and India.

The plant is an occasional adventive on padi-field bunds in N Senegal (1). It is commonly browsed by all stock while green, but not while in seed for the awns on the spikelets are pungent and cause soreness to an animal's mouth. After the awned spikelets have been shed, the dry straw may also be browsed. It is one of the commonest fodder grasses in the Region (1, 2, 5), Kordofan (4); Sudan (10, 11), Somalia (12), Kenya (6, 7, 13, 14) and India (9), etc., but nutritional value is low (3). It is valuable as it is available during the dry season. Camels seem particularly to be attracted to it (3, 5, 10, 13). Herdsmen in Kenya find their animals affected by the sharp 3-awned seed adhering to the fur and wool which even penetrate the skin and cause sores (7) (cf. *A. sieberana* Trin.). The plant produces an indifferent thatching material, but the culms are used in N Nigeria to make mats and basket sieves (8). The plant is also used to stuff saddles (5).

References:

1. Adam, 1960,a: 366. 2. Adam, 1966,a. 3. Adam, 1966,b. 4. Baumer, 1975: 87. 5. Dalziel, 1937: 521. 6. Glover & Samuel 2796, K. 7. Glover & al. 2987, K. 8. Golding 15, K. 9. Gupta & Sharma 1971: 62–63. 10. Harrison 950, 967, K. 11. Harrison 1324, K. 12. Munro 80, K. 13. Mwangangi & Gwynne 1040, K. 14. Veterinary officer 1620, K. 15. Virgo 17, K.

Aristida sieberana Trin.

FWTA, ed. 2, 3: 381. UPWTA, ed. 1, 521, as *A. pallida* Steud.

West African: SENEGAL FULA-PULAAR (Senegal) sirin (K&A) TUKULOR sirin (K&A) WOLOF diarhat (K&A) negéret (K&A) paldinaq (JMD) MALI FULA-PULAAR (Mali) kelbi (A.Chev.) SONGHAI ibirsiagué (A.Chev.) pufo sufo (A.Chev.) TAMACHEK allomoze (A.Chev.) amadzarnɛ (A.Chev.) okras (A.Chev.) NIGER FULA-FULFULDE (Niger) surungeewol (ABM) HAUSA yanta (Bartha) NIGERIA ARABIC-SHUWA gau (JMD) HAUSA datsi *a general term* (JMD) gatsaura (Golding) kas maƙaru *in allusion to the sharp awns which cause injury to horses' mouths* (JMD) katsaura (auctt.) KANURI baya (JMD) súwúlàmè (A&S)

A loosely-tufted perennial with culms to 1 m or more in height, of coastal sand and dry inland sandy locations of the Sahel zone from Mauritania to N and S Nigeria, and across Africa to Somalia, and in Tunisia and Israel.

In the rice-growing area of N Senegal the grass is adventive on padi-field bunds fixing the soil (1). Reports on its value for grazing are contradictory: that cattle in Senegal will not touch it (2, 3), or that all stock will take it during the dry season, and while still young (1, 4). Similarly in Sudan it is said to be unpalatable and scarcely eaten, a situation resulting in it becoming the dominant grass in pasture suppressing other grasses already under pressure from grazing (7). Yet in Somalia it is rated as good for livestock (11), and is readily grazed at Lake Chad (8). In N Nigeria it is recognised as good camel fodder (6). In Niger, herdsmen value it for grazing as it is in flush growth at the middle of the hot dry season (12). Such variation in its quality for fodder is likely to hinge on the plant's phenological state. The long sharp awns, called *kas maƙaru* in Hausa are said to injure a horse's mouth (6). This obviously limits its value after anthesis. Edaptic factors may also be significant.

In Nigeria (13) and in Somalia (11) the culms are used for thatching; in the latter country matting is woven for covering nomadic huts (11).

In Senegal an ointment is made from the powdered roots with butter for topical applications to the site of pain in conjunctivitis, blepharitis and all ophthalmias (9, 10). Analgesic action is also sought by local doctors in the Borno District of Nigeria who prepare an embrocation from the cortex for rheumatic pain (5).

References:

1. Adam, 1960,a: 366, as *A. longiflora*. 2. Adam, 1966,a. 3. Adam, 1966,a, as *A. pallida*. 4. Adam, 1966,a, as *A. longiflora*. 5. Akinniyi & Sultanbawa, 1983. 6. Dalziel, 1937: 521, as *A. pallida* Steud. 7. Harrison 77, 998, K. 8. Hepper 4013, K. 9. Kerharo & Adam, 1964,b: as *A. longiflora* Schum. & Thonn. 10. Kerharo & Adam, 1974: 641–2, as *A. longiflora* Schum. & Thonn. 11. Kuchar 17646, K. 12. Maliki, 1981: 53. 13. Palmer 15, K.

Aristida stipoides Lam.

FWTA, ed. 2, 3: 379. UPWTA, ed. 1, 521.

West African: SENEGAL FULA-PULAAR (Senegal) budel (K&A) MANDING-BAMBARA bébela siraba (JB; K&A) NON makir (K&A) makirö (K&A) SERER diahal dieg (JB; K&A) WOLOF gendarat (JMD; K&A) mpal diinah (JB) paldinaq (auctt.) MALI TAMACHEK telolud (A.Chev.)

185

GRAMINEAE

NIGER HAUSA katchiema (Virgo; Bartha) SONGHAI fóon-sumféy (*pl.* -á) = *tail of the monkey* (D&C) NIGERIA BIROM yùús sɔ́rɔ̀ dùk = *thatch for the roof of the hut* (LB) FULA-FULFULDE (Nigeria) hanhande (JMD) soɗo (*pl.* soɗoji) (JMD) urleho (JMD) HAUSA gárásá (JMD; LB) gwííwàr tsoóhúwáá = *old woman's knee* (auctt.) katsemu (auctt.) tsíntsíyár kòògíí = *brush of the river* (JMD; ZOG) wútsíyàr jààkíí = *tail of the donkey* (JMD; ZOG)

A robust tufted annual with culms to 1½ m high, a ruderal of old cultivations and roadsides in the Sahel across the Region from Mauritania to N Nigeria, and in NE tropical Africa, Tanganyika to SW Africa.

Stock will graze it, but only when it is very young and for want of better browsing (1, 3, 11). The culms are used for thatching and are plaited to form mats (1, 3, 4, 9, 10). The inner nodes are succulent and sweet, and are sucked by children (1, 4, 5).

The plant is frequently galled producing hard lumps in the basal joint of a branch approximately 1½ cm long by 1 cm across. These are known as *makir* or *makirhe* (Nyominka) and *géyi budel* (Fula – Pulaar) in Senegal (7, 8), and *mazarin kyanwa, goron yan, makaranta* and *gudumar biri* (Hausa), and *surun geji* (Fula – Fulfulde) in Nigeria (4). Galled stems are used in Nigeria as toy darts to throw at birds. Nyminka doctors in Senegal consider the galls to be an excellent vermifuge, and Pulaar doctors prescribe the galls powdered and mixed into food or taken by draught in water for anuria (6, 8).

References:

1. Adam, 1954: 88. 2. Adam, 1966,a. 3. Baumer, 1975: 87. 4. Dalziel, 1937: 521. 5. Irvine, 1952,a: 31. 6. Kerharo & Adam, 1964,b: 409.3 7. Kerharo & Adam, 1964,c: 294.3 8. Kerharo & Adam, 1974: 642. 9. Palmer 14, K. 10. Schnell, 1953,b: no. 43. 11. Virgo 16, K.

Aristida sp. indet.

West African: NIGERIA FULA-FULFULDE (Nigeria) dacere ngala (J&D)

An unidentified *Aristida* of the above Fulfulde name in N Nigeria is credited with magical powers – men eat it, money appears (1).

Reference:

1. Jackson, 1973.

Arthraxon lancifolius (Trin.) Hochst.

FWTA, ed. 2, 3: 470.

A delicate trailing annual grass, to 30 cm long; of steep banks and rocky sites; across the northern part of the Region from Senegal to N Nigeria.

No usage is recorded in the Region. In Indonesia it is used as fodder, and has a reported dietetic value of: protein, 3.86%; crude fibre 34.74%; and carbohydrates 46.51% (1).

The pharmaceutical drug, *ergot*, is the source of valuable chemicals causing contraction of the smooth muscles, and of the bladder, and is of use in migraine. The fungus is normally grown on rye. Work in India claims that a promising strain of ergot has been produced on *A. lancifolius* (2).

References:

1. Chadha, 1985: 443. 2. Janardhanan & al., 1982: 166–7.

Arthraxon micans (Nees) Hochst.

FWTA, ed. 2, 3: 470, as *A. quartinianus* (A Rich.) Nash.
West African: SIERRA LEONE YALUNKA tamedi sake-na (Glanville)

A wiry decumbent annual grass to 60 cm long, of shaded clearings, upland evergreen forest and old farmland; across the Region from Guinea to W Cameroons, and extending over tropical Africa, tropical Asia and America and the Caribbean.

Though it does not carry very much foliage, it is deemed in all territories to be useful as a source of cattle fodder (Nigeria: 1, 5; W Cameroons: 3; Uganda: 4; Malawi: 2). In Ethiopia it is a weed of cultivation.

References:

1. Blair Rains 210, K. 2. Jackson 513, K. 3. Maitland 103, K. 4. Snowden 1209, 1239, 1499, K. 5. Tuley 1029, K.

Arundinella nepalensis Trin.

FWTA, ed. 2, 3: 414.

A tufted perennial grass with short rhizome, very variable, culms 60 cm to 3½ m high, usually in the upper part of the range, in moist grassland, margins of marshes and streams of upland sites in Senegal, Guinea and Mali, and E, south tropical and S Africa, and widely dispersed through India to China and Australia.

No usage is recorded for the Region. In Tanganiyika cattle browse it before anthesis, but not when in seed or dry (3). It is said to become bitter during the dry season (4). It is used for thatching (1).

Sotho in Lesotho use the plant to prepare a lotion for washing wounds and compound it into many medicinal prescriptions (2, 4).

References:

1. Greenway 3486, K. 2. Guillarmod, 1971: 411. 3. McGregor 9, K. 4. Watt & Breyer-Brandwijk, 1962: 468.

Axonopus affinis Chase

FWTA, ed. 2, 3: 446.

A stoloniferous sward-forming grass, native of tropical and subtropical America and introduced to many tropical countries, especially for upland locations; recorded present in Sierra Leone where it is used extensively for lawns (1). In Zimbabwe it grows freely in Kikuyu pasture and is preferred by grazing cattle (2).

References:
1. Gledhill 62, K. 2. Mitchell 58262, K.

Axonopus compressus (Sw.) P Beauv.

FWTA, ed. 2, 3: 448. UPWTA, ed. 1, 521.
English: carpet-grass; lawn-grass; blanket-grass; Louisiana grass.
Portuguese: capim-grama; capim-erva-tapete (Feijão).
West African: SIERRA LEONE FULA-PULAAR (Sierra Leone) fɔnyɛ-brurɛ (FCD) GOLA jani (FCD) yani (FCD) KISSI kotokpo (FCD) kɔtokpo (FCD) tandamaneyo (FCD) KONO kpaŋgba (FCD) kpaŋgba-tava (FCD) yani (FCD) LOKO nika-yani (FCD) MENDE boni (NWT)

yani (JMD; FCD) TEMNE *ka*-yan (FCD) YALUNKA kharutu-na (FCD) sɛrkha-na (FCD)
IVORY COAST KWENI tunétunénègo (B&D)

A stoloniferous sward-forming perennial grass with culms to 60 cm high, of pathsides and damp shaded places; native of the Caribbean, and dispersed pantropically; present in the Region from Guinea to W Cameroons, and in E and Central and S tropical Africa.

The grass makes excellent pasture and cattle thrive on it. Chemical analysis indicates good feeding value (2). It is a very useful lawn grass, and withstanding foot-wear it is much used through the tropics for playing fields, golf-courses, etc. Propagation is easy by dibbling in stolon sets. It is valuable for roadside verges and banks to curb erosion. It grows well in light shade and up to upland altitudes, but it will not thrive in very dry conditions.

Kweni of Ivory Coast make a bath and a draught for use against guineaworm (1).

References:

1. Bouquet & Debray, 1974: 92. 2. Burkill, IH, 1935: 276.

Axonopus flexuosus (Peter) Troupin

FWTA, ed. 2, 3: 448.
West African: SIERRA LEONE FULA-PULAAR (Sierra Leone) fɔnyɛ-brurɛ (FCD) GOLA jani (FCD) yani (FCD) KISSI kotokpo, kɔtokpo (FCD) KONO *k*paŋgba (FCD) *k*paŋgba-tava (FCD) yani (FCD) LOKO nika-yani (FCD) MENDE yani (FCD) TEMNE *ka*-yan (FCD) YALUNKA kharatu-na (FCD) sɛrkha-na (FCD)

A robust stoloniferous perennial grass with culms to 1 m high; of damp or swampy sites, usually in the shade; present in Guinea to Fernando Po, and generally dispersed over the rest of tropical Africa.

This grass is not much use for lawns (3). Its robust growth will produce a coarse sward in wet areas with no pronounced dry season. In Zambia it is considered an useful grass in moist pasture (6), though in drier areas, e.g. Fouta Djallon, Guinea, its value for fodder is poor (1). At river wallows in Uganda it is a favourite browsing for hippopotamus (5).

Mende of Sierra Leone apply the name *yani* to this genus. They recognise several forms, of which *A. compressus* is the proper *yani. A. flexuosus* is the 'male' *yani* (2). It has been conjectured that *A. flexuosus* has arisen by secondary speciation from *A. compressus* (4).

References:

1. Adames 346, K. 2. Deighton 2075, K. 3. Deighton 3765, K. 4. Gledhill, 1966: 20–27. 5. Thomas, AS Th.802, K. 6. Trapnell 1701, K. 7. Ward L.101A, K.

Bambusa vulgaris Schrad.

FWTA, ed. 2, 3: 360.
English: bamboo.
French: bambou.
Portuguese: bambu-vulgar.
West African: SENEGAL SERER i n'gol *the bow of a fish-net* (N'Diaye) GUINEA BAGA (Koba) ko-tatami Hovis SUSU tatami (Hovis) SIERRA LEONE BULOM (Sherbro) kana-lɛ (FCD) wus-lɛ (FCD) FULA-PULAAR (Sierra Leone) kɛve (FCD) GOLA sen, seni (FCD) KISSI pilanda (FCD) KONO siminɛ (FCD) KRIO kɛn *i.e.*, *cane* (FCD) LIMBA (Tonko) baräŋ (FCD) LOKO sii (FCD) MANDING-MANDINKA bɔɔ (FCD) MENDE semi (FCD) SUSU tatami (FCD) TEMNE *ka*-sul (FCD) VAI kenye (FCD) seni (FCD) senye (FCD) YALUNKA tatami-na (FCD) UPPER VOLTA GURMA ŋmàlù (Surugue) IVORY COAST KWENI balé (Gregoire) NIGERIA ABUA àgàràbà IJO-IZON (Kolokuma) igbon ịkịrai (KW&T) JUKUN (Wukari) vyo (Shimizu) WEST CAMEROONS BAFUT mfə̀lə̀ (*pl. bi-*) (Mfanyam) NGYEMBOON lefyɔg (Anderson)

A bamboo of medium size, culms quick-growing to 25 m high in open clumps, sympodial, native probably of the Himalayan area, but now widely dispersed through tropical Asia and into America, and present in the Region from Sierra Leone to S Nigeria, and in other African territories.

This is the commonest of the *Bambusa* species to be brought into cultivation (2) and in parts of W Africa it has run wild, e.g. in Sierra Leone (3), Ivory Coast (5) and Ghana (4, 6). The culms are good for all manner of constructional work, tool-handles, weapons, etc. In Sierra Leone it is planted as boundary markers (3). The springy bow of a fish-net called *a kass* used by the Serer on the Senegal River is a cane of this plant (5). The very young new shoots up to 25 cm long are eaten in Asia and plants are cultivated specially to produce them (1). Goats will browse the foliage.

Leaf-sheaths bear short bristles which are urticant. There is record of them being used in food for criminal poisoning.

Species of the genus *Bambusa* are fast growing, easy to propagate and have soil-binding properties. They afford some prospect for reafforestation with a short cycle for harvesting. Potential uses seem to be enormous, e.g., for building and construction-work, paper, plywood, charcoal, especially special purpose charcoal for dry cell batteries, dry distillation to oil, bacteriology, etc. Their potential in W Africa should be examined. The following species are known to have been introduced: *B. arundinacea* Willd., *B. glaucescens* (Willd.) Sieb., *B. longispiculata* Gamble, *B. polymorpha* Munro, *B. tulda* Roxb., *B. tuldoides* Munro, *B. ventricose* McClure, and others.

References:

1. Burkill, IH, 1935: 300. 2. Clayton & Renvoize, 1986: 53–54. 3. Deighton 5517, K. 4. Irvine, 1961: 786–7. 5. N'Diaye, 1964: 116–20. 6. Roberty, 1955: 48.

Bewsia biflora (Hack.) Goossens

FWTA, ed. 2, 3: 397.

A tufted perennial grass, culms to 90 cm high bearing pink to purple inflorescences, in wooded brush of Ivory Coast and central Nigeria, and extensively dispersed between Tanganyika and S Africa.

It is grazed by stock in young growth in Malawi (1), and is probably widely browsed if only for want of better.

Reference:

1. Fenner 303, K.

Bothriochloa bladhii (Retz.) ST Blake

FWTA, ed. 2, 3: 470, in part.
West African: SENEGAL SERER gèrgétièm (JB) gèrkèndièl (JB) WOLOF kumba ndiargan-dal (JB) NIGERIA FULA-FULFULDE (Adamawa) cawkitiningel (J)

A tufted perennial grass, shortly rhizomed, often robust, culms to 1 m high; of damp places, stream-sides, swamp-margins; scattered over the Region in Senegal, Upper Volta, Ghana and N Nigeria, and over E Africa.

The foliage is scented (3) with aromatic substances (1), described as resembling turpentine, which may be why cattle in Ghana avoid it (6). Regardless of its flavour, cattle graze it while still young in Senegal (1, 2a, 2b), but not when it is older. The grass regenerates rapidly after burning. In Uganda (5, 7) and Malawi (4) stock take while it is still young.

References:

1. Adam, 1954: 89. 2a. Adam, 1966, a, as *B. glabra.* 2b. ibid., as *B. intermedia.* 3. Dalziel 257, K. 4. Fenner 314, K. 5. Fiennes 11–?, K. 6. Hall CC.1226, K. 7. Snowden 1170, K.

GRAMINEAE

Bothriochloa insculpta (A Rich.) A Camus

FWTA, ed. 2, 3: 470, as *B. bladhii* (Retz.) ST Blake, in part.

A tufted perennial grass, culms to 2 m high, or decumbent rambling, or stoutly stoloniferous; of grassland and weedy places from lowlands to montane situations; commonly widespread in NE, E and S tropical Africa, and now recognised as occurring in Senegal and Ghana.

No usage is recorded in the Region. In all states of NE and E Africa it is known as a good fodder grass, especially when young, for both domestic stock and wild game. Masai herdsmen of Kenya recognise that the aromatic foliage may cause taint in milk (1).

In Masailand the grass is used for thatching (2, 3).

References:

1. Glover & al. 265, K. 2. Glover & al. 699, K. 3. Glover & al. 2002, K.

Brachiaria Griseb.

West African: UPPER VOLTA FULA-FULFULDE (Upper Volta) pagguri *general for those B.spp. yielding an edible grain* (K&T) NIGER SONGHAI sáari (*pl. -ò*) *general for sand-loving spp. of the Sahel* (D&C)

Brachiaria arrecta (Dur. & Schinz) Stent

FWTA, ed. 2, 3: 443, as *B. mutica* (Forssk.) Stapf, in part.

English: Tanner grass; Joe Tanner's grass (named from the S Rhodesian farmer who brought it from Natal).
West African: MALI ? kussein (Matthes) NIGERIA YORUBA ladikoro DA; Ward

A sprawling perennial grass, prostrate, rooting at the basal nodes, ascending to 1½ m high; of swamps and damp sites by water; heretofore recorded as *B. mutica*, now recognised as separate and known to be in Mali and Nigeria; widely dispersed by introduction in tropical Africa.

It is a promising pasture grass for damp sites and is readily grazed by cattle.

Brachiaria brizantha (Hochst.) Stapf

FWTA, ed. 2, 3: 443. UPWTA, ed. 1, 521.

English: bread grass; large-seeded millet grass (?S Africa, Bos).
West African: NIGERIA FULA-FULFULDE (Nigeria) gawrare (Saunders; JMD) IGBO (Ala) ashama uku (ukwu) =*big ashama* (auctt.)

A robust, tufted perennial grass, culms to about 2 m high; of deciduous woodland, wooded grassland and montane grassland; in Guinea to W Cameroons, and throughout the rest of tropical Africa and into S Africa.

The grass furnishes excellent grazing for all stock, especially whilst young. It is a valuable species for permanent pasture in the guinean zone (3), and it makes good hay (2).

Analysis of hay from Uganda has shown 6–8% protein, 41–42% carbohydrate, crude fibre 27–34% and ash 8–11%. No alkaloid, nor cyanogenetic glycoside was detected (1).

References:

1. Anon, 1924,b. 2. Dalziel, 1937: 521. 3. Foster & Munday, 1961: 314.

Brachiaria comata (Hochst.) Stapf

FWTA, ed. 2, 3: 444, as *B. kotschyana* (Hochst.) Stapf. UPWTA, ed. 1, 522, as *B. kotschyana* Stapf.
West African: NIGER ARABIC (Niger) saba kusku (A.Chev.) NIGERIA IGBO omafè (NWT)

A loosely tufted annual, culms weakly ascending to 60 cm high; of waste places and cultivations, in deciduous bushland, occasionally in Senegal, N and S Nigeria and W Cameroons, and in NE and Central Africa, and Yemen.
The grass is a common weed of cultivation (Ethiopia: 6; Sudan: 5; Central African Republic: 3). It provides good fodder and is said to be probably suitable for making hay (2). Stock browse it at all seasons in Sudan (4, 8). In Ethiopia its abundance around field edges has led to a deduction that it may be host to the parasite *Striga aspera* (Scrophulariaceae) (7).
In Sudan (1, 4) people collect the grain in time of need.

References:

1. Baumer, 1975: 89, as *B. kotschyana* Stapf. 2. Dalziel, 1937: 522, as *B. kotschyana* Stapf. 3. Fay 1630, K. 4. Harrison 935, K. 5. Harrison 1092, K. 6. Parker E.383, K. 7. Parker 4066, K. 8. Peers TO.14, TO.15, K.

Brachiaria deflexa (Schumach.) CE Hubbard

FWTA, ed. 2, 3: 444. UPWTA, ed. 1, 521.
French: fonio à grosses graines (francophone W Africa, Busson).
West African: THE GAMBIA MANDING-MANDINKA jajawo (Fox) GUINEA-BISSAU CRIOULO jégé (RdoF) MALI MANDING-BAMBARA yagué yagué ba (A.Chev.; FB) SONGHAI paguiri (A.Chev.) UPPER VOLTA MOBA kolo rassé (A.Chev.; FB)

A loosely tufted annual grass, culms often weakly ascending to 70 cm high; of cultivated and disturbed land, waste places in deciduous bush and margins of riverine forest with some shade; in the Region from The Gambia to N and S Nigeria, and dispersed across Africa to NE, E and southern Africa, and into Arabia and Madagascar.
The grass is a common weed of cultivation in arable land (15, 16), under plantation crops (14) and in flower gardens (6, 12). In Kenya it is reported in irrigated areas (10). It provides excellent forage, much appreciated by all stock (1, 2; Sudan: 3, 11; Kenya: 9, 17; Tanganyika: 8). The grass is drought-resistant and is thus a valuable forage plant for dry regions (1, 4, 7).
The grain is edible and the grass can be considered as one of the 'kreb' group of grasses (7) – see *Echinochloa pyramidalis* and *Panicum turgidum*. The French name, 'fonio with the large grain', likens it to *Digitaria exilis* the true fonio with which it is often confused. Its growth rate is faster than *D. exilis*, but it requires richer soil and better drainage (5). The wild plant shows some variability, and in the Fouta Djallon area on the Guinea-Mali border a cultivar is grown. This is a food-plant with potential for further selection. The grain provides a flour used in Fouta Djallon to make cakes and batter (Portères fide 5). The grain is also eaten in Senegal (1) and in Zambia (13).

References:

1. Adam, 1954: 89. 2. Adam, 1966, a. 3. Andrews 3210, K. 4. Baumer, 1975: 89. 5. Busson, 1965: 448. 6. Crook P.64, K. 7. Dalziel, 1937: 521. 8. Emson 254, K. 9. Magogo & Glover 814, K. 10. Mallin 297, K. 11. Peers KM.11, K. 12. Phipps 2040, K. 13. Rabson Phiri 16, K. 14. Terry 3042, K. 15. Terry 3116, K. 16. Wiehe N/116, K. 17. Wilson 18, K.

GRAMINEAE

Brachiaria humidicola (Rendle) Schweick.

FWTA, ed. 2, 3: 443.
West African: NIGERIA FULA-FULFULDE (Nigeria) interere (Saunders)

A stoloniferous perennial grass, culms to about 60 cm high, geniculate rooting at nodes; recorded only in swampy upland sites of N Nigeria in the Region, but found widespread in E and S central tropical Africa.

In Nigeria (2), Zambia (3) and Zimbabwe (1) it is reported to be a good pasture grass.

References:

1. Mundy 2825, K. 2. Saunders 35, K. 3. Trapnell 868, 986, K.

Brachiaria jubata (Fig. & De Not.) Stapf

FWTA, ed. 2, 3: 443. UPWTA, ed. 1, 521, as *B. fulva* Stapf.
West African: GUINEA-BISSAU BALANTA lábar (JDES) FULA-PULAAR (Guinea-Bissau) què-el (JDES) MANDING-MANDINKA bondim-ô (JDES) MALI FULA-PULAAR (Mali) sakatéré (A.Chev.) MANDING-BAMBARA ban ngassan (A.Chev.) SONINKE-SARAKOLE handu nkasan (A.Chev.) UPPER VOLTA FULA-FULFULDE (Upper Volta) narukkere (K&T) GRUSI tugau (A.Chev.) GURMA bugmoinu (A.Chev.) dodumuanga (A.Chev.) MOORE kolonkoghse (Scholz) naganionurè (A.Chev.) SAMO kufi moni (A.Chev.) GHANA HAUSA burugu (Ll.W) tumbin jaki = *donkey's belly* (Ll.W) NIGERIA FULA-FULFULDE (Adamawa) huturho (Saunders; JMD) FULFULDE (Nigeria) labunehe (Saunders; JMD) nyeelo (J&D) HAUSA garaji (Taylor) gariji (ZOG) makarin faƙu (JMD; ZOG) sabe (ZOG) IGALA okakakplu (Odoh; RB) YORUBA (Ilorin) ata gbuin gbuin (WG&A)

A tufted perennial grass, culms 60–120 cm high; of valley alluvium, swamp margins and damp places, often on shallow soils suffering seasonal extremes of drought and water-logging; sub-montane to montane across the Region from Senegal to N and S Nigeria, and throughout the rest of tropical Africa.

In all territories the grass produces good pasture grazing readily taken by all stock.

The grass is fibrous-rooting and has potential for strengthening bunds, banks and consolidation of roadsides (3).

The grain is gathered in N Nigeria (2) and in Kordofan (1) for human food in time of scarcity.

References:

1. Baumer, 1975: 89. 2. Dalziel, 1937: 522, as *B. fulva* Stapf. 3. Ward 0062, K.

Brachiaria lata (Schumach.) CE Hubbard

FWTA, ed. 2, 3: 444. UPWTA, ed. 1, 550, as *Urochloa lata* CE Hubbard.
West African: SENEGAL WOLOF kombar-diagandal (JMD) MALI DOGON ányu sámu buguru (CG) TAMACHEK akasof (A.Chev.) ichiban *general for edible grains* (A.Chev.) UPPER VOLTA FULA-FULFULDE (Upper Volta) farduko (Scholz) pagga pucci (K&T) MANDING-BAMBARA bafura (Scholz) MOORE koala (Scholz) GHANA HAUSA garaji (Ll.W) NANKANNI lambusan (Ll-W) NIGERIA ARABIC-SHUWA difera (JMD) umm jigeri (Gwynn; JMD) FULA-FULFULDE (Nigeria) baadeho J HAUSA garaji (JMD; ZOG)

A loosely tufted annual grass, coarse-leaved, culms semi-recumbent, to 60 cm high, of cultivated land and disturbed soil, distributed across the Sahel from Mauritania to N and S Nigeria, and on into Ethiopia and Arabia.

The grass appears as a weed of cultivated crops (Senegal: all crops, 11; The Gambia: ground-nuts, 13; Upper Volta: sorghum, 12; Nigeria: rice, 4: Ethiopia: sorghum, 10), and of formal flower gardens (7). It provides excellent fodder for all stock, and is cut and bundled for sale in W African markets, or is cut for storage as hay (1, 2, 5). It is much valued in Sudan as a palatable cattle-feed (3, 6, 9). In NE Nigeria it is one of the earliest annual grasses to appeear (8). The grain is collected in the Region for human consumption (1, 5).

References:

1. Adam, 1954: 99, as *Urochloa lata* Hubb. 2. Adam, 1966, a. 3. Andrews 3146, 3209, K. 4. Dalziel 477, K. 5. Dalziel, 1937: 550, as *Urochloa lata* CE Hubbard. 6. Evans Pritchard 62, K. 7. Gledhill 795, K. 8. Gwynn 131, K. 9. Harrison 996, 1230, K. 10. Parker E.455, K. 11. Parker 1958, K. 12. Parker 2018, K. 13. Terry 1931, K.

Brachiaria mutica (Forssk.) Stapf

FWTA, ed. 2, 3: 443. UPWTA, ed. 1, 522.

English: Para grass; buffalo grass; water grass; Mauritius grass.

French: herbe de Para; herbe de Guinea (Adam).

West African: MALI ? (Mopti) konya (Matthes) kussein (Matthes) NIGERIA ARABIC-SHUWA birbet (Musa Daggash) HAUSA zarin bauna = *bush cow's rope* (Palmer; ZOG) KANURI zaza (Johnston) YORUBA ladikoro (auctt.)

A sprawling perennial grass, culms prostrate, rooting at lower nodes and ascending to 1.8 m high; of damp and wet sites; of uncertain origin, perhaps of S America and W Africa where it is dispersed throughout the Region, and occurring all over the tropics and sub-tropics by introduction.

It is well-known as one of the best tropical forage plants. It is much relished by all stock, either grazed on pasture, or cut. It makes good quality hay. In trial plantings very heavy yields are reported (2, 3) and nitrogen content over 9% (1). It seeds sparingly and propagation is easiest carried out vegetatively.

References:

1. Adam, 1954: 89. 2. Burkill, IH, 1935: 356–7. 3. Manjunath, 1948: 211.

Brachiaria orthostachys (Mez) WD Clayton

FWTA, ed. 2, 3: 444, as *B. xantholeuca* (Hack.) Stapf, in part.

An annual grass with culms to 1 m high; common on sandy soils and impoverished areas. It is very similar to *B. xantholeuca* (q.v.), with which it was merged in *FWTA*, ed. 2, and is now recognised as distinct; occurring in the Sahel of Mauritania, Senegal and Mali.

It is grazed with relish by stock in Senegal (1, 2).

References:

1. Adam, 1954: 89, as *B. hagerupii* Hitchc. 2. Adam, 1966,a, as *B. hagerupii.*

Brachiaria plantaginea (Link) Hitchc.

FWTA, ed. 2, 3: 443.
West African: GHANA MOORE gyemsa (FRI)

A decumbent annual grass occurring only in Upper Volta, Ghana and W Cameroons in the Region; also in the Congo basin and tropical America.

In N Ghana the grass is cut as fodder for horses (1).

Reference:

1. Irvine 5175, K.

GRAMINEAE

Brachiaria ramosa (Linn.) Stapf

FWTA, ed. 2, 3: 444–5.
Portuguese: pé-de-galinha (Cape Verde, Feijão).
West African: SENEGAL SERER gae rid (JB) WOLOF ndugup i mpiti (JB) NIGERIA FULA-FULFULDE (Nigeria) baadeho (J&D) HAUSA kanarin doki (Golding)

A loosely-tufted annual grass, culms to 60 cm high, of roadsides and waste places across the Region in the Sahel from Mauritania to N Nigeria, and into E Africa southwards to the Transvaal; into Asia as far as Thailand.

The grass is a weed of cultivation. It appears in Ethiopia in irrigated cotton-fields (7) and in Zimbabwe in irrigated sugar-cane (8). It provides good palatable grazing for all stock (1–4). The Fula of N Nigeria value it as horse feed (5). The straw is fed to cattle in India and is readily eaten (6).

The grain is edible, and in parts of S India the grass is cultivated as a short-season crop (3, 6).

References:

1. Adam, 1954: 89. 2. Adam, 1966, a. 3. Gupta & Sharma, 1971: 64. 4. Harrison 980, K. 5. Jackson, 1973. 6. Manjunath, 1948: 212. 7. Parker E.114, K. 8. Taylor 216, K.

Brachiaria reptans (Linn.) Gardner & Hubbard

An annual grass, usually decumbent, culms rooting at the nodes, ascending to 60 cm high; of roadsides and weedy places; native of tropical Asia, and widely introduced throughout the tropics; recorded present in Dahomey.

The plant is a weed of cultivation; reported in irrigated land in Somalia (1, 3), and there browsed by cattle and goats (2).

References:

1. Hansen 6067, K. 2. Munro 77, K. 3. Terry 3408, 3423, K.

Brachiaria ruziziensis Germain & Evrard

A stoloniferous perennial grass, native of Zaïre, and now recorded present in Upper Volta and Ghana, probably introduced. It is known as a forage plant in E and central Africa. It probably has potential in this respect in the Region. It has a distinctly aromatic smell and cattle in Ghana are said to graze it readily (1).

Reference:

1. Rose Innes 32534, K.

Brachiaria serrata (Thunb.) Stapf

FWTA, ed. 2, 3: 443, as *B. brachyolopha* Stapf.
West African: UPPER VOLTA FULA-FULFULDE (Upper Volta) naruce (K&T)
A densely tufted perennial grass with culms to 60 cm high; of deciduous bushland, wooded grassland and upland grass savanna; in Upper Volta to N Nigeria, and widespread elsewhere in tropical Africa. In S central tropical Africa the grass is heavily grazed by stock (Malawi: 2; Zimbabwe: 1) as no doubt it is wherever it occurs.

References:

1. Cleghorn 1345, K. 2. Fenner 202, K.

194

Brachiaria serrifolia (Hochst.) Stapf

FWTA, ed. 2, 3: 444.

An annual grass, culms to 90 cm high, under light shade of deciduous bushland; recorded only in Niger in the Region, but widely dispersed in eastern Africa from Sudan and Ethiopia to Zimbabwe.
All stock graze it during the rainy season in Sudan (2).
In the Mpwapwa district of Tanganyika the seed is collected for eating as a porridge (1).

References:

1. District H.O. EA.15865, K. 2. Peers KO.28, K.

Brachiaria stigmatisata (Mez) Stapf

FWTA, ed. 2, 3: 442. UPWTA, ed. 1, 522.
West African: SENEGAL MANDING-BAMBARA kamimbi (JB) MALI FULA-PULAAR (Mali) niarukého (A.Chev.) MANDING-BAMBARA kamimbi (A.Chev.) kononinimbi = *guinea-fowl grass* (A.Chev.) KHASONKE larba (A.Chev.) SONINKE-SARAKOLE donguénianin tioke (A.Chev.) UPPER VOLTA GRUSI naumé (A.Chev.) MOORE kulumogo (A.Chev.) naganuoyo (A.Chev.) SAMO bassangui (A.Chev.) NIGERIA HAUSA takoure (ZOG) YORUBA (Ilorin) aparum odan (Ward)

A small semi-prostrate annual grass, of disturbed soils in the soudanian zone of the Region from Senegal to N and S Nigeria, and extending into Sudan.
In open sites it forms a turf. It furnishes a good fodder much relished by stock (1, 2). Its dispersal appears to have been furthered by the movement of cattle. About 1940 it first appeared in Sierra Leone attributed to cattle from over the Guinea border (3), and in Zaria by cattle from Lake Chad area (4).
In time of need people collect the grain for food (2).

References:

1. Adam, 1966,a. 2. Dalziel, 1937: 522. 3. Deighton 3981, K. 4. Taylor 29, K.

Brachiaria villosa Vanderyst

FWTA, ed. 2, 3: 444, as *B. distichophylla* (Trin.) Stapf. UPWTA, ed. 1, 521, as *B. distichophylla* Stapf.
West African: THE GAMBIA MANDING-MANDINKA kuling kuling-ô (Pirie) GUINEA BASARI a-kàsy ɓa-yán = *that which waits on the farmer* (FG&G) SIERRA LEONE KORANKO sirilinyaxe (NWT) LOKO òá (NWT) MENDE mbowi (NWT) SUSU funfuri (NWT) NIGER SONGHAI cítòw-cè (*pl.* -cìò) = *foot of the bird* (D&C) NIGERIA HAUSA gagaji (JMD) garaji (auctt.) gariji (JMD; ZOG) ƙasha (ZOG) tufa (Kennedy) KANURI kasha (JMD)

A loosely tufted or creeping annual grass with culms rising to 45 cm high; of waste places and disturbed soils in the Sahel from Mauritania to N and S Nigeria, and in Sudan and central tropical Africa; also in tropical Asia.
The grass is common on waysides. Its semi-creeping habit results in a sort of turf and a pasture which all stock relish for grazing (3; Senegal: 1, 2, 13; Ghana: 3; Nigeria: 6). It is a weed of cultivation invading first year fallows (9) and all arable crops (5, 8). It is particularly troublesome in The Gambia in fields of maize (12), millet (10, 12) and ground-nuts (10, 11).
The grain is collected in the Region (4) and at Cape Verde (7) for human food and sometimes as a famine-food (Sierra Leone: 14).

GRAMINEAE

References:

1. Adam, 1954: 89, as *B. distichophylla* Stapf. 2. Adam, 1966,a: as *B. distichophylla*. 3. Beal 18, K. 4. Dalziel, 1937: 521, as *B. distichophylla* Stapf. 5. Hepper, 1965: 500, as *B. distichophylla* (Trin.) Stapf. 6. Kennedy FHI.8029, K. 7. Malato Belliz & Guerva 370, K. 8. Parker 1974, K. 9. Terry 1872, 3077, K. 10. Terry 1950, K. 11. Terry 3103, K. 12. Terry 3131, K. 13. Trochain, 1940: 194, as *B. distichophylla* Stapf. 14. Thomas NWT.1785, K.

Brachiaria xantholeuca (Hack.) Stapf

FWTA, ed. 2, 3: 444, in part.

West African: THE GAMBIA MANDING-MANDINKA jaje ba = *the large 'jaje'* (Fox) MALI DOGON náá nàma (CG) UPPER VOLTA FULA-FULFULDE (Upper Volta) paggurì (K&T) NIGER HAUSA garza (Bartha) NIGERIA HAUSA garaji (Golding) gariji (Grove; ZOG) ɓasha (ZOG)

A tufted annual grass with culms to 45 cm high, dispersed across the Sahel from Mauritania to N Nigeria, and throughout the rest of tropical Africa.

The grass appears on farmed land and waste places. It is a weed of sorghum in The Gambia (7). It provides a good fodder which stock readily takes (Senegal: 1, 2; Nigeria: 3, 4; Botswana: 6). In eastern Africa it is grazed by game (Zambia, waterbuck: 5; Mozambique, zebra: 8).

References:

1. Adam, 1954: 89. 2. Adam, 1966,a. 3. Golding 8, K. 4. Grove 16, K. 5. Mitchell 4/64, K. 6. Smith, PA 3077, K. 7. Terry 1821, K. 8. Tinley 1855, K.

Brachypodium flexum Nees

FWTA, ed. 2, 3: 371.

A weak-stemmed perennial, matting and producing slender culms to 90 cm high, in shade of upland open or closed-forest in Sierra Leone and N Nigeria.

There is no record of it being put to any use in the Region, but it is noted as being a potentially useful grazing grass at high altitudes in Uganda (1).

Reference:

1. Snowden 1490, K.

Bromus leptoclados Nees

FWTA, ed. 2, 3: 369.

A perennial grass, loosely tufted, culms to 2 m high, of forest clearings in moist or shady sites, in the highlands of W Cameroons and common on highlands throughout tropical Africa and S Africa.

No use is recorded of the plant in W Africa. In Uganda it is recognised as a useful source of fodder (2), and in Kenya, Narok District, it is grazed by all stock (1).

References:

1. Glover & al. 1087, K. 2. Snowden 1444, 1487, K.

Cenchrus biflorus Roxb.

FWTA, ed. 2, 3: 464. UPWTA, ed. 1, 522.

English: bur grass.

French: cram-cram, (from Wolof); kram-kram (Ferry & al.).

West African: MAURITANIA ARABIC (Hassaniya) el gasba *when young* (AN) e'neti Lock gasba Lock initi (AN) initi Lock niti Lock SENEGAL BEDIK kɛbɛ̀ (FG&G) FULA-PULAAR (Senegal) hébbe (Lock) hobbéré (Lock) kébbé (Lock) TUKULOR hébbe (Lock) hobbéré (Lock)

MANDING-BAMBARA gébi (auctt.) norma (JB; K&A) norna (Lock) norolan (Lock) SERER ngoĭ (auctt.) ngoĭin (JB; K&A) WOLOF haham (Lock) hamham (JMD; JB) khakham (Lock) khamkham (JMD; FB) xam xam (K&A) **THE GAMBIA** MANDING-MANDINKA casso (Fox) **GUINEA-BISSAU** FULA-PULAAR (Guinea-Bissau) québè (EPdS; JDES) **GUINEA** BASARI dialángó (FG&G) a-nyalángó (FG&G) KONYAGI u-yalankon (FG&G) **MALI** ARABIC (Mali) gasba (A.Chev.) heskinit *correctly this plant, but applied to others* (A.Chev.) initi (A.Chev.) koreib (A.Chev.) DOGON kolomon (A.Chev.; Lock) konomon (A.Chev.; Lock) kòòlumo ya (CG) MANDING-BAMBARA norna (A.Chev.; Lock) norolan (A.Chev.; FB) KHASONKE khine (Lock) SONGHAI dané (A.Chev.; Lock) dani (A.Chev.; Lock) TAMACHEK uéjag (A.Chev.) uzak (Barth) wadjak (Lock) **UPPER VOLTA** FULA-FULFULDE (Upper Volta) hebbere (K&T) kebbe (K&T) MOORE dani (auctt.) diubiguina (Lock) rani (auctt.) SAMO kinangu (Lock) **GHANA** HAUSA karengia (Krause) **NIGER** FULA-FULFULDE (Niger) gerengyari (ABM) hebbere (ABM) HAUSA karanguia (Bartha) KANURI gobi (Lock) SONGHAI dàanì (*pl.* -ò) (D&C) dané (FB) **NIGERIA** EFIK íkòñ *general term for 'leaf'* (JMD; Lock) nyakkabre (J&D) HAUSA k̃àràngíyáá (auctt.) k̃àràngíyáá gumba *a porridge of the seeds* (JMD) karanguja (auctt.) IGBO ikpọlikpọ (Lock) ikpọlikpọ (FRI; JMD) KANURI ngibbi (JMD; Lock) ngibbi (Lock) njìmì (C&H) njíwì (A&S) njíwì (C&H) NGIZIM ápíiwà (Schuh) TIV kora-kondo (JMD; Lock) YORUBA èèmọ́ (auctt.) èmìmọ́ (auctt.)

An annual grass, culms geniculately ascending to 60 cm high; of old farmland and dry, sandy waste places; dispersed widely across the Region from Mauritania to S Nigeria, and in NE, E and S central Africa, and arid regions of India. The grass grows in dry areas. Its northernmost dispersal is considered to mark the limit of the Sahel (4, 18). It gives a dominant cover on old sand-dunes termed *dior* in Senegal (25). Its presence is understood by nomadic peoples in Senegal to indicate soil suitable for growing millet and ground-nuts (7). It is commonly a pesty weed of cultivation (8; The Gambia: 3, 10). It provides an excellent fodder whilst still young before flowering, and is taken by all stock (7, 8, 24; Mauritania: 20; Senegal: 1, 2; Ghana: 5; Niger: 17; Nigeria: 9, 11, 14, 23; Kordofan: 4; Sudan: 13; India: 12). If cut while still young it makes a good quality hay (4). If cut mature and ensiled, the fermentation process appears to soften the bristles so that stock may consume it without risk of injury (16). It is aggressive in growth and tends to replace other grasses in mixed swards so that it often is the only grass providing fodder in the dry season, when, for want of better, cattle will take it even after flowering and in spite of its bristly, bur-like, spikelets. (1, 5, 8, 17.) Consumption at this stage does present some risk of mouth and stomach injury to stock (4). Cattle-folk in Niger claim that the mature grass has little nutritive value, frequently causes diarrhoea and 'makes the animals' hair stand on end'; it reduces milk-yield, causes arteries beneath the eyes to swell, and animals suffer palpitations and a characteristic muffled cough (17).

The root is said to be used in Mali in aphrodisiac preparations (8, 24), and in Senegal it is used in a medico-magical way as a vulnerary on damage sustained in wrestling contests (15). The reflexed hooks of the bristles on the burs of the inflorescence are pungent and adhere to clothing and pierce the skin. They become entangled in wool fleece and on fur, spoiling them and the quality of leather (4, 7, 8, 12). There is, however, variation of length, presence or absence of these hairs (1), and there may be some advantage where the grass is of special value as fodder to look at the possibility of selection. The hooked bristles, however, do show a beneficial attribute. Locust hoppers passing through belts of this grass become entangled with the hooks and many die, especially those immediately after ecdysis when their chitinous integument is still soft (11, 16).

The seed is pin-head sized and normally remains within the husk; threshing to release it is not easy. One way reported (20) is lightly to fire the straw. The grain is edible and very nutritious. Content of protein is high. Analysis of material from Senegal has been reported to contain 19–21%. Fat content at 8–9% is composed of: *linoleic acid* 42.5%, *oleic acid* 27.2%, *palmitic acid* 22.0%, and traces of others (7). People in areas of marginal subsistence, e.g. Hassaniya in Mauritania (20), Teda in Tibesti (18) and Tamachek of Mali and Niger (1, 7, 8, 24) regularly collect the seed. Elsewhere it is considered a famine-food (Nigeria:

8, 9, 11, 19, 21; Ghana: 5; Kordofan: 4; Sudan: 13; India: 6, 12). In Rajasthan, India, it is regarded as the most nutritious of the famine-foods relied on locally, and regular ingestion is fattening and muscle-building. The usual methods of preparation are as couscous or porridge (see word list above), or ground to flour for baking into cakes or unleavened bread, or eaten raw, or as a part of various culinary recipes. The flour itself is black and is said to be unappreciated (18). The grain is also used to make a cooling drink (7, 8).

In the Cap Vert area of Senegal the grain has undefined magical use (15).

References:

1. Adam, 1954: 89. 2. Adam, 1966, a. 3. Ashrif 24, K. 4. Baumer, 1975: 91. 5. Beal 21, K. 6. Bhandari, 1974: 75. 7. Busson, 1965: 448–51, with chemical analysis of grain. 8. Dalziel, 1937: 522. 9. Daramola FHI.38208, K, FBC. 10. Fox 125, K. 11. Golding 13, K. 12. Gupta & Sharma, 1971: 64–65. 13. Harrison 73, K. 14. Kennedy 8016, K. 15. Kerharo & Adam, 1974: 642–3. 16. Lock, s.d. 17. Maliki, 1981: 50, sp. no. 40. 18. Monod, 1950: 69. 19. Mortimore, 1987. 20. Naegelé, 1958: 876–908. 21. Okiy, 1960: 120. 22. Roberty, 1953: 449. 23. Palmer, 3, K. 24. Schnell, 1960: no. 101. 25. Trochain, 1940: 199, 272.

Cenchrus ciliaris Linn.

FWTA, ed. 2, 3: 464. UPWTA, ed. 1, 522.

English: buffell grass; South African 'Kyasuwa' (Dep. Agriculture, N Nigeria). West African: SENEGAL MANDING-BAMBARA ngolo (JB) zu (JB) WOLOF diam hamham (JB) MALI ARABIC (Mali) heskanit (A.Chev.) TAMACHEK ebanau (A.Chev.) ebeno (A.Chev.) habinni (A.Chev.) labdi (A.Chev.) NIGER HAUSA massinguié (Bartha) NIGERIA HAUSA ƙàràngíyaá (after A&S) ƙàràngíyár Ázbin (JMD; ZOG) KANURI njíbí (A&S)

A perennial grass, mat- or tussock-forming with culms ascending to 1 m high, very polymorphic; of dry, but not very dry, regions across the Sahel of the Region from Mauritania to Niger and N Nigeria, and over NE Africa to the Middle East and India; dispersal by man has been effected throughout tropical Africa and widely in the Old World.

The grass is potentially an useful forage plant for low rainfall areas. All stock graze it (5; Senegal: 1, 2; Nigeria: 4, 6, 18; Kordofan: 3; Sudan: 9, 15; Somalia: 11, 12, 14; Kenya: 7; India: 8, 16), though yield in N Nigeria is reported as low (6). It is considered a good forage in Zaïre, but not particularly good in Mozambique (1). On the Cape Verde Islands it was at one time grown under a little irrigation as the chief forage grass on stock farms (1, 5, 17). Increased yield of milch cows is reported in India which develop a sleek glossy appearance of well-doing. It is considered to be the most nutritious fodder grass in S India and to convert into good quality hay (16). Given the variability of the species, there is scope for selection work to obtain improved yield. Selection has already been attempted in India (8, 16) where a 'black' variety (?grain colour) is considered promising, though all cultivars have shown more-or-less the same food value (8). Selection work should be undertaken in W Africa.

In Kenya elephants are said to be fond of it (13).

The matting habit of the grass makes it a good sand-binder.

The inflorescences are burred. They hook on to animals' fur, and in Kordofan are said to cause damage to hides and sheep fleeces (10).

The grain is edible. It is collected in India in time of scarcity for human consumption (8).

References:

1. Adam, 1954: 89–90. 2. Adam, 1966, a. 3. Baumer, 1975: 91. 4. DA. Samaru/11, K. 5. Dalziel, 1937: 522. 6. Foster & Munday, 1961: 314. 7. Glover & al. 1571, K. 8. Gupta & Sharma, 1971: 65–66. 9. Khalid 63. 10. Hunting Tech. Services, 1964: Kordofan. 11. Kuchar 17250, K. 12. McKinnon S.23, S.63, S.268, S.281, K. 13. Napier-Bax, s.n., K. 14. Peck 54, K. 15. Peers KM.16, K. 16. Sastri, 1950: 115–6, with chemical analysis of hay. 17. Schautz 8, K. 18. Ward 0073, 0090, K.

Cenchrus echinatus Linn.

FWTA, ed. 2, 3: 464.
English: southern burgrass (Bahamas).
French: cram-cram (Roberty).
Portuguese: canapiço (from Crioulo, Cape Verde – Feijão).

A coarse annual grass with culms 30–90 cm high; of roadsides and waste places; native of SE N America, the Caribbean, central tropical America and the Pacific Islands, now generally dispersed throughout the tropics and subtropics and present in W Africa.

This grass resembles *C. biflorus* in having small spiny burs on the spikelets. The French use of *cram-cram* for both species is perhaps confusing. The spiny fruit adhere to the coats of animals. There is no report of it being grazed.

Cenchrus pennisetiformis Hochst.

English: Cloncurry buffel grass; white buffel (Australia, Lock).

An annual grass, occasionally short-lived perennial, culms up to 40 cm high, variable, intergrading with *C. ciliaris*, perhaps not distinct (5), or alternatively a hybrid of *C. ciliaris* and *C. setigerus* (7, 8); of well-drained sand-dunes, dry waste places and salt marsh; native of NE and E Africa, middle East and India; now reported present in Niger.

The grass provides good grazing in dry semi-desert conditions (Sudan: 2, 6; Somalia: 3; Kenya: 1). It has been introduced into Queensland, Australia to give increased carrying capacity of beef cattle ranges with good results (5). In Ethiopia it is reported as a weed of irrigation (4).

References:
1. Adamson 13, K. 2. Khalid 53, K. 3. Kuchar 17466, K. 4. Phillips E.34, K. 5. Lock, s.d. 6. McKinnon S.130, K. 7. Sampson, 1936: 40. 8. Sastri, 1950: 116.

Cenchrus prieurii (Kunth.) Maire

FWTA, ed. 2, 3: 464. UPWTA, ed. 1, 522.
French: cram-cram (Monod).
West African: **MAURITANIA** ARABIC (Hassaniya) gasba Lock initi Lock tilimit Lock **SENEGAL** FULA-PULAAR (Senegal) kébbé buru (Lock) **MALI** ARABIC (Mali) heskanit (JMD) DOGON kòòlumo ána (CG) TAMACHEK tawajjaq (Lock) uazedj (JMD) wadjâk (Lock) wajjag (Lock) wesedj (RM) **NIGER** KANURI ngibbi bulduyé (Lock) SONGHAI-ZARMA dani (Lock) TEDA diger (Lock) mali alyia *the seed* (Lock) **NIGERIA** HAUSA k̀àràngíyár Ázbìn (JMD) k̀àràngíyár hanfoka (Lock) k̀àràngíyár kûra (Lock)

An annual, occasionally perennial grass, decumbent or tufted, ascending culms to 75 cm high; of sandy waste places of the Sahel across the Region from Mauritania to N Nigeria, and extending across Sudan into India.

The grass provides palatable fodder taken before anthesis by all stock in all territories (1–4, 7), and even after flowering if for want of better (6). In Sudan it is noted as good grazing for camels (5).

The grain is edible. For some desert nomads and, for example, Tamachek in Algeria and Niger (2, 3), it is an important staple, and for others it serves as a famine-food in Africa (6) and India (4).

References:
1. Adam, 1966, a. 2. Baumer, 1975: 91. 3. Dalziel, 1937: 522. 4. Gupta & Sharma, 1971: 66. 5. Harrison 962, K. 6. Lock, s.d. 7. Monod, 1950: 65.

GRAMINEAE

Cenchrus setigerus Vahl

FWTA, ed. 2, 3: 464.
English: South African pennisetum (Ilorin, N Nigeria, Ward); 'ajandhaman' pennisetum (Samara, N Nigeria – Dept. of Agriculture); birdwood grass (Australia, after Fd.-Marshall Baron WR Birdwood of Anzac and Totnes (1865–1951) who sent seed from India to Australia; see *The Dictionary of National Biography, 1951–60*, O.U.P., 1971, for biography).

A perennial grass, stoloniferous and clumped, culms bulbous at base, more or less erect to 80 cm high; of sub-desert grassland and arid deciduous grassland; native of NE Africa to India, and introduced to E Africa and present in Ghana and N Nigeria.

It resembles *C. biflorus* but is without spines and burs, and is grown in N Nigeria as a potential pasture grass (2). It is a variable plant and a number of strains have been selected for cultivation in India (4, 7). It will grow on a wide variety of soils except on saline or water-logged sites. In W Australia it has been grown on degraded catchments for restoring a vegetation cover with some measure of success (7). It is palatable to stock at all stages of growth, and in the dry state of Rajasthan in India it is a significant source of fodder (4), though less common than *C. biflorus* (1). In Sudan it is rated as the most important and best source of dry season grazing (5, 6). It is also valued in Kenya (3).

The seed is collected in India and ground to flour for human consumption (1).

References:

1. Bhandari, 1974: 75. 2. DA. Samaru / 18, K. 3. Glover & al. 2772, 2992, K. 4. Gupta & Sharma, 1971: 66–67. 5. Harrison 1316, K. 6. Khalid 40, K. 7. Lock, s.d.

Cenchrus spp. indet.

West African: SENEGAL WOLOF bililar (JB) NIGER ? (Goure) takaaï (Grall) SONGHAI fèejì-ndàano (*pl.* -à) (D&C)

Of the two species cited, *bililar*, given as *C. bilalar* (1), a *nomen nudum*, is an unknown entity. The other, *takoaï*, is said to be bristly and thistle-like, covering sand-dunes in Niger (2). It is perhaps *C. biflorus*.

References:

1. Berhaut, 1967: 395. 2. Grall, 1945: 7.

Centotheca lappacea (Linn.) Desf.

FWTA, ed. 2, 3: 381. UPWTA, ed. 1, 522.
West African: SIERRA LEONE MANDING-MANDINKA naragbadi (FCD) MENDE kulagbi (NWT) nana (FCD) SUSU suisɛxe (NWT) YALUNKA nɔlɔmiŋkɔde-na (FCD)

A forest grass, annual, woody root, culms to 90 cm high, of waste places in forest shade, secondary forest in semi-shaded sites, and under plantation tree crops, commonly throughout the Region from Senegal to W Cameroons and in the western side of central tropical Africa, not occurring in E Africa, but extending far across tropical Asia to Polynesia.

The grain has reflexed spines on the lemma which catch on clothing and the fur of passing animals. This ensures dispersal. In SE Asia, especially by Indian cattle-men on Malayan plantations, the grass is deemed to be good fodder (1).

Reference:

1. Burkill, IH, 1935: 508.

200

Centropodia forskalii (Vahl) Cope

FWTA, ed. 2, 3: 374, as *Asthenatherum forskalii* (Vahl) Nevski. UPWTA, ed. 1, 525, as *Danthonia forskalii* R Br.

West African: MAURITANIA ARABIC (Hassaniya) legsaïbö (AN) MALI SONGHAI zeihife (Jumelle) TAMACHEK AHARAI (JMD) bu-rekkebah (Barth)

A robust, loosely-tufted perennial, with long thin roots, of sand-dunes and arid desert locations in Mauritania, Mali and Niger, and across northern Africa, in Sudan and N Kenya and Angola, and the Middle East and India.

The long stout roots are covered with fine root-hairs to which sand grains adhere (1). The plant probably has some action in sand-binding.

The grass provides good fodder for camels and other domestic stock (2, 3).

References:

1. Bor, 1960: 478, as *Asthenatherum forskalii* (Vahl) Nevski. 2. Dalziel, 1937: 525, as *Danthonia forskalii* R. Br. 3. Maire, H, 1933: 64–65, as *Danthonia forskalei* (Vahl) Trin.

Chasmopodium afzelii (Hack.) Stapf, **C. caudatum** (Hack.) Stapf

FWTA, ed. 2, 3: 506, as *C. caudatum* (Hack.) Stapf. UPWTA, ed. 1, 522, as *C. caudatum* Stapf.

English: cane grass (Sierra Leone, Cole).

West African: SENEGAL DIOLA ésisité (JB) GUINEA FULA-PULAAR (Guinea) kali *from Susu* (JMD) SUSU kali (JMD) SIERRA LEONE BULOM (Sherbro) gban-dɛ (?) (FCD) KONO fa (FCD) fa-mɛsɛ (FCD) KORANKO bɔmie (NWT) kesiowuli (NWT) KRIO kɛn-gras *i.e., cane grass* (FCD) LOKO ŋgara (?) (FCD) waga (NWT) MANDING-MANDINKA gala (FCD) kala (FCD) MENDE ŋgala (*def.* ŋgalei) (auctt.) SUSU kale (NWT) kali (NWT) TEMNE a-bɔbɔruni (NWT) ɛ-thaŋkɛ *applied when the flowering culms have grown up* (FCD) a-, an- wo(h), a-wop *general terms, applied before the flowering culms have appeared* (auctt.) VAI fane-baba (FCD) YALUNKA kal-la (FCD) UPPER VOLTA FULA-FULFULDE (Upper Volta) ŋeloori (*pl.* -ooje) (K&T) NIGERIA FULA-FULFULDE (Nigeria) marorehe (Saunders) HAUSA kamsuvan doki (Saunders) sansari *properly Saccharum spontaneum Linn., but applied here as a group term for similar grasses* (JMD; ZOG) shìnkááfàr dáájii = *wild rice of the bush; applied loosely* (auctt.) shìnkááfàr tsúntsúú = *wild rice of the bird* (JMD)

Robust, coarse, annual grasses, prop-rooted, cane-like culms to $2\frac{1}{2}$ m tall; of the grass savanna. Two very similar species are now recognised, both occurring across the Region, with *C. afzelii* particularly common in Sierra Leone. *C. caudatum* extends across central Africa to Sudan. Vernacular names and attributes are mixed in the records and probably are common to both species.

The grasses provide good cattle and horse fodder (1, 3, 4; *C. caudatum*: Nigeria, 7; *C. afzelii*: Sierra Leone, 5), though on Fouta Djallon, Guinea, there is belief of toxicity (1, 4). Leaves of *C. afzelii* from Ghana have been found to contain *flavonic glycosides* (8).

The pithy stems are commonly cut in Sierra Leone for fencing (*C. afzelii*: 5) and the larger canes serve as arrow-shafts (4).

C. afzelii has been reported on as an indifferent source of paper-pulp (4, 6).

The culms are frequently used in N Sierra Leone to make improvised bee hives in which to take swarming bees (7).

References:

1. Adam, 1966, a. 2. Boboh, 1974. 3. Cole, 1968,a. 4. Dalziel, 1937: 522. 5. Deighton 855, K. 6. Imp. Inst., s.n. K. 7. Saunders 16, K. 8. Rose Innes 3274 (Harborne 6246), K.

Chloris barbata Sw.

FWTA, ed. 2, 3: 400.
West African: GHANA HAUSA káfàr fàráki (Williams)

A loosely-tufted grass, stoloniferous, culms to 60 cm high, recorded as annual the Region, (2), but perennial in E Africa (3); of roadsides and waste places in

Ivory Coast to S Nigeria, and in E Africa, and spreading through the tropics from its home in America.

It is considered to yield a good fodder especially while young (1, 4). Chemical analysis indicates a high food-value (1).

References:

1. Burkill, IH, 1935: 529. 2. Clayton, 1972: 400. 3. Renvoize, 1974: 345–6. 4. Williams 832, K.

Chloris gayana Kunth

FWTA. ed. 2, 3: 400. UPWTA, ed. 1, 522.

English: Rhodes grass.
West African: NIGERIA FULA-FULFULDE (Nigeria) pagamri (J&D) HAUSA garaaji (Golding) kauarin dooki (Golding; ZOG)

A perennial grass, stoloniferous, erect or ascending to 1.2 m high, of riverine woodland, scattered tree grassland or open grassland on light or heavy soils, native of S Africa and introduced into the Region and present throughout in Senegal to Nigeria, and over tropical Africa from NE and E Africa to S Africa; introduced to many tropical territories.

Because of its stoloniferous habit and nodal rooting, the grass is fitted to covering much ground (2). It provides grazing found palatable by stock (1, 3). In N Nigeria it has shown disappointing returns and has proved incapable of withstanding the dry season in the soudanian zone (4) though in Kordofan it thrives exceedingly under irrigation (2). The grass has an inherent variability. Selection should be looked at. Cloning is feasible since it is commonly cultivated in India by propagation by root-stocks. There it gives growth which can be harvested 7–8 times a year and is suitable for silage or hay though it is out-yielded by guinea grass (*Panicum maximum* Jacq.) (9). Horse-dealers consider it to be harmful to horses causing skin-troubles (9).

This grass, as is noted of other *Chloris spp.* has halophytic tendencies. It is reported growing abundantly around a salt-lick in Ethiopia and is heavily grazed, doubtless supplying certain mineral requirements of the game.

Fulfulde cattle-men of N Nigeria note the presence of this grass as a sign of impoverishment of sandy soil (7).

In Iringa Province, Tanganyika, the grass is recorded as growing only on ant-heaps (8), which may perhaps indicate that there is a granary of its seed within.

The longer culms are used in Kenya for thatching roofs (6).

References:

1. Adam, 1966,a. 2. Baumer, 1975: 91–92. 3. Dalziel, 1937: 552. 4. Foster & Munday, 1961: 314. 5. Gilbert & Gilbert 2240, K. 6. Glover & al. 529, K. 7. Jackson, 1973. 8. McGregor 10, K. 9. Sastri, 1950: 130.

Chloris lamproparia Stapf

FWTA, ed. 2, 3: 400. UPWTA, ed. 1, 523.
West African: MALI MANDING-BAMBARA dugu kunsigui (A.Chev.) UPPER VOLTA GRUSI bunagau = *donkey's grass* (A.Chev.) MOORE banga mogho (A.Chev.) NIGERIA HAUSA gergera (Taylor)

A tufted annual grass, culms erect or spreading, 30–60 cm high, on dry bare soil of the savanna in Mali to N Nigeria, and extending to Sudan, Uganda and Tanganyika.

The grass will form a turf (1). It provides good grazing for stock in all territories (Mali: 1; Sudan: 2, 3; Uganda: 4). In Sudan it is said to be taken at all times of the year, and the local Arabic name at Hamadi: *aba malih*, implies that it has a saltiness, and may thus be a source of minerals for browsing stock. The grain is collected for eating by some peoples in Uganda (4).

References:

1. Dalziel, 1937: 523. 2. Harrison, 901, K. 3. Peers T.0.5, K. 4. Wilson 15, K.

Chloris pilosa Schumach.

FWTA, ed. 2, 3: 400. UPWTA, ed. 1, 523.
West African: SENEGAL MANDING-BAMBARA babunsi (A.Chev.) bakoron mbonsi = *goat's beard* (A.Chev.) ngolo ntéguélé (A.Chev.) SERER sivañdan ngoromdom (JB) WOLOF guendjar (JMD) tiokol (JMD) tiokol peul (JMD) THE GAMBIA MANDING-MANDINKA kending na'kurto = *young man's trousers* (Pirie; JMD) LIBERIA MANO duo su (JMD) MALI DOGON kèènie géũ (CG) MANDING-BAMBARA babunsi (A.Chev.) bakoron mbonsi = *goat's beard* (A.Chev.) ngolo ntéguélé (A.Chev.) NIGERIA BURA tiksha digo (Kennedy) FULA-FULFULDE (Nigeria) kerkole (Saunders) kwoyɗe kumare (JMD) HAUSA ƙáfàn gauraka (Kennedy) ƙáfàn pakara (BM) ƙáfàr fàkáráá = *francolin's foot* (auctt.) ƙáfàr gauraka = *crown bird's foot* (JMD; Taylor) ƙáfàr tsúntsúú (JMD) sawun gauraka = *footstep of the crown bird* (JMD) tafin gauraka (JMD) KANURI kila silum (DA) kilasilim (Musa Daggash) YORUBA eéran = *remember* (Verger) eríran = *remember* (auctt.)
NOTE: *Nigerian names apply also to C.pycnothrix Trin.*

An annual grass, culms erect to 1 m high, or geniculately ascending, often rooting at the lower nodes, commong on roadsides, old farmland and disturbed places, throughout the Region from Mauritania to W Cameroons, and into Sudan and Ethiopia, and Zaïre and Malawi.

It is thought to have been a relatively recent arrival in inland Sierra Leone (10) drawn along by the railway (8) and/or road traffic (9). This shows an interesting parallel to the dispersal of *C. barbata* in SE Asia along the lines of communication (6).

It provides good forage and pasturage for all stock at all stages of growth (1–3, 5, 7). Some peoples in the Region (7) and in Kordofan (5) convert it into good quality hay.

Nigerian material has been tested for *alkaloids* and a trace in the leaves is reported (4).

The shape of the infloresence with a few loosely digitate spikes evokes in N Nigeria names alluding to 'birds' feet' and in Senegal and Mali to 'goats' beard', but reference to 'young man's trousers' in The Gambia must surely remain cryptic – see word list above.

Yoruba invoke the plant in an incantation to overcome forgetfulness (11).

References:

1. Adam, 1954: 90. 2. Adam, 1960,a: 367. 3. Adam, 1966,a. 4. Adegoke & al., 1968: 13–33. 5. Baumer, 1975: 92. 6. Burkill, IH, 1935: 529. 7. Dalziel, 1937: 523. 8. Deighton 1462, K. 9. Deighton 1522, K. 10. Deighton 5386, K. 11. Verger, 1967: No. 67.

Chloris pycnothrix Trin.

FWTA, ed. 2, 3: 400. UPWTA, ed. 1, 1.
West African: NIGERIA FULA-FULFULDE (Nigeria) tsawko (Saunders) HAUSA kiri-kiri (Kennedy) tafin gauraka (Saunders) zankon gauraka = *crown bird's crest* (Taylor)
NOTE: *Nigerian names under C.pilosa Schumm. may also refer.*

A stoloniferous perennial grass, culms erect or more often geniculately ascending to 30 cm high, on grassland, fallows, lawns and disturbed situations in Guinea to W Cameroons, and in central and E Africa and S America.

The plant is everywhere considered to be a good grazing grass palatable to stock. It also gives good hay (1). By its stoloniferous habit it may become a weed in the turf of lawns (2).

A cytological report gives its chromosome number as 2n = 40 (3).

References:

1. Baumer 1975: 92. 2. Bumpus Bam./19, K. 3. Gledhill DG.224, K.

Chloris robusta Stapf

FWTA, ed. 2, 3: 400.

West African: NIGERIA HAUSA ƙaasaraa (JMD; 20G) YORUBA (Ilorin) eru lulu do (Ward)

A perennial grass with robust woody culms to 1–2 m high, geniculate and rooting at nodes, of sandy river-beds, sand-banks and river-banks, from Sierra Leone across the Region to N and S Nigeria, and on to Sudan, Zaïre and Uganda.

In the Central African Republic the plant is calcined to make a local salt (1); cf. *C. lamproparia* Stapf, q.v.

Reference:

1. Fay 4409, K.

Chloris virgata Sw.

FWTA, ed. 2, 3: 400.

English: old lands grass; sweet grass (Burkill).

West African: UPPER VOLTA FULA-FULFULDE (Upper Volta) garbere (K&T)

An annual grass, culms erect or geniculate rooting at lower nodes to 60 cm high of scattered tree savanna, bushland, disturbed habitats and waste places on a variety of soils, including brackish and saline, in the Sahel of Mali, Niger and N Nigeria, and in NE, E and south central tropical Africa, and in India.

In Sudan (9), Somalia (1, 6), Kenya (3, 5) and Tanganyika (2, 7) it is held to provide good grazing for all stock, but in Sudan, only in the wet season – it is not a dry season grass. In the arid state of Rajasthan in India its tolerance of salty soil makes it of some importance on saline tracts where it may at times be the only vegetation growing (4, 10). In Ethiopia it has become a weed of cultivation and in particular of irrigated crops (8).

Xhosa of southern Africa use a decoction of the whole plant or of its roots as an additive to a bath for treating colds and rheumatism (11).

References:

1. Boaler B.130, K. 2. Emson 268, K. 3. Glover & Samuel 2764, K. 4. Gupta & Sharma, 1971: 68. 5. Lathbury L.9, K. 6. McKinnon S.18, S.249, K. 7. Marshall 12, K. 8. Parker E.526, E.579, K. 9. Peers K.O.4, K. 10. Sastri, 1950: 131. 11. Watt & Breyer-Brandwijk, 1962: 467.

Chrysochloa hindsii CE Hubbard

FWTA, ed. 2, 3: 402.

West African: UPPER VOLTA MOORE pétrépin ragha (Scholz)

A stoloniferous, sometimes caespitose, annual grass, culms very variable, up to 75 cm high, in seasonally flooded shallow pans and damp clay sites, of the Sahel from Senegal to N Nigeria, and in E Cameroun to Tanganyika and Malawi.

The grass shows a myrmecophilous trait, reportedly occurring in Tanganyika close by anthills (1). It may be that ants store the grain.

Reference:

1. McCallum Webster T.39.K.

Chrysopogon aciculatus (Retz.) Trin.

FWTA, ed. 2, 3: 468.
English: love grass; Port Harcourt grass (Nigeria, Nwanzo).

A perennial sward-forming grass, prostrate stems, nodal rooting, with short erect culms to 30 cm high; native of tropical Asia, and introduced and well-established at various localities in the Region.

It is close-growing and withstands drought rather better than *Axonopus* and *Cynodon.* It is commonly grown as a lawn-grass, but requires frequent mowing (Sierra Leone: 5; Upper Volta: 9; Ivory Coast: 4, 7; Nigeria: 3, 6, 10; India: 8; Malaya: 1). It produces a good turf for roadside verges and consolidates sandy beach foreshores (7). Under higher rainfall conditions it is a good grass for withstanding foot-wear on playing fields.

It will appear as a weed of pasture and cattle browse it for want of better, but is normally avoided if in flower when its long (3–4 mm), needle-like awns cause injury to the mouth of cattle (1, 2, 7, 8). The awns are also a source of damage to the paws of dogs (2).

The inflorescences produce seeds which adhere to objects passing by, - hence the name 'Love Grass.' In Malaya this concept is extended to rubbing the seeds with some lime juice on a decoy female elephant to entice solitary males (1).

In Indonesia the plant has medicinal use as a poison antidote, and in Malaya ash is taken internally for rheumatism (1).

References:

1. Burkill, IH, 1935: 535. 2. Clayton, 1972: 3: 468. 3. Gledhill 628, K. 4. Leeuwenberg 3826, K. 5. Morton & Jarr S.L.1441, FBC. 6. Nwanzo 729, K. 7. Portères 1950,a: 101–2. 8. Sastri, 1950: 151. 9. Scholz 151, K. 10. Wheeler Haines 198, K.

Chrysopogon plumulosus Hochst.

FWTA, ed. 2, 3: 468, as *C. aucheri* (Boiss.) Stapf.

A wiry perennial grass, culms erect or ascending to 90 cm high; of semi-arid deciduous bushland or sub-desert grassland in Niger, and widely dispersed in NE and E Africa.

The grass is rated as providing good grazing for all stock, especially in areas of low rainfall (Sudan: 1, 5; Somalia: 3, 4), and for young weaned pigs (3). Some years ago it was taken under the Somali name *daremo* into a seed merchant's catalogue as a grass of potential improvement of pasture in areas of low rainfall (2).

References:

1. Beshir 14, K. 2. Brookman, ad. not. 18/2/1914, K. 3. Kuchar 17207, K. 4. McKinnon S.10, K. 5. Peers LO.2, K.

GRAMINEAE

Coelachyrum brevifolium Hochst. & Nees

FWTA, ed. 2, 3: 397.
West African: NIGERIA YEDINA magaril kura (Golding)

An annual grass with culms in a loose tuft to 45 cm high, of the desert in Mauritania, Mali and N Nigeria, and in N Africa, Sudan and Arabia. The grass gives good grazing for all stock in Sudan (2). An infusion of this plant with the roots of another known only as *magaril kura* to the Yedina at Lake Chad, NE Nigeria, is taken to purify the blood (1).

References:

1. Golding 37, K. 2. Kalid 30, K.

Coelorhachis afraurita (Stapf) Stapf

FWTA, ed. 2, 3: 509. UPWTA, ed. 1, 523.
West African: SIERRA LEONE MENDE fovo (*def.* -i) (FCD) tomo (*def.* -i) (FCD; Fisher)
NOTE: both general terms.

A tufted perennial grass with culms erect to 2 m tall; of fresh-water swamps and marshy grassland; scattered across the Region from Senegal to N Nigeria, and throughout other parts of tropical Africa.

No specific usage is recorded, but the two Mende names cited above are general terms for tall grasses of use for fencing, thatching, matting, etc., and edible as fodder when young.

Coix lacryma-jobi Linn.

FWTA, ed. 2, 3: 511. UPWTA, ed. 1, 523.

English: Job's tears; Christ's tears; adlai (American, from Bisayan, Philippine Islands).

French: larmes de Job; larmier de Job.

Portuguese: erua-dos-rosários; lágrimas de Job.

West African: SENEGAL BEDIK *ma*-karamba-késé (FG&G) DIOLA balifõ (JB) SERER boror (JB) ñammaket (JB) WOLOF foror (JB) porola (JB) GUINEA-BISSAU SUSU bonco (JDES; RdoF) GUINEA BASARI *a*-mbɛr-kὲsy (FG&G) KISSI fondo (A.Chev.; FB) KONYAGI *wa*-kometa (FG&G) KORANKO forono (A.Chev.; FB) SIERRA LEONE BULOM (Kim) sisig (?) (FCD) BULOM (Sherbro) kpɔklɔ-lɛ (FCD) kpɔkolɔ-lɛ (FCD) FULA-PULAAR (Sierra Leone) fɔrɔndɔ (FCD) GOLA fɔrɔ (FCD) KISSI gbɔlɔ̃ (FCD) gbɔlɔ̃kpɔ (FCD) gbɔlɔ̃ndõ (FCD) KONO gbɔɛ (FCD) gbɔyɛ (FCD) pu-bɔɛ (FCD) KORANKO fɔrɔndɔ-mɛsɛ (FCD) yiri-fɔrɔndɛ (FCD) LIMBA (Tonko) matɔmperega (FCD) LOKO fɔrɔ (NWT; FCD) MANDING-MANDINKA fɔrɔndɔ-tasɛbia (FCD) jina-fɔrɔndɔ (FCD) sankala (NWT) MENDE boboni-vɔlɔ (FCD) fɔlɔ (NWT; FCD) *k*petehu-vɔlɔ (FCD) SUSU bɔhɔri (FCD) bɔŋkɔri (NWT) kali bagi (NWT) TEMNE *am*-pɔlɔ (FCD; M&H) TEMNE (Kunike) ɛ-pɛrɔka (FCD) TEMNE (Port Loko) *ma*-pɔlɔ (FCD) TEMNE (Sanda) *ma*-pɔlɔ (FCD) TEMNE (Yoni) ɛ-pɛrɔka (FCD) YALUNKA gbɛgbɛ-na (FCD) gbɛgbɛ-tasabia-na (FCD) LIBERIA MANO zã (Har.) IVORY COAST ·ANDO· manquassèm *short for.* éwoué man manquam assèmoun· = *I say nothing because of death* (Visser) GHANA AKAN-ASANTE akrokosebia (Bunting fide FRI) FANTE Job n'ani nsuwa (FRI) TWI owu-amma-mankã m'asɛm = *death makes me mute* (auctt.) WASA ahwinie (FRI; E&A) GBE-VHE agu (FRI) VHE (Awlan) agu (FRI) NIGERIA EFIK ŋgkwà eto (Lowe) nkwà ikọ̀t (Lowe; JMD) IJO-IZON (Kolokuma) aká-ịla = *corn bead* (KW; KW&T)

A coarse, erect, annual grass to 2 m tall; of stream-banks and moist places; native of Asia but found wild in E and W Africa, and recorded in the Region from Senegal to W Cameroons; dispersed by man to tropical America and to most temperate countries, often running wild. There are some varieties and cultivated races in Asia, but only var. *lacryma-jobi* is known in Africa, spontaneous and anthropophile around villages and in old cultivation sites.

The mature plant bears shining, pear-shaped fruits about 8–12 mm long which suggest 'tears' that appear in European and Middle Eastern vernacular names. The shell is hard and of various colours: white, light brown, grey to black. Its cultivation in India, Burma, China and Malesia is of great antiquity assuming importance in aboriginal agriculture. It grows in places suited to hill-padi and has served as a fortuitously valuable famine-food when rice-crops have failed. It is amongst the earliest grasses to have been brought into cultivation, a preliminary development being selection of soft-shelled forms. These are grown particularly in the Indo-China region and the Philippine Islands, and in aggregate are known as *ma-yuen* varieties (4, 5, 10, 14, 16).

The silicaceous shell is 30–70% of the whole fruit. It contains a grain not unlike rice and which compares favourably with other grains in respect of protein and fat content. Minerals and fibre are low. The grain contains the following approximations: water, 9.9–10.8%; protein, 13.6–19.1%; fat, 5.7–6.1%; carbohydrate, 58.5–62.7%; fibre, 8.4% and ash 2.2–2.6% (6, 13). Other analyses may be found (5, 15). The grain is, in fact, a good substitute for rice, and in respect of protein is even better. It has been suggested at times for the Philippines and elsewhere as a substitute for rice and wheat (3). It does not, though, store so well, and it requires longer to grow, but is rewarding in needing less labour and is harvestable by machinery (3).

The usual method of preparation for eating is to pound the grain to flour. Trials of milling and baking in the Philippines gave good results with the addition of wheat flour (3, 16), but baking for bread is now not usually done (4). Trials in Brazil indicated that digestibility was found to be much lower than of wheaten flour (20). Powdered flour is taken with water as a drink in India (4) and an infusion of the parched grain is made in Japan (16). Some peoples in Asia take the un-pounded grain to prepare fried titbits, some resembling fried ground-nuts (4).

Cultivation for human consumption in the Region seems to be somewhat limited. The growing of cultivars for their edible seeds is reported in Liberia (1). It may at times be taken when food is short (6), and grown as a garden pot-herb (15).

The grain ferments into a wholesome beer (4, 16, 20). Chinese and Japanese distil a spirit from the seed which is used in rheumatic conditions (20).

The seed-oil is recorded as a mixture of: *oleic acid*, 53%; *linoleic acid*, 30.5%; *palmitic acid*, 16.0% and traces of *stearic* and *linolenic acids* (6). Another report records a predominance of *palmitic acid* (20). The crude oil is said to cause respiratory failure (20).

The principal use of the hard-shelled fruits is as beads for necklaces, rosaries, in decorative work and in rattles for dances (1, 7–9, 16–19, 21). In Ghana they are worn as a sign of mourning for children on religious or ritual occasions (7). The Tenda of SE Senegal put them into petticoats made for girls to wear at the excision ceremony (9). They have a place in magic (6), and at meetings of the Bundu Society in Sierra Leone (12). In S Africa a necklace is put onto an infant to ward off teething troubles, as also elsewhere. In Hawaii a necklace is endowed with curative charms (20).

The fruit is entered in the *US Dispensatory* for use as a tincture and in decoction for catarrhal infection and for inflammation of the urethra (16, 20). The preparation is diuretic (17, 20), depurative and cooling. Chinese use it similarly (20), and the seeds are popularly used in Japapn, Taiwan and China for lung and intestinal cancers (11).

The grass is relished by cattle for grazing (2, 7) and is considered highly by W African graziers (1). It is fed to cattle, horses and elephants in India, and converts into a good silage; it makes good stall-litter, and the mature haulm is usable as thatch (4, 16). Seed of soft-shelled cultivars produce poultry meal (4, 16), and milling offals can be fed to poultry and other farm stock (3).

Sap from the culm is instilled into the eye in Liberia to relieve irritation due to injury (Harley fide 7, 20).

GRAMINEAE

The root is reported to be an excellent anthelmintic (20). In the Philippines it goes into a gonorrhoea remedy, in Malaya into a vermifuge for children, and in India into a medicine for menstrual disorders (20).

References:

1. Adam, 1954: 90. 2. Adam, 1966, a. 3. Anon., 1940, a: 182–4, with references. 4. Arora, 1977: 358–66. 5. Burkill, IH, 1935: 629–31. 6. Busson, 1965: 451, with chemical analyses, 452–3. 7. Dalziel, 1937: 523. 8. Deighton 1167, K. 9. Ferry & al., 1974: sp. no. 107. 10. Grounds, 1989: 68–69. 11. Hartwell, 1969,b: 191. 12. Melville & Hooker 110, K. 13. Purseglove, 1972: 1: 134–7. 14. Quisumbing, 1951: 92–94. 15. Roberty, 1955: 49. 16. Sastri, 1950: 305–6. 17. Vergiat, 1970,a: 84. 18. Viller, 1975: 39. 19. Walker & Sillans, 1961: 186–7. 20. Watt & Breyer-Brandwijk, 1962: 467, with pharmacological detail. 21. Williamson, K. 34,UCI.

Crypsis vaginiflora (Forssk.) Opiz

FWTA, ed. 2, 3: 411, as *C. schoenoides* (Linn.) Lam., in part.

An annual grass, culms prostrate, radiating from base, distally briefly ascending, of damp sites, lake margins subject to flooding in Senegal and across sub-Saharan Africa to Ethiopia and Eritrea. This species is segregated from *C. schoenoides* sensu *FWTA* ed. 2. which is E and S African and N American.

In Kordofan it provides relished dry season grazing for sheep and goats (1).

Reference:

1. Baumer, 1975: 95, as *C. schoenoides* (L.) Lam.

Ctenium elegans Kunth

FWTA, ed. 2, 3: 398. UPWTA, ed. 1, 523.

West African: SENEGAL DIOLA dikãndapali (JB) MANDING-BAMBARA uluku (JB) SERER lab a koy (JB) yagon (JB) WOLOF rèv (JB) rov (JB) GUINEA-BISSAU PEPEL undáte (JDES; EPdS) GUINEA BASARI ε-nókolòmb *perhaps this sp.* (FG&G) MALI DOGON sàmu sáána (CG) MANDING-BAMBARA wolo kaman (JMD) NIGER HAUSA chinaka (Bartha) SONGHAI bàt kàaréy (*pl. -á*) = *box of the white amulet* (D&C) NIGERIA FULA-FULFULDE (Nigeria) sinakaho (JMD) wicco dombru (JMD) wicco pallandi (JMD) wicco wanduho (JMD) HAUSA shinaka (auctt.) wùtsíyàr ɓééráá = *mouse's tail* (JMD; ZOG) wùtsíyàr bírì = *monkey's tail* (JMD; ZOG) wùtsíyàr kádangare = *house-lizard's tail* (JMD; ZOG) wùtsíyàr kúúsùù = *mouse's tail* (JMD; ZOG) KAMBARI-SALKA wẽ̂ẽwę (*pl.* íiw-) (RB)

NOTE: Nigerian names apply also to Ctenium newtonii Hack and Schoenfeldia gracilis Kunth.

A densely tufted annual grass, with wiry culms, 1 m or more high, of dry sandy soils, across the Region from Senegal to N Nigeria, and also into Sudan.

Records of stock grazing are ambiguous. In places it is taken, elsewhere not (2). The foliage is aromatic, described as smelling of citronella or lemon (3) which may account for cattle sometimes shunning it.

The grass is used as thatch for huts (1, 5). Salka people in NW Nigeria make conical covers for beehives (4). In Ghana the grass is woven into baskets plastered over the exterior with cow dung. These are put into trees as beehives (3, 5). It has been tried for producing paper pulp (1, 5).

References:

1. Adam, 1954: 90. 2. Adam, 1966,a. 3. Beal 16, K. 4. Blench, 1985. 5. Dalziel, 1937: 523.

Ctenium newtonii Hack.

FWTA, ed. 2, 3: 399. UPWTA, ed. 1, 523.

West African: GUINEA FULA-PULAAR (Guinea) d'yubali bowal (Adames) SIERRA LEONE BULOM (Sherbro) pui-sa-lε (FCD) LOKO woleŋ (NWT) MENDE fiwa (NWT) foni-gbɔli (FCD) TEMNE kɔ-rin-kɔ-ra-lal (NWT) MALI MANDING-BAMBARA wolo kaman (JMD) UPPER

208

VOLTA FULA-FULFULDE (Upper Volta) laasì dawaadî = *tail of the dog* (K&T) MOORE lamzudu (Scholz) **NIGERIA** FULA-FULFULDE (Nigeria) sinakaho (JMD) wicco dombru (JMD) wicco pallandri (JMD) wicco wanduho (JMD) HAUSA shinaka (JMD; ZOG) wútsíyàr ɓeéráá = *mouse's tail* (JMD) wútsíyàr bíri = *monkey's tail* (JMD; Kennedy) wútsíyàr ƙádangare = *house-lizard's tail* (JMD; ZOG) wútsíyàr kúúsùù = *mouse's tail* (auctt.) KAMBARI-SALKA wɛ̃ɛ̃wɛ (*pl.* iiw-) (RB) YORUBA (Ilorin) abori woroko (Ward)

NOTE: *Nigerian names apply also to Ctenium elegans Kunth and Schoenfeldia gracilis Kunth.*

A tufted wiry perennial grass, culms to 1 m high, of open bushland throughout the Region from Senegal to W Cameroons, and extending to Sudan, Uganda and Angola. In the far north of Nigeria it is used as a thatching grass (2, 5). The leaves are aromatic (4), and it is not much grazed. In places in Sudan it has colonised large areas as a result of over-grazing eliminating other more palatable grasses and on the same premise in N Nigeria its presence is taken as an indicator of over-grazing (5). It is an aggresive pioneer coloniser of shallow stony impoverished soil (3).

References:
1. Harrison 1064, K. 2. Kennedy 8048, K. 3. Rose Innes 31361, K. 4. Rose Innes 31384, K. 5. Ward 0054, K.

Cymbopogon citratus (DC.) Stapf

FWTA, ed. 2, 3: 482. UPWTA, ed. 1, 523.

English: lemon grass.

French: citronelle; fausse citronelle.

Portuguese: citronela; erva Príncipe (Feijão).

NOTE: the true citronella is *C. nardus* (Linn.) Rendle, not this plant.

West African: **THE GAMBIA** MANDING-MANDINKA kanyang yallo (Hayes) **GUINEA-BISSAU** CRIOULO belgata (JDES) **GUINEA** KONYAGI *i-*dɛl tɛgag (FG&G) wa-lɛl wa-rɛgag = *the grass with a scent* (FG&G) **SIERRA LEONE** BULOM (Kim) pei-poto (FCD) GOLA ti (FCD) KISSI bichinyeyo (FCD) KONO pu-dumbi (FCD) pu-lumbi (FCD) KRIO lɛmɔn-gras *i.e., lemon grass* (FCD) MENDE pu-lumbe = *whiteman's citrus* (auctt.) TEMNE an-wo-a-potho (FCD) VAI po-paŋa (?) (FCD) **NIGERIA** EDO ɛti = *thick bush* (DA; Elugbe) isoko (DA) EFIK ikọ̀ñ étí (Lowe; DA) HAUSA tsauri (JRA) IBIBIO myoyaka m̀àkárá = *myoyaka of the Europeans* (DA; BAC) IGBO (Asaba) akwụkwọ = *leaf, in general* (DA) IGBO (Owerri) àchàrà ehi = *cow's coarse grass* (auctt.) YORUBA koóko ọba (DA; Verger) koríko ọba = *lord (king) of the grasses* (DA) koríko òyìnbó = *European grass* (Verger; DA) oko ọba = *lord (king) of the farm* (DA) tíi (DA; Verger) **WEST CAMEROONS** VULGAR ti *used generally; probably the English 'tea'* (DA) BAMILEKE hŭndè (Tchamda) KOOSI beyebe ti (DA) KPE bealibe ti (DA) KUNDU bejaba ti (DA) MABONGE bejaba ti (DA) WOVEA bealibe ti (DA)

A tufted perennial cultigen with culms to 2 m high; known only in cultivation; believed to be of Malesian origin and now found dispersed by man throughout the tropics.

All parts of the plant are aromatic, due to the presence of an essential oil, 0.2–0.5% fresh weight (8, 25), 2.0–2.5% dry weight (14), which is composed of *citral* 63–85%, *myrcene* 12–20%, *dipentene* 3–4% and *methylheptenone* a trace (14, 15). The citral confers the lemon taste. The oil is widely used in perfumery, and its citral is the base from which other chemicals are synthesised. The oil is also a starting point for making *ionones* (artificial violet perfumes) and, through *β-ionone*, to produce vitamin A. (8, 9, 4–16, 18, 28.) Madagascar and Comoro Island are the world's main producing countries, with a small output from Ceylon, Indonesia, Seychelles and Taiwan (18).

An insulin-like material has been isolated from the plant with an insulin value of 440 units orally, or 880 units subcutaneously, per gm (25). Leaf-wax has been shown to contain a ketone named *cymbopogone* and an alcohol *cymbopogonol* (10). A trace of alkaloid is detected in the leaves (2, 14), and an unnamed one in the rhizome (26). Leaf and root contain traces of *hydrogen cyanide* (25). Miscellaneous other principles have been reported (14, 15, 18, 25).

The grass is often grown in gardens, along borders and planted on embankments to check erosion (1, 11, 12). Under forestry in Guinea it is planted between anti-erosion stones set on the contour to fill in the gaps (17), and similarly to check wash on forest pathsides in Ivory Coast and E Cameroun (22). The whole plant has a strong fragrant smell. It is often burnt in houses, either green or dry, to dispel mosquitoes (24). In the early days of the rubber plantation industry in Malaya it was commonly planted near to labourers' quarters in the belief that it would keep away anophaline (malaria-carrying) mosquitoes, a hope not borne out. It has been planted in W Africa in tsetse fly areas as a control to discourage flies breeding (3, 4, 11), but with equally unpromising result. In Sierra Leone the leaves, soaked in water for a week, beaten and dried, are used as a filler for pillows and mattresses in the belief that the scent will repel insects (11, 12). The crushed leaf emits a smell of lime and the exotic origin of the grass is reflected in the Mende name *whiteman's citrus*. Yoruba also recognise its foreign origin in 'European grass', and its superior status in *Lord of the farm* and *Lord of the grasses*. (See word list above.) In all areas the leaves are used in infusions to make a teaïform beverage, commonly shown in Nigerian and W Cameroon names. This infusion is taken as a febrifuge, sudorific and dyspeptic (3, 6, 11–15, 19, 24, 28), and put into hot baths for fumigations (11). Leaves boiled with guava leaves are taken in Nigeria for cough (3). In Trinidad leaf and rhizome infusions are considered pectoral and good for colds, flu, pneumonia, cough and consumption, as well as for fever (27). In the Philippines the leaf is applied to the forehead to relieve headache (25).

The original use of the plant seems to have been for flavouring food. It is widely recorded as used thus in Malesia (9, 20), and in the Philippines for adding to wines, sauces and spices (25). Dispersal originally under Portuguese influence has led to its use in pharmacopoeias as a flavouring rather than as a culinary adjunct. It is, however, recorded that the dried and pounded leaf-tips are used in Zambia to flavour meat (21). It has no place as cattle-fodder, for cattle will not browse it (1, 5, 24), but, however, the residue after distillation of the oil, mixed with 35% cane-molasses and a small quantity of protein concentrates has been found in USA to be an useful cattle-feed (18).

The leaf after distillation of the oil has been tried for paper-making, and deemed unserviceable (9).

The rhizomes are used as tooth-picks, chew-sticks and to rub on teeth for cleansing in W Africa (6, 11, 12, 22, 24). In N Cameroun their use as chew-sticks is said to assuage toothache (22). In Congo (Brazzaville) a tisane of the root is given to children to relieve cough (7), and in Senegal a decoction is considered febrifugal (14, 15).

The Yoruba invoke the grass in a magical incantation *to kill an enemy* (23).

References:

1. Adam, 1954: 90. 2. Adegoke & al., 1968: 13–33. 3. Ainslie, 1937: sp. no. 122. 4. Anon., s.d., f: adnot. K. 5. Baumer, 1975: 95. 6. Boboh, 1974. 7. Bouquet, 1969: 128. 8. Brown & Matthews, 1951: 174–87. 9. Burkill, IH, 1937: 724–6. 10. Crawford & al., 1975: 3099–102. 11. Dalziel, 1937: 523. 12. Deighton 2495, K. 13. Diarra, 1977: 43. 14. Kerharo, 1973: 1. 15. Kerharo & Adam, 1974: 643–4, with pharmacology and phytochemistry. 16. Oliver, 1960: 7. 17. Pouquet, 1956: 11. 18. Sastri, 1950: 412–4, with phytochemistry and oil analyses. 19. Singha, 1965. 20. Soenarko, 1977: 351–3. 21. Trapnell 1460, K. 22. Portères, 1974: 118. 23. Verger, 1967: Sp. No. 131. 24. Walker & Sillans, 1961: 187. 25. Watt & Breyer-Brandwijk, 1962: 471. 26. Willamen & Li, 1970. 27. Wong, 1976: 109. 28. Iwu, 1986: 139.

Cymbopogon commutatus (Steud.) Stapf

FWTA, ed. 2, 3: 482.

A tufted perennial grass with erect culms 1.2 to 1.5 m high; of deciduous bushland in Mauritania and Senegal, and extending into Sudan, Ethiopia and Somalia.

The grass is sweetly scented. It is grazed by cattle in NE Africa but is of low palatability (1–3). In Somalia the culms are made into mats for roofing and as blankets (3).

References:

1. Gilbert & Gilbert 2331, K. 2. Herlocker S-422, K. 3. Peck 339.

Cymbopogon densiflorus (Steud.) Stapf

FWTA, ed. 2, 3: 482.

A tufted perennial grass with culms 1.8 m high; of open spaces along roadsides and wooded grassland; native of central tropical Africa, Gabon to Zimbabwe, and introduced to the Region (Nigeria and possibly other states) and into Brazil.

It is grown in Gabon (7) and in Nigeria (3) as an ornamental and for its aromatic oil.

The leaves are avoided by browsing cattle (Zambia, 2).

In Gabon, the crushed leaves make a treatment for rheumatism (7, 8). In Malawi the flower-head is smoked in a pipe as a cure for bronchial affections (4), and for the same complaints plant-sap is taken in Congo (Brazzaville), where this is also given for asthma and to calm fits (1). In Zaïre it is macerated with *Ocimum basilicum* (Labiatae) and the compound is taken for epilepsy. It is conjectured that any action is due to the camphoraceous volatile oil (9). The plant is also recorded as used as a tonic and styptic (6).

The plant has fetish attributes. In Gabon its inflorescence is burnt in fumigations required in certain rituals, e.g., in making incantations to chase away malign spirits, to cleanse those who have lost their spouse, to rejuvenate and restore the efficiency of a fetish, an amulet or a talisman when the owner has violated a taboo (8). In Tanganyika witch-doctors smoke the flower panicle, either alone or with tobacco, to induce dreams to foretell the future (5). Huntsmen in Gabon use the plant as a fetish lure for taking game (8).

References:

1. Bouquet, 1969: 128. 2. Bullock 2952, K. 3. Clayton, 1972: 482. 4. Fenner 267, K. 5. Newbould & Harley 4319, K. 6. Soenarko, 1977: 328. 7. Walker, 1953,a: 39. 8. Walker & Sillans, 1961: 187. 9. Watt, 1967: 1–22.

Cymbopogon giganteus Chiov.

FWTA, ed. 2, 3: 482. UPWTA, ed. 1, 524.

English: tsauri grass (colloquially from Hausa, Nigeria).

French: beignefala (colloquially from Wolof).

West African: SENEGAL VULGAR begnfala (K&A) beignfala (K&A) BANYUN kala (K; K&A) BEDIK é-dìjìtö (K&A) *gu*-tyám *gu*-pùnà (FG&G) DIOLA éputa (JB) DIOLA (Fogny) ébuk (K&A) ébukay (K; K&A) DIOLA-FLUP ara (K; K&A) FULA-PULAAR (Senegal) dagé (auctt.) gagéli (auctt.) TUKULOR nipéré (K; K&A) MANDING-BAMBARA kiékala (A.Chev.; K&A) tiékala (auctt.) tiékala bilé (A.Chev.) MANDINKA bègnfalo (K) benfalo (K; K&A) kiékala (K; K&A) tiékala (K; K&A) wa (K; K&A) wa kasala (K; K&A) ·SOCE· benfalo (K; K&A) wa (K; K&A) wakasala (K; K&A) NON inak (K; K&A) SERER mbal (K) mbol (JB; K&A) ñak (JB) SONINKE-SARAKOLE konoré (auctt.) WOLOF bègnfala (auctt.) benfala (auctt.) gadié (K; K&A) holl (K) mbönfala (K&A) THE GAMBIA MANDING-MANDINKA kala kasala = *scented reed* (JMD;

211

GRAMINEAE

Pirie) wa *a general term* (JMD) wa serrela = *scented grass* (JMD; Pirie) WOLOF benefalu
(Dawe) **GUINEA** BASARI ɛ-diɛdiyita (FG&G) ɛ-ndɛdiyítá (FG&G) KONYAGI ɔ-samban
(FG&G) **MALI** FULA-PULAAR (Mali) dagé (A.Chev.) gagéli (A.Chev.) MANDING-BAMBARA
kiékala (A.Chev.) tiékala (A.Chev.) tiékala bilé (A.Chev.) SONINKE-SARAKOLE kognioré
(A.Chev.) **UPPER VOLTA** FULA-FULFULDE (Upper Volta) fasuure (*pl.* -uuje) (K&T) gajaalo
(K&T) gajoodo (K&T) GRUSI natamora (A.Chev.) natamoza (K&B) MANDING-BAMBARA
tiékala (K&B) DYULA boborasien (K&B) surugubi (K&B) MOORE kuéré (Geissler) mofogo
(A.Chev.; K&B) SAMO kasseburu (A.Chev.) SAMO (Sembla) sòbwùn (Prost) SENUFO kurukuru
(K&B) **IVORY COAST** AKAN-BRONG wozomo (K&B) BAULE awendé (K&B) KULANGO fimu
(K&B) MANDING-DYULA boborasien (K&B) surugubi (K&B) MANINKA nukian (A&AA)
SENUFO-TAGWANA nuyapien (K&B) **GHANA** ADANGME gbetengã (Bunting; FRI) ADANGME-
KROBO gbetengã = *wolf's grass* (auctt.) AKAN-ASANTE ngkabe (FRI) DAGBANI mopele mogo
(Ll.W) **NIGER** FULA-FULFULDE (Niger) gajaali (ABM) **NIGERIA** ARABIC-SHUWA nal (JMD)
seko (JMD) FULA-FULFULDE (Nigeria) luɓgol (JMD) luɓodi = *strongly smelling* (JMD)
sukkahoreho (Taylor) wajaalo (J&D; JMD) wajande (JMD) GWARI nugbwanu bmagna (JMD)
HAUSA gamba (Kennedy) kyara (BM) mobefa (ZOG) tsabre (auctt.) tsagre (ZOG) tsaure
(auctt.) JUKUN (Wukari) vé (Shimizu) NGIZIM rìyak (*pl.* -àyín) (Schuh) YORUBA ọkà ẹyẹ
(Fahaule & Russell; Verger)

A loosely tufted perennial grass with culms erect, sometimes stilt-rooted, to
2½ m high; of deciduous savanna bushland and wooded grassland; abundant
throughout the Region and in general over all of tropical Africa, with var.
inermis restricted solely to Mauritania and Mali.

This grass is dominant over large regions of the savanna constituting the
major part of the herbaceous flora. It requires good soil and no shade, often
colonising fallows and fire-devastated areas. It prevents soil erosion (30).

The grass has a great reputation in the W African pharmacopoeia, especially
in Senegal, as a panacea. It is credited with both prophylactic and curative
power against fever, yellow fever and jaundice (Senegal: 15, 20; The Gambia: 6;
Tanganyika: 11). A decoction, or infusion, of the inflorescences is taken
internally and in baths, and clumps of the plant reduced to a pulp by pounding
together with other drug-plants are used as a body-friction (1, 4, 5, 9, 14, 15, 18,
20–22, 28, 30). Such treatments have been traditionally prescribed during the
yellow fever epidemics of the end of the last century and the beginning of this by
both African and western medical practitioners (1, 15, 22). Dried panicles are
commonly sold in markets as a fever-remedy, for which it is mixed with lime-
juice, etc., and is one of many folk-medicines for coughs (5). A root-decoction is
taken for cough in Tanganyika (11). Amongst the Wolof it enjoys a reputation
for the treatment of strokes, and Pulaar peoples use it for mental disorders (15,
20). Leaf-sap is taken in Tanganyika for epilepsy while at the same time the
body is embrocated with it and the sap of *Crassocephalum crepidioides* (Com-
positae) admixed (11). The Nyominka give an infusion to children with fever
and to women who have miscarried, who are also bathed in it (15, 19, 20). It has
pulmonary use in the classical remedy of Joal *garap u doala* re-enforced with a
number of other drug-plants (15, 20). It is held to have diuretic properties.
Wolof and Serer in Senegal prescribe a root- and leaf-decoction with bark of
Sclerocarya birrea (Anacardiaceae) for dropsy, and in E Senegal the diuretic
effect is tried on blennorrhoea (15, 17, 20). The plant is made up into a
masticatory for oral troubles: gingivitis, aphthres and stomatitis in children (15,
16, 20, 28). The plant has anodynal and analgesic properties. It commonly
serves as a body-wash or embrocation in fatigue. A leaf-decoction is taken in
Senegal to relieve stomach-ache (9, 19). A nasal instillation of the leaf-sap is
taken in Ivory Coast for migraine (3). For the same purpose a smoke fumigation is used in Nigeria (5, 10). The pain of sore throat is assuaged in Nigeria by
application of a poultice of pounded leaves (5). The 'Soce' of Senegal use
fumigation for lumbago (19).

In Senegal 'Soce' ascribe to the grass fetish value in the case of illness: a piece
is suspended inside the hut (19).

In veterinary medicine, sick horses in Nigeria are subjected to fumigation
from burning roots (5).

The plant enters into human diet in only a small way. In Senegal it is added to the meat of carnivores at the time of cooking as an aid to digestion (15, 16, 20). It is added to meat in The Gambia as a flavouring (27), and in Nigeria to food generally (5). The grain is put into sauces in Dahomey (29). Stock will browse it only while it is still young. It becomes unpalatable when old due to increasing concentration of essential oils (Senegal: 1, 2; Ghana: 12; Niger: 24; Nigeria: 5; Sudan: 26; Kordofan: 4). When cattle do eat it in Sudan their milk is reported to be tainted and to be aerated (8).

Phytochemical examination of the grass shows the presence of 1–1.5% of an essential oil in the flower panicles. By distillation this comes over as a water-soluble essence which can be separated out by the addition of sea salt. It has a strong agreeable smell likened to that of 'ginger grass' (1, 15, 20, 21) whose identity is not disclosed, but probably is *Cymbopogon martinii* (Roxb.) Wats., c.var. *sofia*, an Indian plant yielding ginger oil, with a concentration of ~ 40% *geraniol*. This is an important essence in the perfumery trade (31). Be it noted also that the essential oil of *C. giganteus* is rich in *phellandrine* which is one of the oils of the ginger root (*Zingiber officinale*, Zingiberaceae). The grass rhizome also yields an essential oil assayed at 0.5%. *Glucose* and *rhamnose*, and *glycerides* of *linolenic acid, oleic acid, behenic acid, liquoceric acid* and *arachidic acid* are reported present in the flower panicles amongst other substances (20).

The culms are used in The Gambia in hut-construction (6). They are a common thatching material (1, 5, 10, 13, 23, 29, 32, 33). The culms are worked together to make fencing and palisades (1, 5, 7, 9) and hut-walls and partitions, known as *asabari* in Hausa (5). The culms are commonly used for *zaanaa* (Hausa) mats (5, 7, 10, 25), and are made into beds called *chika* in Hausa and *taras* in Bambara (1, 5). The pithy stems are turned into toy arrow-shafts used by boys (5). In The Gambia hunting arrows are made by tipping the stem with a sharp bamboo (27). Religious teachers (malams) cut the stems for use as pens (5).

References:

1. Adam, 1954: 90. 2. Adam, 1966, a. 3. Aké Assi 12290, K. 4. Baumer, 1975: 95. 5. Dalziel, 1937: 524. 6. Dawe 11, K. 7. Diarra, 1977: 43. 8. Evans-Pritchard 53, K. 9. Ferry & al., 1974: sp. no. 110. 10. Golding 1, K. 11. Haerdi, 1964: 210. 12. Johnson 756, K. 13. Kennedy FHI, 8011, K. 14. Kerharo, 1967: 1391–1434. 15. Kerharo, 1973: 2. 16. Kerharo & Adam, 1963, a: 773–92. 17. Kerharo & Adam, 1964, a: 414. 18. Kerharo & Adam, 1964, b: 434. 19. Kerharo & Adam, 1964, c: 302. 20. Kerharo & Adam, 1974: 644–5, with phytochemistry. 21. Kerharo & Bouquet, 1950: 252. 22. Kerharo & Paccioni, 1974: 345–50. 23. McIntosh 100, K. 24. Maliki, 1981: sp. no. 22. 25. Palmer 35, K. 26. Peers Ka.M.28, K. 27. Pirie 51/33, K. 28. Portères, 1974: 118. 29. Risopoulos 1274, K. 30. Roberty, 1961: 693, 695, as *C. nardus*. 31. Sastri, 1950: 416–8, under *C. martinii* (Roxb.) Wats. 32. Semsei 180, K. 33. Thomas, AS D.95, K.

Cymbopogon nardus (Linn.) Rendle

FWTA, ed. 2, 3: 482.

English: citronella grass; new citronella grass.

French: citronelle.

West African: SENEGAL WOLOF beignefala (GR) MALI MANDING-BAMBARA tiékala-ba (GR) TAMACHEK tiberimt (GR) NIGERIA HAUSA tsaure (JRA)

A tufted perennial grass, culms up to 3 m tall; of deciduous bushland and upland grassland in NE, E and S Africa, and across southern tropical Asia to Burma, and introduced to W Africa, and other tropical countries.

The plant is grown in gardens in the Region as an ornamental and for its aromatic oil (3). Vernacular names cited above are by transference from indigenous *Cymbopogon* species. The English reference to 'new' is in distinction from 'old' citronella grass, or *C. winterianus*.

This species and *C. winterianus* are the world's sources of citronella oil. It is said to be a cultigen derived from *C. confertiflorus* Stapf, native of the uplands of S India and Ceylon (2). It is cultivated in Ceylon. It is hardy and grows on poor soil. The oil is of inferior quality compared with that from *C. winterianus* (q.v.). Oil from Ceylon assays at 55–65% 'total alcohols' which include 7–15% *citronellal*. It is used in perfumery and in trade preparations such as sprays, insect-repellants, detergents, polishes, etc. (1, 10, 11).

In E Africa, stock will browse the young growth (Uganda: 7; Kenya: 5, 6) though horses are said not to like it (6). It is used as thatch on huts (4, 5) and as a floor covering (7). The larger culms are used as arrows in Kenya for teaching boys to shoot (8).

An infusion of the leaves is drunk in Tanganyika (9) and the W Indies (12). During World War 2 citronella oil was prepared as a substitute for chaulmoogra oil, a leprosy drug, and produced similar results (12). *Geraniol*, one of the essential oils present in the form of several of its esters, holds promise of use in perfumery by substitution for expensive essences (10, 12). It is considered vermifugal and febrifugal in S Africa, and refrigerant, stomachic, diaphoretic, diuretic, emmenagogic, antispasmodic and stimulant in India (12).

References:

1. Brown & Matthews, 1951: 174–87. 2. Burkill, IH, 1935: 727. 3. Clayton, 1972, 3: 482. 4. Gilliland 409, K. 5. Glover & al. 1424, 1498, 1753, K. 6. Grant 1042, K. 7. Maitland 25, K. 8. Mwangangi 929, K. 9. Nyaki 78, K. 10. Sastri, 1950: 418–9, with husbandry and phytochemistry. 11. Soenarko, 1977: 349–50. 12. Watt & Breyer-Brandwijk, 1962: 471.

Cymbopogon schoenanthus (Linn.) Spreng.

FWTA, ed. 2, 3: 482. UPWTA, ed. 1, 524, incl. *C. proximus* Stapf.

English: camel grass.

French: chiendent pied de poule (Naegelé).

West African: **MAURITANIA** ARABIC (Hassaniya) afar (AN) iverd (AN) **SENEGAL** MANDING-BAMBARA ñangulé (JB) **MALI** ARABIC (Mali) lemmad (HM) MANDING-BAMBARA tiékala-ni (GR) TAMACHEK taberimt (JMD) tiberimt (JMD) **UPPER VOLTA** FULA-FULFULDE (Upper Volta) buluuje (K&T) buulorɗe (K&T) wuluunde (K&T) wuulorde (K&T) MOORE sompígo *in allusion to rabbits sleeping near it* (AJML) suompiga (Scholz) **GHANA** MOORE sumpiga (FRI) súómpiè (FRI) NANKANNI ku-anka (Lynn) **NIGER** FULA-FULFULDE (Niger) hurdu-dumboore *spp. proximus* (ABM) noobol (ABM) HAUSA nobi (Bartha) tsabre (*pl.* tsaure) (AE&L) SONGHAI gòzów (*pl.* -à) (D&C) **NIGERIA** ARABIC-SHUWA mahareib (JMD) nal (JMD) seko (JMD) FULA-FULFULDE (Nigeria) luɓgol = *strongly smelling* (JMD) luɓoɗi *alluding to the strong smell* (JMD) sukkahoreho (Taylor) wajalo (JMD) wajande (JMD) GWARI jimwi (JMD) nugbwanu bmagna (JMD) HAUSA nobe (auctt.) tsabre *occasionally used* (JMD) tsaure (JMD)

A compact tufted perennial grass with culms 60–90 cm high; on dry stony ground of sub-desert bushland: represented by two subspecies: ssp. *proximus* (Hochst.) Maire & Weiler common throughout the whole Region, and ssp. *schoenanthus*, of N Africa, the Orient and India and occasionally recorded in Mali, Niger and Ghana.

Subspecies *schoenanthus* is one of the traditional drug plants of the ancient Mediterranean philosophers. Hippocrates, born c. 460 B.C., 'the Father of Medicine', is said to have used it (14). It is thought to be the plant referred to by Constantinus Africanus in 1087, by Nicholas of Salemo in the first half of the Twelfth Century and by St Galen in the Ninth Century in treatment for cancerous conditions of the stomach, spleen and liver (8). It has been found amongst the funerary materials of ancient Egyptian tombs (4) and is said to have been used in the toilet and burial preparations of the Prophet Mohammed (14). To some peoples there is an inhibition to burn it on a fire (14).

All parts of the plant are aromatic, wherefor there is a variety of medicinal usages. Distillation of the roots and leaves yields a fragrant oil, 'camel grass oil', but this species is not one of the main sources of cymbopogon oils, and its oil is chemically distinct from citronella oil (6b). The inner core of the rhizome is eaten in the northern part of the Region as an aphrodisiac (6b) and an infusion of the inflorescence is drunk in Ghana for fever (9). In N Ghana and Togo the grass is used for snake-bite (6b, 15), and in Ghana mashed up flowers (9) or ashes of the plant (11) are applied to guinea worm-sores. Smoke from the burning grass is said in Nigeria to dispel temporary maniacal symptoms (6b, 7). The leaves pounded with a little water are used as an embrocation in Ghana for aches in the body (11). The inflorescence is used in the western Sahara to produce abortion (14).

Young shoots are grazed by stock though older material is in general not taken (1, 3, 6b, 13a,b, 14), probably on account of the taste caused by increased concentration of aromatic oils. Herdsmen in Niger recognise between the two subspecies and claim that while ssp. *proximus* gives good forage all the year round, and especially after the rains, ssp. *schoenanthus* is not so readily taken, and when browsed taints the milk. (13a,b) Camels are reported to browse it (Toukonnous, Niger: 2; Hoggar: 12).

Rabbits are said to make their nests near patches of this grass, to which the Moore vernacular name has reference (see list above) (10).

The grass is commonly used for thatching (5, 6b). In Niger, wells dug in sandy soil are revetted with the culms of this grass (13b). In N Nigeria it is chopped up and mixed with clay for building huts, and also used for *zaanaa* matting (6b), and by nomads for covering their huts (6a).

References:

1. Adam, 1966, a. 2. Bartha 5, K. 3. Baumer, 1975: 95, as *C. proximus* (Hochst.) Stapf. 4. Burkill, IH, 1935: 728. 5. Dalziel 486, K. 6a. Dalziel, 1937: 524. 6b. ibid.: 524, as *C. proximus* Stapf. 7. Golding 4, K. 8. Hartwell, 1969,b: 190, as *Andropogon laniger* Desf. 9. Irvine 4605, K. 10. Leeuwenberg 4299, FBC, K. 11. Lynn 1076, K. 12. Maire, H, 1933: 55–56, as *Andropogon laniger* Desf. 13a. Maliki, 1981: 51, as *hurdu-dumboore.* 13b. ibid.: 52, as *noobol.* 14. Monteil, 1953: 13, as *Andropogon schoenanthus* Linn. 15. Saunders 11, K. 16. Scholz 1, 1.a, K.

Cymbopogon winterianus Jowitt

FWTA, ed. 2, 3: 482.

English: old citronella grass; Winter's grass.

A tufted perennial grass with culms 2–2½ m high; known only in cultivation; originally in Ceylon, and now widely distributed in the tropics, especially the Indo-Malesian region, and present too in W Africa. It was at one time considered a variety of *C. nardus,* but is now accorded specific rank.

It is grown in the Region as an ornamental, and for its oil (3). It, together with *C. nardus,* supplies the world with citronella oil. Both are exploited in Ceylon, and *C. winterianus* in Java and Taiwan. It is the less hardy, requiring more rain and more careful husbandry but gives a higher yield of better quality. The haulm is cut above the first node, then dried and subjected to steam distillation. Oil from Java contains 85% 'total alcohols' calculated as *geraniol,* including 35% or more of *citronellal* which is a valuable starting point for making various perfumery chemicals (1, 2, 4, 5).

References:

1. Brown & Matthews, 1951: 174–87. 2. Burkill, IH, 1935: 727, under *C. nardus* Rendle. 3. Clayton, 1972: 3: 482. 4. Sastri, 1950: 418, under *C. nardus* (Linn.) Rendle, with phytochemistry. 5. Soenarko, 1977: 348–9.

Cymbopogon sp.

An unspecified *Cymbopogon* is used by herdsmen in Niger as a medicine for their cattle. It is pounded together with tobacco and kaolin and is administered by smoke fumigation, in ointment or by draught before the illness becomes serious (1).

Reference:

1. Maliki, 1981: 55.

GRAMINEAE

Cynodon aethiopicus Clayton & Harlan

FWTA, ed. 2, 3: 403.

A coarse stoloniferous perennial grass, culms robust to 1 m high, of abandoned cultivations and old cattle pounds, native of E Africa, Ethiopia to S Africa, introduced to other tropical countries, and now found in N Nigeria.
The grass gives good grazing liked by all domestic stock, especially the young new growth after burning. It is cultivated in N Nigeria for fodder cropping (2).
Cytological work on Ethiopian material shows polyploidy, 2n = 18 and 2n = 36 (1).
Medicinal use of the grass in W Lake Province Tanganyika is for stomachache and wounds (3).

References:

1. De Wet OKLA 9218, 9220, K. 2. Scholz 108, a, K. 3. Tanner 4996, K.

Cynodon dactylon (Linn.) Pers.

FWTA, ed. 2, 3: 403. UPWTA, ed. 1, 524–5.

English: Bermuda grass; Bahama grass; Indian couch; Australian couch; Fiji couch; etc.; devil grass (West Indies); dog's tooth grass; dub; dhub; doob (from Hindi: dhub; Bengali: dubh).

French: chiendent; petit chiendent; chiendent des Bermudes; gazon des Bermudes; gazon bleu à Dakar (Adam).

Portuguese: grama; grama de boticas (Feijão).

West African: MAURITANIA ARABIC (Hassaniya) iverd (AN) SENEGAL MANDING-BAMBARA sogosoko (JB) SERER gérédèd (JB) WOLOF harap (JB) kérèf (JB) GUINEA-BISSAU FULA-PULAAR (Guinea-Bissau) bógòbodje (JDES; RdoF) SIERRA LEONE KRIO kroke-gras i.e., croquet-grass (FCD) TEMNE ka-gbatha (FCD) MALI ? (Mopti) ŋghoghon (Matthes) ARABIC (Mali) nedjam (A.Chev.; HM) SONGHAI kiki (A.Chev.) zozobu (A.Chev.) TAMACHEK almès (A.Chev.) UPPER VOLTA FULA-FULFULDE (Upper Volta) cuɓɓe (K&T) suɓɓere (K&T) GHANA HAUSA chiaawar sarki = king's grass - because it is planted at government stations: from sarki: king, chief, leader, man-in-charge, craftsman, etc. (JMD) NIGERIA ARABIC-SHUWA lôh (JMD) BIROM syèsyé (LB) EFIK ńkìménàng (Lowe) FULA-FULFULDE (Nigeria) ɓogol boje = hare's string (JMD) jiriyel (JMD) sirkiyambo (JMD) taagol doneyel (J&D) HAUSA ɓúntùn ƙúdáá from ƙúdáá: fly; a blight of the inflorescence (JMD; ZOG) ɓúntùn shààmúwáá from shààmúwáá: stork; a blight of the infloresence (JMD) jáàjàà mazá (auctt.) karya garma = break hoe: from garma: a sort of fenestrated hoe (JMD) ƙìrí ƙìrì (auctt.) ƙirí-ƙirìí (LB) tsambiya (JMD; ZOG) tsárkíyàr zóómóó, tsìrkiyár zóómóó = hare's bowstring (JMD; ZOG) tsírkiyár dámóó = monitor lizard's bowstring (JMD; ZOG) KANURI kargashi (JMD) karjigu (JMD) kárgásmì (C&H) YORUBA kóoko-igbá (JMD; Verger)

A perennial stoloniferous and rhizomatous mat grass, culms slender rising to 40 cm high, of roadsides, old cultivations, weedy and trampled places, common throughout the Region, and pan-tropics.
It is a good pasture grass and is much valued for grazing. All stock find it palatable (1, 2, 5, 6, 9, 11, 14, 16), but under some dry conditions leading to wilting it produces prussic acid (1, 9, 10, 13, 17). It may become a noxious weed of cultivations invading arable land, rice-padis (8, 9) and irrigation (15). The deep-setting rhizomes make it difficult to eradicate (1, 18). The Hausa name meaning 'break hoe' and the West Indian 'Devil's grass' make plain man's animadversion (6). On the credit side, however, the deep-seated rhizomes help the grass to withstand drought, and, though the top sward is fire-susceptible, they soon throw up new growth after burning. This habit of growth results in the grass being one of the better lawn grasses for drier tropical countries (1, 5, 7, 9, 10). It resists foot-wear and makes a good turf for tennis courts, golf greens and fairways, playing fields, etc., especially under relatively dry conditions. Under wetter conditions more vigorous grasses will choke it out. This, in fact,

discloses a form of husbanding to eradicate the grass from cultivated land by growing a smothering crop of legumes. It is a sand-fixer. It commonly makes a short turf on stream-sides (6, 7) giving protection against wash. In India it is recommended for planting on padi-field bunds consolidating the soil and preventing other plants invading (9). To establish a cover, it is easiest to dibble in pieces of rhizome.

The whole plant is considered in the Philippine Islands to be diuretic and pectoral (13, 17). It is also demulcent, astringent, haemostatic and laxative. The rhizome is official in several pharmacopoeias. In Nigeria it is used as a cardiac tonic, and is considered especially useful for treating dropsy; the action is, however, reported to be more irritant than the use of digitalis (3). In Gabon a root-decoction is used to treat inflamation of the urethra (16) by diuresis.

The roots are edible and might serve as a famine food. They are fed to horses and water buffalo in the Philippines. Besides starch, a substance named *cynodin*, allied to *asparagin*, a good food-substance, is reported (4, 13, 17).

An infusion of the whole plant is used in Kenya as a wash to de-lice camels (12).

The pollen is reported as a hay-fever allergen in the U.S.A. and S America (13).

References:

1. Adam, 1954: 92. 2. Adam, 1966,a. 3. Ainslie, 1937: sp. no. 123. 4. Burkill, IH, 1953: 729. 5. Cole, 1968,a. 6. Dalziel, 497, K. 7. Dalziel, 1937: 524–5. 8. Gill & Ene, 1978: 182-3. 9. Gupta & Sharma, 1971: 71. 10. Hunting Tech. Services, 1964. 11. Maire, 1933,a: 232. 12. Mwangangi & Gwynne, 1024, K. 13. Quisumbing, 1951: 94–95, 1019, 1023. 14. Roberty, 1958: 874. 15. Tewo Ahti 16362, K. 16. Walker & Sillans, 1961: 187-8. 17. Watt & Breyer-Brandwijk, 1962: 471. 18. Williams, RO, 1949: 285-6.

Cynodon nlemfuensis Vanderyst

FWTA, ed. 2, 3: 403.

A stoutly stoloniferous perennial grass, culms robust or somewhat slender to 30–60 cm high, not woody, of two varieties: var. *nlemfuensis* and var. *robustus* Clayton & Harlan, distinguishable on the robustness of the culm, 1–1½ cm and 2–3 cm diameter respectively; of open spaces in bush or forest and on waste places in eastern Africa, Ethiopia to South Central tropical Africa; introduced to other tropical territories, and present in Ghana and S Nigeria. The plant is similar to *C. dactylon*, but the absence of underground rhizomes distinguishes it.

It is an important fodder grass in E Africa, taken by all stock (Sudan: 4; Uganda: 5; Kenya: 2). Var. *robustus* demonstrates an aggressive manner of growth in the Cape Coast area of Ghana and appears to be spreading naturally (3).

Cytological examination shows both varieties to be diploid (2n = 18) but tetraploidy is known in cultivars (1).

References:

1. Clayton, 1974,a: 321. 2. Glover & al. 80, 510, 1982, K. 3. Hall 3451, K. 4. Peers KM23, K. 5. Snowden 1265, K.

Cynodon plectostachys (K Schum.) Pilger

English: (giant) star grass.

A stoloniferous perennial grass, culms stout and woody to 90 cm high by 1–4 mm diameter; of deciduous bushland in disturbed sites and weedy places, of Ethiopia and E Africa and introduced into Nigeria.

It is an useful pasture grass especially in the southern guinean zone (1) providing good grazing for stock generally and for pigs (2). It is grown either alone or in a ley mixture with *Centrosema pubescens* (Leguminosae: Papilionoideae) (3). Propagation is usually done by cuttings, but the use of heat-treated seed is also possible.

References:

1. Foster & Munday, 1961: 314. 2. Modebe, 1963,a: 116–26. 3. Okigbo, 1964: 141–58.

Cyrtococcum chaetophoron (Roem. & Schult.) Dandy

FWTA, ed. 2, 3: 426.
West African: SIERRA LEONE TEMNE *ka*-seth-*ka*-gbel (FCD) IVORY COAST ·NÉKÉ-DIÉ· bikakosiré (B&D)

A delicate lax perennial grass, culms ascending to 1 m high, of forest shade across the Region from Senegal to N and S Nigeria, and to Sudan, Zaïre and Angola.

In Ivory Coast it enters magical application. In a case of a difficult pregnancy due to demoniacal influences, it is necessary to draw four parallel lines from the chin to the navel of the patient with the ash of this grass (1).

Reference:

1. Bouquet & Debray, 1974: 92.

Dactyloctenium aegyptium (Linn.) P. Beauv.

FWTA, ed. 2, 3: 395. UPWTA, ed. 1, 525.
English: crow's foot grass (Adam); comb-fringe grass; finger comb grass (Adam).
French: dactylogène d'Egypte (Naegelé).
Portuguese: pé-de-galinha (Feijão).
West African: MAURITANIA ARABIC (Hassaniya) kra'a l'grab = *crow's foot* (AN) SENEGAL MANDING-BAMBARA ndégélé (JB) SERER ngok (JB) WOLOF ndãnga (JB) tang i mpiteurh (JLT) THE GAMBIA MANDING-MANDINKA kontenterong (Fox) GUINEA-BISSAU BALANTA cúnher (JDES) FULA-PULAAR (Guinea-Bissau) násci (JDES) GUINEA BASARI a-làp ɔ-xèdi FG&G SIERRA LEONE LIMBA kubema (NWT) LOKO taha (NWT; JMD) MENDE pɛtewule (NWT) tugbɛ (FCD) tugbɛlɛ (FCD) SUSU tanse (JMD) tansɛ (NWT) TEMNE mammi (NWT) MALI DOGON kèènie ana (CG) FULA-PULAAR (Mali) burgue boguel = *hare's grass* (A. Chev.) burguel ? = burugal: *swizzle stick* (A.Chev.) ndanguel (A.Chev.) MANDING-BAMBARA ndéguéré (A.Chev.; FB) ntéguélé (A.Chev.; FB) UPPER VOLTA FULA-FULFULDE (Upper Volta) buruugel (*pl.*-uuje) (K&T) GRUSI nebiépélé (A.Chev.) GURMA guananini (A.Chev.) guguni (A.Chev.) MOORE ganaga (FB) guanaga (A.Chev.) uantega (Scholz) NIGER FULA-FULFULDE (Niger) buɗe buɗeeri (ABM) buruugel baali a var. (ABM) guɗe guɗeeri (ABM) guye guyeeri (ABM) HAUSA kutuku (Bartha) NIGERIA ARABIC-SHUWA abu asabe (JMD; Kennedy) koreib, kreb *to several grasses with edible grain* (JMD) um asaba (JMD) BURA fakam (Kennedy) FULA-FULFULDE (Adamawa) falande (Saunders) FULFULDE (Nigeria) burugihi (JMD) pagamri (*pl.* pagamje) (JMD) tumbi (Saunders) HAUSA guɗa-guɗe (JMD; Lowe) gúɗè-gúɗè (auctt.) kurtu (JMD) kutukku (JMD; ZOG) KANURI fàám (C&H; A&S) fagam (JMD) fogam (Kennedy) fòwúm (A&S) TIV ikangel a ika *incl. other grasses with digitate inflorescence* (JMD) YORUBA (Ilorin) ɛwa ɛsin (Ward)

An exceedingly variable loosely tufted or stoloniferous annual grass, prostrate or ascending, culms to 60 cm high, of open waste places, grassland, open woodland and roadsides from sea-level to 2000 m altitude; common throughout the Region from Mauritania to N and S Nigeria, and dispersed across tropical Africa and in warm temperate and tropical regions of the Old World; now also in the New World.

The grass is common around villages, and often in the bush it will form wide expanses of dense cover. It is a weed of cultivated land (The Gambia: 11; N Nigeria: 13). Tenda farmers in SE Senegal attempt to keep it under control by sprinkling salt on the leaves (10). It appears to be well-suited to sandy soil. In Rajasthan, India, it grows on sand with long creeping rhizones putting out deep-penetrating roots, thus serving as a binding agent (14). It has been reported in Sierra Leone growing on sand at the beach-head (15, 22). In S Africa it has been grown as a lawn-turf (27).

In all countries of the Region (8, 29; Mauritania: 23; Senegal: 1, 2, 10; Ghana: 4, 37; Nigeria: 12, 13, 18–20, 24, 28), and elsewhere in tropical Africa and India, the grass provides grazing for all stock. It is sometimes cut in the Region for sale in markets, and for storage as hay (1, 8). Acceptability to stock seems to be variable, perhaps due to edaphic, phenological or even varietal factors. It is said to be a good strengthening horse-feed (1, 8, 18, 28). It serves well in Kordofan being accepted by stock at a time when other fodders are shunned (3), though grazing in N Nigeria is limited to whilst it is still young (34). In Niger, herdsmen recognise two forms of the grass: 1, *mbude-mbudeeri* or *ngyyeeri* which is grazed bare to the ground to the exclusion of other forage plants, on which cows produce thick, rich milk; and 2, *buruugel baali* which is less appreciated, and of which only the upper part is grazed, the lower part being untouched (21). At times and/or places it is reported to produce *cyanogenetic glycosides* (India: 14; southern Africa: 35; Australia: 6, 27), which can be dangerous to stock.

As food value, the grass is below average. Assay of Indian material has given: crude protein 7.25% and fibre 33.74%, dry weight, and of Ghanaian material: protein 16.37%, N-free extract 37.68% and fibre 23.70%, wet weight (8).

Vegetative parts of the plant are applied externally to ulcers in India (14). The plant is reported to be astringent and is used in decoction internally in Africa for dysentery and acute haemoptysis (25).

The value of the grain as food is variously reported upon. In many territories it is not normally an item of diet, but is eaten as a famine-food (7, 8; Mauritania: 23; Senegal: 1, 29, 31; Ghana; 4, 33; Nigeria 12, 24; Jebel Marra: 17; Kordofan: 3, 16; Kenya: 30; India: 5, 27). The taste is reported as being unpleasant and consumption may cause internal disorders. It is said, though, to be eaten in Jebel Marra (36) and in Uganda (9) boiled as porridge or gruel, or baked into cakes. There appears, however, to be a cultivar commonly grown, and not known in the wild, in the Mitengo and Mbugas area of Tanganyika (31), and it probably occurs elsewhere, which produces a platable grain that was commonly eaten as a regular item of diet in olden times before the general availability of maize, rice and tapioca (manioc). Its cultivation required slash and burn of virgin land followed by planting in the residual ash – obviously an ephemeral and destructive form of husbandry. With the introduction of sunn hemp (*Crotalaria juncea*, Leguminosae: Papilionoideae, q.v.) it has been found that this grass will grow quite satisfactorily as a following crop not needing the presence of plant ash, and with the advantage of not calling for destruction of any more forest (26, a, b). It is found to be relatively free from depredation by insects and to be a more reliable crop. The grain stores well for many years. It is a plant of very obvious advantages for regions of irregular rainfall. The grain can be cooked in various ways after grinding. The flower is white but turns red in hot water. The cooked meal does not go sour. It is not indigestible and it has a pleasant smell. It is recorded that for these benefits Wangoni raiders in Tanganyika took this as provisions while out on foray. At the present time the plant is grown for brewing and the product is a very good beer (26, a).

Chemical analysis of grain from Upper Volta shows: protein 15.7% and carbohydrate 65.3%, dry weight (7).

In Tanganyika, alongside the area where the edible-seeded cultivar is grown (see above), there is a wild form, not cultivated, which is used for making baskets (26).

A decoction of seed is used in Africa to relieve kidney pain (25).

References:

1. Adam, 1954: 91. 2. Adam, 1966,a. 3. Baumer, 1975: 96. 4. Beal 2, K. 5. Bhandara, 1974: 76. 6. Burkill, IH, 1935: 747. 7. Busson, 1965: 453, with chemical analysis of the seed, 453/5. 8. Dalziel, 1937: 525. 9. Dyson-Hudson 177, K. 10. Ferry & al., 1974: sp. no. 111. 11. Fox 129, K. 12. Golding 11, K. 13. Grove 18, K. 14. Gupta & Sharma, 1971: 71–72. 15. Harvey 66, K. 16. Hunting Tech. Services, 1964. 17. Hunting Tech. Surveys, 1968. 18. Jackson 1973. 19. JCS 2:518 (UIH 4228), UCI. 20. Kennedy FHI. 8014, K. 21. Maliki, 1981: 47, 50. 22. Melville & Hooker 243, K. 23. Naegelé, 1958,b: 892. 24. Palmer 2, K. 25. Quisumbing 1951: 95. 26a. Rupper 1989,a: in litt., 16/6/1989. 26b. Rupper, 1989,b. 27. Sastri, 1952: 2. 28. Saunders 51, K. 29. Schnell, 1960: no. 104. 30. Sprengler S.48. K. 31. Thomas, L., s.n., 2/11/1950, K. 32. Trochain, 1940: 273. 33. Vigne 4547, K. 34. Ward 0083, K. 35. Watt & Breyer-Brandwijk, 1962: 472. 36. Wickens 1744, K. 37. Williams 604, K.

Danthoniopsis chevalieri A Camus & CE Hubbard

FWTA, ed. 2, 3: 419.
West African: SENEGAL BASARI wandeusse (K&A) SIERRA LEONE MANDING-MAN-DINKA fuadobĩ (FCD) YALUNKA tĕsĕgbe-na (FCD)

A tufted perennial grass, culms to about 2 m high, of grass savanna, known only in E Senegal to Sierra Leone.
In the Musaia area of Sierra Leone it is reported to provide the best thatching grass (2).
In Senegal cattle relish it, even when it is dry (1).
Bedik and Basari people in E Senegal put the plant to undisclosed medicinal use (3).

References:

1. Adam, 1966, a. 2. Deighton 4523, K. 3. Kerharo & Adam, 1964, a: 436.

Desmostachya bipinnata Stapf

UPWTA, ed. 1, 525.
West African: MALI ARABIC (Mali) halfa (JMD) KANURI budaur (A.Chev.)

A tall, tufted perennial grass with stout stolons sending up culms to 2 m high; in dry, hot situations, tussock-forming in sandy desert areas, but also found in low-lying water-logged sites; of the Malian Sahel, and into Sudan and on to India.
Cattle in India do not like it, taking it only for want of better grazing; buffaloes will eat it if it is young (4). In Malian desert it is considered good fodder (2). Analysis of Indian material has given on dry weight bases: crude protein 6.75%, crude fibre 40.30%, fats 1.61%, carbohydrates 42.22% and ash 9.12% (4). Tests for producing paper-pulp resulted in a yield of only 35%. The material was difficult to bleach, and the fibre weak and short. It was usable only as an additive to better pulps (4).
The leafy culms are used for thatching in India (4), and in India (4) and Sudan (1, 3, 5) for roving into a coarse rope.
In India the culms are considered diuretic and are used to treat dysentery and menorrhagia (4).

References:

1. Broun & Massey, 1929: 475, as *D. cynosuroides* Stapf. 2. Dalziel, 1937: 525. 3. Jackson 2354, K. 4. Sastri, 1952: 43. 5. Speke & Grant 63, K.

Dichanthium annulatum (Forssk.) Stapf

FWTA, ed. 2, 3: 471. UPWTA, ed. 1, 525.
West African: MALI TAMACHEK ebastan (A.Chev.)

A densely tufted perennial grass, culms geniculately ascending to 1 m high; of open disturbed places, often by water, stream-channels and irrigated land (var. *papillosum* (A Rich.) de Wet & Harlan); in the soudanian part of the Region in Senegal to Niger, and widespread in NE, E, S tropical Africa and into India, tropical America and Australia.

It is relished by all stock in Senegal (2a,b) and Kordofan (4), and generally in all territories of Africa. It serves as a good desert forage even when dry (1a,b, 6). In India it is esteemed amongst the wild fodder-grasses which stock appear to select in preference (8, 9) when still young or when in flower. Horses, donkeys and buffaloes browse it (3, 7, 8) and various game animals (4, 5).

In India it becomes abundant in old established pastures, yields heavily and stands mowing for hay and for silage if taken before flowering (8, 9). It can be grown as a rough lawn.

References:

1a. Adam, 1954: 91. 1b. ibid.: 91, as *D. papillosum.* 2a. Adam, 1966, a. 2b. ibid., as *D. papillosum.* 3. Amenti 451, K. 4. Baumer, 1975: 96. 5. Blair Rains 93, K. 6. Dalziel, 1937: 525. 7. Godding 9, K. 8. Gupta & Sharma, 1971: 73–74. 9. Sastri, 1952: 54, with chemical analysis.

Dichanthium caricosum (Linn.) A Camus

English: hay grass (Antigua, Burkill); nadi blue grass (Kenya, Muthambi).

A tufted perennial grass, culms geniculately ascending to 1 m high; of open sunny, dry sandy places; native of India and SE Asia and now recorded in francophone W Africa (3) and commonly at sites in N Nigeria (2). It is also present in E Africa.

In the Region it occurs in savanna valley parkland and provides good pasturage (3). In India its fodder value is said to equal that of *D. annulatum* (q.v.), to which it is closely related (4). It is drought-resistant. Its cultivation in several countries has been officially encouraged (1).

References:

1. Burkill, IH, 1935: 802. 2. Ochi 151, K. 3. Roberty, 1961: 694. 4. Sastri, 1952: 55, with chemical analysis.

Dichanthium foveolatum (Del.) Roberty

FWTA, ed. 2, 3: 471, as *Eremopogon foveolatus* (Del.) Stapf. UPWTA, ed. 1, 528, as *E. foveolatus* Stapf.
West African: MALI TAMACHEK okras (A.Chev.) tirichit (A.Chev.)

A densely tufted perennial grass with wiry culms to 60 cm high; of deciduous bushland or sub-desert grassland; from Mauritania to Niger, and across Africa through Kenya and Tanganyika to India.

The grass is a pioneer of shallow sandy soils where conservation practices are necessary (3). It grows on all sorts of soil and produces heavy crops of fodder under favourable conditions (4). It is readily taken by all stock, and is palatable after seeding. It is a valuable desert fodder especially for camels (2, 4). It converts into good hay (3).

221

In Mauritania it has been recorded as a weed of irrigation (1).

References:

1. Arvidsson 68, K. 2. Dalziel, 1937: 528, as *Eremopogon foveolatus* Stapf. 3. Gupta & Sharma, 1971: 81, as *E. foveolatus* Stapf. 4. Sastri, 1952: 183, as *E. foveolatus* Stapf.

Digitaria abyssinica (Hochst.) Stapf

FWTA, ed. 2, 3: 452.

A creeping perennial ruderal grass with wiry rhizomes and culms to 30 cm high; of damp sites; recorded only in N and S Nigeria and W Cameroons in the Region, but widespread elsewhere in tropical and S Africa, and in Madagascar and Ceylon.

No attribute is recorded of the plant in W Africa. It is a pioneer coloniser of disturbed land and in this respect its rhizomatous close-matting growth serves well in Somalia as a soil-binder in anti-erosion work (2). In all eastern and southern Africa it is noted as a pernicious weed of cultivation, and even of irrigation. It is inimical to all arable and plantation crops. In Zambia farmers may abandon land taken over (1), eradication being considered too difficult. Ploughing is impeded and increased traction is said in Somalia to result in injury to the shoulders of draft-cattle (3).

The grass is readily browsed by all stock.

References:

1. Greenway 6222, K. 2. McKinnon S.280, K. 3. Peck s.n., 23/9/1950; s.n., 4/10/1950.

Digitaria acuminatissima Stapf

FWTA, ed. 2, 3: 452.

An annual grass, raising culms to 90 cm, with long stiff racemes, on riparian flats liable to flooding, not common in Mauritania, Mali, Ghana and N Nigeria, and widely dispersed over tropical Africa.

It is found as a weed in padi-fields in N Nigeria (2). It also provides some grazing (1).

References:

1. Dalziel 907, K. 2. Munro 21, K.

Digitaria argillacea (Hitch. & Chase) Fernald

FWTA, ed. 2, 3: 452.
West African: SENEGAL MANDING-BAMBARA foño ni kafini (JB) NIGERIA FULA-FUL-FULDE (Nigeria) sialbo (Saunders)

An annual grass with culms erect to 60 cm high; of waste places in Senegal to S Nigeria.

The grass is a weed of cultivation and invasive of fallows. It furnishes forage for stock.

222

Digitaria argyrotricha (Anderss.) Chiov.

FWTA, ed. 2, 3: 453.

A tufted or trailing annual grass, culms geniculately ascending to 70 cm; of sandy coastal sites, known only in Ghana in the Region, and on the coast of E Africa.

It is recorded in Kenya as a weed of tapioca cultivation (4), and on the coastal dunes as a binder (1). Chickens have been recorded as eating the leaves, especially when they are infected with roundworm (3), and, of unclear significance, rats and mice gnaw the stems (2).

References:

1. Bogdan AB.2527, K. 2. Magogo & Glover 515, 1025, K. 3. Magogo & Glover 668, K. 4. Robertson 3458, K.

Digitaria barbinodis Henr.

FWTA, ed. 2, 3: 454.
West African: NIGERIA ? (Bauchi) ndai *a c.var.* (Philcox) nde *a c.var.* (Philcox)

An annual grass, clumped, slender-stemmed to 60 cm high, recorded only in Mali and N Nigeria.

On Bauchi Plateau, Nigeria, it appears to be occasionally grown under the two cultivar names given above. The collection of the former was mixed with *D. iburua*, the black acha (2), and the latter with *D. exilis*, the white acha (1), instances of mixed cropping recorded under *D. ciliaris*, q.v.

References:

1. Philcox 1005, K. 2. Philcox 1013, K.

Digitaria ciliaris (Retz.) Koel.

FWTA, ed. 2, 3: 453, in part.
English: crab grass; incl. several closely related species (Adam); birds' acha (Philcox).
West African: SENEGAL BEDIK ɓar ɓasyán = *which brings men together* (FG&G) GUINEA BASARI a-xɛrɛtét (FG&G) KONYAGI i-tyit (FG&G) SIERRA LEONE DOGON sáána vòònu (CG) sáána vòònu ya (FB; CG)

An annual grass, decumbent, geniculately ascending to about 60 cm high; of waste places, farmland, roadsides, etc.; common throughout the Region from Cape Verde and Mauritania to N Nigeria, and extending into Sudan and to E Africa, and Asia.

It forms good pasture and is relished by all stock for grazing (Senegal: 2; Nigeria: 6). It is sometimes cut for market sale in Senegal, and also cut for conversion into hay (1). It has been used as a lawn-grass and for tennis courts (1, 5) though it will not stand much foot-wear, nor desiccation.

Birds are keen to seek out the grain (4), hence the English name 'birds' acha'. It is a wild, uncultivated acha (7). It appears vicariously in fields of *D. exilis* in SE Senegal where the Konyagi do not distinguish between the two species till harvest (4). The grain is normally not collected except as a famine-food, though in places, e.g. south of Chad, it may be eaten as a supplementary food (3).

References:

1. Adam, 1954: 91. 2. Adam, 1966, a. 3. Busson, 1965: 455. 4. Ferry & al., 1974: sp. no. 112. 5. Hall 3210, K. 6. Musa Daggash FHI.24891, K. 7. Philcox 1017, K.

GRAMINEAE

Digitaria compressa Stapf

FWTA, ed. 2, 3: 452, as *D. homblei* Robyns.
West African: NIGERIA HAUSA iburun daji = *'iburu'* of the bush (Taylor)

A caespitose perennial grass, culms to 60 cm high; of grassy woodland in N Nigeria, and also in Ethiopia and south tropical Africa.

No usage is recorded in W Africa, but since the Hausa name is a group term for several *Digitaria spp.* which are good fodder plants, it probably qualifies thus (see *D. debilis* and *D. horizontalis*).

Digitaria debilis Willd.

FWTA, ed. 2, 3: 454. UPWTA, ed. 1, 525.
English: finger grass.
West African: SENEGAL SERER rukh (JLT) WOLOF kombardiagandal (JLT) niarh e pic (JLT) THE GAMBIA MANDING-MANDINKA findi bano = *wild 'findi'* (JMD) ja-jeo (JMD) SIERRA LEONE KONO pouvei (FCD) MENDE *n*dewi (JMD) *n*diwi (JMD) pouve (*def.* -i) (FCD) MENDE (Kpa) pouve (*def.* -i) (FCD) TEMNE a-gbel *applied to small-grained grasses: from gbel: fundi bird, or waxbill* (JMD) MALI FULA-PULAAR (Mali) fonio ladde (A.Chev.) musa ladel (A.Chev.) serémé ladde (A.Chev.) MANDING-BAMBARA narkata (A.Chev. fide JMD) SONGHAI bana iddu (A.Chev.) subcoré (A.Chev.) UPPER VOLTA GRUSI banguéré (A.Chev.) GURMA gossolo (A.Chev.) taramanté (A.Chev.) tetemtié-haga (A.Chev.) tetumté (A.Chev.) NIGERIA FULA-FULFULDE (Nigeria) damaliiliho (JMD) damaliiliyel (J&D) HAUSA hàràkiyaa (ZOG) hàrkiíyaa (auctt) ibúrùn dáájìì (JMD; ZOG) karanin dawaki = *'karani'* of horses (JMD; ZOG) IGBO (Agukwu) ilulo egogu ẹzu (NWT) TIV yayaghol (JMD) YORUBA eéran (JMD; Verger)

An annual grass, prostrate at base, geniculately ascending to 60 cm high; of damp sites across the Region from Senegal to W Cameroons, and widespread elsewhere in tropical Africa and to the W Mediterranean.

This grass is one of a group of excellent fodder grasses for both cattle and horses (1, 3). It may be found cut for sale in markets. It also is cut for hay. It can be grown as good pasture, and has use for lawns and on banks, etc.

In time of dearth the seed may be eaten (2–4).

References:

1. Adam, 1966,a. 2. Busson, 1965: 455. 3. Dalziel 254, K; 1937: 525. 4. Okiy, 1960: 120.

Digitaria delicatula Stapf

FWTA, ed. 2, 3: 452.
West African: NIGERIA HAUSA iburun daji (Mayo)

An erect annual grass to 60 cm high, occurring across the Region from Senegal to S Nigeria.

It is regarded as a good fodder in N Nigeria (2) and to be relished by stock in Ghana at all times (1). The Hausa name is a group term meaning 'iburua', or 'acha of the forest', relating the plant to the cereal species *D. exilis* and *D. iburua*. The possibility of this plant providing a supplementary or emergency grain should be borne in mind.

References:

1. Beal 22, K. 2. Mayo s.n., July 1932, K.

Digitaria diagonalis (Nees) Stapf

FWTA, ed. 2, 3: 450.

West African: NIGERIA FULA-FULFULDE (Adamawa) geene sabbere (Pedder)

A robust perennial, culms erect, 1–3 m high with handsome spreading inflorescence; of grass savanna throughout the Region from Senegal to W Cameroons, and generally dispersed over the rest of tropical Africa. A variable species, of which the W African plant is var. *hirsuta* (De Wild. & Th. Dur.) Troupin, merging without clear distinction with var. *diagonalis* and var. *uniglumis* (A Rich.) Pilg. represented in other parts of tropical Africa (1).

No usage is recorded in the Region. In the Central African Republic the leaf is cooked with tapioca for eating; the confection is reported 'to smell very nice' (2). In Malawi (3), Uganda (10) and Kenya (5–8) it is reported as an excellent grazing and fodder grass, much liked by stock and game. Poisoning of cattle, goats, etc., has been recorded in Sudan, the symptoms being white foam at the mouth soon after eating, followed by death. No further detail is given. Elsewhere in Sudan cattle are said to eat it safely (9). In Malawi the grass is valued as one of the earliest to come into growth (3).

The haulm and culm are used in Kenya to thatch huts (7).

Small birds in Kenya seek out the seed to eat (4).

References:

1. Clayton & Renvoize, 1982: 624–7. 2. Fay 4421, K. 3. Fenner 205, K. 4. Glover & al. 587A, K. 5. Leippert 5151, K. 6. McDonald 1017, K. 7. Magogo & Glover 822, K. 8. Magogo & Glover 947, K. 9. Myers 11918, K. 10. Snowden 1251, 1464.

Digitaria exilis (Kippist) Stapf

FWTA, ed. 2, 3: 453. UPWTA, ed. 1, 526.

English: hungry millet; hungry rice; hungry koos; acha grass (Ghana, Irvine), white acha (in distinction from *D. iburua*, black acha).

French: petit mil; millet digitaire (Kerharo & Adam); fonio (from a common vernacular name), fogno; fundi.

West African: SENEGAL VULGAR fonio BEDIK fɔndéŋ *the grain* (FG&G) fɔndéŋ *i*-ɓála *a late form* (FG&G) fɔndéŋ *i*-fɛsyáx = *white fonio, a late form* (FG&G) fɔndéŋ *i* swgɛt = *fonio ripening first; a precocious form* (FG&G) gɛ-pɔndéŋ *the plant* (FG&G) DIOLA eboñay (K&A; FB) efoleb (FB) efoled (K&A) FULA-PULAAR (Senegal) séréné (K&A) MANDING-BAMBARA foño (JB; K&A) MANINKA foño (K&D) MANDYAK findi (FB) WOLOF dekölé (JB; K&A) *n*dengue (FB) find (FB) findi (JMD fide K&A) foño (K&A) THE GAMBIA MANDING-MANDINKA dibong *a short-season c.var.* (JMD) findi, findo (auctt.) findi ba *a late-maturing c.var.* (DA) monyimonyo *a short-season c.var.* (JMD) mormor *a short-season c.var.* (JMD) WOLOF findi (JMD) findi ba (JMD) GUINEA-BISSAU BALANTA fénhe (JDES) BIAFADA bofinhè (JDES) CRIOULO fundo (JMD; JDES) FULA-PULAAR (Guinea-Bissau) fónio (JDES) MANDING-MANDINKA findô (JDES) MANDYAK uante (JDES) urote (JDES) MANKANYA uante (JDES) udote (JDES) PEPEL rote (JDES) urrote (JDES) GUINEA VULGAR fonio (CHOP) BASARI funyáŋ *i*-ɓanàx *a late form* (FG&G) funyáŋ *i*-fɛsyáx = *white fonio;* a late form (FG&G) funyáŋ r̃asy-ɓa-kí = *fonio, twice harvest; a very precocious form* (FG&G) funyáŋ *i*-sɔ́gét = *fonio ripening the first; a precocious form* (FG&G) FULA-PULAAR (Guinea) foignié (FB) foinye (JMD) fongo (FB) fonie (FB) fonyo (JMD) fundé (FB) fundenyo *from Susu* (JMD) fundiune (FB) KISSI kpendo (FB) pende (FB) KONYAGI *i*-farôbt *a precocious form* (FG&G) fayaon (FG&G) funie (FB) punie (FB) LANDOMA pende (FB) LOMA podé (FB) podégui (FB) MANDING-MANINKA foigné (FB) fo(u)nde (FB) foni(-o) (FB) SUSU fonde (FB) funde (FB) SIERRA LEONE VULGAR fundi (auctt.) BULOM peni-lɛ (FCD) BULOM (Sherbro) peni (Pichl) FULA-PULAAR (Sierra Leone) fɔnyɛ (FCD) funyɛ (FCD) GOLA pote (FCD) KISSI kpendô (FCD) KONO fonde (FCD) KORANKO fondiba *long-season, 4-5 months c.vars.* (FCD) fonye (FCD) funa (FCD) fundili *short-season, 2-3 months c.vars.* (FCD) KRIO funde (FCD) fundi (FCD) milɛt (FCD) LIMBA ampindi (FCD) fundili *short-season, 2-3 month c.vars* (FCD; JMD) siragbe *a much-prized short-season c.var with white husk & grain* (JMD; FCD) LIMBA (Tonko) mpende (FCD) LOKO pote (FCD) MANDING-MANDINKA funi (FCD) MENDE fundi (NWT; FCD) pote (auctt.) kputi (FCD) SUSU funde (FCD) fundenyi (NWT; FCD) yele-fui *short-season c.vars.* (FCD) TEMNE ka-enɛ (*pl. pa*-enɛ) *the grain* (JMD) *a*-pende-*pa*-funf *short-season c.vars.* (auctt.) *a*-pende-*pa*-lel *long-season c.vars.* (auctt.) *a*-pende-*pa*-siragbe *a much-prized c.var, with white husk & grain* (FCD) peni (NWT; Mauny) *a*-pɔtɛ (FCD) TEMNE (Kunike) *a*-pɔtɛ (FCD) TEMNE (Port Loko) *a*-pende (FCD) TEMNE (Sanda) *a*-pende (FCD) TEMNE (Yoni) *a*-peni (FCD) VAI pende (FCD) pote (FCD) YALUNKA funde-na (FCD) **MALI** DOGON põ (CG) pon (FB) FULA-PULAAR (Mali) serémé (JMD) MANDING-BAMBARA fani (FB) feni (FB) findi (FB) fini (FB) foni(-o) (auctt.) fundé (FB) MANINKA faïné (FB) fanom (FB) foni(-o) (JMD) funi (Mauny) tau *a pap made of the flour* (FB) SONGHAI fingi (JMD; FB) fodio (FB) foyo (FB) **UPPER VOLTA** BOBO fan (Le Bris & Prost) fan kànpɛnɛ *an early c.var.* (Le Bris & Prost) fen (Prost) FULA-FULFULDE (Upper Volta) sereme (K&T) KURUMBA apendi (Prost) MANDING-MANINKA foni (FB) SAMO (Sembla) fö (Prost) SONINKE-PANA pwẽ (Prost) **IVORY COAST** DAN pohin (FB) pom (FB) KRU-GUERE (Wobe) pohim (FB) KWENI fĩnĭ (Gregoire) **GHANA** DAGBANI kabega (Gaisser fide JMD) HAUSA atcha (Patterson) KONKOMBA epich (Gaisser fide JMD) **TOGO** ANYI-ANUFO ṅfôni (Krass) NAWDM figm (Nicole) kafea (Gaisser fide JMD) pigim (Gaisser fide JMD) TEM (Tschaudjo) tschamma (Gaisser fide JMD) **DAHOMEY** BAATONUN podgi (FB) **NIGER** HAUSA fira (Robin) **NIGERIA** ARABIC-SHUWA difera (JMD) gashish *the grain* (JMD) kreb (JMD) BIROM cùn (LB) déré *a c.var.* (LB) pyẹ̈ng = *the white one; a c.var.* (LB) san *a c.var.* (LB) syinàng = *the red one; a c.var.* (LB) wétẹ̈ = *the dark one; a c.var.* (LB) BURA mili (anon.) FULA-FULFULDE (Nigeria) accari (JMD) sarembe (JMD) GWANDARA-CANCARA caba (Matsushita) GITATA ṅtíya (Matsushita) KARSHI ánea (Matsushita) KORO ṅtíya (Matsushita) NIMBIA ṅtíya (Matsushita) TONI òmbúru (Matsushita) GWARI fulubihi (JMD) HAUSA accà (Lowe) ácca (LB) áccàà (auctt.) akang *a c.var.* (Philcox) burma *a c.var.* (Philcox) chyung *a c.var.* (Philcox) cikarai (anon. fide KW) firo (JMD; ZOG) gumba *a c.var.* (Philcox) imeru *a c.var.* (Philcox) intaya (JMD; ZOG) ndat *a c.var.* (Philcox) salla *a c.var.* (Philcox) IGBO acha ?*loan-word from Hausa* (BNO) òsikapa acha = *accà rice* (BNO; KW) JEN num-mwi (RB) JUKUN (Wukari) cà (Shimizu) KAMBARI-SALKA áccà (RB) KANURI kàshá (A&S) kàshâ (C&H) KORO beenci *a late c.var.* (RB) beentsu *an early c.var.* (RB) LELA pọcho (JMD) MALA ira (RB) irya *general term for fonio* (RB) kọlimọ *a black-seeded form* (RB) NINZAM impuka (RB) SANGA ira (RB) SURA cikarai *a c.var.* (anon. fide KW) sùung (anon. fide KW) zor *a c.var.* (anon. fide KW) YORUBA sùúrù (JMD; Verger)

Products: **SENEGAL** BEDIK féyé *cooked grain* (FG&G) MANDING-MANINKA sanglé *a pap made of the flour* (FB) **GUINEA** MANDING-MANINKA tau *a pap made of the ground flour* (FB) **NIGERIA** BIROM bwrik *a sweet drink* (Hampson) siring *a beer* (Hampson) tuk *a porridge, or gari* (Hampson) HAUSA chehel *the dry flour* (Hampson) giya *a beer* (Hampson) kunu *a sweet drink* (Hampson) tuwo *a porridge, or gari* (Hampson)

An annual grass, culms delicate, to 75 cm high; in the Soudanian-Guinean zone across the Region from Senegal to Lake Chad.

The plant is known only in cultivation confined to rainfall isohyets (500–) 900–1000(–3000) mm p.a. Its grain is an extremely important cereal, staple of many tribes. Its origin is lost in antiquity. It closely resembles *D. longiflora* from which it may be derived (2) and which is locally cultivated on Fouta Djallon in Guinea (q.v.). This is a centre of ancient Manding culture, and philological studies of vernacular names of *D. exilis* indicate that its domestication has hinged essentially on Manding groups (3, 15). To the Dogon it is the source of origin of everything in the world. Ibn Batonta was the first traveller from outside to see the cereal which is recorded in his travels (fide 15).

Though the grain has a low protein content, recorded as 6–8% (3, 12), it is an extremely important cereal over a vast area of W Africa coming into harvest in the 'hungry' months of spring before other crops are available. There are many cultivars, precocious to late maturing, and in colours of grain, white, grey, brown or red (10). Under intensive cultivation three crops of precocious cultivars can be raised each year. In Sierra Leone ɛ *pinde pa funf* (Temne) is applied to short-season (2½–3½-month) cultivars (5, 9). A very precocious one is *yele-fui* (Susu) cropping in 40 days (6). The grain has an agreeable taste and is considered a delicacy. It is normally ground to a flour and eaten as sauce, porridge or pap, seasoned in various ways. It is easily digested and is fed to babies. (3, 4, 7, 8, 11, 13, 14.)

In Senegal the grain has medicinal use as a vehicle for active principles, particularly in treatment of acute intestinal wind and constipation: in such instances powdered root of *Strophanthus sarmentosus* (Apocynaceae) is cooked with it (12). It is also used as a vehicle in diuretic prescriptions (12). The Tenda in SE Senegal add it to sap of *Rauwolfia vomitoria* (Apocynaceae) in couscous for 'those who eat meat of the night' (?prophylactic) (8).

The grain is often fermented to produce a beer, especially in Togo by the Nawdm and Tem (3). In N Nigeria on the Jos Plateau, the earliest harvest is often brewed into beer which serves as payment for the people who help harvest the later ripening cultivars (10).

The grass itself is relished by stock for browsing (1). Straw left after threshing is a valuable fodder (4). It is also used chopped up and admixed with clay for making hut walls (4, 8).

References:

1. Adam, 1966, a. 2. Anon, s.d., e. 3. Busson, 1965: 455, with chemical analysis. 4. Dalziel, 1937: 526. 5. Deighton 1323, K. 6. Deighton 1324, K. 7. Diarra, 1977: 43. 8. Ferry & al., 1974: No. 113–5. 9. Glanville 398, 399, K. 10. Hampson, 1958. 11. Irvine, 1948: 263–4. 12. Kerharo & Adam, 1974: 645–6. with phyto-chemistry. 13. Meek, 1931, 2: 45. 14. Okiy, 1960: 120. 15. Portères, 1955.

Digitaria fuscescens (Presl) Henr.

FWTA, ed. 2, 3: 453.

An annual grass, creeping, stoloniferous, rooting at the nodes; culms ascending to 30 cm; of weedy waste places in Liberia to W Cameroons, and in Tanganyika and generally widespread in tropical Asia.

The plant has some semblance to making a mediocre lawn grass.

Digitaria gayana (Kunth) Stapf

FWTA, ed. 2, 3: 453. UPWTA, ed. 1, 526.

West African: SENEGAL MANDING-BAMBARA nkolu yagé (JB) SERER gèrgétèm (JB) THE GAMBIA MANDING-MANDINKA barto jarge messengo = *small riverside grass* (Pirie) MALI FULA-PULAAR (Mali) debbo daneya (A.Chev.) gague (A.Chev.) MANDING-BAMBARA mussa korui kungue (A.Chev.) ngassa ni kungué (A.Chev.) GHANA DAGAARI mpuru = *old woman* (FRI) MOORE povianga rido = *old woman's grey hairs* (FRI) NIGER HAUSA kalankafoa (Bartha) NIGERIA HAUSA furfurar ba-fillatani = *Fulani man's white hair* (JMD) furfurar gyaatumii = *old man's white hair* (JMD) furfurar tsoohuwaa *name refer to the white appearance* (JMD) gajele (Palmer; ZOG) gaji *incl. other grasses used for planting* (JMD) kálánhúwa (JMD) kaleru (JMD; ZOG) kamfalwa (JMD; ZOG) karan (ZOG) karani (JMD; ZOG)

A loosely tufted annual grass with culms up to 1 m high; inflorescence distinctively silky; on old farmland, fallows and disturbed areas, occurring throughout the Region from Mauritania to N and S Nigeria, and widespread in tropical Africa to Sudan, Zimbabwe and Angola.

It occurs spontaneously on impoverished soils. It is a weed of fallows, especially long-term fallows. Cattle graze it while still young, and horses take it while in grain (1–4).

The culms are used to plait rings and bracelets called *darambuwa*, *garambuwa* or *tafa* in Hausa, and *godogo* in Daura Hausa (4). They are also platted into cheap hats, baskets and a sort of grass quoit.

The fluffy seeds are collected at Sokoto for stuffing cushions (5).

References:

1. Adam, 1954: 91. 2. Adam, 1966,a. 3. Beal 48, K. 4. Dalziel, 1937: 526. 5. Palmer 7, K.

GRAMINEAE

Digitaria horizontalis Willd.

FWTA, ed. 2, 3: 453, in part. UPWTA, ed. 1, 525, under *D. debilis* Willd. and *spp.*

English: finger grass; wild 'findi' (The Gambia, Pirie).

West African: SENEGAL DIOLA ébusé (JB) SERER rukh (JLT) sivândan (JB) WOLOF kumbandiargândal (JLT; JB) niakh e pic (JLT) THE GAMBIA MANDING-MANDINKA ja-jeo (JMD) SIERRA LEONE KISSI yaya (?) (FCD) KONO minãsabine *applies to other small grasses* (FCD) pouvei (FCD) MENDE ndewe (NWT; JMD) pejui (Fisher) pouve (*def.* -i) (FCD) MENDE (Kpa) pouve (*def.* -i) (FCD) TEMNE a-gbel *common for small-grained grasses;* gbel: from *'fundi' bird or waxbill* (auctt.) MALI FULA-PULAAR (Mali) fonio ladde (A.Chev.) mussa ladde (A.Chev.) serémé ladde (A.Chev.) MANDING-BAMBARA narkata (A.Chev.) SONGHAI hana iddu (A.Chev. fide JMD) subcoré (A.Chev. fide JMD) UPPER VOLTA GRUSI banguéré (A.Chev.) gossolo (A.Chev.) GURMA taramanté (A.Chev.) MOORE pɛtrepin (Scholz) tetemtié-haga (A.Chev.) tetumté (A.Chev.) tintinterega (Scholz) NIGER FULA-FULFULDE (Niger) laalowol (ABM) NIGERIA FULA-FULFULDE (Nigeria) damaliliho (JMD) HAUSA harakiyaa (ZOG) hàrƙíyáá (JMD; ZOG) ibúrún dáájìi (JMD; ZOG) karanin dawaki (JMD; ZOG) IGBO (Agukwu) ilulǫ egugo ẹzu (NWT; JMD) TIV yayaghol (JMD) YORUBA eéran (auctt.) YORUBA (Ilorin) eran tapa (Ward)

A creeping or ascending annual grass, culms to 60 cm high; of waste places across the whole of the W African Region, in the islands of the Indian Ocean and in tropical America.

The grass is a weed of coffee plantations in Ivory Coast (5). It is a good pasture grass relished by all cattle, goats, sheep and horses (Senegal: 1; Ghana: 3; Niger: 6; Nigeria: 4, 7). Graziers of Niger claim that it does not withstand the lack of humidity and dries up rather quickly (6).

In Paraguay the whole plant is used in decoction as an abortifacient (2).

References:

1. Adam, 1966, a. 2. Arenas & Azororo, 1977: 298–301. 3. Beal 11, K. 4. Dawodu 16, K. 5. Leeuwenberg 1738, K. 6. Maliki, 1981: 52, sp. no. 51. 7. Ward 0045, K.

Digitaria iburua Stapf

FWTA, ed. 2, 3: 452. UPWTA, ed. 1, 526.

English: black acha.

French: fonio noire; manne noire; ibourou (from Hausa) (Busson).

West African: TOGO ? tchapalo *a fermented drink* (FB) LAMBA afio-uarun (FB) NIGERIA BIROM but a c.var. (LB) cun cérèŋ = *little fonio* (LB) cun yéy = *beautiful fonio cf. D. exilis* (LB) ŋás = *ant-trail: a c.var.* (LB) síŋ a c.var. (LB) wété swit = *dark black: a c.var.* (LB) HAUSA àbúróò (JMD; ZOG) aburu (JMD; ZOG) accà (JMD) alas (JMD; ZOG) dere *a c.var.* (Philcox) ibíròò (JMD; ZOG) ibúróò (JMD; ZOG) iburuu (auctt.) makari (Bargery) ndat *a c.var.* (Philcox) san *a c.var.* (Philcox) wusuwusu *a food preparation* (JMD) JEN nųnghwe (RB) MALA ibulu (RB) NINZAM impwịnci (RB) SANGA utangǫ (RB)

A loosely tufted annual grass, culms to about 1.4 m high, very variable, of wild and cultivar forms; occurring only in limited but disparate localities in Ivory Coast to N Nigeria between 400 to 1000 m altitude with rainfall 900–1000 mm p.a. (1, 10).

The grain is edible, but as the grain integuments are not readily detached in threshing, it is awkward to use and is consumed imperfectly cleaned. Nevertheless it is an important supplementary food to people in the Atakora uplands of Togo and Dahomey, and more especially to the Birom tribe of the Jos Plateau in N Nigeria for whom it serves as a staple (4–6, 8, 10). The plant is very variable. It is a cultigen, but of uncertain provenance. *FTEA* suggests an origin from *D. ternata* (A Rich.) Stapf (3), a widely dispersed species of eastern and southern Africa and tropical Asia. Alternatively it may be derived from *D. barbinodis* Henr. (2), a sub-Saharan species with which it is found growing admixed in a wild state in N Nigeria (9). Cultivars of *D. iburea* have black grain,

228

hence the allusion to blackness in the English and French names. This colour marks them apart from the cultivars of *D. exilis* (q.v.), or white acha, whose grain is white, grey, brown or red (4). They grow taller, to 1.4 m high, whilst *D. exilis* cultivars attain 60–90 cm in height only. In N Nigeria they are generally the last of the acha cultivars to be planted (towards the end of June), and then on the more fertile soil, though it is said they do well on poor and exhausted land (5). Harvesting is in November/December. Yield is usually lower than from cultivars harvested earlier due to adverse effect of the harmattan. Plants may sometimes be grown intermixed with millet (7).

The grain is eaten as porridge or mixed with the meal of other cereals (6). A common confection in Nigeria and Dahomey is couscous, while in Togo it is fermented to a drink called *tchapalo* (1). Chemical analysis of the unmilled grain shows good dietetic value with carbohydrates 77.5% and protein 9.9% (1).

References:

1. Busson, 1965: 459, with chemical analysis of grain, 465–6. 2. Chevalier, 1950: 329–30. 3. Clayton & Renvoize, 1982: 631. 4. Hampson, 1958. 5. Hepper 2867, K. 6. Irvine, 1948: 263–4. 7. Lamb 54, K. 8. Meek, 1931: 2: 45. 9. Philcox 1005, 1013, K. 10. Portères, 1946: 589–92.

Digitaria leptorhachis (Pilger) Stapf

FWTA, ed. 2, 3: 454.
West African: **IVORY COAST** MANDING-MANINKA koronémi (B&D) **NIGERIA** FULA-FULFULDE (Nigeria) geroreje (Saunders) ilanwode (Saunders)

An annual, or short-lived perennial, grass, with decumbent culms, wiry, ascending to about 60 cm; of farmland, roadsides and damp waste places across the Region from Senegal to N and S Nigeria, and extending to the Congo basin and Tanganyika.

The grass provides good forage whilst still young (1, 2, 5). It is a weed of irrigation in Sudan (3). A decoction of the plant is made in Ivory Coast to use as a bath for infants to make them strong (4).

In the Plateau State, Nigeria, Sura collect the seed for food in time of dearth (5).

References:

1. Adam, 1954: 91. 2. Adam, 1966,a. 3. Blair 10, K. 4. Bouquet & Debray, 1974: 92. 5. Saunders 23, K.

Digitaria longiflora (Retz.) Pers.

FWTA, ed. 2, 3: 453. UPWTA, ed. 1, 525, under *D. debilis* Willd. and *spp.*
French: fonio (from Manding); fonio sauvage (i.e., wild fonio).
West African: **SENEGAL** BASARI ε-greb (after K&A) BEDIK *nya*-lèndy FG&G) MANDING-MANDINKA fini, fono (GR) **THE GAMBIA** MANDING-MANDINKA findo (Pirie) projajawo (Fox) **GUINEA-BISSAU** BALANTA cúrè (JDES) iéte (JDES) ura (JDES) BIAFADA buáède (JDES) CRIOULO fundo (GeS; RdoF) fundo bravo (JDES) FULA-PULAAR (Guinea-Bissau) djadje, djadje-maudo (JDES) fónio-tchóli (JDES) MANDING-MANDINKA djádjeô (JDES) MANDYAK guarcam (JDES) obife (JDES) pebife (JDES) MANKANYA upadja (JDES) PEPEL imbilô (JDES) oife (JDES) **GUINEA** BASARI ε-gwòròɓ (FG&G) KONYAGI funyeríti (FG&G) **SIERRA LEONE** KONO fuinε (FCD) MENDE *n*diwi (Fisher; FCD) *n*diwo (FCD) nyina-voni (FCD) puvε *applied generally to small digitarias and other grasses* (FCD) MENDE (Kpa) *n*goka-gbu (?) (FCD) **MALI** DOGON sáána vòònu ána (CG) **UPPER VOLTA** GURMA mòbi (Surugue) **TOGO** KONKOMBA epik (Froelich) impwi (*pl.* ipwi) (Froelich) **NIGERIA** HAUSA harkiya RES harkiyan zomo = '*digitaria' of the rabbit* (Taylor) YORUBA (Ilorin) eran (Ward)

An annual, or short-lived perennial grass, prostrate or upright to 30 cm high; of roadsides, fallows and waste places; common throughout the Region from Senegal to N and S Nigeria, and generally in the Old World tropics, and introduced into the New World.

The grass is everywhere a weed of cultivation. In The Gambia it is a common grass of second and third year fallows (10). Prostrate strains can be cultivated as lawn grass (8, 9, 11), but there are better lawn grasses which it seems capable of suppressing, especially on poor soil (4). In Sierra Leone it is often the first grass to flower after the rains and keeps alive after other grasses have died off in the dry season (4). It provides grazing for stock, but it is said to be bitter in Senegal though classified amongst the good species if grown on fertile soil (Senegal: 1, 2; Mali: 5; Nigeria: 12). Analyses of Indonesian (3) and S African materials (8) indicate good nutritive values.

The grass is reported grown by Mandinka throughout The Gambia (6), but the purpose is not stated. On Fouta Djallon in Guinea the grass is more or less cultivated in extensive fields under the name *fonio* (7) and presumably, therefore, for its grain. The location is said to be sloping dry land, thereby suffering erosion which should be better countered.

There is a form in Indonesia with long stems, and these are used for plaiting (3).

References:

1. Adam, 1954: 91. 2. Adam, 1966, a. 3. Burkill, IH, 1935: 808. 4. Deighton 4016, K. 5. Lecard 252, K. 6. Pirie s.n., 1929, K. 7. Roberty, 1955: 63. 8. Sastri, 1952: 62–64. 9. Small 20, K. 10. Terry 3095, K. 11. Thorold 119, K. 12. Ward 0046, K.

Digitaria milanjiana (Rendle) Stapf

A loosely tufted rhizomatous perennial grass, culms erect or geniculately ascending to $2\frac{1}{2}$ m high; of ubiquitous habitats throughout the eastern and southern part of Africa, and now reported present under cultivation (presumably introduced) in Ivory Coast.

The grass is valued as a superior grazing species taken by all stock. In Tanganyika it is considered amongst the best (1). It has a vigorous growth habit and rapidly colonises open spaces (4), and becomes a weed of cultivation and irrigated land (2, 5). Its presence in Kenya is said to indicate soil good for tapioca (3).

References:

1. Emson 280, K. 2. Makin 298, K. 3. Moggridge 178, K. 4. Rawlins 415, K. 5. Shepphard 282214, K.

Digitaria nuda Schumach.

FWTA, ed. 2, 3: 453, as *D. horizontalis*, in part; and as *D. ciliaris*, in part.
West African: THE GAMBIA MANDING-MANDINKA findi bano = *wild findi* (Pirie) SIERRA LEONE MENDE ndiwi (FCD) GHANA HAUSA harkiya (Ll.W.) NIGERIA HAUSA harakiya (Palmer) KANURI sheshe (anon.)

An annual grass, decumbent, geniculately ascending culms to 1 m high; of waste places, farmland and disturbed sites; common across the Region from Mauritania to Nigeria, and widely dispersed throughout the rest of tropical Africa, and in Mauritius, Brazil and Indonesia.

This species has been confused with *D. ciliaris* and *D. horizontalis* and is now recognised as a separate entity in the Region. With these and others it shares more or less the same attributes and vernacular names. It is a good stock grass and grows into a good pasture. It converts into hay. It is a weed of lawns and is commonly in arable land (1, 2, 5). In Zimbabwe it enters irrigated fields (4).

Along with other digitarias, it is known in The Gambia as *findi bano*: wild findi, and the grain serves as a famine-food (3).

References:

1. Austin 17, K. 2. Parker 1970, K. 3. Pirie 8/33. 4. Shepphard 281865, 282213, K. 5. Terry 1800, 3085, K.

Digitaria patagiata Henr.

FWTA, ed. 2, 3: 454.

A creeping grass of moist soils, recorded only in Senegal. It is reported as a weed of rice-padis (1, 2).

References:

1. Adam 18259, K. 2. Vanden Berghen 3176, K.

Digitaria perrottetii (Kunth) Stapf

FWTA, ed. 2, 3: 452.
West African: SENEGAL SERER siāngan (JB) WOLOF bakèt (JB)

A coarse annual grass with culms erect to 1.3 m high; of bushland and particularly disturbed sites and old farmland; recorded only in Senegal in W Africa, and occurring widely over south tropical Africa and Madagascar.

It is grazed by cattle (2) and sheep (1) in Senegal. Stock is said to take it in Malawi whilst still in young growth (3, 4).

The roots are rubbed on the gums of teething children in Malawi (3).

References:

1. Adam, 1954: 91. 2. Adam, 1966,a. 3. Fenner 145, K. 4. Fenner 317, K.

Digitaria ternata (A Rich.) Stapf

FWTA, ed. 2, 3: 452.
West African: WEST CAMEROONS BAMILEKE fafabo debbo (DA)

A loosely tufted annual grass with culms to 60 cm high; of old cultivations and waste places in the Region from Mali to W Cameroons, and widespread over eastern and southern Africa and the Asian tropics.

In all countries it becomes a weed of cultivation, both arable and in irrigation. It is a good forage grass for all domestic stock.

Digitaria velutina (Forssk.) P Beauv.

An annual grass, culms decumbent, rambling, or geniculately ascending to 80 cm high; of pathsides, farmland and waste places; a species widely dispersed over NE, E, central and S Africa, and in Arabia, and now recorded in N Nigeria.

No usage is recorded in the Region, but in all other territories it gives valuable grazing for all stock in the lowlands to montane altitudes. It is also a weed of arable land and under plantation crops.

231

GRAMINEAE

Diheteropogon amplectens (Nees) WD Clayton

FWTA, ed. 2, 3: 489–90. UPWTA, ed. 1, 520.

West African: SENEGAL FULA-PULAAR (Senegal) garabali (GR) MANDING-BAMBARA bu garabali (JB) MANDINKA garabali (GR) nyânga *a group term for ruderal grasses, esp. on impoverished soil* (GR) wâ (GR) SERER ndâng (JB) WOLOF cicca (JLT) sèl (JB) **MALI** DOGON kéru gŏy (CG) **UPPER VOLTA** FULA-FULFULDE (Upper Volta) garraabal (K&T) garraabe (K&T)

A perennial grass; short rhizome bearing culms 1–2 m high; in poor sandy, stony, compacted and degraded soils; represented by two varieties in Africa of which var. *catangensis* (Chiov.) WD Clayton is found across the Region from Senegal to N Nigeria, and on through central to E and S Africa. The other variety, var. *amplectens*, is of the same distribution excepting W Africa.

When taken at a young stage of growth it is a good forage appreciated by all stock (1–3, 5). It converts into hay (2). Variety *amplectens* in E Africa has longer and pungent awns. Thus cattle less willingly browse it (4, 6).

The mature culms are used in Senegal for thatching (1, 3).

References:

1. Adam, 1954: 87. 2. Adam, 1966, a. 3. Dalziel, 1937: 520, as *D. amplectens* Nees, var. *diversifolius* Stapf. 4. Emson 288, K. 5. MacGregor 42, 43, K. 6. Mwangangi 905, K.

Diheteropogon hagerupii Hitchc.

FWTA, ed. 2, 3: 489. UPWTA, ed. 1, 526.

West African: **UPPER VOLTA** FULA-FULFULDE (Upper Volta) garraabal (K&T) garraabe (K&T) **NIGER** HAUSA kära (Bartha) **NIGERIA** FULA-FULFULDE (Nigeria) dakwumbei (Bargery) galla bar (ZOG) HAUSA galla ɓari (Bargery fide JMD) kàraìràyaú (auctt.) kàrànkàɓaú (JMD; ZOG) shabrai (JMD; ZOG) shamrai (JMD; ZOG) toƙari (auctt.)

NOTE: Nigerian names apply also to some other marsh plants.

An annual grass with culms 1–1½ m high; of dry sandy or gravelly sites; in the soudanian zone across the Region from Senegal to N Nigeria.

No usage is recorded but since it bears the same Hausa name as does *Echinochloa pyramidalis*, it is perhaps recognised as having some value as fodder and other attributes of the latter (q.v.).

Dinebra retroflexa (Vahl) Panz.

FWTA, ed. 2, 3: 393, 395. UPWTA, ed. 1, 526.

West African: **NIGERIA** ARABIC-SHUWA am kachena (A.Chev.) HAUSA firki (Oche) kikiko-masakiya (Golding)

A loosely-tufted annual grass with culms to 1 m high, on wet and humid, or dry, locations in Senegal and N Nigeria, and extending across tropical Africa to S Africa, and eastwards through Egypt and Iraq to India.

The grass is a common weed of cultivated land in all territories. It enters irrigated crops in Ethiopia (cotton, 7) and Somalia (sugar, 11).

It is an excellent fodder and pasture grass readily eaten by all stock (Senegal, 1, 2; Nigeria, 4, 5; Kordofan, 3; Sudan, 8; Kenya, 6, 9; India, 10; etc.). In India the grass is considered especially good for buffaloes, and feeding to stock increases milk-yield, but it is considered unsuitable for making into hay or for ensilage (10).

References:

1. Adam, 1954: 92. 2. Adam, 1966,a. 3. Baumer, 1975: 96. 4. Dalziel, 1937: 526. 5. Golding 40, K. 6. McDonald 1064, K. 7. Parker E.408, E.480, K. 8. Pritchard 25, K. 9. Rucina A/4, K. 10. Sastri, 1952: 67, with chemical analysis. 11. Terry 3343, K.

Echinochloa callopus (Pilger) WD Clayton

FWTA, ed. 2, 3: 443, as *Brachiaria callopus* (Pilger) Stapf.

A tufted annual grass with spongy culms, to 1 m high, of swampy places and shallow pools in the Sahel zone across the Region from Senegal to N Nigeria, and extending to Sudan and Tanganyika.

The grass is greatly relished by cattle in Senegal (1).

Reference:

1. Adam, 1966,a, as *Brachiaria stipitata*.

Echinochloa colona (Linn.) Link

FWTA, ed. 2, 3: 439. UPWTA, ed. 1, 526.

English: jungle rice; sawa millet (India).

West African: SENEGAL FULA-PULAAR (Senegal) burugué (A.Chev.) hu'do belle = *? fat grass* (A.Chev.) hu'do wendu (A.Chev.) yakabré (A.Chev.) yakauré (A.Chev.) TUKULOR baro (A.Chev.) MANDING-BAMBARA bin drima (JB) nteguéré ké (A.Chev.) SERER gadri (A.Chev.; FB) mbakit (JB) mbris (JB) ndiadié (JB) WOLOF *m*baket (auctt.) THE GAMBIA MANDING-MANDINKA manisena (Macluskie) myrameseu (Macluskie) SIERRA LEONE TEMNE suribani (Glanville) MALI ARABIC (Mali) aseral *the grain* (A.Chev.) tadjabar (A.Chev.) taggabar (A.Chev.) FULA-PULAAR (Mali) burugue (A.Chev.) hu'do belle = *? fat grass* (A.Chev.) hu'do wendu (A.Chev.) yakabré (A.Chev.) yakauré (A.Chev.) MANDING-BAMBARA nteguéré ké (A.Chev.; FB) TAMACHEK asseray (A.Chev.) ichiban *the grain* (A.Chev.) UPPER VOLTA FULA-FULFULDE (Upper Volta) pagga pucci (K&T) MOORE kulia mossum = *salt of the marsh* (A.Chev.; FB) NIGER FULA-FULFULDE (Niger) nyereeje (ABM) nyeryaare (ABM) HAUSA guirza (Bartha) SONGHAI báŋg-súbù (*pl.* -ò) = *grass of the pools* (D&C) NIGERIA ARABIC-SHUWA bu-rɛkuba (JMD) difera (JMD) HAUSA baya (Golding) garaji (Golding) sabe (Palmer; ZOG) saben ruwa (Taylor)

An annual grass, culms erect or ascending 60–120 cm high in damp places, in or near water; common throughout the Region from Mauritania to N and S Nigeria, widespread over tropical Africa, and throughout the tropics and subtropics.

The plant is sometimes very abundant, forming extensive prairies on flood-plains. It is an aquatic pasturage on which Niger cattle-men graze their cattle (17). It is particularly common in the Senegal River basin (1, 2) where the grass grows to 1¼ m high and is harvested, tied into bundles and marketed to stock-keepers. The cash return in 1965 was calculated to be twice that obtained from ground-nuts (2). The grass is relished by cattle in all territories, especially in the wet season (7, 25; Senegal: 3; Nigeria: 10; Kordofan: 4; Sudan: 8, 12, 14; Somalia: 15; Kenya: 19; India: 11, 24; Malaya: 5; etc.). It has a good nutritive value which is enhanced when in grain (1, 7). Converted into hay, it is recorded to have 7.1% protein (1). In India hay can be stacked and kept for 5–6 years and when fed to cattle they fatten and come into good condition in a short time (24). In Senegal mineral salts are obtained from the ash of the calcined plant and are used in cooking (1). The tender young shoots are eaten by man in Indonesia (5).

The grass is commonly found in irrigation ditches whence it invades irrigated crops, becoming a weed (Ethiopia: 20; Eritraea: 21, 23). In India it is deemed a noxious pest of rice (11, 26). It is reported as a weed in padi-fields in The Gambia (16), and as serious in the Little Scarcies River area of Sierra Leone (9).

The grain is edible. It is regularly collected in the Niger and Nile river valleys (1, 22). In olden times it was grown in ancient Egypt (6, 7): seed has been determined in the stomachs of predynastic mummies (13). In India it is a poor man's cereal (11), and it is sometimes purposely planted with *Eleusine coracana* or with *Setaria italica*. Indian farmers recognise several races. Some cultivars mature in under two months, even on poor soils (26). It will tolerate soil with up to 2% salt in it (4). This holds promise for further selection and ennoblement for the semi-arid tropics. In all territories it is a famine-food for collection in time of dearth (6, 7, 22, 25; Nigeria: 10; Tanganyika: 18; Kordofan: 13; India: 24). Birds in Kenya seek out the seed (19).

In the Central African Republic the grass is grown for the grain to produce beer (26).

References:

1. Adam, 1954: 92. 2. Adam, 1965: 131. 3. Adam, 1966, a. 4. Baumer, 1975: 96. 5. Burkill, IH, 1935: 888–9. 6. Busson, 1965: 461. 7. Dalziel, 1937: 526. 8. Davies BN.3, K. 9. Glanville 232, K. 10. Golding 9, K. 11. Gupta & Sharma, 1971: 75. 12. Harrison 995, 1051, K. 13. Hunting Tech. Services, 1964. 14. Khalid 19, K. 15. McKinnon S.22, K. 16. Macluskie 12, K. 17. Maliki, 1981: 52. 18. Michelmore M.1587, K. 19. Mwangangi 1365, K. 20. Parker E.221, E.330, K. 21. Popov 1388, K. 22. Raynal-Roques, 1978: 334. 23. Ryding 1340, K. 24. Sastri, 1952: 124–5. 25. Schnell, 1960: no. 105. 26. Wet & al., 1983: 283–91.

Echinochloa crus-galli (Linn.) P Beauv.

English: barnyard millet.

A coarse, tufted, very polymorphic annual grass, culms more or less erect to 80 cm high, of moist or sandy places, native of warm temperate regions and in many parts of the Asian tropics; occuring scattered in NE and E Africa, probably introduced, and now recorded in W Africa in Senegal, Guinea and Ivory Coast (1), and Togo; also in Cape Verde Islands.

In Asia it appears sporadically near swamps and rice-padis (4, 6). It has invaded rice-fields in California where it obstructs mechanical cultivation (4). Effective control is said to be by submerging the fields under 10–15 cm of water and keeping them thus till harvest; also by encouraging germination before the rice is sown, and then ploughing up the weed (7). There has been successful use of it in Egypt in reclaiming brackish land (3, 4, 6), and as a green manure. It is considered a good forage for cattle (4, 5), but not for horses if taken in quantity (4). It is relished by all stock in Senegal (2), but with a very short growth cycle it has a limited value as forage in W Africa (1).

People in Java eat the young shoots as a vegetable (4).

The plant has medicinal use in India for diseases of the spleen and as a haemostatic (6).

The grain is collected as a famine-food in countries subject to recurrent food shortage (4).

References:

1. Adam, 1954: 92. 2. Adam, 1966, a. 3. Anon., s.n., g. 4. Burkill, IH, 1935: 889. 5. Gupta & Sharma, 1971: 75. 6. Sastri, 1952: 125. 7. Wild, 1961: 27.

Echinochloa crus-pavonis (Kunth) Schult.

FWTA, ed. 2, 3: 440.

Portuguese: capim-do-bengo (Feijão).

West African: SENEGAL SERER tiimbéla (JB) GUINEA-BISSAU BALANTA quéo (EPdS; JDES) MANDING-MANDINKA nhamô (EPdS; JDES) SIERRA LEONE SUSU yofoni (NWT) TEMNE ka-fon (FCD)

A robust perennial grass, erect to 1–2 m high; of swamps, stream-sides and shallow water; in the Region from Guinea to W Cameroons, and widespread over the rest of tropical Africa and into S Africa; also in tropical America.

The plant is spreading in Sierra Leone to become a troublesome weed in rice-padis, bordering rivers and creeks (1). It is reported as a weed of rice-fields at Kumasi, Ghana (3). It furnishes grazing for stock in Uganda (4) and Kenya (3).

References:

1. Glanville 228, K. 2. Glover & al. 2625, K. 3. Howes 1000, K. 4. Snowden 1215, 1602, K.

Echinochloa obtusiflora Stapf

FWTA, ed. 2, 3: 439, incl. 3: 444, *Brachiaria obtusiflora* (Hochst.) Stapf.

English: kuskus grass (correctly this refers to *Digitaria exilis* Stapf, but *Echinochloa obtusiflora* is of similar habit).

West African: NIGERIA ARABIC-SHUWA difra (Gwynn)

An annual grass with erect culms to 1 m high, of wet places and shallow pools in the Sahel of Niger and Nigeria; also in E Cameroun and Sudan.

It is a weed of wet rice. The use of *kus-kus* as a name in N Nigeria suggests the probability that the grain is eaten, though this is not recorded. Eastwards in Kordofan, the grain is eaten in time of dearth (1).

Reference:

1. Hunting Tech. Services, 1964: Kordofan.

Echinochloa pyramidalis (Lam.) Hitchc. & Chase

FWTA, ed. 2, 3: 439. UPWTA, ed. 1, 527.

English: antelope grass (E Africa, Dalziel).

West African: SENEGAL DIOLA égil (JB; K&A) FULA-PULAAR (Senegal) hu'du wendu (A.Chev.) TUKULOR didéré (K&A) MANDING-BAMBARA burguké (JB; K&A) lingui *the grain* (A.Chev.) SONINKE-SARAKOLE kimbaka tioké (A.Chev.; K&A) WOLOF ay (K&A) samamgate (A.Chev.) samangate (K&A) sil (A.Chev.; K&A) télâgor (JB; K&A) THE GAMBIA MANDING-MANDINKA jarge ba = *big grass* (Pirie) nyara ba (Macluskie) nyaro ba (Macluskie) MALI ? (Mopti) kussein (Matthes) DOGON pónyô (CG) FULA-PULAAR (Mali) hu'do wendu (A.Chev.) MANDING-MANINKA lingui *the grain* (A.Chev.) SONGHAI farka teli (A.Chev.) fingui (A.Chev.) SONINKE-SARAKOLE kimbaka tioké (A.Chev.) UPPER VOLTA MOORE baukarara = *bush guinea corn* (FRI) kulumodo (A.Chev.) NIGERIA ARABIC-SHUWA um suf (JMD) FULA-FULDE (Nigeria) taagol (J&D) HAUSA gallaɓari (Bargery) gayanshi (Palmer; JMD) geron tsuntsu (DA) gundam cf. Arabic gandum: *wheat* (Lamb) gyaushe (ZOG) iwa (Taylor; Lamb) karairayau (Bargery; JMD) kàrànkàɓáu (JMD; ZOG) kasha *a general term applied to several grass spp. whose grain is collected for food* (JMD) roòbaá *incl. several tall marsh grasses* (JMD; ZOG) shabrai (JMD; ZOG) shamrai (JMD; ZOG) toɍari (Bargery; fide JMD) IJO-IZON (Kolokuma) u̩kású̩ (KW; KW&T) IZON (Oporoma) u̩kású̩ (KW) KANURI kreb *a general term applied to several grass spp. whose grain is collected for food* (JMD) zaza (Johnston)

A reed-like rhizomatous perennial grass, culms robust, erect to $3\frac{1}{2}$ m high, or creeping or floating to $4\frac{1}{2}$ m long; of river-banks, marshes, open water and riverine meadows; common across the whole Region from Senegal to W Cameroons, and widespread in the rest of tropical Africa and into S Africa and Madagascar.

The grass forms great prairies in the flood plains of rivers, especially along the Senegal and Niger Rivers, and in the inundation area of Lake Chad. It thrives in permanent water and withstands seasonal flooding. It is a major constituent of *sudd* in the Niger and Nile Rivers, and is invading some parts of the Volta Lake (11). It is a weed of cultivation. Fula of N Nigeria say that it is a particularly bad weed of sorghum fields, but consider its presence is indicative of soil fertility (28). It is an invader of irrigation (rice-fields – The Gambia: 18, 26; Nigeria: 12; sugar – Nigeria: 14, 15). By its robust habit and deep-penetrating roots it creates a good protection on river-banks against scour and erosion (27). It readily tillers rooting at the nodes even high up the culms, thus propagation from such offshoots is readily established (22).

It is an excellent fodder for all stock (7; Senegal: 1–3, 17; Upper Volta: 13; Ghana: 11; W Cameroons: 20; Kordofan: 6; Sudan: 8; Kenya: 10, 19). The stems tend to be juicy, which makes them the more palatable whilst still green. It makes a good hay, for which in Senegal it is cut twice annually (1). It is reputed to be a particularly good fodder for horses (1, 13, 20), but Hausa believe it makes horses fat, not strong (7). If allowed to grow without cutting, the culms may well become too tall for browsing by any stock.

The culms are used in Kordofan to make enclosures (6), and there (6), as in E Africa (10), they are used as thatch. In places the haulm is incinerated to produce a vegetable salt consisting mainly of sodium carbonate (Senegal: 1; Nupe, Nigeria: 4, 5, 7). This is used as a substitute for normal salt.

The root is said to have a pleasant smell and a decoction is drunk in the Central African Republic as a deodorant (9). Nueyr in Sudan chew the root for cough (24). *Flavones* and *glycosides* have been detected in Zaïrean material (25).

The grass is one of a group occurring in the Sahel, the grain of which is edible and is collected under a general term 'kreb' for human consumption on a regular basis in the Niger and Nile valleys (21), and more especially in time of shortage (1, 4, 7, 12, 16), cf. *Panicum turgidum*, Gramineae. The grain is normally ground to flour. In Sudan the seed is reported eaten as a sort of condiment 'like sugar' (23).

References:

1. Adam, 1954: 92. 2. Adam, 1960, a: 369. 3. Adam, 1966, a: Bull. incl. *E. senegalensis.* 4. Anon., s.d., g. 5. Barter 1156, K. 6. Baumer, 1975: 97. 7. Dalziel, 1937: 527. 8. Evans Pritchard 21, 39, K. 9. Fay 4422, K. 10. Glover & al. 151, 545, K. 11. Hall & al., 1971. 12. Hunting Tech. Services, 1964. 13. Irvine 4677, K. 14. Ivens N.186 (UIH 14747), UCI. 15. Kucera 30, K. 16. Lamb 102, K. 17. Lecard 125, K. 18. Macluskie 10, K. 19. McDonald 1020, K. 20. Maitland 11, K. 21. Raynal-Roques, 1978: 334. 22. Riyom 1, K. 23. Simpson 877, K. 24. Simpson 7104, K. 25. Symoens 14076, K. 26. Terry 1895, K. 27. Ward 0010, K. 28. Jackson, 1973.

Echinochloa rotundiflora WD Clayton

An annual grass, with culms erect or geniculately ascending to 1 m high, of savanna land and swamp in Sudan and Ethiopia, and now recorded in N Nigeria.

In Sudan it provides excellent grazing in the wet season and is considered a very important fodder for all stock (1–4). It is, however, recorded as being slow to regenerate after burning (4).

In Ethiopia the seed is saved for cattle; it is also taken by some birds (5).

References:

1. Andrews 3284, K. 2. Beshir 52, K. 3. Davies BN.15, K. 4. Evans Pritchard 20, K. 5. HADP Soil Survey, s.n., K.

Echinochloa stagnina (Retz.) P Beauv.

FWTA, ed. 2, 3: 439. UPWTA, ed. 1, 527.

French: bourgou (commonly used, from Mali vernaculars); roseau sucré (= *sweet reed*, Vieillard); roseau à miel du Niger (= *honey reed of the Niger*, Dalziel).

West African: SENEGAL MANDING-BAMBARA burgu (JB; K&A) MANDINKA burgu (K&A) GUINEA-BISSAU BALANTA quéo (JDES) MALI ? aluala *a sort of caramel prepared from the sweet sap* (FB) katu *the brown sugar prepared in cubes* (FB) ? (Mopti) burgu (Matthes) gamawa (Lean) kundu-hari *the sweet sap expressed from the submerged stems* (A.Chev.) DOGON nɔmɔ ára (CG) FULA-PULAAR (Mali) borgu (Davey) burdi kamarege (A.Chev.) gamaraho (A.Chev.; Vieillard) gambarawo (A.Chev.; Vieillard) MANDING-BAMBARA burgu (JMD; Vieillard) burgu-lé

(A.Chev.) burgu-ni (A.Chev.) kundu *the succulent submerged stems* (JMD) SONGHAI kundu (A.Chev.) SONINKE-SARAKOLE pérépéré ntioké (A.Chev.) TAMACHEK ekaywod (Duveyrier fide JMD) **UPPER VOLTA** FULA-FULFULDE (Upper Volta) burgu (K&T) GURMA moignima (A.Chev. fide JMD) **NIGER** SONGHAI búrgú (*pl.* -ó) (D&C) **NIGERIA** ARABIC-SHUWA alwa (JMD) helew (JMD) FULA-FULFULDE (Nigeria) burgu (J&D) siseri (JMD) wuruguho (JMD) HAUSA bonekouan (ZOG) burugu (JMD; ZOG) sakera (ZOG) JUKUN (Wukari) swàndzwìn (Shimizu) KANURI alwa (JMD; C&H)

A robust rhizomatous perennial grass, rhizomes stout and often floating, bearing culms to 2 m high (occasionally annual and weaker); of swamps and standing water, often forming floating mats, dispersed across the Region from Mauritania to N and S Nigeria, and generally widespread elsewhere in tropical Africa, and in tropical regions of Asia.

The grass is locally abundant, colonising marshes, and open standing water to form nearly pure stands. It invades rivers causing obstruction of the waterway (Sokoto, Nigeria: 5; Cameroun: 14). It is a major component of *sudd* in several African rivers (6). Its presence is indicative of fresh-water, rather than brackish (13). It is a weed of rice-fields in Madagascar (14) and irrigation channels in The Gambia (12).

The grass produces rich fodder relished by all stock, particularly while still in young growth (5; Senegal: 1; Mali: A Chev. fide 2; Nigeria: 5; Sudan: 8; Zimbabwe: 14; India: 11). Game take it, and the hippopotamus in Zimbabwe browses it (14). The culms are sweet with a sugary sap, said to contain 10% *sucrose* and 7–8% *reducing sugars* (3, 11). Children suck them to obtain the juice. The sap is used to make a drink in Mali which is concentrated by boiling to produce a brown sugar sold in Malian shops, and more especially in Timbuktu to make pastries and a sort of caramel (4) – see word-list above for vernacular names. The submerged rhizomes are particularly rich in sugar and the sweet sap is extraced to form an unfermented drink, but fermentation rapidly sets in resulting in a cider-like beverage, or if distilled to yield alcohol (6, 9). Confectionary and a fermented drink are made from the plant in central Africa (10).

The seed is edible and is one of the 'kreb' group collected for human use, and more especially in time of shortage (6, 10). The grass, in fact, closely resembles wild rice – see *Oryza* – and shattering too easily the panicles are reaped early and dried on a suitable winnowing platform.

The plant is claimed to be the most useful of all wild plants in the Timbuktu area of Mali, providing as well as food and drink, fodder, thatch, caulking for boats, vegetable salt after calcining which is used to make soap and indigo dye (A Chev. fide 2).

The Niger River flood-plain is a recognised locust nursery area, the hoppers growing on the bare soil between tufts of grass exposed after the flood-waters have retreated. Very large areas, especially south of Lake Debo, are covered with pure stands of this species of grass which when the waters have receded is laid and provides excellent pasturage. Trampling, grazing and later firing to promote regrowth leave the soil bare on which locust eggs are laid. Other areas of lesser inundation have admixtures of other grasses which tend to be tufted and thus anyhow have bare ground between the tufts, 20–50% of the ground being bare. *Oryza barthii*, a wild rice, is one that harbours such hoppers (7). If all the land could be clothed with *E. stagnina* and grazing and firing restricted so as not to expose the soil locust egg-laying might be inhibited.

References:

1. Adam, 1966, a, incl. *E. lelievrei.* 2. Anon s.d., B. 3. Baumer, 1975: 97. 4. Busson, 1965: 461. 5. Dalziel 479, K. 6. Dalziel, 1937: 527. 7. Davey & al., 1959: 65. 8. Harrison 910, K. 9. Irvine, 1952, a: 31. 10. Raynal-Roques, 1978: 334. 11. Sastri, 1952: 126. 12. Terry 3257, K. 13. Trochain, 1940: 394–5. 14. Wild, 1961: 25–27. See also: Jacques-Felix, 1964: 557–602.

GRAMINEAE

Echinochloa sp. indet.

West African: NIGERIA FULA-FULFULDE (Nigeria) kayaari (J&D)

Fula of N Nigeria use this plant as horse-fodder (1).

Reference:

1. Jackson, 1973: no. 3.14771.

Eleusine coracana (Linn.) Gaertn.

FWTA, ed. 2, 3: 397. UPWTA, ed. 1, 527.

English: finger millet; African millet; Indian millet; ragi; or ragi millet (from Hindi, India); black millet; korakan; kurkan; marua; tamba millet (Nigeria).

French: eleusine; coracan; millet de Yokohama (Portères).

West African: NIGER DENDI hèènì (*indef., sing.*) (Tersis) NIGERIA BIROM kpáná (LB) kpánáà hós = *of the season of rain: a c.var.* (LB) kpánáà zɛŋ = *of the season of dryness: a c.var.* (LB) FULA-FULFULDE (Nigeria) chargari (JMD; RP) HAUSA tambà (auctt.) támba (LB) támbàà (ZOG) tomba (RP) wanda (Sampson) IGBO oka tamba (BNO) KANURI sarga (JMD; RP) SANGA intɛrr-kum (RB) SURA kùttùng (anon. fide KW)

An erect tufted annual grass, culms to over 1 m high, cultivated in low rainfall zone across the Region from Senegal and especially in Niger and N Nigeria, and extensively across Africa and throughout tropical Asia.

The plant is a crop of importance from E Africa to Japan. It is described as a poor man's cereal being of especial value in poorer countries. Its evolutionary centre is thought to be in the Ethiopian highlands, a cultigen derivative of *E. indica* from which numerous cultivars have arisen. For many people it is a staple, for example, tribes of the upland plateaux of Nigeria (4, 6, 8, 14), Lake Chad (12), Central and E Africa (4), and India (2). Besides being a staple, it has proven a valuable emergency food. During the disruption of normal civilised life in SE Asia by the Japanese invasion of that area in World War II, 1941–45, numerous rural settlement areas owe their survival to being able to grow this plant for food. It crops heavily for a minimum of husbandry effort. It withstands water shortage. The seed is storable for a long period maintaining an edibility and viability for even up to 10 years if stored dry. Folklore in Tripolitania puts it at 100 years! Thus it is an admirable food-stuff for storage against dearth (3, 12), a practice followed by the colonial government in Uganda (13). Furthermore the hard seed testa protects the grain against mould and depredations of insects.

The grain is minute, 400–500 p.gm. (3). It has good nutritional properties for both humans and stock. Protein assays at about 6.7–8.0%, fat 1.3%, carbohydrate 72–86%, etc. (3, 10, 11). Protein is reported to be a good source for all essential amino acids except lysine. The grains can be used to prepare good quality malt, and is good for producing beer (3, 10). White grain is preferred for human consumption. Coloured grain tends to be bitter and is usually reserved for brewing. In many countries the grain is ground and baked into a sort of bread. It is nutritious. In Tanganyika it is prescribed for children and pregnant women (5). The grain ground to a flour is basis for special dietary foods in the USA (15). Alimentary broths are prepared in Kordofan (1). In India flour from the malted grain is used in invalid and infant diets, and is often prescribed for diabetics (7). Millet starch is said to be a prophylactic for dysentery in India (7) and Ethiopia (11).

In Indonesia the plant is eaten as a vegetable (15). Cattle will graze it, but that will be at the expense of the grain. Positive results have been obtained for *hydrocyanic acid* (15), though negative in the seed. Cattle will also take the straw. In Sudan (Jebel Marra) the leaves are woven into string (16).

The plant is susceptible to destruction by locust swarms with critical consequences. It is also attacked by *Striga* (Scrophulariaceae) (9).

References:

1. Baumer, 1975: 97. 2. Bhandari, 1974: 78. 3. Busson, 1965: 463, with chemical analysis, 465–6. 4. Dalziel, 1937: 527. 5. Greenway 3896, K. 6. Hepper 1147, K. 7. Hilu & de Wet, 1976: 199–208. 8. Irvine, 1948: 264. 9. Irvine, 1952,a: 24. 10. Johnson & Raymond, 1964,a: 6–11. 11. Lemordant, 1970: 18–19. 12. Portères, 1951,c: 1–78. 13. Purseglove, 1972: 147–56. 14. Sampson 1, K. 15. Watt & Breyer-Brandwijk, 1962: 472. 16. Wickens 1741, K. See also: Sastri, 1952: 160–6.

Eleusine indica (Linn.) Gaertn.

FWTA, ed. 2, 3: 395. UPWTA, ed. 1, 527.

English: goose grass; wire grass (Adam).

French: eleusine; pied-poule (Berhaut).

Portuguese: pé-de-galo (Feijão).

West African: **SENEGAL** FULA-PULAAR (Senegal) nassi gargagué (A.Chev.) MANDING-BAMBARA gondirima (JB) gondnéma (A.Chev.) guentnéman (A.Chev.) SERER mondon darate (A.Chev.) ratam fa mbé (JB) vodvod (JB) WOLOF budi darate (A.Chev.) budi du khot (JLT) **GUINEA-BISSAU** BALANTA albáli (EPdS; JDES) MANDYAK butchuque (EPdS; JDES) PEPEL blicatchor (EPdS; JDES) **SIERRA LEONE** BULOM (Sherbro) tunkun-dɛ (FCD) FULA-PULAAR (Sierra Leone) sigiri (FCD) KISSI taiyondo (FCD) KONO dutăsa (auctt.) lutăsa (FCD) tugbɛ (FCD) LIMBA gbantama (FCD) LOKO ngitwa (NWT) tese (NWT) MENDE ngetewi (NWT) ngetewuli (auctt.) ngete-wulo (FCD) ngete-wulo-ha (FCD) MENDE (Kpa) ngete-wu (FCD) SUSU tigirinyi (FCD) TEMNE ka-lɔnts (auctt.) YALUNKA tigbiri-na (FCD) **MALI** DOGON sɔ̃ pègu (CG) FULA-PULAAR (Mali) nassi gargagué (A.Chev.) MANDING-BAMBARA gondnéma (A.Chev.) guentnéman (A.Chev.) **UPPER VOLTA** GRUSI gatan (A.Chev.) GURMA garga (A.Chev.) MOORE targanga (A.Chev.; Scholz) **IVORY COAST** ABURE assumoamata (B&D) essuéma (B&D) AKAN-ASANTE kama (B&D) BAULE siganzi (B&D) KRU-GUERE (Chiehn) kpédé, kwédé (B&D) KWENI diridire (B&D) KYAMA n'tena (B&D) **GHANA** GA akpéntè (KD) MOORE tangana = *does not like bush* (FRI) **NIGERIA** BIROM syèsyé (LB) FULA-FULFULDE (Adamawa) tuji (Saunders) FULFULDE (Nigeria) sargande (J&D) seragade (Saunders; JMD) seragalde (Saunders; JMD) tuji (Saunders) GWARI tnatna (JMD) HAUSA ciiyááwar túújíí (auctt.) túújíí = *the greater bustard; the grass is said to be compared with the bird's great strength* (Lowe; auctt.) túújíí (LB) IGALA elade (Odoh; RB) IGBO (Asaba) ẹle NWT IGBO (Okpanam) ile NWT IJO-IZON (Kolokuma) berisọnléí (KW; KW&T) IZON (Oporoma) angóló (KW) TIV jighir (JMD) YORUBA ẹsẹ̀ kannakánná = *crow's foot* (JMD) gbági, gbẹ́gi (auctt.) **WEST CAMEROONS** KPE esinge-singe (JMD)

A tufted annual grass, culms to 60 cm high, slender (ssp. *indica*) or moderately robust (ssp. *africana*), or sprawling prostrate to 1.2 m long; common wayside plant and of disturbed sites; of two subspecies of which ssp. *africana* is the most robust and tetraploid (2n = 36) (18).

The plant is a common weed of cultivation in all territories (The Gambia: 4; Nigeria: 12; etc.). It rapidly colonises newly-cleared sites or abandoned ones (9). In a survey in Sierra Leone in 1960–63 of cattle enclosures, it was found to be the first and dominant colonist of the compacted and puddled earth after cattle were let out for free-range grazing (6). It also is invasive of rice-padis in Bendel State, Nigeria (13). The aggressive resilience of the plant evokes an Hausa saying: '*túújíí*, Thou canst not be hoed up whilst seated', the toughness of the plant being likened to the great strength of *túújíí*, the great bustard (10), and indicating the necessity for the farmer to get up off his back side to do the hoeing. Even power mowers have difficulty in cutting down the plant because of its tough fibrous texture (6). The long-running, deep-seated tough root-system while making the plant difficult to up-root, suggests an use on river-banks and such places where there is a need for soil-binding against erosion (10, 15).

The foliage furnishes good grazing. It is relished by all stock, and is said to be particularly valuable while still young, producing fodder with 8.06–17.12% nitrogen, and under cultivation is the equivalent of good European-standard hay (1, 2). It is furthermore a drought-resistant grass (24). Notwithstanding its good reputation as cattle-fodder, there are reports that under some conditions all parts of the plant show a strong presence of *alkaloids* (3, 11), and *cyanogenetic glycosides* (5, 15, 22, 28, 31) may be generated during wilting, thus rendering it poisonous to stock.

An infusion of the whole plant is used by the Baakpe people in W Cameroons to treat haemoptysis (Santesson fide 10, 20, 28). Sap expressed from the fresh plant is used on sprains in Madagascar (11), and in Ubangi the crushed plant is placed in bathwater or made into decoctions for washes for feverish children (27). Elsewhere (Trinidad: 31; Brazil, Philippine Islands: 28) a decoction is taken as a febrifuge, diuretic and anti-dysenteric. In Ivory Coast sap is instilled into the nostrils to relieve headache, applied to sores as a haemostatic, or in frictions for costal pains, and taken in draught for tachycardia and fainting (8). Ijo of SE Nigeria use the grass for cleaning the ear (30), and Igbo for gunshot wounds (25, 26). In the Philippine Islands, the plant is part of a cure for dandruff and falling hair (28). The Ijo of SE Nigeria use the epidermis of the culm applied topically for ringworm; the treatment is said to sting like application of iodine (30).

The leaves on Jebel Marra are made into a cordage (17) and the culms into hats in the Philippines (23). In eastern and southern Africa baskets, trays and bracelets are made of the fibre of the peduncle (21).

The root has a number of medicinal applications: for tachycardia, in ointment on the forehead for headache, calcined to ash to rub into topical scarification for pain in the sides or kidneys. The ash is said to kill filaria in the mucosae of the eyes (7). The root pulled up and still with some soil adhering is heated in Sierra Leone for fomenting wounds, especially on the soles of the feet (6). Pulped root is made into a plaster in the Ivory Coast for treating adenitis, and given in enema is held to stop immediately abundant and prolonged menstruation (8).

The seed is rich in nitrogen (5, 10). It serves in most territories as a famine-food (10, 14, 16, 17, 19, 21, 23, 28) and in places it may be cultivated (28), but perhaps more relevantly for its value as fodder. An occasional usage of the seed is for brewing beer (Jebel Marra, 17). It is recorded in N Nigeria that if eaten in quantity the seed causes dental troubles (24).

The plant is commonly held to have magical powers. Igbo at Okpanam believe so (26). The Ijo of SE Nigeria will tie a piece of the plant around the waist of a child as protection should that child have to visit a place where an old person has died. They tie some of the plant to a gun to prevent it misfiring, and at *uzíi* ceremony those who have taken *toorú* tie it in their hair so as not to 'disturb' others (29). In Ubangi before one goes out fishing, one soaks the nets in water in which the plant has been crushed, and a gambler carries a piece of the plant for good luck (27).

References:

1. Adam, 1954: 92. 2. Adam, 1966,a. 3. Adegoke & al., 1968: 13–33. 4. Austin 12, K. 5. Baumer, 1975: 97. 6. Boboh, 1974: in litt. 7. Bouquet, 1969: 129. 8. Bouquet & Debray, 1974: 92. 9. Chizea FHI.7873, K. 10. Dalziel, 1937: 527. 11. Debray & al., 1971: 74. 12. Egunjobi, 1969. 13. Gill & Ene, 1978: 182–3. 14. Guillarmod, 1971: 425. 15. Gupta & Sharma. 1971: 76. 16. Hunting Tec. Services, 1964. 17. Hunting Tec. Surveys, 1968. 18. Kennedy O'Bryne, 1957: 65. 19. Okiy, 1960: 120. 20. Oliver, 1960: 25. 21. Purseglove, 1972: 1: 146–7. incl. *E. africana* Kennedy O'Byrne. 22. Quisumbing, 1951: 96–97, 1023. 23. Sastri, 1952: 166. 24. Saunders 26, K. 25. Thomas NWT.1623 (Nig. Ser.), K. 26. Thomas NWT.1804. (Nig. Ser.), K. 27. Vergiat, 1970,a: 84–85. 28. Watt & Breyer-Brandwijk, 1962: 472–3. 29. Williamson KW.1 (UIH 13621), UCI. 30. Williamson K & Timitimi, 1983. 31. Wong, 1976: 104.

Elionurus ciliaris Kunth

FWTA, ed. 2, 3: 505, as *E. pobeguinii* Stapf.

West African: **NIGERIA** FULA-FULFULDE (Adamawa) raneraneho doyonbole (Saunders) FULFULDE (Nigeria) raneraneho (Saunders) HAUSA gámbà (Saunders) gashin fulani (Freeman; Thatcher)

A tufted perennial grass with culms to 2 m tall; of the soudanian savanna from Guinea to N Nigeria.

The grass furnishes some grazing, but the foliage is aromatic, which is in general rendered unpalatable (2, 3).

The root is also aromatic, and in Dahomey it is pulverised and served in decoction with a pimento as a fortifying drink (1).

References:

1. Risopoulos 1288, K. 2. Rose Innes 30,296, 32,306, K. 3. Saunders 18, K.

Elionurus elegans Kunth

FWTA, ed. 2, 3: 305. UPWTA, ed. 1, 528.
West African: **SENEGAL** MANDING-BAMBARA bimbilé (JB) SERER gédétian (JB) WOLOF gèn u diar (JB) **THE GAMBIA** MANDING-MANDINKA koningo nikko (Pirie; JMD) **MALI** FULA-PULAAR (Mali) hu'do fello (A.Chev.) MANDING-BAMBARA bimbilé (A.Chev.) sabi (A.Chev.) SAMO kilaburu (A.Chev.) SONINKE-SARAKOLE kaméré (A.Chev.) komé (A.Chev.) **UPPER VOLTA** DAGAARI naa-nwo (Girault) ner-sar (Girault) MOORE ner-saagha (Girault) **NIGERIA** FULA-FULFULDE (Adamawa) raneraneho doyonkole (Saunders) FULFULDE (Nigeria) ranera-neho (JMD) HAUSA gámbàr kuréégéé = 'gamba' (Andropogon gayanus Kunth) of the squirrel (JMD; ZOG)

A tufted annual grass, 30–60 cm high, in locations across the soudanian savanna from Senegal to N Nigeria.

It is an early grass after annual burning and provides a little indifferent grazing (1, 2).

Dagara people in Upper Volta use the grass in a prescription for treating a victim of snake-bite. The procedure involves fumigation by the strongly aromatic odour of the grass and may last up to 48 hours without break (3).

Women in Mali use the dyed straw to plait into ear-rings (A. Chevalier fide 2).

The grass is parasitised by *Striga*, and may thus act as an indesirable host in the proximity of susceptible crops (4).

References:

1. Adam, 1966, a. 2. Dalziel, 1937: 528. 3. Girault, 1958: 37. 4. Terry 1927, K.

Elionurus hirtifolius Hack.

FWTA, ed. 2, 3: 504–5. UPWTA, ed. 1, 528.
West African: **NIGERIA** FULA-FULFULDE (Adamawa) raneraneho doyonkole (Saunders) FULFULDE (Nigeria) raneraneho (JMD) HAUSA gámbàr kuréégéé = 'gamba' (Andropogon gayanus Kunth) of the squirrel (auctt.)

A wiry tufted perennial grass to 30–60 cm tall, of the grass savanna in Ghana and N and S Nigeria, and in the Central African Republic.

The grass in land subjected to annual burning, is one of the earliest to re-grow (2). In Jebel Marra, Sudan, it is cut for cattle fodder if nothing better is available (1).

References:

1. Hunting Tech. Surveys, 1968. 2. Lily P.157, K.

Elionurus muticus (Spreng.) Ktze

FWTA, ed. 2, 3: 505, as *E. argenteus* Nees.

A densely tufted perennial grass with culms to 60 cm high; of open deciduous bushland; in Guinea to W Cameroons, and extending into NE and E Africa to S Africa, and Yemen; also in subtropical America.

241

It is very quick to come into flower after bush-fires (4). It provides good grazing in Somalia (1), Uganda (5) and Botswana (3) before flowering.

The root of the plant in southern Africa perhaps contains a volatile oil. Sotho chew the root as a toothache remedy (2) and prepare from it a medicine for intestinal upsets (2, 6).

References:

1. Glover & Gilliland 1149A, K. 2. Guillarmod, 1971: 425, as *E. argenteus*. 3. Kelaole 507, K. 4. Lily P.92, K. 5. Snowden 1260, K. 6. Watt & Breyer-Brandwijk, 1962: 473, as *E. argenteus* Nees.

Elionurus platypus (Trin.) Hack.

FWTA, ed. 2, 3: 505.
West African: SIERRA LEONE MENDE leti (Cole) tele (FCD) MENDE (Kpa) leti (FCD) TEMNE ka-gbont (NWT)
NOTE: Mende names refer correctly to Imperata Cyr (Graminae).

A caespitose perennial grass, culms up to 1.8 m tall; of savanna; from Guinea to Ivory Coast.

It provides some fodder, and in Sierra Leone it has been encouraged as a pasture grass (1).

Reference:

1. Cole, 1968.

Elionurus royleanus Nees

FWTA, ed. 2, 3: 505.

A caespitose annual grass with brief culms 5–15 cm high; of dry degraded soils in dry open sites; recorded only in Senegal in the Region, but occurring in Cape Verde Islands and in NE Africa, Arabia and NW India.

In Rajasthan, India, it is said to give palatable grazing at all stages of growth, much appreciated by sheep and goats, but bulk is limited and thus it is relatively unimportant as a fodder (2).

In Kordofan, women plait the dry straw into bracelets (cf. *E. elegans*) and other decorative accessories (1).

References:

1. Baumer, 1975: 97. 2. Gupta & Sharma, 1971: 76.

Elymandra androphyla (Stapf) Stapf

FWTA, ed. 2, 3: 490.
West African: SIERRA LEONE MANDING-MANDINKA fualobĩ (FCD) YALUNKA lɔ̃fu-na (FCD)

A coarse perennial grass with culms to 2½ m tall; of upland distribution on poor sandy soils in savanna; widely dispersed in the Region from Senegal to W Cameroons, and southwards through the Congo basin to Angola.

At Musaia in Sierra Leone it is valued as a thatching grass, considered second best only to *Danthoniopsis chevalieri* (1).

Reference:

1. Deighton 4524, K.

Elymandra subulata Jac.-Fél.

FWTA, ed. 2, 3: 498.
West African: GUINEA FULA-PULAAR (Guinea) chelbi Langdale-Brown

A tufted annual grass with culms up to 2 m tall; of flooded sites and poor drainage, and roadside waste places of the tree savanna in Guinea, Sierra Leone and Ivory Coast.
No usage is reported though the grass is recognised by the Fula in Guinea.

Elytrophorus spicatus (Willd.) A Camus

FWTA, ed. 2, 3: 376.
West African: SENEGAL MANDING-BAMBARA kobin (JB) SERER ndédar (JB) sédar (JB)

An annual grass, culms erect to 60 cm high, of margins of swamps, creeks, water-holes, ditches, drains and pans, dry or drying, common throughout the Region from Senegal to N Nigeria, and dispersed over the Old World tropics.
The plant's preference for water sites that have dried or are drying leads to it becoming a weed of rice-padis, a feature recorded in all territories. Fortunately since the plant does not develop till the padis have been dried off and the rice crop has already been harvested, it is not a pest. In fact, it can be left to grow and later ploughed in as a green manure, or cattle sent to browse over it.

Enneapogon desvauxii P. Beauv.

FWTA, ed. 2, 3: 383.

A small, tufted perennial grass to 23 cm high, of dry rough waste places, in the Region in Mauritania and Niger, and occurring in the Atlantic Islands off W Africa, Algeria to Ethiopia and Arabia to India, and in E Africa southwards to Botswana; also in central and S America.
The grass is palatable to cattle and provides grazing in Kenya (1), Botswana (3) and India (2).

References:

1. Glover & Samuel 2836, K. 2. Gupta & Sharma, 1971: 76–77, as *E. brachystachyus* (Jaub. & Spach.) Stapf. 3. Smith, P.A. 1283, K.

Enneapogon persicus Boiss.

FWTA, ed. 2, 3: 383, as *E. schimperanus* (Hochst.) Renvoize.

A perennial (annual under severest conditions) tussock grass, of dry rocky waste places and limestone hills to 2,500 m altitude, in the Sahel of Mali and Niger, and dispersed across tropical Africa to Somalia and Kenya, and into Pakistan, India and S USSR.
It is palatable to cattle and provides good browsing in Kenya (3, 7), Somalia (1, 5, 6, 9) and India (2, 4), though at times in India it may not be readily taken (8).

References:

1. Boaler B.11, K. 2. Bor, 1960: 610, as *E. elegans* (Nees) Stapf. 3. Glover & al. 281, 838, 1999, K. 4. Gupta & Sharma, 1971: 77, as *E. elegans* (Nees) T. Cooke. 5. Hemming 2258, K. 6. McKinnon S.275, K. 7. Mwangangi & Gwynne 1048, K. 8. Sastri, 1952: 174, as *E. elegans* Cooke. 9. Wieland 4293, K.

GRAMINEAE

Enteropogon macrostachyus (Hochst.) Munro

FWTA, ed. 2, 3: 402.

A caespitose perennial grass with culms to 1 m high, of disturbed sites in tree and bush-savanna; known only in Ghana in the Region, but widely dispersed over NE, E, South Central and SW Africa.

The grass has a termitophile tendency being commonly recorded on termitaria (Ghana: 1, 4; Malawi: 6, 8; Zimbabwe: 7). These reports do not indicate the nature of the affinity, if any. In Ghana it is also found on saline sites (2).

In Sudan (3) and in Kenya (5) it provides good grazing. In the latter territory cane rats are reported to eat it (5). The culms are used in Kenya for thatching (5).

References:

1. Ankrah 20330, K. 2. Baldwin 13613, K. 3. Beshir 18, K. 4. Clayton DC.388, K. 5. Magogo & Glover 888, K. 6. van Rensberg 1424, K. 7. West 7566, K. 8. White 7623, K.

Enteropogon prieurii (Kunth) WD Clayton

FWTA, ed. 2, 3: 400, as *Chloris prieurii* Kunth; and incl. *C. subtriflora* Steud.; and *C. parva* Mimeur, p.402.

West African: **MAURITANIA** ARABIC (Hassaniya) kraa l'grab (AN) **SENEGAL** SERER las a koy (JB) las o fam (JB) las pis (JB) WOLOF gèn u dar (JB) gèn u mbam (JB) **MALI** DOGON pedu kúú (CG) **UPPER VOLTA** FULA-FULFULDE (Upper Volta) ŋaŋarɗe (K&T) ŋaŋargal (K&T) **NIGERIA** HAUSA káfàr gauraka (Golding)

An annual grass, culms erect or geniculately ascending to 45 cm high, of open bush or disturbed ground on poor soils, across the Region in the Soudanian zone from Mauritania to N Nigeria, and in Cape Verde Islands, NE Africa, Arabia and SW Africa.

The grass is deemed to be good forage and is greatly relished by all stock in Senegal (1) and Mauritania (6). It has local importance in Eritrea (5), and in Sudan it is said to be a good rainy season grass (4). In Kordofan (2) and in Rajasthan, India, (3) cattle browse it only till flowering time.

In Mauritania people in time of need raid ant-hill granaries taking the stored grain for their own food requirement (6).

References:

1. Adam, 1966,a, as *Chloris prieurii*. 2. Baumer, 1975: 92, as *Chloris prieurii* Kunth. 3. Gupta & Sharma, 1971: 68, as *Chloris prieurii* (Kunth) Maire. 4. Harrison 76, K. 5. Hemming 1027, K. 6. Naegelé, 1958,b: 892, as *Chloris prieurii* Kunth.

Enteropogon rupestris (JA Schmidt) A Chev.

FWTA, ed. 2, 3: 402.

A caespitose perennial grass with culms to 60 cm high, wiry, of dry locations in scattered tree grassland in Mauritania and Niger; and extending from Cape Verde Islands to the Horn of Africa and in southern Africa.

No usage is recorded in the Region. It is not drought-resistant and value for grazing is low, though when available is taken with relish in Somalia (1) where the culms are used for weaving into mats for roofing (2) and other purposes (1).

References:

1. Keogh 128, K. 2. McKinnon S.124, K.

244

Eragrostis aegyptiaca (Willd.) Link

FWTA, ed. 2, 3: 391.

A tufted annual to about 30 cm high in open waste places in Senegal, Mali and N Nigeria, and in Egypt and Sudan. It is browsed by cattle in Senegal (1).

Reference:

1. Adam, 1966,a.

Eragrostis arenicola CE Hubbard

FWTA, ed. 2, 3: 386.
West African: NIGERIA HAUSA kambura fage = *covers the field* (Taylor)

A tufted annual, culms to 45 cm high, of open dry places, shallow soils and disturbed land, in N Nigeria and W Cameroons, and extending across to NE and E Africa and southwards.

It is a weed of cultivated land reaching sometimes nuisance proportions and evoking at Zaria, N Nigeria, a name meaning *covers the field* (2). In Sudan it invades irrigated cultivations (1).

References:

1. Blair 12, 81, 83, K. 2. Taylor 22, K.

Eragrostis aspera (Jacq.) Nees

FWTA, ed. 2, 3: 387.
West African: THE GAMBIA MANDING-MANDINKA numu ju tio = *blacksmith's grass* (Pirie) GHANA HAUSA buruburwa (Williams) NIGERIA ? yayanga (Dodd) IGBO (Agukwu) odudo ẹzi (NWT) ọgagai (NWT)

A tufted annual, culms erect to 80 cm high bearing large feathery panicles, of disturbed sites and old farmland, in Senegal to S Nigeria, and throughout NE, tropical and S Africa, and extending over the Middle East and India.

It is a common weed of cultivation, and in Sudan of irrigation (2). All stock browse it. In Ghana horses are said to relish it at all times (1), but after anthesis cattle may be put off by the feathery inflorescence. These become dehiscent breaking off entire to be bowled along by the winds (3).

References:

1. Beal 3, K. 2. Blair 8215, K. 3. Welch 150, K.

Eragrostis atrovirens (Desf.) Trin.

FWTA, ed. 2, 3: 390.
West African: SENEGAL MANDING-BAMBARA férala (JB) THE GAMBIA MANDING-MAN-DINKA n'diro (Pirie) SIERRA LEONE MENDE foni *a general term* (NWT) TEMNE kə-mpoeti (NWT) sɔntɔsi (NWT) MALI ? (Mopti) tielé (Matthes) MANDING-BAMBARA ngwose (Lean) GHANA HAUSA tsíntsíyár fàdámàà = *brush grass of the marsh* (Williams) NZEMA atuabo (Thorold) NIGER FULA-FULFULDE (Niger) tapo (Bartlett) SONGHAI dàazi (*pl.* -ò) (D&C) NIGERIA HAUSA burburuwar rafi = '*burburuwa*' *of the stream* (Taylor) karfa (Taylor) tsíntsíiyár fàdámàà = *brush grass of the marsh* (Grove; Palmer) WEST CAMEROONS BAMILEKE saraji (DA)

A variable perennial grass of damp places, or swamps, dispersed throughout the Region from Mauritania to W Cameroons, and in N Africa to Zambia and Angola, and across Asia to the Philippines.

Growing in wet sites, it evokes names in Hausa (see above) connecting it with *marsh* and *stream*. In N Ghana it may produce a good sound ground carpet, and good fodder (3). It is grazed by all stock in Senegal, especially by cattle (1), but there are records elsewhere of inferior utility (N Dahomey: not appreciated by stock, 4; N Nigeria: fodder of poor quality, 5). When culms are produced they reach up to 1 m long and are used for thatching (2).

References:

1. Adam, 1966,a. 2. Kennedy 7280, FBC. 3. Ll. Williams 301, 507, K. 4. Risoupoulos 1253, K. 5. Taylor 3, K.

Eragrostis barrelieri Daveau

FWTA, ed. 2, 3: 393.
West African: MAURITANIA ARABIC (Hassaniya) lehméire (AN)

A loosely-tufted annual grass recorded only on the northernmost limit of the Region at Aïr in Niger. It extends across sub-saharan Africa to NE Africa.
It is a weed of cultivation (3) and irrigation (2), and is browsed by all stock (1).

References:

1. McKinnon S.277A, K. 2. Parker E.104, K. 3. Teuvo Ahti 16,390, K.

Eragrostis barteri CE Hubbard

FWTA, ed. 2, 3: 390.

A caespitose grass, culms to 1 m long, woody, of sand-banks in river-beds and rocky sites in Senegal to W Cameroons.
From river-bed sites the culms left lying by subsiding floods continue to grow and push out flowering shoots. In N Nigeria this provides a little dry season browsing (1).

Reference:

1. Peter & Tuley 5, K.

Eragrostis camerunensis WD Clayton

FWTA, ed. 2, 3: 390.

A densely-tufted perennial grass with culms to 30 cm high, of grassy waste places in Nigeria and W Cameroons.
It provides grazing for stock and for game. It has been found useful at the Gashaka Game Sanctuary (1), the Bambui Farm (3) and the Obudu Cattle Ranch (2).

References:

1. Chapman 4344, K. 2. Haines 142, K. 3. Pedder 23, K.

Eragrostis cenolepsis WD Clayton

FWTA, ed. 2, 3: 389.
West African: GUINEA FULA-PULAAR (Guinea) fonyi choli (Langdale-Brown)

A tufted perennial grass, with culms to 90 cm high, of moist soils in upland localities of Guinea, Sierra Leone and N Nigeria, and known also in E Cameroun.
No usage is recorded, but since it is known to the Fula of Guinea, it perhaps provides some cattle forage.

Eragrostis cilianensis (All.) Lut.

FWTA, ed. 2, 3: 390. UPWTA, ed. 1, 528.
English: stink grass (Dalziel); strongly scented love-grass (Grounds); snake grass (Adam).
West African: MAURITANIA ARABIC (Hassaniya) agmellil (AN) azmelil (AN) tannesmirt (AN) tukurit (AN) SENEGAL MANDING-BAMBARA samba labi (JB) MALI FULA-PULAAR (Mali) burdi koladé (A.Chev.) fitirde (A.Chev.) sarahɔ lékiudi (A.Chev.) MANDING-BAMBARA sama labi (A.Chev.) samba gambi (A.Chev.) wolo kamamba (A.Chev.) SONGHAI subu tufé (A.Chev.) UPPER VOLTA GURMA nuanganu = ox's broom (A.Chev.) MOORE kolo rasé (A.Chev.) NIGERIA HAUSA ámán mússà = sickness of the cat (JMD) budari = zorilla, a skunk-like weasel (auctt.) bulbulin mússà = vomit of the cat (auctt.) bunsurun fàdámàà = he-goat of the swamp (JMD) bunsurun fage = he-goat of the open ground (JMD) kòòmayyàà (Golding) YORUBA akọ yayángán (Dodd; JMD) ẹran awó (Macgregor)

A loosely-tufted annual grass, culms erect or ascending 15–60 cm high, of roadsides, agricultural fallows and weedy overgrazed places, sea-level to montane situations, common throughout the Region from Senegal to N and S Nigeria, widespread over tropical Africa and Old World tropics; also introduced into the western Hemisphere.
The plant is a weed of cultivation. In some parts of the USA, where it has become naturalised, it is considered a pest (1). In Sudan it is even a weed of irrigation (4). By its vigorous habit it has shown capacity in Tanganyika as a pioneer coloniser, along with some other grasses, in reclaiming eroded land (8).
The plant is known in all parts as a source of good forage taken by all stock (5, 15; Mauritania, 11; Senegal, 1, 2; Nigeria, 7; Gabon, 17; Sudan, Somalia, Uganda, Kenya, Tanganyika, India etc.).
In Sudan it is considered a valuable fodder in the dry summer season (10, 14). In Mauritania it is stored as hay for winter and dry season feed (11). Inhibitions have, however, been recorded against feeding it to horses in N Nigeria (6). Inspite of its general acceptance by stock, it has a reputation for having an evil smell (5); in N Nigeria it evokes the unsavory Hausa name meaning cat's vomit (13); cf., also the English name given above.
The culm yields straw used for thatching in N Nigeria, but quality is inferior (6, 7). It is made into palliasses in Kordofan, and plaited into circular covers for pots holding milk or millet mash in process of fermentation (3). People in N Nigeria make the straw into matting (6, 7).
A root-decoction is taken in Ubangi against 'flu in the event of an epidemic' (16).
The grass is edible, and is taken as a normal part of the diet by people in the Lake Chad area (6, 12). People in Mauritania make the grain into cakes (11). It is sometimes eaten in the Cape Verde area of Senegal (1), and in time of dearth Basuto in southern Africa eat it as porridge (9).

GRAMINEAE

References:

1. Adam, 1954: 92. 2. Adam, 1966,a. 3. Baumer, 1975: 97. 4. Blair 14, 32, 99, K. 5. Dalziel 874, K. 6. Dalziel, 1937: 528. 7. Golding 2, K 8. Greenway 10,495, K. 9. Guillarmod, 1971: 425. 10. Khalid 25, K. 11. Naegelé, 1958,b: 894. 12. Okiy, 1960: 120. 13. Palmer 33, K. 14. Peers K.M. 13, K. 15. Schnell, 1960: no. 108. 16. Vergiat, 1970,a: 85. 17. Walker & Sillans, 1961: 188.

Eragrostis ciliaris (Linn.) R Br.

FWTA, ed. 2, 3: 386. UPWTA, ed. 1, 528.

West African: SENEGAL FULA-PULAAR (Senegal) kiu pité (A.Chev.) paguire jaule = 'paguire' of the guinea fowl (A.Chev.) sorgobo (A.Chev.) MANDING-BAMBARA sambu gambi (A.Chev.) wolo gaman (A.Chev.) wolo kaman (A.Chev.) ·SOCE· itimâkoro (K&A) SERER diãmbl'o gôr (JB) diisis gôr (JB) mbañgati (JB) WOLOF salguf (A.Chev.; K&A) salguf u tak (JB) THE GAMBIA MANDING-MANDINKA ndirra, ndirra sina (JMD) n'dyiro (Fox) numu ju tio = blacksmith's grass (Pirie) nyantan (JMD) yiti ma kora from yiti: a sore; ma kora: to clean (Pirie) GUINEA MANDING-MANINKA ologuélé (CHOP) SIERRA LEONE KORANKO sɛrɛlinyaxe (JMD) SUSU funfuri (JMD) sankabesukwi (JMD) TEMNE peni-fafagbi (JMD) peni-pa-gbel (JMD) MALI FULA-PULAAR (Mali) kiu pité (A.Chev.) paguiri jaule = 'paguiri' of the guinea fowl (A.Chev.) sorgobo (A.Chev.) MANDING-BAMBARA sambu gambi (A.Chev.) wolo gaman (A.Chev.) wolo kaman (A.Chev.) SONGHAI fitti-fitti (A.Chev.) subu (A.Chev.) subu furia furia (A.Chev.) subu koré (A.Chev.) TAMACHEK tadjit (A.Chev.) GHANA HAUSA kumburar kama (Williams) NIGERIA ARABIC-SHUWA rauwaj (JMD) FULA-FULFULDE (Nigeria) dutaleho (JMD) sagaje (JMD) saraaho (JMD) saraaho gorko (J&D) sarawal (JMD) tappo (JMD) GWARI aknuse (JMD) HAUSA burburwa (JMD) burburwar fâdámàà = 'burburwa' of marshy land (JMD) gafarfatin (Bargery fide JMD) hàtsàà-hátsàà (auctt.) kàràngíyáá (ZOG) kashe saura = kill the rest (RES) kàtsàà-kátsàà (Golding) kòòmayyàà (JMD; ZOG) kùmbùrà kàmà (Adelodun) kùnbùrà kàmà (ZOG) matsandaka (JMD) matsandaka tsumbe (ZOG) námíjin tsíntsíyáá = male brush grass/broom (auctt.) tsíntsíyáá = a broom (BM; JMD) tsíntsííyár fàdámàà (JMD) tsùmbé (JMD) IGBO (Uburuku) alielie nkuku (NWT) KANURI barata (JMD) gantaska (JMD) kòlàsòlóm (C&H) kòlàsēlēm (A&S) YORUBA àgbàdo-ẹṣin (JMD; Verger) eéran-awó (JMD; Verger) irungbòn ẹfòn from ẹfòn: buffalo (Verger) iwó-awó from awó: guinea fowl (Millen; JMD) ọgbe-agufon after a sp. of bird (JMD) ọkà-ẹṣin = horses' corn: common for fodder grasses (JMD; Verger) yayángán (JMD; Verger)

A tufted annual grass, culms to 60 cm high, a ruderal of roadsides, waste places, cultivations and coastal sandy places, from sea-level to over 1400 m altitude; common throughout the Region in all states, and all over tropical Africa, to Arabia, Madagascar and India, and into tropical America.

The grass grows in all areas as a weed of cultivation. It also grows in sandy locations, even in pure coastal sand (Gabon, 11; Eritrea, 1), and assists in fixing dunes. It provides a good forage for stock and is so used in all territories, stock taking it with relish and at all times of the year. Horses are said to refuse it after seeding time (3). In N Nigeria (6) and Kordofan (2) it is cut for sale as a market produce and for stacking as hay.

The straw is woven into mats for covering food and into a coarse cordage in N Nigeria (6, 9). The longer culms may be bundled for sale as brooms or used as thatch (6). The plant is reduced to ash in The Gambia (10) and Senegal (8) which is spread over cuts, burns and the like, hence the Mandinka name meaning *sore to clean.* An infusion of the plant is taken in Congo (Brazzaville) for stomach-pains (4).

The inflorescence in flower or in seed is reduced to ash in Tanganyika and, like the vegetative parts in The Gambia, is used for curative properties, in this instance mixed with castor-oil on whitlows (7). Maturation is rapid.

The grain is collected in time of scarcity. It has a good nutritive value (5).

'Soce' people in Senegal consider the grass to be a fetish for longevity. Having reached a certain age [? menopause], they deem it good to hang a piece below the bed (8). In Congo (Brazzaville) the plant is considered able to chase away spirits (4).

References:

1. Bally B.6767, K. 2. Baumer, 1975: 97. 3. Beal 4, K. 4. Bouquet, 1969: 129. 5. Busson, 1965: 466–8, with chemical analysis of the grain. 6. Dalziel, 1937: 528. 7. Haerdi, 1964: 210. 8. Kerharo & Adam, 1964, c: 305–6. 9. Palmer 34, K. 10. Pirie 10/33, K. 11. Thompson G.10, K.

Eragrostis curvula (Schrad.) Nees

English: African love-grass (Grounds); weeping love-grass (India, Sastri).

A densely-tufted perennial with culms to 1 m high, native of southern Africa and introduced to many tropical countries as a fodder and ground cover grass. It is now present in W Africa but without report on utility.

Eragrostis cylindriflora Hochst.

FWTA, ed. 2, 3: 391.

A loosely-tufted annual grass, culms to 75 cm high, of weedy overgrazed grassy places in Niger and Ghana, and in Sudan, E Africa to S Africa.

The grass is a weed of irrigation in Sudan (5). It furnishes good grazing for all stock in E Africa (Kenya 3; Tanganyika 1, 2, 4).

References:

1. Gillman 44, K. 2. Greenway 784, K. 3. Leuthold 11, K. 4. Marshall, s.n., 26/2/1932, K. 5. Wickens 1405, K.

Eragrostis domingensis (Pers.) Steud.

FWTA, ed. 2, 3: 391.
West African: SENEGAL SERER dikadika (JB)

A coarse, tufted perennial grass with rigid culms to 1.2 m high, of coastal and inland sandy situations across the Region from Mauritania to N and S Nigeria, and in São Tomé and tropical America.

The plant favours littoral sand-dunes even slightly brackish (1). The roots are conspicuously clothed in a dense sheath of hairs to which sand grains adhere (2). This assists in sand-binding.

References:

1. Adam, 1954: 92. 2. Clayton DC.405, K.

Eragrostis gangetica (Roxb.) Steud.

FWTA, ed. 3: 389. UPWTA, ed. 1, 528.
West African: SENEGAL FULA-PULAAR (Senegal) kiu pité (A.Chev.) paguiri jaule = 'paguiri' of the guinea fowl (A.Chev.) sorgobo (A.Chev.) MANDING-BAMBARA sambu gambi (A.Chev.) wolo gaman (A.Chev.) wolo kaman (A.Chev.) NON ida napen (K&A) idad apen (K&A) SERER diãmbul (A.Chev.) diãmbul a mbèl (JB) WOLOF salguf (A.Chev.) THE GAMBIA MANDING-MANDINKA ndirra (JMD) ndirra sina (JMD) numu ju tio = blacksmith's grass (Pirie) nyantan (JMD) GUINEA MANDING-MANINKA ologuélé (CHOP) SIERRA LEONE KORANKO serɛlinyaxe (JMD) LOKO ruŋgarõhun (NWT) MENDE bɔnji (NWT) leti (NWT) SUSU funfuri (JMD) sankabisuwi (NWT; JMD) TEMNE kupika (NWT) peni-pa-bakbil (NWT) peni-pa-gbel (JMD) MALI DOGON ɛdɛgɛlɛ, sasáá pò ána (CG) FULA-PULAAR (Mali) kiu pité (A.Chev.) paguiri jaule = 'paguiri' of the guinea fowl (A.Chev.) sorgobo (A.Chev.) MANDING-BAMBARA sambu gambi (A.Chev.) wolo gaman (A.Chev.) wolo kaman (A.Chev.) SONGHAI fitti-fitti (A.Chev.) subu (A.Chev.) subu furia furia (A.Chev.) subu koré (A.Chev.) TAMACHEK tadjit (A.Chev.) NIGERIA ARABIC-SHUWA rauwaj (JMD) FULA-FULFULDE (Nigeria) dudaleho (Saunders) dutaleho (JMD) saraaho (JMD) sarawal (JMD) shagaje (JMD) tappo (JMD)

GRAMINEAE

GWARI aknuse (JMD) HAUSA burburwa (auctt.) burburwar fadama (JMD) gafarfatin (Bargery) hàtsàà-hátsàà (JMD; ZOG) kòòmányar rafi = *'koomanya'* of the stream (Maule) kòòmáyyàà (JMD; ZOG) matsandaka (JMD) matsandaka tsumbe (ZOG) námíjìn tsíntsííyáá (JMD) tsìntsííyáá = *broom* (JMD) tsíntsííyár fàdámàà (JMD) tsùmbé (JMD) KANURI barata, gantaska (JMD) YORUBA àgbàdo-ẹ̀ṣin (JMD; Verger) eéran-awó (JMD; Verger) irungbọ̀n ẹfọ̀n from ẹfọ̀n: *buffalo* (Verger) iwó-awó = *guinea fowl* (JMD) ọgbe-agufọn a *sp.* of bird (JMD) ọkà-ẹṣin = *horse's corn* (JMD; Verger) yayángán (JMD; Verger)

A variable loosely-tufted annual grass, culms geniculate slender to 45 cm tall, of weedy waste places, common through the Region from Mauritania to W Cameroons, and in E Africa and India.

It is a weed of cultivation which at Cap Vert has been reported troublesome (1). It provides coarse pasture grazing when young (Senegal, 2; Nigeria, 4; India, 3).

Mende of Sierra Leone make a decoction of the plant for sore feet (5).

References:

1. Adam, 1954: 92, as *E. cambessediana* Steud. 2. Adam, 1966,a. 3. Sastri, 1952: 183. 4. Saunders 28, K. 5. Thomas NWT.21, K.

Eragrostis japonica (Thunb.) Trin.

FWTA, ed. 2, 3: 387, as *E. namaquensis* Nees.

English: Japanese love-grass (Grounds).

West African: MAURITANIA ARABIC (Hassaniya) lehmére (AN) **THE GAMBIA** MANDING-MANDINKA farangtambo (Terry) **NIGERIA** HAUSA bafulatana = *Fulani woman* (Golding) fain rumeya (Kennedy)

A tufted annual grass, culms to 1.2 m high, of sandy locations, stream-sides, and cultivations, dispersed across the Region from Senegal to N and S Nigeria, and in NE, E and S Africa, and across the Middle East to India, Ceylon and Thailand.

The grass is a weed of cultivation. In newly cleared areas for upland rice in The Gambia it is a serious weed, causing, it is said, the rice plants to die in its presence (2). It is grazed by all stock and game at all times. Hausa in N Nigeria feed it together with a shrub named *bera* to cows as a lactogene (1).

References:

1. Golding 39, K. 2. Terry 3209, K.

Eragrostis lehmanniana Nees

English: Lehmann's love grass (India, Sastri).

A tufted perennial grass, culms to 50 cm high, of sandy open wooded savanna, native of southern Africa, and introduced to E Africa and then to Nigeria as a fodder grass, but as yet of unreported utility.

Eragrostis lingulata WD Clayton

FWTA, ed. 2, 3: 393.

A loosely-tufted annual grass to about 30 cm high, of weedy places in Mauritania to Guinea.

Not common in Senegal, but it is recorded there as being browsed by cattle (1).

Reference:

1. Adam, 1966,a, as *E. perbella*.

Eragrostis macilenta (A. Rich.) Steud.

FWTA, ed. 2, 3: 391.

A loosely-tufted annual, culms to 60 cm high, of old cultivations, roadsides, waste places, often in the shade, in uplands of Ivory Coast, N Nigeria and W Cameroons, and extending to Ethiopia and southwards to Zambia. The grass is a weed of cultivation. It gives good grazing in both rainy and dry season in Sudan (1). It is browsed by stock in Uganda (3) and Kenya (3). Birds are said to seek out the grain.

References:

1. Beshir 32, K. 2. Glover & al., 501, 520, 2281, K. 3. Snowden 1225, K.

Eragrostis minor Host

FWTA, ed. 2, 3: 390.
West African: MAURITANIA ARABIC (Hassaniya) azmelil (Kesby)

A loosely-tufted annual grass, culms to 45 cm high, of weedy waste places in Mauritania and Senegal, and also in NE and E Africa, and in temperate and subtropical regions of the Old World; it is found as an introduction also in the New World.

It is grazed by stock in Mauritania (2). Cattle take it in Rajasthan State, India (1).

At times in olden days the grain has been eaten in Mauritania (2).

References:

1. Gupta & Sharma, 1971: 79, as *E. poaeoides* P. Beauv. 2. Kesby 55, K.

Eragrostis mokensis Pilger

FWTA, ed. 2, 3: 389.

A slender annual grass, much tillered at the base, culms to 30 cm high, in montane locations of N and S Nigeria, W Cameroons and Fernando Po, and across equatorial Africa to Kenya.

It is a weed of cultivation. In Kenya stock readily browse it (1, 2).

References:

1. Butler 161, K. 2. Glover & Samuel 3185, K.

Eragrostis paniciformis (A Br.) Steud.

A loosely-tufted perennial grass, culms erect or geniculate to 90 cm high, native of Sudan, Ethiopia, E Africa and Zambia, and now recorded present in Niger.

It is reported as a weed of irrigation in Niger. In Ethiopia and E Africa it is heavily grazed.

Eragrostis pilosa (Linn.) P Beauv.

FWTA, ed. 2, 3: 389. UPWTA, ed. 1, 528.
West African: MAURITANIA ARABIC (Hassaniya) daïfa (AN) lakemera (AN) lekmera (AN) l'hemera (AN) ozeb'an *when young* (AN) tanmaïrit (AN) SENEGAL FULA-PULAAR (Senegal) kiu pité (A.Chev.) paguiri jaule = '*paguiri*' *of the guinea fowl* (JMD) sorgobo (A.Chev.) MANDING-BAMBARA sambu gambi (A.Chev.) wolo gaman (A.Chev.) wolo kaman

GRAMINEAE

(A.Chev.) SERER diãmbul (A.Chev.) salguf (A.Chev.) **THE GAMBIA** MANDING-MANDINKA ndirra (JMD) ndirra sina (JMD) numu ju tio = *blacksmith's grass* (Pirie) nyantan (JMD) **GUINEA** MANDING-MANINKA ologuélé (CHOP) **SIERRA LEONE** KORANKO sɛrɛlinyaxe (JMD) SUSU funfuri (JMD) sankabesuwi (JMD) TEMNE peni-fafahbi (JMD) peni-*pa*-gbel (JMD) **MALI** DOGON sasáá põ bánu (CG) FULA-PULAAR (Mali) kiu pité (A.Chev.) paguiri jaule *from* jaule: *guinea fowl* (A.Chev.) sorgobo (A.Chev.) MANDING-BAMBARA sambu gambi (A.Chev.) wolo gaman (A.Chev.) wolo kaman (A.Chev.) SONGHAI fitti-fitti (A.Chev.) subu (A.Chev.) subu furia furia (A.Chev.) subu koré (A.Chev. fide JMD) TAMACHEK tadjit (A.Chev.) **UPPER VOLTA** BISA zanga bur Prost FULA-FULFULDE (Upper Volta) siiwuko (K&T) si'uko (K&T) **NIGER** SONGHAI géŋgén-sùbù (*pl.* -ò) = *grass of the glade* (D&C) **NIGERIA** ARABIC-SHUWA rauwaj (JMD) BIROM gwés (LB) FULA-FULFULDE (Nigeria) dutaleho (JMD) saraaho (JMD) sarawal (Saunders; JMD) shagaje (Saunders) tuppo (JMD) GWARI aknuse (JMD) bunburwa (RES) HAUSA burburwa (JMD) burburwa fàdámàà (JMD) gafarafa-tin (Bargery) hàtsàà-hàtsàà (JMD) komayya (Saunders; JMD) matsandaka (JMD) námíjin tsíntsíiyáá (JMD) shagaje (JMD) tsíntsíiyáá = *broom* (JMD; LB) tsíntsíiyár fàdámàà (JMD) tsûmbé (JMD) KANURI barata, gantaska (JMD) YORUBA àgbàdo-ẹsin (JMD; Verger) eéran-awó (JMD; Verger) irungbòn ẹfòn *from* ẹfòn: *buffalo* (Verger) iwó-awó *from* awó: *guinea fowl* (JMD) ọgbe-agufọn *a sp. of bird* (JMD) ọkà-ẹsin = *horse's corn* (JMD; Verger) yayángàn (JMD; Verger)

A loosely-tufted annual, culms erect or ascending to 60 cm high, of roadsides, fallow farmland and weedy waste places, throughout the Region from Maurita-nia to N and S Nigeria, and widespread across tropical Africa, and into warm temperate and tropical countries of Asia and America.

The grass is in general a common weed of cultivation in all countries, and is relished by stock (Mauritania, 4; Senegal, 1; Upper Volta, fodder for horses, 5; Nigeria, 2; etc.). It is reported to be drought resistant in Nigeria (6). This species is closely allied to the Ethiopian teff (*E. tef* (Zucc.) Trotter), source of an important staple in that country. The seed of *E. pilosa* is collected by nomads of Central Sahara under the name *asral* (3). Chemical assays show it to contain a fair quantity of protein at 16.1%. People in Mauritania in time of famine raid ant-hill granaries to obtain the seed for cooking into cakes (4).

References:

1. Adam, 1966,a. 2. Dalziel, 1937: 528. 3. Gast, 1972: 51–58, with chemical analysis. 4. Naegelé, 1958,b: 894. 5. Prost s.n., K. 6. Saunders 15, K.

Eragrostis scotelliana Rendle

FWTA, ed. 2, 3: 387. UPWTA, ed. 1.

A loosely-tufted annual grass to 6 cm high, on iron pans and rocky sites in Guinea to W Cameroons, and extending to Zaïre.
The plant is lemon-scented, sometimes strongly so (1).

References:

1. Amshoff 1235, 4265, K.

Eragrostis squamata (Lam.) Steud.

FWTA, ed. 2, 3: 391.
West African: **SENEGAL** SERER diãmbul (JB) firãnd (JB) WOLOF sèlguf u tan (JB) **SIERRA LEONE** MANDING-MANDINKA wunsune (NWT) SUSU funfuri (NWT) TEMNE *ka*-soi (NWT) **NIGERIA** IGBO obumbe (NWT)

A caespitose perennial grass with culms to 1.2 m high, of roadsides and waste places, often sandy, across the Region from Mauritania to N and S Nigeria, and into equatorial Africa to Zaïre.

It is grazed by all stock in Senegal (2), and is recorded as being tolerant of foot-wear and tear (1).

References:

1. Adam, 1954: 93. 2. Adam, 1966,a.

Eragrostis superba Peyr.

A caespitose perennial grass, culms erect to 1.2 m high, of sandy deciduous bushland or wooded savanna, often in disturbed localities, native of eastern Africa from Sudan to S Africa, and now recorded present in the Region in Upper Volta, Ghana and Nigeria.

It is a weed of cultivation. In Sudan it is considered an important fodder-crop for the dry season on account of its plentifulness; it is also an abundant fodder in the rainy season (2). It is under trial as a fodder-grass in Upper Volta (3).

The root is said to be eaten by cane rats (1).

References:

1. Magogo & Glover 698, K. 2. Peers KO33, K. 3. Scholz 133, K.

Eragrostis tenella (Linn.) P Beauv.

FWTA, ed. 2, 3: 386–7.

West African: SENEGAL SERER *n*disis (JB) duyuy (JB) SIERRA LEONE MENDE nyina foni (FCD) NIGER FULA-FULFULDE (Niger) buluuhi (ABM) SONGHAI talaata-kàmbè cirey = *redhands of Talaata (? meaning)* (D&C) NIGERIA FULA-FULFULDE (Nigeria) budo mboju (J&D) HAUSA askan dawaki (DA) IGBO ilulu nza NWT YORUBA iwa igun (Dodd) ori awó (Macgregor) YORUBA (Ilorin) itẹ ẹiye (Ward)

A delicate tufted annual grass, culms to 30 cm high, a ruderal of waste places, roadsides, and on cultivated land, common throughout the Region from Senegal to W Cameroons, and throughout tropical Africa, and tropical Asia.

It is a common weed of arable land. Cattle everywhere graze it (Senegal, 2; Madagascar, 1; India, 5, 7), but cattle-men in Niger say that their cattle do not find it particularly palatable (6). On the campus of University College, Cape Coast, Ghana, it has been grown as a lawn turf (3), self-sown seed maintaining the sward inspite of the grass's annual habit. It is also grown as a covering on the central divider of a roadway (4).

The grain is said in Rajasthan, India, to be nutritious (5). It may perhaps serve as a famine-food.

Igbo of S Nigeria are reported as using the plant as ' "medicine" against bad medicine' (8), a sort of re-insurance.

References:

1. Adam, 1954: 93. 2. Adam, 1966,a. 3. Easterly 1299, K. 4. Easterly 1236, K. 5. Gupta & Sharma, 1971: 79. 6. Maliki, 1981: 47. 7. Sastri, 1952: 183. 8. Thomas, NWT.1632 (Nig. Ser.), K.

Eragrostis tenuifolia (A Rich.) Hochst.

FWTA, ed. 2, 3: 391.

A slender caespitose grass, culms to 60 cm high, of weedy waste places in highland locations of N and S Nigeria and W Cameroons, and dispersed throughout tropical Africa, and in India, Australasia and S America.

The grass is a weed of cultivation in Sudan (2). It occurs abundantly in grassland subject to heavy grazing (1) and is noted as of importance on the Obudu Cattle Range in Nigeria (5). In Uganda (9), Kenya (4, 6–8) and

GRAMINEAE

Tanganyika (3) it is well liked by all stock and grazed at all times of the year. It is said to be also browsed by hares (7).

References:
1. Chapman 4347, K. 2. D.F., Sudan, s.n., K. 3. Emson 278, K. 4. Glover & al., 439, 1549, 1605, K. 5. Harris 361, K. 6. McDonald 988, K. 7. Magogo & Glover 829, K. 8. Maher 3241, K. 9. Snowden 1024, 1408, K.

Eragrostis tremula Hochst.

FWTA, ed. 2, 3: 391. UPWTA, ed. 1, 528.
West African: SENEGAL FULA-PULAAR (Senegal) kiu pité (A.Chev. fide JMD) paguiri jaule = 'paguiri' of the guinea fowl (A.Chev. fide JMD) sorgobo (A.Chev.) MANDING-BAMBARA otokama (JB) sambu gambi (A.Chev.) wolo gaman (A.Chev.) wolo kaman (A.Chev.) SERER diãmbul (A.Chev.) mbèlkèñ (JB) WOLOF salguf (A.Chev.) sèlgue (JB) THE GAMBIA MANDING-MANDINKA ndirra (JMD) ndirra sina from sina: companion, or mate of (Pirie; JMD) numu ju tio = blacksmith's grass (Pirie) nyantan (JMD) GUINEA MANDING-MANINKA ologuélé (CHOP) SIERRA LEONE KORANKO sɛrɛlinyaxe (JMD) SUSU funfuri (JMD) sankabesuwi (JMD) TEMNE peni-fafagbi (JMD) peni-pa-gbel (JMD) MALI FULA-PULAAR (Mali) kiu pité (A.Chev.) paguiri jaule = 'paguiri' of the guinea fowl (A.Chev.) sorgobo (A.Chev.) MANDING-BAMBARA sambu gambi (A.Chev.) wolo gaman (A.Chev.) wolo kaman (A.Chev.) SONGHAI fitti-fitti (A.Chev.) subu, subu furia furia, subu koré (A.Chev.) TAMACHEK tadjit (A.Chev.) UPPER VOLTA FULA-FULFULDE (Upper Volta) ɓuruudi loosely for certain annal spp. (K&T) MOORE sagha (Scholz) GHANA HAUSA komaya (Williams) NIGER FULA-FULFULDE (Niger) saraho (ABM) HAUSA bibirua (Bartha) SONGHAI kullum (pl. -ó) (D&C) sàriiji (pl. -ò) (D&C) NIGERIA ARABIC-SHUWA rauwaj (JMD) FULA-FULFULDE (Nigeria) dutaleho (JMD) saraaho (JMD) sarawal (Saunders; JMD) shagaje (Saunders; JMD) tappo (JMD) GWARI aknuse (JMD) HAUSA burburwa (auctt.) burburwar fàdámàà (JMD) gafarfatin (Bargery) hàtsàà-hátsàà (auctt.) hure (Kennedy) kòòmáyyàà (auctt.) matsandaka (JMD) matsandaka tsumbe (ZOG) námíjin tsíntsíiyáá (JMD) tsíntsíiyáá = broom (JMD) tsíntsíiyár fadama (JMD) tsùmbé (JMD) IGALA ìyō = ram's mane (Odoh; RB) KANURI barata (JMD) gàndàskà (C&H; A&S) gantaska (JMD) kɔ̀làsɔ̀lɔ́m (C&H) kɔ̀làsêlɛ̄m (A&S) NUPE berberinoa (Blaikie) YORUBA àgbàdo-ẹshin (JMD; Verger) eéran-awó (JMD; Verger) irungbòn ẹfɔ̀n from ẹfɔ̀n: buffalo (Verger) iwó-awó from awó: guinea fowl (JMD) ọgbe-agufọn a sp. of bird (Dodd; JMD) ọka ẹshin = horse's corn (JMD; Verger) yayángán (auctt.)

A loosely-tufted annual grass, culms to 90 cm high with attractive trembling panicles, of farmland, roadsides and waste places, commonly throughout the Region from Mauritania to N and S Nigeria, and throughout tropical Africa and on to India.

The grass is weed in the savanna zone which may at times be a little troublesome (1). By contrast, it is an important and valuable source of fodder either grazed in pasture or in the rough, or cut for sale as a market produce, or for converting into hay (9; Senegal, 1–3, 19; Ghana, 6; Nigeria, 7–8, 13, 15, 18; Kordofan, 5; Sudan, 4; etc.). It is much relished by stock either green, or dry, and graziers in Niger aver that their animals produce rich 'heavy' milk after grazing it (14). Indian material is reported to assay at; protein 3.08%, fibre 38.46%, carbohydrate 45–67%, ash 4.84%, fresh weights (17).

The culms, bundled together, are used as hand-brooms for indoor use (9, 12). The culms may also be used for thatching (9). In the Region (9) and in Kordofan (11) they are woven together to make mats and cordage.

In Central African Republic the leaf-blades bundled together and laid on top of a termite nest are used in a cryptic way to catch the insects (10).

The root possesses a strong caramel-like smell (16).

In time of dearth the seed is collected for eating (1, 8, 9), a practice also known in India (17).

References:

1. Adam, 1954: 93. 2. Adam, 1960,a: 369. 3. Adam, 1966,a. 4. Andrews A.522, K. 5. Baumer, 1975: 98. 6. Beal 9, K. 7. Dalziel 271, K. 8. Dalziel 493, K. 9. Dalziel, 1937: 528. 10. Fay 4439, K. 11. Hunting Tech. Services, 1964. 12. Irvine 4615, K. 13. Kennedy FHI.8025, K. 14. Maliki, 1981: 53. 15. Palmer 11, K. 16. Risopoulos 1267, K. 17. Sastri, 1952: 182–3. 18. Saunders 38, K. 19. Trochain, 1940: 270.

Eragrostis turgida (Schum.) de Wild.

FWTA, ed. 2, 3: 387.
West African: GUINEA MANDING-MANINKA ologonelé (CHOP) TAMACHEK diadié (CHOP) NIGERIA FULA-FULFULDE (Nigeria) samereeho (J&D) vutaleho (Saunders) HAUSA alkamar kwaɗi = frog's wheat (auctt.) YORUBA itẹ ẹmọ (Ward)

A loosely-tufted annual grass with culms to 60 cm high, of roadsides, old farmland and weedy places throughout the Region from Senegal to N and S Nigeria, and extending across central Africa to Sudan, Uganda and Zaïre.
The grass provides good grazing (Senegal, 1; Nigeria, 2).
The grain is collected in Guinea in time of dearth (3).

References:

1. Adam, 1966,a. 2. Mayo, s.n., July 1932, K. 3. Pobéguin 1094, K.

Eragrostis viscosa (Retz.) Trin.

FWTA, ed. 2, 3: 386.

A tufted annual grass, culms to 45 cm high, of dry sandy soil, recorded only in N Nigeria in the Region, but extending to E and S Africa and into tropical Asia to Thailand and the Philippines.
It provides grazing in the dry Rajasthan state of India, but its bulk is insignificant (1).

References:

1. Gupta & Sharma, 1971: 80. 2. Sastri, 1952: 183, as E. tenella var. viscosa Stapf.

Eragrostis welwitschii Rendle

FWTA, ed. 2, 3: 393.
West African: SIERRA LEONE KORANKO sɛrɛlinyaxe (NWT)

A loosely-tufted annual grass to 45 cm high, of old cultivations, roadsides and waste places in Guinea to N and S Nigeria, and dispersed across Africa to Ethiopia, Zaïre and Tanganyika.
It is a weed of cultivation. It survives in heavily grazed pasture on the Jos Plateau, Nigeria, and contributes to the fodder available (1).

Reference:

1. Hepper 1174, K.

Eragrostis spp. indet.

West African: NIGERIA FULA-FULFULDE (Nigeria) saraaho debbo (J&D)

The following two plants are known only by the vernaculars cited above.
1. ida napen occurs in Senegal. Nyominka people fear to touch it as contact on the skin may cause erythema with violent itching (2).

2. *saraaho debbo* is used by Nigerian Fula to make brooms, and by sorcerers (*ciiroowo*) to make offensive magic (1).

References:

1. Jackson, 1973: no. 10.12770. 2. Kerharo & Adam, 1964,c: 306.

Eriochloa fatmensis (Hochst. & Steud.) WD Clayton

FWTA, ed. 2, 3: 437, as *E. nubica* (Steud.) Hack. & Stapf. UPWTA, ed. 1, 528, as *E. acrotricha* Hack.
West African: **NIGER** KANURI buteri (A.Chev.) **NIGERIA** HAUSA geron kiyashi = *ant's millet* (Taylor)

An annual grass, culms up to 75 cm high, geniculate and ascending; of damp places, marshes, lake-sides, lowland to montane elevations across the northern part of the Region from Mauritania to N Nigeria, and widely dispersed from Sudan through tropical E and S Africa, and just entering India.

The grass is tolerant of brackish conditions, and calcareous and gypsum-rich soils. It will grow in swamps with floating culms to 1.7 m long. It is widely deemed to be a good grazing grass, especially in the rainy season (Senegal: 1, 2; Mali: 5; Sudan: 6–8; Kordofan: 3; Somalia: 10; Kenya: 11, 12). It is recorded in Sudan as withstanding short dry periods better than many other grasses in overgrazed land (14). When dry as straw it appears to have no attraction to stock (3). The straw in Kenya is used for thatching (12).

In Tanganyika the culms are cooked with a species of lung-fish and the liquid is used as an eye-lotion (4).

Small birds in NE Nigeria seek out the grain (9). Some people of the Malian Sahel collect and eat it (5). The Hausa name meaning 'ants' millet' suggests that ants collect it for storage.

References:

1. Adam, 1954: 93, as *E. nubica* Stapf. 2. Adam, 1966,a: as *E. acrotrichia*. 3. Baumer, 1975: 98, as *E. nubica* (Steud.) Hack. & Stapf. 4. Bullock 2409, K. 5. Dalziel, 1937: 528, as *E. acrotricha* Hack. 6. Davies BN.17, K. 7. Evans Pritchard 15, K. 8. Harrison 11, 1008, 1063, K. 9. Jackson 11.12770, K, UCI. 10. McKinnon S.29, K. 11. Magogo & Glover 696, K. 12. Magogo & Glover 1052, K. 13. Myers 13896, K. 14. Peers K.O.1, K.

Euclasta condylotricha (Hochst.) Stapf

FWTA, ed. 2, 3: 471.
West African: **GUINEA** BASARI a-yèn gekòl (FG&G) **SIERRA LEONE** YALUNKA fuŋfure-khamɛ-na *a general name for certain grasses with large twisted awns* (FCD) **UPPER VOLTA** GRUSI minga (Scholz) **NIGERIA** BUSA kuse togun (Ward) FULA-FULFULDE (Nigeria) ride bara (Ward) HAUSA tusun zake (Ward)

An annual grass with weak slender rambling and geniculately ascending culms to 1½ m high; of partial shade in wooded grassland or deciduous bush-land; common across the savanna zone of the Region from Senegal to N and S Nigeria, and E and S tropical Africa; also in America.

The grass is grazed by cattle in Senegal (1). In Northern Territory, Ghana, cattle are recorded as finding it unpalatable (3).

It is a weed of rice-padis in Sierra Leone (2).

References:

1. Adam, 1966, a. 2. Deighton 4435, K. 3. Ll. Williams 854, 855, K.

Festuca Linn.

FWTA, ed. 2, 3: 367, 369. UPWTA, ed. 1, 528, as *F. gigantea* sensu Dalziel (= *F. mekiste* WD Clayton).

A genus of tufted grasses of montane locations. The following species have been recorded present in W Cameroons and Fernando Po: 1. *F. abyssinica* Hochst.; 2. *F. chodatiana* (St Yves) Alexeev; 3. *F. mekiste* WD Clayton; 4. *F. rigidiuscula* Alexeev; 5. *F. schimperiana* A Rich.; and 6. *F. simensis* Hochst. No usage has been reported for any of these, but *F. chodatiana* (2) and *F. schimperiana* (3) are said to be potentially useful fodders at high altitudes in Uganda. *F. simensis* is grazed by all stock in the Narok District of Kenya (1). The W African species are likely to have value as grazing in highland areas though the amount of foliage may be rather limited.

References:

1. Glover & al. 1455, K. 2. Snowden 1459, K. 3. Snowden 1478, 1482, K.

Guaduella oblonga Hutch.

FWTA, ed. 2, 3: 360.
West African: SIERRA LEONE MENDE kotopɔ (*def.* -i) (FCD) pɔŋgi (NWT) pɔvi-hina = male 'povi' (FCD) pɔvɔ-hina (FCD) TEMNE ɛ-sul-ɛ-ro-kant (NWT)

A herbaceous bamboo, short creeping rhizome, of the floor zone of W African forest from Guinea to Ivory Coast and extending into Angola.

Mende of Sierra Leone confuse this plant with *Olyra latifolia* giving it the same name. It may have similar uses.

Gynerium sagittatum (Aubl.) P. Beauv.

FWTA, ed. 2, 3: 374.
English: uva grass; wild cane (Grounds); white roseau (Caribbean).

A huge cane-like plant, stout woody culms to 8 m high by 3–4 cm diameter bearing a large plumose panicle to 1 m long; native of the Caribbean and S America to Argentina, and introduced into Ghana.

The plant in flower is ornamental. It is used in the Caribbean area for light construction-work (Barbados, 5; Panama, 7) and in Trinidad as laths and the leaves for thatching (3, 4). Throughout S America the canes are used as arrow-shafts (Brazil, 2, 8; Guyana, 1; Venezuela, 10; Bolivia, 9; etc.).

The root ground up is said to have been used in Venezuela on corns and calluses (6).

References:

1. Archer 2314, K. 2. Balée 1074, K. 3. Beard 136, K. 4. Freeman & William, 1928: 83. 5. Gooding & al., 1965: 40–41. 6. Hartwell, 1969,b: 192. 7. Herb. Hook. (Tlayu) 82, K. 8. Prance & al. 10530, K. 9. Renvoize 4724, K. 10. Wurdack & Adderley 43166, K.

Hackelochloa granularis (Linn.) O Ktze

FWTA, ed. 2, 3: 505–6. UPWTA, ed. 1, 528.
West African: SENEGAL MANDING-BAMBARA susan kaba = *hare's maize* (JB) SIERRA LEONE KORANKO fesifesi (NWT) LOKO andande (NWT) MANDING-MANDINKA saŋwanya (FCD) MENDE gungulwi (NWT) TEMNE *ko*-palɛngɛ-*kɔ*-ropetr (NWT) YALUNKA tamedi-sara-na (FCD) MALI FULA-PULAAR (Mali) bambari ladde (A.Chev.) ngoriri (A.Chev.) MANDING-BAMBARA susan kaba = *hare's maize* (A.Chev.) SONINKE-SARAKOLE kanyané maka (A.Chev.)

GRAMINEAE

UPPER VOLTA FULA-FULFULDE (Upper Volta) njadere (Scholz) GRUSI zgrooména (A.Chev.) MOORE soham kamani (A.Chev.) suambu kamana (A.Chev.)

An annual grass reaching 60 cm high; anthropogenic, weed of cultivation, around habitations and waste places; throughout the Region from Senegal to W Cameroons, and widely dispersed over the rest of tropical Africa and throughout the tropics generally.

The grass is a weed of all arable crops and under plantation crops. It gives palatable grazing relished by all stock at all times, but it is short-lived and therefore never abundant (1–5, 10). It is a good fodder for horses (4, 5, 7) and in Bauchi is grown as pasture for them. In India it is deemed to produce good hay (7).

The foliage is covered with hairs which are irritant to the skin (6, 8).

In Central African Republic, where it grows is thought to be good ground for planting sesame (6), yet conversely in Zimbabwe it is an indication of land not suitable for maize (9).

References:

1. Adam, 1954: 93. 2. Adam, 1966, a. 3. Beal 7, K. 4. Dalziel, 1937: 528. 5. Deighton 4429, K. 6. Fay 1982, K. 7. Gupta & Sharma, 1971: 81–82. 8. Langdale-Brown 2396, K. 9. Lawrence 619, K. 10. Magogo & Glover 1082, K.

Helictotrichon elongatum (A Rich.) CE Hubbard

FWTA, ed. 2, 3: 372.

A tufted perennial grass, fibrous-rooted, culms geniculate at base to 1½ m high, of montane forest and open grassland in N Nigeria and W Cameroons, and dispersed in eastern Africa from Sudan and Ethiopia to Zimbabwe and Madagascar.

Abundance of the grass seems to vary, but where present in quantity (Kenya, 1; Tanganyika, 2, 3) it provides grazing for all stock.

References:

1. Glover & al., 1761, K. 2. McGregor 21, K. 3. Vezey-Fitzgerald 4240, 5284, K.

Hemarthria altissima (Poir.) Stapf & CE Hubbard

FWTA, ed. 2, 3: 506.
West African: **NIGERIA** FULA-FULFULDE (Nigeria) burgu = *horse grass* (J; JMD) HAUSA damarage (Palmer; ZOG) manu (Lowe)

A stoloniferous perennial grass with culms decumbent, ascending to 1.6 m high; of damp and wet sites, shallow water, stream-sides, lakesides; in the soudanian zone of Senegal, Mali and N Nigeria, and found widespread in NE, E, south central and S Africa, the Mediterranean and SE Asia.

It is an excellent forage grass, remaining green all the year round and tolerant of dry conditions, growing rapidly. It is readily grazed by all stock (1, 2). The Fula name *burgu* meaning 'horse grass' is a group term applied to several grasses of special value as horse-fodder (4). In Zimbabwe it has been considered to have potential value as a pasture grass (5).

In Lesotho the stems are plaited into ropes used for making screens for courtyard enclosures (3).

Sotho children eat the rhizome raw (3).

References:

1. Adam, 1954: 93. 2. Adam, 1966, a. 3. Guillarmod, 1971: 431. 4. Jackson 9.7770, K. 5. Trapnell 733, K.

258

Heteranthoecia guineensis (Franch.) Robyns

FWTA, ed. 2, 3: 420–1.
West African: **SIERRA LEONE** MENDE alate (NWT) SUSU boxifɔrotai (NWT) TEMNE silɛrɔ (NWT)

A mat-forming annual grass, culms rooting at lower nodes, ascending to 60 cm high, in swamps, shallow streams and pond margins, in Guinea to N and S Nigeria, and extending into E, central and SW Africa.
Dogs are reported in Gabon to eat this plant as a purge when feeling off-colour (1, 2).

References:

1. Walker, 1953,a: 39. 2. Walker & Sillans, 1961: 189.

Heteropogon contortus (Linn.) P Beauv.

FWTA, ed. 2, 3: 473. UPWTA, ed. 1, 529.
English: spear grass; wild oats.
West African: **SENEGAL** MANDING-BAMBARA fila ntaso (JB) **MALI** FULA-PULAAR (Mali) moloko (A.Chev.) niadéré (A.Chev.) MANDING-BAMBARA fila ntaso (A.Chev.) fulanu ntaso (A.Chev.) SONINKE-SARAKOLE guémémé (A.Chev.) **UPPER VOLTA** FULA-FULFULDE (Upper Volta) selɓo (*pl.* celɓi) (K&T) GURMA komango (A.Chev.) MOORE bubongnona sando (A.Chev.) **GHANA** GA akɔsɔfõ (FRI) ananugãi = *spider's arrows* (FRI; Ankrah) HAUSA chiga = *a spike of grass*; applied loosely (Ll-W) **NIGER** SONGHAI bàt círéy (*pl.* -á) = *box of the red amulet* (D&C) **NIGERIA** GWARI kambarahi (JMD) HAUSA bara babba tudu = *lives on a big hill* (Taylor) bunsurun daji = *bush he-goat; applied to several grasses of objectionable qualities* (JMD) buzun kura = *mat of the hyena* (AST; JMD) jan gargan (Kennedy) silka *applied to this spp., but properly for any spike of grass* (JMD; ZOG) tsiígaà, tsiíkaà *applied to any spike of grass* (JMD; ZOG) yartudu (Kennedy) JUKUN (Wukari) sin (Shimizu) YORUBA (Ilorin) ẹru bẹrẹ (Ward)

An untidy tufted perennial grass with culms to about 1 m high; spikelets with awns 5–8 cm long, very pungent; of deciduous bushland and wooded grassland; fairly common from Senegal to N Nigeria, and widespread over the rest of tropical Africa and other warm and temperate regions.
It is an useful fodder relished by all stock up to the time of anthesis, but nutritional value is not high (3, 9). Thereafter it becomes dangerous because of the pungent awns which injure the mouth of grazing animals and if ingested may penetrate the intestines, set up peritonitis and cause death. The awns also entangle with the fur of animals, especially the fleece of sheep. Skin is penetrated and hides are ruined. Severe abcesses may result. (1–3, 6, 9, 11, 12.) The awns give rise to the English name: spear grass. They adhere in masses causing detachment of the spikelet and are blown about by the wind as a 'tumble weed', thus dispersing the plant (4).
The grass is commonly woven into mats (4, 10) and everywhere is used for thatching huts (4, 10, 11, etc.). It has been examined for pulping with poor results (3, 11).
The presence of the grass is understood by the Fulfulde of N Nigeria to indicate fertile soil (7). In India it has been established with success as a protective cover on saline areas (6). It is very drought-resistant, and is hardy so that it may be difficult to eradicate under agricultural operations (9).
In Lesotho the plant is used with *Tribulus terrestris* (Zygophyllaceae) as a remedy for rheumatism in the hands (5, 11). Use of the root is similarly recorded in India (9). The plant is diuretic (9).
In Zambia bees have been observed visiting the plant for pollen (8).

GRAMINEAE

References:

1. Adam, 1966, a. 2. Baumer, 1975: 101. 3. Burkill, IH, 1935: 1143. 4. Dalziel, 1937: 529. 5. Guillarmod, 1971: 435. 6. Gupta & Sharma, 1971: 81. 7. Jackson, 1973. 8. Robson 1029, K. 9. Sastri, 1959: 42–44, with chemical analysis. 10. Thornewill 184, K. 11. Watt & Breyer-Brandwijk, 1962: 473, 493. 12. Williams, RO, 1949: 286–7.

Heteropogon melanocarpus (Ell.) Benth.

FWTA, ed. 2, 3: 473.
West African: SENEGAL SERER balãmbal (JB) WOLOF balãmbal (JB)

A robust annual grass, culms to 2½ m high, stilt-rooted; of abandoned cultivations and path-sides; present in Senegal, The Gambia, Dahomey and N Nigeria, and over the rest of tropical Africa, and into India and tropical America.

Cattle are reported grazing it in Senegal before anthesis (1). The awns are recognised as dangerous to stock, c.f., H. contortus.

Reference:

1. Adam, 1966, a.

Hordeum vulgare Linn.

FWTA, ed, 2, 3: 371. UPWTA, ed. 1, 529.
English: barley.
French: orge.
Portuguese: cevada.
West African: MALI SONGHAI farka subu = donkey's grass (A.Chev.) TAMACHEK timsin (JMD) NIGERIA ARABIC-SHUWA sha'ir (JMD) FULA-FULFULDE (Nigeria) sha'iruri (JMD) KANURI sa'ir (JMD)

An annual herb growing in a clump of culms to about 50 cm high, bearing bearded or not bearded, or bristle-awned seed in 2, 4 or 6 rows in the spike; W African cultivars have 6 rows and are grown in the Sahel of the Region.

This is the common barley of the temperate world. It is thought to have arisen in the Mediterranean area and to have been in cultivation from ancient times. Grain has been found in Neolithic remains dated 4,000–3,200 B.C. and frequently in Egyptian tombs of a millenium later. Barley is more tolerant of dry conditions than is wheat and by the time of the Punic Wars, Third and Second centuries B.C., it was already dispersed and cultivated in the Saharan oases. Its migration southwards into W Africa seems to have been in about the Sixteenth century. It came with its Arabic name: shair into Nigeria and a loan name from the Berbers of the Saharan oases: tumzein into Mali, both words implying a 'grain for food'. It is now grown as a cold season crop, often under irrigation, for its grain. The grain, unlike that of wheat, lacks gluten so that its flour will not produce a risen loaf but baking results in flat biscuits. It is the standard ingredient par excellence for making porridge.

References:

1. Dalziel, 1937: 529. 2. Irvine, 1948: 264. 3. Mauny, 1953: 713–4. 4. Portères, 1958: 14–34.

Hyparrhenia Fourn.

FWTA, ed. 2, 3: 490–1.

West African: **SENEGAL** FULA-PULAAR (Senegal) bowal (GR) MANDING-MANDINKA fuka-bïn (GR) kali (GR) kali ni *group term* (GR) kali-bâ (GR) **GUINEA-BISSAU** MANDING-MANDINKA nhentam-ô, nhentara-ô (JDES) **NIGERIA** IGBO àzàrà = *spear-grass* (BNO; KW)

Essentially an African genus, of the savanna, and dominant there. Of the 55 species of the genus, 28 occur in the Region; mainly used for thatching and providing grazing in the young stages of growth (2, 3). In Sierra Leone the culms of many species are used for making improvised beehives for taking swarms (1).

The vernaculars cited above are group terms.

References:

1. Boboh, 1974. 2. Clayton, 1969. 3. Clayton & Renvoize, 1986.

Hyparrhenia barteri (Hack.) Stapf

FWTA, ed. 2, 3: 492.

West African: **NIGERIA** HAUSA zama (Kennedy)

An annual grass with culms to nearly 2 m tall; on poor soils of roadsides and old farmland; in Togo and N and S Nigeria, and in the Congo basin, Malawi and Zambia.

The culms are used in N Nigeria to weave into Zaria mats and to rove into roping (1).

Reference:

1. Kennedy FHI.8037, K.

Hyparrhenia bracteata (Humb. & Bonpl.) Stapf

FWTA, ed. 2, 3: 494.

A densely-tufted perennial grass with culms 60–150 cm high; on seasonally wet grassland at submontane to montane elevations; found in scattered locations from Ivory Coast to W Cameroons, and in E Africa and tropical America.

On the Vodni Plateau of Nigeria it provides grazing for cattle during the wet season (2), and in W Cameroons it is found in open grassland and has been encouraged in farm pasture (1).

References:

1. Bampus July 46/Bam/26, K. 2. Saunders 9, K.

Hyparrhenia collina (Pilger) Stapf

FWTA, ed. 2, 3: 494.

A loosely-clumped perennial grass with short rhizomes, putting up wiry slender culms to 1.2 m tall; on damp soils of mid-montane grassland; recorded only in N Nigeria and W Cameroons in the Region, and occurring widely in eastern tropical Africa from Sudan to Natal.

In Uganda (2) and Masailand, Kenya (1) it is considered a good grazing grass for all domestic stock. In Masailand it furnishes material for thatching (1).

References:

1. Glover & al. 875, K. 2. Snowden 1469, K.

GRAMINEAE

Hyparrhenia cyanescens (Stapf) Stapf

FWTA, ed. 2, 3: 494.
West African: NIGERIA HAUSA jimpi (Thatcher) timpi (Freeman; Thatcher) YORUBA (Ilorin) fafa (Ward) wafa (Ward)

A robust perennial grass with culms up to 3 m tall; of moist alluvial soils; across the Region from Senegal to N and S Nigeria, and on to Angola.
The grass in Nigeria in its young stage is evanescent, producing inflorescences early. It is thus of little use as fodder (3). In the Zaria area it is used for thatching (2, 3).
Chromosome number is 2n = c.36 (1).

References:

1. Clayton, 1969: 121. 2. Thatcher Samaru/17, K. 3. Thatcher Samaru/19, K.

Hyparrhenia cymbaria (Linn.) Stapf

FWTA, ed. 2, 3: 494.
West African: WEST CAMEROONS BAMILEKE abebe (DF) FULA-FULFULDE (Adamawa) huddo general for 'grass' (DF) HAUSA gamba (DF)

A robust perennial grass, in coarse tufts from slender rambling rhizomes; culms erect, 2–3½ m high, stilt-rooted; of the savanna in uplands and montane situations; in N Nigeria and W Cameroons, and in eastern Africa from Eritrea to Natal and western Africa southwards to Angola; also in Madagascar and Comoros.
The grass is relished by stock while it is still young (Uganda: 6; Kenya: 3, 4, 5). In Kordofan it is said to be heavy yielding and to produce a very palatable hay (1).
In Kenya it is used for thatching (3, 4).
Chromosome numbers 2n = 20 and 30 are recorded (2).

References:

1. Baumer, 1975: 102. 2. Clayton, 1969: 111. 3. Glover & al. 1803, K. 4. Glover & al. 1978, 2314, K. 5. McDonald 947, 1019, K. 6. Snowden 1162, K.

Hyparrhenia diplandra (Hack.) Stapf

FWTA, ed. 2, 3: 496.
West African: SIERRA LEONE KORANKO nyandabinye (FCD) NIGERIA BIROM myèl (LB) HAUSA kíbííyar dááji = bush arrow (LB) tsííkar dááji (LB) WEST CAMEROONS BAMILEKE yemuwe fulako (DA) yemuwel gorko (DA)

A coarse perennial grass with culms 2–3 m tall; of variable habitats, mainly damp places in deciduous bushland and wooded savanna at submontane to montane altitudes; from Senegal to W Cameroons, and dispersed elsewhere throughout tropical Africa, and in SE Asia.
The grass provides good forage for stock when young (1, 6). On the Obudu Plateau of N Nigeria this is one of the three dominant species of the grass community providing important open-range grazing (5, 7). Chemical assay of Kenyan material has given: crude protein 5.66%, fibre 36.42%, and carbohydrate 47.15% on dry weight (4). Protein content is disappointingly low.
In Kenya at Embu, the grass is planted to mark plot borders (2).
Chromosome number is 2n = 20 (3).

References:

1. Adam, 1966, a. 2. Bogdan AB.2688, K. 3. Clayton, 1969: 171. 4. Dougall & Bogdan, 1960: 241–4. 5. Tulley, 1966: 899–911. 6. Walker & Sillans, 1961: 189. 7. Wheeler Haines 370, K.

Hyparrhenia figariana (Chiov.) WD Clayton

FWTA, ed. 2, 3: 494.

An annual grass with culms erect to 2 m high; of waste land, roadsides and in deciduous bush; in upland situations of N Nigeria, and in Zaïre, Sudan and E Africa.

It is reported to be grazed with relish in Uganda by cattle (1), and to be of little use for grazing in Sudan (2), where it serves for thatching (3).

References:

1. Dyson-Hudson 21, K. 2. Harrison 1036, K. 3. Jackson 320, K.

Hyparrhenia filipendula (Hochst.) Stapf

FWTA, ed. 2, 3: 494.
West African: NIGERIA FULA-FULFULDE (Nigeria) wadeho (Saunders) HAUSA tsiikaà (Saunders)

A caespitose perennial grass with culms to 2 m tall; of the savanna in Guinea, N Nigeria and W Cameroons, and a common species of open situations throughout the rest of tropical Africa and S Africa, and into Ceylon and Australia. In Uganda the grass is common in the vicinity of termitaria, some acting as the thicket centre (14). The two varieties of *FWTA*, ed. 2 are now no longer recognised as separate.

The grass provides grazing on the Vodni Plateau of N Nigeria (7), and in all other territories especially while still young (Sudan: 6; Uganda: 10; Kenya: 4, 5; Tanganyika: 3; Zambia: 12; Zimbabwe: 9). In Rwanda it is considered to be the most important highland grass species (8), and in Karamoja district of Uganda its local name means 'tall and good grass' (2); cf. *H. poecilotrichia*. In Sudan it is said to be palatable even when matured in the dry season (6).

In all territories it is used for thatching (Uganda: 'The best thatching grass', 11; Kenya: 4; Tanganyika: 3; Zimbabwe: 9, 13).

Tests for pulping quality have shown it to be suitable (13).

Chromosome number is 2n = 40 (1).

References:

1. Clayton, 1969: 98. 2. Dyson-Hudson 311, K. 3. Emson 292, K. 4. Glover & al. 485, 1900, K. 5. McDonald 986, 1014, K. 6. Peers KO.22, K. 7. Saunders 47, K. 8. Schautz 739, K. 9. Senderayi 3, K. 10. Snowden 1147, 1148, 1149, 1276, K. 11. Thomas AST.4004, K. 12. Trapnell 808, 859, K. 13. Watt & Breyer-Brandwijk, 1962: 493. 14. Wood 955, K.

Hyparrhenia finitima (Hochst.) Anderss.

FWTA, ed. 2, 3: 492.

A tufted perennial grass with robust culms 1–2 m high; a ruderal of roadsides, waste places and farm fallows, deciduous bushland and wooded grassland; recorded only in Sierra Leone in the Region, but widely dispersed on the eastern side of Africa from Ethiopia to S Africa. In Malawi it is recorded as being found in *Brachystegia–Isoberlinia* woodland in association with ant-hills (2).

In Zambia it is used as a thatching grass (1).

References:

1. Duff 1148, K. 2. Jackson 846, K.

GRAMINEAE

Hyparrhenia glabriuscula (A Rich.) Stapf

FWTA, ed. 2, 3: 491.
West African: SENEGAL BASARI ɔ-ndji (K&A) BEDIK gi-ndyi gu-mbara (FG&G) GUI-
NEA BASARI ɔ-nydi (FG&G) UPPER VOLTA FULA-FULFULDE (Upper Volta) bukkaho (K&T)

A tufted perennial grass with culms up to 1½ m high; of seasonally swampy
sites and river flood-plains; across the Region in Senegal, Upper Volta, Ghana
and N Nigeria, and of isolated occurrence in Ethiopia, Malawi and
Mozambique.

No specific usage is recorded, though the grass is evidently recognised by
Basari in Senegal as having medicinal property (1).

Reference:

1. Kerharo & Adam, 1964, a: 436, as *H. amoena* Jac.-Fél.

Hyparrhenia hirta (Linn.) Stapf

FWTA, ed. 2, 3: 492.
French: barbon (Roberty).

A caespitose perennial with dense basal tussock, briefly rhizomed, culms
30–60 cm high; of dry open places, or deciduous bushland and montane
grassland; known only in Niger in the Region, and dispersed over the rest of
tropical Africa and into the Mediterranean; also in Australia and central
America, perhaps introduced.

It is readily grazed by all domestic stock, even camels in Sudan (1), Jebel
Marra (2), Somalia (5, 6, 9), and Kenya (7), but reluctantly in Congo (Brazzaville) (10). It serves well as a thatching grass (7, 8, 11). Tests for pulping have
shown it to be unsatisfactory (11).

In Lesotho the culms are used to weave large baskets (1½–2 m high) for
storage of grain (8).

Chromosome numbers have shown a variation between 2n = 30 and 45 (3, 4).

References:

1. Andrews A 3554, K. 2. Blair 289, K. 3. Celarier A-3026-I, K. 4. Clayton, 1969: 78. 5. Collenette
37, K. 6. Glover & Gilliland 735, K. 7. Glover & al., 2224, 2547, K. 8. Guillarmod, 1971: 435. 9.
McKinnon S.119, S.216, K. 10. Quarré 7839, K. 11. Watt & Breyer-Brandwijk, 1962: 493.

Hyparrhenia involucrata Stapf

FWTA, ed. 2, 3: 496.
West African: GHANA HAUSA shumrayi (Ll.W) NIGERIA BUSA (Bokobaru) fono (Ward)
FULA-FULFULDE (Nigeria) fetinari (Ward) HAUSA kakario (Ward) kyara (Taylor) IDOMA
iganapa (Odoh) IGALA iganapa (Odoh; RB)

A robust annual grass with culms to 2 m high; represented by two varieties:
var. *involucrata* and var. *breviseta* WD Clayton, distinguished on awn length but
of no significance regarding usages; in savanna on dry or shallow soils from
Ghana to S Nigeria, and in Chad and the Central African Republic.

The grass provides thatching material (2, 3), and at Tamale, Ghana, the culms
are woven into coarse matting (2).

Chromosome number is 2n = 40 (1).

References:

1. Clayton, 1969: 159. 2. Ll. Williams 404, K. 3. Ward 0018, 0019, K.

264

Hyparrhenia newtonii (Hack.) Stapf

FWTA, ed. 2, 3: 494.

West African: NIGERIA FULA-FULFULDE (Adamawa) yemuwel debbo (Pedder)

A densely tufted perennial grass with culms 60–120 cm high; on stony hillside locations at submontane to montane elevations; occurring in Guinea to W Cameroons, and in E and S Africa; also widely dispersed over tropical Asia and Malesia.

In W Cameroons (2) and in Uganda (4) it is considered an useful grazing grass. In Zambia it furnishes thatch (3).

Chromosome number is 2n = 40 (1).

References:

1. Clayton, 1969: 149. 2. Hepper 1424, K. 3. Michelmore 334, K. 4. Snowden 1496, K.

Hyparrhenia nyassae (Rendle) Stapf

FWTA, ed. 2, 3: 491.

A tufted perennial grass with culms up to 1½ m high; of savanna woodlands, especially damp sites and swamp edges; known only as a single collecting in Ghana in the Region, and occurring in NE, E, S tropical, and S Africa, and in Indo-china.

The grass is browsed by stock in Sudan (7) and Tanganyika (2, 3) while it is still young and tender. In Zambia the haulm serves as an inferior thatching material (6).

The grain is said to have been eaten in the Upper Nile area in time of famine (4, 5).

Chromosome number is reported as 2n = 20 and 40 (1).

References:

1. Clayton, 1969: 55. 2. Emson 127, K. 3. McGregor 24, K. 4. Speke, 1863: 652, as *Anthristiria ciliata* Retz. 5. Speke & Grant s.n. (Herb. Hook.), K. 6. Trapnell 2542, K. 7. Wickens 1389, K.

Hyparrhenia poecilotrichia (Hack.) Stapf

FWTA, ed. 2, 3: 492.

A perennial grass with culms 60–150 cm high; of savanna and deciduous bushland from lowlands to montane elevations in Guinea to W Cameroons, and in E Africa to S Africa. A variable species, perhaps a hybrid (1).

There is no record of any usage in the Region, but in Uganda it is an useful fodder grass for the dry season and is known in the Karamoja district as *lojokopolon* meaning 'tall grass and good' (2); cf. *H. filipendula*, perhaps a group term.

References:

1. Clayton, 1972: 796–7. 2. Dyson-Hudson 267, K.

Hyparrhenia quarrei Robyns

FWTA, ed. 2, 3: 492.

A tufted perennial grass with culms 1–2 m high; a ruderal of roadsides, clearings and disturbed sites, and deciduous bushland, wooded savanna and evergreen forest margins, on foothills to montane situations; known only in N Nigeria in the Region, and occurring in E Africa, central and S Africa.

In Masailand, Kenya, it is grazed by all domestic stock, and is used for thatching huts (1).

Reference:

1. Glover & al. 2062, 2226, K.

Hyparrhenia rudis Stapf

FWTA, ed. 2, 3: 494.

A coarsely tufted perennial grass, culms 2–3 m tall; of savanna grassland on moist soils; recorded in Ghana and N Nigeria, and widely dispersed elsewhere in tropical Africa from Ethiopia to S Africa.
Cattle in Tanganyika browse it while it is still young (1).

Reference:

1. McGregor 23, K.

Hyparrhenia rufa (Nees) Stapf

FWTA, ed. 2, 3: 492. UPWTA, ed. 1, 529.
West African: SENEGAL MANDING-BAMBARA bin blé (JB) THE GAMBIA KORANKO tikole (Glanville; JMD) MANDING-MANDINKA nyantang furo (Macluskie) YALUNKA khalan-sogo-na (FCD) SIERRA LEONE MANDING-MANDINKA tikolo-yɔ (FCD) UPPER VOLTA GRUSI-KASENA uoh (Scholz) GHANA DAGBANI zankalago (JMD) NIGERIA ARABIC-SHUWA nal al'afin (JMD) BIROM dùúl (LB) BUSA (Bokobaru) fono (Ward) gamare (Ward) FULA-FULFULDE (Nigeria) kalawal (JMD) lemno (JMD) nyanyanga (Saunders; JMD) soɓarla (JMD) wodeho (JMD) GWARI eji (JMD) HAUSA jinfi (Taylor) kyara na fadama (Ward) yamáá (auctt.) KAMBARI-SALKA ĩidyngurùmi (RB) KANURI kinditilo (JMD) TIV ijinga i pupu (JMD) YORUBA (Ilorin) alolo (Ward)

A tufted perennial, occasionally annual, with culms 30 cm to $2\frac{1}{2}$ m tall, variable; typically of the tall savanna of the deciduous bushland and wooded grassland favouring damp sites, but also a ruderal of roadsides and frequented places; common throughout the Region from Mauritania to W Cameroons, and dispersed over the rest of tropical Africa, and in tropical America and introduced to most other tropical countries.

Robust forms are grown in NW Nigeria as grass fencing (7). The haulm when mature is commonly used for thatching (8, 11, 13, 21; Kordofan: 3; Sudan: 2; Kenya: 14). Culms admixed with *Andropogon pseudapricus* (Gramineae) are often puddled with clay for building hut-walls (7, 20). They are also fashioned into *zaanaa* mats (7, 20) and have also shown some potential for converting into paper-pulp (7, 23).

The grass produces one of the best grazing fodders while still young (Senegal: 1; Sierra Leone: 9; Nigeria: 17, 20–22; Gabon: 19; Kordofan: 3; Sudan: 2, 4, 12, 15; Uganda: 18; Zanzibar: 23). There are reports in Kordofan (3) and in Kenya (14) of good resistance to dry conditions, producing fodder which stock will browse when it is older. It is also reported in Kordofan to convert well into hay and silage (3). In Dahomey plants growing on swamp edges remain palatable the year round (16). Cultivation as a pasture grass in N Nigeria has not been successful because of the low quantity of herbage produced (10).

A squat form has been grown in Zambia under frequent cutting to produce a 'lawn' (5). Early grazing will produce the same effect, a thick close sward (6).
Chromosome number is variable, 2n = 30, 36 and 40 being recorded (6).

References:

1. Adam, 1966, a. 2. Andrews A.744, K. 3. Baumer, 1975: JATBA 22: 102. 4. Beshir 22, K. 5. Bullock 3949, K. 6. Clayton, 1969: 63, 64. 7. Dalziel 487, K. 8. Dalziel, 1937: 529. 9. Deighton 4423, K. 10. Foster & Munday, 1961: 314. 11. Glanville 318, K. 12. Harrison 3, 215, 249, K. 13. Jackson, 1973. 14. Kinyua 4, K. 15. Peers KM.9, K. 16. Risopoulos 1258, K. 17. Saunders 2, K. 18. Snowden 1145, K. 19. Walker & Sillans, 1961: 189. 20. Ward 0029, 0030, K. 21. Ward 0050, K. 22. Ward L.152, K. 23. Williams, RO, 1949: 281–2.

Hyparrhenia smithiana (Hook.f.) Stapf

FWTA, ed. 2, 3: 491.
West African: SIERRA LEONE MANDING-MANDINKA fualobī (FCD) YALUNKA lōfu-na (FCD) UPPER VOLTA FULA-FULFULDE (Upper Volta) yantaare (pl. jantaaje) (K&T) NIGERIA FULA-FULFULDE (Adamawa) geene saabal (Pedder)

A tufted perennial grass, endemic to the Region and present in two varieties: 1. var. *smithiana*, culms 30–90 cm high, recorded in Sierra Leone, Upper Volta and W Cameroons; and 2. var. *major* WD Clayton, culms 1.5–2.4 m high, and recorded throughout the Region from Senegal to N and S Nigeria.

Var. *major* is recorded as furnishing a good thatch in N Sierra Leone (2).

Chromosome number is not definite, but is perhaps 2n = 40 (1).

References:

1. Clayton, 1969: 58. 2. Deighton 4522, K.

Hyparrhenia subplumosa Stapf

FWTA, ed. 2, 3: 496. UPWTA, ed. 1, 529.
West African: SIERRA LEONE KORANKO ɡbonkilobon (JMD) nyanda binye (Glanville) MANDING-MANDINKA banbira (NWT) tikolo-yɔ (FCD) MENDE ngale (NWT) SUSU pankasaxi (NWT; JMD) TEMNE ko-bɔŋkolɔ (NWT; JMD) YALUNKA khalansogo-na (FCD) UPPER VOLTA BISA lassa (Scholz) FULA-FULFULDE (Upper Volta) bukkaho (K&T) NIGERIA FULA-FULFULDE (Nigeria) celɓol (pl. celɓi) (JMD) garlabbe (JMD) selbo (Saunders; JMD) woɗeho (JMD) HAUSA buta, butar kurege (JMD) cìkà-dááfi (JMD; ZOG) cìkà-dáájìì = fills the bush (JMD; ZOG) cìkà-dáwà (JMD; ZOG) kíbíar-dáájìì from kíbía: arrow; alluding to the arrow-like reflexed, hard-pointed flower-spikes (JMD; ZOG) lalemo (JMD; ZOG) tsííkàr-dáájìì (JMD; ZOG) tsííkàr-dáwà (JMD; ZOG) tumàà dà gòòbáráá = twitching in conflagration; from tuma: jumping, jerking; da gobara: in conflagration; alluding to the jerking of the dry spikes when burning in a bush-fire (JMD) KAMBARI-SALKA mɔ̀kɔ̀rô (pl. ngk-) (RB)
NOTE: Nigerian names apply to other Hyparrhenia spp. and probably to other genera with similar flower-spikes.

A robust perennial grass with culms 2–3 m high; of guinean savanna woodland; dispersed commonly throughout the Region from Senegal to N and S Nigeria, and extending southwards to Angola, and into Tanganyika.

The grass is palatable to cattle only while it is still young (1, 4, 5, 8). In N Dahomey there is a local vernacular name (not cited) translatable as 'thickets where antelopes are trapped' (7), suggesting that game may assemble there for feeding.

It is used as thatching material generally deemed to be good (4, 5, 8), but at Mamodia in Sierra Leone it is not favoured as there it is too coarse and brittle (6). Salka peoples of NW Nigeria chop up the canes to make building blocks (2) bonded with clay (?). Bristly points on the haulm render it unsuitable for *zaanaa* matting (4). It has been reported usable for paper-pulp (4).

Chromosome number is 2n = 40 (3).

Hausa people hold a prejudice against the canes used as brooms lest the user loses his goods and dies (4).

References:

1. Beal 32, K. 2. Blench, 1985. 3. Clayton, 1969: 164. 4. Dalziel, 1937: 527. 5. Deighton 4465, K. 6. Glanville 329, K. 7. Risopoulos 1290, K. 8. Saunders 3, K.

Hyparrhenia umbrosa (Hochst.) Anderss.

FWTA, ed. 2, 3: 494.
West African: NIGERIA FULA-FULFULDE (Adamawa) shambawal (Pedder)

A perennial grass with stout culms up to 1.8 m high from a slender rhizome; of highland to montane grass savanna by roadsides and on old cultivations; recorded in N Nigeria and W Cameroons, and widely scattered elsewhere in tropical and S Africa.

It is grazed by all stock in Masailand, Kenya, while still young, and when mature is used for thatching (1).

Reference:

1. Glover & al. 1592, K.

Hyparrhenia violascens (Stapf) WD Clayton

FWTA, ed. 2, 3: 492. UPWTA, ed. 1, 529, as *H. soluta* var. *violascens* Stapf.
West African: NIGERIA FULA-FULFULDE (Nigeria) jimfihu (JMD) HAUSA gajiri (JMD) jimfa jimfa (JMD; ZOG) jimfi (JMD; ZOG) KAMBARI-SALKA àəruwâ (RB)

An annual grass with culms erect to 1 m tall; of roadsides and old cultivations; recorded in N Nigeria in the Region, and in E Cameroun and Chad.

It furnishes good fodder when young. When mature it is used for thatching and to make coarse matting and door-screens called *asabari* in Hausa (1).

Reference:

1. Dalziel, 1937: 529.

Hyparrhenia welwitschii (Rendle) Stapf

FWTA, ed. 2, 3: 494.
West African: SIERRA LEONE KORANKO gbonkilobon (Glanville; FCD) MANDING-MANDINKA bintanana (FCD) YALUNKA funfurekhamɛ-na (FCD) kimbigira-na (KM)

A coarsely-tufted annual grass with culms 30–300 cm tall; in the shade of savanna woodland where there is deep and damp soil; distributed from Guinea to N and S Nigeria, and elsewhere generally widespread in E and central tropical Africa to Angola; also in the Comoros. It is recorded in Tanganyika in association with ant-heaps (4).

In Sierra Leone (5) and in Tanganyika (4) cattle browse it while it is still young.

In the Mamodia area of Sierra Leone it serves as a thatching material (3), but at Falaba it is considered to be too brittle (5).

The Mandinka (*buitanana*) and the Yalunka (*funfurekhamɛ-na*) names cited above are both general names given to grasses whose awns are twisted when dry and which untwist when wetted, causing the spikelets to move (2).

Chromosome number is 2n = 40 (1).

References:

1. Clayton, 1969: 142. 2. Deighton 4433, K. 3. Glanville 330, K. 4. McGregor 26, 27, K. 5. Miszewski 17, K.

Hyperthelia dissoluta (Nees) WD Clayton

FWTA, ed. 2, 3: 496. UPWTA, ed. 1, 529, as *Hyparrhenia dissoluta* CE Hubbard.

West African: **SENEGAL** MANDING-BAMBARA dãzo (JB) fodod (JB) mbombod (JB) yégalé (JB) WOLOF cicca (JLT) khat (JLT) sèl a baré (JB) **MALI** MANDING-BAMBARA néanso (A.Chev.) ntaso (A.Chev.) yé-yalé (A.Chev.) **UPPER VOLTA** FULA -FULFULDE (Upper Volta) bodooko (K&T) MOORE gonyamba (A.Chev.) **NIGER** FULA-FULFULDE (Niger) sel-selnnde (ABM) HAUSA jan ɓako (Bartha) SONGHAI sáarí (*pl.* -o) (D&C) **NIGERIA** BIROM càraà vyɛp *from* vyɛp: rat of Gambia (LB) kò a varietal name (LB) FULA-FULFULDE (Nigeria) cingaaji (J&D) cobe (JMD) kalawal (JMD) wuɗeho (JMD) HAUSA gamba (Kennedy; Grove) ƙyáára, ƙyaáraà (auctt.) ƙyáárar bisháára a varietal name; the same as càraàvyɛp of Birom (LB) KAMBARI-SALKA ggyàarâ (*pl.* iigy-) (RB) TIV ijinga i myian (JMD) YORUBA (Ilorin) sɛgɛ (Ward)

A tufted perennial grass with culms 1–3 m tall; a typical savanna grass of waysides and disturbed places from lowlands to montane elevations; common throughout the Region from Mauritania to N and S Nigeria, and generally dispersed elsewhere over tropical Africa and into S Africa, Madagascar, and introduced to tropical America.

The grass is eaten by cattle with relish while growth is young (1, 2a,b, 3, 8, 10, 12, 13). The spikelets are pungently bristled which injure the muzzle of stock and thus inhibit browsing when in flower. Cattle have, however, been observed in Ghana searching for the foliage down below inflorescence level (4). Camels in Niger do not graze it at all (13).

The haulm is ubiquitously used when mature for thatching (1, 5, 7, 8, 12, 14, 16). The culms are commonly used for making matting after the spikelets have fallen (8). They are much used for making *zaanaa* matting to enclose a compound (11, 15), the different stages of which have acquired a special vocabulary (5, 9):

	Hausa	Saka
Cutting the culms	*sharɓee*	*tsùzzàwà*
Sorting, selecting, trimming the culms	*yantaa*	*tsùɗɗàrà*
Bundling culms for weaving	*maayaa*	*tsùkkanalẽ*
Weaving	*saakaa*	*tsùyĩ wookẹ̀tẽ*

An unit piece is approximately 1.8 m wide by 2 m long, the culms being bound together by interwoven culms about 30 cm from the top and bottom. Units are then tied serially to make as long a screen or fence as may be required affixed vertically to a support. Dagaari of Ghana make characteristically square baskets using the younger stems (11).

The culms have some value for producing paper-pulp (1, 8, 16).

Chromosome number is reported as 2n = 40 (6).

References:

1. Adam, 1954: 93, as *Hyparrhenia dissoluta* CE Hubb. 2a. Adam, 1966, a: as *Hyparrhenia dissoluta*. 2b. ibid. 1966, a: as *H. rupechtii*. 3. Baumer, 1975: 102, as *Hyparrhenia dissoluta* (Nees) CE Hubbard. 4. Beal 49, K. 5. Blench, 1985: as *Hyparrhenia dissoluta*. 6. Celarier A-3046, K. 7. Dalziel 265, K. 8. Dalziel, 1937: 529, as *Hyparrhenia dissoluta* CE Hubbard. 9. Furniss, 1990. 10. Harrison 1035, 1042, K. 11. Irvine 4699, K. 12. Kennedy FHI.8015, K. 13. Maliki, 1981: 153, sp. no. 62. 14. Michelmore 692, K. 15. Ward 0061, K. 16. Watt & Breyer-Brandwijk, 1962: 493, as *Hyparrhenia dissoluta* Hubb.

GRAMINEAE

Ichnanthus pallens Doell.

FWTA, ed. 2, 3: 436, as *I. vicinus* (Bail.) Merr.

West African: SIERRA LEONE MENDE aǰe (NWT) SUSU kalɛxɔnɛ (NWT)

A slender perennial grass, rooting at lower nodes, ascending to 60 cm high, recorded only in the mountains of Sierra Leone and Liberia, and extending into E Cameroun, and in Asia and Australia.

In India and Indonesia the grass is readily browsed by cattle and has a satisfactory food value (1, 2).

References:

1. Burkill, IH, 1935: 1221, as *I. vicinus* Merr. 2. Sastri, 1959: 161, as *I. vicinus* Merrill, with chemical analysis.

Imperata cylindrica (Linn.) Raeuschel.

FWTA, ed. 2, 3: 464, 466. UPWTA, ed. 1, 529–30.

English: lalang (from Malay: note – a common spelling, *lallang*, is orthographically wrong); spear grass (cf. *Heteropogon* Pers., Gramineae); Congo Grass(Senegal, JG Adam).

French: herbe (à) baïonnette; herbe à paillottes (Gabon, Walker); imperata.

Portuguese: palha carga (from Crioulo, Cape Verde, Feijão).

West African: SENEGAL BANYUN badied (K; K&A) BEDIK *ma*-diél (FG&G) DIOLA falint (auctt.) FULA-PULAAR (Senegal) sódo (K&A) TUKULOR sódo (K&A) MANDING-BAMBARA dolè (JB; K&B) MANDINKA solim (K&A) MANINKA dolé (K&A) solim (K&A) 'SOCE' solimô (K; K&A) NON idiol (K; K&A) SERER dol (auctt.) WOLOF bodé (JMD; K&A) hada (auctt.) THE GAMBIA MANDING-MANDINKA sɔ (JMD) sɔlingo *from the leaf's cutting edge* (JMD) GUINEA-BISSAU BIAFADA tchumba (JDES) FULA-PULAAR (Guinea-Bissau) sóap (EPdS) sódjô (JDES) MANDING-MANDINKA tumbunsuma (JDES) MANKANYA pêssête (JDES) PEPEL ochête (EPdS; JDES) GUINEA BASARI *a*-lɛréré (FG&G) KONYAGI *wa*-tyagaf (FG&G) SUSU sɔlonyi (NWT; JMD) SIERRA LEONE GOLA tere (FCD) KISSI pulmasa (JMD; FCD) sɔlondo (FCD) KONO suanɛ (JMD; FCD) LIMBA taga (JMD) taga (NWT) LOKO dokin dɔmai (NWT) pɔbɛgɛ (NWT) pɔvɛgɛ (NWT) MANDING-MANDINKA loliŋ (FCD) MENDE bola (NWT) lɛti *a common name for grasses* (auctt.) tele (JMD; FCD) MENDE (Kpa) lɛti (FCD) SUSU sukinyi (NWT) sulunyi (NWT; JMD) sunuŋwuri (NWT) surenyi (JMD) yobainyi (NWT) TEMNE *a*-, *kɔ*-loɛt NWT *a*-lath (auctt.) *a*-wo-*a*-ra-lal (NWT; JMD) VAI tele (FCD) YALUNKA sule-na (FCD) LIBERIA MANO dâh (JMD) MALI FULA-PULAAR (Mali) sɔ'yo (JMD) MANDING-BAMBARA dolé (JMD) KHASONKE gombi (JMD) MANINKA dolé (JMD) SONINKE-SARAKOLE solé (JMD) UPPER VOLTA BISA gay (Prost fide K&B) FULA-FULFULDE (Upper Volta) dolinji (K&T) GRUSI fofo (A.Chev.) MANDING-DYULA lollé (K&B) MOORE pulundi (JMD; K&B) IVORY COAST BAULE aâgni (K&B) WO-WO (K&B) KYAMA nsé (JMD; K&B) MANDING-DYULA lollé (K&B) MANINKA loléhun (A&AA) mangoti (B&D) GHANA ADANGME-KROBO henyu (auctt.) AKAN-BRONG tomene (FRI) GA ŋɛé (KD) ŋeí (KD; FRI) NIGERIA BIROM ekyé (LB) EFIK ásáí (Lowe) FULA-FULFULDE (Nigeria) gasa kigere *from* gasa: *hair* (Saunders; JMD) soo'o (J&D; JMD) soyore (*pl.* soyoji) (JMD) HAUSA toófaá (auctt.) toóhaá (auctt.) zakaran toófaá *the pointed rhizome tips* (JMD) zarenshi (auctt.) IDOMA epe (Odoh) IGALA iwọ (Odoh; RB) IGBO àchàlà *general for any tall or coarse grass, or for straw* (BNO) akata (BNO fide KW) ata (JMD) IGBO (Owerri) àchàrà *general for any tall or coarse grass, or for straw* (BNO) KAMBARI-SALKA àatsùpâ (RB) kàtsùpâ (RB) KANURI fura (JMD; C&H) TIV ihila (JMD) YORUBA ẹ̀kan = *newt* (auctt.) ìṣá (Egunjobi; Verger) WEST CAMEROONS KPE sosongo (Maitland; JMD)

A vigorous, rhizomatous perennial grass, basally tufted with leaves and slender inflorescence rising to over 1 m high; a weed of cultivations, rapidly occupying abandoned farmland and waste places; generally widespread over the whole of the Region, and throughout tropical and S Africa and Madagascar, and over the Middle East to tropical Australasia.

The plant is variable. Five varieties have been recognised, of which *FWTA*, ed. 2, lists two in the African region. They merge into one another, and are more ecoforms than valid separate varieties. By its vigorous, even aggressive habit it forms near pure stands. It collects dead hamper and is very prone to fire, either accidental or intentional. Even the mature green foliage is inflammable. Fire stimulates flowering so that in a few weeks after burning a sheet-lalang area will be covered with flowering panicles. The light fluffy seeds are carried by the wind to new sites. Dispersal may also be vicariously assisted by man. It is recorded that in Sierra Leone with the opening up of the country to agriculture, its spread has been attributed to discarded pillows stuffed with the flower floss that had been used by travellers on the old, now abandoned, railway system (10). The rhizomes are tough and vigorous, and, of course, survive burning. Lalang's troublesomeness in cultivations is that it suppresses seedlings and adversely affects tree crops. It actively competes for nutrients. There may be an element of biotic activity. The leaves with cutting edges are hostile to bare skin and the rhizomes with very sharp pointed ends spike bare feet and injure horses' hoofs. A Fula proverb runs: 'A man needs shoes to walk in *soo'o*', and an Hausa proverb: 'You have to go a long way round to avoid a field of *tóófáá*' (8). Presence of the plant is held by the Fula of N Nigeria to indicate fertility for sorghum fields (14). It is used in India on embankments as a soil-binder (22).

Eradication from cultivations is all important. Noxious chemical sprays may be effective, but hold dangers. Sodium arsenite widely used in rubber estates in Malaya is now discounted because of the risks involved. Frequent cutting to exhaust the rhizomes may serve to check growth, as also frequent mechanical cultivation, but this may result in loss of fertility. Forking and removal of the rhizomes by hand is completely effective but this is only practical if infestation is light and sporadic and the area small. Lalang is a sun-loving plant and it can be shaded out. *Anogeissus leiocarpus* (Combretaceae), *Gmelina arborea* (Verbenaceae) and *Lantana camara* (Verbenaceae) are recorded as being effective, but for lalang control on plantations their presence may not be desirable either where close-planting of the primary crop may be the best solution to obtaining quickly a closed canopy.

The young foliage is tender and stock will browse it. Old leaves become tough and develop razor-sharp edges. They are unpalatable and will be browsed only for want of better. The sharp leaf-margins injure the tender tissue of the mouth of stock (1). They may also cause damage internally.

The older culms are used for thatching in all territories. It is considered good and durable, thought in the Kissi country of Sierra Leone to be the best, and thus was deemed to be the chief's property (10). To give adequate water-proofing the roof pitch has to be steep and the thatch appreciably thick. In Malaya a 20 cm thick thatch may be used, requiring a strong support, and it may last 2 years, or up to 6 years if very steep (7). A life-span is given as 5 years in Lower Dahomey (6). Besides use in a conventional manner, Igala in Nigeria bind the grass in rolls while still on the ground and then place the rolls on the roof (3). A sort of matting, bags, baskets and plates are made in the Region (8). It is used as litter for stall animals in Senegal (1).

Aqueous extracts of the leaves and stems have shown in laboratory experiments some action on tumours (15, 18).

The foliage has been tried for paper. The reports indicated considerable promise but there appears to have been a lack of determination to follow this up (7, 22, 26). W African material should be examined.

The root contains sugars. A sample is reported to hold 18.8%, mainly *saccharose* and *glucose* with traces of *fructose* and *xylose* (15, 18). *Malic, citric, tartaric, oxalic* and *acetic acids* have been reported, and other substances (5, 15, 18). Attempts have been made to ferment the runners into a beer and to extract sugar and alcohol from them but without commercial success (7). The rhizomes are eaten raw by herdsmen in Lesotho (12) and Kipsigi children in Kenya chew them for the sweet flavour (11). The root is held to be galactogenic in Congo (Brazzaville) and is given to suckling women (4). It is a proven diuretic and is given in treatment for blennorrhoea and as an anti-dysenteric (13, 15–18). In Gabon it is given with a mashed banana as a diuretic (25). The pulped-up plant is added to karite butter for use as an embrocation for coughs (1, 9). Temne in

Sierra Leone use it as a cough medicine (23). In southern Africa a root-preparation is used for chest colds in children (12, 26). It is regarded as a specific for hiccups in southern Africa and an indigestion remedy (26). Fula in Senegal take a decoction of the root for schistosomiasis (17). A decoction of the dried plant is taken as a gargle for sore throat, and for neuralgia in Madagascar (9). The ash is alkaline and gets put into certain (not indicated) medicaments in the Region (1, 9). In Indochina and SE Asia there are many medicinal applications (7, 26).

Nine young shoots heated on a metal sheet, powdered and swallowed with palm-oil is claimed in Congo (Brazzaville) to produce an aphrodisiac effect (4).

A trace of alkaloid is reported present in Nigerian material (2). Other reports also record detection of alkaloids (18, 27).

The flossy flowers are collected for stuffing cushions and pillows (8, 10, 23). It has been known in Malaya that areas of sheet lalang are purposely fired to promote flowering to obtain them (7). In Kenya, Digo use the fluff as a substitute for cotton wool in treating sores (20). Chinese consider it haemostatic (26). In the Philippine Islands it is similarly used as a vulnerary to arrest bleeding and taken internally is sedative (21).

The plant is invoked in an Yoruba incantation 'to make a husband fight with his wife' (24).

References:

1. Adam, 1954: 93. 2. Adegoke & al., 1968: 13–33. 3. Blench, 1981–86. 4. Bouquet, 1969: 129. 5. Bouquet & Debray, 1974: 612. 6. Brasseur, 1952: 669. 7. Burkill, IH, 1935: 1228–32. 8. Dalziel, 1937: 529–30. 9. Debray & al., 1971: 74. 10. Deighton 2771, K. 11. Glover & al. 232, K. 12. Guillarmod, 1971: 436. 13. Haerdi, 1964: 310. 14. Jackson, 1973. 15. Kerharo, 1973: 4. 16. Kerharo & Adam 1962. 17. Kerharo & Adam, 1964, b: 552. 18. Kerharo & Adam, 1974: 648–9, with phytochemistry and pharmacology. 19. Kerharo & Bouquet, 1950: 252. 20. Magogo & Glover 381, K. 21. Quisumbing, 1951: 98–99. 22. Sastri, 1959: 169–72. 23. Thomas NWT.3, K. 24. Verger, 1967: sp. no. 92. 25. Walker, 1953: 39. 26. Watt & Breyer-Brandwijk, 1962: 474. 27. Willamen & Li, 1970.

Isachne buettneri Hack.

FWTA, ed. 2, 3: 420.
West African: **SIERRA LEONE** MENDE *n*dewe (NWT; FCD) muli (FCD) muri (*def.* -i) (NWT; Fisher) SUSU sankabesukwi (NWT) TEMNE kə-gbil (NWT) kə-məgbir (NWT)

A herbaceous scrambling perennial grass, of forest glades, in sub-montane situations from Guinea-Bissau to Fernando Po, and in E Cameroon, Uganda and Zaïre.

Sap from the stems is used in Sierra Leone to rub on 'ringworm' patches on the arms (1).

Reference:

1. Deighton 1661, K.

Isachne kiyalaensis Robyns

FWTA, ed. 2, 3: 420. UPWTA, ed. 1.
West African: **SIERRA LEONE** LOKO suewn (NWT) MENDE muli (NWT) nyawule (NWT) TEMNE peni-*pa*-pagbil (NWT) **NIGERIA** IGBO (Agukwu) ilulǫ kele (NWT)

An herbaceous grass weakly ascending, with slender culms to 45 cm long, of marshy places in forest shade in Guinea to S Nigeria, and on to Zaïre and Angola.

Though known in Sierra Leone and Nigeria no usage appears to be recorded.

Ischaemum afrum (JF Gmel.) Dandy

FWTA, ed. 2, 3: 476.
West African: NIGERIA FULA-FULFULDE (Nigeria) ɗadeppure (David) HAUSA kaashin ɓeeraa = *excreta of the mouse* (Palmer; ZOG)

A densely tufted perennial grass, rhizome scaly, culms 60–90 cm high; on black clay soils of impeded drainage; in Niger and N Nigeria, and in NE, E, S central and S Africa, and also in India.

The grass gives useful grazing at all times, but especially while green (Kordofan: 1; Sudan: 2, 5; Somalia: 4; Zimbabwe: 6).

The grain is collected in the Bornu area of N Nigeria for human consumption in time of dearth (3).

References:

1. Baumer 1975, K. 2. Beshir 25, K. 3. Davey FHI.27169, K. 4. McKinnon S.11, K. 5. Peers Ka.M25, K. 6. Trapnell 903, K.

Ischaemum indicum (Houtt.) Merrill

FWTA, ed. 2, 3: 476.

A tufted, erect or decumbent perennial grass; of wet and marshy places; an Asian species, now recorded at scattered points between Sierra Leone and S Nigeria; also in other tropical countries.

In India cattle browse it. In Trinidad taint in milk is attributed to it (1).

Reference:

1. Sastri, 1959: 271–2.

Ischaemum rugosum Salisb.

FWTA, ed. 2, 3: 476.
English: wizard grass (Sierra Leone, Deighton).
West African: SIERRA LEONE BULOM (Sherbro) fofo-bakai-lɛ (FCD) MENDE fovo *general term for several grasses* (FCD) SUSU sanla-yoge (FCD)

A straggling annual grass, culms to 1 m high; of damp and wet sites; an Asian species now found in the Region from Senegal to Upper Volta and Ghana, and in E Africa.

It is a weed of irrigation ditches and wet fields (2, 4). It was first recorded in the Region in 1927 in the Rokel River area of northern Sierra Leone where it has become a serious weed on rice farms in tidal swamps (3). It is also a serious weed of irrigated rice in The Gambia, contamination being carried over in the seed set aside for next year's sowing (6). In the field when young it looks like the rice so its presence is not readily noticed (7). In Fiji, where it also is a nuisance, control is said to be effective by 5 or 6 disc-harrowing of the fallow, and then ploughing in (7).

It is palatable to cattle (7). In India it is considered to be good fodder for cattle and horses (5) and is of fair nutritive value.

The grain is eaten in India in some parts by poorer people (5).

Sherbro in Sierra Leone say that if an ill-wisher throws a piece on the farm of his enemy, it will cover the farm (1). This is perhaps consonant with the English name 'wizard grass' applied in coastal Sierra Leone.

References:

1. Deighton 3541, K. 2. Enti & Hall GC.36002, K. 3. Glanville 226, K. 4. Rose Innes GC.30256, K. 5. Sastri, 1959: 272–3, with chemical analyses. 6. Terry 3249, K. 7. Wild, 1961: 23.

GRAMINEAE

Ischaemum timorense Kunth

FWTA, ed. 2, 3: 476.

A short slender straggling perennial grass to about 50 cm high; an Asian species, now recorded in W Cameroons.
In India it is readily eaten by stock and is considered to be an useful fodder (1).

Reference:

1. Sastri, 1959: 273.

Koeleria capensis (Steud.) Nees

FWTA, ed. 2, 3: 371.

A perennial tussock grass, fibrous-rooted, culms to 25–45 cm high, of montane grassland in W Cameroons, and in eastern Africa from Ethiopia to S Africa.
The plant provides good fodder for all domestic stock and wild game in Kenya (2) and S Africa (1). In Lesotho the culms are woven into hats and baskets (3).

References:

1. Burtt Davy 17734, K. 2. Grant 1247, K. 3. Guillarmod, 1971: 437.

Lasiurus hirsutus (Forssk.) Boiss.

FWTA, ed. 2, 3: 504. UPWTA, ed. 1, 530.
English: tabas grass (Sudan, Bristow).
West African: MALI TAMACHEK guerfis (A.Chev.)

A tufted perennial grass with wiry culms to about 90 cm tall, erect from a woody base; on dry sandy soils of the Sahel in Mali and Niger, and widely dispersed across N and NE Africa, the Middle East and India.
The grass is drought-resistant (6). It provides good nutritious fodder for camels, cattle and sheep (2–4, 7). It is highly valued in the dry regions of India, being grazed at all stages. It forms extensive pastures in Rajasthan on sandy soils and on sand-dunes, which it helps to fix (1, 4, 5). Hay made of it will store, when properly stacked, for 10 years (7).
When the haulm matures it becomes coarse and thick, making good material for thatching (7).
The root yields a fibre which can be turned into brushes used in weaving (7).
The seed is edible. In India it is commonly collected and ground to a flour for human consumption (1, 7).

References:

1. Bhandari, 1974: 75–76. 2. Bristow 14, K. 3. Dalziel, 1937: 530. 4. Gupta & Sharma, 1971: 81–82. 5. Harrison 1402, K. 6. McKinnon S.45, K. 7. Sastri, 1962: 36.

274

Leersia drepanothrix Stapf

FWTA, ed. 2, 3: 367.

A loosely tufted annual or perennial grass, culms to 1 m high from a short rhizome, of marshy places in Senegal to N Nigeria, and in Uganda and Sudan. In Ghana the grass is said to be relished by stock at all times (1).

Reference:

1. Beal 27, K.

Leersia hexandra Sw.

FWTA, ed. 2, 3: 367. UPWTA, ed. 1, 530.

English: rice grass; cut grass (on account of the backward sloping hooks on the leaves which can injure the skin).
French: herbe rasoir.
West African: SENEGAL BANYUN mutut (K&A) MANDING-MANDINKA kamétéô (K&A) GUINEA-BISSAU BALANTA nféndè (EPdS; JDES) unféndè (JDES) BIDYOGO uacúndê (JDES) MANDING-MANDINKA sine-ô (EPdS; JDES) PEPEL olaquicom (EPdS; JDES) NIGERIA FULA-FULFULDE (Nigeria) yaudeho (JMD) HAUSA madariki (Kennedy) YORUBA abẹko (Dawodu; JMD)

A rice-like grass, perennial, culms 30-100 cm high from a rhizomatous base, of shallow water and flooded areas from Senegal to N and S Nigeria and Fernando Po, and commonly dispersed over the rest of tropical Africa into the Transvaal, and across Asia to Australia.

The grass provides very useful fodder in all territories, especially while still young. It is highly esteemed in Australia as the best forage for horses and cattle, either green or as hay (10). Protein content of Indian material is recorded as 5.83%, dry wt., and reports of plants grown in southern USA show 8.6% and a digestibility of dry matter at 44–53% (2). The plant is so prized for its value as fodder that it is in some countries actually cultivated in swamps or under irrigation like rice (3, 10). Plants however have been found in the Philippine Islands to produce traces of hydrocyanic acid in the leaves and stems, and rather more in the roots (9).

The plant is always close by or in water. It may sometimes form floating masses of *sudd*. In Madagascar (11) and in Kenya (7) it is reported to invade rice-padis and to become a troublesome weed. Its presence on swampland is taken in Zambia to be an indication that the land is fit for rice cultivation (6). The leaf midrib develops reflexed hooks which can inflict painful lacerations on bare skin (4, 5). The leaf margin is also unpleasantly razor-like.

The grass is grown in Madagascar to make a turf; it is fire-resistant (1).

In Casamance, Senegal, the plant, part not reported, nor the method of preparation, is given to a patient suffering frequent bouts of coughing accompanied by haemoptysis (8).

References:

1. Adam, 1954: 94. 2. Boyd & McGinty, 1981: 296–9. 3. Burkill, IH, 1935: 1327–8. 4. Clayton, 1972: 367. 5. Hall & al., 1971. 6. Jackson 24, K. 7. Kahurananga & Kibui 2810, K. 8. Kerharo & Adam, 1963,a: 773–92. 9. Quisumbing, 1951: 1023. 10. Sastri, 1962: 59. 11. Wild, 1961: 49.

Leptaspis zeylanica Nees

FWTA, ed. 2, 3: 362, as *L. cochleata* Thwaites.
West African: SIERRA LEONE MENDE hongi (NWT) TEMNE ɛ-suta (NWT) LIBERIA MANO pini gon (Har.)

A sprawling, spreading rhizomatous grass to 1 m high, of ground layer and understorey of shaded places in forest from Guinea to Fernando Po, and across tropical Africa, to Madagascar, Mascarenes, Ceylon, SE Asia and Polynesia.

The plant is capable of extensive spreading to form large pure strands. It produces adventitious cover on coffee plantation in Ethiopia where shade trees have been left standing.

The leaves have an absorbent capacity and are used in the Central African Republic by honey gatherers to mop up spilled honey (2).

In Ivory Coast leaf-sap is massaged onto the neck to reduce enlarged ganglia (1).

References:

1. Bouquet & Debray, 1974: 92, as *L. cochleata* Thw. 2. Carroll 17, K.

Leptochloa caerulescens Steud.

FWTA, ed. 2, 3: 397–8.
West African: SIERRA LEONE SUSU sufe (NWT) NIGERIA HAUSA lale shamuwa (Golding) YORUBA kupuruku (Ward)

An annual grass, culms to 1 m long, sometimes erect but usually decumbent and basally stolon-like, of wet sites in or near water throughout the Region from Senegal to N and S Nigeria, and eastwards to Ethiopia and southwards to Zambia and Angola.

The plant is a rheophyte (2–4), enters swamps (2, 6) and is a weed of irrigation plots and padi-fields (1, 6), reported to be resistant to the weed-killer, *simazine* (7). It is the pioneer coloniser of sand-banks in river creeks (3–5). In this respect it must resist scouring and erosion. Grass-eating fish are reported to browse it in the R Niger (2). It should thus be useful in fish-farming of carp and other grass-eaters.

References:

1. Biswas 103, 305, K. 2. Cook 467, K. 3. Gereau 1253, K. 4. Jones FHI.6889, K. 5. Lean 50, K. 6. Nwauzo 796, K. 7. Vaillant 2769, K.

Leptochloa fusca (Linn.) Stapf

FWTA, ed. 2, 3: 398, as *Diplachne fusca* (Linn.) P Beauv.
English: Lake Chisi grass (Tanganyika, Michelmore).
West African: SENEGAL SERER uk (JB) WOLOF ndibis (JB)

A tussock grass, rhizomatous, perennial, culms moderately stout to $1\frac{1}{2}$ m high, hygrophilous, of swamps and wet sites, on lake and river-margins, across the Region in the Sahel zone from Senegal to N Nigeria, and throughout Africa and into India to China and to Australia.

The plant is a halophyte tolerating brackish, saline and alkaline water. It grows on salt-pans. Its presence indicates a salty soil, but nevertheless it is tolerant of occasional inundation by fresh-water, and also of acidic circumstances arising from oxidation of organic matter (4).

The plant is much relished for grazing by stock and game, browsing in particular young shoots arising after the annual dry season burn (1), but general value as fodder is low (2, 3).

References as *Diplachne fusca* (Linn.) P Beauv.:

1. Baumer, 1975: 96. 2. Burkill, IH, 1935: 834–5. 3. Sastri, 1952: 87. 4. Trochain, 1940: 395.

Leptochloa panicea (Retz.) Ohwi

FWTA, ed. 2, 3: 398.

A tufted annual grass to 60 cm high with slender culms, of wet or damp sites, inundation areas and river-edges of riverain bush, recorded in Ghana, and widely distributed in eastern Africa from Sudan to Transvaal and Natal; native of tropical Asia.

It is a good fodder grass. Stock is reported to graze it in Sudan in both the wet and the dry season (1).

Reference:

1. Peers KaM.31, K.

Leptochloa uniflora A Rich.

FWTA, ed. 2, 3: 397.
West African: NIGERIA IGBO (Agukwu) ẹle ọdodo (NWT) eya (NWT)

A tufted annual grass to 60 cm high of wooded grassland, deciduous bush, often in damp situations, fallow ground and a weed of cultivation in Ghana and S Nigeria, and in E Africa to S Africa, India and Ceylon.

This grass is said to be very succulent and palatable (2) and to be grazed also by game (1). Rats are recorded as eating the base of the tufts (1).

References:

1. Magogo & Glover 447, K. 2. Staples 211, K.

Leptothrium senegalense (Kunth) WD Clayton

FWTA, ed. 2, 3: 413. UPWTA, ed. 1, 530.
West African: MAURITANIA ARABIC (Hassaniya) tigurit (AN) MALI ARABIC (Mali) askanit (A.Chev.) SONGHAI firri (A.Chev.) TAMACHEK ainguiem (A.Chev.)

A short-lived perennial bunch grass, culms to 60 cm long, of sub-desert grassland from Mauritania to Ghana, and across Africa to Egypt and E Africa and into Asia to India.

The grass provides valuable though rather sparse forage to stock in the sub-desert regions (Mauritania: 9; Senegal: 1, 2; Mali: 6; Kordofan: 5; Somalia: 3, 10; India, Rajasthan: 8).

The seed is collected by man for food (1, 6). In Kenya vulturine guinea fowl feed on the grain (4), and in Ethiopia it is eaten by locusts (7).

References:

1. Adam, 1954: 93. 2. Adam, 1966,a. 3. Bally & Melville 15516, K. 4. Bally & Smith B.14592, K. 5. Baumer 1975: 104. 6. Dalziel, 1937: 530. 7. Ellis 37, K. 8. Gupta & Sharma, 1971: 83. 9. Naegelé, 1958,b: 894. 10. Rose-Innes 837, K.

Loudetia annua (Stapf) CE Hubbard

FWTA, ed. 2, 3: 417. UPWTA, ed. 1, 531.
West African: SENEGAL DIOLA é vèt éma (JB) THE GAMBIA MANDING-MANDINKA kereng fenio = *squirrel's tail* (JMD) GUINEA-BISSAU PEPEL ovil (EPdS; JDES) NIGERIA FULA-FULFULDE (Nigeria) cammeya (JMD) nione sule (Saunders) GWARI gwodnu susi (JMD) HAUSA bíndìn kùréégéé = *tail of the squirrel* (JMD; ZOG) búndìn kùréégéé = *tail of the squirrel* (JMD; ZOG) tsíígàà (ZOG) tsííkàà (ZOG) wútsíyàr gwánkií = *tail of the roan antelope* (auctt.) wútsíyàr kùréégéé = *tail of the squirrel* (JMD; ZOG) wútsíyàr zoomoo = *tail of the hare* (JMD; ZOG)
NOTE: *names referring to 'tail' allude to the bottle-brush appearance of the inflorescence.*

277

An annual grass, culms erect to 1.2 m high, of ditch-sides and like humid places in Senegal to N Nigeria, and extending to Sudan and Uganda. The plant offers some browsing to cattle, but only while still young (Senegal: 1; Nigeria: 3–5; Kordofan: 2).

References:

1. Adam, 1966, a. 2. Baumer, 1975: 105. 3. Dalziel, 1937: 531. 4. Rose Innes RRI.318, K. 5. Saunders 51, K.

Loudetia arundinacea (Hochst.) Steud.

FWTA, ed. 2, 3: 417. UPWTA, ed. 1, 531.
West African: SENEGAL WOLOF paldinagh (JMD) SIERRA LEONE KISSI putaka (FCD) KORANKO sɛgɛra (NWT; JMD) sɛra (NWT; JMD) MANDING-MANDINKA sera (FCD) seraŋ (FCD) MENDE čili (NWT) njabɔli (NWT) SUSU filɛsaxe (NWT; JMD) TEMNE kə-bɔnkɔlɔ (NWT) nepelabana (FCD) ɛ-wabe (NWT) ka-walaŋ (auctt.) kə-weliŋ (NWT; JMD) YALUNKA fii-la (FCD) silaŋgi-na (FCD) NIGERIA HAUSA tsíntsííyár duutsee = *brush of the rock* (ZOG) IGBO ajo (NWT; JMD) ata = *grass* (BNO)
NOTE: *other names under L.simplex (Nees) C. E. Hubbard may apply.*

A robust tufted perennial grass with culms to 3 m high, of poor and dry soils especially on rock outcrops, but also sometimes on swampy soil, widespread across the Region from Senegal to W Cameroons, and extending to Ethiopia, and to Angola and Mozambique.

It is a dominant grass in some types of savanna often forming pure colonies (3). It is widely used for thatching (2; Sierra Leone: 3; Nigeria: 7, 10), but it is reported as not lasting (4).

It provides browsing for cattle, especially while it is still in young growth (2; Zaïre: 6; Kenya: 5; Tanganyika: 8). In Kordofan it is said to recover rapidly after a bush-fire when cattle will take the new growth till it reaches about 40 cm high (1). As it gets older the hairs on the mature inflorescence may become troublesome to the eyes of cattle (8).

In Uganda the flowering culms are bundled to make house brushes (9).

References:

1. Baumer, 1975: 105. 2. Dalziel, 1937: 531. 3. Deighton 814, K. 4. Glanville 64, K. 5. Glover & al. 2144, K. 6. Johnston 1099, K. 7. Kennedy 8035, K. 8. Marshall 29, K. 9. Thomas, AS Th.4005. 10. Thomas NWT.1931 (Nig. Ser.), K.

Loudetia coarctata (A Camus) CE Hubbard

FWTA, ed. 2, 3: 417.
West African: GUINEA-BISSAU FULA-PULAAR (Guinea-Bissau) udo-bánorô = *straw of the wolf* (EPdS) udo-bunoro (EPdS) GUINEA FULA-PULAAR (Guinea) kuligi (Adames) laque dawa (Langdale-Brown)

A densely tufted perennial grass on poorly drained soils in Guinea and N Nigeria, and also in Zaïre, Zambia and Sudan.

No usage is recorded, but since it is known to the Fula of Guinea, it has perhaps use as cattle-fodder.

Loudetia flavida (Stapf) CE Hubbard

FWTA, ed. 2, 3: 417.
West African: UPPER VOLTA FULA-FULFULDE (Upper Volta) fegjirdi (Scholz) gommesa-hogo (K&T)

A coarsely tufted perennial grass with culms to $1\frac{1}{2}$ m high, of shallow soils in open savanna woodland from Upper Volta to N Nigeria, and in Sudan, and E to S Africa.

It is specifically recorded as not grazed in Ghana (2), but in Sudan all stock take it in both the wet season and, of especially important value, in the dry season (1).

References:

1. Peers KM.17, KM.18, K. 2. Rose Innes GH.30236, GH.30782, K.

Loudetia hordeiformis (Stapf) CE Hubbard

FWTA, ed. 2, 3: 417. UPWTA, ed. 1, 531.

West African: SENEGAL SERER las a tat (JB) talit (JB) THE GAMBIA MANDING-MANDINKA kereng fenio (JMD) kereng penio = squirrel tail (Perie) UPPER VOLTA GRUSI plumga (Scholz) GRUSI-KASENA plumga (Scholz) NIGERIA FULA-FULFULDE (Nigeria) cammeya (JMD) GWARI gwodnu susi (JMD) HAUSA bíndin kùréégéé (JMD; ZOG) búndin kùréégéé (JMD; ZOG) tsíígàà (ZOG) tsííkàà (ZOG) wútsíyàr gwánkíí = tail of the roan antelope (JMD; ZOG) wútsíyàr kùréégéé = tail of the (ground) squirrel (JMD; ZOG) wútsíyàr zoomoo = tail of the hare (JMD; ZOG)

NOTE: names referring to 'tail' allude to the bottle-brush appearance of the infloresence.

An annual grass, culms to 1½ m high of disturbed sandy soils across the Region from Senegal to S Nigeria, and extending on to Sudan.

In Niger the grass colonises both wind-blown and fixed dunes (3) and may assist in stabilising them.

As fodder for stock it is indifferent but will be browsed while still young and tender (Ghana: 1; Niger: 3; Nigeria: 2).

References:

1. Beal 36, K. 2. Dalziel, 1937: 531. 3. Virgo 2, K.

Loudetia kagerensis (K Schumm.) Hutch.

FWTA, ed. 2, 3: 419.

A tufted perennial grass, culms to 1 m high, wiry, fine-leaved, of open grassy places on shallow soil, sometimes seasonally waterlogged, in Guinea to N Nigeria, and in E Africa, Congo basin and Angola.

In Sierra Leone it is one of the foni community (1). In E Africa it is grazed by cattle and all domestic stock and is deemed to be an important fodder (Urundi: 5; Kenya: 2–4).

The straw is used in Kenya for thatching (2).

References:

1. Deighton 2168, K. 2. Glover & al. 280, 577, 789, 1536, 1855, 1911, K. 3. McDonald 964, K. 4. Maher 2510, K. 5. Schantz 739, K.

Loudetia phragmitoides (Peter) CE Hubbard

FWTA, ed. 2, 3: 417. UPWTA, ed. 1, 531.

English: erapo grass (from Yoruba, S Nigeria, McGregor).

West African: GUINEA-BISSAU FULA-PULAAR (Guinea-Bissau) balabaque (EPdS; JDES) MANDING-MANDINKA sulunhantam-ô (EPdS; JDES) NIGERIA HAUSA tsíntsíyár mázáá (JMD; ZOG) YORUBA èrapò (Dawodu; Verger)

Densely tussocked pampas-like grass, perennial with culms to 4½ m high, of marshy ground throughout the Region from Senegal to W Cameroons, and widespread over tropical Africa.

The grass in flower with its feathery inflorescence is of attractive appearance. The stout pithy culms become hollow and children in Gabon use them to make whistles (3). In Nigeria they are turned into arrow-shafts (1). The Hausa name covers several grasses of this type, but refers particularly to the thickened roots of an E African grass, *Tristachya superba* (De Not.) Schweinf. & Aschers., which are imported and sold in markets as an aphrodisiac (1). It is not clear whether the root of *L. phragmitoides* has this usage.

The leaves are used in the Central African Republic for scrubbing out utensils, e.g. calabashes (2).

References:

1. Dalziel, 1937: 531. 2. Fay 4447, K. 3. Walker & Sillans, 1961: 190.

Loudetia simplex (Nees) CE Hubbard

FWTA, ed. 2, 3: 419. UPWTA, ed. 1, 531.

West African: SENEGAL WOLOF paldinagh (JMD) UPPER VOLTA FULA-FULFULDE (Upper Volta) pisiirɗi (K&T) pisiirgol (K&T) pitiirɗi (K&T) MOORE domsawo (FRI) domshao (FRI) NIGERIA FULA-FULFULDE (Adamawa) sufuwel (Pedder) FULFULDE (Nigeria) burde kalde (Saunders; JMD) yawrondu (Saunders; JMD) HAUSA tsíntsíyár duutsee (auctt.) *NOTE: other names under L.arundinacea* (Nees) C. E. Hubbard may apply.

A tufted perennial grass, culms 30 cm to 1½ m high, in deciduous bushland, rocky sites of shallow soils, occasionally waterlogged, common throughout the Region, and widespread over tropical and S Africa.

It provides grazing only whilst the growth is still young (4; Nigeria: 14; Kordofan: 2; Sudan: 3; Uganda: 6; Kenya: 9; Tanganyika: 7, 11; Malawi: 8; Zimbabwe: 1). It is said to become unpalatable when mature and dry (7) or till seed develops (1). In Upper Volta the dislike is ascribed to a bitterness of the leaves.

In many areas it has become a dominant grass. It is commonly used for thatching in Ghana (13), and on the Kalenda Plain in Zambia it is reputed to be the best of thatching grasses available (12). In Bamako market, Mali, baskets of the culms are traded (5).

References:

1. Appleton 9, K. 2. Baumer, 1975: 105. 3. Beshir 10, K. 4. Dalziel, 1937: 531. 5. Diarra, 1977: 44. 6. Dyson-Hudson 309, K. 7. Emson 89, K. 8. Fenner 320, K. 9. Glover & al. 2325, K. 10. Irvine 4685, K. 11. McGregor 41, K. 12. Milne-Redhead 4402, K. 13. St. Clair-Thompson 1535, K. 14. Saunders 63, K.

Loudetia togoensis (Pilger) CE Hubbard

FWTA, ed. 2, 3: 416. UPWTA, ed. 1, 531.

West African: SENEGAL MANDING-BAMBARA firgala ukasa (JB) MALI FULA-PULAAR (Mali) gombi sogo (A.Chev.) selbéré (A.Chev.) MANDING-BAMBARA firala nkasan (A.Chev.) nkasan (A.Chev.) SONINKE-SARAKOLE ngasan séladé (A.Chev.) UPPER VOLTA DAGAARI bálánplè (FRI) FULA-FULFULDE (Upper Volta) selɓo (*pl.* celɓi) (K&T) GRUSI zaié (A.Chev.) GURMA kiu guma (A.Chev.) MOORE samsogo (A.Chev.) sutu (Scholz) TOGO GURMA (Mango) sangombe (Volkens) NIGERIA ARABIC-SHUWA kaya (JMD) nyamaya *applies to several grasses as for Hausa:* tsìntsííyáá (JMD) HAUSA tsíntsííyáá = *broom, incl. Eragrostis and other grasses* (JMD) KANURI kẹla tsẹlim (JMD)

A tufted annual grass with culms to 1 m high, on dry or shallow soils, common across the Sahel from Mauritania to N Nigeria, and on to Sudan.

The grass provides a little browsing for stock, but only while still young (Senegal: 1; Kordofan: 2).

The flowering culms are bundled into brooms in Ghana (4), and in N Nigeria as is indicated by the Hausa name (see list above). In Upper Volta the stems are plaited into hats (5), and women in N Togo wear armlets made of them (3).

References:

1. Adam, 1966, a. 2. Baumer, 1975: 105. 3. Dalziel, 1937: 531. 4. Hall CC.794, K. 5. Irvine 4727, K.

Loudetiopsis Conert

FWTA, ed. 2, 3: 414.

A genus of 13 species in the Region, most are perennial and with woody reed-like culms to over 1 m tall, of savanna woodland or in pockets in bare rock. No usage is recorded. They appear to be ungrazed, even the young growth after annual burning. The large culms however, must seem to be suitable for thatching.

Loudetiopsis pobeguinii (Jac.-Fél.) WD Clayton

FWTA, ed. 2, 3: 414.
West African: **GUINEA-BISSAU** FULA-PULAAR (Guinea-Bissau) udúbunoro (JDES)

A tufted annual, culms to 60 cm high, of moist places in Senegal to Guinea.

Loudetiopsis tristachyoides (Trin.) Conert

FWTA, ed. 2, 3: 415.
West African: **GUINEA-BISSAU** FULA-PULAAR (Guinea-Bissau) údò-bonóro (JDES) **GUINEA** BASARI a-láfasyèn *to this or related spp.* (FG&G) KONYAGI wɔdion *to this or related spp.* (FG&G) **SIERRA LEONE** KORANKO kulɛbinyi (NWT) MENDE foni *a general term* (FCD) SUSU fainyinyogi (NWT) TEMNE kə-ep (NWT)

A perennial grass, tufted, culms to 1.2 m high, of damp pockets on rock outcrops in Senegal to Upper Volta.

It is not grazed (1). It occurs in the *foni* community in Sierra Leone in the later stages of the succession (2). The Fula name in Guinea-Bissau meaning 'straw of the wolf' (3) suggests inferiority.

References:

1. Adam, 1966, a, as *Danthoniopsis tuberculata.* 2. Deighton 2176, K. 3. Sousa, 1956: 13.

Megastachya mucronata (Poir.) P. Beauv.

FWTA, ed. 2, 3: 381.
West African: **NIGERIA** IGBO ilulu ọkhankwọ (?ọkpankwọ) (NWT) oke ako ọzọ (NWT)

A tufted annual, culms to 90 cm high, of forest shade on damp and wet sites from Sierra Leone to W Cameroons, and widespread over tropical Africa.

Melinis effusa (Rendle) Stapf

FWTA, ed. 2, 3: 457.
West African: **NIGERIA** HAUSA wutsiyar kurege = *tail of the squirrel* (Kennedy)

A straggling perennial grass, culms to 90 cm high; of open hillsides in montane situations of Ghana, N Nigeria and W Cameroons, and in central and SW Africa.

GRAMINEAE

Melinis macrochaeta Stapf & CE Hubbard

FWTA, ed. 2, 3: 455.
West African: NIGERIA FULA-FULFULDE (Nigeria) uringe (Saunders) HAUSA wutsiar zaki
= *lion's tail* (Kennedy)

An annual grass with culms ascending or erect to 1.2 m high, often stilt-rooted at lower nodes; of wooded savanna in Ivory Coast and N and S Nigeria, and generally widespread in tropical Africa.
The leaf is softly hairy, but is not so sticky as those of *M. minutiflora*. It has some use for cattle-grazing in N Nigeria (1).

Reference:

1. Saunders 10, K.

Melinis minutiflora P Beauv.

FWTA, ed. 2, 3: 455. UPWTA, ed. 1, 531.
English: stink grass; Brazilian stink grass; molasses grass; Wynne grass; efwa-takala grass (of several authors, seems to be a corruption of Angolan *evantonkala*, Feijão).
Portuguese: capim melado (= *honey grass*, Feijão); capim gordura (= *fat grass*, Brazil).
West African: SIERRA LEONE KONO minan-sa-bine = *python's lair grass; from,* minan: *boa-constrictor;* sa: *to lie down;* bini: *grass—said to be a favourite sleeping place of pythons* (Dawe; JMD) MENDE fokole = *white sun* (Dawe) GHANA ADANGME-KROBO aketibua (JMD) AKAN-TWI akutuakuru (FRI) NIGERIA YORUBA (Ilorin) ori (Ward)

A perennial grass, culms prostrate basally, ascending or erect to 2 m long; of open hillsides and clearings in Guinea to W Cameroons, and generally wide-spread in the rest of tropical Africa and introduced to most tropical countries.
The fresh plant has a strong smell arising from a viscid secretion of glandular hairs on the leaf and leaf-sheaths. The odour is variously described as like cumin, linseed or molasses. It is said to be strongest in the afternoon (1, 8). The smell is lost on drying. Fresh grass has been analysed as yielding 0.001% of the volatile oil. Because of the smell it has been suggested for planting in tsetse fly isolation belts as a repellant, but the efficacy is open to doubt (2, 9). The whole plant is held to be insecticidal and there has been cultivation for this purpose in Brazil and central Africa (9). In Tanganyika the bruised leaf is rubbed over animals as an insect-repellant and the grass is put in poultry nesting boxes to control vermin (9). Experimental work in Ghana has indicated no deterrent effect on mosquito breeding, and action on tick larvae is mechanical rather than chemical (2).
The grass has some limited use for fodder, and is sometimes grown as a pasture grass (2, 3), but it has been discarded in this respect in N Nigeria for poor persistence and low production (5). If browsed it should be before anthesis, and to be so cut also if for hay. Analysis indicates a fair quality nutrition value (2). Cattle in Brazil are said to fatten quickly on it but lack vigour.
The grass will rapidly take possession of newly cleared land and it will suppress other weeds and give a cover and mulch. In Sierra Leone it is said to be a favourite sleeping place for pythons, hence the Kono name (see above) meaning 'boa-constrictor liedown grass' (4). In the closed-forest country in Ghana, Kwawu farmers know that it is time to prepare their land when it comes into flower (6).
In Tanganyika, the root is used as a purgative (7). In Brazil the plant is a diarrhoea remedy, acting, perhaps, as a carminitive (9).

282

References:

1. Chandler 1498, K. 2. Dalziel, 1937: 531. 3. Dawe 420, K. 4. Dawe 519, K. 5. Foster & Munday, 1961: 314. 6. Irvine 1707, K. 7. Kortischoner 907, K. 8. Snowden 1498,a, K. 9. Watt & Breyer-Brandwijk, 1962: 477.

Melinis repens (Willd.) Zizka

FWTA, ed. 2, 3: 454, as *R. villosum* (Parl.) Chiov. UPWTA, ed. 1, 541.

English: Natal grass; red top; ruby grass.

French: tricholène rose (Roberty).

West African: NIGERIA FULA-FULFULDE (Nigeria) bulule, gawdaho (Saunders) tsaraho (J&D) HAUSA gulbin birai (RES) maí fárín kái = *white headed* (JMD; ZOG) KANURI kelabul (Kennedy) YORUBA eéran ẹyẹ *from* eéran: *general for fodder grasses;* ẹyẹ: *bird* (auctt.) ilosu (Macgregor) sọkọdayà = *make the husband tie the the wife* (Verger; JMD) YORUBA (Ilorin) aboni funfun (Ward) WEST CAMEROONS FULA-FULFULDE (Adamawa) ajarja (Pedder)

An annual, or short-lived loosely-tufted perennial, grass, culms ascending or erect to 1 m high, panicles silvery pink or purple; of old farmland and other disturbed places, often in pure stands; widely dispersed in the Region from Sierra Leone to W Cameroons and generally over other regions of tropical and southern Africa, and in most tropical countries.

The grass is a weed of arable land commonly invading fallows. Use has been made of its vigorous habit in S Africa to revegetate abandoned cultivated land (2). A sheet of this grass in full bloom displaying its pinkish inflorescences is an attractive spectacle, rating it one of the most beautiful grasses in Africa (1). It has been brought into horticultural cultivation in Europe (3). A spontaneous presence, however, is taken as an indication of poor or exhausted soil in N Nigeria (4, 9) and in Tanganyika (7).

It produces good fodder, taken by all stock, especially while still young, and is especially valued for horses (6). In Kenya it is recorded as producing good pasture (5) but cultivation seems in general not to be favoured (3). It converts well into hay and has fair nutritive properties (2, 3). *Hydrocyanic acid* is sometimes produced.

The grass enters an Yoruba incantation 'to make a wife come back to her first husband' (8).

References:

1. Appleton 18, K. 2. Chadha, 1972: 23, with nutritional values. 3. Dalziel, 1937: 541. 4. Jackson, 1973. 5. McDonald 959, K. 6. Saunders 46, K. 7. Strang 2, K. 8. Verger, 1967: sp. no. 170. 9. Ward 0081, K.

Michrochloa indica (Linn. f.) P Beauv.

FWTA, ed. 2, 3: 403. UPWTA, ed. 1, 531.

West African: SENEGAL MANDING-BAMBARA fulabin (JB) MALI VULGAR fukobi (A.Chev.) kulumbi (A.Chev.) FULA-PULAAR (Mali) narkatabéli (A.Chev.) MANDING-KHASONKE dugu konsina (A.Chev.) UPPER VOLTA FULA-FULFULDE (Upper Volta) hondo korlangal (Scholz) kollaanngal (*pl.* kollaade) (K&T) GRUSI-KASENA mbugaga (Scholz) MOORE sutuhanga = *hare's beard* (A.Chev.) NIGERIA FULA-FULFULDE (Nigeria) wicco falandu = *lizard's tail* (JMD) wicco jalando (Saunders)

A loosely tufted annual grass, culms weakly ascending or straggling to 25 cm high of open bare places, shallow and over-grazed soils in bushland or grassland throughout the Region from Senegal to W Cameroons, and dispersed across tropical Africa and pan-tropics.

The grass is grazed by all stock but it is too small to be of much value (1–3).

References:

1. Adam, 1966,a. 2. Dalziel, 1937: 531. 3. Snowden 1467, K.

GRAMINEAE

Microglossa kunthii Desv.

FWTA, ed. 2, 3: 403.
West African: NIGERIA FULA-FULFULDE (Nigeria) wicco sondu (Saunders)

A densely tufted perennial, matting grass, culms to 30 cm high, on open places with bare or shallow soil of bushland or grassland in Ivory Coast to W Cameroons and throughout tropical African and pan-tropics.
In Kenya it is reported grazed by all domestic stock and wild hares (1).

Reference:

1. Glover & al. 579, K.

Monocymbium ceresiiforme (Nees) Stapf

FWTA, ed. 2, 3: 499. UPWTA, ed. 1, 532.
English: oatgrass; wild oat grass (Transvaal, S Africa).
West African: SENEGAL FULA-TUKULOR garlaban (JMD) SERER matifalbène (A.Chev.) WOLOF khat lek (JMD) UPPER VOLTA FULA-FULFULDE (Upper Volta) yantaare (pl. jantaaje) (K&T) NIGERIA BIROM yùús (LB) FULA-FULFULDE (Nigeria) karereyo (Saunders) HAUSA bááyam máárayáá = back of the cob antelope (LB) bááyán mááràyáá (ZOG) ján baújéé applied loosely (JMD; ZOG) jan yámaa (RB) jan yaro (DA) yamaa (auctt.) KAMBARI-SALKA yìssówi (RB) KANURI garbazam applied to several similar grasses (JMD; C&H) WEST CAMEROONS BAMILEKE shamanho (DA)

A tufted perennial grass, culms leafy and weak, 90–120 cm tall, but variable to densely caespitose, dwarf to 30 cm and wiry in southern Africa; on shallow stony hillside sites; common throughout the Region from Guinea to W Cameroons, and generally widely dispersed elsewhere in tropical Africa.
It provides good fodder for stock while still young, i.e., up to about 30 cm high (1, 2, 6, 7, 11). In some places it is the commonest grass of the countryside, e.g. on Loma Mountain. Taller than 30 cm the grass becomes tough, and being weak-stemmed it usually is lodged (12). The leafy haulm is often used for thatching (5, 8, 9). In N Nigeria the straw is chopped up for puddling with clay in house-building (4, 6).
The Hausa names refer to the grass's russet colour, and may include other grasses (Andropogon fastigiatus, Hyparrhenia spp., etc.) which give this colour to the landscape in the dry season (6). In Bornu hunters use it as camouflage concealment (6).
In N Nigeria the grass is used to ward off the evil eye and in forms of oath (Burton fide 6). In Lesotho it is burnt in the autumn in fields of grain to hasten ripening (10).

References:

1. Appleron 33, K. 2. Beal 42, K. 3. DA, Samaru 20, K. 4. Dalziel 284, K. 5. Dalziel 893, K. 6. Dalziel, 1937: 532. 7. Emson 120, K. 8. Fay 4452, K. 9. Glanville 326, K. 10. Guillarmod, 1971: 442. 11. Saunders 33, K. 12. Ward L.147, K.

Olyra latifolia Linn.

FWTA, ed. 2, 3: 362. UPWTA, ed. 1, 532.
West African: SENEGAL BANYUN kisân diabudiof (K&A) DIOLA fu rénum (JB; K&A) ka tira (JB; K&A) GUINEA-BISSAU FULA-PULAAR (Guinea-Bissau) québè-sufo (EPdS) quénè-súfô (JDES) MANDING-MANDINKA suluquemom (JDES) SIERRA LEONE BULOM (Sherbro) kɔkia-lɛ (FCD) KISSI koli-koli (FCD) KONO fonfinɛ (FCD) fũfine (FCD) KORANKO felife (NWT; JMD) LIMBA ntotɛ (NWT; JMD) tutɛ (JMD) LOKO pɔbi (NWT; JMD) pɔvi (FCD) MANDING-MANDINKA fɔnjɔ (FCD) sanbala (JMD) MENDE bɔngi (NWT; JMD) kotopɔ (FCD; JMD) kotopɔvɔ (def. -i) (Dawe; FCD) pɔŋgi (NWT; JMD) pɔvɔ (?) (FCD) SUSU suliai (NWT) TEMNE ɛ-feta NWT ka-sotha (auctt.) YALUNKA tamadikhɛ-khɔŋbei-na (FCD) LIBERIA MANO pinni (Har.) IVORY COAST AKAN-ASANTE drubani (K&B) BRONG dorobéné (K&B) AKYE

284

GRAMINEAE

*n*safo (A&AA) ᴀɴʏɪ doniénié (K&B) ʙᴀᴜʟᴇ féfé (B&D; A&AA) zanfé (B&D) ᴋʀᴜ-ɢᴜᴇʀᴇ (Chiehn) poé B&D) ᴋᴜʟᴀɴɢᴏ fimu (K&B) **GHANA** ᴀᴅᴀɴɢᴍᴇ-ᴋʀᴏʙᴏ adodubɛdsi (FRI) ɔdodobɛŋ (FRI) ᴀᴋᴀɴ-ᴀsᴀɴᴛᴇ doroben (FRI; E&A) ᴛᴡɪ ɔdodobɛn (FRI; OA) doroben (FRI; E&A) ɢᴀ odódobɛŋ (FRI; KD) **NIGERIA** ɪɢʙᴏ amị = *reed* (auctt.) amị milọ̄ (KW) amị olo, amị milō (JMD; KW)

A cane-like grass, thin woody culms to 4 m tall, erect or more or less scandent, from a brief rhizome, perennial; grain conspicuously white shining, hard, 5–6 mm long, shortly bearded, of forest clearings and path sides, common throughout the Region from Senegal to Fernando Po, and in the rest of tropical Africa, and into tropical America.

The leaf, reduced to a pulp, is used as an application to pimples, furuncles and other cutaneous affections to maturate them in Senegal (1, 10) and in Ivory Coast (11).

The plant (? leafy culms) is made into a medicine in Ivory Coast for treating sore throat and into instillations for otitis and nose bleeding (2). Reduced to ash, the ash is rubbed into incisions for hookworms in Tanganyika, and is used in conjunction with medicine taken by mouth (3).

The hollow internodes of the culms are commonly used as straws for drinking beer (4; Kenya: 7, 12; Tanganyika: 13), and as delivery tubes in collecting palm-wine (4; Senegal: 1; Ghana: 9; Nigeria; 14) and in particular in Ghana from felled palms (8). Local names in the Region refer to their use for collecting wine, and in Kenya names allude to *tube* or *pipe* (12). In Ubangi (15) and in the Central African Republic (6) the culms are extensively used as arrow-shafts. Children in Senegal make them into flutes (1), and in Gabon into whistles and toy pop-guns (16). Suitably straight ones may also serve as blowpipes (16).

Pounded root is deemed in Ivory Coast to be haemostatic, and, like the leaf, is made into a drawing poultice to maturate abcesses (2). Sap from the pounded root is instilled into ears for earache in Senegal (1). In Ubangi a root-decoction is used for orchitis and is given to women seeking to become pregnant (15). The liquid in which the roots have been decocted is used in Ghana to make a meal of roasted ground corn to given to a child with diarrhoea (17).

The charred and powdered seed mixed with maleguetta pepper (*Aframomum maleguetta*, Zingiberaceae) is applied in Ivory Coast to wounds and snake-bite (2). Birds are said to eat the fertile florets, and as the seed is probably not digested, owing to its hard testa, dispersal is thus furthered (5, 12).

References:

1. Adam, 1954: 94. 2. Bouquet & Debray, 1974: 92. 3. Culwick 14, K. 4. Dalziel, 1937: 532. 5. Dundas FHI 21479, K. 6. Fay 1653, K. 7. Gardner 1444, K. 8. Irvine 862, K. 9. Irvine 5084, K. 10. Kerharo & Adam, 1974: 650. 11. Kerharo & Bouquet, 1950: 253. 12. Mogogo & Glover 774, K. 13. Proctor 435, 2010, K. 14. Thomas NWT 1706 (Nig.Ser.), K. 15. Vergiat, 1970,a: 86. 16. Walker & Sillans, 1961: 190. 17. Ampofo, 1983: 23.

Oplismenus burmannii (Retz.) P Beauv.

FWTA, ed. 2, 3: 437. UPWTA, ed. 1, 532.

West African: **SENEGAL** ᴡᴏʟᴏꜰ amhay (JB) **GUINEA-BISSAU** ꜰᴜʟᴀ-ᴘᴜʟᴀᴀʀ (Guinea-Bissau) quéuel (EPdS; JDES) ᴍᴀɴᴅɪɴɢ-ᴍᴀɴᴅɪɴᴋᴀ bondim-ô (JDES) bondium (EPdS) **SIERRA LEONE** ᴍᴇɴᴅᴇ yoavi (NWT; JMD) yoyavi (JMD) sᴜsᴜ sunyugi (JMD) ᴛᴇᴍɴᴇ ka-fulu (Glanville; JMD) ka-riŋ *a general term for a creeping grass* (NWT) **IVORY COAST** ɢᴀɢᴜ gbékaople (K&B) ᴋʀᴜ-ɢᴜᴇʀᴇ (Chiehn) bika hakosiré (K&B) bika kosiré (K&B; B&D) ᴍᴀɴᴅɪɴɢ-ᴍᴀɴɪɴᴋᴀ babri (B&D) féyan (B&D) **GHANA** ᴀᴋᴀɴ-ᴀsᴀɴᴛᴇ bogyamono (FRI) **NIGERIA** ɪɢʙᴏ (Okpanam) ọdo olili NWT ʏᴏʀᴜʙᴀ itẹ́-ọká = *python's nest* (JMD)

An annual grass (or perennial, 3) with creeping, trailing culms, aerial-rooted from the lower nodes, to 60 cm long, matting to about 10 cm deep; of shaded sites in forest or tree bushland; widespread across the Region in the guinean zone from Senegal to Fernando Po, and generally throughout the tropics.

285

It is thought to be ungrazed by stock in Senegal (1), but in Sierra Leone it is deemed to be a pasture grass, common under plantation trees (3), and elsewhere to be used as fodder (4). It has affinity to swampy places and in the Bendel State, Nigeria, it is a common weed of rice-padis (5). The 'Chiehn' believe that a panther after eating cleans its claws on this grass (1, 6).

The grass leaves, powdered up and mixed with sap from the palm tree are reputed to be aphrodisiac (1), or they may be mixed with the leaves of *Dienbollia sp.* (?*D. pinnata*, Sapindaceae) to the same end (1, 6). An ointment made with the leaves crushed and compounded with karite butter (*Vitellaria paradoxa*, Sapotaceae) is used in Ivory Coast on guinea-worm sores and snake-bite; it also soothes earache (2).

The inflorescence has a stickiness whereby it adheres to clothing, and for this property is used to catch rats (4).

In the Gagnoa area of Ivory Coast, hunters crush the leaves in water found in the crotch of a tree, and rub the liquid over the face to ensure they will meet game and be the better able to see it (2).

References:

1. Adam, 1954: 94. 2. Bouquet & Debray, 1974: 92. 3. Cole, 1968. 4. Dalziel, 1937: 532. 5. Gill & Ene, 1978: 182–3. 6. Kerharo & Bouquet, 1950: 253.

Oplismenus hirtellus (Linn.) P Beauv.

FWTA, ed. 2, 3: 437.
West African: **SIERRA LEONE** LIMBA bukɛŋgɛ (NWT) MENDE yoavi (JMD) yoyavi *the robust forest form: cf., Streptogyne crinita P. Beauv. with very similar leaves* (FCD) SUSU sumyigi (NWT) sunyugi (JMD) TEMNE ka-fulu (JMD) ka-riŋ *a general term for a creeping grass* (JMD) **GHANA** AKAN-ASANTE bogyamono (FRI) **NIGERIA** EFIK ńtùfíák (DRR) FULA-FULFULDE noppi dombe = *monkey's ears* (Chapman) YORUBA itẹ́-ọká = *python's nest* (JMD)

A perennial grass, with rambling culms up to 1 m long; of forest shade; throughout the Region in lowlands and montane locations, and widely dispersed in tropical Africa, and in the Mascarenes, America and Polynesia.

The grass has a little value as stock-feed in Uganda (2), and at the beginning of this century it was the standard grass cut for stall-feeding horses in Entebbe (1).

References:

1. Mahon s.n., 1901, K. 2. Snowden 1182, 1213, 1230, K.

Oropetium aristatum (Stapf) Pilger

FWTA, ed. 2, 3: 405.
West African: **MALI** DOGON dèdu nà (CG) dèdu yà (CG)

A tiny bunch-grass of numerous thin closely-packed culms 5-8 cm high; of lateritic pans and in rock crevies; Mali to N Ghana, and in Ubangi.

Plants of this species are so unsignificant in stature, and, without vouchers for indentification, it must be conjectured that the vernacular names cited above may more appropriately refer to *O. capense*, a larger plant with some potential interest to the grazier.

Oropetium capense Stapf

FWTA, ed. 2, 3: 405.

A small densely tufted perennial grass with culms to 15 cm high, of open places and shallow soils in deciduous bushland in Mali, and widely dispersed in eastern Africa from Somalia to S Africa.

In Kenya it is said to be eaten by sheep (2), and in Somalia it has a local name *nilo kois* meaning 'makes fat lambs' (1).

References:

1. McKinnon, S.260, K. 2. Veterinary Officer, Isiolo, 1618, K.

Oryza Linn.

FWTA, ed. 2, 3: 365.

English: rice.

French: riz.

Portuguese: orroz.

West African: **SENEGAL** WOLOF diuna *any cultivated rice* (JMD) **GUINEA** KISSI pendekiô (RP) sôbô (RP) KORANKO kini = *cooked rice* (RP) mâlô (RP) suma, sumu = *cooked rice* (RP) MANDING-BAMBARA suma = *food* (RP) MANINKA kini = *cooked rice* (RP) mâlô (RP) MANINKA (Konya) malô (RP) mumu = *cooked rice* (RP) TOMA môlô (RP) **MALI** DOGON ára *loan-word from 'Mande'* (CG) **IVORY COAST** KRU-AIZI saka (Herault) **GHANA** AKPAFU ì-ráâ (Kropp) *à*-rí (Kropp) *à*-rìi (Kropp) AVATIME *kí*-mímí (Kropp) LEFANA ǎǎnimi *cooked rice* (Kropp) *ka*-mù (Kropp) **DAHOMEY** GBE-FON molikun = *about 'li' [Pennisetum glaucum, pearl millet], because of a similarity in appearance of the two grains;* kun: *a suffix for food-plants* (Wigboldus) VHE molu (Wigboldus) **NIGER** DENDI mô *(indef. sing.)* (Tersis) **NIGERIA** ABUA àrùsú (RB) BOKYI eleshe (Osang fide RB) HAUSA lallaki *wild rice spp.* (A&S) HUBA mwalari (RB) IDOMA itsikǎpǎ (Armstrong fide RB) ocikǎpǎ (Armstrong fide RB) JEN nung-maari (RB) JUKUN (Wukari) nggìbyén (Shimizu) KAKA ñzáng (RB) KAMBARI-AUNA íiméḷé (RB) kánàssàarâ *wild rice spp.* (RB) SALKA éèwê wwéḷè *wild rice spp.* (RB) wwéḷé)*pl.* íiwéḷé (RB) KANURI shàngáwá *wild rice spp.* (A&S) shankawa (C&H) KUGBO òróṣì (Wolff) KORO bisika (RB) MALA iṣingkafa (RB) NINZAM ṣingkafa (RB) ODUAL àrùsú (Wolff) OGBIA éróṣì (Wolff) OGBIA-KOLO óróṣì (Wolff) PERO gáppà *a loan-word* (ZF) SAMBA-DAKA boo kem banã *cultivated rice* (RB) SANGA iṣangkapa (RB) SURA àas dì kïnong (anon. fide KW) shìnkápá (anon. fide KW) VERRE maroori (RB) YALA (Ikom) èrèsì (Ogo fide RB) YALA (Ogoja) icikápä (Onoh fide RB) iráis (Onoh fide RB) YATYE iraîs (Armstrong fide RB) YUNGUR marovi *from Fulfulde* (RB) **WEST CAMEROONS** BAFUT ḷésò (Mfanyam) BAMILEKE nkwénndak (Tchamda)

A genus of annual and perennial grasses of some 19–25 species in Africa, Asia, Australia and S America. Six species are recognised as indigenous to the Region of which *O. glaberrima* has arisen by selection from *O. barthii* and has been domesticated. *O. sativa* of Asian origin, the commercial rice, has been introduced. The origins of both species are lost in antiquity. It has been postulated that the widely dispersed and very polymorphic wild rice species, *O. perennis* Moench has given rise by one form to *O. glaberrima* and the closely related West African rice *O. barthii* (*O. breviligulata*), and in Asia by another form to *O. sativa*. At any rate *O. glaberrima* and *O. sativa* are very similar with but minor differences in pubescence of the glume and the size of the ligule. *O. glaberrima* is thought to have originated in the middle and upper Niger River valley some 2–3,000 years ago. There is no knowing the date of its domestication.

By the time of the coming of the European explorers it had acquired a considerable folk-lore and a large number of cultivars for both dry and wet cultivation. Of the Asian rice, *O. sativa*, there is archeological evidence of rice cultivation in China in the third millenium B.C., i.e., some 5,000 years ago. China is not, however, likely to be the cradle of domestication in Asia for the wild species do not occur there, but rather domestication occurred in the Indo-

Chinese region to the south of China proper. The Bengal area appears to be the most likely – and, be it noted, one of the most important and classical areas of rice cultivation in India is the eastern sea-board of Nellore in Andra Pradesh. *Nel* or *nellu* in the Dravidian languages of S India is padi. Cultivation must have been practised considerably before the 5,000 years ago that archeological research has given as the earliest recordable date in India for a rice culture to have spread to and to have been well-established in China by that time. The coincidence of the two datings perhaps implied that man had reached identical stages of social development and that he had similar artefacts of like lasting capacity by that time. As the rice civilisation spread outwards from its Indian epicentre, the grain becoming so esteemed acquired special social attributes. The Chinese emperors engaged in special sowing ceremonies. In northern India it acquired the Sanskrit name *râdjânna*, or 'food of the king', as well as the more pragmatic *dhánga*, or 'supporter (nourisher) of mankind'. Rice is now the staple food of about a half of the human race and any Asian today saying that he is going to feed will say that he is going to take rice. An interesting parallel of the paramount importance of rice is found in West Africa where rice *suma* (Bambara) means 'food'. Rice cultivation in W Africa does not appear to have spread very much from the *O. glaberrima* epicentre in the Niger River basin and a smaller centre in The Gambia. By 1630 it was reported still to be luxury in coastal Dahomey where it acquired the name *moli,* or 'about li', *li* being pearl millet. One must conclude that its general domestication post-dates that of millet. In Asia the rice culture did not reach Persia and the Mesopotamian area till about the Seventh century B.C., though rice must have been known of. Nor did it reach the Nile Delta and the Mediterranean till A.D. 600. So the growth of the two rice cultures, one in eastern Asia, the other in Western Africa, must have occurred in isolation. Yet there are interesting parallels of wet and dry cultivation and the selection of floating cultivars tolerant of flooding, selection against early shattering, and in ethnological, spiritual and animistic practice.

Asian rice has been seriously grown in W Africa only over the past 200 years. In part it has come via the coast working its way inland, and in part perhaps across the continent from Egypt and the east. Whichever way it is now displacing African rice in areas where suitable husbandry can be applied. *O. glaberrima* is, however, the more rugged plant tolerant of poorer conditions and must continue to find a place in W African agricultural economy. The nomenclature and vocabulary which has grown up in relation to African rice is on the whole applicable to Asian rice arriving in the Region unnamed: thus though *suma* (Bambara), *mâto* (Koranko) and *sōbō* (Kissi) refer strictly to *O. glaberrima*, they apply now by extension to all rice generally.

The word *rice* in English (*riz*, French; *arruz*, Portuguese, etc.) is of Aryan origin. Amongst the Sanskrit names is *vriki*, meaning 'that which elevates.' Comparative philologists equate this with *virinzi*, which is old Persian for 'rice.' [In the modern languages of Iran and Iraq the word is *birinj*.] Though rice was not cultivated in the Middle and Near East till much later, it was most certainly known and the Greek sage Sophocles (495–406 B.C.) was writing about ορίνδης (*orindes*) and ορίνδα (*orinda*) demonstrating a common consonantal shift of z to d. Alexander's army (346–326 B.C.) met it and the name for the plant entered classical Greek as όρυζα (*oruza*) and Latin as *aurisa*, or in its more familiar form *oryza*, whence came *riso* (Italian), *riz* (French) and *rice* (English), and adnate words in numerous dead and current European tongues. The Arabs spread rice to the western end of the Mediterranean, and *rozz*, or with its prefix *arrozz*, has entered Portuguese and Spanish as *orroz*.

Reference:

Burkill, IH, 1935: 1592–3. Dufrené, 1887. Grist, 1975. Laufer, 1919: 201. Ogbe & Williams, 1978: 59–64. Portères, 1956,a. Portères, 1956,b: 50. Portères, 1959: 189–233. Portères, 1966. Purseglove, 1972: 161. Wigboldus, 1986: 28.

Oryza barthii A. Chev.

FWTA, ed. 2, 3: 367. UPWTA, ed. 1, 534, as *O. stapfii* Roschev.
English: wild rice; self-sown rice; Mandinka rice (The Gambia, Duke).
French: riz sauvage; riz de marais.
West African: SENEGAL BEDIK màloŋ syɛnór = *rice of the* 'sɛnar' bird (FG&G)
MANDING-BAMBARA khuma malo (JB) SERER day (JB) malo tid (JB) yahas (JB) WOLOF malo
mpit (JB) THE GAMBIA MANDING-MANDINKA kamame mano (Macluskie) mani bano = *wild,
or self-sown rice* (Pirie; JMD) GUINEA-BISSAU BALANTA *n*tanse (JDES) *n*tante (JDES)
untante (JDES) CRIOULO aroz de ganga (JDES) FULA-PULAAR (Guinea-Bissau) marôcúmarè
(JDES) MANDING-MANDINKA cumarô-marô (JDES) PEPEL amanô-mane (JDES) GUINEA
BASARI ɓa-mbínɛkàr (FG&G) a-wínɛkàr (FG&G) FULA-PULAAR (Guinea) maro ladde *from*
ladde: *wild* (RP) maro vèndu *from* vèndu: *depressions, swamps, ponds* (RP) marô-guelode
(JDES) KISSI malo pam'beleya = *rice of the dead; in allusion to the starchy roots in water
providing food for the dead* (FCD; RP) KONYAGI malu wunn = *rice of the sky* (FG&G)
KORANKO djina-suma = *broth of the djinns; made from the grain* (RP) MANDING-BAMBARA kô-
malo = *rice of the rivers* (RP) kuma-malo = *rice of the birds* (RP) malo sina *from* sina: *wild*
(RP) MANDINKA malo-bano *from* bano: *wild* (RP) MANINKA fara loli (RP) kudyô (RP) malo-
mana *from* mana: *wild* (RP) MANDYAK *n*djangantê (JDES) undjangantê (JDES) SIERRA
LEONE BULOM (Sherbro) tɛtɛk-lɛ (FCD) KISSI malumba navileya = *rice of the dead* (FCD)
KONO funyana-kɔɛ (FCD) sambane, sambane pindiki (FCD) LOKO puŋga (FCD) MENDE *n*gafa-
*m*ba (FCD) kpɛkpɛngɛi (FCD; JMD) tɛtɛgɛ *in Sherbro-Krim area* (FCD) TEMNE a-pəla-pa-krifi
(FCD) an-tɛk-tɛk (FCD) *pa*-yar-yar (auctt.) MALI DOGON kumɔ́ ára pílu (CG) UPPER
VOLTA FULA-FULFULDE (Upper Volta) maaro fero (K&T) maaro pooli = *rice of the birds*
(K&T) NIGER HAUSA babo (Bartha) SONGHAI gènji-mò (*pl.*-mòà) (D&C) NIGERIA HAUSA
lállàkíí (ZOG) nanare (ZOG) shinkááfár gyado = *rice of the wart-hog* (ZOG) shinkááfár rishi
= *rice of the wart-hog* (ZOG) shinkááfár-kwààdíí = *rice of the frogs* (ZOG) shinkááfár-
múúgùù-dáwà = *rice of the wart-hog* (ZOG) KANURI fərgàmí (C&H) shàngáwá (A&S)

An annual rice, culms erect or decumbent and rooting at lower nodes, to 1½ m
long, in shallow water throughout the Region from Mauritania to S Nigeria,
and across Africa to Sudan, E Africa and Zambia.
   This rice is believed to be an original African rice and the parent of *O.
glaberrima* (4–6). It is known in the The Gambia (7) that if it is sown in a
mixture of other rices and grain is un-harvested, this species will exterminate the
others in a few years. The grain readily shatters and so ensures a regeneration of
self-sown plants. It is strongly resistant to dessication (2) and is invasive of wet
rice-padis, becoming a noxious weed suppressing cultivated rices.
   It furnishes good grazing for all stock before flowering, after which the scabid
awns may cause injury to the mouth (1, 3).
   The panicle shatters very readily. The plant is not normally cultivated though
the grain may sometimes be collected if plants are present in sufficient abundance, or as a famine-food. For collection it is usual to reap the inflorescence
before it has reached maturity (8), or, if ripe, over a basket or calabash to
contain the falling grain (1).

References:

1. Adam, 1954: 94. 2. Adam, 1960,a: 373. 3. Adam, 1966,a. 4. Dalziel, 1937: 534, as *O. stapfii*
Roschev. 5. Irvine, 1948: 264, as *O. stapfii*. 6. Irvine, 1952,a: 39, as *O. stapfii*. 7. Pirie 46/33, K. 8.
Raynal-Roques, 1978: 334.

Oryza brachyantha A Chev. & Roehr.

FWTA, ed. 2, 3: 365.
West African: SENEGAL MANDING-BAMBARA malo sina (JB)

A stoutly-culmed annual, or sub-perennial, to 50 cm tall, of pools and
swampy location, aquatic but not floating, in Senegal, Guinea, Sierra Leone and
Mali, and in the Congo basin and Sudan.

GRAMINEAE

Oryza eichingeri Peter

FWTA, ed. 2, 3: 365.

A perennial wild rice, culms slender to 60-100 cm high, of damp locations in forest undergrowth. Known only in Ivory Coast in the Region, but widely dispersed in Central and E Africa, and also occurring in Ceylon. In Kenya the plant gives grazing to wild animals, and birds take the grain (1).

Reference:

1. Magogo & Glover 769, K.

Oryza glaberrima Steud.

FWTA, ed. 2, 3: 367. UPWTA, ed. 1, 532.

English: African rice; red rice.

French: riz africain.

West African: SENEGAL BASARI malu (K&A) WOLOF diuna *any cultivated rice* (JMD) tiep (JMD) THE GAMBIA MANDING-MANDINKA deberro (JMD) faton ndure (Haswell) Fula mano = *Fula rice* (JMD) hom(b)o fingo (Pirie; JMD) keberro (JMD) keberro tima (JMD) mama saguia (Haswell) mamading saguia (Haswell) marboe saguia (Haswell) ndure sanneh (Haswell) sanun dingo (JMD) sona sanneh (Haswell) tunkung gunakon (JMD) yakka (JMD) yakka ba (JMD) GUINEA KORANKO djina-suma = *devil's rice, or broth of the djinns; in allusion to the gruel made from the seed* (RP) KPELLE gete (RB) MANO gete (RP) SIERRA LEONE BULOM (Kim) ɛ-kɔ (FCD) BULOM (Sherbro) mantalei (JMD) pɛlɛ (FCD) FULA-PULAAR (Sierra Leone) maro (FCD) GOLA njɔ (FCD) KISSI malô (FCD) malũ (FCD) maru (FCD) KONO koɛɛ (FCD) kwɛ (FCD) KORANKO kɔrɛ (FCD) KRIO rɛs *i.e., rice* (FCD) LIMBA pakalaba (FCD) LIMBA (Tonko) paga (FCD) LOKO mba (FCD) MANDING-MANDINKA suma (FCD) MENDE gobe *a c.var.* (JMD) njangei *a c.var.* (Fisher; JMD) kokovaiye *a c.var.* (JMD) kpakpati *a var., self-sewn, or semi-wild* (JMD) malai *a c.var.* (JMD) mande (JMD) marrai *a c.var.* (JMD) sanganye *a c.var.* (JMD) yake *a c.var.* (JMD) SUSU bomboto male (JMD) male (JMD; FCD) TEMNE a-pala (FCD) ɛ-pasɔr *a second crop* (FCD) pa-timi (JMD) pa-yaka (JMD) VAI kɔrɔ (FCD) YALUNKA mal-la (FCD) MALI DOGON ára géũ (CG) *also some floating rices of the Upper Niger River* (Camus & Viguier) SONINKE-MARKA barakanaï mobéri barkanaï kossa bussadiamu coïbane kossa daïa kossa karosa bura *bouyant but withstanding immersion for 3-4 days* karosapla pogo simoburu NIGERIA HAUSA babu rashi = *there is no loss* (Hardcastle) biya gero = *follow the millet* (JMD) ɗán ciso (JMD; ZOG) ɗán kashin shanu = *product of cattle manure* (Hardcastle) ɗán mai zudan (Hardcastle) ɗán tako (ZOG) ɗán tanko (JMD) jan irin gari (Hardcastle) shinkááfáá shééfè (ZOG) shinkááfár biyau (JMD; ZOG) shinkáfáá *applies generally* (JMD; ZOG) tamba (Jenkins) tanko (Jenkins) IGBO òsikapa ɔ̀bàra ɔ̀bàrà = *red rice* (BNO; KW)

Floating forms (Camus & Viguier): MALI MANDING-BAMBARA simoba simobaledio simober simoranéo simouadéo
NOTE for Hausa and Nigeria - c.var names mostly qualified by place names, colour or other features.

A robust annual grass, culms stout, 1-1½ m high, very variable, grain panicles black or white, of swampy and wet sites in Senegal to N Nigeria, and present in Central tropical Africa.

The plant produces an edible grain for which it has been grown. It is an indigenous rice species probably derived from *O. barthii* on which the W African rice culture between the Senegal River and N Nigeria grew up. This culture was established by Proto-Manding peoples, after millet and *Digitaria* spp. had been brought into the service of mankind, at c. 1,500 B.C., and flourished till the Sixteenth century when Asian rice cultivars reached W Africa (1). The species shows immense diversity, and some 1,500 varietal cultivars are known. The greatest quantity is to be found west of longitude 4°W in Ivory Coast (4-6). A curious parallel with Asian rice is to be found in this diversity with cultivars adapted to hill (dry), swamp, flood and floating conditions. Some will grow in water to 2½ m deep and withstand immersion for 3-4 days, and produce culms 4-5 m long (2).

As with the other indigenous rices, the grain tends to shatter requiring reaping of the entire panicle before final maturity. Asian rice strains do not suffer this defect. Nevertheless *O. glaberrinia* is still sometimes cultivated (7) though its interest as a cultivated crop is being displaced by *O. sativa.* The production of black grain seems also to be aesthetically undesirable. As a wild crop, the grain may still be collected.

In the Central African Republic the root is eaten raw for diarrhoea (3).

References:

1. Busson, 1965: 4. 2. Camus & Viguier, 1937: 201–3. 3. Fay 4484, K. 4. Portères, 1950, b: 484–507. 5. Portères, 1956,a: 341–84, 541–80, 627–700, 821–56. 6. Portères, 1962: 195–210. 7. Raynal-Roques, 1978: 334.

Oryza longistaminata A Chev. & Roehr.

FWTA, ed. 2, 3: 365. UPWTA, ed. 1, 532, as *O. barthii* A. Chev., in part.

English: wild rice.
West African: SENEGAL FULA-PULAAR (Mali) bulguré (A.Chev.) maro laddé (A.Chev.) maro vèndu (A.Chev.) PULAAR (Senegal) bahuré (A.Chev.) MANDING-BAMBARA ko malo (A.Chev.) kuma malo = *crested crane rice* (A.Chev.) SONINKE-SARAKOLE sakuru malo = *flamingo rice* (A.Chev.) WOLOF ndomgoduane (A.Chev.) tiep walo (A.Chev.) SIERRA LEONE BULOM *an*-tetchek (Stocks; JMD) BULOM (Sherbro) teteki (Stocks; JMD) TEMNE *pa*-kin-kin =*devil rice* (JMD) an-tek-tek (FCD) *a*-tetchek (Stocks) an-tetek (Glanville; JMD) *pa*-yariare =*streamside wild rice* (FCD; JMD) MALI ˙BAOMI˙ phenl (Davey) FULA-PULAAR (Mali) bahuré (A.Chev.) PULAAR (Senegal) bulguré (A.Chev.) maro laddé (A.Chev.) maro vendu (A.Chev.) MANDING-BAMBARA ko-malo (A.Chev.) kuma malo = *crested crane rice,* (A.Chev.) ndiga (Davey) UPPER VOLTA FULA-FULFULDE (Upper Volta) burgu (K&T) hudendia (Scholz) IVORY COAST MOORE bangé saga (A.Chev.) kolkodo (A.Chev.) TOGO GRUSI bugau (A.Chev.) NIGERIA FULA-FULFULDE (Nigeria) naanaare (J&D) HAUSA bau (JMD) bawu (Lamb) lállàkí (JMD) nanare (BM; JMD) shinkáàfár gyado, sh. mugun-dawa, sh. rishi = *wart-hog's rice* (JMD) shìnkáàfár kwaɗi = *frog's rice* (JMD) shìnkàáfàr watsu = *self-sown rice* (JMD) KANURI karagala (von Duisberg) karala (von Duisberg fide JMD)

A robust perennial with extensive rhizomes, culms to $1\frac{1}{4}$ m high by 1 cm diameter, soft and spongy, rooting at the nodes, of swamp, shallow ponds and flooded grassland, common throughout the Region from Senegal to N Nigeria, and extending all over tropical Africa to SW Africa, the Transvaal and in Madagascar.

The plant may colonise inundated meadows to form a blanket cover. Some sheltered parts of Volta Lake in Ghana have thus been taken over, even when the water is $1\frac{3}{4}$ m deep (4). In N Nigeria it is recorded as growing in water as much as 1 m deep (7). Thick stands provide excellent grazing for cattle (1, 6). In parts of Sudan it is greatly valued by graziers during both rainy (3) and dry seasons (5) for its excellence as fodder. In wet-rice cultivation it is a noxious weed suppressing cultivated-rice strains (1, 6, 10). Hausa in NW Nigeria refer to such weeds as *bau* (1). It is occasionally cultivated (8), and if self-sown is called *shinkafar watsu* (1).

The seed has an appearance similar to Asian rice. It is eaten here and there, is said to have a good flavour and may sometimes be found available in country markets (1, 3, 5, 6, 9). It is recognised as a famine-food (1, 7). Harvesting, however, poses difficulties. The grain shatters extremely readily. Temne of Sierra Leone have been noted to consider that the plant is not a kind of rice at all since it 'never' fruits (2). It is a practice, however, to reap the entire panicle just before coming to maturity (8), or to shake the ripe panicle over a basket or a calabash into which grain falls (1, 6). Another disincentive to handling the panicle is the long scabrid awns (1).

The straw which may attain considerable length if the plant has grown in flood water is valued for thatching (1).

Warthogs will eat the roots (1).

GRAMINEAE

References:

1. Dalziel, 1937: 532, as *O. barthii* A. Chev. 2. Deighton 1543, K. 3. Evans-Pritchard s.n., 12–16/9/ 1935, K. 4. Hall & al., 1971. 5. Harrison 247, K. 6. Irvine, 1952,a: 39, as *O. barthii*. 7. Lamb 100, K. 8. Raynal-Roques, 1978: 334. 9. Stocks s.n., 14/2/1921, K. 10. Wild, 1962: 22–23, as *O. perennis* Moench.

Oryza punctata Kotschy

FWTA, ed. 2, 3: 365.

An aquatic rice, stout-culmed, 60–90 cm high, spongy, of swampy streams in Ivory Coast to S Nigeria and central, E and SW Africa, Madagascar and Thailand.

The grain is recorded as eaten as a famine-food in Kenya (1).

Reference:

1. Taylor 1211, K.

Oryza sativa Linn.

FWTA, ed. 2, 3: 365. UPWTA, ed. 1, 533–4.

English: rice; Asiatic or Asian rice; padi (from Malay) or paddy (Anglo-Indian), especially for wet-rice in cultivation, and for the unhusked grain.

French: riz; riz asiatique.

Portuguese: arruz.

West African: **SENEGAL** BEDIK màlóŋ (FG&G) màlóŋ bandyul a *c.var.* (FG&G) màlóŋ *i*-swgɛt a *c.var.* (FG&G) DIOLA émano (JMD; JB) MANDING-BAMBARA malo (JMD; JB) SERER malo (JB) WOLOF diuna a *general term* (JMD) malo (JMD; JB) tiep (JMD) **THE GAMBIA** MANDING-MANDINKA mano (Williams; JMD) tubal mano = *foreign rice* (JMD) **GUINEA** BASARI malú ɓandyúl a *c.var.* (FG&G) malú syísyèt a *c.var.* (FG&G) malú-*i*-ɓánàx a *c.var.* (FG&G) malú-*i*-sɔ́gét a *c.var.* (FG&G) FULA-PULAAR (Guinea) malo (JMD) KISSI dixi, dixô, dixiô *Indian introductions* (RP) malô (RP) selegbo (RP) KONYAGI malu (FG&G) KORANKO mâlô (RP) KPELLE halemoni (RP) selemonu (RP) MANDING-MANINKA malo (JMD) MANO halemoni (RP) selemonu (RP) SUSU dixi, dishi, disi *Indian introductions* (RP) malé (JMD) mereke, meleke, merke, merkeni *American introductions* (RP) **SIERRA LEONE** BULOM pɔlè (Pichl) pɛlɛ (FCD) BULOM (Kim) ɛ-kɔ (FCD) BULOM (Sherbro) kɔ̀kɔ̀vaya a *c.var.* (Pichl) FULA-PULAAR (Sierra Leone) maro (FCD) GOLA ɲjɔ (FCD) KISSI malô (FCD) malũ (FCD) maru (JMD; FCD) KONO koɛɛ (FCD) kwɛ (JMD; FCD) KORANKO fara kɔrɛ *padi (wet) rice* (JMD) gbilema kɔrɛ *dry (hill) rice* (JMD) kɔrɛ (JMD; FCD) KRIO rɛs *i.e., rice* (FCD) LIMBA pagala ba ka tité *dry (hill) rice* (JMD) pakalaba (JMB; FCD) LIMBA (Tonko) paga (FCD) LOKO mba (FCD) mali (JMD) MANDING-MANDINKA suma (FCD) MENDE jonge a *second crop* (FWHM) mba-gale *the grain* (FWHM) mba-wui *the ear* (FWHM) mbei (*def.*) (FWHM) SUSU kharima male *dry (hill) rice* (JMD) male (FCD) meri male *padi (wet) rice* (JMD) TEMNE ka-ɔla (*sing*) a grain (NWT) ka-ɔs (*pl.* tsoe-ɔs) *grain in husk* (NWT) ɛ-pasər a *second crop, an aftermath* (NWT; FCD) a-pɔla (*pl.*) (NWT; FCD) a-pɔla-*pa*-ro-bat *padi (wet) rice* (NWT; JMD) a-pɔla-*pa*-ro-gban *dry (hill) rice* (NWT; JMD) ka-yaka (*pl.* tsoe-yaka) (NWT) VAI kɔrɔ (FCD) YALUNKA kharima-mal-le *dry (hill) rice* (JMD) mal-la (JMD; FCD) meri mal-le *padi (wet) rice* (JMD) **LIBERIA** KRU-DEWOIN mɔ̃ (Marchese) GUERE (Krahn) kɔ̃ (Marchese) KUWAA kɔ̃lɔ̃ (Marchese) BASA mɔ̃ (Marchese) MANO bu (JMD) ·MPWESE· djɔ (JMD) **MALI** MANDING-BAMBARA malo (JMD) **UPPER VOLTA** FULA-FULFULDE (Upper Volta) maaro (K&T) GURMA mùùli (Surugue) KURUMBA amwi (Prost) MOORE mwi (Prost) SAMO (Sembla) mããn (Prost) **IVORY COAST** KRU-AIZI saka (Marchese) BETE sika (Marchese) DIDA sáká (Marchese) DIDA (Vata) saká (*pl.* -a) (Marchese) GODIE sũké (*pl.* sukáa) (Marchese) GREBO bla (Marchese) GREBO (Jrewe) kòwɔ (Marchese) GREBO (Tepo) gblã (Marchese) GUERE kɔ̃ (Marchese) GUERE (Chiehn) kòò (Marchese) GUERE (Krahn) dī (Marchese) GUERE (Wobe) kō (Marchese) KLAO kɔ (Marchese) KOYO sáká (Marchese) NEYO saka (Marchese) NIABOUA kɔ̃ɓɔ̃ (Marchese) KWENI sáá *from Baule*: sáká (Gregoire) **GHANA** ADANGME-KROBO omõ (FRI) AKAN-ASANTE ɛmõ (JMD; E&A) ɛmu (JMD) FANTE omõ (auctt.) TWI ɛmõ (auctt.) ɛmu (JMD) mõ (auctt.) DAGBANI sinkafa *from Hausa* (JMD) sunkafa (Gaisser) GA omõ (auctt.) GBE-VHE mɔli, molũ, mɔlu (auctt.) VHE (Pecí) mɔli (FRI) GRUSI mumuna (JMD) KONKOMBA imul (Gaisser; Froelich)

292

MAMPRULI sinkafa *from Hausa* (JMD) MOBA mori (JMD) NANKANNI muie (JMD) NZEMA azan (JMD) SISAALA miiriŋ (Blass) **TOGO** BASSARI imogule (Gaisser) KABRE mau (Gaisser) ungau (Gaisser) NAWDM mi (Gaisser) miirbe (Nicole) miri (Gaisser) TEM (Tschaudjo) mau (Gaisser) ungau (Gaisser) **NIGER** SONGHAI mò (*pl.* mòà) (D&C) **NIGERIA** BATA hoyyanga (Barth) BIROM cun bikwòók = *fonio with envelopes* (LB) egì ŋàs = *egg of the ant trail* (LB) EDO izɛ (JMD) EFIK èdêsì (auctt.) FULA-FULFULDE (Nigeria) burungo (JMD) maaroori (J&D; JMD) GWANDARA-CANCARA hyiŋkapa (Matsushita) GITATA ǹšikapa (Matsushita) KARSHI šiŋkapa (Matsushita) KORO šiŋkafa (Matsushita) NIMBIA ciŋkafa (Matsushita) TONI šikafa (Matsushita) GWARI shewi (JMD) HAUSA gume (JMD) shánshèèráá *unhusked grain* (JMD; ZOG) shéfè (JMD; ZOG) shinkááfáá (auctt.) ICHEVE élísì (Anape fide KW) IGALA òchikápā *from Hausa:* shinkaáfaá (RB) IGBO òsikapa (auctt.) ISOKO osikapa (JMD) JUKUN (Gwana) among (Barth) KANURI esmalle (Barth) pergami (JMD) MARGI pirgami (Barth) NUPE chenkafa, yokofa (JMD) NZANGI morori (Barth) TIV chingapa (JMD) tsinggapa (JMD) WANDALA-GAMARGU fergamye (Barth) YORUBA rɛsì, ìrɛsì (JRA; JMD)

Products: **GUINEA** KISSI mumu *cooked rice* (RP) KORANKO kini, suma, suma *cooked rice* (RP) **NIGERIA** HAUSA a chi da mai *a cooked preparation* (JMD) betso, buza *a beer made from rice and honey* (JMD) ƙuƙus *a small hard-grained var. uncookable* (JMD) rauno *straw, chopped and mixed with mud, used for building* (JMD) zazzaɓaɓɓiya *steamed rice in the husk* (JMD) NOTE: see (1.) Portères, *J. Agr. trop. Bot. appl.* 1956, 1966, for a detailed study of *Oryza* in cultivation in West Africa and for Guinean cultivar names. (2) Espírito Santo, *Nomes vernaculos de Algumas plantas da Guine Portuguesa,* 1963, for cultivar names in Guinea-Bissau.

An annual, many-stemmed grass to about 50 cm high, but some floating swamp forms to over 3 m long, extremely polymorphic, bearing an open nodding panicle of grain; a cultigen originating in Asia and introduced to W Africa in the last 500 years, and to many tropical and sub-tropical countries.

Asian rice has assumed major importance as a world cereal, a staple for a half of the world's population. Its great plasticity and variation lends it to adaptation to numerous sets of conditions. Primarily there are strains which will grow on dry land provided they receive a minimum of 700 mm rainfall during the growing period. The principle variation is found in cultivars selected for cultivation in water. Optimal depth is variable for specific strains. Some can grow under flooding, the basal end of the culm elongating allowing the culm to float on the rising water, and then the culm roots nodaly when the water level falls. Other strains will grow in saline or brackish water. Certain strains are daylight length sensitive, even plus or minus a quarter hour being significantly critical. All such forms of Asian rice have been introduced to the Region. Parallel development is found in African rice, *Oryza glaberrima*, and strains of the two species are interfertile producing hybrids and great problems for agrobotanical taxonomy. The principal rice growing areas of W Africa are Senegal to W Guinea and N Sierra Leone, the Upper Niger river basin and Nigeria, but with changing social attitudes rice is no longer a perquisite of the chiefs and headmen, but is eaten by all the population so that wet rice is commonly cultivated wherever adequate water is available. Asian strains are tending to displace the native African rice.

The growing plant makes good fodder for cattle, but at the expense, of course, of obtaining any grain.

The paramount purpose of growing the plant is for its grain. The grain in its husk, or padi, is good fodder for farm animals, especially poultry. Removal of the husk exposes the grain covered by a pink or mauve aleurone integument. In this form the grain is edible, but it is usual for aesthetic purposes to mill it off leaving a white kernel. Another method of arriving at the kernel is to parboil the seed, i.e. to soak the seed in cold, or perhaps warm water for three to four days. The water is run off and the grain heated by steam to boiling point. The husk is then milled off leaving a slightly gelatinised more or less transparent grain. An analogous process of parboiling is reported practice in the Region (7). The grain is said to be more easily digested.

Rice is not a complete diet. Milled rice has about 86–90% carbohydrate and 7–8% protein. Amino acids are naturally low, and as a proportion is present in the aleurone layer, milling leads to further imbalance. Without the addition of other suitable foodstuffs to the diet, especially those rich in protein, malnutrition results usually in the form of beriberi (4, 5, 9).

Milling produces a small proportion of broken grain, rather less under parboiling than purely mechanical cleansing. In Italy this is commonly ground down to flour, and sometimes it is mixed with wheat flour to make a high class bread (9). A form of parboiling is done by Mandinka and Susu peoples to produce a sort of bread or baked cake with honey added taken as food or used in ceremonial rituals (7). Bran, the aleurone layer, has a high food value with 75% digestibility. Its addition to the diet helps to offset symptoms of vitamin B deficiency. It is recommended for milch-cows. Rice oil can also be extracted. It is not edible but can be used in soap-making (9).

Highly polished rice is a good source of industrial starch used, for example, in the laundry industry. It has cosmetic use in Asia. The grain is commonly fermented to make wine, spirits, beer and industrial alcohol. A beer is made from rice in Senegal, Nigeria and doubtless elsewhere in the Region.

The husk is of no value as food, but will burn and can be used as fuel in rice-mill engines. It is a common commodity for packing fragile goods. The husk can be successfully used as an aggregate in preparing lightweight concrete with a density of not greater than 100 lb/cu ft when dry (6).

Rice straw has about 32–33% carbohydrates, 3.7% proteins and 1.4% fat. Digestibility is low, but it can be admixed with other fodders for milch-cows and bullocks. It can also be ensiled. It is suitable for turning into fibre-board. Some 300,000 tons were said to be available in Nigeria in 1973 when there were proposals for converting the straw into boards (3). Chopped up and mixed with mud it is favoured for building hut-walls in Nigeria (7) and Senegal (1) where it is also used to weave into baskets and mats. In China it forms a substrate in mushroom cultivation (9).

Medicinally water in which rice has been cooked is taken as a beverage and as a vehicle for other medicines, and the fine meal is rubbed onto skin eruptions and rashes, and used as a poultice (7). Rice water is taken as a bland diuretic in Nigeria in urinary complaints (2). In Gabon it is considered a strengthening food for children and helpful in cases of diarrhoea and dysentery (12). Rice powder is used in Senegal against itch (1). Cooked rice, rice flour and rice paste are variously used in Asian countries for certain cancerous conditions (10).

In Zanzibar a black dye is made from pounded roasted grain mixed with soot (8).

Examination of plants in Sierra Leone has shown alkaloids to be absent from the root and saponin to be present (11).

References:

1. Adam, 1954: 94. 2. Ainslie, 1937: sp. no 256. 3. Anon., 1973. 4. Burkill, IH, 1935: 1593–610. 5. Busson, 1965: 468, with chemical analyses. 6. Chittenden & Flaws, 1964: 187–99. 7. Dalziel, 1937: 533–4. 8. Greenway, 1941: 222–45, sp. no. 185. 9. Grist, 1975. 10. Hartwell, 1969,b: 195. 11. Taylor-Smith, 1966,a: 538–41. 12. Walker & Sillans, 1961: 190–1.

Oxytenanthera abyssinica (A. Rich.) Munro

FWTA, ed. 2, 3: 360. UPWTA, ed. 1, 534.

English: bamboo; West African bamboo (Percival); savanna bamboo (E Africa, FRI).

French: bambou.

West African: SENEGAL BASARI o-kadjié (K&A) BEDIK u-hátyè (FG&G) ma-kátyè (FG&G) DIOLA bubul (JB; K&A) fugil (JB; K&A) FULA-PULAAR (Senegal) kéwé (K&A) MANDING-BAMBARA bó (auctt.) kévé (JB) MANINKA bó (K&A) 'SOCE' bó (K&A) SERER giol (JB; K&A) ingol = the bow of a fishing bow (N'Diaye Samba) WOLOF wa (auctt.) THE GAMBIA FULA-PULAAR (The Gambia) kebe (DRR) kewal (DAP) MANDING-MANDINKA bo the cut culms (JMD) boho (DRR; DAP) bongo (def.) (auctt.) WOLOF wah (DRR; DAP) GUINEA-BISSAU BALANTA sougué (JDES) BIAFADA djame (JDES) BIDYOGO edjô (JDES) CRIOULO bambú (JDES) FULA-PULAAR (Guinea - Bissau) québè (JDES) quéné (JDES) PULAAR (Guinea-Bissau)

djambarlam (EPdS) MANDING-MANDINKA bô (JDES) djambátam-ô (def.) (EPdS; JDES) MAN-
DYAK najane (JDES) udjame (JDES) MANKANYA djame (JDES) PEPEL djamá (JDES) **GUINEA**
BAGA (Koba) kô-tatami (Hovis) BASARI ɔ-xàtyé (FG&G) KONYAGI wa-diag (FG&G) u-ryag
(FG&G) SUSU tatami (Hovis) **SIERRA LEONE** BULOM thong (JMD) wus-lɛ (FCD) BULOM
(Sherbro) kana-lɛ (FCD) FULA-PULAAR (Sierra Leone) kɛvɛ (FCD) GOLA sen (FCD) seni (FCD)
KISSI bɔho (JMD) pilanda (FCD) KONO siminɛ (FCD) KCRANKO bɔ (JMD) KRIO kɛn (FCD)
LIMBA ko-ai (pl. ba-wai) (JMD) LIMBA (Tonko) barāŋ (FCD) LOKO sii (FCD) MANDING-
MANDINKA bəə (FCD) MENDE semi (JMD; FCD) SUSU bomi as a pole used as a mast [?boom]
(JMD) tatami (JMD; FCD) TEMNE ka-sul the cut pole (JMD; FCD) ka-thong (JMD) ka-tɔn
(JMD) VAI kenye (FCD) seni (FCD) senye (FCD) YALUNKA tatami-na (JMD; FCD) **LIBERIA**
LOMA temui (JMD) **MALI** MANDING-BAMBARA bɔ (A.Chev.) SONGHAI dianacaré (A.Chev.)
SONINKE-SARAKOLE koré (A.Chev.) **UPPER VOLTA** BOBO buna (Le Bris & Prost) FULA-
FULFULDE (Upper Volta) lebooji (K&T) GURMA mia (A.Chev.) MOORE tanhuisi (A.Chev.)
SAMO (Sembla) báalé (Prost) **IVORY COAST** KRU-BETE kɔle (Pageaud) **GHANA** ADANGME
pamploo FRI ADANGME-KROBO pamplo (Bunting fide FRI) pāplo (Bunting fide FRI) AKAN-
FANTE nkampon (FRI; E&A) prampru (auctt.) TWI mpampro (FRI; E&A) mprampuro (FRI)
AKPAFU ɔ́-bébéī (pl. sɪ-) (Kropp) ANYI-ANUFO anohwere FRI kyɛmponyi FRI mbaramboro FRI
AOWIN anohwere (FRI) krɛmponyi (FRI) mpampro (FRI) SEHWI mbaramboro FRI AVATIME
pámpró (Kropp) GA pampló (Auctt.) pamploo (FRI) pāplo (Bunting fide FRI) GBE-VHE
(Awlan) pamplo (FRI) VHE (Pecí) pampru (FRI) GRUSI miu (FRI) GUANG-ANUM nkampro
(Bunting fide FRI) GONJA nkampro (FRI) HAUSA gora (FRI) LEFANA mpàmpró (Kropp)
MAMPRULI nyoringa (Arana & Swodesh) **DAHOMEY** BAATONUN téma (JMD) **NIGER** DENDI
káálá (indef. sing.) (Tersis) **NIGERIA** AKPA òcyácyō (Armstrong) ARABIC-SHUWA gana (JMD)
BIROM syɛ̀ (LB) BURA kida (anon.) EDO oyo (JMD) EFIK ńnyányángá ḿbàkárá (auctt.) FULA-
FULFULDE (Nigeria) kewal (pl. kewe) (JMD) GWARI kawu (JMD) HAUSA góora (JMD; LB)
góora di a large pole (JMD; ZOG) IDOMA aco (Armstrong fide RB) ọcacō, ọcácọ̄ (Armstrong
fide RB) IGBO àchàlà oyibo = European achara (JMD) àchàrà oyibo = European achara
(JMD) ḿkpọ àchàrà a bamboo pole (Lowe) ọhọloibo (JMD) ọtọsị (auctt.) IJO-IZON (Egbema)
bọmọụn (Tiemo fide KW) KANURI gàmáré (C&H) halwa (JMD) kava (Benton fide JMD) LOKE
eman (Berry & Guy) ketitahng (Berry & Guy) NGIZIM tákárwá (pl.-áwín) (Schuh) ODUAL ạligua
(Gardner) PERO bálbàl (ZF) gónrò a bamboo pole (ZF) SURA ràas (anon. fide KW) YALA
(Ikom) àtàng (Ogo fide RB) YALA (Ogoja) ɔkpɔ̀kpɔ̀ (Onoh fide RB) YORUBA apàko (JMD)
aparun (JMD) ọpa a pole (JMD) ọparun general for canes or giant grasses (JMD) pàko (JMD)
WEST CAMEROONS NGYEMBOON lekwê a sp. of bamboo (Anderson) ndyung (Anderson)
nkâ´ a split bamboo (Anderson) shyǔ´ a pole (Anderson)

A stout, woody bamboo; culms to 10–15 m high by 8–10 cm diameter, thick-
walled, scarcely hollowed; in clumps, rhizomatous, sometimes immense, forming impenetrable thickets which in Uganda may spread over several square
miles; outer culms becoming recumbent and flowering, then dying; widespread
in dry forest and soudanian woodland across the Region from Senegal to N
Nigeria, and throughout tropical Africa.

The leaf is browsed by cattle in Senegal (1), and Kordofan (2), even in a dry
state. Chimpanzees are said to eat the leaf and the pith, especially from young
new shoots (21). A leaf-decoction is valued in Senegal for urinary problems,
being prescribed for lack of urine, as well as for too much urine, particularly in
diabetes (14–16). Leaf-decoction is also administered in Senegal for generalised
oedemas and albuminurea (16). Undefined medicinal application is reported in
Mali, where in Bamako market leaves are an item of trade (8). In Ghana leaves
rubbed (?) on house-walls are said to keep away lice (Twi: sommɔe) (13).

Young leafy shoots are edible and are said to be taken in Uganda in time of
famine (Eggeling fide 13).

The strong woody culms are the most valued part of the plant as building
material for huts, furniture and fencing, for splitting to weave into baskets and
panniers, for spears, bows and arrows, musical instruments, xylophones, and
tambourines, etc. (4, 8, 10, 12a,b, 13, 20). In Jebel Marra, Sudan, walking-sticks
are commonly made by cutting suitably-sized culms with a piece of the basal
rhizome (12). In the Region a charcoal is made of the culms (6).

The leaf and culm are reported to contain an (unnamed) alkaloid (23). Culm
sheaths are hairy, and the hairs are rubbed off in Tanganyika for use as a
wound-dressing which is applied with the rhizome pith cooked to a mash and
bandaged on (11).

Bobo people of Upper Volta are reported to use the rhizome in treatment for dysentery (17). It is said in Zimbabwe that on flowering, the flower-heads are frequently covered with honey-dew (3). On ripening the spiky seed heads may cause wounds which are refractory to healing (12). The seed is collected on Jebel Marra for grinding to flour and to use in brewing in time of famine (12, 22). In Tanganyika the seed is ground into a meal with other grain to make a tonic for small children (11). The plant, which can be conveniently propagated by root-division, is commonly cultivated in the Njombe area of the Tanganyika Southern Highlands for its seeds to make an alcoholic drink (9, 18, 19).

References:

1. Adam, 1966,a. 2. Baumer, 1975: 108. 3. Crook 807, K. 4. Crook 921, K. 5. Dalziel 502, K. 6. Dalziel, 1937: 534. 7. Deighton 5581, K. 8. Diarra, 1977: 44. 9. Eggeling 6077, K. 10. Ferry & al., 1974: sp. no. 125. 11. Haerdi, 1964: 210. 12a. Hunting Tech. Services, 1964. 12b. Hunting Tech. Surveys, 1968. 13. Irvine, 1961: 787–9. 14. Kerharo & Adam, 1963,b: 853–70. 15. Kerharo & Adam, 1964,b: 564. 16. Kerharo & Adam, 1974: 650–1. 17. Le Brist & Prost, 1981. 18. Lynas DC 159, K. 19. Pitt 559, K. 20. Snowdon 1051, & letter, K. 21. Tutin 26, K. 22. Wickens 2925, K. 23. Willaman & Li, 1970.

Panicum Linn.

FWTA, ed. 2, 3: 426–9. UPWTA, ed. 1, 534.
West African: SENEGAL WOLOF baket *edible grain spp.* (anon.) salguf (JMD) SIERRA LEONE MENDE faditi (JMD) faliti (JMD) fɔni *short grasses of open flats* (JMD) muli (JMD) muri (JMD) ndewe (JMD) ndiwi (JMD) ngalei-hei = *female 'ngalei'; ngalei is general for tall grasses, esp. Chasmopodium caudatum (Hack.) Stapf* (JMD) nyina-fɔni (JMD) yowo *chiefly creeping grasses* (JMD) SUSU funfuri (JMD) fure-funfuri (JMD) sankabesukwi (JMD) sunyugi (JMD) TEMNE kɔ-gbel (JMD) a-pende-*pa-ma*-gbel (JMD) peni-fafagbi (JMD) peni-*pa*-gbel (JMD) *ka*-sul-*ka*-robat (JMD) NIGERIA FULA-FULFULDE (Nigeria) gerorohe (*pl.* -oje) (JMD) HAUSA geron tsuntsaye = *birds' millet* (JMD) IGBO ìkpò milī = *water ikpò* (JMD; KW) ilulo (JMD) ilulo kputu (JMD) ilulo mgbada (JMD) ilulo ñkịtā = *ilulo of the dog* (JMD; KW) ilulo okili mbawisi (JMD) ilulọkele (JMD) YORUBA ìtẹ̀àparò = *francolin's nest* (JMD; Verger) motisan (JMD)

A large pan-tropical genus of ∼ 470 spp., of which some 50 are recorded present in the Region, mostly indigenous, but some introduced; annual or perennial, woody shrubs to low herbs; of desert, savanna, forest and swamp. Some are domesticated, e.g. *P. maximum* Jacq. (guinea grass), and *P. coloratum* Linn. (buffalo grass). Some ought to be domesticated, e.g., *P. turgidum* Forssk. and *P. laetum* Kunth. Many provide useful fodder, and probably almost all provide more or less some browsing. The grain of several is collected for human food, and perhaps the grain of others, so far unrecorded, might be eaten in time of great want.

Panicum afzelii Sw.

FWTA, ed. 2, 3: 433.
English: magbel's millet (Sierra Leone, after the small black bird called *magbel* in Temme, Deighton).
West African: SIERRA LEONE MENDE nyina-fɔni, nyina-voni (FCD) SUSU funfuri (NWT) TEMNE a-pende-*pa-ma*-gbel *from* ma-gbel: *a small black bird which eats the grain* (FCD) *ka*-seth-*ka*-gbel (JMD)

A delicate grass to 30 cm high, of shallow or disturbed soil of moist places, especially near *foni* flats in Sierra Leone, and occurring from Senegal to Ghana.

Panicum anabaptistum Steud.

FWTA, ed. 2, 3: 432. UPWTA, ed. 1, 535, in part.

West African: SENEGAL BASARI a-ïlutu (K&A) BEDIK a-ïlutu (K&A) MANDING-BAMBARA guo (JB; K&A) ngâ (JB; K&A) NON idadapen (K&A) idanapen (K&A) GUINEA BASARI a-yílútú (FG&G) KONYAGI a-ngay (FG&G) MALI MANDING-BAMBARA paguiri mayo (A.Chev.) suebee (Davey) yufane (A.Chev.) UPPER VOLTA FULA-FULFULDE (Upper Volta) siiwuko (K&T) si'uko (K&T) NIGER SONGHAI génsi (pl. -ò) (D&C) NIGERIA FULA-FULFULDE (Nigeria) burɗi, buwirɗi (JMD) HAUSA tafartso (JMD; ZOG) tsíntsííyáá = a brush (JMD; Palmer) tsíntsííyár tafartso (ZOG)

A robust perennial grass with culms to 1½ m high, on flood-plains, common across the Region from Mauritania to W Cameroons, and on to Sudan.

The grass provides a good fodder when young (1, 3). It is commonly used as a broom, hence the Hausa name of that meaning (3, 7) and for thatching and matting (2–4). Tenda people of the Senegal–Guinea border use it for filtering millet beer (4).

In contact with the skin it provokes an erythema and violent itching (6), and has unspecified medicinal usage in E Senegal (5).

References:

1. Adam, 1966,a. 2. Dalziel 274, K. 3. Dalziel, 1937: 535. 4. Ferry & al., 1974: sp. no. 126. 5. Kerharo & Adam, 1964,a: 436. 6. Kerharo & Adam, 1964,c: 316. 7. Palmer 16, K.

Panicum antidotale Retz.

English: blue panic (India, Deshaprabhu).

A coarse perennial grass, with creeping root-stock throwing up culms to 2 m high, native of India, the Middle East and Australasia, now introduced into Nigeria.

It has been grown on the Ilorin stock farm of N Nigeria (2), but there is no report of its utility. In India (1) it is drought-resistant and winter-hardy. It produces a fodder best utilised by cutting to keep it in active growth. When old it acquires a bitter taste. As hay, quality is good. In America it is used for erosion control on flood-plains and as a wind-break. Smoke from the burning plant is a fumigant for wounds and disinfectant in small-pox. The plant is used for throat infections and as an antidote for rabies.

References:

1. Deshaprabhu, 1966: 222–3, with chemical analysis. 2. Ward 0069, K.

Panicum brazzavillense Franch.

FWTA, ed. 2, 3: 431, as *P. pubiglume* Stapf.
West African: NIGERIA FULA-FULFULDE (Nigeria) ramafada (Saunders)

A tufted perennial grass, culms to 60 cm high, on seasonally wet or swampy soil across the Region from Senegal to S Nigeria, and in Congo (Brazzaville).

It is a pasture grass, known to Fula in N Nigeria (1), and presumably therefore a source of cattle-fodder.

Reference:

1. Saunders 24, K.

GRAMINEAE

Panicum brevifolium Linn.

FWTA, ed. 2, 3: 429.
West African: SIERRA LEONE KISSI deno-yeno *probably general for grasses growing along bush paths* (FCD) LIMBA nimbo (NWT) LOKO kɔibɔre (NWT) nyanbile (NWT) MENDE muli (NWT) muri (NWT) SUSU fumfuri (NWT) kurudɛrafumfuri (NWT) sankabesugi (NWT) yane (NWT) TEMNE kə-bil (NWT) kuseta-*ka*-gbil (NWT)

An annual grass, culms rambling, rooting at the nodes to 1 m long by 30–60 cm high, of forest shade throughout the Region, and dispersed over tropical Africa and the Old World tropics.

Cattle (1) and game (2) graze it readily, thus it seems to be a good fodder. It is a common weed of rice-padis in Bendel state of Nigeria (3).

References:

1. Burkill, IH, 1935: 1655. 2. Magogo & Glover 575, K. 3. Gill & Ene, 1978: 182–3.

Panicum calvum Stapf

FWTA, ed. 2, 3: 433.

A perennial grass, slender rambling culms to 1½ m long, of upland and montane locations in forest shade and margins in Ivory Coast, Nigeria and W Cameroons, and across Africa to E Africa.

It occurs as a weed of cultivated and waste lands in W Nigeria (1). In Uganda it is reported to be much liked as fodder by stock (2).

References:

1. Egunjobi, 1969. 2. Snowden 1231, K.

Panicum coloratum Linn.

FWTA, ed. 2, 3: 434.

A tufted perennial grass, culms to 1.2 m high, very polymorphic, native of NE, E and southern Africa, and introduced to N Nigeria, and other tropical countries.

The plant is drought-resistant and in all territories it provides good grazing for all stock.

Panicum congoense Franch.

FWTA, ed. 2, 3: 432.
West African: SIERRA LEONE BULOM (Sherbro) peniŋbon-lɛ (FCD)

A densely tufted perennial grass, culms 30–60 cm high, of marshy places in Senegal to Ghana, and in the Congo basin and Angola.

Panicum dinklagei Mez

FWTA, ed. 2, 3: 431.
West African: SIERRA LEONE KISSI yeno (FCD) KONO wetɛ (FCD) LOKO njekbɛvɛ (NWT) MENDE folimbe (NWT) muli (FCD) nyina-fɔni (FCD) sukule-voni (FCD) SUSU furefumfuri (NWT) TEMNE a-kol-saba (Glanville) kə-yapa (NWT) **LIBERIA** MANO pini ti (Har.)

A robust perennial grass with thin cane-like culms scrambling over other vegetation, in Guinea to Ghana.

Panicum dregeanum Nees

FWTA, ed. 2, 3: 432.

Densely tufted perennial grass, culms 60–90 cm high, on seasonally damp soil and marsh in lowlands and montane situations, scattered here and there in the Region from Senegal to Liberia and N Nigeria and W Cameroons, and throughout tropical Africa.
It is recorded as being good forage in Zimbabwe (1), and taken by stock in Malawi while in young growth (2).

References:

1. Appleton 8, 21, K. 2. Fenner 242, K.

Panicum ecklonii Nees

FWTA, ed. 2, 3: 429.
West African: **WEST CAMEROONS** BAMILEKE bajiho (DA)

A densely caespitose perennial grass, with bulbous base, culms to 30 cm high of montane grassland on sandy rocky soil in Sierra Leone to W Cameroons, and in Zaïre, Tanzania and S Africa.

Panicum fluviicola Steud.

FWTA, ed. 2, 3: 432. UPWTA, ed. 1, 534, as *P. aphanoneurum.*
West African: **SENEGAL** DIOLA oselléta (JB) FULA-TUKULOR synk (JLT) MANDING-BAM-
BARA guo, ngan (JB) SERER dundi (JB) lundi (JB) **MALI** ? (Mopti) ghonya (Matthes) **GHANA**
DAGBANI mgannikago (Ll.W) **NIGERIA** YORUBA (Ilorin) gọgọwu obo (Ward) pinkun (Ward)

A reed-like grass, perennial, culms 1–2 m high, of flood-plains, moist soils and riparian woodland, common throughout the Region from Mauritania to N and S Nigeria, and scattered over tropical Africa.
The plant is grazed when young (1). In Kordofan it is said to be eaten by all stock at all times except during flowering, and to be much appreciated as straw (2). As it is a grass of semi-swampy places, it has potential for grazing and hay in such locations (4).
Some tribes in N Nigeria collect the grain as food (3).

References:

1. Adam, 1966,a. 2. Baumer, 1975: 108. 3. Dalziel, 1937: 534, as *P. aphanoneurum.* 4. Ward 0017, K.

Panicum gracilicaule Rendle

FWTA, ed. 2, 3: 432.
West African: **SENEGAL** DIOLA éburèy (JB)

An annual grass with culms to 60 cm high, on roadsides and weedy places, in Senegal (1) across the Region to N Nigeria, and extending to Zimbabwe and Angola.

Reference:

1. Berhaut 1967: 414, as *P. tambacoundense* Berh.

GRAMINEAE

Panicum griffonii Franch.

FWTA, ed. 2, 3: 433.
West African: SIERRA LEONE MENDE timbe (NWT) NIGERIA HAUSA geron tsuntsu
(Saunders)

An annual grass, straggling or weakly ascending to 30 cm high, of shallow
rocky soils across the Region from Senegal to W Cameroons, and extending to
Zaïre and Angola.
On plateau land in N Nigeria it serves as a grazing grass (1).

Reference:

1. Saunders 1, K.

Panicum heterostachyum Hack.

FWTA, ed. 2, 3: 429.

An erect or ascending annual grass, culms to 60 cm high, in shade on
deciduous bushland or wooded savanna to sub-montane elevations; recorded
only in Niger in the Region, but occurring widespread throughout tropical
Africa, and also in Guyana and Trinidad.
No usage is reported in Niger. In Tanganyika Sukuma add the leaves to the
brew in making beer to increase its strength. It is said that if the leaves which
fall from the culm are used 'those drinking will stagger' (1).

Reference:

1. Tanner 682, K.

Panicum hochstetteri Steud.

FWTA, ed. 2, 3: 433.
West African: WEST CAMEROONS BAMILEKE nikih (DA)

A perennial grass, loosely tufted, culms weak-stemmed to 1 m high, in
montane forest shade and savanna above the tree-line in Sierra Leone to W
Cameroons, and across Africa to Ethiopia and E Africa.
In Uganda it is reported to provide useful fodder (1).

Reference:

1. Snowden 1603, K.

Panicum laetum Kunth

FWTA, ed. 2, 3: 434. UPWTA, ed. 1, 535.

English: wild fonio.

French: haze (Busson); fonio sauvage (wild hungry rice, Kintz & Toutain).
West African: SENEGAL MANDING-BAMBARA diadié (JB) MALI FULA-PULAAR (Mali)
balbaldi (A.Chev.) fardulo debbo (A.Chev.) paguiri (A.Chev.) paguiri lendia (A.Chev.) MAND-
ING-BAMBARA diadié (A.Chev.) jajié (A.Chev.) yague missi *the grain* (A.Chev.) SONGHAI paguiri
(A.Chev.) TAMACHEK ichiban (A.Chev.) UPPER VOLTA FULA-FULFULDE (Upper Volta)
pagguri (K&T) MOORE tinga ponoga (A.Chev.; FB) NIGER FULA-FULFULDE (Niger) sabeeri
(ABM) SONGHAI génsi (*pl.* -ò) (D&C) NIGERIA BIROM cun ǹnòn = *fonio of the birds* (LB)
FULA-FULFULDE (Nigeria) sa ɓeri (JMD) GWARI fuliyaho (JMD) HAUSA bachakura (JMD)
báina (LB) báyáá (auctt.) báyaá sa're (ZOG) bundigi *the chaff* (auctt.) garaji (ZOG) gariji
(ZOG) jaàtaú (ZOG) saɓe (JMD; ZOG)

300

An annual grass, tufted with culms ascending or erect to 70 cm high, on damp soils in the Sahel zone from Mauritania to N Nigeria, and extending to the Horn of Africa, and in Tanganyika.

In parts of the Soudano-Sahel it forms immense meadows, even into the desert along the banks of water-courses. It is much appreciated by stock for grazing (1–3, 6, 10, 11).

The grain is a very important commodity, especially in time of scarcity. It is often sold in markets, and is eaten in the form of cakes, porridge, etc. The ripe grain shatters and it is collected by a sort of basket, called *akaimi* in Hausa, which is swept over the inflorescences in such a manner that the grain is knocked off, falling inside. Protein has been assayed at 11.1% and aminoacids are reported to be somewhat low. The plant is, however, of considerable significance in areas of marginal subsistance and steps ought to be taken seriously to select improved strains for cultivation (1, 4–9, 11).

The plant has a valuable potential for restoring over-grazed desert pasture, and, as for its grain, selective programme should be carried out for improved fodder production (8, 12).

References:

1. Adam, 1954: 94. 2. Adam, 1966,a. 3. Andrews 3159, K. 4. Busson, 1965: 475–6. 5. Dalziel 262, K. 6. Dalziel, 1937: 535. 7. Gast, 1972: 51–55, with chemical analysis. 8. Irvine, 1952: 40. 9. Kintz & Toutain, 1981. 10. Maliki, 1981: no. 52. 11. Schnell, 1960: no. 116. 12. Williams, JT & Farias, 1972: 15.3

Panicum laxum Sw.

FWTA, ed. 2, 3: 429.

An annual grass, culms 30–60 cm high, of roadsides and open spaces, especially on damp soil and in seasonally flooded land, native of tropical America, introduced to W Africa and now naturalised in all but the driest areas from Mauritania to N and S Nigeria.

Over a space of some 20 years it has spread more or less throughout Sierra Leone. It makes a good lawn-grass mowing well, tillering abundantly; on damp sites it suppresses other grasses, and in rice-padis it is liable to become a troublesome weed (2). On parts of Loma mountain it forms dense pure stands. It has value as cattle-fodder and merits promotion in this respect (1).

References:

1. Adam, 1972: 59–62, with chemical analysis. 2. Deighton 5439, K.

Panicum maximum Jacq.

FWTA, ed. 2, 3: 429. UPWTA, ed. 1, 535.

English: Guinea grass.

French: herbe de Guinée.

Portuguese: erva-da-Guiné.

West African: SENEGAL DIOLA bu silita, o pipi (JB) GUINEA-BISSAU MANDING-MANDINKA siluntentam-ô (JDES) GUINEA SUSU mengui serhé (A.Chev.) SIERRA LEONE MENDE ngalei-hei = *female 'ngalei'*; ngalei *is general for tall grasses, esp. Chasmopodium caudatum (Hack.) Stapf* (FCD) GHANA ADANGME-KROBO go (auctt.) AKAN-FANTE nkyekyer (FRI) GA nto (FRI) GBE-VHE kogbe (FRI) NIGERIA EFIK ńnyányángá énàng = *grass of the cow; incl. other large grasses* (auctt.) IGBO agarama (JMD) ìkpò mìlī = *water ikpò* (auctt.) okẹ àchàlà (KW) okẹ àchàrà (JMD; AJC) IGBO (Umuahia) oke àchàrà (JMD; AJC) IJO-IZON (Kolokuma) pẹrẹ osí, pẹrẹ usí = *chief elephant grass; cf.* izɔn usí: *the true elephant grass* (KW; KW&T) YORUBA ikin (JMD; Verger) YORUBA (Ilorin) iran akun (Ward) WEST CAMEROONS BAFOK makok (DA) DUALA èkòro a bɔlɔ (Ittmann) KPE makoko (DA) KUNDU makoko (DA) LONG makok (DA) WOVEA makoko (DA)

A densely tufted robust perennial grass with culms to 3 m high from a stout rhizome, variable, of shady or open situations in woodland or deciduous bushland, roadsides, river-banks, sealevel to montane, across the Region, and African tropics to S Africa and Madagascar; widely dispersed by introduction throughout the tropics.

The plant generally requires humidity though there are some drought-resistant ecotypes able to grow on rocky, stony or sandy soil (3). It provides a valuable fodder utilised in all territories. It is suitable for permanent pasture (8) but management is difficult by grazing alone (10, 14). It gives a good return when cut three to four times a year (1, 2, 6) for stall or at stake feeding. It is often cut for market trade in W Africa (4). Kept cut and in vigorous growth the young shoots make good hay. There is a suspicion, however, that it, along with other suspect plants, may at some stages cause a sheep illness called 'dikoor' in S Africa (1, 4, 13). The plant is laxative and traces of *hydrocyanic acid* have been detected (2, 6, 11).

The straw is useful for thatching (12). The culms serve as brooms and are used in Central African Republic for basket weaving (7). Sap from the crushed fresh plant is used in Madagascar as a cicatrisant on wounds and sores (5). In Uganda the grass is tied around the head for relief of headache (13).

In Zambia, Senga people collect the seed for food in time of scarcity (9). The seed can serve as grain for feeding birds. It was for this purpose rather than as a source of stock fodder that the plant was first introduced into the W Indies two centuries ago (2).

References:

1. Adam, 1954: 95, with chemical analysis. 2. Burkill, IH, 1935: 1656. 3. Clayton & Renvoize, 1982: 471–2. 4. Dalziel, 1937: 535. 5. Debray & al., 1971: 74. 6. Deshaprabhu, 1966: 223–5, with chemical analyses. 7. Fay 1695, 4456, 4459, K. 8. Foster & Munday, 1961: 314. 9. Mitchell 6146, K. 10. Ogor & Hedrick, 1963: 146. 11. Quisumbing, 1951: 1024. 12. Schnell, 1953,a: no. 24. 13. Watt & Breyer-Brandwick, 1962: 480–2. 14. Williams, RO, 1949: 282–3.

Panicum pansum Rendle

FWTA, ed. 2, 3: 434. UPWTA, ed. 1, 534, under *Panicum* Linn. as *P. kerstingii.*
West African: SENEGAL MANDING-BAMBARA tobaku (A.Chev.; JB) MALI MANDING-BAMBARA tobaku A. Chev. fide JMD NIGERIA IGBO ilulo kputo (NWT) ilulo ṅkàtà *from* ṅkàtà; *a kind of round basket* (NWT; KW) ilulo okelemgbada (NWT) ilulo okili mbawisi (NWT)

An annual grass, culm to 90 cm high of waste places and wooded grassland, over the Region from Senegal to N and S Nigeria, and extending across the Congo basin to Tanganyika and Angola, and into Sudan.

The grass occurs as a weed of cultivation in Sudan (3) and in millet and ground-nut fields in Upper Volta (4). It is grazed in Senegal whilst still young (1).

In Mali the grain is occasionally collected as food (2).

References:

1. Adam, 1966,a. 2. Dalziel, 1937: 534 as *P. kerstingii.* 3. Myers 7024, K. 4. Scholz 77,a, K.

Panicum parvifolium Lam

FWTA, ed. 2, 3: 433.
West African: SIERRA LEONE BULOM (Sherbro) lomɔtheŋ-lɛ (FCD) MENDE bendɛva (NWT) muli (NWT) TEMNE a-ruma-ra-pompo (Glanville) NIGERIA IGBO ilulo agilị ṅkịtā = *ilulo of dog's hair* (NWT; KW)

A perennial grass, creeping, stoloniferous, culms up to 50 cm long, of swamps, dispersed across the Region from Senegal to S Nigeria, and over tropical Africa and Madagascar; also in tropical America.

The grass invades padis on tidal fringes in the Scarcies River area of Sierra Leone (1) and may be a weed of rice-cultivation.

Reference:

1. Glanville 225, K.

Panicum paucinode Stapf

FWTA, ed. 2, 3: 433.
West African: NIGERIA HAUSA tsíntsííyar fadama (Turnbull) wútsíyàr bòòdáríí = *tail of the zorilla* (Taylor; ZOG)

A robust annual grass, culms to 90 cm high on shallow or disturbed places of moist soil, known only from Guinea, Ghana and N Nigeria, and in E Cameroun.

The Hausa name (see above) meaning 'broom of the swamp' suggest its use as a broom, but this is not recorded.

Panicum phragmitoides Stapf

FWTA, ed. 2, 3: 432. UPWTA, ed. 1, 535, as *P. anabaptistum* Steud., in part.
West African: NIGERIA FULA-FULFULDE (Nigeria) fuda tumbi (Saunders) HAUSA hunda tumbi (ZOG) korukoru siawa (Freeman) tafartso (ZOG) tsíntsííyáá = *a broom* (auctt.) WEST CAMEROONS BAMILEKE geme saliho (DA) ngyanle (Gillett)

A robust, reed-like perennial grass, culms to 2 m high, in savanna on dry soils, across the Region from Guinea to W Cameroons, and into the Congo basin to Malawi and Angola.

The grass is planted in Sokoto, NW Nigeria, as a field boundary marker (1, 2). The culms are commonly cut to use as brooms, hence the Hausa name (1–3). In Bamenda the grass is used in thatching if nothing better is available (4).

References:

1. Dalziel 480, K. 2. Dalziel, 1937: 535, as *P. anabaptistum* Steud., in part. 3. Diarra, 1977: 44. 4. Gillett 13, K.

Panicum porphyrrhizos Steud.

FWTA, ed. 2, 3: 434.
West African: GHANA HAUSA sisia (FRI) MOORE sara = *broom* (FRI)

A robust perennial, caespitose culms to 1 m high, on moist soils, near water, by ditches, swamps and rivers in Senegal, Ghana and Niger, and widely dispersed in NE, E and S central Africa.

The Moore name in Ghana (see above) suggests use as a broom (2). In Kenya it provides good grazing for all stock (1).

References:

1. Glover & al. 2971, K. 2. Irvine 5027, K.

Panicum praealtum Afzel

FWTA, ed. 2, 3: 431.
West African: GUINEA BASARI a-yilútú (FG&G) SIERRA LEONE LOKO baikɔ (NWT) MENDE muli (NWT) pagwɛ (NWT) TEMNE peni-*pa*-gbel (NWT)

A robust perennial grass, culms 60–120 cm high, of sandy soil in savanna across the Region from Senegal to N and S Nigeria, and also in E Cameroun.

GRAMINEAE

Panicum repens Linn.

FWTA, ed. 2, 3: 434. UPWTA, ed. 1, 535.

English: torpedo grass (Adam).

Portuguese: escalracho (Feijão).

West African: SENEGAL DIOLA é kéna (JB) é sélèk (JB) MANDING-BAMBARA bama subu (JB) GUINEA-BISSAU BALANTA uncanda (JDES) PEPEL otigna (JDES) SIERRA LEONE MENDE ŋgŋka-kpo (FCD) piso (FCD) yolo (FCD) yowo (FCD) SUSU sumfe (NWT) TEMNE an-gbalɛt (FCD) ka-waya (FCD) VAI sɔ-mɛsɛ-mɛsɛ (?) (FCD) wasa (FCD) MALI MANDING-BAMBARA bama subu (A.Chev.) SONGHAI buga subu (A.Chev.) farka teli (A.Chev.) NIGERIA HAUSA ƙàfíífiyàà (auctt.) YORUBA epoṣe (Onochie; Verger) YORUBA (Ilorin) ẹkuro imado (Ward)

A perennial grass with long rhizomes throwing up culms to about 60 cm, on sandy soil in the proximity of water, fresh or brackish, river-banks, and foreshore, dispersed throughout the Region from Senegal to W Cameroons, and over tropical Africa, and generally pan-tropical.

The vigorous rhizomatous growth in sand acts as a good sand-binder (1, 5). It has been put to effective use in Malaya fixing river silt caused by tin-mining (4). It produces a good mat for turf and lawns, resistant to foot-wear (6). In cultivation it can however become a pernicious weed difficult to eradicate (4, 6, 7). It produces good fodder readily taken by all stock (1–3, 5). It has a high nutritive value said to be above average (4, 6).

The root and seed are reported to be slightly cyanogenic (6). In Indonesia the rhizome is given as medicine for abnormal menstruation (4).

References:

1. Adam, 1954: 95. 2. Adam, 1966,a. 3. Baumer, 1975: 108. 4. Burkill, IH, 1935: 1657. 5. Dalziel, 1937: 535. 6. Deshaprabhu, 1966: 228, with chemical analyses. 7. Ward 0057, K.

Panicum sadinii (Vanderyst) Renvoize

FWTA, ed. 2, 3: 429.

West African: SIERRA LEONE LOKO soiwu (NWT) MENDE gɔnyenyi (NWT) ĵoki (NWT) SUSU kalekame (NWT) sunyugi (NWT) TEMNE pɔgina (NWT) ɛ-sulɛ-ro-bat (FCD; Glanville) ko-sulu-ko-sin (NWT) sut (NWT)

A perennial grass, culms basally decumbent and rooting, ascending to 60–90 cm high, in shady locations in Guinea to Liberia, and in Zaïre, Zambia and Angola.

Panicum strictissimum Afzel.

FWTA, ed. 2, 3: 431.

West African: SIERRA LEONE LOKO poteŋgu (NWT) yuwɛ (NWT)

A perennial grass with erect or ascending culms 60–90 cm high, of swamps, and known in Guinea, Sierra Leone and S Nigeria; also in Congo (Brazzaville).

Panicum subalbidum Kunth

FWTA, ed. 2, 3: 434. UPWTA, ed. 1, 535, as *P. longijubatum* Stapf.

West African: SENEGAL FULA-PULAAR (Senegal) gambéré (A.Chev.) kidi (A.Chev.) MANDING-BAMBARA diimina (JB) guimena (A.Chev.) guimi (A.Chev.) SERER diàngh a mbèl (JB) WOLOF mpal (A.Chev.) THE GAMBIA MANDING-MANDINKA barto jarge ba = *big riverside grass* (Pirie) barto kinto = *river corn* (Pirie; JMD) kinti bora = *bearded corn* (auctt.) kununding mano (Macluskie) GUINEA BASARI diàmbɛ ɛ-ngéla (FG&G) ɛ-nyàmb ɛ-ngèlá (FG&G) SIERRA LEONE LIMBA boroboro (NWT) SUSU foni (NWT) MALI ? (Mopti)

304

populdja (Matthes) FULA-PULAAR (Mali) gambéré (A.Chev.) kidi (A.Chev.) MANDING-BAMBARA guimena (A.Chev.) guimi (A.Chev.) SONGHAI tietiɛsu (A.Chev.) **UPPER VOLTA** FULA-FULFULDE (Upper Volta) pagga pucci (K&T) MOORE mofogo (A.Chev.) ponjanga sugpio logho (Scholz) rudu (A.Chev.) tonso naba konjiodu (Scholz) **NIGERIA** BIROM gwés (LB) FULA-FULFULDE (Nigeria) bordi (Kennedy) HAUSA gora-gora *from* gora: *bamboo* (Taylor) haikin fadama (Grove) kollogi (Kennedy) macara (JMD; ZOG) shinkafa berewa (Thatcher) tsíntsíiyáá (LB) KANURI kaya (Kennedy)

A robust annual, or short-lived perennial grass with spongy culms up to 2 m high, of swamps and wet locations throughout the Region from Mauritania to N and S Nigeria, and widespread over tropical Africa.

The plant invades cultivations. It poses serious control problems in rice-padis in Sudan (11) and The Gambia (10). It is considered to be one of the most noxious weeds of cereal fields (9). It is difficult to uproot.

Cattle find it highly palatable, especially before the stem hardens (1–3, 12; Sudan: 13; Kenya: 5, 8). In Zimbabwe it is used to feed *Tilapia melanopleura*, a farmed vegetable-eating fresh-water fish (7). The fish relishes the succulent stems.

The stems are used by children in N Nigeria to make whistles (3, 4) and as straws for drinking (4). The leaf-blades, 20–50 cm long by 7–15 mm wide, are plaited by Fula women into fans (4).

Birds in Central African Republic very much like the grain (6). In W Africa it is sometimes collected for human food (4).

References:

1. Adam, 1954: 95, as *P. longijubatum* Stapf. 2. Adam, 1966,a. 3. Dalziel 289, K. 4. Dalziel, 1937: 535, as *P. longijubatum* Stapf. 5. Edwards 2947, K. 6. Fay 4462, K. 7. Goldsmith 173/68, K. 8. Marshall s.n., 26/2/1932, K. 9. Terry 1867, K. 10. Terry 3156, K. 11. van Eyck 2, K. 12. Ward 0015, K. 13. Wickens 2104, K.

Panicum turgidum Forssk.

FWTA, ed. 2, 3: 433. UPWTA, ed. 1, 535.

West African: **MAURITANIA** ARABIC (Hassaniya) abukar *the green panicle* (Williams & Farias) az, aze *the grain* (Williams & Farias; AN) morkba *the plant* (AN) mrekba *colloquially* (Le Riche) murkba (Williams & Faries) nnshe *the dry culm; also a porridge of the grain* (Williams & Farias) tishilat *the grain* (Williams & Farias) umm rekba = *that which has knees; after the markedly articulate culm* (Le Riche) **MALI** ARABIC (Mali) bu-rɛkuba (JMD) mrokba (HM) SONGHAI afoajo (JMD) afodio (Williams & Farias) afodjo (Williams & Farias) foyo (JMD; Williams & Farias) TAMACHEK afazo (JMD) afezu (HM fide JMD) **NIGER** ARABIC (Niger) markuba (Grall) FULA-FULFULDE (Niger) gajalol (ABM) SONGHAI afodio (Williams & Farias) afodjo (Williams & Farias) foyo (Williams & Farias) TAMACHEK afeza (Williams & Farias) afezu (Williams & Farias) fadhik (Williams & Farias) TEDA gumchi (Grall)

A perennial tussock-grass with thick root-stock, smooth solid culms attaining 1½ m height, of dry sandy soil in Mauritania, Senegal, Mali and Niger; through N Africa to Somalia, and into India.

The plant produces long deep-penetrating and quick growing-roots which on sand-dunes act to bind the sand (4, 5, 10, 15–17). It is common on the Sahel steppes (2), and in parts of Niger (Agedes area) it forms near pure stands on sand-patches (6). In Sudan it is dominant on locust laying-ground and serves as food for the young insects (8).

The stiff straw is commonly woven into mats and used for thatching (3, 15, 16).

In the Aïr district, Tamachek people weave the stiff straw, as the weft, with thin threads of leather, as the warp, into mats which can be rolled only one way (Rodd fide 3, 16). The culms are also woven into baskets in the western Sahara (13).

The grass is grazed, but it is palatable only while still young (1, 2, 13, 15, 16) and its utility depends on whether there is other better grazing available. Its nutritional value is average but it is rich in phosphorous (2). It is said to be good camel-fodder (3) and is taken by them and donkeys when in a dry state (4, 5, 11, 16). In Rajasthan State, India, sheep only browse it when old (5). Cattle herdsmen in Niger say that within 2–3 days of milch cows grazing it their milk becomes foul-smelling (12).

The culms are used in desert areas as firewood (15), and old culms, dried and powdered, are used in the Hoggar as a wound-dressing (11, 16).

The root is carried by female marabouts (religious teachers) in Mauritania for corporal punishment of wayward pupils (9). Calcined roots produce a sort of soda which Tamachek in the Hoggar add to a tobacco quid (11).

Around the Sahara borders, peoples collect the grain for food, eaten as a soup, a sort of bread or ground to a flour as a porridge. It is said to be anti-diabetic. The grain may also be stored against a time of scarcity (3, 7, 13, 14, Chevalier fide 15, 16). Between 1900 and 1906, the French administration imposed taxes on the people of the Lake Chad area in the form of 'krebs', i.e. the grain of wild grasses including *P. turgidum*, to feed occupation troops (16). In time of dearth, people raid ants nests to obtain the grain the ants have stored. The plant is extremely drought-resistant, and with its clear value to peoples on marginal subsistence it ought to be subjected to selection for improved strains to be brought into cultivation.

In the Hoggar area the plant is held to have mystical properties (16).

References:

1. Adam, 1966,a. 2. Adam, 1966,b: 342. 3. Dalziel, 1937: 535. 4. Deshaprabhu, 1966: 231. 5. Gupta & Sharma, 1971: 85–86. 6. Guile s.n., April, 1966, K. 7. Irvine, 1952,a: 39. 8. Johnston 3014, K. 9. Leriche, 1952: 980. 10. McKinnon S.147, K. 11. Maire, H, 1933: 57–58. 12. Maliki, 1981: 49. 13. Monteil, 1953: 21. 14. Naegelé, 1958,b: 894. 15. Schnell, 1953,b: no. 46. 16. Williams & Farias, 1972: 13–20. 17. Yeates 14, K.

Panicum walense Mez

FWTA, ed. 2, 3: 433.
West African: **SENEGAL** SERER muya muy (JB) **SIERRA LEONE** MENDE ndiwi (Fisher) **NIGERIA** HAUSA gidan durwa (Maule)

A slender annual grass, culms to 60 cm high, often forming dense cover over shallow or disturbed sites on moist soil across the Region from Senegal to N and S Nigeria, and extending to Sudan and Zambia; also in India, China and Malaysia.

The grass is a weed of cultivated land and of padi-fields in India (3).

It is grazed by all stock in Senegal being especially relished by cattle (1). In India it furnishes excellent fodder for cattle (3), and also in Indo-China (2).

Poor people in Punjab in India collect the grain for food (3).

References:

1. Adam, 1966,a, as *P. humile*. 2. Burkill, IH, 1935: 1655–6, as *P. humile* Nees. 3. Deshaprabhu, 1966: 230, as *P. austroasiaticum* Ohwi.

Panicum spp. indet.

West African: **GUINEA-BISSAU** BALANTA bucansóle (JDES) PEPEL iúfo-iúfo (JDES)

Parahyparrhenia annua (Hack.) W D Clayton

FWTA, ed. 2, 3: 498.

West African: SENEGAL MANDING-MANINKA kali-ni (GR)

An annual grass, loosely tufted slender culms up to 1½ m tall; on lateritic pans and in shallow water-logged pools and in rice-padis; common in Senegal to N Ghana, and in Cameroun to Sudan.

It is dominant in grassland on Fouta Djallon in Guinea, and beautifully conspicuous when in flower by its scarlet anthers. No particular attribute is recorded for it, but by presence in padis-fields it must be a weed of rice-cultivation.

Paratheria glaberrima CE Hubbard

FWTA, ed. 2, 3: 457.

A prostrate perennial grass, aquatic; endemic to Sierra Leone.

The grass is mat-forming in tidal padi-fields, colonising and consolidating the bunds (1).

Reference:

1. Deighton 4337, K.

Paspalidium geminatum (Forssk.) Stapf

FWTA, ed. 2, 3: 440. UPWTA, ed. 1, 535.

West African: MALI FULA-PULAAR (Mali) marbéré (A.Chev.) TAMACHEK baugassongau (A.Chev.) NIGERIA HAUSA geron tsíntsííyáá (BM) hákóórín kàréé = dog's tooth (JMD; ZOG) mákòòrín kàréé (ZOG) tùmbín kúúsúù = mouse's stomach (Palmer; ZOG) KANURI angarago (JMD: C&H)

A perennial grass, rhizomatous or stoloniferous, rhizomes spongy, sometimes floating, culms rooting at the base, ascending to 1 m high; of wet sand-banks, sites of temporary inundation, marshes and in water to 2 m deep, sometimes brackish; throughout the Region from Senegal to N and S Nigeria, and widespread over the rest of tropical Africa, and into Madagascar, Egypt and India.

The grass gives a good forage taken generally by all stock (3; Senegal: 1, 2; Nigeria: 7, 11; Somalia: 6: Uganda: 10; India: 4, 5). The succulent culms seemingly render the grass most palatable. In Uganda it is grazed by the hippopotamus (9).

By its rheophytic nature, it is recorded as contributing to blocking of smaller streams (3). In Ethiopia it is found in irrigation channels (8).

No record has been seen of the seed being collected for human use, but one might expect the seed to be edible and collectable in time of want. The seed of the related P. flavidum (Retz.) A Camus is so treated in India (4).

References:

1. Adam, 1954: 95. 2. Adam, 1966,a. 3. Dalziel, 1937: 535. 4. Deshaprabhu, 1966: 269. 5. Gupta & Sharma, 1971: 86–87. 6. Munro 66, K. 7. Palmer 21, K. 8. Parker E.573, K. 9. Purseglove P.3512, K. 10. Snowden 1647, K. 11. Tuley 1131, K.

Paspalum conjugatum Berg.

FWTA, ed. 2, 3: 445. UPWTA, ed. 1, 536.

English: buffalo grass; sour grass (Jamaica); Hilo grass (American, from Hawaii).

GRAMINEAE

Portuguese: capim-gordo (Feijão).
West African: **SIERRA LEONE** KISSI kpoŋgbo-piando (?) (FCD) KONO wowɛgbinɛ (FCD) MANDING-MANDINKA yanɛ (NWT; JMD) MENDE kapie (JMD) yani *applied also to Axonopus with which it is confused* (auctt.) SUSU alekɔre (JMD) balekɔre (NWT) yanɛ (NWT; JMD) TEMNE ka-gbata (JMD) ka-lant (Glanville; JMD) ka-yan (FCD) YALUNKA kharatu-na (FCD) **IVORY COAST** BAULE kama (B&D) KYAMA dianderika (B&D) **GHANA** AKAN-TWI nsɔhwea (FRI) NZEMA asamo akwanta (FRI fide JMD) **NIGERIA** EFIK efɔk ngkùkù (auctt.) IGBO ikute àlà = *grass of the ground: i.e. creeping* (auctt.) IGBO (Uzuakoli) oji-ikpere-eje (FRI) IJO-IZON (Kolokuma) dŭwęi berisǫnlei (KW&T) IZON (Oporoma) dǔęí (KW)

A stoloniferous perennial grass producing small tufts with culms to about 60 cm high in open places of the forest; recorded from Sierra Leone to Fernando Po, and generally dispersed over the rest of tropical Africa; native of tropical America and now widespread in the tropics.

The vigorous stoloniferous growth results in close-matting, often to the exclusion of other plants. The grass invades forest clearings and may impede forest regeneration. It is suitable, but somewhat coarse, as a lawn-grass withstanding mowing and foot-wear (5, 6, 11). It is drought-resistant, remaining green long into the dry season, and has been used for the turf of all the lawns in the Victoria Botanic Gardens, W Cameroons (8). By its vigour, it can to some extent suppress lalang, *Imperata cylindrica* (Gramineae) (4).

The grass provides good grazing for cattle and horses (6), taken preferably before seed-set as the seed is reported to stick in the throat and cause choking (7). Cats and dogs are said to eat the leaf as a purgative (10, 11).

The leaves have a variety of medicinal uses. In Gabon they are cooked and eaten with ground-nuts for stitch, and pounded with the leaf of *Desmodium salicifolium* (Leguminosae: Papilionoideae) an ointment is made to apply in frictions for heart-troubles (10, 11). A similar prescription as the latter is applied as a compress for contusions, sprains, dislocations, etc. (10, 11). Analgesic action is also sought in Congo (Brazzaville) by using leaf-sap in palm-oil for application to points of pain after scarification in cases of headache and on the sides of the body (2). The leaf with those of *Macaranga sp.* (Euphorbiaceae) and *Renealmia sp.* (Zingiberaceae) is used in a vapour bath in Congo (Brazzaville) for fever (2), and also the sap expressed from leaf softened over a fire mixed with a little salt is instilled to relieve eye-injury: if used immediately it is said to be very effective (2). Leaf-infusions are taken in Trinidad for fever, flu, pneumonia, pleurisy and fatigue (12). A decoction by draught and in a bath is used in Ivory Coast as a tonic to counter emaciation in adolescents and to hasten onset of puberty (3).

Dietetic value of the grass is fair (4). Malayan material is reported to contain 2.9% *protein* before flowering, and appreciable amounts of *carotene* and *ascorbic acid* (7). The grass has some haemostatic property ascribed to the presence of a glycoside, *paspaloside* (7, 12).

A decoction of fresh root is taken in the Philippine Islands for diarrhoea (9).

The stolons evoke amongst the Efik (see word-list above) a name for the grass meaning 'web of the grasshopper', and Efik children use them to make the framework of playthings (1).

References:

1. Adams, RFG, 1943. 2. Bouquet, 1969: 129. 3. Bouquet & Debray, 1974: 92. 4. Burkill, IH, 1935: 1673–4. 5. Cole, 1968, a. 6. Dalziel, 1937: 536. 7. Deshaprabhu, 1966: 269. 8. Maitland 6, K. 9. Quisumbing, 1951: 104. 10 Walker, 1953, a: 39. 11. Walker & Sillans, 1961: 191. 12. Wong, 1976: 109

Paspalum deightonii (CE Hubbard) WC Clayton

A perennial grass present in Guinea to Nigeria, hitherto regarded as a variety of *P. scrobiculatum*, but with marginal distinctions enough to consider it a separate species (1).

No usage is reported but it should be examined within the *P. scrobiculatum* complex (q.v.).

Reference:

1. Clayton, 1975: 101–5.

Paspalum dilatatum Poir.

FWTA, ed. 2, 3: 445.
English: Dallis grass.

A tufted perennial grass, culms robust, to 180 cm high, native of S America, and known to be present in Ghana; introduced to most tropical countries. The grass produces excellent pasture and hay. It has been strongly recommended for establishment in tropical and subtropical countries. It is reported as a weed of cultivation in E Africa (1), and that it may be infected with *Claviceps* and thus be toxic to stock grazing it (2).

References:

1. Clayton & Renvoize, 1982: 608. 2. Watt & Breyer-Brandwijk, 1962: 483.

Paspalum lamprocaryon K Schum.

FWTA, ed. 2, 3: 446, as *P. auriculatum* sensu FWTA, non JS Presl.
West African: **SIERRA LEONE** KORANKO biŋgodi (NWT) LOKO načawaŋge (NWT) MENDE kapie (NWT) SUSU alɛkɔre (NWT) yani (NWT) **WEST CAMEROONS** BAMILEKE fafabo fafroko (DA)

A perennial grass, culms stout, geniculate and rooting at lower nodes, ascending to 1½ m high, of marshy and damp sites, not abundant but dispersed across the Region from Senegal to W Cameroons, and in E and S Central tropical Africa.
It produces abundant good green forage utilised in Kordofan (1).

Reference:

1. Baumer, 1975: 109.

Paspalum notatum Fluegge

A rhizomatous mat-forming perennial grass; culms 15–50 cm high, native of S America and now recorded in Ghana (2) and in E and Central Africa.
No usage is recorded in Ghana. It has been introduced to E Africa for grazing and erosion control (1). It is reported in Gabon to make an excellent lawn and good pasture (3).

References:

1. Clayton & Renvoize, 1982: 608–9. 2. Jackson 2510, K. 3. Walker & Sillans, 1961: 191.

Paspalum paniculatum Linn.

FWTA, ed. 2, 3: 446.
Portuguese: capim grama-da-Guiné (Feijão).

A perennial growing in coarse clumps to 1.2 m high, native to tropical America, introduced and present in the Region in Liberia, W Cameroons and

309

Fernando Po; also in Gabon southwards to Angola and Uganda; generally dispersed pan-tropically.

Uses in Gabon are as for *P. conjugatum* (q.v., 2). In E Africa it is used as ground-cover in upland plantations (1).

References:

1. Clayton & Renvoize, 1982: 608. 2 Walker & Sillans, 1961: 192.

Paspalum scrobiculatum Linn.

FWTA, ed. 2, 3: 446, as *P. orbiculare* Forst. ; and *P. polystachyum* R.Br. UPWTA, ed. 1, 536.

English: bastard millet; ditch millet; koda (millet); or kodra (millet) – the cultivated form in India from Hindi: *koda, kodra.*

West African: SENEGAL FULA-PULAAR (Senegal) parkatari (A.Chev.) MANDING-BAMBARA barabudiaba (K&A) baraburyaba (A.Chev.; K) barabuyaba (JB) barobia (A.Chev.) diadié (A.Chev.; K&A) tiéku (auctt.) SONINKE-SARAKOLE dara koré (auctt.) WOLOF gargâda (auctt.) ndugupfit (auctt.) THE GAMBIA MANDING-MANDINKA barankato (Fox) falising-ô (Macluskie) fatang-ô (Macluskie) GUINEA-BISSAU FULA-PULAAR (Guinea-Bissau) djábi-maudo (JDES) SIERRA LEONE KISSI maloninda (FCD) KONO minasabinɛ (FCD) pendiki (FCD) KORANKO lɛfɛbuiyɛ (NWT) MANDING-MANDINKA binkolo (FCD) MENDE kapia (auctt.) kapika (auctt.) yanee (FCD) zimi (Fisher) SUSU alekore (JMD) balekore (JMD) TEMNE *k*-pika (auctt.) YALUNKA kharatu-na (FCD) LIBERIA MANO duo su (Har) MALI ? (Mopti) kussein (Matthes) DOGON bɔlɔ ánala = *big buttocks* (CG) FULA-PULAAR (Mali) parkatari (A.Chev.) MANDING-BAMBARA baraburya ba, borobia, diadié (A.Chev.) tiéku (A.Chev.; FB) KHASONKE laruha (A.Chev.) SONGHAI nkungurumo (A.Chev.) SONINKE-SARAKOLE dara koré (A.Chev.) UPPER VOLTA BISA hiburu (Scholz) GHANA AKAN-ASANTE chesimbri (DA) DAGBANI bamrog (Ll-W) MOORE gonera (FRI) NIGER HAUSA tùmbin g̃àaku (AE&L) tumbin jaki (AE&L) SONGHAI nkungurumo (FB) NIGERIA FULA-FULFULDE gauri cholli = *birds' guinea-corn* (Chapman) HAUSA tamban tsuntsu (Kennedy) tùmbín jà3kíí = *belly of the donkey* (auctt.) IGALA òwú (Odoh; RB) IGBO ìkpò ntà = *small ikpo* (NWT) ìkpòo (JMD) IGBO (Umuahia) ikute àlà = *grass of the ground*; *general for grasses of this shape of inflorescence* (AJC) YORUBA ɔkànli (auctt.) WEST CAMEROONS BAMILEKE fafabo gorko (DA)

A short-lived perennial grass, or cultivars annual; very variable, stoloniferous tufted or culms geniculate and rooting or ascending 10–60 cm high, of moist, damp or wet sites throughout the Region, and elsewhere in tropical Africa, and throughout the Old World Tropics. Authorities have divided it into a number of species and subspecies which are considered together here as an apomictic swarm (fide *FWTA* ed. 2, 3: 446).

The grass provides good forage for all stock before anthesis (1, 3, 7, 13, 23, 28, 30; Gabon: 39; Kordofan: 6; Uganda: 36; Kenya: 27, 29; etc.), but productivity is said to be low in N Nigeria and is thus discarded as a poor pasture grass (19). Grazing after seed-set should be viewed with caution. Some forms produce toxic seed (see below) which cause digestive troubles (1). In India there has been report of elephants dying as a result of eating the grain (17).

The grass is an aggressive coloniser of disturbed and cultivated land. It commonly invades both hill- and padi-plantings of rice (10, 13, 17; Senegal: 2; The Gambia: 4, 5, 20, 37; Upper Volta: 32; Nigeria: 21, 26; Ethiopia: 31, 34; etc.). For its vigorous stoloniferous habit, it is planted in Ghana on drain-banks to curb erosion (11). In Nigeria the grass has been reported used as thatch (13, 38).

A few medicinal usages are recorded of the plant in Asia. A decoction of the root and rhizome is used in childbirth (35, 40), and in the Philippine Islands sap expressed from the stem is held to be useful for corneal opacity (35). A tranquilising substance is obtained from water in which the grain of some forms has been decocted (8, 25, 42). An undisclosed medicinal use is reported in Kumasi, Ghana (12).

The grain of the wild perennial plant is in general not edible on account of toxicity. In India in ancient times the plant has been subjected to a degree of ennoblement and domestication. The perennial life-cycle has been changed into an annual cycle and non-toxic strains selected (9). This is *kodo millet* which is a minor grain crop grown throughout India. Toxicity is, however, not entirely lost for it is recommended that kodo grain be stored for six months before consumption as immature or newly-gathered grain is thought to be poisonous (18), a tantalising prospect should there be shortage of food. In N Nigeria the grain is considered unwholesome, and the Hausa name may indicate digestive disturbances arising from ingesting too much of it (13). Nevertheless, a selection process similar to that in India seems to have been in train in W Africa where it serves as a supplementary or emergency food. As a common weed of rice-fields it is usual to leave it to grow along with the rice crop, to reap both together, but to cook and eat the two grains unseparated. This is particularly so in N Sierra Leone (14, 16). The best *Paspalum* grain is held to come from hill-rice plantings rather than from padi-fields (14). There are reports of it being grown in that area also as a pure crop (16). Zabrama pastoralists and nomads in Upper Volta collect the seed which is cooked like rice (23). There is here an obvious case for further enquiry, tapping the knowledge of the folk who eat this grain without poisoning themselves, and, since the grain may serve as an emergency food, for selection of superior cultivars.

Superior strains of grain are reported to have 6–10% protein, fat 1.4–4.2% and carbohydrates 54–59% (1, 10, 18, 25). Toxicity of poisonous strains is thought to lie in the testa and pericarp (24, 25, 40, 41) and may be due to ergot. It affects man and animals alike. Symptoms which may be manifest within 20 minutes are loss of consciousness, delirium, trembling, dilation of the pupils, and weak pulse (17, 24). To prevent poisoning, the grain needs to be carefully removed from the surrounding parts (17). Alternatively without such threshing the seed can be boiled, the water extracting the poisonous substances (8).

References:

1. Adam, 1954: 95. 2. Adam, 1960, a: 374. 3. Adam, 1966, a. 4. Ashrif 20, K. 5. Austin 193, K. 6. Baumer, 1975: 109. 7. Beal 5, K. 8. Bhide & Aimen, 1959: 1735–6. 9. Burkill, IH, 1935: 1674–5. 10. Busson, 1965: 476, with chemical analysis of grain. 11. DA (Ghana) Herb. No. 1, K. 12. DA (Ghana) Herb. No. 2, K. 13. Dalziel, 1937: 536. 14. Deighton 811, K. 15. Deighton 4427, 4428, K. 16. Deighton 5288, K. 17. Wet & al. 1983: 159–63. 18. Deshaprabhu, 1966: 270–3, with copious detail. 19 Foster & Munday, 1961: 314. 20. Fox 144, K. 21. Gill & Ene, 1978: 182–3, as *P. orbiculare* Forst. 22. Glanville 73, K. 23. Irvine 4507, K. 24. Kerharo, 1973: 5. 25. Kerharo & Adam, 1974: 651–3, with phytochemistry and pharmacology. 26. Lawlor & Hall 654, K. 27. McDonald 950, K. 28. MacGregor 3, K. 29. Maher 3240, K. 30. Melville & Hooker 212, K. 31. Mooney 9254, K. 32. Parker 2008, K. 33. Parker 2050, K. 34. Parker E.155, E.363, E.610, K. 35. Quisumbing, 1951: 104–5. 36. Snowden 1109, 1367, K. 37. Terry 1855, K. 38. Thomas NWT.1890 (Nig. Ser.), K. 39. Walker & Sillans, 1961: 192. 40. Watt, 1967: 1–22, as *P. commersonii* Lam. 41. Watt & Breyer-Brandwijk, 1962: 482, as *P. commersonii* Lam. 42. Bouquet & Debray, 1974: 93.

Paspalum urvillei Steud.

FWTA, ed. 2, 3: 445.

A tufted perennial grass to 2 m high, native of S America, and now established in Liberia, and introduced to most tropical countries.

It is a potential fodder grass.

Paspalum vaginatum Sw.

FWTA, ed. 2, 3: 445. UPWTA, ed. 1, 536.

English: silt grass (Malaya, Henderson).

West African: SENEGAL WOLOF hey (JMD; K&A) xérof (K&A) THE GAMBIA MAND-ING-MANDINKA niro (Fox) SIERRA LEONE BULOM (Sherbro) kɛnkɛn-dɛ (FCD) MENDE gbonje *applies in S Mende areas to several grasses* (FCD) pisui *applies to Acroceras and other creeping grasses* (JMD) TEMNE k-kire-kire (auctt.) GHANA GBE-VHE (Awlan) gbeklẽ (FRI)

A stoloniferous perennial grass, extensively creeping over tidal flats, foreshore and inland marshes, culms to 60 cm high, around the whole Region from Senegal to Fernando Po, and in E Africa, and generally pan-tropical and sub-tropical.

The plant is a halophyte. Its presence indicates a salty soil (13). In the mangrove community it holds a niche populating the mud above the *Avicennia* pioneer zone (10). It readily colonises littoral mud-flats and tidal estuaries and lagoons (1, 5, 11). At Lagos the Yellow Fever Commission reported in 1930 that it was creating breeding sites for the mosquito *Anopheles gambiae* in shallow littoral water (3). Besides salt-water, it occupies sites of brackish and fresh-water. It is a troublesome weed in tidal padi-land in Sierra Leone (7) and The Gambia (12). It is very difficult to eradicate (5, 6). It generally has a sand-binding capacity, and on the credit side in the Rokupr area it is observed in an useful role consolidating padi-field bunds. On drier sites it forms a turf and sometimes can successfully be grown as a lawn which withstands foot-wear (Senegal: 1).

The grass provides good pasturage for stock (2, 5, 9, 14). Cattle are said to fatten on it (4). This is not altogether without danger (16): in Natal there has been suspected *Claviceps* infection arising in cattle showing ergot-like symptoms after grazing it (15).

In the lagoon area of S Ghana Vhe farmers cut the haulm and bury it with mud and sand, and after turning it one to two months later use the composted material as an agricultural manure (8).

In Java it has use in external application for rheumatism and eye trouble (15).

References:

1. Adam, 1954: 95. 2. Adam, 1966,a. 3. Barber s.n., 5/11/1930, K. 4. Burkill, IH, 1935: 1675. 5. Dalziel, 1937: 536. 6. Deighton 977, K. 7. Glanville 229, K. 8. Irvine 2780, K. 9. Irvine 4934, K. 10. Marchal, 1959: 185. 11. Morton & Gledhill SL. 897, K. 12. Terry 3268, K. 13. Trochain, 1940: 395. 14. Walker & Sillans, 1961: 192. 15. Watt & Breyer-Brandwijk, 1962: 483. 16. Kerharo & Adam, 1974: 653.

Paspalum virgatum Linn.

A clumped perennial grass, culms erect to 2 m; native of tropical S America, now reported present in Upper Volta where it is being used on stream-banks to prevent erosion (1).

Reference:

1. Scholz 98, K.

Pennisetum clandestinum Hochst.

FWTA, ed. 2, 3: 460.
English: Kikuyu grass.

A creeping perennial grass with stout rhizomes, mat-forming, culms brief to 15 cm high; native of E African highlands, and introduced to montane areas of N and S Nigeria and W Cameroons; generally dispersed to tropical uplands and subtropics throughout the world.

The plant is robustly turf-making. In Bermuda it has been grown for anti-erosion measures (2). In hill-stations it is grown as lawns. It makes a good pasture with grazing much relished by stock. On the Obudu Plateau it has been grown with *Trifolium baccarinii* Chiov. (Leguminosae: Papilionoideae) to give an excellent sward during the wet season with the grass remaining green in the dry season (4). Its cultivation appears to be finding wider acceptance displacing other grassland communities, though in places in Tanganyika sheep farmers are

said to dislike it, claiming that its dense matting growth harbours sheep parasites (3).
In Tanganyika, the whole plant (1) or the bruised leaf (5) are used as a styptic.

References:

1. Bally, 1937: 10–26. 2. Brunt 1082, K. 3. Gillett 17779, K. 4. Tulley, 1966: 903. 5. Watt & Breyer-Brandwijk, 1962: 483. See also: Purseglove, 1972, 1: 203.

Pennisetum divisum (Forssk.) Henr.

FWTA, ed. 2, 3: 463.
West African: MALI ARABIC (Mali) mekhamla (HM) oum khamela (HM)

A glabrous bushy perennial grass with stout woody stems of the dry Saharan area of Aïr in Niger, and in N Africa, Arabia and India.
In olden times in Egypt and Arabia it was commonly grown for its grain, known as 'kasheira', as a favoured food (1). The herbage is eaten by most domestic stock in Hoggar (2).

References:

1. Daniell, 1852: 398. 2. Maire, H, 1933: 58.

Pennisetum glaucum (Linn.) R Br.

FWTA, ed. 2, 3: 463, as *P. americanum* (Linn.) K Schum.; incl. *P. dalzielii* Stapf & CE Hubbard; *P. stenostachyum* (Klotzsch) Stapf & CE Hubbard; *P. violaceum* (Lam.) L Rich.; *P. senegalense* Steud.; and *P. fallax* (Fig. & De Not.) Stapf & CE Hubbard. UPWTA, ed. 1, 538–40 as Grain Pennisetums; incl. p. 536: *P. dalzielii* Stapf & Hubbard; p. 537: *P. mollissimum* Hochst.; *P. ochrops* Stapf & Hubbard; and *P. perrottetii* K Schum.; & p. 538: *P. stenostachyum* Stapf & Hubbard.
English: pearl millet; bulrush millet; spiked millet.
French: mil; petit mil; mil chandelle; mil à chandelle.
West African: SENEGAL BEDIK bendah (K) gi-ŋà *P. gambiense race* (FG&G) DIOLA balut (K) balutabu (K) dugup (AS fide JMD) FULA-PULAAR (Senegal) gauri (K; K&A) TUKULOR mutil (JMD) mutiri (K) nutil (K) MANDING-BAMBARA kuya (A.Chev.) nunkuru (JB) MANDINKA suno (K) 'SOCE' suno (K) NON tio tande (JMD) tioh *an early c.var.* (JMD) tomak *a late or bristly c.var.* (JMD) SERER diimb (JB) gatiah (JB) mati (JB) pod *P. gambiense race* (JB) WOLOF deguerem (A.Chev.; JMD) diembu (A.Chev.; JB) dora diemb (JB) dugup *early c.vars.* (AS; K) sanio, sanyɔ *late c.vars.* (auctt.) seguerem (A.Chev.) suna *early c.vars.* (auctt.) **THE GAMBIA** MANDING-MANDINKA majo *late c.var.* (JMD) sanyo *late c.vars.* (DA; JMD) suno *early c.vars.* (DA; JMD) **GUINEA-BISSAU** CRIOULO midjo (JMD) midjo preto (JDES) milho preto (EPdS) FULA-PULAAR (Guinea-Bissau) madja (JDES) **GUINEA** BAGA (Koba) kö mãk (Hovis) BAGA (Sitemu) kö mẽk (Hovis) BASARI syɔ̀ngɔ̀ (FG&G) FULA-PULAAR (Guinea) mutiri (CHOP fide JMD) KONYAGI u-suri (*pl.* wa-tyuri) (FG&G) LANDOMA kō-mãk (Hovis) MANDING-MANINKA suno (CHOP fide JMD) sunan (CHOP fide JMD) SUSU tengué (CHOP fide JMD) TEMNE kö-mãk (Hovis) **SIERRA LEONE** BULOM (Kim) nyɔmui (?) (FCD) BULOM (Sherbro) so-lɛ (FCD) su-lɛ (FCD) FULA-PULAAR (Sierra Leone) mutiri (FCD) GOLA dida (FCD) KISSI soamdawomdô (FCD) soandlawomdô (FCD) KONO sɛnɛ (FCD) KORANKO sanyɔ̃ (JMD; FCD) soɔ̃ (FCD) KRIO kus-kus *properly this refers to the prepared food, and to Sorghum* (JMD) LIMBA tafeya (FCD) tefeye (JMD) LIMBA (Tonko) tehɛr-kpɔsuma (FCD) LOKO kpe-nyɔ (FCD) MANDING-MANDINKA sanyɔ̃ (FCD) MENDE kpele-nyɔ = *rough maize; in allusion to the silicaceous hairs* (FCD; JMD) SUSU tenge (JMD; FCD) TEMNE ta-sor (JMD; FCD) ta-sur (JMD; FCD) VAI ŋwɔnyɛ-nyɔ (FCD) YALUNKA tengi-na (JMD; FCD) **MALI** ARABIC (Mali) bechna (JMD) MANDING-BAMBARA bishen (JMD) heni (Barth) sanyŏ *late c.vars.* (JMD) suna(n) *early c.vars.* (JMD) TAMACHEK abora (Foureau fide JMD) ebeno (A.Chev.) eneli (JMD) tabenhaut (A.Chev.) **UPPER VOLTA** BOBO gbègélù (Le Bris & Prost) mànà-furu (Le Bris & Prost) FULA-FULFULDE (Upper Volta) gawri (K&T) gawri ndaneeri (K&T) mutiri (K&T) GRUSI-LYELA mèla (Nicolas) GURMA diwe (Surugue) KURUMBA ayam-pumõ (Prost) SAMO (Sembla) gméè kàn (Prost) mõn (Prost) **IVORY COAST** ? gnon (A&AA) KRU-AIZI jho (Herault) jo (Marchese)

GRAMINEAE

BETE ŋ (Marchese) nyɔɔ́ (Marchese) DIDA kȯkwé (Marchese) GODIE nyōɔ́ (Marchese) GREBO (Jrewe) gbújȯ (Marchese) GREBO (Tepo) sȯghlȧ (Marchese) GUERE kɛ̃ɛ̃ (Marchese) GUERE (Wobe) kplȧȧ (Marchese) NEYO wȋ (Marchese) **GHANA** VULGAR nara *early c.vars.* (Appa Rao & al.) shibras *wild forms* (Appa Rao & al.) zia *late c.vars.* (Appa Rao & al.) ADANGME-KROBO ŋma (FRI) AKAN-ASANTE eujo *a 17th century name; an allusory reference to the sun:* egwju (Wigboldus) ɛwio (auctt.) AKPAFU *ka* - kpȧȧ (Kropp) AVATIME ȧdzȧgȯ (Kropp) DAGBANI isa lochɔ *late c.vars.* (Gaisser fide JMD) isa-nyi *a late c.var.;* *short form* (Gaisser fide JMD) nara *an early c.var.* (FRI) za (FRI) za-lia *big-grained c.vars.* (FRI) za-nyan *small-grained c.vars.* (FRI) GA ngmȧa *from* ng: *a prefix;* mȧa: *an abbreviation of the Portuguese* macaroea, maja, mays *or* maiz, *ignoring differentiation between the various known cereals* (FRI; KD; Wigboldus) GBE-FON likun *from* li: *millet;* kun: *a suffix for crop names* (Wigboldus) VHE gbekui (FRI) lu (FRI; Wigboldus) GRUSI chara *early c.vars.* (Ll.W) mupona *late c.vars.* (Ll.W) GUANG kȕdȕrbȋ (Rytz) KONKOMBA nyu (*pl. ỉ*yu) (Gaisser fide JMD; Froelich) LEFANA ȧ-tȕkɔ̀ (Kropp) MAMPRULI nara *early c.vars.* (FRI) za (FRI) za-lia *big-grained c.var.* (FRI) za-nyan *small-grained c.var.* (FRI) MOBA yoii (FRI) NANKANNI nara *early c.var.* (Ll.W.) zia *late c.var.* (Ll.W.) **TOGO** ANYI-ANUFO nyepe (JMD) BASSARI iyɔ *late c.var.* (Gaisser fide JMD) KABRE amala *late c.var.* (Gaisser fide JMD) NAWDM amale *late c.var.* (Gaisser fide JMD) dowili *late c.var.* (Gaisser fide JMD) nȧȧdȕ (Nicole) nara *early c.var.* (Gaisser fide JMD) TEM (Tschaudjo) adalla *late c.vars.* (Gaisser fide JMD) mise *early c.var.* (Gaisser fide JMD) **NIGER** HAUSA hatchi *applied generally to wild grasses with seed resembling millet* (Marchal) SONGHAI hèenȋ (*pl.-ȯ*) (D&C) SONGHAI-ZARMA haïni *a general term* (Marchal) hamo (Robin) hanyi kirey *a short season c.var.* (Robin) **NIGERIA** ? ȋhȋnmȋghȧfȋȧnmȋ (Elugbe) AKPA ȋdvȕ (Oblete fide RB) ANGAS mȧngȕn (Kraft fide KW) moor (Hoffman) ARABIC-SHUWA dukhn (JMD) dukhn (JMD) liji *a short season c.var.* (JMD) mattiye *maiwa type* (JMD) BADE ȧawȕn (Kraft fide KW) BARAWA-GEJI cȋlȋhwȏ̀ (Kraft fide KW) BATA-BACAMA lȧmȧtȯ (Kraft fide KW) lȧmųtȋ (Kraft fide KW) BETTE-BENDI ȧ-bȗng (Ugbe fide KW) BIROM gȇy *maiwa type* (LB) gey sunang = *red dwarf millet; geero type* (LB) gyȇrȯ *geero type* (LB) gyȯrȯ *geero type* (LB) BOLE maɗȯ (Kraft fide KW) BURA yȧrȋ (Kraft fide KW) CHIP mȧr (Kraft fide KW) CHOMO lȧhwe (Shimizu) DERA mȯɗȯ̀ *geero type* (Newman; Kraft) nyȏm *maiwa type* (Kraft) DGHWEDE wȋrȧ (Kraft) DOKO-UYANGA ebing (Cook) FALI mųxųrȉn (Kraft fide KW) xamzȋ̌kȗ (Kraft fide KW) xȧmzȗ (Kraft fide KW) yɛɗȋ (Kraft fide KW) FULA-FULFULDE (Nigeria) muri (JMD) mutiri (JMD) muuoi (J&D) GA·ANDA sȅkȩtȩ (Kraft fide KW) shȅkįtȧ (Kraft fide KW) GA·ANDA (Boka) shȇtȧ`ȧ (Kraft fide KW) GALAMBI mȧrzȋ (Schuh) GEERUM mȯɗȯ (Schuh) GERA mȧrɗȧ (Schuh) mȧrɗȧ (Kraft) GLAVDA mȧdȧrȋyȧ (Kraft) GOEMAI maar (Hoffmann) shȏng (Kraft) GUDE wȩedetsų́ (Kraft) xȋnȧ (Kraft) GUDUF mȧdȋȧwȧ (Kraft) mȧdȧrȇy (ȧ) (Kraft) GWANDARA-CANCARA joro (Matsushita) mȇwȧ (Matsushita) GITATA gyora (Matsushita) mȋ̃wȧ (Matsushita) KARSHI gyero (Matsushita) KORO gyoro (Matsushita) mȇywa (Matsushita) NIMBIA gyoro (Matsushita) mįyȯngo (Matsushita) TONI gyɔlo (Matsushita) yȏ̃wȧ (Matsushita) GWARI mawi (JMD) sawi (JMD) HAUSA damrȯȯ (JMD; LB) daurȯȯ (JMD; LB) dȧwrȯȏ (Matsushita) gȇȇrȯn dȧn kȧȧrȕwȧ *probably a general name* (auctt.) geȇrȯȯ (auctt.) k̇yaasuuwaa (Lowe) mȧiwȧȧ (auctt.) shibra (JMD) shura (JMD) HUBA iyȧdȋ (Kraft fide KW) iyeɗȋ *geero type* (KB) HWANA wȕshȯhȧrȧ (Kraft) IDOMA adlȧ (Armstrong fide RB) eyȇ (Armstrong fide RB) IGALA ȯkȏdȕ (RB) IGBO ọkȧ ȋnarị *pennicilary type* (BNO) ọkȧ mịlȇtȋ *maize millet, bulrush type* (BNO; KW) IVBIE ȧdȯ (Elugbe) IZAREK izȕk (Kraft fide KW) JARAWA-BANKAL mɔ́r (Kraft fide KW) JEN įnyȅ (RB) JIRU hul (RB) JUKUN zumya (JMD) JUKUN (Pindiga) mȅnȅ (RB) JUKUN (Wukari) jȋmȋ (Shimizu) mbe (Shimizu) KAJE dzȕk (Cook) KAMBARI-AUNA ȋȋyȇȇnjȋ *geero type* (RB) ȗȗtȇwȧ *maiwa type* (RB) SALKA vɔ̂ɔ́jȋ (*pl.* yȏɔ́jȋ *geero type* (RB) yȋttȧ̂wȧ *maiwa type* (RB) KAMWE gȯrwa (Kraft fide KW) mȧɛxƁi (Kraft fide KW) xȧmzȅ (Kraft fide KW) xȧmzȩ̀ (Kraft fide KW) xȧnzȯ̀ (Kraft fide KW) KAMWE (Kiria) 'yȧrȋn (Kraft fide KW) KANURI ȧrgȏm (C&H) ȧrgȏm mȇtȧ *white grain* (C&H) ȧrgȏm mȯrȯ *reddish grain* (C&H) ligi (Lamb) matia *maiwa type* (JMD) metia (JMD) moro *geero type* (JMD) nzaimo (C&H) KAREKARE mȕɗȋyȧ (Kraft fide KW) KATAB-KAGORO zuk (Cook) KOFYAR maar (Hoffmann) KYIBANNU yȧrȋ (Kraft fide KW) LENYIMA akpe (Cook) LEYIGHA nsange (Cook) MALA igilọ (RB) ịshina (RB) mewa *maiwa type* (JMD) MARGI mȋgȧ (Kraft fide KW) MARGI (South) ɗȕwȧtȕ (Kraft fide KW) tȕmbȕsȕ (Kraft fide KW) yȧɗȋ (Kraft fide KW) MARGI-PUTAI mȧtȋyȧ (Kraft fide KW) mųtȋyȧ (Kraft fide KW) MATAKAM gȧgȧr (Kraft fide KW) MBEMBE (Tigong) zȧ (RB) zȧ (RB) NGAMO sȧwȧ (Kraft fide KW) NGGWAHYI mȧtȋyȧ (Kraft fide KW) NGIZIM mȧrɗȗ (*pl.* -ɗȧɗȋn) *geero type* (Schuh) mȯɗȋyȧ (*pl.* -ayȋn) *maiwa type* (Schuh) NINZAM amar (RB) NUPE kapai (JMD) mai (JMD) NZANGI mȅɗȋkȋcȋ (Kraft fide KW) OKPAMHERI (Emarle) ȏvȯfȩ̀ (Elugbe) OLULUMO (Ikom) ǹ-dȕk (Cook) PERO Ɓwọrọng (Kraft fide KW) mwȯɗȯ̀ (Kraft fide KW) POLCI yȯghȧ (Kraft fide KW) POLCI-ZUL dȧwrȯ (Kraft fide KW) SAMBA-DAKA maka *geero type* (RB) LEEKO yȩtura *maiwa type* (RB) SANGA iyo (RB) SURA gyȇwȕrȯ (Kraft fide KW) TANGALE mȯɗȧ (Kraft fide KW) TERA mȇrȇ *geero type* (Newman) shegȧ *maiwa type* (Newman) TERA-PIDLIMDI yȯmdȋ (Kraft fide KW) TIV abaffi *maiwa type* (JMD) agasse *maiwa type* (JMD) aminne *geero type* (JMD) UHAMI-IYAYU gegebo (Elugbe) UMON akpoi (Cook) UNEME ȯkȧbȧbȧ (Elugbe) VERRE sȩȩtȕ (RB) WANDALA mȧgȯyȅ (Kraft fide

KW) WOM mìsa (Meek) YALA (Ikom) èsià (Ogo fide RB) YALA (Ogoja)) īkpēē (Onoh fide RB) YORUBA ẹmẹyẹ̀ (JMD; Verger) YORUBA (Ilorin) mayi (JMD) YUNGUR muula *geero type* (RB) ZAAR gyóró (Kraft fide KW)

Parts and products: SENEGAL FULA-PULAAR (Senegal) sano *the bran* (K&A) WOLOF tiéré = *cuscus* (AS fide JMD) MALI MANDING-BAMBARA dalo *a beer* (JMD) dègè *steamed sifted flour with salt, piment and milk added* (Diarra) di dègè *a fritter of sifted flour with honey and chili* (Diarra) dolo *a beer* (JMD) moni *a gruel* (Diarra) niondolo *a beer made from the grain, usually maiwa type* (JMD) takula *oven-baked balls made from the flour-paste* (Diarra) UPPER VOLTA DAGAARI sāb *a pap made from the grain with seed of gombo: Abelmoschus esculentus (Malvaceae)* GHANA VULGAR koko *a thin porridge* (Appa Rao & al.) marsa *a deep-fried pancake* (Appa Rao & al.) pito *beer; a corruption of* poitouw, *Portuguese from Angola* (Appa Rai & al.; Wigboldus) tô *a stiff porridge* (Appa Rao & al.) AKAN-TWI ahei *beer* (Wigboldus) GBE-AJA li *conjecturally a reduction of* lili: *what is made small by grinding, hence something fine-grained* (Wigboldus) VHE ahali *germinated grain for brewing* (Wigboldus) DAHOMEY GBE-AJA liha *beer* (Wigboldus) FON ahali *beer* (Wigboldus) VHE ahei *beer; from* ahali, *Fon: germinated grain* (Wigboldus) liha *beer* (Wigboldus) NIGER HAUSA kunu (AE&L) kùnuu *a sort of pap* (AE&L) làalumee, làalàmee (AE&L) lalame *a sort of pap* (AE&L) lalume *Ader dialect* (AE&L) NIGERIA FULA-FULFULDE (Nigeria) baudi (pl.) *the grain head before harvesting* (JMD) bumangal *the inflorescence* (JMD) damana *young plants; applies also to guinea corn* kitikwa, kitiku *a bundle of the straw plaited with the spikes outward for oxen transport* (Bargery fide JMD) lifafi *leaves or stunted plants used for fodder* (Bargery fide JMD) samfoyi *self-sown plants* (Bargery fide JMD) yabana *young plants* (Bargery) HAUSA arauye *a food-preparation* (Bargery fide JMD) areye *a food-preparation* (Bargery fide JMD) azuƙuƙu *a food-preparation* (Bargery fide JMD) chankwama *a cooling drink, or food-preparation* (Bargery fide JMD; JMD) dauro *the bran* (JMD) fito *from pito, corruption of* poitouw, *Portuguese from Angola* (Wigboldus) gaban kabare *a food-preparation, so-called in allusion to the colour of the yellow-breasted weaver bird:* kabare (Bargery fide JMD) giya *a sort of beer, usually made from maiwa type* (JMD) gumba *a food-preparation of pounded grain mixed with water and sugar to promote lactation* (Bargery fide JMD; Schuh) nanaye *the flour* (Bargery fide JMD) sanganche *a food-preparation* (Bargery fide JMD) tsuge *a food-preparation* (Bargery fide JMD) JEN nungye *the grain* (RB) KANURI àrgə̀m métá *threshed maiwa type grain* (JMD; C&H) àrgə̀m móró *threshed geero type grain* (JMD; C&H) NGIZIM ɓátlâ (pl. -átíín) *fermented grain for brewing beer; perhaps also applies to Sorghum grain* ráfíiwà (pl. - àwín) *the chaff* sə̀náasə̂n (pl. -sásin) *a deep-fried cake of the grain* támbàazai (pl. -azín) *pounded grain mixed with water and salt* tlárám (pl. - mámín) *the dried sheaves on the culms* tlátlámáawà (pl. -àwìn) *the dried leaves* WEST CAMEROONS VULGAR cus-cus *general throughout francophone territories, for the grain usually reduced to a meal, granulated and flavoured* tô *the pap* FULA-FULFULDE (Adamawa) jaduri (Thillard fide JMD)

Pathology: SENEGAL sikin *'Green ear disease'; from loose bearded appearance, cf. Serer,* sikin: *a beard* (JMD) MANDING-BAMBARA dunuli *'Sugary disease'* (JMD) WOLOF benat *'Sugary disease'* (JMD) NIGERIA HAUSA domana *a black smut fungus* (JMD) kuturta *a blight* (JMD) WEST CAMEROONS FULA-TUKULOR diuman *'Sugary disease'* (JMD)

The genus *Pennisetum* is large and diverse. The perennial species are treated separately here species by species. These are mainly fodder grasses. The grain pennisetums have evolved through millenia of human selection. They are annual grasses and fall into a number of cultivated races and degenerated or weedy, and wild forms of more or less continuous characters. Some authorities have maintained them as separate species with several in W Africa (22) with cultivated races arising independently. But all are interfertile and current thinking is that the swarm is one species, *P. glaucum, sensu lato* (5, 7, 9, 21).

Bulrush millet as a grain crop has probably evolved in W Africa where the greatest concentration of wild forms exist. It is very tolerant of dry conditions and poor sandy soil. Some very fast-growing precocious forms complete their growth cycle on 250 mm rainfall. A form in the Lake Chad area grows without rain, subsisting on subsoil water (16). In times past before the introduction of that other arid zone cereal, *Sorghum*, it was a staple (18, 20). Even so, *Sorghum* requires a minimum of 375 mm rainfall and people living below this isohyet still depend on bulrush millet. Its vital importance is reflected in a form of totemism practised by the Lyela living in the Sahel of Upper Volta whose tribal name is consonant with their word for millet: *mèla*, and their use of patronymics *bemèla* (male) and *émèla* (female) (19).

In the very arid areas only the fastest growing strains can be cropped, but southwards in the Region to areas of longer wet-season and greater rainfall slower growing and later maturing forms can be cultivated. Thus it is a common practice to intermix the strains, or even to interplant with sorghum to extend the harvest period (21). Such practices grew up in N Ghana whither it is reported bulrush millet arrived as early as 1,250 B.C. The early strains were commonly referred to as 'hungry millet', or *nara*, and late strains as *zia*. Intermixed with cultivar forms, there often occur weedy or degenerate forms, perhaps hybrids in the seed-corn, known as *shibras* (4). Along the coast it appears that Portuguese navigators spread bulrush millet. Pre-colonial Europeans were dubbed by the Asante 'People of the Sun', and the grain they were said to be distributing became called *eujo*, or *εwio* after *egwju*: the sun. By 1637 it had become a major food-crop in the coastal area of Dahomey (24). From W African centres of development, bulrush millet spread some 2,000 years ago to eastern Africa, and thence into India where it has assumed great importance in the dry regions (9, 12). In modern times it has gone to the USA and other countries.

Though bulrush millet is an arid zone, dry land crop, it is in places cultivated under irrigation (13). On dry land, cultivation required is minimal, but improved husbandry is rewarding in producing bigger returns (9, 21), as also breeding and selection for precociousness and other specific factors. Pests can be a nuisance, especially birds. Strains which have bearded seed-heads are less susceptible to depredation by small birds such as finches (8a). It is a practice of some people in Senegal to reflex the flowerhead-stem to a drooping position shortly before harvesting to reduce bird-damage (Trochain fide 8a). A number of pathological conditions is listed above. 'Sugary Disease' is said to cause toxic symptoms when the grain is prepared for eating, it tasting sweet and causing distension of the stomach and a drunken sensation (8a).

Nutritional quality of the grain is variable depending on strain, edaphic factors and water supply. Carbohydrate content is 67–80% (6, 9, 13), of which the main component is starch, white and smoothe equating to commercial corn starch (13). Protein lies between 9–14% (6, 9, 13). The grain is a good source of all essential amino acids. Grain is normally threshed though it is a tedious process under African village conditions, occupying much of the housewife's time (6). The grain may be ground to flour, and eaten cooked into a stiff porridge or baked into unleavened bread, or in other manner of preparations (4, 8a, 13). A malt can be made for admixture to other foodstuffs to give balance to diet. Selection has been exercised for grain for specific end-products. The Tenda reserve the lower quality grain for women (10). Dark coloured grain goes for brewing into beer (4). In Dahomey such is the popularity of *liha* beer that it replaces palm-wine (24). So great is the importance of bulrush millet in African human economy that a diverse vocabulary for forms and products has grown up (see word-list above).

The grass is grown in the Region as a good green manure and for forage (1a–c, 2a,b, 4, 8b,c), though use as fodder for stock must inevitably be at the expense of raising grain. It is extensively cultivated in India as a fodder grass (12). It must be conjectured that given a great variability such usage is made of the weedier forms. There are reports in Senegal that young seedlings or young culms (15) may be toxic to cattle. There is also belief in Senegal that the roots of precocious forms are poisonous causing a sort of 'sleeping sickness' (1e, 2a, 8a, 14). The cause may be due to the presence of an alkaloid, perhaps *agroclavine* and associated substances (14). There is, however, no certain proof of poisonousness (8a). Millet is known to be sometimes infected with ergot and some forms to produce *hydrocyanic acid* (9). It may be that either of these is the source of toxification. In India stubble is commonly fed to stock, while the green crop is turned into silage (12). The leaves have been tested in Nigeria for toxcity towards the fresh-water snail, *Bulinus globulus*, a vector of schistosomiasis, and have been found completely benign (1).

316

The culms are used to make screens and divans (1c,d), fencing, roofing and fuel (4, 8a). Though in general *Sorghum* culms are preferred for basketry, *Pennisetum* culms can be used, and then they are split and the pith is removed before working (4). They also serve as fire-sticks (8a).

Red and purple flowered forms are the source of a dye used on leather and wood (1c,d, 8a, 23). Similar use is recorded in Algeria (8a). Craftsmen at Agadez in Niger and at Timbuktu in Mali making leather boxes for containing domestic and personal bric-à-brac use a red paint made from millet husks for decoration. *Bata*: Hausa, is a general name for the containers, but variants in design have their own separate names (17).

The grain has medical uses. It has been applied for chest disorders, and ground up given to children as an anthelmintic (23). Bran is used to make poultices (8a) and in Senegal after heating as a massage for kidney pains (14, 15). Also in Senegal, Wolof take the grain or the ear to treat leprosy, blennorrhoea and poisonings (14).

The Dagaari of Upper Volta ascribe ritual significance to the grain. The bat is an important figure in local philosophy as the bearer of illnesses and souls of the departed. As a part of a rogation ceremony held in the dry season (January–February) a small bat is stuffed full with millet pap: *saab*, till it is so full it cannot fly (11), and thus cannot spread disease nor remove the spirits of the dead. In Niger seed from dark purple plants is purposely mixed with grain for sowing for magical purposes to ensure a good crop (23).

References:

1a. Adam, 1954: 96, as *P. mollissimum* Hochst. 1b. ibid., 96, as *P. rogeri* Stapf & Hubbard. 1c. ibid., 96, as *P. pycnostachyum* Stapf & Hubbard. 1d. ibid., 96, as *P. typhoideum* Stapf & Hubbard. 1e. ibid., 96–97, as *P. aff. violaceum*. 2a. Adam, 1966, as *P. violaceum*. 2b. ibid., as *P. mollissimum*. 3. Adewunmi & Sofowora, 1980: 39: 57–65, as *P. americanum* K Schum. 4. Appa Rao & al., 1985: 25–38, as *P. americanum*. 5. Brunken, 1977: 161–76. 6. Busson, 1965: 477, with chemical analyses, 479–83. 7. Clayton & Renvoize, 1982: 672–3. 8a. Dalziel, 1937: 538–40, as *Grain Pennisetums* 8b. ibid., 537, as *P. mollissimum* Hochst. 8c. ibid., 537, as *P. perrottetii* K Schum. 9. Deshaprabhu, 1966: 296–308, as *P. typhoides* (Burm.f.) Stapf & Hubbard, with chemical analyses. 10. Ferry & al., 1974: No. 130, as *P. gambiense*. 11. Girault, 1960: 120. 12. Gupta & Sharma, 1971: 87, as *P. typhoides* (Burm.) Stapf & Hubbard. 13. Johnson & Raymond, 1964, a: 6–11, as *P. typhoideum* Richard, with many references. 14. Kerharo, 1973: 18. 15. Kerharo & Adam, 1964, b: 567. 16. Lamb 109, K (*P. gibbosum* Stapf & Hubbard). 17. Lhote, 1952: 919–55. 18. Meek, 1931, 2: 45, as *P. spicatum* 19. Nicolas, 1953: 836, as *P. spicatum*. 20. Okiy, 1960: 119–20, as *P. typhoideum*. 21. Purseglove, 1972, 1: 204–14, as *P. typhoides* Stapf & Hubbard. 22. Stapf & Hubbard, 1934. 23. Watt & Breyer-Brandwijk, 1962: 483, as *P. typhoides* Stapf & Thunb. 24. Wigboldus, 1986: 299–383.

Pennisetum hordeoides (Lam.) Steud.

FWTA, ed. 2, 3: 461. UPWTA, ed. 1, 537.

West African: SENEGAL BEDIK sikili (FG&G) WOLOF bara (A.Chev.) THE GAMBIA MANDING-MANDINKA barato barra (Pirie fide JMD) farato barra = *swamp grass* (Pirie) GUINEA BASARI a-sigini (FG&G) FULA-PULAAR (Guinea) hulhuldé (A.Chev.) wolondé (A.Chev.; Adames) KONYAGI wa-ryuy (FG&G) SUSU kuli (A.Chev.) YALUNKA puki (A.Chev.) SIERRA LEONE LIMBA foi (NWT) ningolobubeyuwe (NWT) LOKO k'pabalu (NWT) nguɔgɔi (NWT) ngwɛgwe (NWT) MANDING-MANDINKA turunyã (FCD) MENDE mumiyami (NWT) SUSU kuli (NWT) TEMNE kɔ-balingi (NWT) kɔ-ep (NWT) a-linki (NWT) YALUNKA kul-la (FCD) MALI DOGON sapa (A.Chev.) FULA-PULAAR (Mali) bogo dolori (A.Chev.) hulhuldfé (A.Chev.) wolonde (A.Chev.) MANDING-BAMBARA ngolo (JMD) UPPER VOLTA GRUSI yakalo (A.Chev.) GURMA hihangon (A.Chev.) MOORE kénibédo (A.Chev.) kimogo (A.Chev.) kim-ubogo (A.Chev.) SAMO dansa (A.Chev.) NIGERIA BUSA bushi (Ward) FULA-FULFULDE (Nigeria) buludi (Ward) HAUSA kansuwa (Ward) ƙyasuwa (JMD; ZOG) IGBO (Owerri) eru (AJC; JMD) KAMBARI-AUNA kàkàsɔ̃wũ̀ (RB) SALKA àapù (*pl.* ụ̀mù) (RB) YORUBA ilósùn (Verger) tòlò (Verger)

An annual grass, culms to 1.2 m high; of disturbed areas throughout the Region from Senegal to W Cameroons and Fernando Po, and on to the Congo basin; also occurring in India.

It is an abundant annual in fields and waste places, becoming a persistent weed in cultivation which is difficult to eradicate, especially on damp sites (4–6). It is a good fodder grass and much relished by stock while still young (1, 3). It is suitable to establish in pasture leys (3), and converts into a good quality hay (7).

Medicine-men in Congo (Brazzaville) take sap from the crushed plant for topical application for pain in the lumbar and costal regions with or without scarification (2).

References:

1. Adam, 1966, a. 2. Bouquet, 1969: 129. 3. Dalziel, 1937: 536–7. 4. Deighton 4551, K. 5. Ferry & al., 1974: No. 131. 6. Pirie 5/33, K. 7. Ward 0021, K.

Pennisetum macrourum Trin.

FWTA, ed. 2, 3: 461, as *P. glaucocladum* Stapf & CE Hubbard and *P. giganteum* A Rich.
West African: **WEST CAMEROONS** BAMILEKE gotonga (DA)

A stout reed-like perennial grass, with creeping rhizome, culms to 5 m high; of stream-banks; scattered in the Region in Guinea, Togo and N Nigeria, and in Sudan, E Africa and southwards to Botswana.

While the plant is still young it is a good fodder much appreciated by cattle (1–3).

References:

1. McDonald 961, K. 2. Snowden 1144, K. 3. Thomas, AS Th.2105, K.

Pennisetum pedicellatum Trin.

FWTA, ed. 2, 3: 460. UPWTA, ed. 1, 537.

English: barra grass (The Gambia, from Mandinka, Pirie); matting grass (The Gambia, Pirie).
West African: **SENEGAL** FULA-PULAAR (Senegal) ulunde (JMD) wolonde (JMD) MAND-ING-BAMBARA ngolo (auctt.) SERER bob (JB) dan (JB; K&A) faf (K&A) fayfay (JB; K&A) fof (JB) ga (JLT fide JMD) mbop (JMD; K&A) SONINKE-SARAKOLE bara (K&A) WOLOF bara (auctt.) bob (JB) mbop (K&A) **THE GAMBIA** MANDING-MANDINKA barra (Pirie) **MALI** FULA-PULAAR (Mali) ulunde (JMD) wolonde (JMD) MANDING-BAMBARA ngolo (JMD) **UPPER VOLTA** FULA-FULFULDE (Upper Volta) bogodollo (K&T) bogodollooji (K&T) MOORE ngolo (Scholz) kimbogo (Scholz) **GHANA** DAGBANI china (Ll.W) **NIGER** HAUSA kissana Bartha SONGHAI hárgéy (*pl.* -à) (D&C) **NIGERIA** ? (Nigeria) suroja (Saunders) vichu zeen (Saunders) ARABIC-SHUWA umm dufufu (JMD) FULA-FULFULDE (Nigeria) wuuluunde (*pl.* buuluuɗe) (J&D; JMD) HAUSA hura *generally for the plant, but in Kano for the flowering head only* (auctt.) huran giwa (Golding) kaá-fi-riímií = *better than kapok: in allusion to the floss of the mature spike* (JMD; ZOG) kámsuwa (JMD; A&S) ƙan suwaa (auctt.) ƙya suwaa (auctt.) tsat suwaa (BM) IGALA ìkpàkpàlà (Odoh; RB) KANURI férá (JMD; C&H) fàrá (A&S) YORUBA ęsu (Macgregor) YORUBA (Ilorin) ilosun (Ward)

An annual grass, rarely perennial, culms to 1½ m high; of dry savanna, dispersed throughout the whole of W Africa, and in E Cameroun to Sudan and India. Two subspecies occur in The Region: ssp. *pedicellatum* and ssp. *unispiculum* Brunken. The former is represented throughout, the latter is Ghanaian and occurs on disturbed sites. Both are considered together here.

The grass provides good forage for cattle and horses before flowering (1, 2, 4, 5, 7, 10, 12). It converts into good quality hay when cut early (5, 13). In Kordofan it has been successfully ensiled (4). It is considered promising for laying down in pasture (5) to provide dry season fodder (6). Plants show a variability in leafiness and earliness which offers scope for undertaking selection (6).

The culms are woven into mats in The Gambia – hence the English name (5, 11). They are commonly used for thatching in the Region (1, 5, 7, 8) and in Kordofan (4); also for daubing with clay to make hut-walls.

In Borno, a decoction of the whole plant is considered diuretic (3), and in Senegal it is used internally for the same purpose and externally as a haemostatic (9).

References:

1. Adam, 1954: 96. 2. Adam, 1966, a. 3. Akinniyi & Sultanbawa, 1983: 101. 4. Baumer, 1975: 109. 5. Dalziel, 1937: 537. 6. Foster & Munday, 1961: 316. 7. Golding 10, 41, K. 8. Jackson, 1973. 9. Kerharo & Adam, 1974: 653–4. 10. Macgregor 1, K. 11. Pirie 81/33, K. 12. Saunders 68, K. 13. Ward 0013, 0060, K.

Pennisetum polystachion (Linn.) Schult.

FWTA, ed. 2, 3: 460–1, incl. *P. subangustum* (Schumach.) Stapf & CE Hubbard and *P. atrichum* Stapf & CE Hubbard. UPWTA, ed. 1, 537, and 538, as *P. subangustum* Stapf & Hubbard.

English: mission grass (Fiji, Dalziel); barra grass (from Mandinka, Pirie); matting grass (The Gambia, Pirie); golden grass (ssp. *atrichum* (Stapf & CE Hubbard) Brunken, The Gambia, Pirie).

West African: SENEGAL DIOLA ésulâg (JB; K&A) FULA-PULAAR (Senegal) buludé (K&A) WOLOF ardièmba (JB) bara (A.Chev.) THE GAMBIA MANDING-MANDINKA barra (Pirie) sano barra = *golden grass* (Pirie) GUINEA-BISSAU BALANTA feéta (JDES) BIAFADA mambinro (JDES) GUINEA BASARI a-tyéɓ laŋét = *which the dog drinks [sic]* (FG&G) FULA-PULAAR (Guinea) wolonde (A.Chev.) KONYAGI saɓire (FG&G) SUSU kuli (A.Chev.) YALUNKA puki (A.Chev.) SIERRA LEONE BULOM (Sherbro) goŋgo lɛ (FCD) KISSI sɛnsɛnde-pɔ mbolɛŋ (FCD) KONO sɛnsɛnɛ-musuma (FCD) KORANKO yɛrɛmɛ (NWT) LOKO nguague (FCD) LOKO (Magbile) ngɔbɔina (NWT) MANDING-MANDINKA fafwiya (NWT) turunyã (FCD) MENDE gbana-lɛvu *the plant in flower*(FCD) ngaile (Dawe) gbanalɛvu *when in flower*(FCD) gbana-lɛvu (*def.*-i) *the inflorescence* (FCD) ngoŋgo (*def.* -i) (auctt.) ngoŋgo-lɛvu, ngoŋgo-lɛvu-ha *when in flower; female-cf. Rottboellia exaltata, male*(auctt.) gɔngɔ-levu-he *when in flower; female*(FCD) ngugu (*def.* -i) (FCD) ngungu (*def.* -i) (auctt.) kpanalɛuu *when in flower*(FCD) kpana-lɛvu *the infloresence*(JMD) MENDE (Kpa) gɔngɔ-levu (*def.* -i) (JMD) MENDE (Up) fovo (*def.* -vui) *when not in flower*(FCD) SUSU kuli (auctt.) TEMNE a-balingi (NWT) a-kɔp-ka-bɛra = *female 'kɔp'* (FCD) a-linki (NWT) YALUNKA kul-la (FCD) kul-la-khɔngbena (FCD) panyirakul-la (FCD) MALI DOGON sapa (A.Chev.) FULA-PULAAR (Mali) bogo dolori (A.Chev.) hulhuldé (A.Chev.) wolonde (A.Chev.) MANDING-BAMBARA ngolo (A.Chev.) nkolo (A.Chev.) KHASONKE bara (A.Chev.) SAMO dansa (A.Chev.) UPPER VOLTA FULA-FULFULDE (Upper Volta) bogodollo (*pl.* -ooji) (K&T) GRUSI yakalo (A.Chev.) GURMA hihangon (A.Chev.) MOORE kimbogo (Scholz) kimogo (A.Chev.) kim-ubogo (A.Chev.) GHANA HAUSA kyasuwa (Ll.W) NIGERIA ARABIC-SHUWA umm dufufu (JMD) BUSA bushi (Ward) FULA-FULFULDE (Nigeria) wulunde (*pl.* buludé) (JMD; Ward) HAUSA hura *used generally, but in Kano for the flowering head only* (auctt.) káá-fi-ríímíí = *better than kapok: the floss* (JMD; ZOG) kámsuwa (JMD; A&S) ƙan suwaa (auctt.) ƙya suwaa (auctt.) ƙya suwar fadama (DA; Thatcher) ƙyamsúwáá (ZOG) IGBO (Ala) ugbene jinni (NWT) IGBO (Umuahia) àchàrà nwankịtā = *grass of the dog*(AJC) KANURI fàrá (JMD; A&S) YORUBA ilosùn (auctt.) inásua (JMD; Verger)

A polymorphic grass, annual or perennial, culms 30 cm to 1.8 m high, of varying stiffness and colouring of inflorescence; of two subspecies: ssp. *polystachion* and ssp. *atrichum* (Stapf & CE Hubbard) Brunken, the latter distinguished by involucral bristles scaberulous, not ciliate; on fallow or disturbed sites, throughout the whole Region, and ssp. *polystachion* generally over tropical Africa and the tropics, and ssp. *atrichum* to central and S tropical Africa.

The grass gives good browsing for all stock at least till it begins to flower (1, 2a,b, 7a,b). It is often planted around villages and in fields and roadsides in Senegal (1). It makes good horse-feed (4, 7a,b) and Hausa horse-boys are said to prefer feeding this grass if it can be got (14). Pigs in S Nigeria will eat it (6). Some forms cease growth early in the dry season and are therefore of little value as pasture grass (9a). Other forms are more free-growing, more precocious, leafier and convert into good quality hay, providing feed for the dry season (3,

319

9b). Hay grown in Guinea is recorded as having 8.9% nitrogenous material (1). Strains grown in Ilorin, Nigeria, can produce two cuts a year (18). Examination of variability in habit and utility should be made to select superior forms.

In Kenya, the culms are used for thatching (13) and in The Gambia to make matting (15).

At Musaia, Sierra Leone, the presence of the grass in the bush is held to indicate potentially good farm land (8).

In Ubangi the grass is calcined to yield a vegetable salt, and sap from young culms is applied to cuts and wounds to promote healing (16). In Congo (Brazzaville) plant-sap is used to cleanse sores. Application is said to be painful, but nevertheless some medicine-men claim it to be useful in treating conjunctivitis (5). A root-decoction is taken in Tanganyika as an antemetic, and water in which a flower-spike has been pulped is instilled into ears to quell earache (10). The grain appears to have similar analgesic action. A poultice of the grain may be applied topically for pain in the sides, to the shoulder for dislocation and for internal pain (5).

The panicles of some forms are attractive. They are used in Gabon to decorate ceremonial drums (17).

Peul of Senegal ascribe medico-magical powers to the plant, particularly with a corn cob and *Tapinanthus* (Loranthaceae) in treating impotence (11, 12).

References:

1. Adam, 1954: 96. 2a. Adam, 1966, a. 2b. ibid., as *P. subangustum*. 3. Baumer, 1971: 109. 4. Beal 13, K. 5. Bouquet, 1969: 130. 6. Carpenter AJC.819 (UIH 4361), UCI, as *P. subangustum*. 7a. Dalziel, 1937: 537. 7b. ibid., 538, as *P. subangustum* Stapf & CE Hubbard. 8. Deighton 4525, K. 9a. Foster & Mundy, 1961: 314. 9b. ibid., 316, as *P. subangustum*. 10. Haerdi, 1964: 211. 11. Kerharo & Adam, 1964, b: 567, as *P. subangustum* Stapf & CE Hubbard. 12. Kerharo & Adam, 1974: 655, as *P. subangustum* Stapf & CE Hubbard. 13. Magogo & Glover 603, 824, 1072, K. 14. Maitland 153, K. 15. Pirie 81/33. 16. Vergiat, 1970, a: 86. 17. Walker & Sillans, 1961: 193, as *P. setosum* Rich. 18. Ward 0012, K.

Pennisetum prostrata Griseb.

FWTA, ed. 2, 3: 457.

A prostrate perennial grass, aquatic; in Senegal to S Nigeria, and to central and SW Africa, Madagascar and tropical America.

The grass is abundant in freshwater, or occasionally brackish, swamp areas, quickly colonising areas empoldered for rice-cultivation in Sierra Leone. It grows on the bunds and consolidates them.

Pennisetum purpureum Schumach.

FWTA, ed. 2, 3: 461. UPWTA, ed. 1, 537–8.

English: elephant grass; Napier grass; Napier's fodder.

French: herbe à éléphants; fausse-cane à sucre.

Portuguese: capim-elefante (Feijão).

West African: **SIERRA LEONE** BULOM (Sherbro) ŋà` (Pichl) FULA-PULAAR (Sierra Leone) tambɛn (FCD) KISSI cheŋgio (FCD) KONO fa (FCD) fa-wa (FCD) KORANKO molike (FCD) KRIO bush-shuga-kɛn *i.e., bush sugar cane* (FCD) LOKO ŋgara (FCD) MANDING-MANDINKA moloko (FCD) mɔlɔkɔ-yɔ (FCD) MENDE mbowi-hei = *female 'mbowi'*(FCD) ŋgala (FCD) ŋgalɛ *applied generally but mainly to this sp.* (auctt.) ŋgɔŋɔi (Migoed; JMD) TEMNE a-, aŋwo *a general name*(FCD) an-lal (FCD) *ka-*staf *for the red variety, sometimes cultivated and of which the culms are cut for walking-sticks* (FCD) YALUNKA kulon-na (FCD) kulu-na (FCD) **IVORY COAST** ANYI chelié (K&B) BAULE nè (B&D) KWENI dia (K&B) dian voli (B&D) **GHANA** AKAN-ASANTE akokɔ ani = *fowl's eye* (FRI) anan hwerew = *cow's sugarcane* (FRI) hwediɛ (FRI; E&A) GA adaí (KD) glã (FRI; KD) NZEMA élanke-akanla = *cow's sugarcane* (FRI) **TOGO** GBE-VHE adá (JMD) **NIGERIA** BIROM eromo (LB) DERA yiwó (Newman) EDO ǫghǫdóǧbǫ (JMD; Elugbe) EFIK ṁbíit (AJC fide JMD) ṁbǫ̀kǫ̀k ékpò = *sugarcane (of) demon*

320

(auctt.) FULA-FULFULDE (Adamawa) toll ore = *hollow elephant's grass* (Pedder) FULFULDE (Nigeria) gawri ngabbu = *hippopotamus's corn* (JMD) toloore (J&D) HAUSA daáwàr kádàá (JMD; ZOG) kyambama (JMD; ZOG) yambámáá (auctt.) IBIBIO ṁbọ̀kọ̀ èkpò = *devil's cane* (JMD; BAC) ICEN fe (RB) IGBO àchàlà (JMD) àchàrà (auctt.) àchàrà mili *from* mili: *water* (JMD) IGBO (Asaba) ùkpò ukwu = *big ukpo* (JMD) IJO-IZON (Kolokuma) izáí *the infloresence* (KW) ịzọ́n úsí = *ijo, or true elephant grass* (KW) osï (KW; KW&T) usï (KW; KW&T) IZON (Oporoma) epírí (KW) KAMBARI-SALKA iiliinâ (RB) ODUAL ikpu (Gardner) SURA cár (anon. fide KW) TIV awo (JMD) uwua nor (JMD) YORUBA eèsún (auctt.) eèsún funfun *the white form* (auctt.) eèsún pupa *the red form* (auctt.) YORUBA (Ijebu) ikẹn (JMD) **WEST CAMEROONS** BAFOK sosom e nyak (DA) KOOSI sosom (DA) KPE bekoko (DA) KUNDU bekoko (DA) LONG besong (DA) LUNDU makoko (DA) MBONGE bekoko (DA) TANGA makoko (DA) WOVEA makoko (DA)

A robust perennial grass, often clumping, culms bamboo-like to 2.5 cm diameter at the base by up to 8 m high, but less in upland situations at ~ 2000 m; riverine, valley bottoms, forest margins on rich soil, often cultivated as elephant grass or Napier grass; common across the Region from The Gambia to W Cameroons and Fernando Po, and throughout the rest of tropical Africa; widely introduced in the tropics.

It is the dominant grass of the fertile region around Lake Victoria in Uganda where the rainfall is 1,000–1,500 mm p.a. (22). It is said to have been first brought into cultivation in Zimbabwe (10), or Transvaal (9) from whence the practice has spread throughout Africa and elsewhere. It can be grazed, but will not stand intensive grazing for a long period. It becomes unpalatable on flowering because of the bristly awns (5). It does better as fodder and pasture grass, responding well to cutting and maintaining a good return for 5–6 years when replanting by stem stools has been recommended (12, 22). Since it is drought-tolerant and gives a return even during severe periods, it is economically valuable (18). If cut while still young, say, 1 m tall, it converts into good hay, or is suitable for ensiling (10, 12, 22, 27). It has a chromosome number 2n = 28, and crossing with *P. typhoides*, 2n = 14, has produced a sterile triploid, 2n = 21, more productive of fodder than its parents (12, 21). Under some conditions cut fodder of *P. purpureum* has been reported to produce *hydrocyanic acid* (28).

For human food Igbo, particularly at Umuahia in Nigeria, put the young leaves rich in protein, carbohydrates, fat and vitamins in soup (10, 16, 19). This is mildly laxative (16). In The Gambia young shoots are collected during the rains for cooking in ground-nut stew (23). The culms reduced to ash are lixiviated to produce a vegetable salt in Gabon (25).

The vigorous compact habit of growth lends the plant to the establishment of stands on river-banks to prevent erosion and scour (21, 22). The culms are used to make fences (6, 10, 22, 26), huts and screens, and are planted to form windbreaks. They are used for thatching (W Cameroons: 14), but elsewhere lack favour (28). In Sierra Leone the grass is planted as a boundary-marker between garden plots (11). The culms are a suitable source of paper-pulp (4, 20, 28) producing good quality paper (12). In Gabon the dried culms serve as pipe stems (25). Culms, cut to a sharp point, are used as spears in mimic battle in some religious rites in N Nigeria (Meek fide 10), and in Gabon culms may be laid across a door threshhold to prevent entry of evil spirits; or for the same purpose clumps of the grass are grown near to houses (25). On a similar line of thought sap enters into a Tanganyikan formulation with a number of other drug-plants to be taken by mouth as a tranquilliser for someone possessed by spirits (15). Spiritous association is found also in the Efik name for the plant: 'sugar-cane of the demon' (1).

The plant enters into a number of medical applications. Extracts are strongly diuretic and are used as such in Nigeria (20). In Rio Muni an infusion of both foliage and culms is used in anuria (28), and in Ivory Coast–Upper Volta a root-decoction is given for blennorrhoea (17). A leaf-infusion in Congo (Brazzaville) constitutes a gargle and mouthwash for buccal affections, gingivitis and thrush (7). Sap expressed from young shoots which have been heated over a fire is mixed with a little salt and instilled in Ivory Coast into the eyes for cataract; the sap is also considered healing on wounds (8). Sap taken from the top of the culm and mixed with that from other drug-plants is used in Tanganyika on fresh wounds, and pith taken from the ends of young culms softened in a fire is a dressing for contusions (15). Stem-sap is used in Gabon for ear-trouble (25).

Ash from the calcined culms is put up in Gabon as an ointment with palm-oil or false shea butter (*Lophira lancolata*, Ochnaceae) as base for treating herpes and other skin-complaints, and ash added to slices of the large green banana is put on ulcers on the soles of the feet (24, 25). Also in Gabon, culms of the grass reduced to ash together with parts of a number of other plants go into a complex prescription for treating blennorrhoea (24, 25).

Examination of Nigerian material has shown a trace of *alkaloid* in the leaf (2). In S Nigeria the seed is reported used to cure headache (3).

References:

1. Adams, RFG, 1947: 23–24. 2. Adegoke & al., 1968: 13–33. 3. Ainslie, 1937: sp. no. 266. 4. Anon., 1913. 5. Baumer, 1975: 109. 6. Blench, 1985. 7. Bouquet, 1969: 130. 8. Bouquet & Debray, 1974: 92. 9. Burkill, IH, 1935: 1688. 10. Dalziel, 1937: 537–8. 11. Deighton 2919, K. 12. Deshaprabhu, 1966: 294–6, with chemical analysis. 13. Foster & Munday, 1961: 314. 14. Gillett, S 14, K. 15. Haerdi, 1964: 211. 16. Iwu, 1986: 143. 17. Kerharo & Bouquet, 1950: 253. 18. Ogor & Hedrick, 1963: 146. 19. Okiy, 1960: 120. 20. Oliver, 1960: 33. 21. Pouquet, 1956: 12. 22. Purseglove, 1972, 1: 203–4. 23. Tattersall, 1978: 13. 24. Walker, 1953, a: 39–40. 25. Walker & Sillans, 1961: 192. 26. Ward 0047, K. 27. Ward L.107, K. 28. Watt & Breyer-Brandwijk, 1962: 483.

Pennisetum ramosum (Hochst.) Schweinf.

FWTA, ed. 2, 3: 460. UPWTA, ed. 1, 538.
West African: **NIGERIA** HAUSA géérón kádàà = *millet of the crocodile* (auctt.)

A tough caespitose annual grass with culms to 1.2 m high, on sites of heavy soil, swamps, riversides and wet places in N Nigeria, and across Africa to NE and E Africa.

In Sudan it becomes a weed of cultivation (2). In all areas it provides excellent grazing while still young (Nigeria: 3, 5; Sudan: 4, 7; Tanganyika: 6; Kordofan: 1), but the flowers bear harsh bristly awns which put cattle off from taking it after anthesis.

The culm is said to be used in Sudan to chastise naughty boys (4).

References:

1. Baumer, 1975: 109. 2. Beshir 57, K. 3. Dalziel, 1937: 538. 4. Evans Pritchard 37, K. 5. Golding 27, K. 6. Marshall 18, s.n. 26/2/1932, K. 7. Peers K.O.6, K.

Pennisetum setaceum (Forssk.) Chiov.

FWTA, ed. 2, 3: 461.
West African: **MALI** DOGON avú (CG)

A densely tufted perennial grass, culms up to 1.3 m high, leaf-blades, harsh and rigid, of stony and dry open places, rare in the Region in Senegal and Ghana (?introduced), and in N Africa, Tunisia to Israel, and in E Africa to Tanganyika.

It is grown in many warm countries as an ornamental, and is so recorded in Ghana (4).

It provides forage for cattle, sheep and goats in Sudan (1), but the harshness of the foliage is evidently not benign to the mouth of animals. In Somalia it has a name meaning 'cuts tongues' (2), or *arap jeb*: tongue sore, *jeb* being a sliver of wood inserted into a young camel's tongue to wean it from suckling (3).

References:

1. Andrews A.2717, A.3593, K. 2. Burne 27, K. 3. Collenette 18, K. 4. Scholz 103, K.

Pennisetum thunburgii Kunth

FWTA, ed. 2, 3: 463, as *P. glabrum* Steud.

A loosely tufted rhizomatous perennial grass, culms to 90 cm high; of damp places, old pasture and cultivations, and roadsides, recorded only in N Nigeria in the Region, but extending to Sudan, Ethiopia, E Africa and Zambia.
The grass occurs as a weed of cultivation, especially in cereal fields. It is reported to become a weed of lawns in Ethiopia where its stoloniferous habit produces an acceptable turf (1). It is grazed by all stock.

Reference:

1. Friis & al 1397, K.

Pennisetum trachyphyllum Pilger

FWTA, ed. 2, 3: 461.

A perennial grass, culms ascending, rooting at lower nodes, to 2 m high; of damp, semi-shaded sites or path-sides and evergreen forest glades in W Cameroons only, but also in NE, E and S tropical Africa.
It is a leafy grass and providing good fodder and grazing when growing in suitable situations; stock take it readily (1, 2).

References:

1. Lyne Watt 1190, K. 2. Snowden 1445, K.

Pennisetum unisetum (Nees) Benth.

FWTA, ed. 2, 3: 459, as *Beckeropsis uniseta* (Nees) K Schum. UPWTA, ed. 1, 521, as *B. uniseta* K Schum.
West African: **SIERRA LEONE** MANDING-MANDINKA tɔŋfɔ-jɔ̃ (FCD) YALUNKA foɲfo-folesɛkhɛ-na (FCD) foɲfoɲfole-na (FCD) **UPPER VOLTA** BISA himikom (Scholz) **NIGERIA** BUSA (Bokobaru) ali liya (Ward) FULA-FULFULDE (Nigeria) ali liya (Ward) file (Ward) HAUSA fafewa (JMD) furo kogo (Ward) garangautsa (JMD; ZOG) ƙara(n) ƙauji *from* ƙauji: *roughness* (JMD; ZOG) ƙaran ƙausa (JMD; ZOG) korkoro (JMD) kwarkwaroo (ZOG) IDOMA ufie (Odoh) IGALA ùkpáfēlē (Odoh; RB) IGBO (Ala) ar'o amị amị: *reed* (JMD; KW) ar'o anị ànị: *ground* (NWT; KW)
NOTE: *Nigerian names cover several species of Pennisetum as well and perhaps other genera with rough leaf-edge and hollow stems.*

A tufted perennial grass with robust culms to $3\frac{1}{2}$ m high by $1\frac{1}{2}$ cm diameter; of deciduous bushland and wooded savanna in shaded positions of lowlands to montane elevations; widespread across the Region from Senegal to W Cameroons, and throughout tropical Africa generally.
In all countries the grass provides fodder for cattle, but usually only whilst still young and tender (1–4, 6, 7, 10). It withstands mowing and cutting for feed, regenerating quickly and giving several cuts in the year (3). Its hay is a popularly given horse-fodder on the Obudu Plateau at 500–600 m altitude (11).
The hollow culms are used in Igala country to make an end-blown flute (5). In Sudan they are made into whistles and drinking straws (9), and in Ghana are used for syphoning (4).
The root compounded with the rhizome of ginger (*Zingiber officinale* Roscoe, Zingiberaceae) and cooked, the decoction is drunk in Tanganyika for sore-throat and tonsilitis (8).
The grain is collected in Ethiopia for human consumption as a cereal and for the production of a beer (1, 6).

GRAMINEAE

References:

1. Adam, 1954: 88, as *Beckeropsis uniseta* K Schum. 2. Adam, 1966, a: as *B. uniseta*. 3. Baumer, 1975: 88, as *B. uniseta* (Nees) K Schum. 4. Beal 30, K. 5. Blench, 1981–82. 6. Dalziel, 1937: 521, as *B. uniseta* K Schum. 7. Deighton 4432, K. 8. Haerdi, 1964: 209, as *B. uniseta* (Nees) Stapf. 9. Jackson 484, K. 10. Saunders 4, K. 11. Tulley, 1966: 905, as *B. uniseta*.

Pennisetum villosum Fresen.

FWTA, ed. 2, 3: 463.

A perennial grass with handsome plumose involucres; an Ethiopian species, recorded in Senegal (2).

It is grown as an ornamental in Senegal, and also in France (1). It is perhaps cultivated in Eritrea and appears as a weed of irrigated land (4). Cattle browse it, trampling the culms, producing a mat (3), from which the plant continues to grow in a semi-decumbent manner covering the ground.

References:

1. Adam, 1954: 96, as *P. longistylum* Hochst. 2. Berhaut, 1967: 393. 3. Hemming 1046, K. 4. Ryding & Ermias K.1140, K.

Pennisetum spp. indet.

West African: SENEGAL BEDIK *gi*-ŋà-ɛŋ *mɔ*-ɓɔ́y (FG&G) GUINEA-BISSAU CRIOULO midjo madja (JDES) GUINEA BASARI madià (FG&G) KONYAGI madya (FG&G) NIGERIA HAUSA dan gariya (ZOG) dán-bàrnó = *the one from Borno* (ZOG)

Perotis hildebrandtii Mez

FTWA, ed. 2, 3: 411.
West African: NIGERIA YORUBA zkoyi (Maitland)

A loosely-tufted annual, culms geniculately ascending to 45 cm high, a ruderal of roadsides, lawns and waste places in Sierra Leone to S Nigeria and in E Africa and Seychelles.

Perotis indica (Linn.) O Ktze

FWTA, ed. 2, 3: 411. UPWTA, ed. 1, 540, excl. Senegal.
West African: NIGERIA HAUSA búndin kureégé = *tail of the squirrel* (JMD; ZOG) wútsíyàr kùréégéé = *tail of the squirrel* (JMD; ZOG)

A loosely-tufted annual, culm semi-erect to 30 cm high, a ruderal of dry sandy roadsides and waste places in Sierra Leone and S Nigeria, and across equatorial Africa to Tanganyika and in India and SE Asia.

The grass is relished for grazing but it lacks quantity. It has fair nutritional value. Indian material is reported as containing protein 4.03%, fibre 30.56% and carbohydrate 49.41%, dry wt.

Reference:

1. Deshaprabhu, 1964: 314.

Perotis patens Gandoger

FWTA, ed. 2, 3: 411.
West African: GHANA HAUSA gééróó susu = *early millet* (Ll.W) NIGERIA HAUSA wútsíyàr kùréégéé (Kennedy) YORUBA èrò yéwa (Verger) iru ẹgbara (anon.; Ward)

324

A loosely-tufted annual, or short-lived perennial grass, culms up to 60 cm high, graceful, a ruderal of waste places, roadsides, and a weed of cultivation, often on dry sandy soils, in Ivory Coast to S Nigeria, and widespread throughout tropical and S Africa and into Madagascar.

In all territories it is browsed by cattle with appreciation. In Sudan it is reported to produce a sort of turf but indifferent as a lawn grass (2). An infusion of the leaves is drunk for colds in Uganda and the same preparation is is given to cattle (1).

References:

1. Dyson-Hudson 255, K. 2. Myers 6343, K.

Phacelurus gabonensis (Steud.) WD Clayton

FWTA, ed. 2, 3: 504, as *Jardinea congoensis* (Hack.) Franch. UPWTA, ed. 1, 530, as *Jardinea congoensis* Franch.

West African: **TOGO** KABRE loɛ (Volkens) TEM (Tschaudjo) loku (Volkens) **NIGERIA** FULA-FULFULDE (Nigeria) iware (JMD) HAUSA dĩwa (JMD) iwa (JMD; FRI) shada (JMD) IDOMA onji (Odoh) IGALA ẹ̀jì (Odoh; RB) KAMBARI-SALKA iiliinâ (RB) YORUBA (Ilorin) ige (Ward) ire (Ward)

A coarse perennial grass with cane-like culms to 3 m tall; of swampy places; occurring from Ghana to N and S Nigeria, and extending into Sudan and southwards through the Congo basin into Zambia and Angola.

The culms are commonly made into baskets, fish-traps, screens, coarse matting, sleeping-mats, grain-storage containers and covers, etc. (1–4, 9). In Adamawa it is often used for thatch (7). There (7) and in Togo (3) it enters into a form of semi-cultivation to provide material for these requirements. In N Nigeria, Kontagora Emirate, the culms are woven into a framework which is plastered with a filler to make beehives. For the filler, cow dung is preferred as it does not crack (6). In Ubangi (8) and the Central African Republic (5) the canes are used as splints in binding bone-fractures, and in the former territory they serve as arrow-shafts (8).

A root-decoction is given to small children in Ubangi for colic; it is also taken for blennorrhoea. Fresh root pulped to a mash is applied as a poultice to the stomach for intestinal pain, and sap from the pounded root is put up the nostrils to stop nose-bleeding (8).

The root is worn by women in Ubangi as a perfume with the added satisfaction that it is held to repel evil spirits (8).

References:

1. Broun & Massey, 1929: 440, as *Jardinea congoensis* Franch. 2. Dalziel 898, K. 3. Dalziel, 1937: 530, as *J. congoensis* Franch. 4. Fay 4445, K. 5. Fay 4446, K. 6. Irvine, s.d., a, as *J. congoensis*. 7. Peter & Tuley 51, K. 8. Vergiat, 1970: 85–86, as *J. congoensis* Franchet. 9. Ward 0091, L.104, K.

Phalaris minor Retz.

West African: **MAURITANIA** ARABIC (Hassaniya) tassala (AN)

An annual with slender culms 20–100 cm long; of waste places; originally of the Mediterranean, now worldwide, spread probably by shipborne ballast and cargo; present in Mauritania (1).

The grass, when fresh, is very poisonous and dangerous to stock (2, 3), but is safe when dried (2).

References:

1. Anderson, DE, 1961. 2. Monteil, 1953: 23. 3. Naegelé, 1958: 894.

GRAMINEAE

Phragmites australis (Cav.) Trin., **ssp. altissimus** (Benth.) WD Clayton

FWTA, ed. 2, 3: 374. UPWTA, ed. 1, 540, as *P. vulgaris* Druce.

English: common reed; reed grass.

French: roseau.

Portuguese: caniço.

West African: **SENEGAL** DIOLA ba gigimb (JB; K&A) MANDING-MANDINKA bakinto (JMD; K&A) SERER diundi (JB; K&A) gati (JB; K&A) mbeleg-kumpa (JB; K&A) WOLOF barah (JB; K&A) portöl (JB; K&A) **GUINEA-BISSAU** PEPEL oncôco (EPdS; JDES) **NIGERIA** ARABIC-SHUWA bus, buzzam (JMD) FULA-FULFULDE (Nigeria) golßi, golßol = *the tubular mouthpiece of a pipe* (JMD) wicco nyiwa (JMD) HAUSA bùshááràà (ZOG) bushaßi (JMD; ZOG) bushara (JMD) gabara (JMD; ZOG) kárán bùshááràà *the jointed culms* (JMD; ZOG) kárán macara *a rattle made from the hollow culms* (JMD; ZOG) kárán mashai *a rattle made from the hollow culms* (JMD; ZOG) kashala (Golding) wûtsíyàr gíiwáá = *tail of the elephant* (JMD; ZOG) YORUBA ifù (JMD; Verger)

A perennial reed, creeping rhizome, culm jointed, hollow, to 3½ m high or more bearing a plumose panicle, on banks and in shallow water of rivers and lakes, and swampy areas, across the northern part of the Region in Senegal, The Gambia, Niger and Nigeria, and extending through northern Kenya to Arabia and Iran, and in Spain. The type is widespread in the temperate regions of both hemispheres of the Old and New Worlds.

The plant is decorative with its feathery inflorescence and is suitable as an ornamental plant (7). It accepts a wide range of soil conditions from pH 3.8 to 7.5 (1). It invades rice-padis in N Senegal and may become a troublesome weed if not efficiently removed (2). The tough roots are good for stopping coastal erosion (ssp. *typica*) (4). When young cattle will graze it a little (3, 8), but on the whole it is not regarded as a fodder-plant, and it may be injurious (5) causing tympanites and fungal infection (12). Ssp. *typica* is a source of a manna in western USA obtained as a sort of dried sugar from sap exuded through punctures (10, 12). W African material should be examined.

The culms are used to make thatching, matting, hut-walls and partitions, screens, enclosures, pallisades and arrow-shafts (1, 4, 59). In Nigeria the hollow jointed stem is commonly used to make pipe stems, flutes and the mouthpiece of the Hausa bagpipe called *algaita*, also a kind of rattle (see vernacular name-list above) (5). Ssp. *typica* has been used in some countries for paper pulp but only as an additive to other pulps (4, 12).

The plant has been found very efficient in taking up noxious substances from drainage effluent. It has thus a well-proven and successful potential for use in treating domestic sewage (4, 14).

The root-stock is said to be edible (11, 12). It is incorporated into diuretic and diaphoretic medicines (4, 12). Alkaloids and saponins are recorded present in small amounts (6, 13).

References:

1. Adam, 1954: 97, as *P. communis* Trin. 2. Adam, 1960,a: 374, as *P. vulgaris*. 3. Adam, 1966,a, as *P. vulgaris*. 4. Cruz, 1978: 46–50, as *P. communis* Trin. 5. Dalziel, 1937: 549, as *P. vulgaris* Druce. 6. Debray & al., 1971: 74, as *P. communis* Trin. 7. Deighton 2873, K. 8. Golding 36, K. 9. Guillarmod, 1971: 446, as *P. communis*. 10. Harrison, 1950: 407–17, as *P. communis* Trin. 11. Raynal-Roques, 1978: 334. 12. Watt & Breyer-Brandwijk, 1962: 483, as *P. communis* Trin., 484 as *P. vulgaris* B.S. & P. 13. Willamen & Li, 1970. 14. Wolverton, 1982: 373–82, as *P. communis* Trin.

Phragmites karka (Retz.) Trin.

FWTA, ed. 2, 3: 374.

West African: **NIGERIA** FULA-FULFULDE (Nigeria) golbi (J&D) golbiho (J&D) HAUSA wûtsíyàr gíiwáá = *tail of the elephant* (JMD) YORUBA ifu (Millen)

A perennial reed, creeping rhizome, culm erect to 7 m high bearing plumose panicle, in or near water, rivers and marsh lowlands and into upland locations

326

throughout the Region from Senegal to N and S Nigeria, and widely dispersed across northern tropical Africa into India, Malesia and N Australia.

The plant's presence is deemed by Fulfulde of N Nigeria to indicate fertile soil (2).

The culm is used in Nigeria for arrow-shafts (4), pipe tubes (1), and to make flute mouthpieces (2). In India the plant is used to make paper, and its harvesting is a valuable village industry. Fibre is short, 0.3–3.2 mm long, so the pulp is used admixed with longer-fibred pulps, but the product is suitable for writing paper and printing. The reed is also suitable raw material for making chip-board of fair quality. It is used for thatching, matting, basketry and fish-traps, furniture, smoking pipes, flutes, etc. Culms with the flower-panicle make good brushes, and flowering stalks yield a fibre suitable for cordage. The reed is a suitable material for producing *furfural*, a valuable industrial solvent (3).

The young growth is eaten by cattle. Material from Kashmir is reported to contain 6.6% protein, 43.6% carbohydrates and 41.7% fibre (3).

The root has use in India in healing broken bones. The rhizome and root are antemetic and diaphoretic, and are used to treat diabetes. The whole plant has application against rheumatic complaints (3).

References:

1. Barter 1031, K. 2. Jackson, 1973. 3. Krishnamurthi, 1969: 33–34. 4. Millen 52, K.

Phyllostachys aervea Carr. ex A & C Rivière

English: fish-pole bamboo.

A bamboo with culms 5–8(–10) m tall by 3–5 cm diameter, lower nodes often strangely shaped; green while young, and turning grey with age; native of China, and widely dispersed in Japan, Europe and America; very hardy; now reported present at Buea, W Cameroons (3).

It is commonly cultivated in the Far East, and in China is deemed to be one of the most important garden bamboos (5), becoming feral in Japan (4). The culms have ornamental value and are used to make fishing-rods, walking-sticks, umbrella-handles, fan-handles and pipe-stems (1, 2). In Hong Kong besides those uses, it is used for furniture and handicrafts; the young shoot is eaten as a vegetable (1).

At Buea it is grown as a hedge, and its cultivation in W Cameroons has been postulated for export of the canes to Europe for garden-stakes (3).

References:

1. But & al., 1985: 69. 2. Chen & Chia, 1988: 64. 3. Dundas FHI.13903, K. 4. Suzuki, 1978: 72. 5. Wang & Shen, 1987: 53.

Poa Linn.

FWTA, ed. 2, 3: 369.

A genus of small straggling grasses, annual or perennial up to 60 cm high, of the mountains of W Nigeria, W Cameroons and Fernando Po.

P. annua Linn., *P. leptoclada* Hochst., *P. migoedii* Melderis and *P. schimperana* Hochst. are recorded from the Region, but without usage. In temperate regions *P. annua* is a weed of cultivation. *P. leptoclada* is a weed of cultivation in Kenya (3), an is grazed by all stock in the Narok District (2) and in Somalia (1). *P. schimperana* is reported to be potentally a good grazing grass in Uganda (4).

The two last-named grasses may perhaps provide browsing in the montane area of E Nigeria and W Cameroons.

References:

1. Bally & Melville 15985, K. 2. Glover & al. 1041, K. 3. Greenway 6005, K. 4. Snowden 1480, K.

Pogonarthria squarrosa (Licht.) Pilger

FWTA, ed. 2, 3: 397.
West African: NIGERIA FULA-FULFULDE (Nigeria) lammulammugel = *sheep's salt grass* (JMD)

A densely tufted perennial grass to 90 cm high of grassland, open woodland and open waste places on poor soil in Ghana to N and S Nigeria; a common grass throughout eastern and southern Africa.

Reports on the value of grazing are ambivalent. In Tanganyika cattle are said to eat it at all stages, even when dried up (7), and conversely that it is not liked by stock (5). In Zimbabwe it is reported as 'probably a good forage' (1) and taken by cattle when young (10), but also to be of little value as fodder (2). In the Transvaal it provides excellent autumn and winter grazing (4), but in the Orange Free State to be seldom touched by stock (3). Such contradictions must imply a presence or an absence of alternative more palatable species, a situation inferred in a report from Zimbabwe of the grass's presence in cattle areas as an indicator of over-grazing (8), i.e., it has become dominant over more acceptable species browsed out. Thus cattle may condition the grass biota.

The plant has unspecified medicinal use in Zambia (9). Fula in N Nigeria crush the leaves which are cooked in porridge taken as a hepatitis medicine (6). The Fula vernacular name meaning 'sheep's salt grass' suggests that the grass might be a source of a vegetable salt.

References:

1. Appleton 12, K. 2. Appleton 25, K. 3. Brandmullen 116, K. 4. Burtt-Davy 1694, K. 5. Emson 257, K. 6. Jackson, 1973. 7. McGregor 30, 31, K. 8. Mullin 54573, K. 9. Rabson Phiri 207, K. 10. Senderayi 25, K.

Polypogon monspeliensis (Linn.) Desf.

FWTA, ed. 2, 3: 374.
English: annual beard grass.
West African: MAURITANIA ARABIC (Hassaniya) gemh elfar (AN)

A tufted annual grass, culms geniculate or erect to 80 cm, of damp sites, cosmopolitan from northern Europe to India and China, introduced and naturalised in most warm temperate countries; now reported from Mauritania and Niger.

The grass has an attractive silky panicle for which it is sometimes cultivated in gardens as an ornamental (4). It is a weed of cultivation (Ethiopia, 1, 8; India, 3), even in irrigation (Ethiopia, 9). In moist localities growth becomes very lush and affords rich grazing for stock (2). Material from Jammu in India assayed at protein 11.8% dry weight, carbohydrate 38.0%, fibre 24.9% plus beneficial amounts of minerals (7), but in the arid environment of Rajasthan the grass is described as being of little value as fodder (3). Stock is recorded as grazing it in Ethiopia (5) and in Kenya (6).

References:

1. Ash 657, K. 2. Bor, 1960: 403. 3. Gupta & Sharma, 1971: 88. 4. Hubbard, 1954: 285. 5. IECAMA J-46, K. 6. Kinua 6, K. 8. Krishnamurthi, 1969: 203. 9. Mooney 6375, K. 10. Parker E-1, K.

Polytrias diversiflora (Steud.) Nash

FWTA, ed. 2, 3: 468, as *P. amaura* (Büse) O Ktze.

A perennial mat-forming prostrate grass, native of SE ASia, commonly introduced to tropical countries, and found in W Cameroons.
It is used in tropical countries as a lawn-grass (1) and is grown in this capacity at Victoria, W Cameroons (3, 4).
In Indonesia it is reported to produce fodder for cattle of moderate nutritional value (2).

References:

1. Bor, 1960: 202. 2. Burkill, IH, 1935: 675–6, as *Eulalia praemorsa* Stapf. 3. Keay FHI.28661, K. 4. Maitland 98, K.

Pseudechinolaena polystachya (Kunth) Stapf

FWTA, ed. 2, 3: 436.
West African: **SIERRA LEONE** MENDE nane (*def.* -i) (FCD)

An annual grass, basally prostrate with slender ascending culms to 30 cm high, matting, in damp and shady places in the forest at submontane to montane elevations in the Region from Guinea to W Cameroons and Fernando Po, and widely dispersed in NE, E, S central and S Africa, and throughout the tropics.
The decumbent habit of the grass results in thick matting. In Ethiopia it adventitiously forms natural ground cover under coffee (1). All stock will browse it though the quantity of leaf is sparse (Uganda: 7; Kenya: 5; Asia: 3, 6).
In Congo (Brazzaville), a decoction of the plant is put into hip-baths and vaginal douches to promote conception and to combat female sterility, and it is used in enemas for diarrhoea and dysentery (2).
The glume at maturity is armed with a bristly hook. Mende of Sierra Leone commonly apply the name *nane* to plants with prickly fruit (4). The sophistry of the use in Congo (Brazzaville) to promote fertility may perhaps be sympathetic magic.

References:

1. Amshoff 5256, K. 2. Bouquet, 1969: 130. 3. Burkill, IH, 1935: 1812. 4. Deighton 416, K. 5. Glover & al. 1884, K. 6. Krishnamurthi, 1969: 283. 7. Snowden 1183, K.

Puelia olyriformis (Franch.) WD Clayton

FWTA, ed. 2, 3: 362.

A perennial sub-woody herbaceous grass of the forest floor of the closed evergreen forest, to about 60 cm high, in Sierra Leone and Liberia, and in Central Africa from E Cameroun to Tanganyika and Angola.
The plant has a short rhizome with some roots ending in a tuber. In Congo (Brazzaville) the roots are eaten to relieve pain in the kidneys, and as a stimulant in impotence; they are also boiled with papaya roots and the liquid is drunk morning and evening, one glassful, to relieve stomach-ache (1).

Reference:

1. Bouquet, 1969: 128.

GRAMINEAE

Rhytachne gracilis Stapf

FWTA, ed. 2, 3: 511.

A slender annual grass with culms 15–60 cm tall; of sandy or shallow soil; scattered in Senegal to Ivory Coast.
It provides a little grazing for cattle in Senegal while young (1).

Reference:

1. Adam, 1966, a.

Rhytachne rottboellioides Desv.

FWTA, ed. 2, 3: 511. UPWTA, ed. 1, 541.
West African: **GUINEA** FULA-PULAAR (Guinea) sodorko (A.Chev.) **SIERRA LEONE** TEMNE a-kek (auctt.) **GHANA** KONKOMBA lipomale (JMD) **NIGERIA** HAUSA iwa (Abrahams)

A densely caespitose perennial grass with culms to 60 cm high; on shallow soils over iron-pan or in swamps and seasonally wet grassland from lowlands to montane locations throughout the Region, and generally widespread over tropical Africa, and into Brazil and the Caribbean.
In Malawi it becomes a weed of padi-field margins (4). When young it is grazed by cattle in Senegal but it is of little value (1). In northern Sierra Leone and in Guinea people make fancy baskets from it (2, 3) and Konkomba in Togo use it for making fine light straw hats (Volkens fide 2).

References:

1. Adam, 1966, a. 2. Dalziel, 1937: 541. 3. Glanville 132, K. 4. Wiehe N/140, K.

Rhytachne triaristata (Steud.) Stapf

FWTA, ed. 2, 3: 511. UPWTA, ed. 1, 541.
West African: **SIERRA LEONE** MENDE foni yambei *misapplied as this is not a true 'foni' grass* (FCD) **MALI** DOGON tɔrɔ béɛ náa yába ána = *if you can split them open take a cow, male* (CG) **NIGERIA** HAUSA iwa (Abrahams)

A slender annual grass with culms to 60 cm tall; of shallow soils over rock outcrop, disturbed sites and stream-sides; across the Region from Senegal to W Cameroons; and in Sudan and Zambia.
It is commonly a weed of arable fields. It provides palatable grazing relished by stock at all stages of growth (1–4).
In Sierra Leone, the stems are plaited into finely worked ornamental coverings, along with leather, for cigarette-tins, bottles, etc. (3).

References:

1. Adam, 1966, a. 2. Beal 23, K. 3. Dalziel, 1937: 541. 4. Ll. Williams 860, K.

Rottboellia cochinchinensis (Lour.) WD Clayton

FWTA, ed. 2, 3: 506, as *R. exaltata* Linn.f. UPWTA, ed. 1, 541, as *R. exaltata* Linn.f.
West African: **SENEGAL** MANDING-BAMBARA vaga sian (JB) SERER falèmbal, kañañar (JB) WOLOF pellen (JB) **THE GAMBIA** MANDING-MANDINKA safala (Macluskie) Saloum barra = *Saloum grass; from the area of that name in Senegal* (Pirie; JMD) **SIERRA LEONE** BULOM (Sherbro) fofo-lε (FCD) KISSI sεnsεndeŋ (FCD) KONO sεnsεnε-kaima (FCD) LOKO hahã (FCD) nyɔɔgɔ (NWT) MANDING-MANDINKA saŋwanya-musuma (FCD) MENDE fovo (*def.* fovui) *general for tall grasses* (Fisher; JMD) ngoŋgo, ngungu *said to be the 'white gungui': cf.Pennisetum subangustum (Schumach) CEHubbard*, the *'red gungui'*, (auctt.) goŋgo lεvu *when in flower, or very tall* (FCD) goŋgo lεvu-hina *male* (JMD; FCD) MENDE (Kpa) gungu (*def.* -i) (Fisher)

330

GRAMINEAE

SUSU kale, kali (auctt.) TEMNE *a-*, *anwo* (NWT; JMD) *a-kɔp-ka-runi* = *male 'kɔp'* (auctt.)
YALUNKA kalo-na (FCD) khalo-na (FCD) **MALI** FULA-PULAAR (Mali) niélo (A.Chev.) niélo
yélori (A.Chev.) MANDING-BAMBARA sian (A.Chev.) wanga (A.Chev.) SAMO kuono (A.Chev.)
SONINKE-SARAKOLE gambé (A.Chev.) **UPPER VOLTA** FULA-FULFULDE (Upper Volta) ŋeloori
(*pl.* -ooje) (K&T) GRUSI-KASENA gandjanga (Scholz) MANDING-BAMBARA sian (K&B) MOORE
kaliniaga (A.Chev.; K&B) kalinjango (Scholz) karyaga (A.Chev.; K&B) SENUFO-NIAGHAFOLO
lawula (K&B) **GHANA** DAGBANI nyehin (Ll.W) MOORE kalinyada = *horse corn* (FRI)
NIGERIA BIROM cilá (LB) FULA-FULFULDE (Nigeria) loyo (JMD) nyalo (JMD) HAUSA dààwà-
dááwàà (ZOG) dáddáwáá *applied loosely,* (JMD) gyaazámáá (auctt.) IDOMA agahama (Odoh)
IGALA àgàhámà (Odoh; RB) IGBO agumbogo (NWT; JMD) IGBO (Agukwu) àchàlà nkịtā =
elephant grass of the dog (JMD; KW) YORUBA hólo (auctt.)
NOTE: Sierra Leone names are often the same as for *Pennisetum subangustum* but are distinguished
by suffixes meaning 'male' or 'female' (FCD).

An annual grass with culms of variable height upto 2½ m tall, stilt-rooted at
the base and basal sheaths painfully hispid with pungent hairs; partially
anthropogenic near habitation, of fallows, abandoned farmland and disturbed
ground; often forming extensive prairies in the soudano-guinean zone across the
Region from Senegal to W Cameroons, and throughout the Old World tropics,
and recently introduced to the Caribbean.

The grass is a troublesome weed of cultivation, particularly on fertile well-
cultivated land. It invades rice-padis in Bendel State, Nigeria (12), and is
reported as a weed of irrigation in Somalia (13). It provides on the whole
valuable fodder, particularly while still young, which is relished by all stock
(Senegal: 1, 2; Sierra Leone: 5; Ghana: 4; Nigeria: 9, 19; Gabon: 21; Kordofan:
3; Sudan: numerous collectors; Zambia: 20), and is notably valued for horses
(Senegal: 1; Sierra Leone: 10; Ghana: 15; Nigeria: 7, 8; Gabon: 21; Sudan: 16;
Zambia: 20). In Ghana the Moore name refers to 'horse corn' (15). It will
appear in cultivated pasture and may become a nuisance, but subjected to
mowing to keep it short suitable herbage results. When mown it does in fact
produce a good quality hay (22), and since regrowth is rapid mowing is
repeatable after 6–8 weeks to give an aftermath cut as good as the first (1, 3, 9).

In The Gambia the culm is used to make matting (9, 18), and in Senegal the
haulm is a filling for mattresses (1).

A macerate of the whole plant is taken as a drink for hernia (1, 17).

The grain is plump like wheat. It is harvested in Kordofan in time of dearth
(14) and has been suggested as a potentially useful crop in Niger (11).

The plant fodder is rich in protein and is considered amongst the best of the
Gramineae (3); material from Zaïre is reported to contain *phenolics – quercetin 3*
derivatives (6).

References:

1. Adam, 1954: 97, as *R. exaltata* Linn.f. 2. Adam, 1966, a: as *R. exaltata*. 3. Baumer, 1975: 111–2,
as *R. exaltata* Linn.f. 4. Beal 20, K. 5. Cole, Ayodele, 1968, a: as *R. exaltata*. 6. Compere 2083
(Harborne 699), K. 7. Dalziel 509, K. 8. Dalziel 870, K. 9. Dalziel, 1937: 541, as *R. exaltata* Linn.f.
10. Deighton 859, K. 11. Eden Foundation 73, K. 12. Gill & Ene, 1978: 183–3, as *R. exaltata* Linn.
13. Hansen 6062, K. 14. Hunting Tech. Surveys, 1968, as *R. exaltata*. 15. Irvine 5025, K. 16. Ismail
A.3272, K. 17. Kerharo & Bouquet, 1950: 253, as *R. exaltata* Linn. 18. Pirie 49/33, K. 19. Thomas
NWT.532, K. 20. Trapnell 840, K. 21. Walker & Sillans, 1961: 193, as *R. exaltata* Linn. 22.
Williams, RO, 1949: 283–4, as *R. exaltata*.

Rottboellia purpurascens (Robyns) Jac.-Fél.

FWTA, ed. 2, 3: 506, as *Robynsiochloa purpurascens* (Robyns) Jac.-Fél.

A robust grass of swamp and deep water locations, culms to 1.6 m tall,
rooting at lower nodes; present in Guinea and Sierra Leone, and in the Congo
basin.

In the rice-growing area of Sierra Leone it is a weed of padi fields (1).

Reference:

1. Jordan 598, K.

GRAMINEAE

Saccharum officinarum Linn.

FWTA, ed. 2, 3: 466. UPWTA, ed. 1, 541.

English: sugar-cane.

French: canne à sucre.

Portuguese: cana-de-açúcar; cana sacarina.

West African: SENEGAL BEDIK sukara (FG&G) WOLOF bānta (JB) bantu u sukar (JMD) diamb u sukar (JMD; JB) GUINEA BASARI a-nyísá (FG&G) KONYAGI sukuru (FG&G) SIERRA LEONE BULOM igbáŋ (Pichl) BULOM (Kim) gbaŋ-poto (FCD) BULOM (Sherbro) gban-dɛ (FCD) FULA-PULAAR (Sierra Leone) gawudɛ (FCD) GOLA kuma (FCD) KISSI chɛŋgima-kpandeaū (FCD) KONO fa-nyɔ-kenɛ (FCD) nyɔ-kenɛ (FCD) KRIO shuga-kɛn *i.e.*, *sugar cane* (FCD) LIMBA (Tonko) gboga (FCD) LOKO sugaken (FCD) MANDING-MANDINKA likala (FCD) MENDE nyɔkɔ (FCD) SUSU khemun-yi (FCD) TEMNE ka-gbokaŋ (*pl. tsoe-*) (JMD) ka-suka-ken (FCD) VAI gbā (FCD) kpā (FCD) kpaŋ (Heydorn) UPPER VOLTA BOBO neme-wẹlo Le Bris & Prost IVORY COAST 'ANDO' ahrana Visser KRU-BETE jībɔ̀dɔ̀ Pageaud KWENI gɔɔ̀ *also meaning a warri-board* (Gregoire) GHANA ADANGME afunu (FRI) ADANGME-KROBO afungu (FRI) ahleu (FRI) AKAN-ASANTE ahwereɛ (FRI; E&A) FANTE ahwere (E&A) TWI ahwedie (E&A) ahwerew (FRI; E&A) AKPAFU ī-gbārá (*pl. ā-*) (Kropp) AVATIME ɔ̀-sɔ́ (*pl. è-*) (Kropp) GA ʃẽ (FRI) shē (KD) GBE-VHE boɣlẽ (FRI) boɣlẽ biri = *black sugar cane* (FRI) boɣlẽ fe (FRI) boɣlẽ yibɔ = *black sugar cane* (FRI) fofoŋu (FRI) VHE (Awlan) fofoŋu (FRI) fonfon (FRI) fonfongu (FRI) VHE (Pecí) boɣlẽ (FRI) boɣlẽ-biri (FRI) boɣlẽ-fe (FRI) boɣlẽ-yibɔ = *black sugar cane* (FRI) GUANG tàkàndá (Rytz) LEFANA à-ʃywérè (Kropp NZEMA akanla (FRI) SISAALA ahiria (Blass) TOGO ANYI-ANUFO tàkàndá (Krass) NIGERIA ABUA úgwẹ̀ (Wolff) AKPA ànggbó ibó (Oblete fide RB) BIROM rà'ke (LB) BOKYI bonkwunkwo (Osang fide RB) BURA lake (anon.) EBIRA-ETUNO úji (Moomo) EDO okwere (JMD) ukhuire (JMD; Elugbe) EFIK m̀bɔ̀kɔ̀k (auctt.) óbúbít m̀bɔ̀kɔ̀k *a dark coloured form* (NWT) FULA-FULFULDE (Nigeria) chife mayo (JMD) lamarudu (JMD) GWARI dakwohi (JMD) HAUSA àràkké (auctt.) dalimi (ZOG) gwalagwaji (ZOG) gyaùrón ràkéé *stool-shoots* (JMD; ZOG) kárán sárkíí = *cane of the chief* (JMD; ZOG) kuburu *a form with many short joints* (JMD; ZOG) ƙyarno (ZOG) ráake (LB) ràkéé (auctt.) IBIBIO m̀bɔ̀kɔ̀ (JMD) ICHEVE èwó ùgènyí = *European's 'ewo' (a sort of grass)* (Anape fide KW) IDOMA ẹ́mɔ̀ (Armstrong fide RB) IGALA ilèchè (RB) IGBO àchàrà mmako (NWT; KW) okpètè (auctt.) IJO-IZON (Egbema) ukpú (Tiemo fide KW) IZON (Kolokuma) izághái *the inflorescence* (KW&T) KAMBARI-SALKA àràkè (RB) KANURI rèkè (C&H; A&S) KUGBO ókpɔ̌gh (Wolff) NGIZIM sɔ̀kàr (*pl.*-ràrín) (Schuh) ODUAL ukpɔ̂gh (Gardner; Wolff) OGBIA ókpò (Wolff) OGBIA-KOLO ɔ̀lukpó (Wolff) PERO kántù (ZF) rèkè (ZF) sùgá *the product: from English* (ZF) SANGA magang (RB) SURA jèer (anon. fide KW) TIV iyelegh (*pl.* ayer) (JMD) URHOBO akele (JMD) YALA (Ikom) mangkɔng kɔng (Ogo fide RB) YALA (Ogoja) imàngkɔ̀ (Onoh fide RB) YORUBA irèkè (auctt.) WEST CAMEROONS BAFOK sosom (DA) BAFUT asïsáng (*pl. e-*) (Mfanyam) DUALA mukòkẹ (Ittmann) KOOSI nkoge (DA) KPE mukoku (DA) KUNDU ekoko (DA) LONG sosong (DA) MBONGE dikoko (DA) NGYEMBOON nkungkû' (Anderson) WOVEA mukoko (DA)

Pathology: NIGERIA HAUSA dalimi *sugar cane blight* (JMD) gwalagwaji *sugar cane blight* (JMD)

A perennial clumped grass, very polymorphic with canes of various dimensions, 1.5 to 6 cm diameter, 5 to 25 cm node-length, and 2½ to 6 m high; a domesticated cultigen grown everywhere throughout the Region, and worldwide as the commercial sugarcane.

The original species probably came from SW Pacific and from an early date has been subjected to selection from an enormous range of forms derived from bud mutations, man picking out those clones capable of yielding sweeter sap, then raw molasses and finally crystalline sugar. Plant geneticists and breeders have added to this process by introducing the genes of related species, especially *S. spontaneum*, *S. barberi* and *S. sinense*. The resultant thick canes have become known as 'noble canes' to distinguish them from the thin canes of these other species. The reader is referred to the voluminous literature on sugar-cane research (5, 6, 14).

The English word: sugar, is derived from Arabic: *sukar*, or *sukwar*, and this root appears in all European languages (11). The plant is thought to have arrived in Egypt under Arab trade from India about a millenium ago. It was known in the Greeko-Roman world, but not for making sugar (12). The cane was chewed for its sweet sap, an use still prevalent in W Africa (7, 9, 10, 13) and many parts of the world. It is commonly grown in the Region here and there in domestic gardens for casual cutting and chewing, and any surplus taken to market for sale (7). The plant is suitably adapted and could be cultivated on a larger scale for the production of sugar. Chewing the cane promotes salivation with a cleansing action synergistic with friction of the cane as a chew-stick (13). It is laxative, and Igbo of Nigeria use the plant in poultice (10). The sap enters into cooking for making sweetmeats and sweet drinks, and sometimes alcoholic drinks. It is a sweetening agent put into medicines, and in Nigeria it is given in large quantity for wasting diseases, cancer and other illnesses (2). The plant, sap or molasses, has use in a number of countries for various cancerous conditions (8). In Gabon it is added to violent purgatives (16) and is recognised as an antidote against ordeal-poisoning by *Strychnos ijaca* (Loganiaceae), for which it has to be taken admixed with human faeces! (17, 18).

A trace of *alkaloid* has been detected in the stem of Nigerian material (1). Under some conditions *hydrocyanic acid* is produced (18). Flavonoids and vitamins are present in the leaves (10).

There is a wax on the surface of the cane which at one time was extracted from the bagasse after the sugar had been taken out. This wax should have many industrial applications if available in adequate quantity, and of improved colour, light colours being preferred. Yield appears to be about 2 lb per ton of cane (3, 4).

A variety with red leaves is credited in Gabon with magical powers of protection when planted alongside village paths (17). Yoruba invoke it for the sweet satisfaction of being victorious over an enemy (15).

References:

1. Adegoke & al., 1968: 13–33. 2. Ainslie, 1937: sp. no. 306. 3. Anon., 1942, b: 11–12. 4. Anon., 1943, c: 86–90. 5. Burkill, IH, 1935: 1925–40. 6. Chadha, 1972: 103–65. 7. Dalziel, 1937: 541. 8. Hartwell, 1969, b: 195. 9. Irvine, 1948: 265. 10. Iwu, 1986: 144. 11. Lemordant, 1971: 144. 12. Mauny, 1953: 694–5. 13. Portères, 1974: 136. 14. Purseglove, 1972: 215–56. 15. Verger, 1967: sp. no. 120. 16. Walker, 1953, a: 40. 17. Walker & Sillans, 1961: 193–4. 18. Watt & Breyer-Brandwijk, 1962: 484.

Saccharum spontaneum Linn.

FWTA, ed. 2, 3: 466. UPWTA, ed. 1, 542.
West African: SIERRA LEONE TEMNE *ka*-fon (FCD) NIGERIA BIROM mòs (LB) sáá kwɛt = *partner of the arrow* (LB) FULA-FULFULDE (Nigeria) kaulewol (*pl.* kauleji) (JMD) GWARI jibwi (JMD) HAUSA abóókin kíbíiya = *partner of the arrow* (LB) àbóókin kíbíyàà (JMD; ZOG) falfoli *applies generally to pithy-stemmed grasses* (JMD) kíbiyàà (ZOG) kyámróó (auctt.) kyaúrón kíbíyaá = *arrow reed* (JMD; ZOG) kyáúróó (LB) sansari (JMD; ZOG) shasari (JMD; ZOG) sheme (JMD; ZOG) JUKUN (Wukari) kwekyàn (Shimizu) KAMBARI-KIMBA kàkyàndá (RB) SALKA àakùtsû, àakùtsû (RB) TIV igomough ki ikaregh (JMD) YORUBA eèsú (auctt.) esùsú (JMD)
NOTE: Hausa names probably refer to the culm and its uses rather than as a designation of the plant itself.

A perennial, rhizomatous, stool-forming culms bearing large white silky-plumed inflorescences; usually of river sand-banks; in Upper Volta to N and S Nigeria, and extending to N and NE Africa along the Mediterranean into Spain, and eastwards into E Africa, the Middle East, India and eastern Asia.

The plant is very polymorphic. Small forms may have culms only 35 cm high; large forms attain 8 m. The culms are hardy, pithy, often hollow and with little juice. Bud-sporting is known. Chromosome number varies from 2n = 40 to 128. It has been used by plant breeders as valuable gene material in the ennoblement of sugar-cane (2, 3, 8).

The plant is an efficient soil stabiliser. It is commonly used in N Africa (Morocco, Algeria, Libya, Sahara: 1, 4) and in India (5) to fix shifting sand-dunes. It makes a good wind-break and screen for plantations and re-afforestation in sandy areas (4). The canes are used in the soudanian zone to make arrow-shafts; also to make beds and fish-traps, and sometimes for roofing and thatching (4, 6). Cuts by the sharp leaf-edges as a result of handling may lead to dermatitis (3). In Uganda in the days before the general availability of imported salt, the leafy canes were calcined and a vegetable salt was extracted from the ash (7). The canes are a potential source of paper-pulp (5).

Cattle will eat the foliage if it is still tender, and in India it is considered to be a favourite fodder for buffaloes (5). In Kordofan grazing is restricted to canes not over 1 m high, and the danger of feeding them at certain times is recognised because of the liability of fermentation (1).

The tender shoots are cooked and eaten as a vegetable in Java (9).

The grain is collected in Uganda for human consumption (10).

References:

1. Baumer, 1975: 112. 2. Burkill, IH, 1935: 1925–40, under *S. officinarum*. 3. Chadha, 1972: 103–. 4. Dalziel, 1937: 542. 5. Gupta & Sharma, 1971: 89. 6. Hunting Tech. Services, 1964. 7. Maitland 19, K. 8. Purseglove, 1972: 214–5. 9. Raynal-Roques, 1978: 327–43. 10. Wilson 235, K.

Sacciolepis africana CE Hubbard & Snowden

FWTA, ed. 2, 3: 425. UPWTA, ed. 1, 542, as *S. interrupta* sensu Dalziel. West African: THE GAMBIA MANDING-MANDINKA nyara fingho (Macluskie) GUINEA-BISSAU PEPEL umpôlpôl (JDES) SIERRA LEONE KORANKO birikolonosɛfine (NWT) LOKO hɔmbɔ (NWT) kintibo (NWT) MANDING-MANDINKA kɔbɔrɔ (FCD) MENDE be (NWT) SUSU kuli (NWT) TEMNE ɛ-linke (NWT) YALUNKA tamedi-sara-na *applies to Acroceras and probably to similar fodder grasses* (FCD) LIBERIA MANO bu su (Har.) MALI ? (Mopti) niépoto (Matthes) UPPER VOLTA MOORE kulugu savandé (Scholz) NIGERIA HAUSA babaci (auctt.) bubuci (JMD; ZOG)

A rhizomatous perennial grass, with thick spongy culms, decumbent, rooting at lower nodes in swamps and shallow water, across the Region from Senegal to N and S Nigeria, and throughout tropical Africa.

The habit of the plant suggests an use in mud-binding in alluvial swamps (2). It furnishes good fodder relished by stock at all times (1, 2). In Sierra Leone it is fed in the dry season mixed with *Acroceras spp.* (Gramineae) to horses (3). Though people in N Sierra Leone recognise this grass and *Aroceras spp.* as being different yet similar and growing in the same sort of habitat, they use the same names (see above). It is a weed of rice-padis in the Sokoto River area of NE Nigeria (5), and in Sudan (8) where it is said to appear late after most other weeds. It is a weed of corn-fields in Ethiopia (7).

The green spikelets produce a lather when rubbed in water, and are thus used as soap in Malawi (9).

The grain is collected in N Nigeria for consumption in time of dearth (2, 5). A similar report from E Africa (6) regarding *S. interrupta* Stapf, an Asian species, may also refer here. Like certain cultivars of *Oryza*, the culm can extend to as much as 2 m with rising flood water (5).

Examination for improved grain and fodder yield is recommended (4).

References:

1. Adam, 1966,a: as *S. interrupta*. 2. Dalziel, 1937: 542, as *S. interrupta* Stapf. 3. Deighton 4541, K. 4. Irvine, 1952,a: 40. 5. Lamb 101, K. 6. Raynal-Roques, 1978: 337, as *S. interrupta* Stapf. 7. Stewart 88, K. 8. Van Eyck, 4, K. 9. Wiehe N/497, K.

Sacciolepis cymbiandra Stapf

FWTA, ed. 2, 425.
West African: SIERRA LEONE MENDE yɔli (NWT)

An erect perennial grass with a spongy culm to 1 m high of marshy places and in water from Guinea-Bissau to N and S Nigeria.
It is a waterside grass grazed by stock at all times (1).

Reference:

1. Beal 45, K.

Sacciolepis indica (Linn.) A Chase

FWTA, ed. 2, 3: 425, as *S. auriculata* Stapf.

A decumbent or ascending annual grass, culms rooting at the nodes to 1 m high, of wet places up to montane altitudes in Sierra Leone to N and S Nigeria, and in NE, E, S tropical and SW Africa, and across the Indian sub-continent and SE Asia.
In Asia it is recorded as forming part of good grazing land and that cattle are fond of it, but the yield is small and it is thus rarely used though dietetic analysis is good (2, 3).
In Kenya it is reported to be a weed in maize-fields (1), and in Zambia in rice-padis (4).

References:

1. Bogdan AB.5309, K. 2. Burkill, IH, 1935: 1940–1. 3. Chadha, 1972: 165. 4. Mortimer 68079, K.

Sacciolepis micrococca Mez

FWTA, ed. 2, 3: 425.

An annual grass, tufted culms to 70 cm high, of damp and swampy places at mid-montane elevations from Senegal to N and S Nigeria, and widespread across tropical Africa.
The plant is a weed of cultivation, commonly in rice-padis.

Sacciolepis myosuroides (R Br) A Camus

A slender, erect or creeping perennial grass of wet sites up to 1,800 m altitude in the Indian sub-continent, and SE Asia, now reported introduced to Senegal (1).
It is a good fodder relished by cattle (1, 2), and also for buffaloes (2).

References:

1. Adam, 1966,a. 2. Chadha, 1972: 166.

335

GRAMINEAE

Sacciolepis typhura (Stapf) Stapf

An erect tufted, rhizomatous grass, culms to 1½ m high, of wet sites, and occurring from Senegal to W Cameroons, and widely dispersed over tropical Africa especially towards S Africa.

In the Pankshin District of Nigeria it provides some grazing for cattle (1).

Reference:

1. Saunders 12, K.

Sacciolepis sp. indet.

West African: NIGERIA FULA-FULFULDE (Nigeria) tappo (J&D)

An unidentified species, known to the Fula cattle-men of N Nigeria, and thus likely to be of value as cattle-feed.

Schizachyrium brevifolium (Sw.) Nees

FWTA, ed. 2, 3: 478–9. UPWTA, ed. 1, 542, under *S. platyphyllum* Stapf.
West African: SENEGAL WOLOF hat u blek (JB) GUINEA-BISSAU PEPEL enchèrè (EPdS; JDES) NIGERIA HAUSA jan rauno = *red 'rauno'* (Taylor) takrumo (Freeman) zomo (Freeman)

A delicate straggling annual grass, culms to 60 cm high; usually found growing beneath taller grass species, or in shaded sites on damp ground; a variable species of var. *brevifolium* dispersed generally over the whole Region and widespread in the tropics, and var. *flaccidum* (A Rich.) Stapf in Guinea, N and S Nigeria and W Cameroons, and into other parts of tropical Africa and India.

It is a weed of abandoned cultivations and a roadside ruderal. It provides grazing at all stages of growth, but is lacking in bulk (1, 2, 5). Nutritional value is said to be good in Senegal (1) but indifferent in SE Asia (3, 4).

The leafage is used in the Central African Republic as a stuffing for beds, having a local name meaning 'keeps chest agreeable' (6). Whether this is as a physical comfort or for a medicinal benefit is not clear.

References:

1. Adam, 1954: 97. 2. Adam, 1966, a. 3. Burkill, IH, 1935: 1975. 4. Chadha, 1972: 249. 5. Dalziel, 1937: 542. 6. Fay 4479, K.

Schizachyrium exile (Hochst.) Pilger

FWTA, ed. 2, 3: 479. UPWTA, ed. 1, 542.
West African: SENEGAL MANDING-BAMBARA holo (JB) SERER mãmbil hon (JB) mbati o gay (JB) ndiébèl (JB) pèrlãnd (JB) UPPER VOLTA FULA-FULFULDE (Upper Volta) hudo wodeeho = *red grass: applies to several grass spp.* (K&T) hudo-nai (Scholz) GRUSI-KASENA uonga (Scholz) MOORE bomodo (Scholz) NIGER SONGHAI sùb cìréy (*pl.* -à) (D&C) NIGERIA ARABIC-SHUWA geishsh an naar (JMD) BUSA don'si (Ward) FULA-FULFULDE (Nigeria) tidingho (JMD) HAUSA farin shibei = *white 'shibei'* (Kennedy) gamba (BM) geronu (Golding; ZOG) jan bako (auctt.) ján-baíyé (ZOG) ján-baújéé (JMD; ZOG) ján-rámnóó (auctt.) ján-ráno (JMD; ZOG) ján-raúnóó = *red 'rauno'* (auctt.) kùmbùrà tsuúlí = *swell the anus* (ZOG) kyauria (Ward) KANURI ngármásàm (C&H; A&S) ngirmasan (JMD) NUPE dogó (JMD) TIV acho *a general term* (JMD) YORUBA (Ilorin) lẹlẹuriẹ (Ward)

A slender annual grass, culms erect, 60–120 cm tall; on poor, dry, disturbed, gravelly soils, open places and bordering bush and woodland; throughout the soudanian zone of the Region from Mauritania to N and S Nigeria, and over the rest of tropical Africa and into India and Thailand. The grass is a common weed in fields of ground-nuts and sorghum in The Gambia (11). It is abundant on roadsides and field-edges where it is believed in Ghana to be a possible host of *Striga aspera* (Scrophulariaceae) (10). It provides indifferent grazing for stock, taking it only while it is still young (1–5, 8, 9). The mature grass is widely used for making coarse matting (5, 7, 8), for thatching (6, 8, 9, 12) and, being chopped up and mixed with mud, for use as a building material (1, 3, 6, 7). Such material is known in Hausa as *datse* in Katsina, and *datsi* at Kano. The term *rauno* also is loosely applied, but is properly the analagous product made from rice or wheat straw, or *Digitaria exilis* (Gramineae), q.v. (7).

The grass turns a russet-red in the dry season, evoking the Hausa *jam* or *jan*: red, in the local name. Other grasses of similar appearance are also so called (7).

References:

1. Adam, 1954: 97. 2. Adam, 1966, a. 3. Baumer, 1975: 112. 4. Beal 8, K. 5. Chadha, 1972: 249. 6. Dalziel 259, K. 7. Dalziel, 1937: 542. 8. Golding 5, K. 9. Kennedy FHI.8022, K. 10. Parker 2030, K. 11. Terry 3166, K. 12. Ward 0027, K.

Schizachyrium maclaudii (Jac.-Fél.) ST Blake

FWTA, ed. 2, 3: 479.
West African: SIERRA LEONE MANDING-MANDINKA fulufala (FCD) TEMNE kirin-kə-ro-lal (NWT) *ka*-soe Glanville YALUNKA funfure-na (FCD)

An annual grass, similar to *S. brevifolium*; dispersed in the Region from Guinea to W Cameroons, and occurring also in French Guiana and Colombia. It provides good pasture in Sierra Leone (1).

Reference:

1. Glanville 65, K.

Schizachyrium nodulosum (Hack.) Stapf

FWTA, ed. 2, 3: 479.
West African: UPPER VOLTA FULA-FULFULDE (Upper Volta) hudo-nai (Scholz)

An annual grass to about 30 cm high; of rocky and shallow gravelly soil; in the soudanian zone across the Region from Senegal to N and S Nigeria.

It is similar to *S. urceolatum*, and Fula herdsmen in Upper Volta do not distinguish them apart.

Schizachyrium platyphyllum (Franch.) Stapf

FWTA, ed. 2, 3: 478. UPWTA, ed. 1, 542.
West African: SENEGAL WOLOF khat diambor (JMD) khat lek (JMD)

An annual or perennial grass with culms to 1.2 m high; as an understorey of taller grass species in marshy places; across the Region from Senegal to W Cameroons, and in NE, E and central Africa.

It is grazed by cattle providing fairly good fodder in the wet season (1, 2).

References:

1. Adam, 1966, a. 2. Dalziel, 1937: 542.

GRAMINEAE

Schizachyrium pulchellum (Don) Stapf

FWTA, ed. 2, 3: 481.
West African: SENEGAL MANDING-BAMBARA bin blé (JB)

A prostrate creeping perennial grass with rampant stolons, culms to 1½ m high; of coastal dunes and sandy sites; around the coastline from Senegal to S Nigeria, and on to the Congo estuary.

The vigorous stolons are nodal-rooting and spreading over coastal sand-dunes help to fix them. It is capable of covering large areas, even down the beach to high-water level (1–5, 8). It seems to act in concert with other sand-binding plants. In the Lagos area it is reported to come in as a follow-up coloniser after "convolvulus" (6), probably *Ipomoea pes-caprae*. On the Zaïre coast it penetrates established *Scaevola plumeri* (Goodeniaceae) coppices anchoring the sand between the plants (7).

References:

1. Adam, 1954: 97–98. 2. Akpabla 1879, K. 3. Deighton 5670, K. 4. Killick 259, K. 5. Morton A.1781, K. 6. Stubbings 11, K. 7. Vanden Berghen, 1979: 185–238. 8. Wheeler Haines 152, K.

Schizachyrium ruderale WD Clayton

FWTA, ed. 2, 3: 481.
West African: UPPER VOLTA BISA hutarda Scholz GRUSI-KASENA uonga Scholz

A loosely tufted annual grass, culms to 2 m high; of roadside spoil-heaps and new fallow land; of the soudanian zone from Senegal to Dahomey.

It provides indifferent browsing for stock, and then only when young (1).

Reference:

1. Beal 43, K.

Schizachyrium rupestre (K Schum.) Stapf

FWTA, ed. 2, 3: 481.

A caespitose perennial grass, culms to 1 m high; on stony slopes; across the Region from Senegal to N and S Nigeria.
Cattle in Senegal graze it while it is still young (1).

Reference:

1. Adam, 1966, a.

Schizachyrium sanguineum (Retz.) Alston

FWTA, ed. 2, 3: 479. UPWTA, ed. 1, 542, as *S. semiberbe* Nees.
West African: UPPER VOLTA BISA hutarda (Scholz) FULA-FULFULDE (Upper Volta) yantaare (*pl.* jantaaje) (K&T) GRUSI-KASENA uonga (Scholz) NIGERIA BUSA gadega (Ward) HAUSA ján datse (ZOG) kyaure (Ward) YORUBA (Ilorin) pasi (Ward)

A tufted perennial grass, culms to 3 m tall, very polymorphic; of a wide range of habitats from stony hillsides to wet stream-sides; common throughout the whole Region, and in E Africa and throughout the tropics.

The grass is in general considered to give good fodder for cattle at all times (5; Senegal: 1; Ghana: 2; Nigeria: 7; India: 4), especially the wet season, or where it grows near streams and there remaining evergreen. But under certain unspecified conditions cattle are recorded as refusing it, even as tender young growth (Ghana: 6; Kaiama, N Nigeria: 9; Malaya, indifferent grazing: 3). With its robust culms, it is commonly used as a thatching material (5, 8–11). On the

338

Vodni Plateau, Nigeria, matting and *zaanaa* screens are made of them (7). They are suitable for paper-pulp (5, 11).

References:

1. Adam, 1966, a, as *S. semiberbe*. 2. Beal 50, K. 3. Burkill, IH, 1935: 1975, as *S. semiberbe* Nees. 4. Chadha, 1972: 249–50, as *S. semiberbe* Nees. 5. Dalziel, 1937: 542, as *S. semiberbe* Nees. 6. Rose Innes 30365, 30384, 31598, K. 7. Saunders 11, K. 8. Vigne 4643, K. 9. Ward 0026, K. 10. Ward L.146, 0059, K. 11. Watt & Breyer-Brandwijk, 1962: 493, as *S. semiberbe* Nees.

Schizachyrium scintallans Stapf

FWTA, ed. 2, 3: 481.
West African: SIERRA LEONE MENDE foni *an ecological term*(FCD)

A slender annual grass to 60 cm high; on ironstone pans; in Sierra Leone. The Mende name, *foni*, is properly applied to the community of plants growing on the wet stony pans, but this grass is the first coloniser of such habitats and thus is given the name (1).

Reference:

1. Deighton 2175, K.

Schizachyrium urceolatum (Hack.) Stapf

FWTA, ed. 2, 3: 479.
West African: UPPER VOLTA FULA-FULFULDE (Upper Volta) hudo-nai (Scholz)

An erect annual grass to 30 cm high; of shallow stony soil; in the soudanian zone across the Region from Senegal to N Nigeria.
It is similar to *S. nodulosum*, and Fula herdsmen in Upper Volta do not distinguish them apart.

Schmidtia pappophoroides Steud.

FWTA, ed. 2, 3: 383.

A tussock, rhizomatous grass, culms to 90 cm high, of dry stony and sandy places of open bushland in Mauritania and Senegal, and widespread in NE to S Africa.
It is reported to afford, grazing in Somalia (2), Uganda (1), Kenya (4) and Botswana (3), but it is of little importance.

References:

1. Eggeling E5819, K. 2. Fausset 45, K. 3. Smith, PA 3116, K. 4. Wilson 22, K.

Schoenefeldia gracilis Kunth

FWTA, ed. 2, 3: 402. UPWTA, ed. 1, 543.
West African: SENEGAL FULA-PULAAR (Senegal) suarelalale (A.Chev.) MANDING-BAMBARA furabã (JB) ulukumissé (JB) urga (A.Chev.) SERER a las ndiad (JB) golombale (A.Chev.) ndiad (JB) salombale (A.Chev.) WOLOF gèn u gold (JB) pellen (A.Chev.) rov (JB) MALI DOGON sáána yà (CG) FULA-PULAAR (Mali) burdi (A.Chev.) guare lalale (A.Chev.) lakio wanduho (A.Chev.) MANDING-BAMBARA furala (A.Chev.) ulukumissé (A.Chev.) urga (A.Chev.) UPPER VOLTA FULA-FULFULDE (Upper Volta) ɓuruuɗi *loosely for certain annual spp.* (K&T) nyomre (K&T) nyoobe (K&T) sarao (K&T) selɓo (*pl.* celɓi) (K&T) GRUSI tugu légan (A.Chev.) MOORE lamyondon (A.Chev.) silazuré (A.Chev.) NIGER FULA-FULFULDE (Niger) geenal = *hay;* when dried up (ABM) huɗo rimo (ABM) SONGHAI sùb káaréy (*pl.* -à) (D&C) NIGERIA HAUSA kakuma (ZOG) rumayza (ZOG) rumaza (ZOG) sáwún kádàfkaràà = *footprint of the*

bustard (auctt.) shinaka (auctt.) tafa ɗuwa (JMD; ZOG) wútsíyàr ɓééráá = *mouse's tail* (auctt.) wútsíyàr bírìi = *monkey's tail* (auctt.) wútsíyàr ƙádangare = *house lizard's tail* (JMD) wútsíyàr kúúsù = *mouse's tail* (JMD)

A tufted annual grass, culms slender to 1 m high, of sandy areas, soils and hard pans of the Sahel/Soudanian zone across the Region from Mauritania to N Nigeria and eastwards to the Red Sea, and also in India.

The digitate spikes of the inflorescence and the conspicuous braiding of the spiklet awns evoke fanciful names (see list above) in Nigeria.

The grass provides indifferent browsing taken mainly whilst growth is still young (1, 2, 5, 10, 14) and especially by sheep and goats (3, 7). Cattlemen in Niger, however, claim that it makes excellent forage for cattle during the rains. At that time while the culm remains soft it is known as *huɗo rimo*, but after the last rains it dries up quickly and is then known as *geenal*, terms which generally designate any fresh grass or any dried hay respectively (12). In India it is considered good fodder (9) and especially for camels.

The grass is commonly used for thatching huts (N Nigeria: 6–8, 13, 15; Kordofan: 3; India: 4). It is also roved into cordage (4; 6), being used in Kordofan for tying bundles of forage and fastening loads to pack-animals (3). In this area, also, the culms are bundled to make brushes.

References:

1. Adam, 1954: 98. 2. Adam, 1966,a. 3. Baumer, 1975: 112. 4. Chadha, 1972: 254. 5. Chevalier 33898, K. 6. Dalziel 506, K. 7. Dalziel, 1937: 543. 8. Golding 7, K. 9. Gupta & Sharma, 1971: 89–90. 10. Harrison 104, K. 11. Hunting Tec. Services, 1964. 12. Maliki, 1981: 51. 13. Palmer 6, K. 14. Roberty, 1953: 452. 15. Taylor 5, K.

Sehima ischaemoides Forssk.

FWTA, ed. 2, 3: 476. UPWTA, ed. 1, 543.
West African: **MALI** TAMACHEK allomoze (JMD) **NIGER** SONGHAI sùb káaréy (*pl.* -á) (D&C)

A tufted annual grass, culms to 60 cm high; on dry soils in deciduous bushland or sub-desert grassland, sub-montane to montane; in Mali and N Nigeria, and into Sudan and E Africa and S tropical Africa, and eastwards to Pakistan and India.

The grass is a good fodder in sub-desert sandy areas, growing during the brief wet period (1, 2). In Kordofan it is browsed at all stages. It grows on saline soils in Rajasthan, India, and is utilized for soil conservation (3). It is palatable to cattle.

References:

1. Baumer, 1975: 113. 2. Dalziel, 1937: 543. 3. Gupta & Sharma, 1971: 90.

Setaria barbata (Lam.) Kunth

FWTA, ed. 2, 3: 424. UPWTA, ed. 1, 542.
English: bristly foxtail grass (general for *Setaria*).
West African: **SENEGAL** MANDING-BAMBARA diièyba (JB) **GUINEA-BISSAU** BALANTA ntchacufalo (JDES; EPdS) untchancufalo (JDES) FULA-PULAAR (Guinea-Bissau) udetchoboi (EPdS) udetcholói (JDES) MANDING-MANDINKA bara (JDES) **SIERRA LEONE** KORANKO kwɔfiɛ (NWT) sibule (NWT) LOKO yarangi (NWT) **NIGERIA** FULA-FULFULDE (Nigeria) kosabere (Saunders; JMD) HAUSA gambadawari (Kennedy) IGBO ele ạdodo (NWT; JMD) iloloriwanga (NWT)

A loosely tufted annual, culms ascending to 1½ m high, weed of disturbed land, shade-loving, from sea-level to montane situations throughout the Region from Senegal to Fernando Po; primarily a W African species, but occurring in NE, E, central and SW Africa, and perhaps introduced elsewhere.

The grass is an acceptable fodder for all stock, especially horses and donkeys (1–3, 5, 6). In the savanna it tolerates light shade (1). In Sudan (7) and Kordofan (3) it is held to be an important dry season grass; stock will take it in the wet and the dry season even though the spikelets bear irritant bristles.

In Congo (Brazzaville) the plant-sap is applied in massages to epileptics (4).

The Ejagham people reckon that the dry season starts from the time of flowering of this grass (Talbot fide 5).

References:

1. Adam, 1954: 98. 2. Adam, 1966,a. 3. Baumer, 1975: 113. 4. Bouquet, 1969: 130. 5. Dalziel, 1937: 543. 6. Kennedy FHI. 8017, 8026, K. 7. Peers KO.32, K.

Setaria incrassata (Hochst.) Hack.

FWTA, ed. 2, 3: 423, as *S. ciliolata* Stapf & CE Hubbard.

A tufted perennial, briefly rhizomed, culms to 60 cm tall, on heavy clay soils but also in a variety of habitats from savanna to stony hillsides and margins of evergreen forest; from lowlands to upland situations in N Nigeria, and widespread to Ethiopia, E Africa and southwards to S Africa.

No usage is recorded in the Region. In Sudan, E and S central tropical Africa it is commonly recorded as grazed in all territories, often being suitable all the year round and deemed very valuable fodder. The culms in eastern Africa grow appreciably larger, to 2 m long, than in W Africa, and are used in Sudan in hut-building (2) and to make rope (1). They are generally used for thatching in Sudan (4), Kenya (5, 6) and Tanganyika (7), and in the last-named territory for basket-making (3).

In Sudan (8) and in Kenya (6) birds seek out the grain to eat.

References:

1. Andrews 3276, K. 2. Beshir el Nayer 61, K. 3. Bullock 2543, K. 4. Evans-Pritchard 13, 18, K. 5. Glover & al. 1556, 1993, 2245, K. 6. Magogo & Glover 771, K. 7. Michelmore O.1589, K. 8. Simpson 7052, K.

Setaria longiseta P. Beauv.

FWTA, ed. 2, 3: 423. UPWTA, ed. 1, 543.

West African: **NIGERIA** FULA-FULFULDE (Nigeria) shabalaho (Saunders; JMD) IGBO ilulo okweliokwokwo (NWT) odu lili (NWT) YORUBA àse-olongo *from* olongo: *a species of bird* (JMD)

A loosely tufted perennial grass, culms 60–90 cm high, of shady places on pathsides, fallow land and riverine locations, from Guinea to N and S Nigeria, and to E Africa.

It is appreciated for grazing by stock (1, 2), and is suitable for conversion into hay (1).

Igbo at Onitsha, S Nigeria, prepare from it a drink given to children in the dry season against fever (3).

References:

1. Dalziel, 1937: 543. 2. Snowden, 1093, 1143, K. 3. Thomas NWT.1861 (Nig. Ser.), K.

Setaria megaphylla (Steud.) Dur. & Schinz

FWTA, ed. 2, 3: 424 incl. *S. chevalieri* Stapf. UPWTA, ed. 1, 543 as *S. chevalieri* Stapf.

English: horse grass (Sierra Leone, Cole); buffel grass (S Africa).

GRAMINEAE

West African: **SIERRA LEONE** KISSI fɔyɔndo (FCD) KONO bobo, bobo-yamba (FCD) KRIO hɔs-gras *i.e., horse grass* (FCD) LOKO bobo, *m*bobo (NWT; JMD) *m*boworo (FCD) ndɔgɔbeni (NWT) tira (NWT) MANDING-MANDINKA tukɔdɔbĩ (FCD) MENDE *m*bovi (Cole) *m*bowi (auctt.) *m*bowo (FCD) *m*bowo-la *the leaf, but usually meaning the whole plant,* (FCD) njɔpo-bowi *from* njɔpo: *fallow ground* (JMD) SUSU foni *a general term* (NWT) furudevakali (NWT) koseaxuli (NWT) xɔriexuli (NWT) TEMNE kə-(g)bil (NWT) kə-bil-kə-leŋ (NWT) *ka*-fonte (FCD) *an*-fonte (Glanville) kə-roi (NWT) YALUNKA wogowaga-na (FCD) **LIBERIA** MANO ka (Har.) **IVORY COAST** ? djuaya (A&AA) ABURE moya moya (B&D) AKAN-ASANTE aboigna (B&D; E&A) AKYE hintsun (A&AA) kotsin-té (A&AA) BAULE abobonia (B&D) maka (B&D) KYAMA aboya (B&D) aguan (B&D) MANDING-MANINKA denzenbré (B&D) **GHANA** AKAN-ASANTE awaha (FRI) TWI awaha (E&A) awaha (FRI; E&A) GBE-VHE wadjere (Thompson) **NIGERIA** EDO ọkẹ́shín *corn of horses* (Kennedy; Elugbe) IJO-IZON (Kolokuma) akáráká (KW; KW&T) YORUBA ọkà-ẹṣin (JMD; Verger) **WEST CAMEROONS** DUALA èkòko enùmb'a pwèèpwèè (Ittmann)

A coarse perennial grass, clumped, culms to 3 m high, stout, in shady and damp sites on forest margins in all territories of the Region, and common throughout tropical and S Africa, in tropical America and occasionally found in India.

This is the commonest of the tall grass species of Africa. Authorities have previously recognised *S. chevalieri* Stapf as a separate species with a drooping panicle and flexuous lower branches distinct from a stiff erect panicle in *S. megaphylla*. But these conditions intergrade and the two entities are now treated as conspecific (5). Reported usages are more or less the same.

The grass in general is a good forage appreciated by all stock (1a, b, 7–11, 14–17). There are reports though of it causing scouring (1a). It is good for cutting for stall-feeding, yet such use requires care as there are reports of wilted grass producing *hydrocyanic acid* (21b). Clearly careful study is necessary for a full evaluation as fodder (1b). In Kenya thickets of the grass are reported to harbour tsetse fly. Thus stock browsing therein are at risk of contracting trypanosomiasis (10).

The woody culms are used in hut-building (1b–7).

The whole plant reduced to ash produces a vegetable salt used in Ubangi (18). In Ivory Coast the ash is applied to sores, and to assuage the pain caused by the spit venom of *Naja nigricollis*, the spitting, or black-necked cobra, said to be the most dangerous snake of Africa (4a).

The plant has anodynal and analgesic properties. Zulu in S Africa apply crushed leaves to bruises (21a). In Congo (Brazzaville) sap is massaged into areas of pain. For more vigorous action the affected part may be scarified by rubbing with the rough leaf, and ash of the calcined plant applied (3a, b). A leaf-decoction is put into a bath or given by mouth in Ivory Coast to babies suffering convulsions or fits of epilepsy (4b). Leaf-sap together with that of *Dracaena steudneri* (Dracaenaceae (Agavaceae), an E African species, not in W Africa) is given in Tanganyika for mental derangement (12). A leaf-decoction is sedative on cough, and is indicated for oedema (4b). Ijo in SE Nigeria rub leaves crushed with salt on the forehead for headache, and squeeze the sap on to a sore after it has been cleaned (22).

The grass has a reputation for beneficial action on urino-genital troubles. A leaf-decoction is given in Ivory Coast for amenorrhoea and blennorrhoea (4b). In Gabon crushed leaves macerated with some chips of bark of a *Croton* species (Euphorbiaceae) is used in draught or douch for blennorrhoea (19, 20). In Tanganyika the root decocted in palm-wine or pineapple juice or water with sometimes *Senna occidentalis* (Leguminoseae: Caesalpinioideae) or *Ficus sp.* (Moraceae) added is taken 1 glassful x 3 daily for blennorhoea, while the same preparation is given to a pregnant woman to ease delivery (3a). The root in Ubangi is held to be abortifacient (18). The plant with *Cissus aralioides* (Vitaceae) and *Selaginella sp.* (Selaginellaceae) is used in Congo (Brazziville) to prepare a bath for someone with fever, especially if the fever is due to meeting an evil spirit (3a). In cases of listlessness and sleepiness a root-decoction is given in Ubangi as a pick-me-up, and after downing it, the patient's face is washed with a leaf-decoction (18).

The broadly pleated linear leaves are used in Ghana to wrap (? for cooking) plantains (13).

The grain in S Africa has been reported as toxic to small birds (1b, 7, 21a, b).

References:

1a. Adam, 1954: 98. 1b. Adam, 1954, as *Setaria chevalieri* Stapf. 2. Adam, 1966, a, as *S. chevalieri* Stapf. 3a. Bouquet, 1969: 130. 3b. Bouquet, 1969: 130 as *S. chevalieri* Stapf. 4a. Bouquet & Debray, 1974: 93. 4b. Bouquet & Debray, 1974: 92, as *S. chevalieri*. 5. Clayton & Renvoize 1982: 540. 6. Cole, 1968,a, as *S. chevalieri*. 7. Dalziel, 1937: 543, as *S. chevalieri* Stapf. 8. Glanville 56, K. 9. Glover & Samuel 2883, K. 10. Glover & al. 409, K. 11. Glover & al. 1987, K. 12. Haerdi, 1964: 212, as *S. chevalieri* Stapf. 13. Irvine, 1586, K. 14. Kennedy, 1748, K. 15. Lane-Poole 451, K. 16. Magogo & Glover 787, K. 17. Snowden 1152, 1216, K. 18. Vergiat, 1970,a: 87. 19. Walker, 1953,a: 40. 20. Walker & Sillans, 1961: 194. 21a. Watt & Breyer-Brandwijk, 1962: 485, as *S. sulcata* Raddi. 21b. Watt & Breyer-Brandwijk, 1962: 484, as *S. chevalieri* Stapf. 22. Williamson, K, s.d.,a: as *S. chevalieri*. Stapf.

Setaria palmifolia (Koen.) Stapf

FWTA, ed. 2, 3: 424, as *S. chevalieri* sensu Clayton, in part. UPWTA, ed. 1, 543, as *S. chevalieri* sensu Dalziel, in part.

West African: **SIERRA LEONE** SUSU kalixɔnɛ (NWT) TEMNE kɛ-ben NWT

A loosely clumped perennial grass, creeping rhizome, culm to $1\frac{1}{4}$ m high, recorded from Sierra Leone to W Cameroons in the Region, and seemingly not elsewhere in Africa, but occurring in Asia, India to China, and in the Caribbean.

No usage is recorded of the plant in the Region. The *Herb. K* holding has been held as *S. chevalieri* Stapf (= *S. megaphylla* (Steud.) Dur. & Schinz.), and now recognised as a separate entity (2). Usages, if any, are perhaps as recorded for *S. megaphylla*, q.v.

In Malaya, the leaves are part of a decoction taken for irregular menses (1), and the young shoots are eaten as a vegetable (3). The grain is eaten in parts of the Philippines as a substitute for rice (3).

Traces of *hydrocyanic acid* have been found in the roots (3).

References:

1. Burkill, IH, 1935: 1999–2000. 2. Clayton, 1982: 540. 3. Quisumbing, 1951: 109–10, 1024.

Setaria poiretiana (Schult.) Kunth

FWTA, ed. 2, 3: 424, as *S. caudula* Stapf.
West African: **WEST CAMEROONS** BAMILEKE puchuho (DA)

A tufted perennial grass forming large clumps, culms to 1.8 m high, of shady places in and around the forest in sub-montane to montane locations of N and S Nigeria and Fernando Po; and in NE, central and E Africa.

The stems are covered with fine bristly hairs (3) and the leaves are recorded as being slightly urticant (2). It is nevertheless occasionally grazed in Sudan (1), and is much liked by stock in Uganda (4).

References:

1. Andrews A.1958, K. 2. Lewalle 879, 2150, K. 3. Mooney 5695, K. 4. Snowden 1245, K.

GRAMINEAE

Setaria pumila (Poir.) Roem. & Schult.

FWTA, ed. 2, 3: 423, as *S. pallide-fusca* (Schumach.) Stapf & CE Hubbard.
UPWTA, ed. 1, 543, as *S. pallidifusca* Stapf & Hubbard.

English: bristly fox-tail grass; cat's tail grass (Fiji); dove corn (The Gambia, Pirie).

West African: SENEGAL MANDING-BAMBARA mbasambi uluka (JB) SONINKE-SARAKOLE baga ndioke (A.Chev.) WOLOF vul-vul (JLT) THE GAMBIA MANDING-MANDINKA nyakurou (Macluskie) pura barra = *dove corn* (Pirie); = *doves' grass* (JMD) MALI FULA-PULAAR (Mali) laki davangel = *dog's tail* (A.Chev.) MANDING-BAMBARA mbassambi uluku (A.Chev.) KHA-SONKE ulu ndenku (A.Chev.) UPPER VOLTA FULA-FULFULDE (Upper Volta) safuure (*pl.* -uuje) (K&T) GRUSI nazogau (A.Chev.) GURMA bayori (A.Chev.) MOORE basuré (A.Chev.; Scholz) kinsuré, ku-sugo (A.Chev.) GHANA DAGBANI fukagli (Ll.W) HAUSA geron darli (Ll.W) NIGERIA ARABIC-SHUWA danabal kalb = *tail of the dog* (Kennedy) gundul (JMD) HAUSA dùùsáá (JMD; ZOG) duza (JMD; ZOG) geron darli (ZOG) geron tsunsaye = *bird's millet* (auctt.) ƙyás (ZOG) ƙyasuwar rááfíí = *Pennisetum (ƙyasuwa) of the pool* (JMD) ƙyasuwar tá fàdámà (JMD; ZOG) wútsíyàr ɓééráá (Kennedy) NUPE esaggi luko (JMD)

A loosely tufted annual grass, culms ascending to 60 cm tall, very polymorphic, of pathsides, weedy places and sites of cultivation, common throughout the Region in all states and widespread in tropical and warm temperate regions of the Old World; now present in N America.

It is a weed of cultivation. On suitable sites it forms pasture good for grazing acceptable by all stock (1, 2, 5, 7–9). In all NE, central and E African territories it is considered a good grazing grass though its utility may be limited to the early rainy period. It is drought-resistant (5).

The plant is sometimes used as thatch (5). In Lesotho it is twisted into cords for binding sheaves of sorghum (6).

The grain is edible. Khasonke of Mali are reported to collect it for food (5). It is collected in Kordofan in time of shortage (3) and in Ethiopia to make a sort of bread (Schimper fide 5).

References:

1. Adam, 1954: 98. 2. Adam, 1966,a. 3. Baumer, 1975: 113. 4. Broadbent 104, K. 5. Dalziel, 1937: 543. 6. Guillarmod, 1971: 453. 7. Ll. Williams 831, K. 8. Taylor 30, K. 9. Trochain, 1940: 272.

Setaria sphacelata (Schumach.) Stapf & CE Hubbard

FWTA, ed. 2, 3: 423, incl. *S. anceps* Stapf; as *S. sphacelata* var. *torta* (Stapf) WD Clayton; and *S. aurea* Hochst.; as *S. sphacelata* var. *aurea* (Hochst.) WD Clayton. UPWTA, ed. 1, 543.

English: golden timothy grass (S Africa); Rhodesian timothy grass.

West African: SENEGAL MANDING-BAMBARA ka kumbélé, ngolo ké (JB) SONINKE-SARAK-OLE bara yuguné (A.Chev.) buré bur (JB) mati a koy (JB) volvol (JB) WOLOF diem bu (A.Chev.) MALI FULA-PULAAR (Mali) laki davangel, lakiérande, mbihuri (A.Chev.) MANDING-BAMBARA kekumbilé (A.Chev.) ngoloké (A.Chev.) ngolokumbilé (A.Chev.) uluniku = *little dog's tail* (A.Chev.) UPPER VOLTA BISA gimun (Scholz) gimun (Scholz) GRUSI kukuré dabilé (A.Chev.) HAUSA doronkayé (A.Chev.) MOORE bazuré (A.Chev.) konsuga (Scholz) uanzonguri (A.Chev.) GHANA MOORE bazure = *dog's tail* (FRI) bazure = *dog's tail* (FRI) NIGERIA FULA-FULFULDE (Adamawa) bishebadi (Pedder) bishebadi (Pedder) FULFULDE (Nigeria) wicco wandoho *loosely applied* (JMD) FULFULDE wicho panduhi = *monkey's tail* (Chapman) HAUSA babaci (auctt.) geron tsuntsu (auctt.) wútsíyàr bírìì (JMD; ZOG) YORUBA pekunodo (Ward) YORUBA (Ilorin) luludo (Ward) pekun odo (Ward)

A tufted perennial grass, shortly-rhizomed or occasionally creeping, culms up to 2 m high, on moist soils, but occurring on a wide range of habitats, stony hillsides, bush/wooded savanna or swamps and river-banks; variable with 3 varieties in the Region: var. *sphacelata*, caespitose with culms to 1 m high from a short rhizome occurring from Senegal to N Nigeria, and generally widespread in

344

tropical and S Africa; var. *torta*, tufted to 2 m high on moist soils, common throughout the Region in all territories, and in central and S Africa; and var. *aurea*, densely caespitose, culm to 1.2 m high, known only from Ivory Coast to W Cameroon in the Region, and in Sudan, E Africa and Zaïre.

The grass is good for grazing and is readily taken by all stock, especially while the growth is still young. It is drought-resistent and succeeds on poor and shallow soil, and responds to cutting to promote new green growth, but it becomes woody on flowering (3, 9). Var *aurea*, introduced from S Africa, is particularly esteemed as Rhodesian timothy grass. Var *torta* is reported as an important constituent of the highland ranch grass community (8). There are strains of the grass, however, which do give a poor return. These merit discarding (5). The grass has potential for conversion into hay and silage, but this seems to be little practised (Kordofan: 2; Zambia: 1).

In Zambia the grass is dug in as a green manure (6). In many areas, especially in E Africa, the grass is used for thatching (7). In southern Africa the flowering culms are bundled to make brooms (10).

The leaf sheaths can be manipulated to make a sort of screeching sound which fascinates cane rats and is used in the Central African Republic in trapping them (4).

The grain is suspect of being toxic, but Zulu of S Africa are reported to eat it in time of shortage; they boil it before grinding into meal (10).

References:

1. Appleton 18, K. 2. Baumer, 1975: 113. 3. Dalziel, 1937: 543. 4. Fay 1609, K. 5. Foster & Munday, 1961: 314. 6. Fraser 8, K. 7. Glover & al. many nos., K. 8. Tulley, 1966: 899–911, as *S. anceps*. 9. Ward 0014, K. 10. Watt & Breyer-Brandwijk, 1962: 485.

Setaria verticillata (Linn.) P Beauv.

FWTA, ed. 2, 3: 421. UPWTA, ed. 1, 544.

English: rough bristle grass.

French: epi collant (Berhaut).

Portuguese: milhã-verticilada (Feijão).

West African: **MAURITANIA** ARABIC (Hassaniya) lesséig (AN) **SENEGAL** MANDING-BAMBARA sé norna (JB) SERER ñadãg rèv (JB) ngok o mbèp (JB) WOLOF ñãmbãg (JB) **MALI** FULA-PULAAR (Mali) kebbe tioffé *from* kebbe: *bur grass* (A.Chev.) MANDING-BAMBARA nornaba (A.Chev.) sé norna = *adhering to the foot* (A.Chev.) **UPPER VOLTA** FULA-FULFULDE (Upper Volta) hudél (Scholz) GRUSI-KASENA gadjané (Scholz) MOORE suntu (A.Chev.) SONINKE-SARAK-OLE khine mésséni (A.Chev.) **NIGERIA** ARABIC-SHUWA amlisego (Musa Daggash) FULA-FULFULDE (Nigeria) nyakkabre (J&D) HAUSA ɗánkáɗáfi (auctt.) maɗaɗɗafi (Palmer; ZOG) YORUBA éémọ́ ẹyẹ = *bird's gum* (Verger); *bird's bur* (JMD; Macgregor)

A loosely tufted annual grass, culms ascending to 1 m high, a ruderal of waste and weedy places often in damp and shady sites across the Region from Mauritania to S Nigeria, and generally throughout in warm temperate and tropical regions of Africa and Asia.

The grass is a common adventive on farmland, around cattle pounds and such anthropogenic sites. It is a weed on cultivated land (1). In Uganda the whole plant is calcined to produce ash used as a cooking salt (20). It provides good forage while still young (5; Mauritania: 16; Senegal: 2; Kordofan: 4; Kenya: 9, 15; Tanganyika: 6, 13, 17; Malawi: 7). It is said to be rich in protein (1). It may also be eaten by stock when dry as hay (1, 4, 5), which is of good nutritional quality, but during flowering flower spikelets with many bristles inhibit browsing, and, indeed, it is recorded that if taken in large quantity death may result (9). There is also a time at the start of the rains before seed has been set cattle ingesting it suffer bloat (8). Furthermore, thickets of the tall grass harbour tsetse fly, so that cattle browsing therein are at risk of contracting trypanosomiasis.

The leafy culms are used in Lesotho (10) and southern Africa (20) to weave into hats.

The flower panicle is bur-like, characterised as mentioned above, by having many bristles which stick to the clothing of passing persons and the fur of animals. The Nigerian Fula name: *nyakkabre* is applied to this and similarly bristled plants, e.g., *Cenchrus biflorus* (Gramineae), *Pupalia lappacea* (Amaranthaceae), and *Triumfetta spp.* (Tiliaceae), with this habit and which make barefoot walking uncomfortable (12). The inflorescences are applied by small boys in N Nigeria to trapping birds (14), and even quite large beetles are entrapped in Uganda (18). An ingenious usage is reported in the Shinyanga District in Tanganyika. The openings of bins used for storage of grain are blocked by a wadge of the flowering spikes of this grass, and the local rat population, learning that contact has unpleasant consequences, does not attempt to raid (3).

Nomads in the Sahara sometimes collect the seed for food (5). Poorer people in Rajasthan, India, are reported to eat the grain (11). In Kordofan it is a famine food (4).

Yoruba invoke the grass for its adhesive property in an incantation to keep a miscarried baby on earth (19).

References:

1. Adam, 1954: 98. 2. Adam, 1966,a: 450–537. 3. Agric. officer, Shinyanga, H1272/1928. 4. Baumer, 1975: 113. 5. Dalziel, 1937: 544. 6. Emson 121, K. 7. Fenner 201, K. 8. Glover & al., 440, 688, K. 9. Glover & al. 2012, K. 10. Guillarmod, 1971: 453. 11. Gupta & Sharma, 1971: 91. 12. Jackson, 1973. 13. Marshall SA.26/2/1932, K. 14. Musa Daggash FHI.24899, K. 15. Mwangangi 1366, K. 16. Naegelé, 1958,a: 293–305. 17. Newbould 5810, K. 18. Thomas, AS Th.3921, K. 19. Verger, 1967: No. 90. 20. Watt & Breyer-Brandwijk, 1962: 485.

Setaria viridis (Linn.) P. Beauv.

FWTA, ed. 2, 3: 424.
English: green bristle grass.
Portuguese: milhã-verde (Feijão).

A loosely tufted annual grass, culms to 50 cm high, of sandy places, native of Europe and warm temperate regions, now recorded at Aïr in Niger.
The plant is in general a weed in cultivation.

Setaria sp. indet.

West African: NIGERIA FULA-FULFULDE (Nigeria) lesleddeho (J&D)

An unidentified grass found growing under trees, known to Fula graziers of N Nigeria.

Sorghastrum bipennatum (Hack.) Pilger

FWTA, ed. 2, 3: 468.
West African: SIERRA LEONE KORANKO ninki-bafoen (Glanville; FCD)

A weak-stemmed annual grass with bunched culms to 1.2 m high; of disturbed sites on damp soil, dispersed across the Region from Guinea-Bissau to W Cameroons, and common over E, central and S central tropical Africa and Madagascar.
The grass is said to provide some grazing but only while still young before anthesis (1, 2).

References:

1. Beal 40, K. 2. Risopoulos 1257, K.

Sorghastrum stipoides (Kunth) Nash

FWTA, ed. 2, 3: 468, as *S. trichopus* (Stapf) Pilger.

West African: SENEGAL DIOLA hu hèt è kobol (JB)

A tufted perennial grass, polymorphic, with culms to 1½ m high from a brief rhizome; of low-lying situations liable to flooding, scattered across the Region from Senegal to N Nigeria, and over E and S tropical Africa and tropical S America.

It provides some grazing for cattle, and Masai in Kenya burn it to promote young growth (1).

The plant has use in Kenya for personal adornment: young men stick flower-heads in their hair and Kipsigi girls push pieces of the thick culms through slits in their ears (1).

Reference:

1. Glover & al. 663, K.

Sorghum Moench

FWTA, ed. 2, 3: 466.

The genus has a wide distribution throughout the warmer regions of the world, with the largest number of species in the NE quarter of Africa, which is probably the evolutionary centre. Considerable taxonomic study has been carried out which broadly recognises three weedy species and one hybrid in the W African region, and the grain sorghums with an over-classification into numerous species, now relegated to one species, *S. bicolor*, composed of many races.

References:

De Wet, 1978: 477–84, with many references. Doggett, 1970. Purseglove, 1972, 1: 259–87. Snowden, 1936.

Sorghum arundinaceum (Desv.) Stapf

FWTA, ed. 2, 3: 467, incl. *S. aethiopicum* (Hack.) Rupr.; *S. virgatum* (Linn.) Pers.; *S. lanceolatum* Stapf and *S. vogelianum* (Piper) Stapf. UPWTA, ed. 1, 548.

English: Kamerun grass (Dalziel).

French: mil sauvage.

West African: SIERRA LEONE KORANKO hula murutuna (FCD) kɛnde *as in general for guinea corn* (NWT; JMD) LIMBA taloa (NWT) taŋki (NWT) MANDING-MANDINKA wada-kende (FCD) MENDE kete *as in general for guinea corn* (NWT; JMD) ngɔgbɔ-getei = *bush guinea corn* (FCD) SUSU tamidi (NWT) TEMNE ta-gbɔyɔ *as in general for guinea corn* (NWT) kɔ-tanbe (NWT) YALUNKA buruna-murutu-na (FCD) hula murutu-na (Glanville) muruntu-na (NWT) LIBERIA MANO gea kpai (Har.) GHANA DAGAARI bàki (FRI) TOGO KONKOMBA gaman daé (Froelich idè, idi (*sing.* n'dé) (Froelich) NIGER FULA-FULFULDE (Niger) gabaraari (ABM) HAUSA daonan daji karkara (Bartha) SONGHAI gènji-hàmà (*pl.* -à) (D&C) NIGERIA FULA-FULFULDE (Nigeria) dakwumbey (JMD) ngaiori (Jackson) HAUSA dááwàr dààmínáá = 'dawa' *of the rainy season* (ZOG) dááwàr dòòrínáá = 'dawa' *of the hippopotamus* (JMD; ZOG) dááwàr kádàà = 'dawa' *of the crocodile* (JMD; ZOG) dááwàr rààfíí = 'dawa' *of the stream* (JMD; ZOG) dawardan kurana (auctt.) firki (Davey; Oche) IGBO (Umuahia) akpụkpa enyi mirĩ = *corn of the elephant of water, i.e., hippopotamus* (auctt.) IJO-IZON (Kolokuma) ịkịráị *the culm* (KW; KW&T) ịmbẹbẹlẹ (KW; KW&T) IZON (Oporoma) omụngẹlẹ (KW) KANURI kara-kara (JMD; C&H) YORUBA ọkà-iye (anon.; Verger)

Tufted annual plants or weak biennials, very variable, of several varieties intergrading together without clear demarcation as to be better identified as ecotypes (5); culms up to 4 m high; of swampy sites, stream-banks, disturbed places and old farmland, or of desert conditions (race *aethiopicum*), throughout

347

the W African region, and generally over tropical Africa and tropical Asia into Australia, and introduced into subtropical and tropical America.

All races give grazing for stock: *arundinaceum* (4); *lanceolatum*, Sudan: donkeys (2), all stock (10); Somalia: camels, cows (9); *virgatum*, Sudan: horses, donkeys, camels (1); *aethiopicum*, Niger: cattle (8). Cattle-folk in Niger recognise that race *aethiopicum* has a high nutritive value and that a small quantity of the grass is adequate to satisfy an animal (8). However, though an useful dry season fodder, it is poisonous under certain conditions of rain in Sudan (7).

Race *lanceolatum* is grown in N Nigeria as a field boundary (4), and in Sudan it is used to make camp wind-screens (6). The mature dried stems of race *vogelianum* are used by the Ijo of SE Nigeria as rush tapers for light (11).

Leaf-sheaths and blades of race *arundinaceum* are sometimes tinged red like those of the cultivated varieties of *S. bicolor* (*S. caudatum* var. *colorans* sensu Snowden), known as *kárán dáfìi* (Hausa) which are used for dyeing, and are picked for admixing (3, 4).

The grain is collected in E and W Africa for eating in time of scarcity (race *arundinaceum* (4); race *lanceolatum*, Sudan (6)).The grain of race *vogelianum* is fed to poultry by the Ijo of SE Nigeria, and other birds will come to seek it out (11).

References:

1. Andrews A.2763, A.2772, K, as *S. virgatum* (Hack.) Stapf. 2. Andrews A.2771, K, as *S. lanceolatum* Stapf. 3. Dalziel 1322, K, as *S. arundinaceum* Stapf. 4. Dalziel, 1937: 548. 5. De Wet, 1978: 481–2. 6. Evans Pritchard 32, K, as *S. lanceolatum* Stapf. 7. Harrison 981, K, as *S. aethiopicum* (Hack.) Rupr. 8. Maliki, 1981: 48, sp. no. 19, as *S. aethiopicum*. 9. Munro 74, K, as *S. lanceolatum* Stapf. 10. Peers s.n., 31/8/53, K, as *S. lanceolatum* Stapf. 11. Williamson, K. 75, UCI, as *S. vogelianum* (Piper) Stapf.

Sorghum bicolor (Linn.) Moench

FWTA, ed. 2, 3: 467. UPWTA, ed. 1, 544–6.

English: grain sorghums; great millets.

French: mil; gros mil; sorgho; guernotte (18th century, Mauny).

Portuguese: sorgo; miglio zaburro (early Portuguese, Mauny).

West African: SENEGAL BASARI dagave (K&A) BEDIK be-ndab after (K&A) gi-ndáb (FG&G) gi-ndáb gi-mbara (FG&G) ty-ndap after (K&A) DIOLA basit-bantabu (K; K&A) NON bo (AS) kotj *large-grained races* (AS) pim *ordinary size grain races* (AS) WOLOF bassi (JMD) tiń (K&A) THE GAMBIA MANDING-MANDINKA basso (JMD) kinti, kinto (JMD) kous, koos *general for cereals except maize* (JMD; Williams) nyŏ *corn in general* (JMD WOLOF bassi (JMD) GUINEA BASARI dɛfàx (FG&G) dɛfàx *i*-waráx (FG&G) dɛxâf (FG&G) ɛ-ndéfáx (FG&G) KONYAGI kombo va-tɛfary (FG&G) u-limp (FG&G) vya-ntimp (FG&G) SUSU mengi (JMD) SIERRA LEONE KORANKO kenele (JMD) LOKO gete (JMD) kete (JMD) MENDE kɛti (JMD) kiti (JMD) SUSU mengi (JMD) TEMNE ta-gbɔyɔ (JMD) kɔ-kbɔyɔ (JMD) YALUNKA meni-na (JMD) LIBERIA MANO ding (JMD) VAI kende (Heydorn) MALI ARABIC (Mali) bechna (JMD) DOGON ɛmé (CG) DOGON (village dialects) de (Bertho) isey (Bertho) ndza (Bertho) ngu (Bertho) nguyu (Bertho) sie (Bertho) tchye se (Bertho) yo (Bertho) yu (Bertho) FULA-PULAAR (Mali) gauri (JMD) MANDING-BAMBARA kende (JMD) nion *any corn* (JMD) MANINKA kende (JMD) nion *any corn* (JMD) SONGHAI hame (Barth) saba (Barth) TAMACHEK abora (JMD) abura (JMD) tafsut (JMD) UPPER VOLTA BOBO mà-pɛnɛ (Le Bris & Prost) mùkô (Le Bris & Prost) yáfa *a small white-grained var.* (Le Bris & Prost) FULA-FULFULDE (Upper Volta) mbayeeri *white-grained* (K&T) mbayeeri mboɗeeri *red-grained* (K&T) gawri mbayerri *white-grained* (K&T) magaaji ɓodeeji *red-grained* (K&T) mojonoori *red-grained* (K&T) GRUSI-LYELA yā, yala *a white-grained c.var.* (K&T) SAMO (Sembla) bïrï mɔ̀ɔ̀ndè (Prost) gmɛ́ɛ̀ sen (Prost) IVORY COAST KRU-BETE nyɔ̀ɔ̀ (Pageaud) MANDING-MANINKA bemberi-gué, bessekre *any dry season sorghum crop, but especially 'durra'* (CHOP) GHANA ADANGME-KROBO koko (FRI) AKAN-ASANTE atokɔɔ (FRI; E&A) TWI atoko (FRI) kokɔte (FRI) DAGBANI bammatica, kapielle (Coull) kulia *sorghum used only for beer* (JMD) kumpesi (JMD) tschi (Gaisser) worsuli = *horse's tail* (Coull) GA àkoko (FRI; KD) coco *perhaps because of the reddish grain; a word used in central Ghana for red or reddish colour of some other crops* (Wigboldus) GBE-VHE fo (FRI) vo (JMD) VHE (Pecí) fo (FRI) GRUSI yara (Coull) GUANG kùtúwè *a general term* (Rytz) kùtúwé pèpèr *a red guinea-corn*

348

GRAMINEAE

(Rytz) KONKOMBA idɛ (Gaisser) MAMPRULI katapielle (Coull) ka-ziaa (Coull; Arana & Swodesh) za *a late c.var.* (Arana & Swodesh) MOBA apargu, demoni (Coull) NANKANNI chi, kyi (Coull) SISAALA kadaaga *a red guinea-corn* (Blass) mɛŋpɛ *a white guinea-corn* (Blass) miaa, miiŋ *general terms; properly used for millet* (Blass) miipulima *ordinary millet* (Blass) nɔruŋ *an early crop* (Blass) nyuŋso *a white guinea-corn* (Blass) TOGO ANYI-ANUFO ajulûwá *a c.var.* (Krass) akpaluku *a c.var.* (Krass) ǹggání (Krass) BASSARI idi, machu (Gaisser) GBE-VHE VO (JMD) GURMA dïimón (Tersis) KABRE mela (Gaisser) NAWDM cóŋú (Nicole) mela (Gaisser) tschonde (Gaisser) TEM néu (Gaisser) TEM (Tschaudjo) mela (Gaisser) DAHOMEY GBE-AJA vo (Wigboldus) FON abokun *a loan word* abo *from Aja;* kun: *a suffix for crop names* (Wigboldus) VHE vo (Wigboldus) NIGER SONGHAI hàmà (*pl.* -à) (D&C) NIGERIA ? ìhiǹmì (Elugbe) ABÕN agwi (Meek) AKPA īkwù (Oblete fide RB) AKPET-EHOM (Ubeteng) ekpoi-deunderim (Cook) ANGAS kás (Kraft) shwé (Kraft) ARABIC-SHUWA dorra (JMD) himeirun (JMD) BADE mùujín (Kraft) BANDAWA-MINDA siɛ (RB) BARAWA-GEJI wô (Kraft) BATA (Malabu) zumwe (Meek) BATA (Zumu) gweo (Meek) BATA-BACAMA gwóyó (Kraft) zùmwέy (Meek; Kraft) BETTE-BENDI ɨ́-kúlátàng (Ugbe fide KW) ɨ́-kwól á-tàm = *maize of Atam; after the Ibibio name for the Bette* (Ugbe fide KW) BIROM tàkàndá *after Hausa* (LB) yara (LB) yara dìk *a c.var.* (LB) yara pyɛng = *white Guinea corn; a c.var.* (LB) yara syinàng = *red Guinea corn; a c.var.* (LB) BITARE ajonggo (Meek) BOGHOM ó (Kraft) BOLE kuté (Benton fide JMD) lὲkítὲ (Kraft) BURA bangtankir (anon.) banjanga (anon.) mtî (Kraft) yeri (anon.) CHIP shû (Kraft) CHOMO léme thaũ (Shimizu) DERA kúrè (Newman; Kraft) wayàati *a c.var.* (Newman) ́nyàm (Newman) DGHWEDE xíyà (Kraft) DOKO-UYANGA ansam-dkara (?) (Cook) EBIRA-ETUNO àkụ́ (Moomo) EDO ọ́kẹ́shín *from Yoruba;* = *horse corn* (JMD) ESAN imoghioi (Elugbe) FALI khui (Meek) xá (Kraft) zìyìkwî (Kraft) zùkú (Kraft) zúkụ̀n (Kraft) FULA-FULFULDE (Adamawa) ndammungeri *a red form* (JMD) FULFULDE (Nigeria) gaw (*pl.* gaweje), gawri (*pl.* gawrije) *general for corn* (JMD; J&D) lacceri (JMD) lakawal *a c.var.* grown for its straw (J&D) mbayeri (JMD) GA'ANDA hwέrmà (Kraft) kwέrmà (Meek) xwέrmé (Kraft) GA'ANDA (Boka) xwɔ̀rmá (Kraft) GENGLE som (Meek) GERA sὲwɔ́ (Kraft) GLAVDA xíyà (Kraft) GOEMAI sowa (Kraft) swaa (Hoffmann) GUDE súkúngwá (Meek; Kraft) GUDU gɔ́:wà: (Meek; Kraft) GUDUF xáyá (Kraft) xíyà (Kraft) GWANDARA-CANCARA dá' (Matsushita)GITATA dáã̂ (Matsushita) KARSHI dá̂' (Matsushita) KORO dá' (Matsushita) NIMBIA dá' (Matsushita) TONI dã' (Matsushita) GWARI ewi (JMD) HAUSA bakin rakumi (AS) dáawàa (auctt.) gabjin (Abrahams) ganda gaura (Abrahams) igwu abakpa (Armstrong fide RB) takanɗáa (LB) zangare *the stripped head* (Mortimore) HUBA ubi (RB) úhì (Meek; Kraft) HWANA kwârmá (Meek) xwârmá (Kraft) IBIBIO akpakpa (Cook) ICEN fafa (Meek) iyī̄ (RB) kwi (Meek) zə (RB) ICHEVE ɨ́-kùlé (*pl.* ɨ́-kùlé) (Anape fide KW) IDOMA igwu (Armstrong fide RB) IGALA ókīlī (RB) ókòlī (RB) IGBO okà okìrì =*maize guinea corn* (auctt.) okìlì (JMD; KW) IGBO (Ezinehite) ukhòorū (Armstrong) IVBIE ádóòkhò (Elugbe) IZAREK nyór (Kraft) JAKU mísá (Kraft) JANJO mọ (Meek) JARAWA-BANKAL gù (Kraft) JEN nɛ̃mọ (RB) JIRU yi (RB) JUKUN zadzòn (Shimizu) zajíkwìn (Shimizu) JUKUN (Abinsi) zuku (Meek) JUKUN (Donga) yera (Meek) JUKUN (Gwana) za (Meek) JUKUN (Kona) za (Meek) JUKUN (Pindiga) zaà (RB) JUKUN (Takum) za (Meek) JUKUN (Wase) za (Meek) JUKUN (Wukari) aza jukwi (Meek) za (Shimizu) zafyen (Shimizu) JUKUN-JIBU iza (Meek) KAJE nywât (Cook) KAKA nzang (RB) nza-sẹ (RB) KAMBARI-AUNA iishìnâ (RB) SALKA mmɔ̃́ɔ̃́ (*pl.* iimɔ́ọ́) (RB) KAMWE kha (Meek) lkha (Meek) xà (Kraft) KAMWE (Kiria) ikha (Meek) xà (Kraft) KANURI ajagama (JMD) gabeli (JMD) kiliram (JMD) mari, mariya *white forms* (JMD) ngaberi (JMD) ngafeli (JMD) ngàwúli *Guinea corn in general* (C&H) KAREKARE wàtè (Kraft) KATAB suwat (Meek) KATAB-KAGORO swálák (Cook) KOFYAR-MERNYANG suwaa (Hoffmann) KORO bekpu (RB) KUGAMA som (Meek) KUMBA sori (Meek) KYIBAKU 'wùxì (Kraft) LAAMANG khia (Meek) LENYIMA akpe (Cook) LEYIGHA nsange (Cook) LIBO kwǫma (Meek) LONGUDA kwanla (*pl.* kwama) (Meek; RB) MALA idawn (RB) MAMBILA ya (Meek) yiẹ (Meek) yiir (Meek) yirẹ (Meek) MARGI okhi (Meek) ukhi (Meek) wúxí (Kraft) MARGI (South) wúxī, wúxì (Kraft) MARGI-PUTAI kwishì (Kraft) kwúshì (Kraft) MATAKAM dáwn (Kraft) MBEMBE (Tigong)-ASHUKU ẹkẹ (Meek) kèn (RB) NAMA kwiri (Meek) ǹdjì-kùnè (RB) MBOI kama (Meek) koma (Meek) MBULA misa (Meek) MUMBAKE tura (Meek) MUMUYE ze (Meek) zie (Meek) MUNGA mom (Meek) NDORO iga (Meek) NGAMO kútè (Kraft) NGGWAHYI úxì (Kraft) NGIZIM gavərka (Schuh) NINZAM ikpu (RB) NUPE ekpan (JMD) eyì *a general term* (JMD) eyì-kpan (JMD) NZANGI kwama (Meek) kwɔ̀mɔ́ (Kraft) NZANGI (Holma) maudzen (Meek) OKPAMHERI (Emarle) îjì (Elugbe) OLULUMO (Ikom) è-gû (Cook) jɔ́rɔ́ (Cook) PEREMA gbera (Meek) POLCI cáw (ZF) cə́n (Kraft) gbóròn (ZF) móɗɗò (ZF) PITI idar (Meek) POLCI ûu (Kraft) POLCI-BULI 'ó (Kraft) ZUL 'ɔ̀ (Kraft) SAMBA-DAKA yiiri (RB) yiri (Meek) LEEKO yẹra (RB) SANGA ida (RB) SUKUR mbwaran (Meek) SURA kàs (Kraft; anon. fide KW) papira (anon. fide KW) shwaa (anon. fide KW; Hoffmann) TANGALE sὲw (JMD) TEME kǫom (Meek) TERA gàwà (Newman) TERA-PIDLIMDI gúdì (Kraft) TIKAR-NKOM nzan (Jeffreys) TIV bange (JMD) gbenge (JMD) gɔ̀wə̀nggɔ́ (KW fide RB) uwua (JMD) UHAMI-IYAYU gegebọ baba (Elugbe) UMON akpoi (Cook) URHOBO ọki-ighan (JMD) VERRE kowọp (Meek) réérù (RB) WAJA khria (Meek) WAKA kǫng (Meek) WANDALA khiya (Meek) xíyá (Kraft) WOM che (Meek) YALA (Ikom) ìgù *cf. Zea mays* (Ogo fide RB) YALA (Ogoja) igwīanya (Onoh fide RB) YANDANG

GRAMINEAE

kǫwg (Meek) YEDINA kanggeli (Lukas) mazakwa *dry season Guinea corn; this language?*
(Golding & Gwynn) míau (Lukas) miyo (JMD) phíau (Lukas) YORUBA bàbà (JMD) bomǫ́
(JMD; Verger) ǫkà ẹ̀sin = *horse corn* (JMD; Verger) ǫka-bàbà (JMD; Verger) ǫkà-iṣí *with
vars. designated by colour* (JMD; Verger) YORUBA (Oyo) ǫkà *cf. Yoruba (Egba): maize* (JMD)
YUNGUR kwoma (RB) ZAAR wâ (Kraft) **WEST CAMEROONS** BATA (Bacama) akuthinda
(Meek) TIKAR-NKOM nzan (Lowe)

Cultivars: **SENEGAL** ? samé (JLT) sevel (JLT) **GUINEA-BISSAU** FULA-PULAAR (Guinea -
Bissau) basse-bássi (JDES) nhame-quinto (JDES) quinterim (JDES) PULAAR (Guinea-Bissau)
baieri (JDES) **NIGERIA** ARABIC-SHUWA mereya *a red form, small grain* (JMD) sambul *a white
form* (JMD) FULA-FULFULDE (Nigeria) ajagamari, ɓalwa-cololari, cakalari, jarmari, kuberi,
malleri, mukka jeɗa, yaldari, murmureri, ndunguri, ngobari, njigana-mbaya, njigari, njolomri,
salambata, same, sudumari, teleri, umpu, wunwundi, yeleri (Taylor) gawri jigaari = *short
season gawri* (J&D) gawri mbayeeri (J&D) gawri mbayeeri (J&D) gawri njolomri (J&D) gawri
njolomri (J&D) gawri yolobri (J&D) gawri yolobri (J&D) mbayeri daneeri *a white form* (JMD)
mbayeri mboɗeeri *a red form* (JMD) IGBO okili īnyarī, okà-ajata (JMD; KW) IGBO (Awka)
ǫka-nwaijẹta, ebwafẹle (JMD) IGBO (Onitsha) inyari, inari (JMD) okili mme *a red form* (JMD)
okili ǫcha *a white form* (JMD) KAMBARI-SALKA àkùnkwáshìlî *a c.var grown for beer production,
and to use in arrow-poison* (RB) karandafi *a c.var grown for beer production* (RB) YORUBA
bòromǫ́ *a red-grained form* (JMD; Verger) ṣoṣoki *a brown-grained form* (JMD; Verger)

Parts of the plant: **SENEGAL** BEDIK bambaràlì *grain hard to pound* (FG&G) ɓi-nwm
grain which is tender, i.e., easy to pound (FG&G) **GUINEA** BASARI ɔ-diamb *reaped millet lying
on the ground* (FG&G) ɛ-nágà *plant in flower* (FG&G) a-ndyɛrá *inflorescence* (FG&G) ɓa-
nyelɛsyɛtɛn *bran* (FG&G) ɓa-nyewa *threshed straw* (FG&G) ɛ-pálangà *a head of millet*
(FG&G) **GHANA** AKAN-FANTE awi *the grain* (JMD) TWI awi *the grain* (FRI; JMD) **NIGERIA**
FULA-FULFULDE (Nigeria) bafuri, ndalturi *stool shoots from roots left at harvest* (JMD)
chingirri, gabara, hambana *when young*, karkara, katsakatsa, kusumburwa, magargera *culms of
self-set plants* (JMD) mbafu *young plants* (JMD) sammere (*pl.* chammaje) *the panicle* (JMD)
HAUSA baho, gyauro, gyamro *stool shoots from roots left at harvest* (JMD) buniya, bununi,
binini *the pollen* (JMD) damana, yabanya *young plants* (JMD) garera, soshiya *the bare fruiting
head after removal of the grain* (JMD) karmami *leafy shoots or badly-grown plants used only as
fodder* (JMD) kona, kono *the chaff* (JMD) KAMBARI-AUNA karaatòkìishìnà *a head of grain* (RB)
kàssúngù *a culm* (RB) kɔ́rẹ̀ẹ̀rẹ̆ *new leaf* (RB) SALKA áakúnkú *a culm* (RB) ɔ̀əgɔ̀zɔ̂ *a head of
grain; also applies to rice* (RB) ɔ̀rẹ̀ẹ̀rẹ̆ *leaves* (RB)

Products: **SENEGAL** MANDING-BAMBARA dolo *beer* (JMD) WOLOF puh *beer* (JMD)
GHANA ? (Northern Territory) pito *beer; a corruption of* poitouw, *Portuguese from Angola*
(Sodah Ayenor & Matthews; Wigboldus) AKAN-TWI ahavo *beer* (Wigboldus) atoko-sã *beer*
(FRI) DAGBANI dam *beer* (JMD) GBE-VHE liha *beer* (JMD) **DAHOMEY** ? dama *beer* (JMD) ?
(North) chapalo *beer* (JMD) **NIGER** HAUSA aleewà, alewa *molasses or sweetmeats* (AE&L)
NIGERIA FULA-FULFULDE (Nigeria) cananari *the culms of self-sown plants used to make flutes*
(JMD) HAUSA alewa *molasses or sweetmeats made from sweet-stemmed forms* (JMD) fito *beer;
from* pito, *a corruption of* poitouw, *Portuguese from Angola - cf.* giya: *beer from any source*
(JMD; Wigboldus) KANURI mbal *beer* (JMD) KATAB akan *beer* (Meek) sywât *a food-stuff*
(Meek) KORO mme *beer* (RB) MARGI komil *beer* (JMD) PERO dìlà *a red wine* (ZF) YORUBA ǫti
bàbà, pito *beer* (JMD) **WEST CAMEROONS** NGYEMBOON mesangá *flour* (Anderson)

Pathology: **NIGERIA** FULA-FULFULDE (Nigeria) mbumari, ɓuntu, ƙatsu-ƙatsa *a loose,
black, dusty grain-smut caused by Sphacelotheca reiliana, (Kühn) Clinton* (JMD) turɓushi *a
plant affected by Sphacelotheca reiliana (Kühn) Clinton* (JMD) HAUSA birtuntuna, burtuntuna *a
loose, black, dusty grain-smut caused by Sphacelotheca reiliana (Kühn) Clinton* (JMD) darɓa *a
sticky 'honeydew' blight caused by aphids or plant-ticks known as manɗo* (JMD) kuturta *a blight
attacking both Sorghum and Pennisetum* (JMD)

   Annual or perennial grasses with stout culms to 4 m high or more by up to
3.5 cm diameter, basally bearing a panicle of loose or dense grain, erect or
sometimes goose-necked, generally very variable; a very important dry area
cereal, particularly in parts of Africa and India where rainfall is under 1,000 mm
p.a. It shares with bulrush millet, *Pennisetum glaucum*, Gramineae, q.v., the
position of a major staple in the W African Soudano-Sahel, yielding place to the
latter only in the driest areas. It has been conjectured that domesticated
sorghum originated from selections of *S. arundinaceum* before 1,000 B.C. in the
Ethiopian area with distribution across the Soudan to Upper Niger and the
Mande civilisation. Other selections may have occurred independently elsewhere

350

(8, 12, 15). Such was the great utility for the plant and its variability as to usage and habitat, that innumerable different strains have been selected by man, presenting to taxonomists an endless web for classification. Snowden recognised 31 species, 157 varieties and 571 forms (6, 16). All these units are more or less completely interfertile and are now regarded as cultivars or races of a single species, *S. bicolor*, as here recognised, with 5 basic races (10, 12).

Young foliage may be dangerous to stock. From germination there is present a cyanogenetic glycoside, *dhurrin*, which under certain conditions of growth and weather builds up to toxic proportions. It comes and goes and appears to be a link in the plant's metabolism. Drought also promotes its formation. Tiller shoots after harvesting are often very poisonous (1, 6, 11, 14). An antidote for poisoned animals is said to be large quantities of molasses or milk (11). Salka people of N Nigeria use the plant in arrow-poisons (3). Older plants seem to be free of toxicity, and the foliage is fed to livestock (1, 2, 11) after the grain has been harvested.

Some tribes in N Nigeria macerate the stems in water and bind them together to make a sort of mat; leaf-sheaths are similarly used (11). Such large mats called *màɓàgi* in Salka are used for covering doorways (3). The culms are commonly used for fencing and hut-building, especially for the radiating bends of conical hut-roofs (11). The Fulfulde of N Nigeria sometimes grow the grass specially for the stems (13). In N Sierra Leone they are used to make improvised beehives for taking swarming bees (4). As a source of paper-pulp, the product is moderately strong, but short-fibred, which could be used as a filler with longer-fibred pulps (6, 9, 11). The pulp is deemed unfit for viscose manufacture in production of artificial silk (9). The stems contain a relatively high amount of *pentosan* (27.8%) which offers possibility for production of *furfural*, as industrial solvent (9). Ubiquitously the stems serve as fuel and the ash is used as a manure (11).

Cultivated sorghum is grown chiefly for its grain. Amongst the world's grain crops it ranks fourth in importance after wheat, rice and maize. There are bearded and unbearded strains. In Africa, where bird depredation is serious, the former are preferred as the awns obstruct the birds. Even so, in the absence of unawned strains, birds, particularly quelea (*Quelea quelea*), will attack any strain and farmers may harvest only such fields which he can protect by noise or even laying the crop (5). There are strains of marked precocity and delayed maturity and all grades between with life cycles 80–200 days. Human consumption is normally in the form of gruel, porridge or cake. Some forms, particularly those with dark-coloured bitter grain, are used for fermenting to beer. Grain also serves as poultry and cattle-feed as, for example, in the USA where breeding and selection has been advanced to raise commercial strains. Dietetic value is poor. The grain is conspicuously deficient in minerals except phosphorus and magnesium. In threshed and polished grain vitamins A and B, and protein are depleted. (7, 8, 11, 15).

References:

1. Adam, 1954: 98, as *S. vulgare* Persoon. 2. Adam, 1966, a, as *S. sp.* (cult.) 3. Blench, 1985. 4. Boboh, 1974. 5. Bourke, 1963: 121–32. 6. Burkill, IH, 1935: 2056–9, as *S. vulgare* Pers. 7. Busson, 1965: 484, with chemical analyses, 486–8. 8. Chadha, 1972: 436–9. 9. Chittenden & al., 1951: 299–305. 10. Clayton & Renvoize, 1982: 726–7. 11. Dalziel, 1937: 544–6. 12. De Wet, 1978: 477–84. 13. Jackson, 1973. 14. Kerharo, 1973: 7. 15. Purseglove, 1972M 1: 259–87. 16. Snowden, 1936.

Sorghum bicolor (Linn.) Moench, race bicolor

UPWTA, ed. 1, 547, as *S. membranaceum* Chiov. var. *baldratianum* Chiov.; and *S. notabile* Snowden var. *notabile.*

English: kafir-corn.

West African: **SENEGAL** BEDIK faɓák *a c.var. with grey grain* (FG&G) sangare *a c.var. with flat bulky grain* (FG&G) **THE GAMBIA** MANDING-MANDINKA bassi qui (Williams) bassi

351

GRAMINEAE

wulima (Williams) **GUINEA-BISSAU** CRIOULO midjo cabalo (JDES) MANDING-MANDINKA bambaram-bassô (JDES) **GHANA** DAGBANI kaziegu *a red form* (Coull) **NIGERIA** HAUSA jan dawa *a red race* (Schule) IDOMA abám (Armstrong) acǫ (Armstrong) NUPE ekpan bokungi *a white form* (JMD) ekpan djúrúgi *a red form* (JMD) ekpan emigi *a yellow form* (JMD) ekpan etswa gútági *a 3-months crop* (JMD) ekpan gbadza, ekpan guduchi (JMD) eyi takungi *a dwarf form* (JMD) gbagù, màyì gbagu (JMD) kpayi *a long-headed form* (JMD) kuyi, kuyi beyigi (JMD) màyì *a long-headed form* (JMD) màyì chintara ebe = *monkey's tail; a long-headed form* (JMD) shurugi (JMD) surù *a very small form* (JMD)

Kaoling sorghum, S. membranaceum sensu Snowden: **NIGERIA** HAUSA makaa-too-da-waayoo *a soft sweet form* (JMD; ZOG)

Sugar sorghum, S. notabile sensu Snowden: **NIGERIA** HAUSA léébèn raakúmìì = *lips of the camel* (JMD; ZOG) wútsíyàr gíìwáá = *tail of the elephant* (JMD; ZOG)

A heterogeneous race, low-yielding but grown for its grain which is eaten, brewed into beer, and fed to stock. Baking trials under an United Nations programme using a 50:50 mixture of wheaten and millet flour produced an acceptable loaf, and bread containing 40% millet flour is now commercially sold in the Chad area. It is well-received, long keeping and with a good flavour (1, 3).

Many strains have sweet stems which are chewed. Attempts to grow such strains as a commercial source of sugar have not been successful because of the difficulty of crystallising the product. The press-cake after extraction of the sugar contains about 10% hard wax (4).

The Tenda of SE Senegal consider that the plant is proprietary to men. Its cultivation is precise and a specific terminology has been developed (2). Most of the crop is fermented for beer, the balance being pounded to flour for consumption as gruel or sauce (2). The flour and beer enter into ritual uses (2).

References:

1. Anon., 1971, a: 215–17. 2 Ferry & al., 1974: sp. no. 135, *S. vulgare*. 3. Raymond & al., 1954: 152–8. 4. Watt & Breyer-Brandwijk, 1962: 486–8, as *S. dochna* Snowden.

Sorghum bicolor (Linn.) Moench, race caudatum

UPWTA, ed. 1, 546.
West African: **NIGERIA** HAUSA abantako (JMD) aje bichinka (JMD) bahon buka (JMD) ba-kadaniya (JMD; ZOG) bakar kona (JMD) balasana (JMD) balasaya (ZOG) banda (JMD) bankun (JMD) ɓargon doki = *marrow bone of the horse* (JMD) ɓargwan doki (JMD) chi da gero (JMD) dan karuna (JMD) farfara (JMD) farfari (JMD) fasa kundu = *burst gizzard* (JMD) gagare (JMD) gayaji (JMD) giwar kamba (JMD) giwar kumbo (JMD) gobe ya gurbi *a late-maturing form* (JMD) jar dawa (JMD) jengere (JMD) kárán dáfìì (ZOG) karya koshiya = *break spoon* (JMD) kaura *a collective term, referring mostly to yellow and red forms* (JMD; ZOG) kújèèrár, kujerar kúrcíyáá = *chair of the dove* (JMD; ZOG) kur kura (JMD) kwanbar-bada (JMD) kwar biyu (JMD) lafiya (JMD) maisalati (JMD) murmura (JMD) shèèkár kúrcíyáá = *nest of the dove* (JMD; ZOG) tagwaye = *twins* (JMD) takan chi da dawa (JMD) takan da (JMD) tsuwen jaki (JMD) yaddara (JMD) yar biyu (JMD) zabakon farfara (JMD) zaùnàà ínúwà (JMD; ZOG)

Kafir corn, S. caudatum sensu Snowden: **MALI** FULA-PULAAR (Mali) kellori (A.Chev.) mbayeri mbo'deri = *red corn* (A.Chev.) MANDING-BAMBARA diélica nion = *non-agriculturists' corn* (A.Chev.) moigne (A.Chev.) oinaca (A.Chev.) MANINKA fara oro (A.Chev.) SAMO malanga (A.Chev.) **UPPER VOLTA** BOBO péné sogo, téni soko (A.Chev.) MOORE monom (A.Chev.) **GHANA** DAGBANI molesche (Volkens & Gaisser fide JMD) paga *a general term* (JMD) **TOGO** ANYI-ANUFO sotemondi (Volkens & Gaisser fide JMD) BASSARI idikantui (JMD) KABRE gbanjinga, kpalejinga, panyinga (Volkens & Gaisser fide JMD) NAWDM kpanj-inga (Volkens & Gaisser fide JMD) TEM (Tschaudjo) fulgani (Volkens & Gaisser fide JMD) furgani *a dark red form* (Volkens & Gaisser fide JMD) palenyina, kpaleyinga (Volkens & Gaisser fide JMD) umbonu (JMD) TOBOTE enanantjo (JMD) **NIGERIA** ARABIC-SHUWA gardi-binohi (JMD) kuludu *the dark red dye from the plant* (JMD) HAUSA farin janjare, janjare *red forms* (JMD) karan dafi (JMD) karan dafi baki *mature leaf sheaths yielding a yellow dye for palm fibre, etc.* (JMD) karan dafi jaje *not-mature leaf sheaths* (JMD) KANURI masogó (JMD) muji (JMD) shìmkùwì (C&H) NUPE dindorogi, dundorogi (JMD)

352

kafir corn, S. caudatum f. colorans, sensu Snowden: **GHANA** KONKOMBA gamandae (Volkens & Gaisser fide JMD) **TOGO** BASSARI ikamaudi (Volkens & Gaisser fide JMD) **NIGERIA** FULA-FULFULDE (Nigeria) yambe (JMD) NGIZIM dàashà a *red race* (Schuh)

Culms stout, 1–4 m high by 1–3 cm in diameter near the base; panicle contracted, sometimes goose-necked (3a); an old race of grain sorghum with remains found in archeological deposits of a millenium ago; now closely associated with peoples of Shari - Upper Nile, and widely grown in Chad, Sudan, NE Nigeria and Uganda (2, 3b).

This is an important race for its grain, very variable in form and in habitat. Grain tends to be soft and not storing well. Some strains mixed with wheatflour produce a palatable well-risen loaf. Other strains have bitter grain and are fermented to produce beer.

Several strains have leaf-sheaths and adjoining parts of the stem coloured deep red-purple. These are often grown as a source of dye for mats, cloth, calabashes and body pigments. Used alone red colour results, or mixed with natron and other pigments black is produced (1).

References:

1. Dalziel, 1937: 546. 2. De Wet, 1978: 480. 3a. Snowden, 1936: 161. 3b. ibid.: 181.

Sorghum bicolor (Linn.) Moench, race durra

UPWTA, ed. 1, 546, as *S. cernuum* Host var. *orbiculatum* Snowden; and 547, as *S. durra* Battand. & Trab. var. *niloticum* (Koern.) Snowden.

English: durro, after Arabic: *dorra.*
West African: **IVORY COAST** MANDING-MANINKA bemberi-gué, bessekré *also applied to any dry season sorghum crop* (CHOP fide JMD) **NIGERIA** ARABIC-SHUWA berbere (Taylor fide JMD) FULA-FULFULDE (Nigeria) maskuwari (JMD) murel, murerri, kandiri (Taylor fide JMD) ndaneri (JMD) purdi (Taylor fide JMD) HAUSA másakuwa (JMD; ZOG) mázakuwa (JMD; ZOG) KANURI keriram *a sweet form* (Taylor fide JMD) masakwa (Taylor fide JMD) zàrmà (C&H)

Milo sorghum, S. cernuum sensu Snowden: **SENEGAL** MANDING-BAMBARA maka nǒ (JB) tamari (JB) SERER fèla (JB) ngok (JB) WOLOF dahnat (JB) **THE GAMBIA** MANDING-MANDINKA bambara-basso (JMD) gajabo *straight and goose-necked forms* (JMD) manio (Williams) **MALI** MANDING-BAMBARA amadi bubu (Dumas fide JMD)

Guinea corn, S. margaritiferum sensu Snowden: **SENEGAL** MANDING-BAMBARA haussa kala (JB) SERER tèñ (JB) tiñ (JB) **THE GAMBIA** MANDING-MANDINKA bassi wulengo Jalla (JMD) bassiba kinto (JMD) kingi koyo (JMD) kingi wulengo (JMD) nyaro kinto *from nyaro, said to be a corruption of Portuguese:* senhora (JMD) samba Jabbo (JMD) sou fenio = a *horse's tail,* (JMD) WOLOF ditinge (JMD) tinge (JMD) **GUINEA** MANDING-BAMBARA kende missé, kende ni-ulé (JMD) kendi bilé (JMD) MANINKA lomba (JMD) tayim (JMD) tayula (JMD) SUSU mengi béli *a red grained form* (JMD) mengi firé (JMD) **SIERRA LEONE** FULA-PULAAR (Sierra Leone) baheri (FCD) GOLA gbadi (FCD) gbali (?) (FCD) KISSI chɔndɔ (FCD; JMD) KONO kende (FCD) saine (JMD) KRIO kus-kus (FCD; JMD) LIMBA serikiti (FCD; JMD) LIMBA (Tonko) taŋgi (FCD) LOKO kete (FCD) MANDING-MANDINKA kende (FCD) MENDE kɛti (FCD) SUSU mɛŋgi (FCD) TEMNE ta-gbɔyɔ (FCD) ta-gbɔyɔ-ta banko *black-seeded form* (JMD) ta-gbɔyɔ-ta-gbalkanta *white-seeded, red glumed form* (JMD) ta-gbɔyɔ-ta-gbotho (JMD) ta-gbɔyɔ-ta-yerike (JMD) ta-gbɔyɔ-ta-yulu (JMD) ta-lasoi (JMD) ta-merike (JMD) ta-yante (JMD) kɔ-yule (JMD) VAI gende (?) (FCD) kende (FCD) YALUNKA murutu-na (FCD) **NIGERIA** HAUSA baakin raaƙúmii = *mouth of the camel; var. tremulans, sensu Snowden* (JMD; ZOG) gagara biri = *defeats the monkey* (JMD) girar burtuu = *eyebrow of the hornbill; a form with a drooping panicle* (JMD; ZOG) ƙyááràà (ZOG) ƙyaram (JMD) ƙyarama (JMD) ƙyaran (JMD) ƙyaranaa (ZOG) leébèn raaƙúmii = *lips of the camel; var. tremulans sensu Snowden* (JMD; ZOG) mallen kabi = *'malle' of Kebbi* (JMD; ZOG) mallen mama (JMD; ZOG) mallen zamfara = *'malle' of Zamfara* (JMD; ZOG)

Culms stout, up to 4½ m high by 2 cm wide, rarely sweet. This race has close association with Islamic people along the southern Sahel from Sudan into W Africa (2, 3). It is a dry season crop grown after flood-waters have retreated.

The grain is usually white and soft, but other colours occur, brown and reddish. It can be eaten without grinding, and it can also be mixed with wheat for making bread. The seed is long-lived and may be stored for years buried in puddled clay. (1)

References:

1. Dalziel, 1937: 546. 2. De Wet, 1978: 480. 3. Snowden, 1936: 187-9.

Sorghum bicolor (Linn.) Moench, race guinea

UPWTA, ed. 1, 547, as *S. exsertum* Snowden var. *amplum* Snowden; *S. gambicum* Snowden; *S. guineense* Stapf var. *involutum* Stapf; *S. margaritiferum* Stapf and *S. mellitum* Snowden var. *mellitum*.

English: guinea-corn; sweet sorghum (*S. mellitum* var. *mellitum* sensu Snowden); Sierra Leone guinea-corn (*S. margaritiferum* sensu Snowden).

French: gros mil; sorgho; tigne; tègne (*S. margaritiferum* sensu Snowden).

West African: SIERRA LEONE BULOM tòl-lɛ (FCD) BULOM (Sherbro) thòl (Pichl)

Guinea corn, S. caudatum, f. colorans, sensu Snowden: GUINEA KONYAGI vyɛntɛli ve-ndyax (FG&G) vye-ntɛli vyempalax (FG&G) SIERRA LEONE FULA-PULAAR (Sierra Leone) gauri (FCD) MALI ? (Lake Faguibine) djibi a *black-glumed form* (JMD) DAHOMEY ? obo (Newton fide JMD) NIGERIA GWARI dakwuhi (JMD) ewi dakwuhi (JMD) NGIZIM ángàabà (*pl.* -àbín) (Schuh)

Guinea corn, S. exertum, var. amplum, sensu Snowden: TOGO TOBOTE edyipempɛ (JMD) NIGERIA HAUSA zabakon ƙaura (JMD; ZOG)

Guinea corn, S. gambicum sensu Snowden: THE GAMBIA MANDING-MANDINKA basse ba kinto (JMD) bassiba (JMD) bassoba (JMD) ferdu basso (JMD) gajabo *straightnecked form* (JMD) kinto (JMD) GUINEA MANDING-MANINKA nion-ni-fing = *small black* (JMD) nion-ni-gélé (JMD) SIERRA LEONE FULA-PULAAR (Sierra Leone) gari (FCD) KORANKO bimbɛri (FCD) MALI MANDING-BAMBARA bemberi (JMD) bemberi ba (JMD) keniki (JMD) sanko (JMD) suku *late forms* (JMD) MANINKA nion-ni-fing = *small black* (JMD) nion-ni-gélé (JMD) NIGERIA HAUSA kaso ba ni (JMD) kaso bammi (ZOG)

Guinea corn, S. guineense sensu Snowden: SENEGAL BEDIK gi-nìyál (FG&G) gi-nìyál ge-pɛsyà (FG&G) MANDING-BAMBARA bãgéné (JB) SERER kõngosan (JB) GUINEA BASARI delí fɛsyàx (FG&G) delí wáràx (FG&G) TOGO NAWDM dgefɛlu (JMD) nyiamune (JMD) TEM kansina (JMD) tyetya kulma (JMD) NIGERIA HAUSA abantako (JMD) asniya (JMD) basharamba, bazamfara (JMD) doran zabo = *hump of the guinea-fowl* (JMD) doro = *the hump* (JMD) dungogiya a *goose-necked form* (JMD) fara-fara (JMD) farfara (JMD) haƙorin karuwa = *tooth of the prostitute; a form with both red and white grain in the same panicle* (JMD) janare (JMD) janhauya (JMD) masaba (JMD) tan ƙwasau *goose-necked forms* (JMD) tankware *goose-necked forms* (JMD) wútsíyàr tunku = *tail of the mongoose*(JMD) yalan (JMD)

Guinea corn, S. mellitum, var. mellitum, sensu Snowden: SENEGAL WOLOF diahnât a *goose-necked form* (AS) THE GAMBIA MANDING-MANDINKA sucar kalo (JMD) NIGER DENDI kááná (*indef. sing.*) (Tersis) NIGERIA FULA-FULFULDE (Nigeria) badal (*pl.* bade) (JMD) cibal (*pl.* cibe, cife) (JMB) cibal ngesa *from* ngesa: *of the field* (JMD) cife takanɗari (JMD) lakaawal (*pl.* lakaaje) (JMD) lakatblel (JMD) HAUSA kárán tálàkà = *poor man's cane* (JMD; ZOG) tàkàn ɗaá (auctt.) IDOMA ɛmjigwu (Armstrong) KAMBARI-SALKA ttàkámù (*pl.* iita-) (RB) KANURI kàngálè sawadə (C&H) ngáwúrà (A&S) sabadɛ (JMD) SURA ewùng (anon. fide KW) YEDINA kanggeli aali (Lukas)

Culms stout, 2–5 m high by 2–3 cm diameter, long loose panicle, often goosenecked; a race particularly of W Africa, widely distributed in The Gambia, Guinea, Sierra Leone and N Nigeria.

A strain, *S. gambicum* sensu Snowden, is the native guinea-corn of The Gambia and Mali. A black-glumed form of it is grown like rice in standing water near Timbuktu (1). The grain of another strain, *S. margaritiferum* sensu Snowden, is cooked and eaten like rice (2). Strains of *S. mellitum* have sweet culms. They are used in Ghana to make a sweetmeat called *alewa*: Hausa (4). A commercial strain developed in the USA is source of a thick golden brown syrup. The crushed culms after expression of the sweet sap are called 'pummies' and are considered a good soil conditioner and mulch (3).

A strain grown in N Nigeria (*S. exsertum* var. *amplum* sensu Snowden) has red culm sheaths from which a red dye is extracted for dyeing leather (1).

References:

1. Dalziel, 1937: 547. 2. De Wet, 1978: 480–1. 3. Hemmerly, 1983: 406–9. 4. Irvine, 1948: 265.

Sorghum bicolor (Linn.) Moench, race kafir

UPWTA, ed. 1, 548, as *S. nigricans* Snowden var. *peruvianum* Snowden.

English: sweet sorghum; kafir-corn.
West African: **NIGERIA** HAUSA a chi duka = *eat all; var. of* takanda (JMD) ayango (JMD) ayangwa (JMD) badankama (JMD) ba-fillatana = *Fulani woman* (JMD) ba-gobara = *Gobir woman* (JMD) chi da dawa (JMD) chi da gero (JMD) dalimi *several-headed* (JMD) danduwa (JMD) garangatsa (JMD) kafar shamuwa = *foot of the stork* (JMD) kaso ba ni (JMD) kuburu *short-jointed* (JMD) kundukundu *var. of* takanda (JMD) marangwadiya (JMD) takanda (JMD) wutsiyar saniya = *tail of the cow* (JMD) yar gwanki = *daughter of the antelope* (JMD) yar rani (JMD) yar wuri *an early form* (JMD) KANURI mbio (JMD)

Culms stout, 1–3 m high by 2–3 cm diameter, sap insipid or sweet, racemes more or less compact; a race of importance principally in eastern and southern Africa of the savanna from Tanganyika to S Africa (2, 3), and grown also in N Nigeria.

The name 'kafir' is derived from Arabic: *unbeliever,* referring to Bantu who are main growers of this race and for whom it serves as a staple (2).

Like some strains of other races, this race has many strains which are sweet-stemmed and are grown purposely for the stem to chew or to make molasses and sweetmeats termed *alewa* in Hausa: cf. race *guinea.* The stems can be fed to stock even after drying. Some sweet-stemmed strains also yield grain (1).

References:

1. Dalziel, 1937: 548, as *S. nigricans* sensu Snowden. 2. De Wet, 1978: 480. 3. Snowden, 1936: 126.

Sorghum X drummondii (Steud) Millsp. & Chase

English: Sudan grass.
French: sorgo d'Alep (Roberty).

An annual grass with stout culms up to 4 m high; a natural hybrid between *S. bicolor* and *S. arundinaceum* found in E Africa. Though the hybrid in the wild is unstable, reverting by successive back-crossing to one or other of the parents, a segregate called 'Sudan grass' has been cultivated (4) and is introduced to several countries as a fodder grass. It is recorded present in Sierra Leone (5).

It is reported to serve well in USA for pasture, hay and silage (6). In Kordofan growth is rapid, particularly under irrigation or during the rains and cuts can be made about every 40 days (2). In Sudan all stock readily take it (1). However, it is possible that the grass may contain sufficient hydrocyanic acid to be dangerous to stock (2, 3). Content of *dhurrin,* the poisonous cyanogenetic glycoside, is not constant, being increased when the plant is under stress of drought or other stunting factors. A manurial regime ensuring adequate amount of available phosphate tends to lower the glycoside content. Slow drying also reduces it. Nutritional value in other respects is good (3).

In Uganda ash from the calcined plant is used as cooking salt (3).

References:

1. Andrews A.3846, K. 2. Baumer, 1975: 114, as *S. sudanense* (Piper) Stapf. 3. Chadha, 1972: 448–9, as *S. sudanense* (Piper) Stapf. 4. Clayton & Renvoize, 1982: 726. 5. Deighton 6025, K. 6. Purseglove, 1972, 1: 260, as *S. arundinaceum* (Desv.) Stapf var. *sudanense* (Stapf) Hitchc.

Sorghum halepense (Linn.) Pers.

FWTA, ed. 2, 3: 467.

English: Johnson grass.

A stout perennial with extensive creeping rhizomes; an autotetraploid arising from a natural hybridisation between a diploid African *S. arundinaceum* and a diploid Asian *S. propinquum* followed by doubling of the chromosomes. The original mother plant was found in Turkey in 1830, taken to the USA and cultivated by a Col. Johnson, whose name the plant bears, as a fodder plant. It is found wild in the Mediterranean region, and under cultivation in most of the warmer countries of the world. In some countries it has become a troublesome weed. Though domesticated as a forage plant it may contain dangerous amounts of *hydrocyanic acid*. It has been used in hybridisation with other sorghums. (2–6)

The grain is eaten by Indians in Rajasthan State admixed with grain of *Pennisetum glaucum* (Gramineae) (5). For others it serves as a famine-food (2). The cyanogenetic glycoside, *dhurrin*, present in other parts of the grass, is absent from the grain (2), and in Rajasthan the grain is eaten as an antidote against hydrocyanic acid poisoning (5).

The grass is halophytic and very strongly drought-resistant (1), and it is when stunted by drought that death of cattle due to grazing it is reported (2).

References:

1. Bhandari, 1974: 77. 2. Chadha, 1972: 446–7. 3. De Wet, 1978: 478. 4. Doggett, 1970. 5. Gupta & Sharma, 1971: 91–92. 6. Purseglove, 1972: 1: 261.

Sorghum purpureo-sericeum (A Rich.) Aschers. & Schweinf.

FWTA, ed. 2, 3: 467.

A robust annual grass, culms to 1 m high; of clay soils near water, sub-montane in N Nigeria, and E Africa, NE Africa and in India.

In India it grows on poor land and yields heavily. The grass is relished by cattle. It is drought-resistant (1). In Sudan all stock take it but it is not reckoned as an important dry season grass (3). Stock take it in Uganda (2).

Women in Uganda bundle the culms to make brooms (2).

References:

1. Chadha, 1972: 448. 2. Dyson-Hudson 282, K. 3. Peers K.07, K.

Sporobolus africanus (Poir.) Robyns & Tournay

FWTA, ed. 2, 3: 410.

West African: NIGERIA FULA-FULFULDE pagame (Chapman) MAMBILA goor (Chapman)

A tufted perennial grass, culms 30–60 cm high, of disturbed sites in montane grassland in Nigeria and W Cameroons, and across Africa to Ethiopia and through E Africa to S Africa; and of disjuncted worldwide distribution in the Indic and Pacific basins.

The grass is grazed in N Nigeria (1, 8). On the ranching grasslands of the Obudu Plateau it occurs in near pure stands, tolerant of heavy grazing and is thus a good cropper (7). Its presence is, in fact, taken as an indication of intensive grazing. It is recognised as a useful fodder grass in Uganda (5) and Kenya (2). The root has a faint smell of vetiver (4). In E Africa the plant (part not specified) has topical application to wounds and to snake-bite (9). Birds eat the grain (3) which by inference may be fit for human use in emergency. The whole plant is used in thatching in Kenya (3). In Zimbabwe it is plaited into baskets, hats, etc. (6).

References:

1. Chapman 3242, K. 2. Glover & al. 782, K. 3. Glover & al. 1752, K. 4. Michelson 1184, K. 5. Snowden 1175, 1493, K. 6. Trapnell, 1342, K. 7. Tulley, 1966: 899–911. 8. Veterinary officer 2502, K. 9. Watt & Breyer-Brandwijk, 1962: 488, as *S. capensis* Kunth.

Sporobolus cordofanus (Steud.) Coss.

FWTA, ed. 2, 3: 407.

A tufted annual grass, culms to 30–60 cm high, of dry sandy soil in the Sahel from Senegal to N Nigeria and extending from Sudan to Zimbabwe.

It is recorded a good grazing grass taken by all stock in Sudan (1), Uganda (2, 4) and Kenya (3).

References:

1. Beshir 13, K. 2. Dyson-Hudson 261, K. 3. Glover & al. 2977, K. 4. Liebenberg 1761A, K.

Sporobolus dinklagei Mez

FWTA, ed. 2, 3: 406.
West African: SIERRA LEONE KONO saia (FCD) LOKO tai (NWT) MENDE kɔli-gbiti (FCD)

A caespitose perennial grass, culms to 10 cm high, of sandy soils and roadsides, known only from Sierra Leone to Ivory Coast.

Sporobolus festivus Hochst.

FWTA, ed. 2, 3: 410. UPWTA, ed. 1, 548.
West African: SENEGAL MANDING-BAMBARA dugu kunsigi (JB) THE GAMBIA MANDING-MANDINKA kunindingyamo (Fox) MALI DOGON dèdu sasáá pŏ̃ (CG) FULA-PULAAR (Mali) fitirde jaule (A.Chev. fide JMD) MANDING-BAMBARA kafini, kononi, kononi mbi (A.Chev. fide JMD) SAMO bama ningui = *bird's broom* (A.Chev.) UPPER VOLTA FULA-FULFULDE (Upper Volta) balbalɗe (K&T) *m*balbaldi (K&T) ɓurundi *loosely for certain annual spp.* (K&T) GRUSI lulisacé (A.Chev.) MOORE lulisaga, lusaga (A.Chev.) NIGER FULA-FULFULDE (Niger) ndiriiri (ABM) SONGHAI táláatá-kàmbè ciréy = *the red hands of Tallaata*; meaning? (D&C) NIGERIA FULA-FULFULDE (Nigeria) huɗo mboju = *hare's grass* (JMD; J&D) nalle wainaɓe = *herds-men's henna*; *loosely applied to grass with purplish inflorescence or leaves* (JMD) HAUSA hakin furtau *from* hakin: *a kind of grass* (JMD; ZOG) lállèn birii = *monkey's henna* (JMD) lállèn shamuwa = *henna of the stork* (Taylor) tsintsiyar fadama *from* tsintsiya: *a stout grass of which brooms are made*; fadama: *swamp, marsh* (Grove) YORUBA irun awó *from* awó: *guinea fowl* (JMD)

A densely caespitose perennial grass, culms to 60 cm high, on well-drained open places, shallow sandy soils, especially in crevices of rock outcrop, wide-spread across the Region from Mauritania to W Cameroons, and throughout tropical Africa and into S Africa, SW Africa and Madagascar.

It is sometimes very abundant and forms a sort of turf (5). The foliage is a good fodder-grass relished by livestock at all times (Senegal: 1, 2; Ghana: 3; Uganda: 6). Cattlemen in Niger say that it increases milk production of their

357

stock and makes the milk tasty (10). In Sudan, however, it is considered to be only of limited value (8) restricted to the beginning of the rains (4). Hares and rabbits are reported to eat it (1, 5).

The grass mixed with other grasses is used in the Region for thatching (1, 5). In the Central African Republic it is used to wrap earthnuts as protection from animal damage – hence the Suma name meaning 'mice cannot eat it' (7).

The grain is collected in Jebel Marra for human need in time of dearth (9).

References:

1. Adam, 1954: 99. 2. Adam, 1966,a. 3. Beal 37, K. 4. Beshir 404, K. 5. Dalziel, 1937: 548. 6. Dyson-Hudson 254, K. 7. Fay 4492, K. 8. Harrison 903, K. 9. Hunting Tech. Surveys, 1968. 10. Maliki, 1981: 48.

Sporobolus helvolus (Trin.) Dur. & Schinz

FWTA, ed. 2, 3: 408. UPWTA, ed. 1, 548.

West African: **MALI** FULA-PULAAR (Mali) sakatéré (Davey) MANDING-BAMBARA schakatee (Davey) TAMACHEK afer (A.Chev.) **NIGER** FULA-FULFULDE (Niger) ngarziiri (ABM) **NIGERIA** HAUSA màndà-mábdàà (Golding; ZOG)

A tufted perennial grass with slender stolons, culms to 25 cm high, in open sandy situations, seasonally flooded, across the Sahel from Mauritania to N Nigeria, and extending to the Horn of Africa and E Africa; into Arabia, India and Burma.

It is a good fodder grass (4), relished by cattle in Senegal (1). It is well liked by all livestock throughout the year in Niger where herdsmen claim that it adds flavour to the milk and increases production (8). In Sudan it is held to be a valuable dry season fodder (7, 10), though in Kordofan it is said not to be grazed in a dry state (2). In Kenya it is grazed by all stock and wild game (5, 9, 11). In India it is recognised as a good fodder for sheep and goats (6) and for camels and horses as well as cattle, and that it makes a good hay (3). Burmese material has been analysed as containing: protein 4.18%, carbohydrates 59.62% and fibre 26.54% (3). It is a grass which dogs in Kenya will eat to promote vomiting (9).

The crushed plant is added to an infusion of tea and the liquid is drunk in India for malaria (3).

The grass serves as thatching for huts in Kenya (9).

References:

1. Adam, 1966,a. 2. Baumer, 1976: 114. 3. Chadha, 1976: 24–25. 4. Dalziel, 1937: 548. 5. Glover & al. 2969, K. 6. Gupta & Sharma, 1971: 93. 7. Harrison 982, K. 8. Maliki, 1981: 56. 9. Mwangangi 1462, K. 10. Peers M.O.6, K. 11. Stewart 486, K.

Sporobolus ioclados (Trin.) Nees

FWTA, ed. 2, 3: 407.

A caespitose perennial grass, sometimes stoloniferous, culms 30–60 cm high, of dry clay soil and hard pans, usually alkaline, seasonally moist, of the Sahel in Mauritania to Niger and widely dispersed throughout tropical Africa, and into India.

It is grazed in all African territories by stock and game (Somalia: 6 (except for camels), 9; Kenya: 3, 14; Tanganyika: 1 (game on the Serengeti Plain), 7, 10 (gazelle), 11; Zimbabwe: 2; Botswana: 5; etc.). Growing on salt pans it is said to have a salty tang (4). In Zambia it is reduced to ash, lixivated and evaporated to yield a salt (8, 12). It has been recorded that commando units operating for months in the South African bush during the Boer War were able to obtain sufficient salt minerals to remain healthy by cooking meat over a fire made up of this grass (Viljoen, fide 8).

The presence of the grass is taken in Zambia to indicate soil unfavourable for cultivation (12).

In Sudan the long flowering culms are bundled to make brooms (13).

References:

1. Brooks 64, K. 2. Davies D.1356, K. 3. Glover & al. 2976, K. 4. Jackson 1206, K. 5. Kelaole s.n., K. 6. Kuchar 17689, K. 7. Leippert 5597, K. 8. MacOwen s.n., 12/4/1903, K. 9. McKinnon S.13, S.233, K. 10. Oteke 236, K. 11. Proctor 3162, K. 12. Scudder 35, K. 13. Simpson 7144, K. 14. Veterinary Officer, Isiolo, 1615, K.

Sporobolus infirmus Mez

FWTA, ed. 2, 3: 410.
West African: SIERRA LEONE KORANKO babwiye (NWT) MENDE munyi (NWT) NIGERIA YORUBA irun awo (Macgregor)

A variable slender annual grass, culms to 30 cm high, of rocky outcrops in Guinea to W Cameroons.

Though known to peoples in Sierra Leone and Nigeria, no usage seems to be recorded.

Sporobolus jacquemontii Kunth

FWTA, ed. 2, 3: 410.
West African: LIBERIA MANO ni mo (Har.) GHANA DAGBANI nàgsàa (FRI) NIGERIA HAUSA tsintsiyar Kwaro *from* tsintsiya: *a brush*; Kwaro: *a tribe in Zaria* (Taylor) IGBO (Okpanam) ǫdo nlili (NWT)

A tufted perennial grass, culms 50–70 cm high by 1–2 mm in diameter, similar to *S. pyramidalis* (q.v.), even perhaps a depauperate form, native of tropical America and introduced into W Africa and recorded in Sierra Leone to W Cameroons.

At Ibadan, Nigeria, it is reported to be a noxious weed of lawns (3). It has value as fodder being grazed by stock on the Benue Plateau (2) and cut for feeding to horses in Ghana (1).

References:

1. Irvine 4728, K. 2. Kennedy 8027, K. 3. Townrow 19, K.

Sporobolus microprotus Stapf

FWTA, ed. 2, 3: 407.
West African: UPPER VOLTA MOORE liudi sagha (Scholz) lussa (Scholz) NIGERIA BURA babatu (Kennedy)

A loosely-tufted annual grass, culms erect or ascending to 30 cm high, of roadsides and waste places of the Sahel and Soudanian zone from Senegal to N Nigeria, and extending across Africa to Eritrea and Kenya.

In Somalia (4) and in Uganda (3) it is said to be a good pasture grass eaten by all stock. In Kordofan it is taken at all stages of growth (1). It is a weed of cultivation in N Nigeria (2), and in Eritrea a prostrate form of rosette growth helps in fixing sand-dunes (5).

References:

1. Baumer, 1975: 14, as *S. humifusus* var. *cordofanus* Stapf. 2. Kennedy FHI.8005, K. 3. Liebenberg 176, K. 4. McKinnon S.83, K. 5. Stower & al. 4026, K.

GRAMINEAE

Sporobolus molleri Hack.

FWTA, ed. 2, 3: 408.
West African: GUINEA FULA-PULAAR (Guinea) sigiri (Adames)

A tufted annual grass, culms to 30 cm high, of cultivated land and pathsides, in Guinea to N and S Nigeria and Fernando Po; across Central Africa in Zaïre, Tanganyika and Zimbabwe.
The plant is a weed of cultivated land.

Sporobolus montanus Engl.

FWTA, ed. 2, 3: 407.

A densely caespitose perennial grass of montane locations in NE Nigeria and W Cameroons.
On the Mambila Plateau this is recorded as being the first grass to regenerate to provide browsing after burning (1).

Reference:

1. Chapman 2744, K.

Sporobolus myrianthus Benth.

FWTA, ed. 2, 3: 410.

A loosely-tufted perennial grass, culms slender to 90 cm high, of open weedy places in wooded montane savanna in N Nigeria, and in E Africa and to Angola.
It is a common member of grass pasture in E Africa and Malawi where it is grazed by stock (Kenya: 1, 4; Malawi: 2, 3).
In Malawi the mature panicle culms are bundled into brushes (2).

References:

1. Bogdan AB.1734, 1806, K. 2. Fenner 206, K. 3. Fenner 208, K. 4. Leippert 5170, K.

Sporobolus natalensis (Steud.) Dur. & Schinz

A tufted perennial grass, culms to 120 cm high, of roadsides and disturbed places, native of E Cameroun to Ethiopia and to S Africa, now recorded in Nigeria and W Cameroons.
The culms are used in Ethiopia in basket-making (1).

Reference:

1. Tadesse Ebba 747, K.

Sporobolus nervosus Hochst.

FWTA, ed. 2, 3: 410.

A densely caespitose perennial grass, culms 15–60 cm high, of open places with shallow soil in bushland, recorded only in Mauritania in the Region, but occurring widely in eastern Africa from Ethiopia to Tanganyika, and in SW Africa and Arabia.

It is grazed by all stock in Somalia (1, 2). It dies down rapidly at the onset of the dry season but quickly produces regeneration immediately after rain, so conspicuously so that it has a Somali name meaning: 'Why should I wait?' (1).

References:

1. Keoch 129, K. 2. McKinnon S.229, K.

Sporobolus panicoides A. Rich.

A loosely tufted annual grass, culms to 1 m high, on dry sandy soil in shade of deciduous bushland, or on seasonally wet soils, native of E and S tropical Africa, now recorded in Niger.

In Uganda (4) and Malawi (2) it is grazed by all stock at all stages of growth. It is said to be good food for guinea fowl and other birds in Zimbabwe (3), where the grain itself is collected for human consumption (1, 3).

References:

1. Crook 2364, K. 2. Fenner 291, K. 3. Mshasha 36, K. 4. Wilson s.n., 7/1954, K.

Sporobolus paniculatus (Trin.) Dur. & Schinz

FWTA, ed. 2, 3: 407.
West African: NIGERIA FULA-FULFULDE (Adamawa) huɗo mbujo (Saunders) FULFULDE (Nigeria) finde (Saunders) HAUSA lallen biri = henna of the monkey (Saunders) wichiyar zoomoo = tail of the rabbit (BM)

An annual grass, solitary or loosely-tufted, culms erect or ascending to 60 cm high, of rocky sites, roadsides, and waste places in open bushland or wooded grassland, in the whole Region from Senegal to W Cameroons, and widespread throughout tropical Africa, and into Madagascar and Mexico.

The plant is a weed of cultivated land (1). It forms a low spreading turf in Sierra Leone giving good pasture grazing (1, 2).

References:

1. Cole, 1968. 2. Deighton 3564, K.

Sporobolus pectinellus Mez

FWTA, ed. 2, 3: 410.
West African: IVORY COAST ? kubégnin (A&AA)

A delicate tufted annual with culms to 45 cm high, widely dispersed on stony outcrops over the Region from Senegal to S Nigeria and in Zaïre and Uganda.

Sporobolus pellucidus Hochst.

FWTA, ed. 2, 3: 410.

A densely caespitose perennial grass, culms to about 60 cm high, of open places of deciduous bushland, recorded only in Niger, but occurring widely in eastern Africa from Ethiopia to Zambia.

It is heavily grazed by stock in Somalia (3) and Uganda (2, 4). It is reported subject to attack by army worms in Kenya (1).

References:

1. Brown 1952.G, K. 2. Liebenberg 1754, K. 3. McKinnon S.134, K. 4. Thomas, AS Th.3510, K.

GRAMINEAE

Sporobolus pyramidalis P. Beauv.

FWTA, ed. 2, 3: 408. UPWTA, ed. 1, 548.

English: rat's tail grass.

West African: SENEGAL MANDING-BAMBARA ménu (JB) tura basa (JB) **SIERRA LEONE** LIMBA pɛlis (NWT) LOKO kɔ-bwɛrɛ (JMD) koebwɛre (NWT) mendo (NWT) MENDE mbowi-hei = *female 'mbowi'*, *a loose term for certain grasses* (FCD) foni *a general term* (NWT) kɔli-gbiti (auctt.) SUSU filirasaxai (NWT) kuradagi (NWT) TEMNE sonta tusip (NWT) **LIBERIA** MANO nĩ mo (Har.) **MALI** FULA-PULAAR (Mali) burdi (A.Chev.) ganségui (A.Chev.) MANDING-BAMBARA tura basa (A.Chev.) wolo kaman (A.Chev.) **UPPER VOLTA** GRUSI nama nazan (A.Chev.) GURMA moki piégu (A.Chev.; K&B) MANDING-BAMBARA tura baoa, wolo kaman (K&B) MOORE ganga (Scholz) gansacé (A.Chev.) gansaga (Scholz; A.Chev.) sompiga (A.Chev.; K&B) **GHANA** NANKANNI saga (JMD) **NIGERIA** FULA-FULFULDE (Adamawa) cekelgol (J) FUL-FULDE (Nigeria) gerorce (JMD) HAUSA jaja karfi (RES) tsintsiya (RES) tsuntsiar gero (J) IGALA iyà-òkólō (Odoh; RB) IGBO ilulọ enyi nnono (JMD) MAMBILA jiru (Chapman) YORUBA motisan (JMD)

A densely tufted perennial grass with culms 90–160 cm high, 2–5 mm diameter at base; a waterside grass and of disturbed land and occurring over most of the Region from Senegal to W Cameroons, and present widely over Central and E Africa, and tropical S Africa, in Yemen and perhaps in NE S America. This species is very closely related to *S. jacquemontii* (q.v.) with which it intergrades.

The plant is of a graceful habit and is suitable in garden borders as an ornamental (7). It is a fetish plant to some W African people, and in Adamawa, N Nigeria, Mambila take their tribal oaths on it (17).

It is reported as a weed of rice-padis in Bendel State, Nigeria (14). It provides forage for stock especially when young and tender (7; Senegal: 1; Sierra Leone: 26; Ghana: 5; Zaïre: 25; Sudan: 24; Ethiopia: 13; Uganda: 20; Kenya: 15, 21, 23; Tanganyika: 22; Malawi: 11; Zambia: 2; etc.). In Kordofan besides being browsed in a green state, cattle will eat the straw (4), and in Tanganyika it is grazed at all stages of growth (22).

In Mali the plant is plaited to make bowls and baskets (7). Women's skirts are made of it in the Central African Republic (10). In Sudan it is roved into cord (8) and rope (16). There was at one time an interest in it for paper manufacture (7). It has been used as a filling material for matresses, and for mixing with mud in house-building (28). In N Nigeria the culms are bundled together to make brooms (18). Fula plait it into *dekko*-matting (29).

In the Central African Republic, the plant is reported as being burnt to obtain a vegetable salt (9). It has medicinal applications. The whole plant is used in E Africa as a styptic (3), and in Tanganyika as a snake-bite antidote (27). There also it is made in a cryptic presciption given with other medicines and applied with charms, amulets and magic 'to make babies strong' (6). Medicinemen in Sudan use it in an unspecified way (8). In the Man region of Ivory Coast, a decoction of the whole plant is given in draught as a counter-poison (19).

The grain is edible and is collected in time of famine (7). It is regularly collected for consumption in the Omo Valley in Ethiopia (12). Birds eat the seed in Kenya (15) and the Central African Republic (9).

References:

1. Adam, 1966,a. 2. Appleton 20, K. 3. Bally, 1937: 10–26, as *S. indicus*. 4. Baumer, 1975: 114. 5. Beal 38, K. 6. Culwick 21, K. 7. Dalziel, 1937: 548. 8. Evans-Pritchard 11, K. 9. Fay 4487, K. 10. Fay 4489, K. 11 Fenner 237, 250, K. 12. Fukui & Fukui 1163, K. 13. Gilbert & Gilbert 1557, K. 14. Gill & Ene, 1978: 182–3. 15. Glover & al. 137, 421, 698, 1866, 1906, 1991, 2064, 2221, K. 16. Harrison 2, K. 17. Hill s.n., ad not. 6/12/1926, K. 18. Kennedy 7296, K. 19. Kerharo & Bouquet, 1950: 253. 20 Lock 68/28, K. 21. McDonald 989, K. 22. McGregor 35, K. 23. Magogo & Glover 503, 688, K. 24. Peers KaM.24, K.O.19, K. 25. Schautz 737, K. 26. Scott-Elliot 5250, K. 27. Tanner 4704, K. 28. Wiliams, RO, 1949: 453, as *S. indicus*. 29. Jackson, 1973, as *Sporobolus verticillatus*.

Sporobolus robustus Kunth

FWTA, ed. 2, 3: 408. UPWTA, ed. 1, 548.

West African: **SENEGAL** DIOLA dămbur (JB) dămbus (JB) haiselle (A.Chev.) WOLOF dămbur (JB) darak (JB) **THE GAMBIA** MANDING-MANDINKA wontokado (Fox) **SIERRA LEONE** BULOM (Sherbro) boncho (FCD)

A stoloniferous perennial grass, culms 60–90 cm high, of saline soils, sand-dunes and estuaries from Senegal to W Cameroons and extending along the whole Atlantic seaboard to Angola.

The plant is a grass of the littoral occupying a niche to landward of the sand-binding pioneer species *S. virginicus* on lagoon and estuary margins and on brackish soil (1, 3, 4, 6, 8) becoming dominant in salt-marsh (2) and in salt-spray zone (7). Its presence is indicative of salty soil (5, 9).

References:

1. Adam, 1954: 99. 2. Broadbent 26, K. 3. Clayton, 1965: 296. 4. Dalziel, 1937: 548. 5. Deighton 3543, K. 6. Deighton 4116, K. 7. Rose Innes GH.30358, K. 8. Rose Innes GH.30888, K. 9. Trochain, 1940: 395.

Sporobolus sanguineus Rendle

FWTA, ed. 2, 3: 406.

English: crucifer grass (Ghana: Rose Innes).

A tufted perennial grass, culms to 60 cm high, of rocky outcrops in Guinea to N and S Nigeria, and dispersed throughout tropical Africa.

The branchlets of the infloresence are borne in cruciform tiers, hence the English name. No usage is recorded but it probably provides forage for stock.

Sporobolus spicatus (Vahl) Kunth

FWTA, ed. 2, 3: 408. UPWTA, ed. 1, 549.

English: salt grass (Kenya: Livingstone & Delamere).

West African: **MAURITANIA** ARABIC (Hassaniya) akrich (AN) izigzig (AN) **SENEGAL** WOLOF ndamséki (JB) **MALI** ARABIC (Mali) beurgu (A.Chev.) TAMACHEK beurgu (A.Chev.) **NIGERIA** HAUSA wútsíyàr ɓééráá = *rat's tail* (auctt.)

A shortly-tufted perennial, stoloniferous and matting, culms 30–60 cm high with pungent leaves, on alkaline or saline soils of the Sahel from Mauritania to N Nigeria, and throughout the drier parts of Africa from the Mediterranean to the Transvaal, and through the Middle East into India.

The plant is said to be the most alkali-tolerant grass in Kenya (9) and in Tanganika at Lake Rukwa it grows to a pure stand over soda-impregnated soil (11). With its stoloniferous habit it is a coloniser and important fixer of sand-dunes in the sandy littoral of Senegal (1, 3, 15). Inspite of its pungent leaves, its decumbent creeping habit can produce a soft turf which might have scope for making a good lawn (Ethiopia: 6; Somalia: 7).

Camels will browse it (5; Senegal: 1, 2; Kenya: 13), but cattle may or may not take it, presumably according to the stage of growth. They do not graze it in Senegal (2), and a vernacular name for it in Somalia means 'Mouth sore through gnawing' (4). Other reports are of good grazing (Mauritania: 14; Kenya: 9; Tanganyika: 10; Somalia: 8; etc.). The foliage has a salty taste.

In Tibesti the grain is eaten (12). The plant, presumably the grain, is reported in Tanganyika, Lake Rukwa area, to be known as a famine-food (11).

GRAMINEAE

References:

1. Adam, 1954: 99. 2. Adam, 1966,a. 3. Broadbent 36, 175, K. 4. Collenette 21, K. 5. Dalziel, 1937; 549. 6. Gilbert 2553, a, K. 7. Gillett 4326, K. 8. Gillett 4737, K. 9. Livingstone & Delamere 81–1, K. 10. Michelmore 718, K. 11. Michelmore 1165, K. 12. Monod, 1950: 70. 13. Mwangangi & Gwynne 1258, 1263, K. 14. Naegelé, 1958,b: 895. 15. Vanden Berghen, 1979.

Sporobolus stapfianus Gandoger

FWTA, ed. 2, 3: 410.

A densely caespitose perennial grass, culms slender, to 40 cm high, of sandy, or shallow soils, or rocky locations in wooded grassland or deciduous bushland in N Nigeria, and across Africa to Ethiopia, and E Africa south to Natal, and in Madagascar.

It grazed by stock in Uganda (3) and Kenya (1, 4), but it has insufficient bulk to be of importance as a fodder. Rabbits and hares eat it in Kenya (2).

References:

1. Glover & al. 262, 287, K. 2. Glover & al. 93, 122, 262, 711, 1990, K. 3. Snowden 1331, 1498, K. 4. Stewart 315, 440, K.

Sporobolus stolzii Mez

FWTA, ed. 2, 3: 407.
West African: SENEGAL SERER ndiisis (JB) nduyuy (JB)

A tufted annual grass with culms to 60 cm high, of roadsides and waste places in the Sahel from Senegal to Niger, and dispersed from Ethiopia southwards to Zimbabwe.

In Senegal it is very short-lived but provides browsing on sandy and calcareous soils (1). It is described as a good pasture grass in Tanganyika (3) readily taken by all stock (2).

References:

1. Adam, 1954: 99, as S. granularis Mez. 2. Emson 81, K. 3. Rounce C.11, K.

Sporobolus tenuissimus (Schrank) Kuntze

FWTA, ed. 2, 3: 411.

A weakly-culmed annual, 30–90 cm tall; a ruderal of roadsides and waste places, recorded from Senegal, Guinea, Liberia, S Nigeria and Fernando Po, and across Africa to Tanzania, to Arabia, Mauritius and India; also in tropical America.

There is no record of its value as forage, but it is occasionally cultivated as an ornamental (1).

Reference:

1. Adam, 1954: 99, as S. minutiflorus Link.

Sporobolus tourneuxii Coss.

FWTA, ed. 2, 3: 408.

A tough perennial grass, culms to 30 cm high, almost woody at base, of saline soils in the Sahel of Mauritania, Tunisia, Somalia, Arabia and Pakistan. In Somalia it is rated as a good grazing grass (1–4).

References:

1. Boaler B.101, K. 2. Gillett 4096, K. 3. McKinnon S.102, K. 4. Peck 217, K.

Sporobolus virginicus(Linn.) Kunth

FWTA, ed. 2, 3: 408.
West African: MAURITANIA ARABIC (Hassaniya) alech (AN) SENEGAL SERER géredièdi (JB) tèh (JB)

A perennial grass with long slender rhizomes, culms to 60 cm high, of sandy seashores, and occasionally on saline soils inland, from Mauritania to Dahomey, and in E Africa, and on tropical seashores generally.
It is an important pan-tropical pioneer sand-binder colonising beach-heads down to high-water level and coastal sand-dunes (5), commonly recorded along the W African coastline (Senegal: 1, 4, 9; The Gambia: 3; Sierra Leone: 2, 6; Ghana: 8; Liberia, Dahomey, etc.) Its value as forage is described as being in Senegal only mediocre and this while young (1), but in Mauritania it provides good pasturage for all stock (7).
People in Mauritania remove grain from ant-hill granaries in time of dearth and grind it for cooking into cakes (7).

References:

1. Adam, 1954: 99. 2. Adames 35, K. 3. Blair-Rains 1, K. 4. Broadbent 178, K. 5. Clayton, 1965: 295. 6. Deighton 2680, K. 7. Naegelé, 1958,b: 896. 8. Rose Innes 32700, K. 9. Vanden Berghen, 1979.

Sporobolus sp. indet.

An unidentified species with odouriferous roots is used in the construction of a fish-lure called *codio-codio* used in the Senegal River (1).

Reference:

1. Busnel, 1959: 350.

Stenotaphrum dimidiatum (Linn.) Brongn.

FWTA, ed. 2, 3: 435.

A robust, stoloniferous perennial grass, dense sward-forming with culms to 30 cm high, native of the Indic basin on sandy foreshores, now introduced to the Region, present on the coast of Ghana, and inland in Upper Volta; also in E Africa and in Brazil.
The grass grows into a coarse mat. It gives a good succulent fodder (1, 3). By its vigorous growth it has use as a ground-cover and an ability to smother out weeds. In Zanzibar it is said to be useful on contour strip planting on

embankments and bunds (3). It withstands even fairly dense shade. In Kenya it is reported to be an useful lawn grass though a little bit coarse (2).

It has medicinal use in Brazil as a diuretic and sudorific (3).

References:

1. Chadha, 1976: 41. 2. Gillett 18692, K. 3. Williams, RO, 1949: 287–8.

Stenotaphrum secundatum (Walt.) O Ktze

FWTA, ed. 2, 3: 435. UPWTA, ed. 1, 549.

English: buffalo grass (Australia); pimento grass (Jamaica); St Augustine's grass (USA).

French: gross chiendent (Berhaut).

Portuguese: capin-de-Santo-Agostinho (Feijão).

A strongly stoloniferous perennial grass of the coastal region of W Africa from Senegal to W Cameroons, and introduced inland, and widely dispersed to the east coast of N and S America, the Pacific basin and Australasia.

By its vigorous rhizomatous growth, it is a good sand-binder especially near the sea (3, 5, 7). When growing in shade the rhizomes are long-jointed, but in the sun they are short-jointed, thus making a dense compact sward, and a good turf for lawns (1, 2, 4–6, 8, 9). It is said to be of a pleasing green colour, but requires frequent mowing, and unfortunately is not hard-wearing.

It produces a good cattle-feed when young; and is widely grown in several countries (1, 2, 4–6).

The plant has medicinal use in S Africa for urinary stones, gall stones and stomach pains especially due to acidity and wind. The root serves as a diuretic in Brazil. Plant extracts have given negative anti-biotic tests (10).

References:

1. Adam, 1954: 99. 2. Burkill, IH, 1935: 2075, with reference to chemical analysis. 3. Cole, Ayodele 638, K. 4. Duffey 1/49, K. 5. Dalziel, 1937: 549. 6. Maitland 25C, K. 7. Morton A.298, K. 8. Scholz 113, K. 9. Walker & Sillans, 1961: 195. 10. Watt & Breyer-Brandwijk, 1962: 488.

Stipagrostis acutiflora (Trin. & Rupr.) de Winter

FWTA, ed. 2, 3: 376.

West African: MAURITANIA ARABIC (Hassaniya) asserdun (Bayard) sfar (Bayard)

A slender perennial grass with culms to 45 cm high, of dry waste places in Mauritania, Mali and Niger, and in Egypt and Sudan.

Stipagrostis hirtigluma (Steud.) de Winter

FWTA, ed. 2, 3: 371.

A tufted annual grass, culms to 45 cm high, of arid waste places in Mali and N Nigeria, and extending eastwards through Sudan, Ethiopia and Kenya to the Middle East and India, and southwards in Africa to Angola, Botswana and S Africa.

No usage is recorded of the grass in the Region. In Sudan all stock graze it and it is said to be the best camel grass for dry season grazing (2, 3). In Rajasthan, India, under dry conditions it gives a palatable fodder before seeding (1).

References:

1. Gupta & Sharma, 1971: 62, as *Aristida hirtigluma* Edgew. 2. Harrison 914, 955, K. 3. Peers M.O.3, K.

Stipagrostis plumosa (Linn.) Munro

FWTA, ed. 2, 3: 376.
West African: **NIGER** ARABIC (Niger) nsil (Grall) TEDA mali (Grall)

A densely caespitose perennial, or annual under very arid conditions, in Mauritania, Senegal, Mali and Niger, and in the Sahara to Sudan, Arabia, Middle East to India.
No usage is recorded in the Region. In Algeria (1) and the western Sahara (2) it provides excellent forage for camels and horses.
The grain is edible (2).

References:

1. Maire, H, 1933: 61. 2. Monteil, 1953: 17.

Stipagrostis pungens (Desf.) de Winter

FWTA, ed. 2, 3: 376. UPWTA, ed. 1, 521, as *Aristida pungens* Desf.
West African: **MAURITANIA** ARABIC (Hassaniya) sbot (Bayard) **MALI** ARABIC (Mali) drinn (JMD; HM) TAMACHEK tulult (JMD) **NIGER** ARABIC (Niger) sbat (Grall) KANURI (Kanem) madiugu (JMD) TEDA madjiugu (Grall)

A stiff coarse grass of sand-dunes and dry waste places in Mauritania to Niger, and in N Africa and the Sahara.
It produces good forage for camels and other stock (1–4), especially while still green. The culms are used in the western Sahara for weaving into mats, bags, etc. (4). Rope and pack saddles are also made (4). Plaited leaves produce an inferior cordage that is used in the Hoggar region for camel harness (3). The green culms are used as probes by local doctors in the western Sahara in treating wounds (4).
The grain is edible and is eaten by Moorish people of the Sahara region (2–6). Nutritionally it is claimed to be the equivalent of barley (3). In the Hoggar it is fed to horses (3).

References:

1. Adam, 1966. 2. Dalziel, 1937: 521. 3. Maire, H, 1933: 62. 4. Monteil, 1953: 18. 5. Monod, 1950: 69. 6. Williams, JT & Farias, 1972: 13.

Stipagrostis uniplumis (Licht.) de Winter

FWTA, ed. 2, 3: 376.

A caespitose perennial grass, culms to 60 cm high, of arid bushland or semi-desert scrub in Senegal, Mali and Niger, and throughout much of tropical Africa and S Africa.
It is grazed with relish by cattle in Senegal (1), and generally in the drier areas of the Region (3). In Kenya it is taken by all wild game and domestic stock (2) and birds use the straw for nest-building. Grazing, however, requires some care as losses, especially of young sheep, have been recorded as a result of ingesting the panicles from which seed-hairs form masses in the stomach causing obstruction (4). Under some conditions the grass may produce hydrocyanic acid (4).

References:

1. Adam, 1966,a. 2. Mwangangi & Gwynne 1094, K. 3. Schnell, 1953,b: no. 44. 4. Watt & Breyer-Brandwijk, 1962: 465.

Streptogyna crinita P. Beauv.

FWTA, ed. 2, 3: 362, 365. UPWTA, ed. 1, 549, as *S. gerontogaea* Hook.f.
West African: **SENEGAL** BANYUN damordégad (K&A) DIOLA èmokok (JB) **SIERRA LEONE** KISSI dilag-boinyondo, malomboenda (FCD) KONO tuabiya (FCD) KORANKO sibile

GRAMINEAE

(NWT) sibire (NWT) LIMBA naganage (NWT; JMD) LOKO buragai, kongeŋgi (NWT) naga-naga (FCD) nanaka (JMD) nanaŋa (NWT) MENDE kaiyowo, kayowo *when not in seed* (FCD) yoyavi *esp. when in seed* (auctt.) TEMNE ε-, *ka*-sarɔk (auctt.) ε-sul (NWT) YALUNKA fɔtom-boŋsεkhεdekhɔle-na (FCD) **LIBERIA** MANO gie kia ku (Har.) **IVORY COAST** AKAN-ASANTE atiéréfos (B&D) BAULE dontrè (B&D) KRU-GUERE (Chiehn) kwamossé, kwamurè, pamussè, yoè (B&D) 'NÊKÉDIÉ' kpa mussé (B&D) **GHANA** ADANGME-KROBO detshεngã (FRI) wondu wondu (FRI; AS Thomas) AKAN-ASANTE etwã (FRI; E&A) FANTE kyeretwẽ = *catch antelope* (FRI; E&A) **NIGERIA** IGBO ako ọzọ (NWT; JMD)

A perennial grass, culms 30–75 cm high, erect, rhizomatous, flower spike to 20 cm long with spikelets 20–30 mm bearing awns to 25 mm long, strongly barbed; of the ground layer of forest throughout the Region from southern Senegal to Fernando Po, and across tropical Africa and into India and Ceylon.

The grass occurs along forest tracks through clearings and persists on cacao plantations, orchards and cultivations where there is shade. It is considered a noxious pest particularly at the fruiting stage as the strongly barbed awns catch in the hair of legs of passers-by, and are difficult and painful to remove (3–5, 7, 10, 11). They catch also on clothing and on the fur of animals. The plant is commonly used to catch mice and rats: a number of inflorescences is rolled up into a ball which is placed on the rodents' runs, and the animals become so entangled as to be unable to rid themselves of the attachment (2, 4, 13). In Gabon they are put out on lagoons as anchored inducement to water-birds to settle upon, thus becoming caught and captured (13). Sophistry arising from such practical applications leads to the use of the plant in Gabon (13) in a charm mixture to capture the confidence of influential personages, and in Congo (Brazzaville) (1) as a love philtre. In Nigeria Igbo use the plant in an unexplained manner to make an *idaii* charm (12).

In Sierra Leone the leaf is boiled in water which is then used to cook rice to be given to dogs as a vermifuge (6). A leaf-decoction is used in Ivory Coast as an efficacious lotion on certain skin-eruptions (2). Plant-sap is also used on sores and fractures (2). The leaf and rhizome are cooked in Tanganyika for taking for pneumonia (8). The plant, with other drug plants, is taken [probably as a decoction in draught] in Congo (Brazzaville) for various localised pains, e.g., costal pain, kidney pain, and rheumatism, and as an emmenagogue and abortifacient (1). Pith is used in Casamace, Senegal, for dental caries and all sorts of toothache (9).

References:

1. Bouquet, 1969: 131, as *S. gerontogaea* Hook.f. 2. Bouquet & Debray, 1974: 93, as *S. gerontogaea*. 3. Chipp 53, K. 4. Dalziel, 1937: 549, as *S. gerontogaea* Hook.f. 5. Dawkins D.459, K. 6. Deighton 838, K. 7. Fisher 1, K. 8. Haerdi, 1964: 212. 9. Kerharo & Adam, 1974: 657–8, as *S. gerontogaea* Hook.f. 10. Lane-Poole 344, K. 11. Morton & Gledhill SL. 295, FBC. 12. Thomas NWT. 2010 (Nig. Ser.), K. 13. Walker & Sillans, 1961: 195, as *S. gerontogaea* Hook.f.

Tetrapogon cenchriformis (A Rich.) WD Clayton

FWTA, ed. 2, 3: 399. UPWTA, ed. 1, 549, as *T. spathaceus* Hack.
West African: **MALI** TAMACHEK tadjemait (A.Chev.) **NIGER** ARABIC (Niger) bené ekoissé (A.Chev.) bu areiba (A.Chev.)

A tufted annual grass, to 60 cm tall, of dry sandy or rocky situations in the Sahel from Mauritania to Niger, and in Cape Verde Island to Somalia and Arabia, and E Africa.

The grass appears on sand-dunes during the rains (4), and provides grazing for cattle in Senegal (1). In Kordofan (2) and in Sudan (3) it is recorded as producing mediocre and good grazing respectively in both the rainy and dry seasons. Sheep take it readily in Kenya (Masailand), especially while it is young (5).

References:

1. Adam, 1966,a, as *T. spathaceus.* 2. Baumer, 1975: 116, as *T. spathaceus* (Hochst.) Hack. 3. Beshir 24, K. 4. Dalziel, 1937: 549, as *T. spathaceus* Hack. 5. Glover & Samuel 2769, K.

Thelepogon elegans Roth

FWTA, ed. 2, 3: 473. UPWTA, ed. 1, 549.

West African: SENEGAL BASARI ɔ-ndeuss (after K&A) ɔ-ngén ɔ-tyérún (FG&G) NIGERIA FULA-FULFULDE (Nigeria) hanhande (JMD) labaho (Saunders) tagarawal (Taylor) HAUSA daàtà-dáátàá (ZOG) dáátanniyáá (ZOG) dáátanniyár dáácíí (ZOG) dáddáátàà (JMD; ZOG) dandata (JMD) ɗata-ɗata (JMD) ɗatanniya (auctt.) ɗatarniya (JMD; ZOG) ɗwaatanna (auctt.) ɗwaatarniya (JMD; ZOG) gíshírín dáwaakíí = *salt of the horse* (JMD; ZOG) KANURI kagera kagum *from* kagera: *bustard* (JMD)
NOTE: Hausa names allude to the bitter taste.

A coarse annual grass, culms prop-rooted, erect to 1 m high; of disturbed sites across the Soudanian/Sahel zone of the Region from Senegal to Niger and N Nigeria, and generally throughout tropical Africa and into India and Malesia.

The grass makes good pasturage on light Sahel soils (11). It is particularly valued as horse-feed (1, 3, 4, 7). In N Nigeria it is gathered at harvest time or after the rains, chopped up and fed to horses for about ten days to counteract the effects of their being too long on green fodder (5). It is distinctly bitter, a character giving rise to the Hausa name meaning 'horses' salt.' The other Hausa roots derive from *ɗachí:* bitterness (8). (See word list above.) Cattle are said to take it, either green or in a dry state (Nigeria: 5; Kordofan: 2; Sudan: 9) but seemingly not everywhere for stock in Dahomey (10) and in Tanganyika (6) are reported as refusing it.

Chemical analysis of Indian material shows crude protein 5.6%, crude fibre 28.2% and N-free extract 29.8% (3). Alkaloids *thelepogine* and *thelepogidine* and three unidentified bases are reported present (3, 12).

References:

1. Adam, 1966, a. 2. Baumer, 1975: 116. 3. Chadha, 1976: 207. 4. Dalziel 284, K. 5. Dalziel, 1937: 549. 6. Marshall 49, K. 7. Moiser 132,a, K. 8. Palmer H.73.35, K. 9. Peers Ka.M36, K. 10. Risopoulos 1256, K. 11. Roberty, 1961: 694. 12. Willamen & Li, 1970.

Themeda triandra Forssk.

FWTA, ed. 2, 3: 471. UPWTA, ed. 1, 549.

English: rooi grass (S Africa, India); red grass (India).

A polymorphic species, perennial, tufted, culms erect, 30 cm to 2 m high; of open deciduous bushland, often dominant; widespread across the Soudanian zone of the Region from Senegal to N Nigeria, and in E and central Africa, and tropical parts of the Old World.

The grass provides valuable fodder while still young and before anthesis. It is taken with relish by all stock (1, 3, 4, etc). It is commonly cut for hay in India (3). In Zaïre it is put into seed mixtures for pasture leys in farms (8). Recovery from cutting and grazing may be a bit slow (2). Analysis of grass grown at Bombay, India, gave on a dry weight basis: protein 2.52% and crude fibre 32.23%, with a variety of minerals (3). Wilted culms contain *hydrocyanic acid* and are a danger to livestock (3, 9, 10).

The presence of this grass is considered in S Africa to be an indication of the suitability of the land for growing pineapples (3).

The culms are used in Uganda for hut-building (5). Ubiquitously they serve as thatch, but in Lesotho the quality is said to be poor and not lasting (6). Culms have been found suitable for producing paper-pulp (3, 10).

A decoction of the roots is drunk in Tanganyika for dysmenorrhoea (7).

The seed is collected in India as human food in time of famine (3). Sheep farmers in southern Africa find the seed a nuisance as it becomes entangled in sheep's fleece, causing serious injury known as 'traumatic purulent dermatitis' (10).

References:

1. Adam, 1966, a. 2. Baumer, 1975: 116. 3. Chadha, 1976: 209, with chemical analysis. 4. Dalziel, 1937: 549. 5. Dyson-Hudson 427, K. 6. Guillarmod, 1971: 455. 7. Haerdi, 1964: 212. 8. Johnston 1097, K. 9. Quisumbing, 1951: 1024. 10. Watt & Breyer-Brandwijk, 1962: 465.

Themeda villosa (Poir.) A Camus

A tall tufted reed-like perennial grass, culms stout, to 1.8 m tall; native of tropical Asia in open spaces, and introduced to other tropical countries; now present on the Mambila Plateau of Nigeria.

The flowering plant is graceful and is sometimes grown as an ornamental.

The young shoots are sweet with a significant concentration of sugar. They are eaten in Malaya as a salad (1), and in India are grazed by stock (2). Later the leaves become hard and are worthless as fodder. The culms yield good quality paper-pulp except for difficulty in bleaching (1).

References:

1. Burkill, IH, 1935: 2147. 2. Chadha, 1976: 209.

Trachypogon spicatus (Linn.f.) O Ktze

FWTA, ed. 2, 3: 473.
West African: NIGERIA FULA-FULFULDE (Nigeria) celbiiho (J&D) juda (Saunders) HAUSA yandaya (Freeman)

A tufted wiry perennial grass with culms to 1.2 m high; of upland grassland and open bushy or woody grassland; from Ivory Coast to N Nigeria, and generally over tropical Africa and in tropical America.

The grass is palatable to animals while it is still young (2, 4, 6). It is used as thatch on houses (1, 3, 6). It has some capacity for pulping (6).

It is said to be a weed of plantations in Tanganyika but it is easy to control (5). The root secretes a substance which stimulates germination of the seed of *Tagetes minuta* (6).

The foliage has been used in China as a tea-substitute (6).

References:

1. Jackson, 1973. 2. McGregor 37, K. 3. Magogo & Glover 330, K. 4. Saunders 64, K. 5. Strang 8, K. 6. Watt & Breyer-Brandwijk, 1962: 448.

Tragus berteronianus Schult.

FWTA, ed. 2, 3: 413.
West African: UPPER VOLTA FULA-FULFULDE (Upper Volta) kebbe baali (K&T) kebbel baal (K&T)

A loosely tufted annual grass, culms erect or ascending to 60 cm high, a ruderal of weedy trampled places on poor sandy soils from Mauritania to Upper Volta, and widely dispersed over tropical Africa, and into Arabia, Afghanistan and China, and in warmer parts of America.

The grass is grazed by all stock (Kordofan: 1; Kenya: 3; Tanganiyika: 5). It is said to be an excellent sheep's grass in Tanganyika (2). The glumes with hooked spines cluster the spikelets together into little burs. They are troublesome in sheep's fleeces (3) and difficult to remove from any sort of woollen clothing (4).

References:

1. Baumer, 1975: 116. 2. Emson 234, K. 3. Glover & al. 514, K. 4. Goldsmith 48836, K. 5. Marshall 9, K.

Tragus racemosus (Linn.) All.

FWTA, ed. 2, 3: 413. UPWTA, ed. 1, 549.
West African: MAURITANIA ARABIC (Hassaniya) tinismirt (AN) MALI TAMACHEK abugur nekli (A.Chev.) tafado fodo (A.Chev.) UPPER VOLTA FULA-FULFULDE (Upper Volta) kebbe baali (K&T) kebbel baal (K&T)

A loosely spreading annual, culms to 30 cm high, a ruderal of roadsides and dry sandy waste places in Mauritania to Niger, and extending to Ethiopia and Somalia into Arabia, and in S Europe and N and S Africa.

The plant is a good fodder for all stock while young, (Mauritania: 7; Senegal: 1; Mali: 3; Kordofan: 2; Somalia: 5, 6; Zimbabwe: 8). After flowering the glumes develop hooked spines so that the harsh prickly flower-spike then makes the grass unpalatable. The hooked glumes attach themselves to bird's feathers and thus disperse the seed.

In Ethiopia the grass has been reported as a weed of irrigated cotton (4).

References:

1. Adam, 1966,a. 2. Baumer, 1975: 116. 3. Dalziel, 1937: 549. 4. Drake-Brockman 68, K. 5. Gillett 3929, K. 6. McKinnon S.17, K. 7. Naegelé, 1958,b: 896. 8. Smith, PA 1280, 3613, 3898, K.

Tricholaena monachne (Trin.) Stapf & CE Hubbard

FWTA, ed. 2, 3: 455.

A short-lived perennial, culms decumbent, rooting at lower nodes, ascending to 60 cm; on sandy sites, often foreshore within salt-spray zone, known only in Ghana in the Region, but widely dispersed elsewhere in tropical Africa and in Madagascar and the Mascarenes.

It provides grazing, said to be well liked by stock in Tanganyika (1). It is a common weed on old cultivations and colonises areas of sandy soil. In Zambia its tough creeping habit clothes areas of Kalahari sand (3) helping in stabilisation. The culms are used in Zimbabwe to make grain bins (2).

References:

1. Emson 256, K. 2. Taylor 136, K. 3. Trapnee 1035, K.

Trichoneura mollis (Kunth) Ekman

FWTA, ed. 2, 3: 393.
West African: NIGER HAUSA tchirki n'zomo (Bartha) SONGHAI bèr-káar bii (*pl.* -iò) (D&C) NIGERIA HAUSA garajin fadama BM

A slender tufted annual to 30 cm tall of dry bushland in the Sahel from Mauritania to Niger and NE Nigeria, and across Africa to Somalia and Arabia.

The grass is grazed by all stock in Senegal (1) and in Kordofan (2), but quantity is limited.

References:

1. Adam, 1966,a. 2. Baumer, 1975: 116.

GRAMINEAE

Tripogon minimus (A Rich.) Hochst.

FWTA, ed. 2, 3: 393.
West African: UPPER VOLTA FULA-FULFULDE (Upper Volta) bahel (K&T) bahel maccuɗo (K&T) mangel mangurla (K&T) NIGER HAUSA bubukuwa (AE&L) buubuuƙùwaa (AE&L) NIGERIA HAUSA búbbúúkùwáá (JMD; ZOG) buƙai (auctt.) buƙe (Bargery; JMD) ɗakesa (auctt.) ɗaskara (auctt.) diskara (ZOG) sabko búbbúúkùwáá *from* búbbúúkùwáá: *the short-legged pelican; alluding to the grass from the supposed habitat where the pelican alights; including other short grasses* (JMD)

A small densely-tufted perennial to 15 cm high, of shallow soils on rocky outcrops across the Region from Mauritania to N and S Nigeria, and throughout tropical Africa to S Africa.

It provides grazing for cattle in Senegal (1), and Kordofan (2) but limited in quantity by its diminuitive stature. It is valued because it precedes all other fodder grasses in producing new growth before the dry season ends (2, 3).

References:

1. Adam, 1966,a. 2. Baumer, 1975: 117. 3. Dalziel, 1937: 549.

Triraphis pumilio R. Br.

FWTA, ed. 2, 3: 376. UPWTA, ed. 1, 549.
West African: MALI TAMACHEK akachakar (A.Chev.)

A small tufted annual to 25 cm high of dry sub-desert locations in Mauritania to Niger, and in NE Africa and S Africa.

It is recorded as being a poor *gizu* grass in Sudan, growing in the desert in the early part of the dry season drawing on soil moisture said to be brought to the surface by dry cold north wind (3). It furnishes limited browsing in Mali (2) and the Kordofan (1) for donkeys and sheep.

References:

1. Baumer, 1975: 117. 2. Dalziel, 1937: 549. 3. Harrison 942, 968, K.

Triticum aestivum Linn.

FWTA, ed. 2, 3: 371. UPWTA, ed. 1, 550, as *T. durum* Desf. and *T. vulgare* Vill.
English: wheat.
French: blé.
Portuguese: trigo.
West African: MALI ARABIC (Mali) gemah (JMD) MANDING-BAMBARA halkama (JMD) SONGHAI halkama (JMD) TAMACHEK ihered, ired (JMD) timzin *but more usually refers to* Hordeum (JMD) TOGO KANURI luwaidu *a food preparation* (JMD) NIGERIA ARABIC-SHUWA gameh (JMD) k'ameh, el k'ameh (JMD) BIROM álkámà (LB) FULA-FULFULDE (Nigeria) alkamari (JMD) HAUSA álkámàà (auctt.) KANURI láàmà (A&S) legamma, shegar (JMD) PERO àlkámà (ZF)

Products: TOGO HAUSA alfintar, alkaki, alkubus, bashi, dashishi, dawúde, fankaso, funkaso, kuskus *food preparations* (JMD) algaragi *cakes boiled in oil and seasoned with tamarind and natron* (JMD) algaragis *a food without oil* (JMD) gurasa *baked buns with seasoning* (JMD) kirikirino *a product like macaroni* (JMD) rauno *the straw chopped up and mixed with clay for building* (JMD) tuwon ɓaure *a finely ground flour with butter* (JMD)

An annual herb sending up a clump of culms to about 50 cm high bearing bearded or not bearded spikes, of the soudanian zone in Mauritania, Mali and N Nigeria, and cultivated as the common bread wheat in all temperate countries of the world.

Wheat has evolved in the NE African/Mesopotamian evolutionary area. It was perhaps being grown in Egypt by 6,000 B.C. by which time a system of agriculture had been well-established. By 3,000 B.C. a knowledge of cereal cultivation was distributed throughout the Sahara via the 'Saharan Fertile Crescent', i.e., the chains of highlands forming a migration route between the Nile valley and W Africa (9). Wheat has entered N Nigeria under the Hausa: *álkámàà*, from Old Egyptian: *qemhu* meaning 'nourishment', with the definitive Arabic prefix (7), and similarly into adnate W African languages.

Wheat is grown for its grain in N Nigeria as a wet season crop, or under irrigation in the dry season. Indigenous cultivars ripen in 50–55 days and yield is low. It is traded in markets being no longer a rich man's crop owing to imports of flour and extended cultivation (1–6, 8). A variety of food preparations is commonly made by Hausa – see above for vernacular names.

Wheat straw is compounded with clay as a standard building material (1).

References:

1. Dalziel, 1937: 550, under *T. durum* Desf. & *T. vulgare* Vill. 2. Diarra, 1977: 45, as *T. vulgare* L. 3. Golding & Gwynne, 1939: 631–43. 4. Irvine, 1948: 264–5, as *T. vulgare*. 5. Mauny, 1953: 691, as *T. vulgare/T. durum*. 6. Miège, 1963: 335. 7. Portères, 1958: 311–45. 8. Zeven, 1974: 137–44. 9. Zeven, 1980: 40–41. See also: Watt & Breyer-Brandwijk, 1962: 488.

Urelytrum agropyroides (Hack.) Hack.

FWTA, ed. 2, 3: 504, as *U. auriculatum* CE Hubbard; *U. pallidum* CE Hubbard; *U. gracilius* CE Hubbard and *U. semispirale* WD Clayton.
West African: NIGERIA FULA-FULFULDE (Nigeria) ande (Saunders)

A perennial grass, tussock-forming, with culms 90–120 cm tall; in open submontane wooded grassland; recorded only in Ghana and N Nigeria, and sporadically in E Africa to S Africa and Madagascar.

The root is burnt in Malawi and the ash is applied to septic wounds (2) and under supervision of a Malawi assistant in Harare Government Hospital in Zimbabwe, the roots are collected for use against general stomach complaints (1).

References:

1. Biegel 4159, K. 2. Fenner 239, K.

Urelytrum digitatum K Schum.

FWTA, ed. 2, 3: 504, as *U. fasciculatum* Stapf.

A perennial tussock grass with culms nearly 2 m tall; of rough short savanna; occurring in the uplands of N Nigeria and W Cameroons, and also in E Cameroun and Central African Republic.

The culms are used on the Mambila Plateau of N Nigeria to make *zaanaa* mats (1).

Reference:

1. Chapman 5085, K.

Urelytrum giganteum Pilger

FWTA, ed. 2, 3: 502. UPWTA, ed. 1, 550, as *U. thyrsioides* Stapf.
West African: NIGERIA FULA-FULFULDE (Nigeria) soɗornde (*pl.* coɗorɗe) (JMD) soɗornde mayo (JMD) zemako (JMD) HAUSA jéémàà (JMD; ZOG)

A robust perennial grass with culms to $2\frac{1}{2}$ m tall; of marshes, river-banks and similar swampy places; recorded only in N and S Nigeria and W Cameroons in the Region, and occurring also in Sudan, Uganda and the Congo basin.

In N Nigeria the grass serves as fodder and is used to make *zaanaa* mats (2). The culms in Central African Republic are cut for low quality arrow-shafts (1).

References:

1. Fay 1694, K. 2. Kennedy 8041, K.

Urelytrum muricatum CE Hubbard

FWTA, ed. 2, 3: 502.
West African: NIGERIA HAUSA rumaya (DA) rumiya (DA)

A tufted perennial grass, culms erect to $2\frac{1}{2}$ m high; of open or wooded savanna; across the Region from Senegal to N Nigeria.
The grass is lightly grazed in the Bimbila area of Ghana (4), doubtless for want of better, for the culms are bitter-tasting (3) and in general are not taken (1).
The haulm is tough and has potential for roving into rope (2).

References:

1. Ankrah 20405, K. 2. DA. Shika/P.H. 49, K. 3. Rose Innes 30760, 31535, 32624, K. 4. Rose Innes 32196, K.

Urochloa mosambicensis (Hack.) Dandy

A tufted or stoloniferous perennial grass, culms to $1\frac{1}{2}$ m high; of wooded grassland and deciduous bushland, in E Africa, S central and S Africa, Burma and other tropical countries by introduction, and now recorded in Ghana.
The plant is an useful pasture grass in all territories where it is indigenous, for which quality it has been introduced elsewhere. Its stoloniferous habit leads to formation of turf and it has been recorded as being grown in Zimbabwe as a lawn (1).
In Zambia it is valued as one of the more important edible grasses (3), the grain being a normal supplementary food to some people. It may be collected wild, but sometimes it is cultivated in gardens alongside maize (2).

References:

1. Bingham 631, K. 2. Mitchell 2938, K. 3. Scudde 27, K.

Urochloa trichopus (Hochst.) Stapf

FWTA, ed. 2, 3: 440. UPWTA, ed. 1, 550.
West African: MALI TAMACHEK saakat (A.Chev.)

A coarsely tufted annual grass with culms rising to 60 cm high (to 1.7 m in E Africa); of wooded grassland and bushland, usually on sandy soil in the Sahel from Mauritania to N Nigeria, and extending to NE, E and S central tropical Africa and Mozambique.
The grass provides good fodder for stock in sub-desert regions (4; Senegal: 1; Kordofan: 2; Jebel Marra: 5). It is a weed of cultivation in Ethiopia (7; in irrigation, 8).
The grain is sometimes garnered for food in the Soudanian region (4) and is commonly collected at places in Zimbabwe (3, 9), where it has been under experimentation by the Department of Agriculture as a potential plant for cultivation. In Tanganyika it is considered a valuable famine food at Uwanda

and south of Lake Rukwa (6). It is sometimes collected in Kordofan in time of shortage (2).

References:

1. Adam, 1966, a. 2. Baumer, 1975: 117. 3. Beattie 251208, K. 4. Dalziel, 1937: 550. 5. Davies BN.9, K. 6. Michelmore 993, 1171, K. 7. Parker E.277, K. 8. Parker E.601, K. 9. Yalala 419, K.

Vetiveria fulvibarbis (Trin.) Stapf

FWTA, ed. 2, 3: 470.
West African: SENEGAL MANDING-BAMBARA kongo dli (JB) MALI MANDING-BAMBARA gongo-dili *the root* (Diarra) NIGERIA FULA-FULFULDE (Adamawa) cawkitiningel (J) isirko (J)

A robust perennial grass, culms to 2 m high; of the flood-plain; disjunctedly dispersed from Senegal to W Cameroons.

The grass is grazed in Senegal by cattle while still tender (1), but is said to be ungrazed in Northern Territory of Ghana (3).

In Mali the root is an article of trade in Bamako market for perfuming drinking water (2).

References:

1. Adam, 1966, a. 2. Diarra, 1977: 45. 3. Ll. Williams 840, K.

Vetiveria nigritana (Benth.) Stapf

FWTA, ed. 2, 3: 470. UPWTA, ed. 1, 550.

English: vetiver.

French: vétiver.

West African: SENEGAL BEDIK ga-ndèny (FG&G) FULA-PULAAR (Senegal) sâban (K&A) sêban (K&A) TUKULOR sâban (K&A) sêban (AS; K&A) MANDING-BAMBARA bangasa (JB; K&A) ngokoba (JB; K&A) NON tul (AS) SERER siñ (JB) siñti (JB; K&A) WOLOF sep (auctt.) tiep (auctt.) GUINEA-BISSAU FULA-PULAAR (Guinea-Bissau) cudoendo (JDES) GUINEA BASARI a-ndéy (FG&G) SIERRA LEONE MENDE pindi (NWT) susu barewali (NWT) TEMNE an-wo-ŋa-ro-baŋ (FCD) MALI ? (Mopti) ghana (Matthes) FULA-PULAAR (Mali) dayi (Lean) dimipallol (A.Chev.) kiéli (A.Chev.) MANDING-BAMBARA babin (A.Chev.; Lean) gongo-dili *the root* (Diarra) ngokoba (A.Chev.) ngɔngɔn (A.Chev.) SONGHAI diri (A.Chev.) SONINKE-SARAK-OLE kamaré (A.Chev.) UPPER VOLTA BOBO nyen (Le Bris & Prost) FULA-FULFULDE (Upper Volta) ciidol (*pl.* ciidi) *the culm* (K&T) codorďe (K&T) sidiiho (K&T) sidiire (K&T) sodorko (K&T) GURMA kulkadéré (A.Chev.) MOORE rudum (A.Chev.) GHANA AKAN-ASANTE sansan *the fibrous roots sold for use as a wash-scrubber* (JMD) DAGBANI kulikarili (Ll.W) NIGER HAUSA ǧeemà *Ader dialect* (AE&L) SONGHAI díirí-ñá (*pl.* -ñóŋó) (D&C) NIGERIA DERA jámáli (Newman) FULA-FULFULDE (Nigeria) ngongonari from Bambara (Mali) (JMD) sekko *matting made from the fibrous roots* (J) sodornde (*pl.* codorde) (JMD; J&D) somayo, zemako (JMD) HAUSA darambuwa *armlets and girdles made from the fibrous roots* (JMD) jema (auctt.) kambu *armlets worn by hunters for protection* (JMD) IDOMA aganya (Odoh)

A robust perennial grass, clumped, with culms to over 2 m high; of stream-sides and swampy flood-plains; throughout the Region from Mauritania to N and S Nigeria, and in NE, E, S tropical Africa, and disparately in Ceylon, Thailand, Malaysia and the Philippines.

The presence of the grass in general indicates a wet locality, temporarily swampy, even slightly brackish, but usually fresh (1, 21b). Sometimes it is grown as an ornamental (1). Its value as fodder is variously reported, perhaps according to the stage of growth. In young growth cattle take it (Senegal: 1, 2; Nigeria: 6). It is said to be readily grazed in Kordofan, cattle entering flood-water to reach it (3), but in Sudan to be not touched (8). It is planted at Katsina, NW Nigeria, to mark field boundaries (17) and around the edges of fields in Malawi (14) but this latter is of unstated significance. Clumps of the grass have sand-binding property, and it is planted for this purpose on the banks of the Jur

River in Sudan (20). There is a claim to the plant repelling termites: an old brick-and-timber house at Badagri, Nigeria, now used as a mission post, has four large clumps of this grass, one at each corner, planted to repel termites from attacking the timbers of the house. Apparently this has been successful (25).

The culms are used for thatching (6; Senegal: 1; Mali: 15; Nigeria: 5, 11, 18), matting (1, 6), known to the Nigerian Fulfulde as *sekko* (10) and for plaiting into armlets and rings known to the Hausa as *darambuwa* (6). Stems are split for twisting into headbands worn by youths at marriage (6). In Ghana the stalks are woven into straw hats (16). In Senegal the culms and foliage are wrapped around the panicles of sorghum to protect them against the depredation of grain-eating birds (21a). In Nigeria the culms are woven into wicker fish-traps (24).

The roots are varyingly aromatic, probably according to edaphic conditions (6), but in places the grass is grown for this property. The Tenda place dried roots amongst clothing to add a scent (9), and women of Senegal soak them in water to produce a perfume (1, 21a). In Gabon the roots are dried and reduced to a powder to make an ointment and cosmetic (23). Nigerian women put them in sachets or plait them into necklaces to provide a body-perfume (6). Hausa hunters tie them up in leather armlets (Hausa: *kamba*) as a charm against injury (6). Roots are commonly soaked or infused in drinking water to add a pleasing flavour (6; Senegal: 9; Mali: 7; Gabon: 22, 23). They are added to water to purify it (Nigeria: 17). There is belief in Kano that water so purified cannot be drunk by a witch (19). At Dakar root added to soda-impregnated water acts as a desodorant (19). A little root in a calabash of water taken on a long march in Sudan is believed to keep it sweet (8). In Senegal the root is used medicinally as an anti-diarrhoetic for children. It is considered to be a stimulant and is used with other drugs in treating mental conditions (13). The root enters veterinary medicine in an infusion used in Ghana to treat certain animal diseases, e.g. *garli* (1, 6, 16). In the Niger River the root is used in the construction of an acoustic fish-trap known as *semu*. The scent is deemed to be vital to success (4, 12). Fibrous roots at one time sold in Kumasi market under the name *sansan* as a scrubber are held to be of this grass (6).

References:

1. Adam, 1954: 99–199. 2. Adam, 1966, a. 3. Baumer, 1975: 81–116. 4. Busnel, 1959: 346–60. 5. Dalziel 273, K. 6. Dalziel, 1937: 550. 7. Diarra, 1975: 45. 8. Evans Pritchard 49, K. 9. Ferry & al., 1974: No. 137. 10. Jackson, 1973. 11. Kennedy 7283, K. 12. Kerharo & Adam, 1964, b: 585. 13. Kerharo & Adam, 1974: 658–9, with phyto-chemistry. 14. Lawrence 2, K. 15. Lean 9, K. 16. Ll. Williams 722, K. 17. Mortimore, 1987. 18. Palmer 23, K. 19. Samia al Azharia Jahn, 1977. 20. Simpson 7639, K. 21a. Trochain, 1940: 102. 21b. ibid.: 398. 22. Walker, 1953, b: 275. 23. Walker & Sillans, 1961: 198. 24. Ward S.701. 25. Wheeler Haines 147, K.

Vetiveria zizanioides (Linn.) Nash

FWTA, ed. 2, 3: 470. UPWTA, ed. 1, 551.

English: vetiver; cuscus grass; khus-khus (India).

French: chiendent odorant.

Portuguese: capim-de-boma; capim-vetiver (Feijão).

West African: SIERRA LEONE MANDING-MANDINKA sumare (FCD) SUSU kale (FCD) TEMNE *ka*-benis (FCD)

A densely tufted, wiry, perennial grass, rhizomatous, culms up to 2 m high; native of India and introduced to many tropical countries. It is to be found at various points in the Region, between Senegal and Nigeria.

The plant is grown for its aromatic roots, source of an aromatic oil known as Oil of Vetiver, and valued in high-class perfumery where its persistent odour makes it of great value as a fixative in admixture with other perfumes. A hot climate, ~ 25°C, is necessary; a firm sandy soil is preferred for ease of harvesting and to obviate loss of the finer roots wherein the higher concentration of oil lies (3). Java used to be the chief producing country before World War II and a yield of 3% from the dried root is recorded (18). Present day supplies mostly come from Haiti, La Réunion and India, with yields averaging 1.5% dried root and 80 Kg per Ha per annum (12). Examination of material for Ghana has given 2.25% of volatile oil (6), and 1% from Zaïre. (18) Production in India shows edaphic responses (5). There is plainly scope for selection of improved strains and methods of husbandry under W African conditions.

The root is used in India on abdominal tumours (8). Roots are also reduced to powder for use as a perfumed body cosmetic (7). In Mauritius the root has been used as an abortifacient (18). Infusions are made in Trinidad for fever, flu, pleurisy and yellow fever (19). In Gabon (16, 17), and in India (5), roots are laid between clothing to convey the scent and to keep away insects. The grass has been grown on plantations in Malaya as an anti-malarial measure but it seems ineffective in repelling mosquitoes.

In W Africa (6) and many other countries the grass is grown along roadsides (9), to mark garden boundaries (10), as a hedge plant (1), to bind soil against erosion, especially on contours (2, 12, 15, 17), and as a possible mulch on coffee plantation (11). In W Africa it is planted along field boundaries to prevent encroachment of *Cynodon dactylon* (Gramineae) (6). The roots are woven into screens in India to hang over doors and windows in hot weather, when, sprinkled with water, the air is cooled and perfumed (5, 7); also for mattresses and underbedding (5). At Kano, Nigeria, the root is used for water-purification with the presumed advantage by magical belief that a witch will not be able to drink such water (14).

The leaves are without scent. Cattle will graze them while still young (6, 13). They can be cut for stall litter. As a source of paper-pulp they are assessed as second rate (4).

References:

1. Akpabla 565, K. 2. Anon. G.5901, K. 3. Brown & Matthews, 1951: 174–87. 4. Burkill, IH, 1935: 2228–30. 5. Chadha, 1976: 451–7, with physico-chemical characteristics of the oil. 6. Dalziel, 1937: 551. 7. Gupta & Sharma, 1971: 95. 8. Hartwell, 1969, b: 197. 9. Irvine 1638, K. 10 Jordan 475, K. 11. Nicoll 267119, K. 12. Purseglove, 1972, 1: 297–8. 13. Roberty 1961: 695. 14. Samia al Azharia Jahn, 1977. 15. Trochain, 1940: 102. 16. Walker, 1953, b: 275. 17. Walker & Sillans 1961: 196. 18. Watt & Breyer-Brandwijk, 1962: 489. 19. Wong, 1976: 109.

Vossia cuspidata (Roxb.) Griff.

FWTA, ed. 2, 3: 502. UPWTA, ed. 1, 551.
West African: SENEGAL WOLOF hay (JB) MALI ? (Mopti) temboro (Matthes) UPPER VOLTA FULA-FULFULDE (Upper Volta) burgu (K&T) NIGERIA FULA-FULFULDE (Nigeria) lingurehu (Bargery) lonyo (Bargery) HAUSA barugu (ZOG) burugu (Palmer; ZOG) damba (JMD; ZOG) daura (auctt.) falfoli (JMD) lambami (auctt.) milla (auctt.) tattakiya (auctt.) tsamba (JMD)
NOTE: Nigerian names apply generally to swamp-grasses.

A perennial hydrophilous grass with spongy culms submerged or floating, up to 7 m long, rooting at submerged nodes and putting up aerial stems 1–2 m above water; of lakes, rivers and streams; from Senegal to N Nigeria, and throughout the rest of tropical Africa and in India.

The grass is a major component of sudd vegetation in the Upper Nile (5). It occupies positions of periodic flooding (5; Senegal: 16; Sierra Leone: 9; Kordofan: 2; Uganda: 10) where its copious rooting is helpful in reclaiming inundated areas. It is present in the lagoons of Ivory Coast and Ghana (8) and in the Volta Lake, establishing itself along the margin in the draw-down area exposed during the dry season when the lake level falls. On return of the rains when the level rises it floats, furnishing harbour for snails which further the spread of schistosomiasis (12). It becomes a troublesome weed of canals and river systems (18) forming dense floating masses which obstruct the movement of canoes (7, 8) and is difficult to clear. Even motor-propelled boats are said in Zambia to suffer from straggling stems catching in propellers (14).

The grass provides good fodder in a green state and is readily grazed by cattle (Senegal: 1; Nigeria: 5, 6, 13; Kordofan: 2; Tanganyika: 3; Zambia: 15; Zimbabwe: 18). The green leaf is fed to buffaloes in India (4). It is a favourite food of hippopotamus (3, 15, 17, 18), and is also grazed by elephant (11).

The older stems are used for making a coarse matting (5, 6) and stems with leaves still attached serve to make mats, mattresses and thatching (2).

The stems have been examined as a source of paper-pulp and have been found unsuitable for paper requiring strength, e.g. wrapping paper. Its pulp has, however, a possible use when strength is not so critical as, for example, for newsprint or writing paper as a filler for bulking stronger long-fibred pulps (4, 6).

References:

1. Adam, 1966, a. 2. Baumer, 1975: 117. 3. Bullock 2977, K. 4. Chadha, 1976, a: 560. 5. Dalziel, 1937: 551. 6. Furlong, 1944: 149–53. 7. Hall & al., 1971. 8. Hepper, 1984: 47. 9. Jordan 600, K. 10. Langdale-Brown 2324, K. 11. Nicholson 46, K. 12. Odei, 1973: 57–66. 13. Palmer 19, K. 14. Seagrief 2294, K. 15. Trapnell 953, K. 16. Trochain, 1940: 394–5. 17. Walker & Sillans, 1961: 196. 18. Wild, 1961: 18–19.

Vulpia bromoides (Linn.) SF Gray

FWTA, ed. 2, 3: 369.

A very slender annual grass to about 25 cm high in montane waste places of W Cameroons, and of Europe, the Mediterranean and mountains of tropical Africa; widely introduced elsewhere.

No usage is recorded for the Region. In Sudan it provides browsing for sheep and goats (1), and on Jebel Marra it is a weed of irrigated wheat fields (2), and of wheat in Ethiopia (3).

References:

1. Andrews 3526, K. 2. Blair 292, K. 3. IECAMA RS 124, K.

Zea mays Linn.

FWTA, ed. 2, 3: 511. UPWTA, ed. 1, 551–3.

English: maize (an anglicisation of Arawak-Carib *mahiz* recorded by Columbus); corn (USA); Indian corn (i.e. American Indian corn, not Asian Indian); mealie (S Africa, from *milje*, Cape Dutch); also Egyptian corn; Turkish corn(in mistaken notion of its original home).

French: maïs.

Portuguese: milho; milho grande; milho grosso; milho-maeês.

West African: **SENEGAL** DIOLA (Fogny) ekôntibaba (Pasch) ékuntubaba (*pl.* sikutumbara) (K&A) husit *from* basit: *millet* (Pasch) sitikon = *Italian millet* (Pasch) DIOLA-FLUP kumorha (K&A) FULA-PULAAR (Senegal) maka (K&A) makarbodiri (K&A) makari (K&A) mala (K&A) TUKULOR gwari makka (Pasch) makarbodiri (K&A) makka (auctt.) mala (K&A) sataba (Pasch) MANDING-BAMBARA maño (K&A) MANDINKA maño (K&A) MANINKA maño (K&A) ˈSOCEˈ maño (K&A) MANDYAK búmaagi (Pasch) SERER mumbáawo (Pasch) púmaidsi (Pasch) pursin (Pasch) SONINKE-SARAKOLE maka (K&A) WOLOF maka (Pasch) makandé (auctt.) mbogi (JMD) mboha (auctt.) mbox(a) (K&A) morha (K&A) wende (JMD) **THE GAMBIA** MANDING-MANDINKA manyɔ (JMD) tubah-nyɔ (JMD) **GUINEA-BISSAU** BIAFADA ntubanyo (Pasch) CRIOULO midjo bassil (JDES) FULA-PULAAR (Guinea-Bissau) cába (JDES) tubanhô (JDES) PEPEL bumaadsa (Pasch) bumbaawa (Pasch) **GUINEA** BADYARA tubanyo (Pasch) BAGA (Kalum) kènkáabe (Pasch) BAGA (Koba) k-akabe (Hovis) BAGA (Sitemu) kö-bay (Hovis; Pasch) ts-akabe (Hovis) BASARI ɛ-mákà *i*-sɔ́gét (FG&G) *i*-rundù (FG&G) BEDIK *gé-maka (Pasch) FULA-PULAAR (Guinea) butali (JMD) kaba (JMD; Pasch) FULU-PULAAR (Guinea) maka, makaré (Dumas & Renoux fide JMD)* KISSI sòàng (Pasch) KONO nyóde (Pasch) KONYAGI u-tĕflɛf (FG&G) KPELLE woloma kpway cf. kpway: *millet* (Pasch) LANDOMA kö-babo (Hovis) MANDING-MANINKA bura-gué (CHOP) cissé-nion (CHOP) diokoroni (CHOP) kaba (JMD) maño (K) sagada (CHOP) MANO kpèy (Pasch) SUSU kabè (auctt.) TEMNE kémank cf. kémank kèsóor: *Sorghum guineense* (Pasch) **SIERRA LEONE** BULOM nkang-ntol (JMD) nkuskus (Pasch) BULOM (Kim) kã (FCD) kã-moɛ (FCD) BULOM (Sherbro) kaŋ-dɛ (FCD) khàŋ (Pichl) nkan (Pasch) nkison (Pasch) FULA-PULAAR (Sierra Leone) kaaba (FCD) GOLA di (FCD) diomɔkɔ (FCD) kedï (Pasch) KISSI soã (FCD) swahu (JMD) KONO nyuɛ (FCD) KORANKO nyõ (FCD) KRIO kɔn (FCD) LIMBA kutanki (Pasch) taŋki (FCD) LIMBA (Tonko) teher-baŋwuridi (FCD) LOKO dahɔyɔ (NWT) nyɔ (FCD) MANDING-MANDINKA kama (FCD) MENDE nyɔ (FCD) nyoo (Pasch) SUSU kaabɛ (FCD) kabe (NWT; FCD) TEMNE a-mank (NWT; FCD) VAI nyóoroo (Pasch) nyɔrɔ (FCD) nyuɛ (FCD) tama-nyɔ (FCD) YALUNKA kabé-na (JMD) **LIBERIA** DE baai (auctt.) KPELLE gbado *a fetish name* (Pasch) gbai (auctt.) ˈKRUˈ gbuu (Pasch) pamu (Johnston fide JMD) KRU-GREBO yibo (Pasch) yubwɔ (Johnston fide JMD) BASA gbu (Johnston fide JMD) LOMA gbáazi (Pasch) MANDING-MANINKA nyɔ (Johnston fide JMD) MANO kpai (Johnston fide JMD) VAI nyoru (Johnston fide JMD) **MALI** MANDING-BAMBARA manyò = *big millet* (auctt.) KHASONKE mako (Pasch) **GHANA** ADANGME-KROBO kroju (Pasch) AKAN-ASANTE aburo (FRI; Pasch) awiaburo = *suncorn; the second crop planted in the season of less rain* (FRI) FANTE eburo (Pasch) TWI aburoo (FRI; E&A) aburow (FRI; E&A) AKPAFU efiita (Pasch) *i*-titã (*pl. a-*) (Kropp) AVATIME ò-dóólí (Kropp) DAGBANI kaloana (Gaisser) kaluwana (JMD) káwoana cf. ka: *red Sorghum* (Pasch) GA abelé (KD) able (auctt.) blafo (Pasch) GBE-VHE akplẽ (FRI) VHE (Pecí) kpledzi (FRI) kple-ti *the plant* (JMD) **DAHO-MEY** GBE-FON (Gũ) gbado (Adandé gbli (Pasch) FON (Maxi) gbadye (Pasch) GEN ebli (Adandé) PHERA (Popo) abirri (JMD) birri (JMD) ebli (Adandé) GURMA bayuri *from* yuri: *millet* (Pasch) maana-sore *from* sore: *millet* (Pasch) sege-yore *from* yore: *millet* (Pasch) TEM-DOMPAGO ámeláamelá (Pasch) YOM manzo (Pasch) YORUBA-NAGO agbado (Adande) **NIGER** DENDI kòtòkòalí (*indef. sing.*) (RB) FULA-FULFULDE (Niger) makkari (Pasch) massaru (Pasch) SONG-HAI kólgótí (*pl. -ó*) (D&C) kolkoti (Robin) **NIGERIA** ? ókà (Elugbe) ABŌN agu mana (Meek) AGOI ansam (Cook) AGWAGWUNE ejama (Cook) AGWAGWUNE (Etuno) esut (Cook) AKPA ĩkpãngkpà (Armstrong; Oblete fide RB) AKPET-EHOM (Ubeteng) ekpoi (Cook) EHOM (Ukpet) ekpai (Cook) AOMA ǫka (Elugbe) ARABIC-SHUWA dura shami (JMD) masar (JMD) umm abât (JMD) umn abat (Pasch) BAKPINKA anjam (Cook) BANDAWA sɔ̃ko (RB) BANDAWA-MINDA siɛk (RB) BATA (Garua) daené (Pasch) BATA (Malabu) dawai (Meek) BATA (Zumu) mapinawo (Meek) BATU aku kwan = *guinea corn of the Jukun* (Meek) BETTE-BENDI ù-kúl (*pl. i̱*-kúl), ù-kwál (*pl. i̱*-kwál) (Ugbe fide KW) BIROM yara kàpas (LB) BITARE ajo kwana = *guinea corn of the Jukun* (Meek) BOKYI nkurung (Osung fide RB) BOLE damasar (Pasch) damasr (Meek) BURA pinau (Meek) pino (anon.) CHAWAI dir kwozak (Meek) CHAWAI-JANJI masiri (auctt.) KURAMA imasarim (Meek) CHOMO mikî̃ (Shimizu) DERA ápónò (Newman) DOKO-UYANGA ansam (*pl. arasham*) (NWT; Cook) EBIRA-ETUNO aagba (Pasch) aagwa (Pasch) agwawa (Pasch) EDO ka (Pasch) ǫka (auctt.) EFIK ibòkpòt (auctt.) ibòkpòt úmón (NWT) EJAGHAM nchamm (auctt.) EJAGHAM-EKIN nsamm (auctt.) ENGENNI áka (Elugbe) EPIE àkán (Elugbe) ERUWA ǫkà (Elugbe) ESAN ǫka (Elugbe) FALI nggulia (Meek) FULA-FULFULDE (Nigeria) butaali (*pl.*) (JMD; J&D) kaba *at Sokoto* (JMD) GAˈANDA puno (Meek) GBIRI-NIRAGU pimisire (Meek) pitigadin (Meek) GENGLE som kiva (Meek) GHOTUO ǫkà (Elugbe) GUDE àmbàbàt (Pasch) nggulę (Meek) GUDU gau buza (buzy) cf. buzy: *Sorghum* (Meek; Pasch) zakzak (Meek) GWANDARA-CANCARA dákú-hye (Matsushita) GITATA dàwúšè (Matsushita) KARSHI dákúše (Matsushita) KORO dákúši (Matsushita) NIMBIA dáza (Matsushita) TONI dákúše (Matsushita) GWARI agbado (Pasch) nyawi (JMD) nyiawie (Pasch) HAUSA agwado *from Yoruba* (JMD) dááwàr Másàr (JMD; ZOG) dawaa baaMásàráá (JMD; ZOG) HUBA hi buku (Meek) hibəku (RB) HWANA panu (Meek) HYAM gupara (Meek) IBIBIO àkpà-akpa (Ilonah) akpakpa (Cook) akpakpa (Cook) àkpàkpà (NWT) ICEN amirkpa (Meek) makpa (Meek) mkpà (RB) ICHEVE i̱-kúléghí (*pl.* é-kúléghí) *white or yellow forms* (Anape fide KW) IGBO ikpapka (Iwu) ǫgbàdù (KW) ǫgbàdụ (Pasch) ǫkà (auctt.) ǫkà (Pasch) IJO-IZON (Kolokuma) jzǫ́n áká = *the people's (Izon) corn* (KW&T) ISOKO ǫkaa (Elugbe) IVBIE àlàkpà (Elugbe) JANJO ihwę (Meek) JARA likam (Meek) JIRU acim (RB) JUKUN za kwa (Meek) JUKUN (Donga) kpankara = *guinea corn of the Akpa* (Meek) zakpa (Meek) JUKUN (Gwana) zakim (Meek) JUKUN (Kona) zakeim (Meek; RB) JUKUN (Pindiga) zaakim (RB) JUKUN (Wase) zakpa (Meek) JUKUN (Wukari) azankpa (Meek) zamkpà (Shimizu) JUKUN-JIBU amaû (Meek) KAJE nywat épat (Pasch) nywât épàt (Cook) yakpat (Meek; Pasch) KAKA gǫmbi (RB) KAMBARI-AUNA khàrèbì (*pl.* àkàrèbì) (RB) SALKA kkárábú (*pl.* íikárábú) (RB) KAMUKU limasára (Pasch) KAMWE khauwa *from Kanuri: Sorghum* (Pasch) khavwa (Meek) KAMWE (Kiria) khavwa (Meek) KANURI àrgêm (A&S) másàr (C&H) masarmi (JMD) KAREK-ARE damasar (Pasch) damasr (Meek) KATAB suwa kpat = *guinea corn of the Hausa:* okpat (auctt.) KATAB-KAGORO shwa pa (Pasch) silok akpat = *guinea corn of the Akpa* (Meek) solak akpat (Meek) swá pa (Cook) SHOLIO tsaa kpat (Meek; Pasch) KORO beŋkpa (RB) KOROP nkwi (NWT) KUDA-CHAMO aakalaaba (Pasch) KUGAMA som kiva (Meek) KUMBA sopa (Meek) KYIBAKU khiya masere (Pasch) masar (Meek) LAAMANG babir (Meek) LENYIMA akpe (Cook)

LEYIGHA nsaŋe (Cook) LIBO kwang ufa (Meek) LOKE esahma (Berry & Guy) é-saamã̂ (pl. ń-) (Elugbe) LONGUDA apenwa (Meek) LUBILO nsam (Cook) MAGU fuan (Meek) MALA idãnyagọ = Guinea corn that carries a child (RB) MAMBILA kọọm (Meek) tap (Meek) MAMBILA-KAMKAM fan (Meek) MARGI apanau (Meek) khiya masere (Pasch) masar (Meek) MBEMBE ọ́ghàk-kpà (Cook) MBEMBE (Tigong)-ASHUKU kâ k'pa (Meek) kùmkpà, kũkpa (RB) NAMA kâ-g'ba (Meek) kumkpa (RB) MBOI fatuma (Meek) MBULA misakono (Meek) MUMBAKE izitura (Meek) MUMUYE (Gnoore) zagin (Meek) MUMUYE (Lankaviri/Saawa) za ki (Meek) MUMUYE (Yaa) zakin (Meek) MUMUYE (Zing) zagin (Meek) MUNGA mom kwaẹ = Guinea corn of Kona (Meek) NDE ẹ̀-gú (pl. à-) (BAC) NDORO akwana (Meek) NGAMO haigim (Meek) masar (Meek) NGIZIM másármì (pl. -màmín) from Kanuri (Meek; Schuh) NINZAM isangkpar (RB) NKUKOLI nshamm (NWT; Pasch) NUPE káawa (Pasch) kàba (Banfield; JMD) NUPE (Kupa) aakaaba (Pasch) NYAMNYAM kwon ga = guinea corn of the Jukun (Meek) NZANGI mapinawe (Meek) NZANGI (Holma) mapinawin (Meek) OBULOM ọ̀bị̀àkà (Cook) OGONI-GOKANA kpákirà (Pasch) OKPAMHERI (Emarle) úbaakpâ (Elugbe) OLULUMO ẹ̀-gû (Cook) OLULUMO (Ikom) ì-gù (Cook) PEREMA diptura (Meek) PERO kóomò (ZF) PITI idal tibok from idar: guinea corn; tibok: hat (JMD) ROBA kilbokta (Meek) komberi ma (Meek) SAMBA-DAKA ka yiri (Meek) kááda (Pasch) kai (Meek) kaii (RB) SANGA idi-mansẹri (RB) SOMYEWE mu buba (Meek) SUKUR khlabir = guinea corn of the Pabir (Meek) SURA béng shwàa (anon. fide KW) TEME kofa (Meek) TERA likám (Meek; Newman) TERA-PIDLIMDI jẹrldi (Pasch) pinodi (Meek) TIKAR-NKOM gombie (Jeffreys) TIV ikuleke, ikuleko (Pasch) ikureke (JMD) UHAMI-IYAYU ọgbado (Elugbe) ɔɔgbadoo (Elugbe) UKUE ɔɔka (Elugbe) UKUE-EHUEN igbadoo (Elugbe) UMON akpoi (Cook) UNEME ɔ̀ɔ̀kpà (Elugbe) URHOBO ọka (JMD; Elugbe) YORUBA àgbàdo (auctt.) àgwàdo (Pasch) ìgbàdo (JMD) YORUBA (Egba) àgbàdo (Pasch) ọkà (JMD) óoka (Pasch) YORUBA (Ife) egbáado (Pasch) YORUBA (Oshogbo) yángán (JMD) YUNGUR kụl bọkta (RB) WEST CAMEROONS BAFOK ngui (DA) BAFUT ánsáng (pl. b-.) (Mfanyam) DUALA mbàsì (Pasch; Ittmann) KOOSI ngun (DA) KPE mbasi (DA) mukala (Pasch) KUNDU ngui (DA) ngwi (Pasch) LONG ngui (DA) LUNDU begbabo (DA) MBONGE mgbi (DA) NGEMBA n̄čừò, n̄čwiɔ́, ǹgèsaŋɔ́ (Chumbow) NGYEMBOON kwata (Anderson) ngesáng (Anderson) TANGA mbwe (DA) TIKAR-NKOM gombie (Lowe) WOVEA mbasi (DA)

Cultivars: SENEGAL BASARI maka (Pasch) BEDIK gi-máka from a-mák: to be a scar, in allusion to the serrated seeds on the cob (FG&G) GUINEA FULA-PULAAR (Guinea) fula kabe a precocious form, eaten unripe (JMD) kaba dialo (Dumas & Renoux fide JMD) MANDING-MANINKA bura-gué lisse-nion (Dumas & Renoux fide JMD) diskorobé small cobs eaten unripe (Dumas & Renoux fide JMD) gogo-ulé red c.var. (Dumas & Renoux fide JMD) kaba-ba a large form, high yield (CHOP, & Dumas & Renoux fide JMD) kabalé-diua very precocious form, eaten unripe, small cobs (CHOP, & Dumas & Renoux fide JMD) kaba-ulé red c.var (CHOP fide JMD) sataba a large c.var., late maturing with several small spikes (CHOP, & Dumas & Renoux fide JMD) susu kabé a late maturing form (CHOP fide JMD) MALI MANDING-MANINKA kabalinké with shining white seed (Dumas & Renoux fide JMD) mańo (K&T) sara-gué white-seeded (Dumas & Renoux fide JMD) sara-ulé red-seeded (Dumas & Renoux fide JMD) sataba a tall c.var., 5-6 months crop, pale yellow-seeded (Dumas & Renoux fide JMD) semonio-ni a dwarf c.var., grain small, round, white (Dumas & Renoux fide JMD) SONINKE maka (Pasch) SONINKE-BOZO manyimo (Pasch) manyumani (Pasch) UPPER VOLTA BISA kampana (Pasch) körgare (Pasch) BOBO bàmàkà (Le Bris & Prost) dugu (Prost) vaghaka (Prost) DAGAARI kaman (K&B) FULA-FULFULDE (Upper Volta) mbammbaari (Pasch; K&T) kamanaari (Pasch; K&T) makka (Pasch) GRUSI kamana (K&B) GURMA a-gbaata (Pasch) kókòdlì (Surugue) KIRMA dyu-dyalle (Pasch) KURUMBA atɔre-yana (Prost) NANKANNI kám ááne (Pasch) SAMO siê (Pasch) SAMO (Sembla) gyɛɛ̈ỹ (Prost; Pasch) SENUFO-TUSIA sankɔdo, sankokɛ (Pasch) IVORY COAST ABE agbɔ (Pasch) ABURE dabo (Pasch) ADYUKRU kokotu (Pasch) AKYE ngkwa (Pasch) ALADYĀ dudu (Pasch) ANYI able (Pasch) ARI dodo (Pasch) AVIKAM dudu (Pasch) BAULE able (Pasch) DAN kièn (Pasch) kpè (Pasch) ·EOTILE· lɔbɔ (Pasch) GWA gɔmutu (Pasch) KRU-AIZI dodo (Marchese) dɔdo (Pasch) dụdụ (Herault) BETE gògò (Pageaud) gùgoo (Marchese) guugoo (Marchese) DIDA dìdí (Marchese) GODIE dìdí (Marchese) GREBO (Jrewe) gbòsú (Marchese) GREBO (Plapo) yube (Pasch) GREBO (Tepo) gbòluu (Marchese) GUERE kpaaú (Marchese) GUERE (Wobe) kpooú (Marchese) KOYO dòdó (Marchese) NEYO dyò-bwo (Pasch) jɔjoo (Marchese) NIABOUA gò (Marchese) KULANGO sɔyɔzug (Pasch) KWENI gbaĩ (Pasch) gòò (Pasch) nyo (Pasch) yee (Gregoire) KWENI-BE zinin (Pasch) GBAN kpoto (Pasch) NWA dyogo (Pasch) gbongan (Pasch) TURA kpan (Pasch) kpi (Pasch) KYAMA mũntu (Pasch) LIGBI kaara (Pasch) MANDING-DYULA gbìsì (Pasch) gbògò (Pasch) kàbà (Pasch) nyo (Pasch) MOORE kaamaana (Pasch) kamaande (Pasch) SENUFO padégué (Pasch) SENUFO-BAMANA bèrèguè (Pasch) FOLO nandege (Pasch) TAGWANA nàdeng (Pasch) GHANA ADANGME blafo (Pasch) ADANGME-KROBO blɛfo (FRI) jubja (Pasch) kokodibɔ in allusion to the red and white seeds on the same cob (FRI) AKAN-FANTE asuwuntsem (DA) churamba (DA) kokonko (DA) mpagua (DA) TWI aburow pa = good corn; white-seeded (FRI) DAGBANI kawalog (JMD) molikom (JMD) ulinga (JMD) GBE-VHE blɛfo a commonly grown c.var. with white seeds (DA)

380

bli (JMD; Pasch) kokodibo *a red seeded c.var.* (DA) kpeli (Pasch) numele blɛfo *a smaller form of blɛfo* (DA) oloto *a quick maturing white seeded c.var.* (DA) VHE (Awlan) akplẽ (FRI) blé (JMD; Pasch) VHE (Pecí) ada kple (FRI) bli-kple (FRI) kple (FRI) watsi kple *white-seeded* (FRI; DA) GRUSI kamana (JMD) GRUSI-BUILE kiumbéna (Pasch) GUANG abló (Pasch) adzembé (Pasch) kpoli (Pasch) kúboyo (Pasch) kùbòyù (Rytz) woyu (Pasch) GUANG-NKONYA kpolíi (Pasch) KONKOMBA nkalma (Gaisser) nkalma mɛ *red-seeded* (Gaisser fide JMD) nkalma pi *white-seeded* (Gaisser fide JMD) LEFANA ɔ-ntá (Pasch) ɔ́-ńtá *(pl. lé-.) (Kropp)* LOWILI agbáno *(Pasch)* MAMPRULI *kaluwana (JMD) kawalog (JMD) ka-wana(-h) (Arana & Swodesh) molikom (JMD) ulinga (JMD)* MOBA *kaluwandi (JMD)* NANKANNI *karuwena (JMD)* SAN-TROKOFI eííita (Pasch) **TOGO** AHLO *ayíta (Pasch)* ANYI-ANUFO *àbùé (Pasch; Krass)* AVATIME *adebii (Pasch)* BASSARI *emerge (Gaisser) idalɛ (Gaisser)* GURMA *akokora (Pasch) kpábol (Tersis) likokólé (Pasch)* KABRE *samela (Gaisser)* KEBU *uutuubó (Pasch)* KONKOMBA *ima dè, inkalmâ, n'kalma (Froelich)* KPOSO *edeefa (Pasch) yooklo (Pasch)* NAWDM *baragenarede (Gaisser) mala'a (Pasch) tschinjamla (Gaisser)* TEM *(Tschaudjo) oamela, uamela (Gaisser)* TOBOTE *itàrende (Pasch)* YORUBA-IFE *(TOGO) baafoo (Pasch)* **DAHOMEY** BUSA *aagbado (Pasch) agbado (Pasch)* GBE-FON *agogokomé, agogodo komé white-seeded c.var., dwarf plants, early maturing* (Henry; Adandé) gbade (auctt.) gbade asanmignan *yellow-seeded* (Henry fide JMD) gbade wéwé *white-seeded c.var.* (Henry) gbadevovo *red-seeded c.var.* (Henry) gbagî *seed yellow, very hard and resistant to weevils* (Adande) gbaguen *seed deep yellow* (Henry fide JMD) gbo *white-seeded c.var., long season crop, seed soft, susceptible to weevils, but much valued for the ease of grinding and quality of the flour* (Henry; Adandé) gboli *cobs of mixed yellow and white grain* (Henry) goéku(-n) *commonly grown, long season (3-5 months) crop, grain resistant to weevils, good for grinding and to make akassa* (Henry; Adandé) hûvé, hunvé *blood red-seeded c.var., long season crop* (Henry; Adandé) kévé *yellow-seeded, long season crop, grain good for grinding* (Adandé) khévet *yellow-seeded* (Henry fide JMD) kîto *yellow-seeded, short season crop, grain small and eaten in soup or baked* (Adandé; Henry) ligbo nuku, lingbo nuku = *sheep's eye; good for flour* (Henry; Adandé) niali, nioli *white grain, long season c.var., soft seeds susceptible to weevils, but greatly valued for its ease of grinding and quality of the flour* (Henry; Adandé) toga *yellow-seeded, late maturing c.var.* (Henry) FON (Gũ) agbade (Pasch) agbadekũ (Pasch) **NIGERIA** ARABIC-SHUWA ligi *a dwarf c.var.* (Golding & Gwynn) EDO ivibowɛn *white-grained c.var.* (JMD) ɔka-azɛn *yellow-grained c.var.* (JMD) FULA-FULFULDE (Nigeria) masariel *small-grained c.var.* (Westermann fide JMD) masariri (JMD) mbayeri Masar (JMD) HAUSA baƙaa *black-grained c.var.* (JMD) burudi (ZOG) damin uku, gudus, rawaya, ba-gwariya, baƙar *másàraá = black 'masara'; yellow-grained c.vars.* (JMD) ɗán Másáráá (auctt.) duna *black-grained c.var.* (JMD) fara *white c.var.* (JMD; ZOG) garáá (ZOG) géémùn másàráá = *beard of the maize* (ZOG) góóyón másàráá = *backpack of the maize* (ZOG) gwáárí (ZOG) gwari *any c.var. with large grain* (JMD) háƙórín káárùwà = *teeth of the prostitute; in allusion to the dark and light-coloured seeds* (JMD; ZOG) jinin kare = *blood of the dog; dark red-grained c.var.* (JMD) kán másáráá = *head of the maize* (ZOG) ƙwad-da-gayya *with a loose fruiting head* (JMD) kwatakwali (JMD; ZOG) leƙa kogi = *peeping over the stream; a c.var. grown under irrigation* (JMD) maburaki (ZOG) maburkaki (ZOG) mádáráá (JMD; ZOG) másaráá (LB) másàráá (auctt.) másáráá kwona *a tall c.var. with large white grain* (JMD; ZOG) másàráá wàádáá = *maize of the dwarf; a dwarf c.var.* (JMD; ZOG) másàrár fúlààníí = *maize of the Fulani; a c.var. reaching to a man's height* (JMD; ZOG) ràáwáyáá (ZOG) rara *white-grained c.var.* (ZOG) yag gulumbe *a small c.var. with whitish-red grain* (JMD; ZOG) IDOMA àníwó (Armstrong fide RB) anyɔ̀lígá (Armstrong fide RB) igbañgkpa (Armstrong) IGALA aakaagwa (Pasch) àákpà *by elision from àkákpà* (RB) àkákpà (RB) IGBO akpukpa (AJC) ɔka amiacha, ɔka apuakampo, ɔkà irabo, ɔkà mme *c.vars. probably named after places* (JMD) ɔkà ntà = *small maize, white-grained c.var.* (JMD; KW) ɔkà ofigbo *yellow-grained c.var.* (JMD) ɔkà-uku = *big maize* (JMD; KW) okoru mmano nto (JMD) wokuvu (AJC) IGBO (Awo-Idemmiri) ɔkhà (Pasch) IGBO (Ezinehite) ɔkhà (Pasch) ukhòorū (Armstrong; Pasch) ukhwòorū (Armstrong; Pasch) IGBO (Ohuhu) ukwhà orū (Armstrong) IGBO (Umuahia) ukoru (JMD) IGBO-UKWUANI ɔkâ (KW) okaà (Armstrong; Pasch) IJO-IZON agbodo *from Yoruba* (Pasch; KW) IZON (Egbema) agbodo (Tiemo fide KW) IZON (Kolokuma) aká (KW&T) NUPE kaba ejégi (JMD) kaba èkpo (JMD) kaba liákpiagi *long-headed c.var.* (JMD) URHOBO ɔka-uweni *yellow-grained c.var.* (JMD) UVBIE óka (Elugbe) VERRE resara (Meek) ripung (RB) vesara (Pasch) WAJA babir (Meek) WAKA jɛki (Meek) WANDALA khiya masere (Meek; Pasch) wom muro (Meek) YALA (Ikom) ìgù *cf: Sorghum* (Ogo fide RB) YALA (Ogoja) igu-mãakpa (Onoh fide RB) YANDANG sikon (Meek) YEDINA mahar (JMD) mahara (Pasch) masarmi (Pasch) YESKWA uuguza (Pasch) YEKHEE (Auchi) óka (Elugbe) YEKHEE (Avianwa) óka (Elugbe) YORUBA ɛlépɛ, ɛlɛpɛ̀rɛ́ *soft-grained c.var.* (JMD) igbado funfun *white-grained c.var.* (JMD) igbado pupu *yellow-grained c.var.* (JMD)

Parts and products: **SENEGAL** 'CASAMANCE' fulem futa ekuntubébu *the dried tassel* (K) SONINKE-SARAKOLE makaturantu *the bran* (K) **SIERRA LEONE** TEMNE a-(g)bəfəl *a fruiting* *head* (JMD) **GHANA** AKAN aburo-pata *a corn store* (FRI) aburow guane *ripe cobs of corn* (FRI) akassa *ripe grain ground to flour partially fermented made into a paste and wrapped in a leaf* (JMD) buro-fua *a grain of corn* (FRI) buro-hono *the husks* (FRI) kenki, komi, ɔdɔkõno, banku *leavened corn dough boiled as balls or in masses, wrapped in leaves* (JMD) klaklo *unleavened dough mixed with ripe plantain rolled into balls and fried* (JMD) AKAN-TWI ahei *maize spirits* (JMD) GA abólòó, agidi *foodstuffs made from the grain* (KD) àhai *a non-alcoholic beverage made from the grain* (KD) ahwánya *the flowers* (KD) akudono, ngm da *maize beer* (JMD) baŋkũ *a porridge of slightly fermented dough* (KD) trema-sugbɔ *a sweet maize pap* (KD) GBE-VHE (Pecí) a-kplẽ *the grain or meal* (JMD) **DAHOMEY** GBE-FON (Gũ) ahavo *beer; originally applied to Sorghum beer; possibly a loan-word from Gbe-Vhe* ahali: *germinated millet* (Wigboldus) **NIGERIA** FULA-FULFULDE (Nigeria) sawtere *the tassel* (JMD) waundu *the fruiting head* (JMD) HAUSA burudĩ, kan masara, gemun masara, maburkaki *the tassel or silk* (JMD) dawa da mudu *unripe grain, roasted and eaten on the cob* (JMD) gara *the tassel, or the male inflorescence* (JMD) goyon masara *the fruiting head* (JMD) kututu *cobs used to make pipe-bowls* (JMD) toton masara *the cob proper, or the naked axis free of grain* (JMD) zakankan *a sort of potash obtained after burning maize stems: as also for Sesamum* (JMD) IDOMA ácígbangkpa *the cob* (Armstrong fide RB) IGBO ijali ọgbedu, iraga, ǹzà *tha male inflorescence* (NWT) isisi ọkà *the cob* (NWT) mba *the fruiting head* (NWT) ǹzàlì = *the tassel* (KW) IJO-IZON (Kolokuma) aká apʉra *the sheath* (KW&T) aká bọlọʉ *tips of the cob* (KW&T) aká ịwẹsị *the silk* (KW&T) aká pomu *leaf* (KW&T) aká sele *grain* (KW&T) aká tuu *base of cob* (KW&T) aka umgbou *empty cob* (KW&T) aka warị *cob* (KW&T) ẹwẹsị *the silk* (KW&T) ịwẹsị *the silk* (KW&T) YORUBA àdùn *parched corn, ground and mixed with honey or oil* (JMD) agidi *ripe grain ground to flour, partially fermented and made into a paste and wrapped in a leaf* (JMD) eginrin àgbado *the fruiting head* (JMD) ejẹ̀rẹ *the flower* (JMD) èkuru àgbado *corn meal* (JMD) erínkà, erinigbado *corn on the cob* (JMD) ọti-àgbàdo, sẹ̀kẹtẹ *maize beer* (JMD; Verger)

NOTE: *throughout the foregoing lists there is frequent reference to* masar *or* masara, i.e., Egypt, *and to* maka *or* makari i.e., Mecca, *in allusion to a mistaken belief that* Zea mays *originated from either of these two countries presumably under Islamic influences.*

A robust annual grass, usually single-stemmed, occasionally tillering, with stout culm, sometimes stilt-rooted at the basal nodes, to 1–4 m high, even to 6 m, and 3–4 cm in diameter; monoecious and diclinous with male inflorescences, called a 'tassel', terminal, and female flowers, called the 'ear', a modified spike on a short lateral axillary branch borne down the stem; very variable and of numerous cultivars; anthropogenic and dispersed over the whole Region; a central American plant, now cultivated throughout the tropics, subtropics and into temperate regions.

Palynological evidence has dated the grass's presence in Mexico to at least 80,000 years ago, well before man's appearance in the Americas. Maize is, however, now found growing in a wild state. The pregenital ancestor is extinct, perhaps as a result of early man cultivating the strains he selected in the localities where the ancestral plant had grown, thus displacing it. Archeological research has been an epic of botanical detection and has shown maize's centre of domestication to have been in southern Mexico around 5,200–3,400 B.C. The advanced civilisations of Central America, Aztec, Inca, Maya, grew up on this cereal as a staple (4, 9, 15, 19, 20). Columbus on his first Voyage of Discovery in 1492 met with it, and he and his crew ate the grain with relish. By this time it had been spread in the New World from the Great Lakes in the north to Chile in the south, limited only by the available length of the warm summer season and rainfall. Columbus and other explorers carried seed back to Europe and its worth was quickly recognised in southern Europe and N Africa. Maize is extremely variable and adaptable to a wide range of climatic and soil conditions. Seven races, not of botanical status, but convenient divisions for usage, are generally recognised: pod corn, *tunicata*; popcorn, *everata*; flint maize, *indurata*; dent maize, *indentata*; soft, or flour maize, *amylacea*; sweetcorn, *saccharata*; and waxy maize, *ceritina*, and a few horticultural curiosities (4, 19, 20). To these races the applied skill of the plant breeder and geneticist has turned maize into a plant suitable for cultivation in very extensive areas of the world. It is now, after wheat and rice, the third ranking cerial.

It seems probably that maize arrived in W Africa under Portuguese influence. It was certainly there by the beginning of the Sixteenth Century, i.e. within 8 years of its arrival in Europe (16). Its cultivation in the Region is limited by the acceptance of other staple foods and climatic and edaphic conditions. It is to some extent complementary to the grain sorghum (*Sorghum bicolor*) and bulrush millet (*Pennisetum glaucum*) (q.q.v.), being cultivated largely in the south and in riverine situations inland, whilst the other two cereals are in the drier northern region. It is subject to food preference in areas where rice cultivation has been developed during the past century, e.g. Senegal to Guinea/ Sierra Leone and the Upper Niger Basin to N Nigeria, and in the region of the Guinea yam culture of Ivory Coast to S Nigeria. In some areas, however, it has found ready acceptance where predators make other food crops not attractive to grow, e.g. yams by yam beetle attack and Guinea-corn by *Aphis sorghi* (7).

Dietetically the grain has a chemical composition of about: starch 71–77%, sugar 2%, oil/fat 5%, and protein 6–15% (5, 20). The protein contains a substance named *zein* but lacks *lysine* and *tryptophane* amongst the amino acids, substances which prevent pellagra. *Gluten* content is so low as to make the flour unusable for bread-making. The oil, 80% of which lies in the germ, is composed of: *linoleic acid* 50%, *oleic acid* 39%, *palmitic acid* 7%, *stearic acid* 3.0%, and traces of *arachidic* and *lignoceric acids* (5).

The grain is eaten in the Region mainly as pap, which is sometimes let down to produce various drinks, and it is fermented into a beer (5, 7, 19, 23). Tenda people use it only if millet is not available, and then it is deemed proprietary to men (8). The cob is commonly roasted for eating (7). As a food it is in general deficient in protein necessitating the consumption of other foods to make a balanced diet. In the world at large the grain is eaten in numerous confections, and turned into beer and distilled into spirits. It is a very important feed for livestock and poultry, high in energy, low in fibre and easily digested (19). The plant itself is also an excellent forage (1).

The grain yields many industrial products. The main one is starch which can be used as such or converted into dextrin, syrups or sugars. It has valuable medicinal use for application to areas of infection and suppuration conferring protection against bacteria, and as a simple dusting powder against chafing and excoriations (3). It is emolient and helpful in poultices; it is a suitable base for tablets (12, 17). The seed contains *allantoin* which is found to accelerate cell proliferation in sluggish wounds, and is particularly helpful in osteomyelitis (12, 24). In America the seed has been used in many treatments for tumours and warts (10). Dagaari in Upper Volta prepare an ointment from the calcined tassel in karite butter for application to oedema, a treatment used in a Government dispensary and found to be effective (14). The oil is good for soap or glycerine manufacture, or after refining for human use (4, 6, 19). It is a substitute for olive oil in certain pharmaceutical formulae official in the *British Pharmaceutical Codex* and the *British Veterinary Codex* (17). An oestrogen factor is present in the seed, principally in the oil from the embryo, which has had beneficial use in cases of chronic eczema and other skin complaints and asthmatic conditions (12).

The plant may have some value as a bee-plant; bees are reported to collect the pollen (11).

The whole plant is used in N Nigeria as a horse medicine for mucal diarrhoea (7, 21). Urino-genital problems are commonly treated with prescriptions using the whole or part of the plant (7, 21, 24). The culm, converted to charcoal, is put into medicines for gonorrhoea. In Guinea (18) and Gabon (23) a diuretic tisane is made from maize, and an infusion of the stigmas is used for all diseases of the urethra. A young cob and stalk made into a decoction is a treatment in Java and China for bladder and kidney troubles (17, 24). The fresh and dried styles and stigmas are official in several pharmacopoeas for diuretic properties in urethritis and cystitis (17, 25). A high concentration of *potassium salts* is present and this excites the renal epithelium; *salicylic acid* is also present which is beneficial in use for rheumatism (12, 25). Bekik of Senegal obtain a salt substitute from the ash of the burnt culm (8), as also the Hausa in Nigeria (7).

The sheath of the cobs (corn shucks) is often used as a food wrapper, or of exported fruit. It can be made into a sort of cloth and used to make matting (7). The cob-axis freed of grain is used to make pipe-bowls (7). In Gabon the sheaths are put into mattresses as a filling (23). The incinerated cob is included in a snuff used in southern Africa (24).

The cob has a phallic appearance and because of the numerous seeds fixed on the central axis, it naturally has symbolic significance. The cob grilled and powdered is used in Senegal for impotence and sterility (13). Red-seeded cobs are considered especially effective (12). Its productive power is reflected in an Odu incantation in Yorubaland which runs that àgbàdo (maize) has good luck: it goes into a field naked and returns home with 200 children and 200 sets of clothes (22). In Lower Dahomey, maize, like millet, is said to have magical origins and powers, and is used for religious purposes (2). In the word-lists above, there is frequent reference to *Masar* (Egypt), and *Maka* (Mecca) in the belief of an origin from these countries.

References:

1. Adam, 1954: 100. 2. Adandé, 1953: 220–82. 3. Ainslie, 1937: sp. no. 362. 4. Burkill, IH, 1935: 2285–92. 5. Busson, 1965: 489, with chemical analysis, 491. 6. Chadha, 1962: 13–37. 7. Dalziel, 1937: 551–3. 8. Ferry & al., 1974: sp. no. 138. 9. Galinat, 1977. 10. Hartwell, 1969, b: 197–8. 11. Howes, 1945. 12. Kerharo, 1973: 18. 13. Kerharo & Adam, 1964, b: 587. 14. Kerharo & Bouquet, 1950: 254. 15. Mangelsdorf, 1974. 16. Mauny, 1953: 684–730. 17. Oliver, 1960: 10, 91. 18. Pobéguin, 1912: 39. 19. Purseglove, 1972, 1: 298–333. 20. Robinson & Treharne, 1988: 199–207. 21. Singha, 1965. 22. Verger, 1967: No. 14. 23. Walker & Sillans, 1961: 196. 24. Watt & Breyer-Brandwijk, 1962: 489. 25. Wong, 1976: 109.

Zoysia Willd.

FWTA, ed. 2, 3: 413.
English: zoysia grass.

A genus of ~ 10 species of tropical and sub-tropical Asia and Australasia, mat-forming and perennial, of which *Z. matrella* (Linn.) Merr. and *Z. tenuifolia* Trin. have been introduced to the Region. Both give a very fine, rather bristly deep-piled sward of superior appearance.

Gramineae spp. indet.

West African: **SENEGAL** FULA-TUKULOR symbane *used to bind fishtraps; it does not rot* (Busnel) **GHANA** AKAN-TWI oseprempre (OA) **NIGERIA** ARABIC-SHUWA kalasihum *a salt-bearing species* (Sikes) kanido *a salt-bearing species* (Sikes) pagam *a salt-bearing species* (Sikes) HAUSA badashi *at Birnin Gwari* (RES) farin kyamroo *a strong-culmed sp. to make arrow-shafts* (RB) IGALA áfǫ̀tǫ̀ *a heavy swamp grass of Ife country* (RB) JUKUN (Wukari) bwá *for fodder* (Shimizu) cògbon *a thatching grass* (Shimizu) fò *indicates good farmland* (Shimizu) 'gòn *for mat-making* (Shimizu) gótò gèngèn *for fodder* (Shimizu) kán (Shimizu) kpó *for fodder* (Shimizu) kyinvò zènkù *for fodder* (Shimizu) tsu *for mat-making* (Shimizu) KAMBARI-SALKA ɔɔpɔ̀sù = Hausa: farin kyamroo; *a strong grass for making arrows* (RB) wìizû *for brooms* (RB) KANURI jáurálá *a broom grass* (C&H) kalasihum *a salt-bearing species* (Sikes) kanido *a salt-bearing species* (Sikes) pagam *a salt-bearing species* (Sikes) súndòk *a broom grass* (C&H) TIV dŭl (KW fide RB) ikanggɔrɔka (KW fide RB)

During the course of this work a number of grasses identified only as to local vernacular name, listed above, has been met with. Besides the notes given besides each, the following require further comment:

1. *kalasihum, kanido* (Arabic-Shuwa, Kanuri, in Nigeria) are endemic grasses found near Lake Chad. They are salt-bearing (3).

2. *pagam* (Arabic-Shuwa, Kanuri in Nigeria) in the vicinity of Lake Chad is salt-bearing, but elsewhere appear to be not so (3). The grass is perhaps *Dactyloctenium aegyptium* (Linn.) P Beauv., cf. *pagamri*, Fula.

3. *symbane* (Fula-Tukulor in Senegal) refers to a grass used in an acoustic fish-lure. The grass does not readily rot (2).

4. *oseprempre* (Akan-Twi in Ghana) is a grass whose roots are ground to a paste in water with a seed or two of *Piper guineense* (Piperaceae) for nasal instillation treatment of headache (1).

References:

1. Ampofo, 1983. 2. Busnel, 1959: 347. 3. Sikes, 1972: 92.

GUTTIFERAE

Guttiferae Juss.

West African: **SIERRA LEONE** MENDE *ŋ*jole-dɛli (*def.* -i, -i) (S&F) *ŋ*jole-lɛli (*def.* -i, -i) (S&F) sɔle-dɛli (*def.* -i, -i) (S&F) sɔle-leli (*def.* -i, -i) (S&F)

Allanblackia floribunda Oliv.

FWTA, ed. 2, 1: 291. UPWTA, ed. 1, 89, and 90, as *A. parviflora* A. Chev.

English: tallow tree.

Trade: timber – lacewood (Liberia, Cooper & Record).

West African: **SIERRA LEONE** KISSI njailiŋ (S&F) KORANKO mbrembra (S&F) MENDE *ŋ*jolei-dɛlii (S&F) *ŋ*jolei-lɛlii (S&F) sɔle (*def.* -i) (FCD; DS) sɔlɛ-deli (*def.* -i, -i) (FCD; S&F) sɔlɛ-lɛli (*def.* -i, -i) (S&F) **LIBERIA** KRU-BASA gbar-chu (C; C&R) **IVORY COAST** ABE otéro laha = *male 'otero'; cf. Pentadesma butyracea, Guttiferae - female 'otero'* (PG) otéro lakya = *long 'otero'; in allusion to the fruit being x3 or more longer than wide; cf. P. butyracea* (PG) utéra (Aub.; FB) wohotelimon (A.Chev.) ABURE ewotébo (A.Chev.) wotobé (A.Chev.) AKAN-FANTE akumasé *cf. Carapa procera (Meliaceae)*; (A.Chev.) AKYE bissaboko (A.Chev.) pissalé boko (Aub.; FB) ANYI alabénun (A.Chev.) banummabi (Aub.; FB) essuinduku (Aub.; FB) KYAMA tiabokebihia (Aub.) NZEMA akumasé (A.Chev.) **GHANA** AHANTA anane (auctt.) AKAN-ASANTE bohwe (auctt.) *okisidwe* = *rat's nut; the seeds* (FRI) okusie-dua (JMD; FRI) sonkyi from Nzema (auctt.) FANTE sonkyi (E&A) TWI bohwe (JMD) sonkyi (E&A) WASA sonkyi ()E&A) suenkyi (auctt.) sunkyi = *elephant's back; meaning unknown* (CJT) ANYI banummabi (FRI) esonye duku (FRI) ANYI-SEHWI sunkyi = *elephant's back; meaning unknown* (CJT) NZEMA anane (CJT) soin (CJT) sonkyi (auctt.) sonye (TFC) suein (TFC; BD&H) **NIGERIA** BOKYI mia (auctt.) EDO ízéni = *elephant rice* (auctt.) izǫkhain = *grass-cutter rice* (JMD; Elugbe) EFIK édíáng *cf. Treculia africana, Moraceae* (auctt.) IGBO egba (KO&S) IGBO (Awarra) ǫcha (DRR) IGBO (Isu) ocha (DRR) IGBO (Umuakpo) ala-enyī = *elephant's breast* (DRR) IGBO-IZI alenyi (NWT) IJO-IZON obobiobo (JMD; KO&S) KAKIBA (Ibani) balanono (LAKC) KAKIBA (Kalabari) balanono (LAKC; JMD) YORUBA orógbó erin = *elephant's bitter kola* (auctt.)

An evergreen tree to 30 m high or more, of the rain-forest in wet situation, common, from Sierra Leone to W Cameroons, and on into Zaïre and Uganda.

The bole is straight, not buttressed, but sometimes fluted. The heart-wood is pale red or brown and fairly hard, resinous, moderately heavy; sap-wood is whitish. The wood is said to be resistant to termites but is not particularly durable. It is fairly easy to work and finishes well but it is of little commercial importance though it has appeared on the market in Liberia as 'lacewood'. (2, 7, 8, 10, 14, 15.) It has an attractive figure when quarter-sawn, and is suitable for carpentry (8, 17). The tree has other attributes (see below) which make it potentially of some value. Cultivation appears desirable and is certainly practicable though in managed forestry it is considered a weed-species and is eradicated (14).

The wood is used in Nigeria in hut-building for making walls (11), doors and window-frames (8) and in Liberia for planks (7). In Ghana small trees are cut for poles (16) and find use as mine pit-props and bridge-piles (9). There is unspecified use of the wood in Ivory Coast (12). The twigs are used in Ghana as candlesticks (9), and the smaller ones as chew-sticks and tooth-picks in Ghana (8) and Gabon (17).

The inner bark contains a sticky yellow resin. The bark has anodynal properties. In the Region it is pounded and rubbed on the body to relieve painful conditions (8). In Gabon a decoction is taken for dysentery and as a mouthwash for toothache (17) and in Congo (Brazzaville) for stomach-pains (3). In Congo a decoction of the bark or the leaves is taken for cough, asthma, bronchitis and other bronchial affections while the lees from this preparation are rubbed over areas of pain after scarification (3). Sap expressed from a crushed up mixture of bark and that of *Mammea africana* Sab. (Guttiferae), maleguetta and sugar-cane is taken in Congo for urethral discharge (3). Congo material has been reported to contain abundant *flavonins* in the bark and roots, some *tannins*, and traces of *steroids* and *terpenes*(4).

Trees are usually highly fructiferous. The fruit is large, up to 30 cm long by 10 cm in diameter with upward of 100 seeds borne within in a translucent mucilage. There is no recorded usage of this mucilage, but a decoction of the whole fruit is used in Ivory Coast to relieve elephantiasis of the scrotum, though this may simply be based on the Theory of Signatures because of the size and shape of the fruit (5). However, an alkaloid has been reported in the fruit-sap, a derivative of *tryptophane* and related to *eseroline* found in the Calabar bean, *Physostigma venenosum* Balf. (Leguminosae: Papilionoidae) and is mildly stimulatory (13, 17).

Animals eat the fallen fruits (9, 15), which are used in Ghana as bait in rat-traps; hence the Asante name meaning 'rat's nut' (9).

The seeds are oily. Some races in Gabon eat them after cooking, especially in time of dearth (17). The seed coat is thin, brittle and easily removed. The kernel is about 62% of the seed by weight, and the fatty fraction of it is about 71.5% which can be obtained either by boiling for 30 minutes in water or by solvent extraction. The fat is a white solid consisting of 52.5% *stearic acid*, 46.9% *oleic acid* and traces of *palmitic* and *linoleic acids*. Stearic acid is an important industrial substance, and this appears to be a promising source for both pure and technical grades of it. Some selection work has already been done in Zaïre. (1, 6, 18.) The residual cake is too bitter for cattle-feed. The fatty substance is mildly purgative.

References:

1. Anon., 1936. 2. Aubréville, 1959, 2: 330. 3. Bouquet, 1969: 131–2. 4. Bouquet, 1972: 27. 5. Bouquet & Debray, 1974: 93. 6. Busson, 1965: 211–3, with seed analysis. 7. Cooper & Record, 1931: 39–40, as *A.* *parviflora* A. Chev., with timber characteristics. 8. Dalziel, 1937: 89–90. 9. Irvine, 1961: 143–4. 10. Keay & al. 1960: 177. 11. King-Church 47, K. 12. Pellegrin, 1959: 216–30. 13. Resplanda, 1955: 542–6. 14. Savill & Fox, 1967, 130–1. 15. Taylor, CJ, 1960: 171. 16. Vigne 886, K. 17. Walker & Sillans, 1961: 197, incl. *A. klainei* Pierre. 18. Watt & Breyer-Brandwijk, 1962: 494. See also: Portères, 1974: 112.

Calophyllum inophyllum Linn.

FWTA, ed. 2, 1: 291.

English: Alexandrian laurel (Ghana, Irvine; a transferred and inappropriate name, Burkill).

Portuguese: loureiro-de-Alexandria (Feijão).

A largish tree reaching to 13 m high with a spreading crown, a plant of the seashore and coastal dunes of the Indic and Pacific basins from E Africa to Tahiti. It has been dispersed by man to many tropical countries including the W African region.

The tree has been successfully grown as an avenue roadside shade-tree in Nigeria (4). It stands clipping and will make a good hedge. It has been grown in Ghana as a wind-break (2, 3).

The wood is reddish brown, hard, close-grained, and usually in short lengths. Throughout its natural range as a maritime plant, it is put constantly to use in boat-construction. It has proved useful as railway-sleepers, and taking a good polish it is used in cabinetry and high-class carpentry. At one time it was sold in London as 'Borneo mahogany.'

The bark is medicinal in Asia: in India for orchitis; and in Indonesia after childbirth, for vaginal discharges, passing of blood and in gonorrhoea. *Tannin* is present in it.

Leaves left to soak in water impart a pleasant smell to it. This is used in Indonesia as an eye-lotion, and in the Philippines as an astringent for piles.

The fruits are more or less poisonous and only the endosperm of the still immature fruit is safe to eat. Mature fruit have been crushed and used as rat-bait. The embryo which becomes large on maturity contains oil and a resin. The oil is one of the oldest known to be used for illumination in India and across the Pacific. It is used secondarily on dermal troubles and is an ancient treatment for leprosy.

References:

1. Burkill, IH, 1935: 408–11. 2. Burtt Davy & Hoyle 1937: 53. 3. Irvine, 1961: 144–5. 4. Keay FHI.22476, K. 5. Quisumbing, 1951: 617–20.

Endodesmia calophylloides Benth.

FWTA, ed. 2, 1: 287.
West African: NIGERIA IJO-IZON bonason (KO&S)

A small riverain tree to 17 m high with a thin bole, of swamp-forest in S Nigeria and W Cameroons, and in E Cameroun, Gabon to Cabinda.

The wood is pinkish, hard, heavy and durable (2), but not of any recorded usage. Leaf-sap is used in Congo (Brazzaville) in eye-instillation for filaria (1).

References:

1. Bouquet, 1969: 137. 2. Keay & al., 1960: 171–2.

Garcinia afzelii Engl.

FWTA, ed. 2, 1: 295. UPWTA, ed. 1, 91, as *G. mannii* sensu Dalziel, in major part.

English: bitter kola (Scott Elliot, Sierra Leone).
West African: GUINEA KISSI kollopello (A.Chev.; LC) MANDING-MANINKA kouru (LC) SIERRA LEONE KONO sokiti(-i) *the roots* (S&F) KORANKO wurumamalal (S&F) MANDING-MANDINKA wurukuna (FCD) MENDE *n*denyani(-i) (auctt.) *n*doneani (LAKC) sokitii (S&F) TEMNE *ta*-sagbe (S&F) YALUNKA kɔre-na (FCD) LIBERIA KRU-BASA kar (C) IVORY COAST ABE siébé (A.Chev.) tiokué (Aub.) MANDING-MANINKA ko-oworo, koworo = *water kola* (A.Chev.) GHANA AKAN-ASANTE *n*soko-dua (BD&H) BRONG *n*sokɔ, *n*soku (BD&H; E&A) TWI *n*soko (auctt.) *n*soko-dua (auctt.) WASA *n*soko-dua (LC; E&A)
NOTE: Vernaculars recorded in Guinea and Sierra Leone for G. mannii are probably correctly ascribed here.

A tree to 18 m high, short bole, bushy spreading crown, of the evergreen forest often in damp situations, or of dry forest understorey from Guinea to S Nigeria.

The sap-wood is whitish, turning to yellow in the air; heart-wood is pinkish to deep yellow or olive-brown at the centre. The wood is hard, heavy, fine-grained and takes a good polish. It is suitable for carpentry and general construction-work though liable to fungal attack. It is resistant to teredo worm and so is used for wharves and bridges. (1, 3, 4, 9.) In Sierra Leone it furnishes a pole-crop from forest reserves (7).

The roots particularly, but sometimes the twigs, are used as chew-sticks. They are commonly cut up into pencil-lengths and traded in markets in Sierra Leone (5, 7–10), Ivory Coast (1, 2) and Ghana (3, 6, 11). Use of them is held to strengthen the gums and to prevent dental caries. Phytochemical tests have shown the presence of a high amount of flavonic substances in the chew-sticks and of *tannins* and *flavones* in root- and stem-barks (2), also of *gamboge* which is haemostatic, antiseptic and vulnerary (11). *Raphia* wine in which root-bark has been soaked is drunk in Sierra Leone as an aphrodisiac, and the seeds are acknowledged to be highly erogenic so that they 'should not be eaten when brother and sister are left alone' (9). Bark is chewed in Sierra Leone for cough and stomach-ache, or a decoction of it is taken (5).

The fruits are brownish-yellow, to 2.5 cm in diameter and contain 2–4 seeds embedded in an acidulous pulp which is edible and much relished.

References:

1. Aubréville, 1959: 2: 338. 2. Bouquet & Debray, 1974: 94. 3. Chalk & al., 1933: 49, as *G. mannii.* 4. Dalziel, 1937: 91. 5. Deighton s.n. 18/12/1946, K. 6. Irvine, 1961: 145. 7. King 185, K. 8. King-Church 6/1922, K. 9. Savill & Fox, 1967: 131. 10. Scott-Elliott 4841, K. 11. Portères, 1974: 124, as *C. mannii* Oliv.

Garcinia elliotii Engl.

FWTA, ed. 2, 1: 294. UPWTA, ed. 1, 90.
West African: **SIERRA LEONE** MENDE gwɛtewuli (NWT) kewe (GFSE) SUSU kufuri (NWT) TEMNE *ta*-sagbe (S&F)

A shrub to 60 cm high of the forest zone in Sierra Leone.
The seeds have medicinal use for indigestion (1).

Reference:

1. Scott Elliot 5094, K.

Garcinia epunctata Stapf

FWTA, ed. 2, 1: 295.

English: bitter kola (Sierra Leone, Smythe); chew-stick (Sierra Leone, Smythe; Liberia, Cooper and Record).
West African: **SIERRA LEONE** MENDE sɔkiti (DS) TEMNE *ta*-sagbe (S&F) **LIBERIA** KRU-BASA kar (C; C&R) **IVORY COAST** ABE siébé (Aub.) 'ANDO' alikpadjo (Visser) MANDING-MANINKA ka woro (A.Chev.)

A tree to 30 m high of the dense evergreen semi-deciduous, swamp and galleried forest in Sierra Leone to Nigeria and extending to Angola.

The timber is hard, heavy, tough, strong and coarse-textured (5). In Sierra Leone it is split to make chew-sticks (6), while the Ando of Ivory Coast use the twigs for that purpose (7). There appears to be no other recorded usage.

There is a sticky yellow resin present in all parts of the tree. In Liberia the roots are crushed and soaked in palm-wine to make 'male medicine' which is strongly aphrodisiac (4). The bark is eaten in Congo (Brazzaville) to soothe cough and the leaves are made into a soup for upset tummy (2). The leaves are prepared in decoction by the Ando of Ivory Coast for troubles of the teeth (7).

The fruits are up to 4 cm in diameter and contain an edible pulp (1, 4, 5).

Phytochemical analysis has shown abundant *flavonins* in the bark and roots, and some *tannins* (3).

References:

1. Bamps, 1970: 65–66. 2. Bouquet, 1969: 132. 3. Bouquet, 1972: 27–28. 4. Cooper 272, K. 5. Cooper & Record, 1931: 40–42, with timber characters. 6. Smythe 221, K. 7. Visser, 1975: 54.

Garcinia gnetoides Hutch. & Dalz.

FWTA, ed. 2, 1: 294.

English: false chew-stick tree (from the Wasa name, Irvine).
West African: IVORY COAST AKYE oropupati (Aub.) GHANA AKAN-FANTE ablari
(TFC) WASA kyapiakwa (CV) tweapia-akoa (E&A) NZEMA brɔmabene (FRI) ehurike (CV)
NIGERIA BOKYI osun ojie (Catterall)

A tree to 17 m high with straight bole to 60 cm in girth, of the forest zone
from Liberia to W Cameroons.
Several species form a group with this. It shares the same vernaculars with *G.
smeathmanii* in Ivory Coast, but foresters reserve them for the latter (1). The
Wasa name in Ghana (5) and the Bokyi in Nigeria (2) refer to several other
species as well, but the reference to *G. granulata* Hutch. & Dalz. in Ghana (4)
belongs to this plant.
The bark yields a little yellow resin. A decoction of the leaves is a purgative
(3).

References:

1. Aubréville, 1959, 2: 334. 2. Catterall 11, K. 3. Chipp 14, K. 4. Irvine, 1961: 146. 5. Vigne 222, K.

Garcinia kola Heckel

FWTA, ed, 2, 1: 294. UPWTA, ed. 1, 91.

English: bitter kola; false kola; male kola; orogbo kola nut (from the Yoruba
name).
French: faux (fausse) colatier; petit kola; cola amère; cola mâle.
West African: SENEGAL WOLOF bitikola (K&A) SIERRA LEONE KONO sagbe (S&F)
KRIO bita-kola *i.e.*, *bitter kola* (FCD; JMD) MENDE *n*denyanie (Joru) sagbe (*def.* -i) (FCD;
S&F) TEMNE *ta*-sagbe (FCD; S&F) LIBERIA KRU-BASA swa-meh (C; C&R) MENDE kofě
(C&R) IVORY COAST ABE auolié (Aub.; JB) ABURE atuékwe (B&D) AKAN-ASANTE sundi
(B&D) ANYI tiampia (Aub.; JB) KYAMA haingre (Aub.) GHANA VULGAR tweapea (FD) AKAN-
ASANTE sĕfufudua (FRI) tw(e)apĕa (auctt.) tweapia (E&A) WASA twiapia (CJT) ANYI tiampa (FRI) ANYI-SEHWI twiapia
(CJT) NZEMA suapea (CJT) NIGERIA BETTE-BENDI *ù*-gyè (*pl. ì-*) (Ugbe fide KW) BOKYI oje
(auctt.) EDO ɛ̃dun (auctt.) EFIK èfìàrí (auctt.) EJAGHAM-EKIN efrie (McLeod; JMD) HAUSA cida
goro (Etkin) gooro (ZOG) IBIBIO efiat (auctt.) ICHEVE èmìàlè (Anape fide KW) IDOMA ígólígó
(Armstrong fide RB) IGBO àdù (auctt.) agbuilu (Iwu) akara-inu *from* inu: *bitterness* (DRR;
Singha) aku-ilu (Iwu) ùgolò (auctt.) IGBO (Awka) ugugolu (DRR) IGBO (Owerri) akụ ilū =
bitter palm kernel (DRR; KW) akuruma (DRR) IJO-IZON (Kolokuma) akaǎn (auctt.) ISEKIRI
okain (Kennedy; JMD) YORUBA orógbó (auctt.) WEST CAMEROONS BAFUT ngyà (*pl. bî-*)
(Mfanyam)

An evergreen tree to 30 m high, but usually to about 12–15 m, of the dense
rain-forest understorey, bole straight, unbuttressed, girth up to 1.80 m, spreading, heavy crown, often in wet situations, riverain and swamp, and up to 1200 m
altitude, occurring naturally from Sierra Leone to S Nigeria and on into Zaïre
and Angola, but distributed further by man and often found cultivated around
villages. Plantation cultivation can be successfully carried out: three year old
stump-planting under shade is recommended, but seed-at-stake is possible (21).
The specific name *kola* is taken from the genus *Cola* Schott. & Endl. (Sterculiaceae) and that in turn is derived from Temne and other kindred languages for
the true kola nut. *G. kola* is the source of the false kola.
The sap-wood is creamy white. Heart-wood is yellow, darkening to brown at
the centre, hard, close-grained, finishing smoothly and taking a good polish. It
is durable and fairly resistant to termites (10, 11, 15, 20, 21). The wood has had
unspecified use in Nigeria (11). Its principal application is for chew-sticks. In
Liberia they are said to whiten the teeth (10). In Ghana it is the smaller trees
which tend to be felled for this purpose and the wood is cut and split into pencil-

sized pieces (8, 21).It is bundles of these which are a common market commodity throughout the Region, for the chew-sticks of this species are considered superior to any other (13). The roots are also used, sometimes in preference (Sierra Leone: 20; Ghana: 13; Nigeria: 11, 28), and in the Ibadan area of S Nigeria they are thought to prevent dental caries. Tests have, however, shown no anti-biotic activity (17). In Sierra Leone the root is chewed to clean the mouth (20). In Igbo (Nigeria) pharmacology extracts of stems, roots and seeds have shown strong anti-hepatotoxic and hepatotropic activity. Petroleum ether and acetone extracts were found to be markedly anti-microbial (23).

The bark contains an abundant sticky resinous gum. It has water-proofing property. Certain people in W Cameroons use it to protect powder in the priming pans of flinklock guns from rain. It is incendiary and twigs burn brightly and can be used as tapers (11). It is used on skin-infections in Liberia (9, 10), and Congo (Brazzaville) (5). The powdered bark is applied in Nigeria to malignant tumours, cancers, etc., and the gum is taken internally for gonorrhoea, and externally to seal new wounds (4). In Congo a bark-decoction is taken for female sterility and to ease child-birth, the intake being daily till conception is certain and then at half quantity throughout the term (5). The bark is added to that of *Sarcocephalus latifolius* (Rubiaceae), a tisane of which has a strong reputation as a diuretic, urinary decongestant and for chronic urethral discharge (5). The bark is also thought to be galactogenic (5), whilst in Ghana the bark is used with *Piper guineense* (Piperaceae) and sap from a plantain stalk (Musaceae) to embrocate the breast for mastitis (29). In Ivory Coast a decoction of the bark is taken to induce the expulsion of a dead foetus, and seed and bark are taken to treat stomach-pains (7). Bark and leaves are used in Congo for pulmonary and gastro-intestinal troubles (5). Root and bark are administered in Sierra Leone as a tonic to men 'to make their organs work well' (12) and in that country too bark is added to palm-wine to improve its potency (20, 22). Bark is administered in Ivory Coast as an aphrodisiac (7). In S Nigeria a cold-water extract of root-bark with salt administered to cases of *ukwala* and *agbo-or*, identified as bronchial asthma or cough, and vomiting, is said to promote improvement (14, 27). The bark is used in tanning in Ghana (13), and during the 1939–45 War thousands of tons were exported as a tanning material (24). *Tannins*, a *reducing sugar* and traces of an *alkaloid* have been detected in the bark; *flavonins* are also present (2, 6, 13, 14), the whole being extremely bitter, resinous and astringent.

The leaves have a bitter taste (28). They are used in Congo as a deterrant to fleas (5). A leaf-infusion is purgative (4, 13).

The fruits are edible, orange-sized, and contain a yellow pulp surrounding four seeds. The fruits are eaten in Nigeria as a cure for general aches in the head, back, etc., and as a vermifuge (4). Igbo medicine-men prescribe the fruit for arthritic conditions (26). Wild animals go for them and the elephant is particularly partial, coming from afar to trees in season (21). The seeds are an important article of commerce being traded well beyond the distribution area of the tree. They are the false kola nut, or the bitter kola, as distinct from the true kola which is from *Cola nitida* (Vent.) Schott. & Endl. and *C. acuminata* (P. Beauv.) Schott. & Endl. (Sterculiaceae). The true kola nut has separate cotyledons. The false kola nut can be distinguished in not so separating. They are used as a masticatory, having a bitter, astringent and resinous taste, somewhat resembling that of the raw coffee bean. This is followed by a slight sweetness (or lingering pepperiness, 1). The false kola nut is relished as an adjuvant rather than a substitute for the true kola, increasing the user's enjoyment of the latter and allowing consumption of larger quantities without indisposition. Similarly they enhance the flavour of local liquor. The residue after chewing is white. They are eaten raw and not in prepared food (11). They have medicinal attributes. Mastication will relieve coughs, hoarseness, and bronchial and throat troubles (11, 18). They are taken dry as a remedy for dysentery (4). They are said to provide an antidote against strophanthus poisoning (11, 21). They are vermifugal (19). In Senegal (16), on Mt. Nimba, Liberia (1), in Ivory Coast (3) and in Congo (5) they are considered aphrodisiac. In Liberia the seeds are chopped up and steeped in water or better still in beer, while in Congo they enter into many medico-magical remedies taken with palm-wine 'to cleanse the stomach and to give strength in love' (5). In Nigeria the simple act of mastication of the seeds is held to be as effective (4).

The active principle, or principles, in the nut remain enigmatic. Caffein, which is present in the true kola, is absent. A trace of *alkaloid* has been reported in Nigerian materials (2), but absent in other samples (16, 18). *Tannins* are present which may contain the anti-bacterials *morellin* and *guttiferin* (25). The seeds have been shown to have four fluorescent substances of an undetermined nature (7, 16). Activity may also lie in resins which are as yet unidentified.

References:

1. Adam, 1971: 374–5. 2. Adegoke & al., 1968: 13–33. 3. Adjanohoun & Aké Assi, 1972: 149. 4. Ainslie, 1937: sp. no. 164. 5. Bouquet, 1969: 132. 6. Bouquet, 1972: 28. 7. Bouquet & Debray, 1974: 94. 8. Burtt Davy & Hoyle, 1937: 54. 9. Cooper 216, K. 10. Cooper & Record, 1931: 40, with timber characters. 11. Dalziel, 1937: 91. 12. Herbalists, Joru, Nov. 1973. 13. Irvine, 1961: 146–7. 14. Iwu, 1986: 140. 15. Keay & al., 1960: 185–6. 16. Kerharo & Adam, 1974: 338–9. 17. Lowe: 1976. 18. Oliver, 1960: 27, 65. 19. Pellegrin, 1959: 225–6. 20. Savill & Fox, 1967: 132. 21. Taylor, CJ. 1960: 171. 22. Unwin & Smythe 50, K. 23. Iwu, 1984. 24. Portères, 1974: 123–4. 25. Etkin, 1981: 84. 26. Iwu & Anyanwu, 1982: 263–74. 27. Akubue & Mittal, 1982: 357. 28. Isawumi, 1978, b: 114. 29. Ampofo, 1983: 8.

Garcinia livingstonei T Anders

FWTA, ed. 2, 1: 294. UPWTA, ed. 1, 90, as *G. baikieana* Vesque.
West African: SENEGAL KONYAGI pratone, pratond (JB) MANDING-BAMBARA kôkumô, sumésunsu (JB) MANINKA mädähô (JB) sungala (JB) GUINEA MANDING-MANINKA sungala (Aub.) MALI MANDING-BAMBARA sumé sunsu (Aub.) MANINKA *ko* rokna *ko*: tree (A.Chev.) sungala (Aub.) UPPER VOLTA MANDING-BAMBARA sumé sunsu (Aub.) GHANA DAGAARI kulitia (AEK; FRI) GBE-VHE abriva (auctt.) KONKOMBA lepenabul (Kersting) MOORE kulitia (AEK; FRI) WALA kombauadai (AEK; FRI)

A shrub or tree to 17 m high, of fringing forest by stream-banks and in swamp of the savanna, occurring sparingly across the Region from Senegal to N Nigeria, in Cameroun and widely distributed in eastern Africa from Somalia to South Africa.

The wood is yellowish white. It is said to be susceptible to borer but to be of unspecified use in Botswana (5). The branches make stirring sticks in Kenya (4).

The bark yields a yellow gum which hardens to a substance like gamboge (1, 5). Leaf and flower-extracts have given positive anti-biotic tests (5). In the Meru district of Kenya the flowers are much visited by bees (3).

The fruit is bright yellow and 1.25 to 2.5 cm across and contains a pleasantly acid-sweet pulp which is edible. In Uganda, Karamojo area, it is considered a 'herdsman's' food for plucking and eating on the spot, not for taking home (2). In Mozambique the fruit is fermented to produce an alcoholic drink (5).

References:

1. Burtt Davy & Hoyle 1937, 53. 2. Dyson-Hudson 403, K. 3. Gardner 18570, K. 4. Van Someren 79, K. 5. Watt & Breyer-Brandwijk, 1962: 495.

Garcinia mangostana Linn.

FWTA, ed. 2, 1: 295.
English: mangosteen (from olden Malay, *manggusta* or *manggistan*, Burkill).
Portuguese: mangostão (Feijão).
West African: GHANA AKAN-TWI tweapea-akoa (FRI; E&A)

A tree to 7–8 m high with dense heavy profusely branched crown, known only from cultivation in SE Asia and thence taken by man to other parts of the tropics. Successful introduction has not always been achieved, and dispersal has been limited. The tree is in cultivation in W Africa. A constantly humid climate is required.

The timber is dark brown, rather hard and heavy. It has been used in SE Asia for cabinetry, house-building, rice-mortars and pounders, etc. In Ghana it is said to be used for chew-sticks (2, 3). But since the tree is grown primarily for its fruit, the wood is of greater utility in a live state. The fruit is the mangosteen, rated one of the most delectable of the tropics. Good fruits may attain 6–7 cm in diameter and contain about 5–7 seed surrounded by a white, sweet and succulent flesh. The flesh is present even if the seeds have aborted or atrophied. Furthermore the fruit is parthenocarpic. There may be scope for developing seedless clones. The most serious drawback to this plant is that it rarely fruits younger than 15 years of age.

The fruit-shell contains 7–13% *tannin*. It can be used for tanning and might be worth collecting for this purpose if in abundance near to a tannery. In Malaya the shell is used to obtain a black dye, and also to prepare astringent medicines for use in dysentery, enteritis, etc.

The seeds contain 3% oil, and the kernels can be ground to produce a vegetable butter.

References:

1. Burkill, IH, 1935: 1052–5. 2. Irvine 1003, K. 3. Irvine, 1961: 147. 4. Sastri 1956: 103–5.

Garcinia mannii Oliv.

FWTA, ed. 2. 1: 295. UPWTA, ed. 1, 91, in minor part.
West African: **NIGERIA** BOKYI osun ojie (Catterall; JMD) EDO opahan (AHU; JMD) EFIK ọkọk (KO&S) IGBO akụ ilụ = *bitter palm kernel,* or *bitter kola* (auctt.) YORUBA (Ondo) aromọ (Kennedy; JMD) okuta = *stone; from the hardness of the wood* (JMD)

A tree to about 12 m high by 60–100 cm in girth, short bole, compact crown, of the dense forest and know only from S Nigeria and W Cameroons in the Region, and occurring also eastwards to Gabon and Equatorial Guinea.

The wood is very hard, meriting a name meaning 'stone' in Ondo. The bark contains a thick orange sap which is not of recorded usage, but the bark itself is bitter. The root-bark may be dried, powdered and taken with spices or condiments for serious forms of diarrhoea and dysentery (1). The root soaked in palm-wine or gin is taken in Nigeria as an aphrodisiac (1). The root-wood is know also as an aphrodisiac in E Cameroun (2).

References:

1. Dalziel, 1937: 91. 2. Mildbraed 10693, K.

Garcinia ovalifolia Oliv.

FWTA, ed. 2, 1: 295. UPWTA, ed. 1, 92.
West African: **GUINEA** MANDING-MANINKA guessé-guessé (A.Chev.) **NIGERIA** EDO ọviedun (Ross; KO&S) HAUSA goorò (Lowe)

A shrub or tree to 10 m high, of the fringing forest in the savanna zone from Guinea and Mali to S Nigeria, and extending onward to Ethiopia, Uganda and Angola.

No usage is recorded in the Region. In Zaïre the wood is used to make canoes (2). Some *flavonins* have been reported in the bark and roots, and also traces of *saponins* and *tannins*; in the fruits, leaves bark and roots *steroids* and *terpenes* are reported (10).

References:

1. Bouquet, 1972: 28. 2. Dalziel, 1937: 92.

Garcinia punctata Oliv.

FWTA, ed. 2, 1: 295.

A tree to 12 m high by 30 cm in girth, of dense and galleried forest, often in wet situations and then with stilt-roots, in S Nigeria and W Cameroons, and extending to Malawi and Angola.

The powdered bark is applied to snake-bite in Congo (Brazzaville), and sap from the bark or bark-decoction is taken by draught for costal pain and cough; the painful areas are scarified and embrocated with leaf-sap of this species to which has been added gunpowder and charcoal made of *Schwenckia americana* Linn. (Solanaceae) and *Dichrostachys glomerata* (Forssk.) Chiov. (Leguminosae: Mimosoideae) (1).

The seed kernel is eaten in W Cameroons (2).

References:

1. Bouquet, 1969: 133. 2. Gartlan 2, K.

Garcinia smeathmannii (Planch. & Triana) Oliv.

FWTA, ed. 2, 1: 294–5, incl. *G. polyantha* Oliv. UPWTA, ed. 1, 90, as *G. barteri* Oliv. and 92, as *G. polyantha* Oliv.

English: false chew-stick tree (Ghana, Burtt Davy & Hoyle, Irvine).

West African: SENEGAL DIOLA-FLUP o sénom kāngem (JB) MANDING-MANINKA makakandi (JB) GUINEA-BISSAU BIDYOGO n-tchocodó (auctt.) untchócodó (Rdof) FULA-PULAAR (Guinea-Bissau) macacundje (EPdS) GUINEA MANDYAK macacundje (JDES) macadunje (RdoF) SIERRA LEONE KONO nguoka-kɔnɛ (S&F; FCD) MENDE gbɔwi, kɔndoi, kɔndui (NWT) ngolo-kala (*def.* -kalei) (FCD; S&F) nguo-kala (*def.* -kalei) (FCD; S&F) sagbe (*def.* -i) (FCD; S&F) sɔle (*def.* -i) (Aylmer; SKS) SUSU kureteburaxe, kurutuburaxe (NWT) TEMNE ta-sagbe (S&F; FCD) IVORY COAST ABE oropupati (auctt.) AKYE hinnépo (auctt.) ANYI mamiékini (A.Chev.) maniékimi (auctt.) KYAMA ahiohué (Aub.; RS) aniaya (B&D) GHANA VULGAR tweapea-akoa (FD) AKAN-ASANTE bɔhwe (auctt.) TWI bɔhwe (BD&H; FRI) WASA kyapiakwa (auctt.) tw(e)apeakwa = *false chew-stick* (FRI) tweapia-akoa (E&A) AKPAFU duamawipe (FRI) ANYI maniekini (FRI) TOGO TEM tchifomura (Gaisser; JB) TEM (Tschaudjo) tsche-fonuvia (Gaisser) tyifomire (Kersting) NIGERIA EDO ọbiẹn-edun (JMD) ọviẹdun (auctt.) YORUBA afon (KO&S) ẹrú orógbó (KO&S; Verger) ozoda (AST) WEST CAMEROONS BAMILEKE leny (ATJohnstone)

A tree of the evergreen forest, to 23 m high with straight clear bole to over 1 m in girth, horizontal whorled branching, usually near streams in lowlands and uplands to 1300 m elevation, from Guinea-Bissau to W Cameroons, and widely dispersed across Africa to Malawi, Zambia and Angola.

The wood is hard and durable. It has been used for posts in Ghana (3). The Wasa name, and the English, doubtless drawn from it, suggests use for chew-sticks, if only as a substitute.

The bark contains a bright yellow, sticky latex which is used as a wound-dressing (2). In Congo (Brazzaville) skin-infections are treated by rubbing with a bark-decoction followed by an application of the latex (1). The latex is also instilled into the eyes for ophthalmias, and the bark (? decoction) is taken by mouth as a purgative in treatment for female sterility and as a poisoning antidote (1).

393

GUTTIFERAE

The fruit-pulp is edible and is taken in time of scarcity (2). The seeds are used more or less like those of *G. kola*, the false kola nut.

References:
1. Bouquet, 1969: 133, incl. *G. polyantha* Oliv. 2. Dalziel, 1937, 90, 92. 3. Irvine, 1961: 147, as *G. polyantha* Oliv.

Garcinia staudtii Engl.

FWTA, ed. 2, 1: 294.
West African: NIGERIA BOKYI oje (KO&S)

A tree to 13 m high, bole straight to 60 cm in girth, an undershrub of the evergreen forest, recorded in S Nigeria and in E Cameroun.
The wood is pale brown and very hard (1, 2).

References:
1. Catterall 1K/18/1934, K. 2. Keay & al., 1960: 182–3.

Harungana madagascariensis Lam. ex Poir.

FWTA, ed. 2, 1: 290. UPWTA, ed. 1, 87–88.
English: dragon's blood tree.
French: guttier du Gabon (Walker).
West African: SENEGAL BALANTA nidièl (JB) psèr (auctt.) BANYUN kilé éińn (K&A) kilé éñimé (JB; K&A) su budèr (auctt.) 'CASAMANCE' biné (Aub.; RS) fusimègue (Aub.; RS) nguel (Aub.; RS) uliyolo (Aub.; RS) CRIOULO po de faya (JB) DIOLA biné (auctt.) bu kakāndia (JB) fu simed, fu timet, fu simat, fu simodi (JB) fusimeg (A.Chev.; K&A) DIOLA (Oussouye) étóno (K&A) DIOLA-FLUP é toño, é tóno (K&A; JB) FULA-PULAAR (Senegal) sumbala (JB; K&A) sungala (JB) u liéli (JB) MANDING-BAMBARA sāndala giri (JB) sumbala (K&A) MANDINKA ulo yélo, uliélu (auctt.) MANINKA subala giri, vobé (auctt.) MANDYAK be ñakat, bi ñañak (JB) MANKANYA be nakenten, be vana (JB) 'SUSU' vobe (JB) THE GAMBIA FULA-PULAAR (The Gambia) sumbula (DAP) MANDING-MANDINKA wuliyolo (DAP) WOLOF ngwel (DAP) GUI-NEA-BISSAU BIDYOGO uómnhé (JDES) CRIOULO pó de faia (EPdS; JDES) FULA-PULAAR (Guinea-Bissau) súngala (JDES) uliéli (EPdS; JDES) uluélo (EPdS) MANDING-MANDINKA ulielô (EPdS; JDES) GUINEA BALANTA umpátè (JDES) FULA-PULAAR (Guinea) sumbala (auctt.) sungala (CHOP) KONO léré (RS) KPELLE lolo (RS) LOMA podo, podo gui (RS) MANDING-MANINKA (Konya) sungala (RS) MANDYAK binhanhaque (FD) MANO lulu (RS) luru (RS) SUSU uobé (Aub.; RS) wobé (CHOP) TEMNE *k*-pel (CHOP) SIERRA LEONE BULOM (Sherbro) pel-lɛ (FCD; S&F) GOLA yoŋgo (FCD) KISSI howa (S&F; FCD) yo (S&F; FCD) yuleŋ (S&F; FCD) KONO kuŋgbali-kɔne (S&F; FCD) suŋgbali (S&F; FCD) KORANKO sumbale (NWT) toubale (NWT) KRIO blɔd-tri *i.e., blood tree* (S&F; FCD) pam-ɔil-tik *i.e., palm-oil tree* (S&F; FCD) LIMBA kutoso (NWT) LOKO *m*be(-i) (S&F; FCD) mbɛli (NWT) mbema (JMD) mbe-nyɛgɛ (NWT) ndeli (NWT) MANDING-MANDINKA sumbalaŋ (FCD) MENDE *m*bɛli (*def.* -i) (auctt.) yoŋgoe (*def.* -i) (S&F) MENDE (Up) yoŋgo (FCD) SUSU kobekume (NWT) wavesive (NWT) wobe (NWT; JMD) wobewobe (NWT) wubi (JMD) TEMNE *a*-pil (auctt.) VAI suŋgba (FCD) YALUNKA webe-na (?) (FCD) wobe-na (FCD) YORUBA-AKU igi ekpó (JMD) LIBERIA BANDI mbeli (Konneh) KRU-BASA giĕ (Barker) MANO lolo (JMD; RS) UPPER VOLTA DAGAARI kāl (Girault) IVORY COAST ABE uombé (auctt.) wombe (A.Chev.) ABURE vossobo (B&D) AKAN-ASANTE kossuwa (K&B) AKYE ŋgoa (auctt.) ngoe (RS) ANYI kossa (K&B) BAULE fadosa (K&B) fosona (K&B) gugru (K&B) kossua (K&B) FULA-FULFULDE (Ivory Coast) sumbala (Aub.) GAGU kpébo (B&D) lolo (K&B) 'KRU' blakaï (K&B) dozero (K&B) kehéré baïgri (K&B) KRU-BETE bubréi (K&B) GUERE soro (RS) sorowe (K&B) sro (RS) GUERE (Chiehn) bugulu (B&D) gugru (K&B; B&D) GUERE (Wobe) tora (RS) KWENI golibénuré (B&D) nia iri, nian iri (B&D) nien iri = *blood tree* (K&B) KYAMA akuin (auctt.) kossoa (B&D) MANDING-MANINKA laguiri (Aub.; K&B) suba (K&B) subalaguiri (Aub.) sunaï (B&D) 'NÉKÉ-DIÉ' goglu (B&D) GHANA AKAN-ASANTE bofua = *yellow velvet; from the use of the wood to dye this material* (FRI) ɔkɔsoa (auctt.) kusua (auctt.) BRONG ɔkɔsoa (FRI; E&A) FANTE ngodua, nngodua = *oil tree* (auctt.) ɔkɔsoa (auctt.) TWI bofua = *yellow velvet; from the use of the wood to dye this material* (FRI) oduben = *sulphur; alluding to the colour of the wood* (FRI; JMD) ɔkɔsoa (auctt.) WASA ɔkɔsoa (FRI) NZEMA kɔsoa (auctt.) NIGERIA ANAANG ébánănàń (KW)

EDO itue (auctt.) itue-ẹgbo (JMD) EFIK òton (auctt.) ESAN uruarua (KO&S) HAUSA alíllìbár ràảfíí (auctt.) IBIBIO atón (Singha; KO&S) IGBO ọtọrọ (AJC) útùrù (auctt.) IGBO (Ila) nyemu nyemu (NWT) IGBO (Okpanam) ọmasika (NWT) IGBO (Owerri) eterē (JMD) IGBO (Uburuku) ọmasika (NWT; JMD) IGBO (Umuahia) oturu (AJC; JMD) IJO-IZON bou pulóú (auctt.) IZON (Kolokuma) bou pulóú (KW) IZON (Oporoma) abó (KW) YEKHEE uyaruerue (JMD) YORUBA adindix (IFE) amùjè̩ = *drinker of blood; from* mu: *to arrest*; èjè: *blood* (auctt.) egbo re *the roots* (IFE) elépo (auctt.) legun soko *the roots* (IFE) **WEST CAMEROONS** DUALA tolongo (JMD) KPE morolongo (JMD) otolungo (JMD) wutulongo (AJC; JMD)

A much-branched shrub or small tree, occasionally reaching 12 m high on a straight trunk, with a heavy canopy, of clearings in the forest zone and on forest margins of the savanna region, locally abundant in secondary forest, occurring right across the Region from Senegal to W Cameroons, and widespread in tropical Africa, Madagascar and the Mascarene Islands. The generic name is drived from Malagassy vernacular name.

The plant is a vigorous coloniser. It is amongst the first pioneers to occupy savanna land after destruction of the vegetation by bush fires, thus affording protection and cover for the soil (5). It also invades grassland thereby accelerating reversion to secondary bush-savanna. In Sierra Leone, at least, the tree is often ant-infested (29).

The wood is orange-red to yellow, and is particularly attractive. It works well and finishes easily, but unfortunately timber in sufficiently large enough pieces is not commonly available. The wood is light in weight and durable in contact with the ground, though in exposed positions it is liable to insect-attack. It is nevertheless widely used in hut-construction for poles, roof-joists, planks etc. throughout its range in Africa. In western Ghana it is used to make thwarts and seats in canoes (14, 20), and in S Nigeria for yam-sticks and the construction of store-houses (12). The branches diverge at an angle of 60–80° from the main stem, and suitable pieces have been cut for use as hockey-sticks (12). The wood is useful as a fuel in local metalurgy for softening metals (14).

The bark is easily peeled off in long strips. It is rather wet and turgid. The main inner part yields a brilliant orange gum which turns red on exposure, but the innermost layer of bark and the outer layers of the wood yield a yellow sap from which a more delicate dye or stain is obtained which is used in Ghana for dying velvet – hence the Asante and Twi name *bofua*. The light colour is also likened to sulphur in the Twi name *oduben*. The gummy latex is used in all areas to furnish a yellow dye also for use on cloths, matting and many other articles, the usual practice being to chop up pieces of the bark which are boiled in water with the thing to be dyed. The Ijo of the Niger delta area use the latex as a yellow paint (38). The exudate thickens and a balsam has been recovered from it. It makes a good stain for wood and in Tanganyika has been used as a sealing-wax (6). The yellow dye can be made brown by mixing in powdered camwood (14). At Ohuhu-Ngwa in S Nigeria, Igbos boil the inner bark in water and the resultant yellow liquid is added to pounded cassava to form a stiff, gluey paste. This is applied to newly fired pots while they are still hot as a sort of finish. The significance of this is not clear, but one result is to stain them a dark yellow (42).

The resinous sap is used for all manner of cutaneous complaints: sores, itch, scabies, ringworm, craw-craw, mange, ?prickly heat: mycoses, often after the affected area has been scarified to draw blood with, for example, the rough leaf of *Ficus exasperata* Vahl (Moraceae), the sand paper tree (14; Sierra Leone: 8; Liberia: 24; Ivory Coast: 10, 19; Nigeria: 13, 25; Gabon: 34, 35; Congo: 9; Ubangi: 41; Zambia: 23; Zanzibar: 37; Madagascar: 36). It is used on leprous areas in Liberia (14), Ivory Coast (10, 19) and Congo (Brazzaville) (9). Ash (? of the bark) is applied in Sudan to areas of scabies (4). The gum is held to be styptic and haemostatic. Some vernacular names refer to blood, and the Yoruba specifically refer to its quality of arresting the flow of blood (see above). For this it is applied to cuts, and in Liberia (7, 33) to fresh circumcision wounds. The dried gum is also used in Nigeria as a wound-dressing (2). In extension of this line of thought, a bark-decoction, or of the root, is widely deemed helpful as a remedy for troubles in which blood is manifest: for dysentery and piles (2, 14, 31); in Ivory Coast as a placental ecbolic (19) and emmenagogue (10); in Gabon for painful menstruation and as an ecbolic (34, 35); in Congo for gynaecological conditions: dysmenorrhoea, menstrual irregularity, miscarriage and sterility, and also for haematuria (9); in Ubangi for female sterility (41); in Tanganyika to arrest menstruation (21, 36); in Madagascar for cough with bloody sputum (15). Also on the Theory of Signatures, the yellow colour of the gum evokes usage of the bark, roots or the gum itself for jaundice (Sierra Leone: 29; Ivory Coast: 10, 19; Nigeria: 25). Then, likening the expissation of the gum to milk-flow, the bark or root is used in a treatment in Tanganyika to stimulate breast-development (30, 36). Healing quality of the gum is sought in Gabon (34, 35) and in Tanganyika (39) in application to ulcers, on which in the latter territory the gum is allowed to solidify.

The bark or the gum is held to have purgative properties. In Senegal the bark, and the leaves, are used for stomach-ache (18). Sap expressed from the inner bark is taken slightly warmed in Tanganyika as a purgative (16), and water in which bark has been boiled is given to babies suffering from constipation and wind (27). Similar infant treatment is recorded in Malawi (20). It is taken for dysentery in Ivory Coast (10), Nigeria (31) and Congo (9). Sap washed out from bark taken from the east and west sides of the tree trunk (a relic of sun-worship?) has been recorded used in Liberia as a remedy for tape-worm (14). In the Central African Republic also the bark has vermifugal use (26).

The bark, or the gum, has application in various countries for chest and breathing difficulties. A decoction of it is taken in Congo for bronchial affection, and cough, and as an expectorant and emetic (9), in Madagascar for cough (15), in Gabon for asthma, liver complaints and as an emetic (35). In Ivory Coast the plant (? bark, gum) has a reputation as an aphrodisiac, and is sometimes used for toothache and oedemas and is a constituent of various arrow-poison mixtures (19).

In Casamance (Senegal) there is veterinary use of the bark which is crushed and scattered in water where poultry goes to drink as a preventative for poultry illnesses (17, 18).

In Gabon the bark-sap is added to fermented beverages for colouring (14). The bark is sufficiently fibrous to be used sometimes in Ghana for tying roofs (20).

The twigs, leaves and leaf-buds find similar medicinal uses. The leaves are used in treatment for chest-complaints in Nigeria (2), asthma in Sierra Leone (29) and on skin-troubles in Sudan (3, 11); for haemorrhage, diarrhoea, gonorrhoea, sore-throat and fevers in Madagascar (15, 36); and for stomach-ache and menstrual troubles in Ubangi (41). The leaf, and the roots are also considered febrifugal and anti-malarial in Tanganyika (16). The leaf-sap is taken in Tanganyika for amenorrhoea (16), and in Congo for heart-troubles (9). The leafy shoots are deemed stomachic in Nigeria (31) and laxative in Gabon, where they are also chewed as a masticatory with kola nut for urethral discharge (34, 35). The two terminal leaves of a shoot are deemed in Congo to be anti-tussive (9). Puerperal infection is treated in Liberia by eating the unopened buds beaten up with palm-oil (Harley in 14).

Food-poisoning of vegetable origin, symptomised by a swollen stomach, is treated in Ubangi with a preparation made of the root (41).

The fruits are edible, though stomachic and mildly laxative, and in excess emetic (14). They are used in local cooking in Guinea and fermented to produce a sort of cider which taken on an empty stomach may sometimes cause vomiting (28). The fruits and the young leafy shoots are taken in Nigeria as a laxative for stomachic and intestinal troubles (14). An ointment made from the fruit in animal fat is used in Ivory Coast on inflamed ganglia (19). In Sierra Leone the fruit is occasionally taken following an abortion in the belief, as indicated above for the bark-gum, that the red sap will staunch bleeding (14, 29). Resin from the flower is used in southern Africa for colic, puerperal infection, roundworm and as a rubefacient (36). The seed serve as bait in Ubangi for trapping small birds (41).

Analysis of Nigerian material has shown the presence of a trace of *alkaloids* in the bark (1). In Madagascar alkaloid has been reported absent, but abundant *saponin* in the leaves, and the presence of *flavones, leucoanthocyanins* and *tannins* in the stem (15). The yellow colouration has been ascribed to a phenolic pigment named *harunganine*. A number of related substances is present and an extract called 'harongan' is being subjected to examination for stomach and pancreas disorders (10, 18). Examination of Nigerian material for anti-biotic activities has shown action on Gram +ve *Sarcina lutea* and *Staphylococcus aureus*, no action against Gram −ve organisms, and no fungistatic action (22).

Because of the reputation as a haemostatic and the red colour of the gum, the plant has the fanciful name of *dragon's blood tree*, and under the Yoruba name *amújẹ̀*: arrester of blood, the plant features in an Yoruba incantation for the purification of blood (32). The simile to the oil-palm (see the vernacular names above) also is against meeting snakes while climbing the palms to tap them. A young leaf taken from a terminal bud is carefully placed under the tongue. It must not touch the teeth. If the leaf remains unbitten the tapper will be protected, but should he damage the leaf the protection will be lost (8). Whether this plant has some property as a snake-repellant should be examined. In Upper Volta it enters into rituals for supplication (40).

References:

1. Adegoke & al. 1968: 13–33. 2. Ainslie, 1937: sp. no. 176. 3. Andrews A.1563, K. 4. Anon. 17, K. 5. Aubréville, 1950: 144. 6. Bancroft 184, K. 7. Barker 1278, K. 8. Boboh, 1974. 9. Bouquet, 1969: 137. 10. Bouquet & Debray, 1974: 95. 11. Broun & Massey, 1929: 117, as *Haronga madagascariensis* Chois. 12. Carpenter 67, UCI. 13. Carpenter 104, UCI. 14. Dalziel, 1937: 87–88. 15. Debray & al., 1971: 75. 16. Haerdi, 1964: 101. 17. Kerharo & Adam, 1963, a: 773–92. 18. Kerharo & Adam, 1974: 484, with phytochemistry and pharmacognosy. 19. Kerharo & Bouquet, 1950: 55–56. 20. Irvine, 1961: 140–1. 21. Koritschoner 648, K. 22. Malcolm & Sofowora, 1969: 512–7. 23. Milne-Redhead 4550, K. 24. Okeke 2, K. 25. Oliver, 1960: 28. 26. Pellegrin, 1959: 218. 27. Pirozynski P.81, K. 28. Pobéguin, 1912: 40, as *Harunga paniculata*. 29. Savill & Fox, 1967: 138. 30. Semsei 651, K. 31. Singha, 1965. 32. Verger, 1967: no. 33. 33. Voorhoeve, 1965: 107. 34. Walker, 1953, b: 277. 35. Walker & Sillans, 1961: 204–5. 36. Watt & Breyer-Brandwijk, 1962: 495. 37. Williams, RO, 1949: 291–2. 38. Williamson KW.18, UCI. 39. Wilson 22, K. 40. Girault, 1960: 120. 41. Vergiat, 1970, a: 88–89, as *H. paniculata* Loes. 42. Murray, 1972: 167–8.

Hypericum lalandii Choisy

FWTA, ed. 2, 1: 287.

A perennial herb to 1 m high, of damp grassland, recorded only from the Bauchi Plateau of N Nigeria, but occurring widespread in central, eastern and southern Africa and across Asia into SW China.

There is unspecified medicinal usage of the plant in Lesotho (1, 2).

References:

1. Guillarmod, 1971: 435. 2. Watt & Breyer-Brandwijk, 1962: 495.

Hypericum peplidifolium A. Rich.

FWTA, ed. 2. 1: 287.

A perennial herb of montane grassland, often in wet situations, recorded only from W Cameroons and Fernando Po in the Region, but occurring widely in central, eastern and south-central Africa.

The plant is grazed by all stock in Kenya (3). Kipsigis of Kenya eat the fruits and say that like the leaves they have an acid taste (2). The Chagga of Tanganyika use the leaves as an indigestion remedy (1; 5), and in the Iringa region the plant (part not stated) is used for conjunctivitis and probably for trachoma (4).

References:

1. Bally, 1937: 10–26. 2. Glover & al. 812, K. 3. Glover & al. 1035, K. 4. Hobbs 6A, K. 5. Watt & Breyer-Brandwijk, 1962: 496.

Hypericum revolutum Vahl

FWTA, ed. 2, 1: 287, as H. lanceolatum Lam. UPWTA, ed. 1, 88, as H. lanceolatum Lam.

English: giant St. John's wort (Kenya, Dale & Greenway).

West African: **WEST CAMEROONS** KPE mosoni (AJC)

A shrub or tree to 12 m high, of montane situations in W Cameroons and Fernando Po, and distributed across central Africa to eastern Africa.

The timber does not attain much size, but in Kenya is large enough to be used by the Masai for hut-building (2). The foliage is strongly aromatic, and the plant has been a source of a balsam (3). The bark and rotten timber have been reported to be eaten by gorillas in Rwanda (1). Masai herdsmen in Kenya feed the foliage to goats (2), but the plant is a cause of photo-sensitisation in sheep in southern Africa though it is slower-acting and the result is less intense than allergy caused by other *Hypericum spp.* (3). In Ethiopia the leaves are cooked and eaten with meat against stomach-upsets (4).

References:

1. Fossey B/3, K. 2. Glover & al. 930, K. 3. Watt & Breyer-Brandwijk, 1962: 495, as *H. lanceolatum* Lam. 4. Lemordant, 1971: 159, as *H. leucoptychodes* Steud.

Hypericum roeperanum Schimp. ex A. Rich.

FWTA, ed. 2, 1: 287, as *H. riparium* A. Chev.

A shrub or small tree to 5 m high, in bush and riverain montane situations in Guinea, Nigeria and W Cameroons, and widely dispersed over central, eastern and south tropical Africa.

The roots are recognised in Tanganyika to be toxic (2) and are used in an undetailed manner to treat infertility (1).

References:

1. Koritschoner 888, K. 2. Koritschoner 1392, K.

Mammea africana Sab.

FWTA ed. 2, 1: 293. UPWTA, ed. 1, 92, as *Ochrocarpus africanus* Oliv.

English: the fruit – African apple; mammee apple; African mammee-apple; mammy supporter (Sierra Leone from 'mammea sapota'), African apricot; the timber – bastard mahogany; Calabar mahogany.

French: the fruit – abricotier d'Afrique; the timber – obota (from a Gabonese vernacular).

West African: **SENEGAL** DIOLA-FLUP é dagalésa (JB) WOLOF zaué (FB) **SIERRA LEONE** KONO kaikɔmba (S&F) MENDE kaikɔmbe (*def.* -i) (auctt.) kaikonbe (*def.* -i) (Joru) TEMNE bakum, mammee (LC) **LIBERIA** DAN mö(-n) (GK) mung (AGV) KRU-BASA bahn (auctt.) GUERE (Krahn) gó-ai (GK) kunfe (AGV) MENDE kaikumba (C&R; AGV) **IVORY COAST** ABE djimbo (auctt.) ABURE okué (B&D) ADYUKRU kotoé (A.Chev.; JB) AKYE agnuhé (Aub.; JB) animbéré (A.Chev.) ANYI bélétui (Aub.; JB) blétuné (Aub.) kélipé (A.Chev.) BAULE pépia (A.Chev.; JB) KRU-GUERE (Wobe) zawé (A.Chev.) KULANGO (Bondoukou) kélipé (A.Chev.) quelipeketipe (LC) KYAMA moya (Aub.; B&D) NZEMA bletuné, sumapia (A.Chev.) **GHANA**

AKAN-ASANTE abrompesowa (CJT) bɔmpagya *the most usual name* (auctt.) duforokoto (CJT; E&A) TWI bompegya (auctt.) WASA bɔmpagya (auctt.) pegya (CJT) ANYI bletui, bletune (FRI) ANYI-AOWIN pasêê (FRI) SEHWI pasêê (auctt.) paseng (CV) pasin (CJT) NZEMA abɔtɔabiri (FRI) apegya (CJT) **NIGERIA** ABUA odọ (JMD) BOKYI bobuat (Catterall fide JMD) okut (KO&S; LC) ótông (LC; JMD) EDO ogi-otien (JMD) otien-ogioro (auctt.) ótíẹn-wáre (auctt.) udegbu (Ross; JMD) EFIK édéng (auctt.) EJAGHAM (Keaka) njenku (LAKC; JMD) otun (AHU; JMD) ovep (LC) EKPEYE odudu (DRR; JMD) ENGENNI agba (JMD) IGBO ekpili (KO&S) ekpiri (BNO) IGBO (Awka) ekpili (Kennedy; JMD) IGBO (Owerri) ukutu (DRR; JMD) IGBO-IKWERE (Isiokpo) okakilo (DRR; JMD) IJO-IZON (Kolokuma) bọ́lọ́ (KW&T) KAKIBA (Ibani) bọlọ (LAKC) NEMBE bọlọ, borlor (LAKC) IṢEKIRI borọlọ, owosokotọ (LAKC; JMD) OKOBO ọdùdù (DRR; JMD) URHOBO loolo (Kennedy; JMD) urherame (auctt.) YORUBA ologbomodu (auctt.)

A tree to over 30 m high, trunk without buttress, but sometimes flared, and to 3.75 m in girth, in wetter sites of the forest zone from southern Senegal to W Cameroons and on to Zaïre and Angola.

The tree is easy to cultivate and has been successfully grown in plantations but growth is slow and overhead shading is necessary (23). It has also been established by underplanting, and conversely the mature trees are good shade-bearers (24). The new leaf flushes are deep red making an attractive, if but ephemeral, show.

The sap-wood is light brown. When first cut the heart-wood is deep red and turns a mahogany-brown. It is hard and fairly heavy, with an attractive grain, coarse but even-textured, durable, but not termite-proof (10, 13, 14, 17, 22–24). The wood contains resin-canals and mineral-deposits. It has been exported from Gabon (26) but the gumminess limits its value as an export article (24) and mineral cells blunt cutting tools (17). Though on the hard and heavy side, it might serve as a substitute for mahogany. In Sierra Leone it substitutes *Entandrophragma* (Meliaceae) in local furniture construction (22) and is con-sidered as a primary structural timber (11). In Ghana it is listed amongst timbers of possible economic value (15) and is currently recommended for general joinery (3). It is a timber of general utility for carpentry, joinery, interior decoration, planks, beams, bridge piles, and heavy construction-work (8, 10, 13, 14, 17, 21, 24, 26). Canoe dugouts are cut from the trunk in SE Nigeria (1, 18, 27).

The bark exudes a yellow resinous sap. This, or the bark in decoction, is widely and commonly used for dermal infections, craw-craw, itch, etc. (2, 7, 10, 14, 16, 20, 23). In Sierra Leone the extracted latex is mixed with clay and is put on to heal scratches and is applied to skin-diseases, and water in which the bark, or the fruit, has been boiled is used to draw jiggers from the feet (22). Similar treatment for jiggers is reported in Liberia (24). Water in which the roots have been boiled is used as a wash to treat children with an itch known as *boro*, Twi, in Ghana (17). The bark is an occasional remedy for syphilis in Liberia (14) and Nigeria (20). A decoction is administered in Ivory Coast as an anodyne for fever-pains (7). In Gabon simlar treatment is sought to assuage rheumatic pain; also the decoction is used to cleanse ulcers, and bark-scrapings are rubbed over cutaneous eruptions and onto dogs for mange (25, 26). In Congo (Brazzaville), bark crushed with maleguetta peppers and sugar-cane is macerated for 24 hours and the liquor is taken for ovarian and stomach troubles, blennorrhoea and as an ecbolic; this prescription is put into hip-baths and vaginal douches for metritis and vaginitis; a bark-decoction is drunk for whooping-cough and put into baths for children with fever (5). Bark is used as a fish-poison in Liberia (12) and in Nigeria (2), but it is evident that its potency is not great for the fish are only stunned and use is limited to small streams (13) and to dammed creeks (24) where the volume of water is restricted.

The bark is a source of potash, and is burnt along with dried plantain-skins to make soap with palm-oil or other fats (10, 14). This soap is said in Nigeria to be particularly helpful to people suffering from minor skin-complaints (2). In Guinea an edible salt is obtained from the bark (21). Congolese root and bark material has been reported to contain *flavonins, tannins, steroids* and *terpenes* (6).

The fruit is subspherical, 7.5 to 10 cm in diameter. It has wrongly acquired in Sierra Leone the name of 'mammy supporter', mis-applied from the sapota, *Ponteria sapota* (Jacq.) HE Moore & Stearn, an important, and closely related fruit, of the humid lowlands of Mexico and Central America, which is known in Caribbean Creole as 'mammee sapota' or 'mamey sapota'. The fruit-pulp is sweet smelling, fibrous but edible and is commonly consumed (4, 14, 19, 22, 24, 26). The rotting pulp is very attractive to large forest snails which gather by the trees in season (14, 23) and can then be collected up for human consumption. The fruits have medical application in Nigeria: pulped and mixed into a bark-and root-infusion to make a thin paste to paint onto itch and other skin-afflictions and allowed to dry (2).

Each fruit has 3–4 or more seeds. They contain about 9.6% oil consisting of a mixture of *oleic acid* 34.9%, *palmitic acid* 32.3%, *linoleic acid* 22.9%, *linolenic acid* 6.1% and *stearic acid* 3.8%. A quantity of *amino acids* is present with unusually high amounts of *arginine* and *glutamic acids* (9). The seeds are edible, and in some areas the oil is extracted for culinary purposes (10, 13, 14). Some wild animals also seek out the seeds for eating (24, 26). The seed-husk is taken in Gabon as a gunpowder measure (26).

The tree is held in Gabon to have magical potency for acquiring riches, and for warding off accidents and epidemics (26).

References:

1. Adams, RFG, 1943: 292. 2. Ainslie, 1937: sp. no. 246, as *Ochrocarpus africanus*. 3. Anon., s.d. a. 4. Aubréville, 1959, 2: 328. 5. Bouquet, 1969: 134. 6. Bouquet, 1972: 28. 7. Bouquet & Debray, 1974: 94. 8. Burtt Davy & Hoyle, 1937: 54. 9. Busson, 1965: 213–5, with seed analysis. 10. Chalk & al., 1933: 54. 11. Cole, 1968, a. 12. Cooper 223, K. 13. Cooper & Record, 1931: 41–42, with timber characters. 14. Dalziel, 1937: 92. 15. Foggie, 1957: 131–47. 16. Herbalists, Joru Village, Nov. 1973. 17. Irvine, 1961: 148–9. 18. King-Church 43, K. 19. Kunkel, 1965: 138. 20. Oliver, 1960: 30. 21. Portères. s.d. 22. Savill & Fox, 1967: 132-3. 23. Taylor, CJ. 1960: 174. 24. Voorhoeve, 1965: 108. 25. Walker, 1953, b: 276. 26. Walker & Sillans, 1961: 199. 27. Williamson, K. & Timitimi, 1982.

Mammea americana Linn.

FWTA, ed. 2, 1: 293.

English: mammee apple; mammy apple; apricot of San Domingo (West Indies).

French: abricotier d'Amérique.

West African: SIERRA LEONE KRIO mami sɔpɔta (LP; FCD) sɔpɔta (FCD)

A tree of the West Indies and the northern part of S America which has been taken by man to many parts of the tropics. It is often cultivated in Sierra Leone for its fruit which is eaten fresh or prepared in various ways. It is wrongly named 'mammy supporter' in mistaken identity with the true sapota – see comments under *M. africana. Mammee*, or *mamey* are Caribbean Creole group-names and 'mammee apple' is an appropriate English name. In the West Indies besides consumption of the fruit, it is fermented into a wine. The scented flowers are distilled into a liquer called Eau de Créole. A resin is obtained from the bark which is used in foot-baths to extract jiggers from the feet, and is made into washes for ticks. (2, 5, 6.) The seeds also contain an insecticide effective on a large number of insect species affecting plants and animals. The toxic substance, *mammein*, is analogous to *pyrethrins* (1, 4, 7). Examination of the leaves has detected no action against avian malaria (3).

The wood is hard and durable. It takes a polish and is suitable for cabinetry (2).

References:

1. Bezanger-Beauquesne, 1955: 570. 2. Burkill, IH, 1935: 1399. 3. Claude & al., 1947: 145–74. 4. Heal & al. 1950: 121. 5. Uphof, 1968: 329. 6. Usher, 1974: 375. 7. Wong, 1976: 132.

Mesua ferrea Linn.

FWTA, ed. 2, 1: 291.

English: ironwood; Indian rose chestnut.

A small tree, of India and SE Asia, and widely dispersed by man in the tropics and present in the W African region.

The tree is an ornamental with attractive white showy flowers and waxy leaves. In open positions it retains its branches to the ground. The timber is good and where the tree thrives it becomes of large enough size for railway-sleepers and for construction-work. It is durable in contact with the ground.

The seeds are oily. Content in the kernels is about 75% and it can be used for illumination. The cake after expression is bitter and not edible. A resin can be obtained from the young fruits and the bark that may be suitable for varnishes.

Reference:

1. Burkill, IH, 1935: 1458–60.

Pentadesma butyracea Sab.

FWTA, ed. 2, 1: 291–2. UPWTA, ed. 1, 93–94.

English: butter; butter tree; tallow; tallow tree; candle tree; candle butter tree; black mango tree.

French: arbre à beurre; lami (from Susu of Guinea); beurre de lami.

West African: GUINEA-BISSAU FULA-PULAAR (Guinea-Bissau) boncom-hadje (JDES) boncom-háe (EPdS) MANDING-MANDINKA boncom-ô (JDES) SUSU lami (JDES) GUINEA LOMA namu (JB) MANDING-MANINKA krinda (auctt.) SUSU lami (auctt.) SIERRA LEONE KONO sɔɛ-kɔne (S&F) MENDE kɔndi (JMD) mbeke-wa (S&F) njolei-dɛlii, njolei-lɛlii (S&F) nɔmɔli (NWT) sokai (S&F) sole (FCD) sɔle (def. -i) (JMD; Cole) sɔlɛ-dɛli (def. -i, -i) (JMD; S&F) sɔlɛ-lɛli (def. -i, -i) (S&F) SUSU lami (NWT) TEMNE a-mut (JMD) an-yot, an-yoth (auctt.) LIBERIA DAN u-wee-ti (AGV) weh-ti (GK) KRU-BASA gbah *the young tree* (C) waye, waye-kpay (C; AGV) MANO lē (Har.) lõn (JMD) MENDE mdayen (C&R) IVORY COAST ABE otéro-lemu = *short 'otero';* *in allusion to the fruit being round or nearly so; cf. Allanblackia floribunda, Guttiferae* (PG) otéro-lesī = *female 'otero'; cf. Allanblackia floribunda - male 'otero'* (PG) utélimon (auctt.) utéra (auctt.) AKYE pichalé-boko (auctt.) piché aboko (auctt.) pissalé boko (A.Chev.) BAULE akpan-ni-uaka = *fruit-bat tree* (Henry; FB) KYAMA tiakukebia (A.Chev.) MANDING-MANINKA krinda (Aub.) SENUFO lorokiéré (Henry fide JMD) GHANA AHANTA agyiapa (Bunting; FRI) ajiappa (TFC) bromabine (auctt.) paegya (Brent fide FRI) AKAN-ASANTE abotoma-sɛbie (E&A) ehukei (BD&H) kisidwe *the nuts* (JMD) paegya (BD&H) FANTE pagya (FRI) peja (JMD) pija (TFC; BD&H) WASA abotasebie (DF; E&A) abotoasebie = *butter fat* (auctt.) abotomasɛbeɛ (auctt.) ANYI-AOWIN asuaindokum (auctt.) ehurike (FRI) ɛsoedɔkuno (FRI) NZEMA bromabine (CV; CJT) ehuka (CJT) ehukei (TFC) ehurike (FRI) paegya (JMD) paeja (TFC) pegya (FRI) seowu (auctt.) soinankaw (CJT) suen (Reece; FRI) TOGO TEM (Tschaudjo) agbete (Gaisser) budyonu (Volkens) YORUBA-IFE (TOGO) akutu (Volkens) NIGERIA BOKYI bofuakechi (Catterall; JMD) EDO eduɛni (JMD) izeni (KO&S) izéniɛ́zɛ (auctt.) EFIK ùdíá ébìòng (JMD) EJAGHAM-ETUNG oran-atanki (Catterall; JMD) IGBO òzè (KO&S; KW) IGBO (Awka) òzè (Kennedy; JMD) IGBO (Umuakpo) ala-enyi-milī = *Allanblackia of the water (mili)* (DRR; JMD) IGBO-UKWUANI aghe (auctt.) IJO-IZON akanti (KO&S) NUPE lòlobì (Yates; JMD) YORUBA grogboerin (Lowe) orógbó erin (auctt.) YORUBA (Ijebu) iroro, oro (Kennedy; JMD) YORUBA (Ikale) ɛkuso (JMD) YORUBA (Ondo) ekuso (JMD)

An evergreen tree to 28 m high by 2.75 m girth, of the forest zone often in swampy situations from Guinea to W Cameroons, and extending into Gabon.

A tree is easy to grow from seed and is quick to mature. It suckers freely from stumps. In primary forest it is rather sparsely scattered but in disturbed secondary jungle it may be locally abundant. It is considered suitable for reafforestation purposes (10, 11), but the timber appears to be of limited economic importance. It has been grown successfully in Ghana under plantation conditions producing 9 m high trees in 10 years (15).

Sap-wood is whitish, yellowish or pale brown; heart-wood is reddish. The wood is fairly hard and heavy, easy to work but of coarse texture (1, 9–11, 14, 16), yet it may finish to an attractive appearance (13). In Sierra Leone it is described as a secondary structural timber (6) and is occasionally logged (14). In Liberia it may be sawn into planks and used in rough construction-work (7, 9) and for making canoes (8). The slender boles are used for boat-masts and cut into oars in Guinea and Sierra Leone (1), and similarly and for mine-props in Ghana (11, 15). The wood is said to be resistant to termites. In Benin (Nigeria) combs are cut from it. It also serves generally as a fuel.

The bark contains a yellow, resinous latex which is more abundant towards the inner side. It dries to an orange-red colour. Bark-decoction is purgative and is so used in Ivory Coast (4) and in Gabon (17, 18), and a macerate is used in Gabon against cutaneous parasites (17, 18). The bark serves in Ghana as a fish-poison (10, 11). In Congo (Brazzaville) it is sometimes taken as an aphrodisiac (2). Traces of *flavonins, saponins, tannins, steroids* and *terpenes* have been reported in the bark and roots of Congo material (3).

The roots are used Liberia as chew-sticks (9), and a root-decoction as a vermifuge (Harley fide 10).

The leaves after roasting and crushing are given to children in Sierra Leone to relieve constipation (14).

The fruits are broadly ovoid, 15 cm long by 10 cm across, pointed. Each bears 3–10 or more seeds in a yellowish pulp. A copious yellow thick juice is exuded when the fruit is cut. This is similar to the bark-exudate. The fruit is said to be edible when immature but is too hard on ripening (5, 9).

The seeds are brown with flattened sides, vinous red in cross-section. Before ripening they are said to be edible, but they become very bitter ('sour', 7) on maturing. They have been passed off as an adulterant for true kola and can be distinguished by an examination of the base: true kola has radiating lines indicating the separation into cotyledons, while the seed of *Pentadesma* is uniformly solid and is composed of a large embryo without the cotyledons differentiated, and no endosperm (10, 16). Air-dried seeds carry 32–42% of solid fat which can be extracted by pounding and cooking. It varies from pale to dark brown, is almost tasteless and has a pleasant smell. It is edible and resembles to some extent shea butter (*Vitellaria paradoxa* Gaertn., Sapotaceae), and has the advantage of being free from any bitter latex, and it does not turn rancid. It is used for cooking and is made into unguent for skin and hair; it is applied to kill lice, jiggers. etc. It is suitable for turning into soap, candles and margarine (1, 4, 5, 10, 11, 12). Composition is: 53.6% *oleic acid*, 45.2% *stearic acid*, 1.2% *palmitic acid*, with an assortment of *amino acids*. The cake after extraction of the fat is almost free of carbohydrates and has little nitrogen; tannin and other substances present render it unsuitable for feeding purposes; nor is it of value as manure (10, 11).

Porcupines eat the seeds dropped by forest trees (10, 11). In Gabon they serve as a bait for trapping these animals (18).

In the Mano area of Liberia the fruit has magical power to prevent poisoning when going amongst strangers (10).

References:

1. Aubréville, 1959, 2: 326. 2. Bouquet, 1969: 134. 3. Bouquet, 1972: 28. 4. Bouquet & Debray, 1974: 94. 5. Busson, 1965: 214–5, with seed analysis. 6. Cole, 1968, a. 7. Cooper 80, K. 8. Cooper 273, K. 9. Cooper & Record, 1931: 42, with timber characters. 10. Dalziel, 1937: 93–94. 11. Irvine, 1961: 149–50, with kernel cake analysis p. lxxxii. 12. Kerharo & Bouquet, 1950: 56. 13. Punch, s.n. 19/2/1900, K. 14. Savill & Fox. 1967: 133–4. 15. Taylor, 1960: 176. 16. Voorhoeve, 1965: 111. 17. Walker, 1953, b: 276. 18. Walker & Sillans, 1961: 200.

Psorospermum Spach

FWTA, ed. 2, 1: 288.
West African: **THE GAMBIA** FULA-PULAAR (The Gambia) jankumo (DAP) kokuno (DAP) MANDING-MANDINKA jankumo (DAP) kajijankumo (Hallam) kokuno (DAP) WOLOF

jankumo (DAP) kokuno (DAP) **MALI** MANDING-BAMBARA diurasungalani (Aub.) MANINKA carédianguma (Aub.) karidia kuma (Aub.) kitidiancuma (Aub.) **IVORY COAST** LIGBI vio (Aub.)

A genus of small trees and shrubs, not always distinguished apart, and the vernacular names cited above refer to all species occurring in the respective language area. Infusions of leaves of each of the species in the Kita region of Mali are taken as diuretics and strong febrifuges in preference to the preparations of *Combretum micranthum* G Don (Combretaceae), the general panacea *kinkeliba* (1).

Reference:

1. Aubréville, 1950: 147.

Psorospermum alternifolium Hook. f.

FWTA, ed. 2, 1: 289. UPWTA, ed. 1, 88.
West African: **GUINEA** FULA-PULAAR (Guinea) keti (A.Chev.) **SIERRA LEONE** MANDING-MANDINKA funkui (NWT) MENDE b'atue (NWT) giji (NWT) TEMNE ε-cintr (NWT) kisui (NWT) runko (NWT)

A shrub or tree to 4 m high occurring in Guinea, Mali and Sierra Leone. The wood (? stems) is used in hut-construction (2). In Ivory Coast the plant (part not indicated – ? bark) is used to treat fevers and skin-troubles (1).

References:

1. Bouquet & Debray, 1974: 95. 2. Thomas NWT.44, K.

Psorospermum corymbiferum Hochr.

FWTA, ed. 2, 1: 289, 762. UPWTA, ed. 1, 89 as *P. kerstingii* Engl. and *P. thompsonii* Hutch. & Dalz.
French: millepertuis velu (Bailleul).
West African: **SENEGAL** BEDIK *gi* komoñir (JB) FULA-PULAAR (Senegal) kati diãnkuma (JB) koti diãnkuma (JB) MANDING-BAMBARA diura sungalani (JB) kari diakuma (JB) MANDINKA kato diãnkuma = *mange of the cat* (K&A; JB) MANINKA kiti diãnkuma (JB) 'SOCE' kating diãnkumõ, kurkutumãndi (JB) SERER dadran (JB) **THE GAMBIA** MANDING-MANDINKA katijankumo (Hallam; Bojang) wollo koyo (Frith) **GUINEA-BISSAU** FULA-PULAAR (Guinea-Bissau) catidjancuómo (JDES) codidjancuma (JDES) MANDING-MANDINKA catidjancuómo (JDES) codidjancuma (JDES) **GHANA** GUANG kàlùwúyà **TOGO** ? akpalami (Von Doering) nina deyu (JMD) **NIGERIA** FULA-FULFULDE (Nigeria) cawaiki (JMD) HAUSA cidakara (RES) kaskawami (JMD) kiskawali (JMD) kiskawoli (Lely) YORUBA légun oko (auctt.)

A shrub or tree to 5 m high of the savanna zone from The Gambia to S Nigeria and in Uganda.
Leaves and twigs are reported to be boiled yielding a supernant oil which is skimmed off and used in Togo for craw-craw and similar skin-conditions (Kersting in 2) – cf. *P. senegalensis*. An analogous pattern of treatment is recorded from The Gambia where the roots are boiled in palm kernel oil. The oil is then mixed with milk for application to boils on the skin and to the eruptions of a disease called *kato* which is identified as small-pox (3). The leaves are also boiled in water in The Gambia and the water is then taken for cough and stomach-ache, and the water in which leaves and roots have been boiled is taken for 'body sickness' (?debility) and loss of blood (3).

Nigerian material has been screened and a trace of *alkaloid* found in the roots (1).

References:

1. Adegoke & al., 1968: 13: 13–33. 2. Dalziel, 1937: 89. 3. Hallam, 1979: 41.

Psorospermum febrifugum Spach

FWTA, ed. 2, 1: 290. UPWTA, ed. 1, 88.
West African: **THE GAMBIA** MANDING-MANDINKA katwan-kumo (JMD) **SIERRA LEONE** KORANKO kiliɔŋgume (NWT) LOKO nyobalia (NWT) MENDE mbeibamba (NWT) tɛli (NWT) TEMNE ɛ-naŋka (NWT) ɛ-pil a-runi (NWT) **IVORY COAST** ANYI wanzokoroma (K&B) **NIGERIA** FULA-FULFULDE (Nigeria) sowike (FNH) URHOBO owegba (Cole) YORUBA légun oko (IFE; Verger)

A shrub or tree to 4 m high, variable and represented in the Region by two varieties, var. *febrifugum* and var. *ferrugineum* (Hook. f.) Keay & Milne-Redhead, of merging and unclear differentiation, of the savanna from Guinea to W Cameroons and generally widespread in tropical Africa.

The plant in Uganda is said to show some resistance to fire (8). When burnt the wood produces much smoke, and the Gbaya of Central African Republic hold it to be unusable on this account (21).

The bark, and more especially the root-bark, is used for all manner of skin-troubles. It is reported used in Sierra Leone on craw-craw (18), and on skin-troubles in Ivory Coast (5). The powdered root is used topically on parasitic skin-diseases in Nigeria (12). In Tanganyika the root ground up and mixed with oil is said to be a good remedy for pimples, eruptions and wounds (2, 20).

The plant, and especially the bark, has been regarded as possessing anti-leprous properties and to be of value against insect-bites (20). Another Tanga-nyikan treatment for scabies is to mix the bark-ash with castor-oil for application to the area after scarification, and the sediment after decocting the bark is used on dry eczema of allergic or fungal origin (10). In Congo (Brazzaville) the root-bark is recognised as efficaceous against skin-itch and jiggers, and in infusion is used as a wound-treatment and on skin-eruptions (1). The pulped plant or the bark in decoction is also used in Congo on furuncles and leprosy, and is taken internally for dysmenorrhoea, dysentery, tuberculosis and whooping cough (3). Miscellaneous other uses are: in Kenya the root pounded with water is taken as a dose to cure an unspecified malady called 'tongue' disease (11); in Tanganyika as a 'baby remedy' (14); to relieve constipation (13); sap is expressed to apply on sores (15, 17) to treat salpingitis (9); as an enema for ankylostomiasis (16) and in a formulation the decoction is taken for syphilis (10).

The plant has a reputation for febrifugal action (Ivory Coast: 25; Nigeria: 12; Gabon: 19; Angola: 7, 20), as its scientific name might imply but tests on the bark have given negative anti-malarial results (20) and similarly against avian malaria (6). Root-bark has been analysed as containing *catechuic tannins* 9.5%, and a toxic red fluorescent anthraquinonic pigment, *corymbiferin*, which is related to *hypericin* and has a photosensitising capacity: it is toxic to kidneys and intestines (12). *Steroids* and *terpenes* are also present in the bark and roots (4).

The dried flower-stalks are powdered and are added to *Acalypha ornata* Hochst. & A. Rich. (Euphorbiaceae) for use in Tanganyika as a wound-dressing following circumcision (10).

The fruits are eaten in Tanganyika (16, 17).

References:

1. Aubréville, 1950: 146. 2. Bally, 1937: 10–26. 3. Bouquet, 1969: 138. 4. Bouquet, 1972: 28–29. 5. Bouquet & Debray, 1974: 95. 6. Claude & al., 1947: 145–74. 7. Dalziel, 1937: 88. 8. Dawkins P.513, K. 9. Gaetan 102, K. 10. Haerdi, 1964: 101. 11. Mwangangi 1287, K. 12. Oliver 1960: 35, 79. 13. Pirzynski P.170, K. 14. Semsei S.1228, K. 15. Tanner 5226, K. 16. Tanner 6043, K. 17. Tanner 6045, K. 18. Thomas, NWT.2238, K. 19. Walker & Sillans, 1961: 206. 20. Watt & Breyer-Brandwijk, 1962: 498, 500. 21. Roulon, 1980: 224.

Psorosperum glaberrimum Hochr.

FWTA, ed. 2, 1: 289.
West African: GUINEA-BISSAU FULA-PULAAR (Guinea-Bissau) catidjancuómo (JDES) codidjancuma (JDES) GUINEA BASARI a-ngwày ɓi̱-yil = *genet of the spirits* (FG&G) MANDING-MANDINKA catidjancuómo (JDES) codidjancuma (JDES) SIERRA LEONE MAND-ING-MANDINKA funsari bankɔ (NWT) TEMNE ɛ-turibwerakantr (NWT)

A much-branched shrub to 4 m high recorded from The Gambia to Dahomey.

Psorospermum lanatum Hochr.

FWTA, ed. 2, 1: 289. UPWTA, ed. 1, 88, as *P. guineense* Hochr., in part.
West African: GUINEA FULA-PULAAR (Guinea) keti diankuma (CHOP) MANDING-MANINKA kokunu (A.Chev.) SUSU loli (CHOP)

A shrub to about 50 cm high, recorded only from Guinea.
The boiled bark yields a soapy product which in Guinea is mixed with oil to put on the skin, or onto the sores of animals as a fly-repellant. A decoction of the plant is also taken in Guinea for neuralgia (1).

Reference:

1. Dalziel, 1937: 88.

Psorospermum senegalense Spach

FWTA, ed. 2, 1: 289.
West African: SENEGAL BALANTA sukus (K&A) BEDIK gi-komoniir *after* (K&A) gi-ngòr ɓa-yidi = *the chase of the spirits* (FG&G) FULA-PULAAR (Senegal) katidiâkuma (K&A) MANDING-BAMBARA diurasumgalani (JB; K&A) makarakun koyoté (K&A) MANDINKA kati-diâkuma (K&A) katodiâkuma = *mange of the cat* (K&A) MANINKA katidiâkuma, katodiâ-kuma = *mange of the cat* (K&A) MANINKA (Yassine) katidiemkumo (K&A) 'SOCE' katêndâk-omô (K&A) kunkutu (K&A) kurkutumâdi (K&A) mâdiô (K&A) WOLOF êklen, êklèn (K&A) inklen (K&A) GUINEA-BISSAU FULA-PULAAR (Guinea-Bissau) catidjancuómo (JDES) codid-jancuma (EPdS; JDES) MANDING-MANDINKA catidjancuómo (JDES) codidjancuma (JDES) GUINEA BASARI a-ngwày ɓi̱-yil = *genet of the spirits* (FG&G)

A bush of tortuous branches to 2.5 m high, of the bush and wooded savanna of the soudanian zone, recorded only from Senegal to Sierra Leone.
The plant has a general usage in Senegal for all skin-affections. Several vernacular names relate it to 'mange of the cat' (1). A bark-decoction, of the roots, is used in washes and baths for common dermal troubles and for herpes, mange, eczema, etc., and to leprous and syphilitic conditions (2, 4). In Dakar a general application is by powder for colics, by decoction for blennorrhoea, and by macerate for leprosy (4). In the Ferlo region a decoction of bark from the main stem and of the roots is added to baths and is made up into draughts for pains in the joints in attacks of fever (2, 4). The pulped bark and pulped root is used topically in Guinea on dematoses generally, and a decoction of leafy twigs is given by draught as a diuretic and febrifuge (5). A filtrate from a prolonged boiling of the leaves is deemed in Senegal to alleviate respiratory trouble and is

taken for leprosy. An oily film comes to the surface of this preparation which can be separated off on cooking. This is used externally for skin-troubles (3, 4).

Pieces of the plant are attached to millet by the Tenda to discourage bees from nesting on it (60).

Analysis of Guinean material has shown the presence of *tannins*, *sugars* and a fluorescent *anthroquinonic pigment*, the last of which is toxic and causes photo-sensitivity and acute irritation to the kidneys and intestines (4, 5).

The plant enters into a Fula magical treatment to confer protection against evil: bark burnt over live embers makes much smoke in which the naked patient is fumigated (2). Protective properties are attributed to the plant by the Tenda; it is made into an amulet to protect hunters from meeting a lion, and *Loranthus* growing on it is tied to a post near a house to ward off sorcerers (6).

References:

1. Kerharo & Adam, 1962. 2. Kerharo & Adam, 1964, b: 570. 3. Kerharo & Adam, 1964, c, 318. 4. Kerharo & Adam, 1974: 485–6. 5. Planche, 1948: 546–65. 6. Ferry et al., 1974: sp. no. 213.

Psorospermum tenuifolium Hook. f.

FWTA, ed. 2, 1: 290.

A much-branched shrub to 3 m high of swamp-forest, often coastal, in S Nigeria and W Cameroons and extending into Zaïre.

No usage is recorded for the Region. In Congo (Brazzaville) it takes the place in the forest zone of *P. febrifugum* which is a savanna species, and is used there in the same way for treatment of all skin-ailments (1).

Reference:

1. Bouquet, 1969: 138.

Psorospermum spp. indet.

UPWTA, ed. 1, 88, as *P. guineense* Hoch. in part.

West African: IVORY COAST MANDING-MANINKA kédiofuin (B&D) NIGERIA ALAGO oñmogu = *kills fowls* (Hepburn; JMD) HAUSA huda tukunya = *pierce the pot* (JMD) hùndà túkúnyá (ZOG) kàshè kaàjí = *kills fowls* (auctt.) kaskawani (ZOG) maganin tsuntsaye = *medicine against birds* (JMD) maganin tsuntsu = *bird medicine* (Hepburn)

An unnamed species, probably of this genus, is used in Liberia for boat-keels because of its hardness, toughness and durability (1).

A species recorded in the Nassawara Province of N Nigeria is known by Hausa and Arago names which indicate a toxicity towards birds. The Arago prepare an infusion by soaking the bark in which they place their guinea-corn seed before it is sown. Birds are said not to touch the treated seed (2).

References:

1. Cooper & Record, 1931: 38 (as Cooper 382). 2. Hepburn, SUM24, K.

Symphonia globulifera Linn. f.

FWTA, ed. 2, 1: 293. UPWTA, ed. 1, 94, as *S. gabonensis* Pierre.

English: hog gum tree; hog plum.

Trade: the bark gum – hog gum, doctor's gum, mani wax or karamani wax (from a Guyanan vernacular name); the timber – osol (Gabon).

West African: GUINEA-BISSAU CRIOULO mundela (RdoF) SIERRA LEONE MENDE njolei-dɛlii, njolei-lɛlii (S&F) njore (Aylmer) sɔlei-dɛlii, sɔlei-lɛlii (S&F) torlei (DS) LIBERIA DAN hué, wué (GK) KRU-BASA waye-pu (C; C&R) IVORY COAST ABE beu (auctt.) oropopati (A.Chev.) siébé (A.Chev.) AKYE bon (Aub.; RS) GHANA AKAN-TWI asoduru (FRI) WASA ahɔke (auctt.) NZEMA ahurke (FD) ehulike (CJT) ɛhurɛkɛ (E&A) **NIGERIA** ABUA ojeme (DRR; JMD) BOKYI edueni-eze (DRR; JMD) édún-eze (auctt.) obanefufu (Catterall; JMD) ohun-eze (?) (DRR; JMD) oviedun eze (?) (DRR; JMD) ovien odun eze (?) (auctt.) EKPEYE agba (DRR; JMD) ENGENNI agbaba (DRR; JMD) IGBO agbā (Singha; KW) IJO-IZON (Kolokuma) okilolo (auctt.) okolólo (KW&T) ókolólo (KW&T) IZON (Oporoma) okilolo (KW) ongóyówéi (KW) ISEKIRI eyin-odan (Kennedy; JMD) MAMBILA geeb (Chapman) URHOBO arhohen (Kennedy; JMD) YAMBA beek (Chapman) YORUBA agérigbédé = *we lift the head and know the language* (auctt.)

A tree to over 30 m high with a straight bole, sometimes buttressed and sometimes with stilt-roots, by nearly 2 m in girth, branches whorled and at right angles producing a small crown, of river-sides and swampy locations in the rain-forest from Guinea-Bissau and Sierra Leone to W Cameroons, and on into Tanganyika and Angola, and widely dispersed in south and central tropical America.

The wood sinks when fresh (8, 19). Sap-wood is pale yellow or nearly white and is sharply demarcated from the reddish or yellowish brown heart-wood (11, 12, 19, 25). It is liable to pin-hole borer and removal on felling is recommended (19). The grain is straight but inclined to be coarse-textured, and there may be a little warping in seasoning (19, 25). It is said to work well taking staining and polishing, and finishing with an attractive lustre. It is a good wood for cabinetry, turnery and general carpentry (11, 12, 19, 20, 29). It can be used as a substitute for mahogany, and there has been use of it in Europe for veneers and plywood (14). Besides its resistance to insect and fungal attack, a relative abundance is an asset. It is locally sawn into planks and squared timber (7, 11). Planks are made into canoes and boats and in S Nigeria and in Central Africa the trunks are cut into small dug-out canoes. In Ghana it is currently recommended for general purposes (2) and in Liberia it is used for cheap furniture (11). All manner of household utensils are commonly made of it. The Ijo of the Niger Delta know the tree as the 'paddle tree' and make paddles from the wood (30). In Uganda the tree attains a greater size and is deemed an useful timber tree (24). In Guyana trunks are hollowed out for dug-out canoes (3). In Brazil it provides timber for nautical purpose (15) and in Belize railway-sleepers (12). No medicinal usage is recorded for the wood in Africa, but in Belize it is known as 'Waika chew-stick' (16, 18).

The bark contains a deep yellow resin which blackens on exposure to air. It is of the 'bassora' or hog gum type and is insoluble in water (17). It is usually collected as an expissated lump at the foot of trees and partial purification is done in hot water. It is strongly adhesive and water-resistant. In Liberia it is a general glue (10). In Sierra Leone it is mixed with bees-wax and used for fastening arrow-heads to the shaft (4). An interesting parallel to this is a similar application in Guyana but using charcoal as well as bees-wax as additives; this mixture is also used for jointing wood (26, 27). On the Amazon at Para, boats are water-proofed with the resin (13). Caulking of canoes and the mending of cracked calabashes with the resin is carried out in Gabon (29), Cameroun and Zaïre (12), Congo (Brazzaville) (5) and Central African Republic (20). There is also general use in attaching fletches to arrow-shafts, pipe-bowls to stems, knife-blades to handles, etc.

The resin has vulnerary properties. It is used in Gabon on sores, as a protective against jiggers and as a cure for dermal infections; it is also added to a prescription for treating chancres (6). The resin is taken internally in Nigeria for gonorrhoea and as a diuretic and externally as a wound-dressing (1), and for application to craw-craw (23).

Suitable materials can be impregnated with the resin for burning as torches (12).

The bark-macerate is considered in Ghana (19) and Nigeria (23) to be a stomachic tonic and in Gabon as emetic and is given for chest-complaints (28, 29). In Congo a bark-decoction is taken for blennorrhoea, haematuria, heart-troubles, stomach-disorders in women and as an ecbolic; a bark- or leaf-decoction is furthermore used in vapour-baths to relieve localised pain, rheumatism and oedema (6). Some *saponins* and a trace of *tannins* have been detected in Congolese bark-material (7).

For nose-bleeding sap from the leaves is sniffed up the nose in Congo (6).

The flowers, usually profuse, are pinkish red to deep crimson, shortly-branched on the main branches giving the tree in flower a very attractive appearance. In Congo 'to obtain pleasant dreams' one burns on the hearth in the evening some flowers with other sweet smelling plants (16).

References:

1. Ainslie, 1937: sp. no. 328. 2. Anon., s.d., a. 3. Archer 2401, K. 4. Aylmer 55, K. 5. Bamps, 1970: 36–38. 6. Bouquet, 1969: 134. 7. Bouquet, 1972: 29. 8. Burtt Davy & Hoyle 1937: 55. 9. Cooper 105, K. 10. Cooper 333, K. 11. Cooper & Record, 1931: 42–43, with timber characters. 12. Dalziel, 1937: 94. 13. Ducke 7968, K. 14. Eggeling & Dale, 1952: 155. 15. Froes 1791, K. 16. Gentle 2915, K. 17. Howes, 1949, a: 51. 18. Hummel 19, 178, K. 19. Irvine, 1961: 150–1. 20. Pellegrin, 1959: 229. 21. Savill & Fox, 1967: 134–5. 22. Schipp 466, K. 23. Singha, 1965. 24. Styles 173, K. 25. Taylor, CJ, 1960: 177. 26. Uphof, 1968: 506. 27. Usher, 1974: 562. 28. Walker, 1953, b: 276. 29. Walker & Sillans, 1961: 200–1. 30. Williamson 127, UCI.

Vismia guineensis (Linn.) Choisy

FWTA, ed. 2, 1: 288. UPWTA, ed. 1, 89, as *V. leonensis* Hook. f. and *V. leonensis*, var. *macrophylla* Hutch. & Dalz.

West African: **GUINEA** FULA-PULAAR (Guinea) bélendé (CHOP) SUSU loli (CHOP) **SIERRA LEONE** GOLA duma-yoŋgo (FCD) KISSI cholompɔmbɔ (S&F; FCD) KONO suŋgbali-kaima (S&F) KORANKO konfure (S&F) kɔnfuri (NWT) LOKO mbɛli-mbamba (S&F; FCD) ngumbi (NWT) MENDE *m*baibambe (JMD) *m*bɛbambe (NWT) *m*bɛimbamba (*def.* -ei) (FCD) bele-*m*bambei (SKS) *m*bɛli (NWT) *m*bɛli-*m*bamba (*def.* -ei) (auctt.) bɛli-*m*bambe (JMD) *m*bɛli-heni (FCD) *m*bɛli-hina (*def.* -ei) (auctt.) SUSU wobe (NWT) TEMNE am-pel-pelaŋ (aucct.) a-pil (NWT) a-pil-a-runi cf. a-pil: *Harungana madagasceriensis* (Jordan) **LIBERIA** KRU-BASA ge-ahn, ge-ahn-de-pay *from* ge-ahn: *grave, or spirit world; from which no living person returns* (C&R) MANO lolo mia (JMD) **IVORY COAST** ABE uombéhiapi (Aub.) AKAN-ASANTE titinondra, titinuera (B&D) AKYE nguamo (A.Chev.) onhonpéhiapi (A.Chev.) KRU-GUERE (Wobe) gbloglè (A&AA) KYAMA tanompi (Aub.) **GHANA** AKAN-ASANTE ɔkɔsoa-nimmaa (auctt.) ɔkɔsoa-nini (FRI; E&A) GBE-VHE agbɔti (auctt.) VHE (Awlan) gbɔti = *goat's bitter leaf* (Williams; FRI) **NIGERIA** EDO ovitue *from* ovie: *small;* itue: *Harungana* (JMD; KO&S) HAUSA kiska wali (ZOG) IGBO oke oturu = *male, or big oturu* (KO&S)

A tree to about 16 m high, but often shrubby, of secondary jungle from Guinea to W Cameroons, and extending to Zaïre.

The wood is reddish, fairly hard and strong, of medium texture, finishing smoothly and taking a high polish. It is not very resistant to decay. The poles are used for hut-posts. (5–8, 11).

The bark exudes on cutting a reddish or yellowish resinous gum. It may be pounded and made into an ointment for craw-craw, or the oily resin extracted by boiling with palm-kernels producing an ointment for the same use (5; Harley in 6; 8). In Ubangi the gum is rubbed on the body for fleas and itch (10), while in Ivory Coast it is used for fevers and general skin-complaints (3). In the lagoon area of Ivory Coast, a healing lotion is made from the bark for applying to sores caused by scabies (13). In Sierra Leone the Mende allow the resin from cut young stems of 5–7 cm diameter to drip onto the freshly-made wound in the circumcision ceremony. This is followed by the operator chewing alligator-pepper seeds (*Aframomum melegueta* K. Schum., Zingiberaceae) and blowing the cud onto the wound. This is deemed to be both haemostatic and cicatrisant (2, 8).

Young leaves, probably of this plant, are made into an infusion in Guinea for treating gleet, and as a purgative, and for women with abdominal trouble (8, 9). They are also made into a preparation with *A. melegueta* in both Guinea and Sierra Leone for throat-inflammation (6, 11). In Ivory Coast sap expressed from the young leaves is used in washes on infants for jaundice, and for adults the young leaf-macerate is added to palm-wine and taken by mouth (1). Another treatment in Ivory Coast for jaundice is to rub the body with the young leaves made into pellets with very little water (1). In Liberia the leaf-buds are held to have anodynal effect if one crushes them and inhales the vapour in cupped hands for the relief of vertigo: from this attribute comes the Mano name meaning 'small pain-killer' (12).

In the coastal part of Ivory Coast the twigs serve as chew-sticks (13).

The tree is important in Liberia in juju, to which the Basa name owes reference (5), and is one to which the believer will turn to obtain advice (4). It is also an ordeal plant: a pot containing the inner-bark macerated in raw palm-oil, to which some of the resin is added, is placed on a tripod formed of three branches of the tree, and the accused immerses his hand in the boiling mixture as an ordeal. Guilt is indicated by scalding (5).

References:

1. Adjanohoun & Aké Assi, 1972: 150. 2. Boboh, 1974. 3. Bouquet & Debray, 1974: 95. 4. Cooper 348, K. 5. Cooper & Record, 1931: 38–39, with timber characters. 6. Dalziel, 1937: 89. 7. Deighton 668, K. 8. Irvine 1961: 143. 9. Pobéguin, 1912: 20. 10. Portères, s.d. 11. Savill & Fox, 1967: 139. 12. Watt. 1967: 1–22. 13. Portères, 1974: 140, as *V. leonensis* Hook. f.

Vismia rubescens Oliv., var. rubescens

FWTA, ed. 2, 1: 288.

A scrambling shrub to about 6 m high, recorded only in S Nigeria in the Region, and occurring also from Gabon to Zaïre.

The stems exude a yellowish red gum which is used in Gabon as a wound-dressing and which is also a dye-stuff for cloth.

References:

1. Walker, 1953, b: 277. 2. Walker & Sillans, 1961: 206.

HERNANDIACEAE

Gyrocarpus americanus Jacq. spp. pinnatilobus Kubitziki

FWTA, ed. 2, 1: 60, as *G. americanus* Jacq.

A species of many subspecies, of which ssp. *pinnatilobus* occurs in Guinea, Mali and the Ivory Coast as a shrub or small tree to 15 m high. The species is widely distributed in tropical and South Africa, tropical Asia, Polynesia, Australia and South America.

The leaves when crushed emit a disagreeable strong smell (2). The bark of Ivory Coast material has been found to contain alkaloids (1). Leaf, stem and fruit of Madagascan material have been found to have an alkaloid (4), and Australian material has been shown to contain the alkaloids, *phaenthine* and *d-magnocurarine*, the latter belonging to a group of ganglion-blocking agents which may prove useful in cardiology (3).

References:

1. Bouquet & Debray, 1974: 95. 2. Jaeger, 1964: 123. 3. Watt & Breyer-Brandwijk, 1962: 520. 4. Willaman & Li, 1970.

Hernandia beninensis Welw.

FWTA, ed. 2, 1: 59.

A forest tree to 12 m high, recorded only on Fernando Po in the Region, and at São Tomé.

In Gabon the bark has been reported as used in enemas and fumigations for its supposed capacity to cure madness (2, 3), but since this species is now said not to occur in Gabon (1), the identity of the material concerned must remain in doubt.

References:

1. Kubitziki, 1969: 148. 2. Walker, 1953,b: 277. 3. Watt, 1967: 1–22.

Illigera pentaphylla Welw.

FWTA, ed. 2, 1: 59.
West African: IVORY COAST ABURE furignama (B&D)

A forest liane occurring from the Ivory Coast to W Cameroons, and in E Cameroun to Uganda and Angola.

The leaf-sap is used in the Ivory Coast for ophthalmia. The sap seems to be a good remedy for shingles (2). In Congo (Brazzaville) leaf-sap is taken in draught for breathlessness and difficult respiration (1).

References:

1. Bouquet, 1969: 135. 2. Bouquet & Debray, 1974: 95.

HOPLESTIGMATACEAE

Hoplestigma kleineanum Pierre

FWTA, ed. 2, 2: 16.
West African: IVORY COAST ·KRU· niéuetu (Aub.) KRU-BETE béli (Aub.)

A tree to 25 m tall, straight bole, to 40 cm diameter, strongly shouldered at the base; of secondary jungle in Ivory Coast and Ghana in the Region, and also in Gabon (1–3).

The slash is fibrous, and moist, quickly darkening to brown, and with a walnut-like smell (2). The wood is a clear yellow marbled rosy pink. It finishes nicely and is good for joinery and carpentry (3).

References:

1. Aubréville, 1958, 3: 155. 2. Hall & N'Da s.n., K. 3. Walker & Sillans, 1961: 202–3.

HUMIRIACEAE

Sacoglottis gabonensis (Baill.) Urb.

FWTA, ed. 2, 1: 355. UPWTA, ed. 1, 134.

English: bitter bark tree; cherry (Liberia, but not to be confused with African cherry of timber trade, auctt.); mahogany (Liberia, but not to be confused with African mahogany of timber trade, Cooper & Record); cherry mahogany (Sierra Leone, Deighton); ozouga (Ghana, Liberia, timber trade); atala; ntala; tala (Nigeria, timber trade, Dalziel).

West African: SIERRA LEONE KONO gbɔkɔne (S&F) MENDE kpewin (Joru) kpɔ-wuli, kpɔ-wulo = excrement tree (auctt.) LIBERIA DAN doh (GK; AGV) GOLA kwuo (AGV) KPELLE wuwi (AGV) KRU-BASA dauh (C; AGV) GUERE (Krahn) déwe (GK) toboe (AGV) IVORY COAST ABE akuapo (Aub.; JMD) ABURE aomon (B&D) AKAN-ASANTE ofana (B&D) FANTE amuan (A.Chev.) AKYE hué (A.Chev.) ANYI efeuna (Aub.) effuinlin (Aub.) KYAMA akaya (B&D) akohia (Aub.) NZEMA amuan (A.Chev.) GHANA VULGAR tiabutuo (DF) AKAN-WASA afamkokoo (CJT) fawere (E&A) tiabutuo (auctt.) NZEMA fawire (CJT) NIGERIA ABUA okia (DRR; JMD) BOKYI edat (KO&S) EDO úgu (auctt.) EFIK èdát (auctt.) idat (LAKC) n̈dât (auctt.) EJAGHAM ẹdat (JMD) ndat (JMD) IBIBIO edat (KO&S) IGBO nche (auctt.) ntala (LAKC) IGBO (Umudike) nche (Herington) IGBO-UKWUANI okpi-uta (Kennedy fide JMD) IJO-

410

IZON (Egbema) itala (JMD) tala (JMD; Tiemo fide KW) ISEKIRI itala (KO&S) YORUBA atala (auctt.)

A tree of lowland rain-forest beside water, in fresh-water swamp or other damp situations, to 40 m or more high, a large crown and emergent, often the largest tree of the forest; bole to 60–80 cm diameter, irregularly and often profoundly fluted; recorded from Sierra Leone to Nigeria, and southeastwards into Zaïre and Angola. This is the one extra-territorial species of a S American genus (10).

The wood is hard and heavy. Sap-wood is white to a dull yellow-brown. Heart-wood is brown and durable. Owing to the poor shape of the bole the timber is not commercially exploitable, but it does have local domestic use sawn into beams, planks and boards (6, 9, 10, 21).

It is recorded as frequently used by the Public Works Department of Sierra Leone for heavy outdoor construction-work (19). It is recommended in Ghana for making waggons and carriages (2). Poles are used in that country in hut-building (20) and the wood for furniture and interior decoration (14) with an attractive appearance like mahogany. The timber is said to warp and shrink (8, 9). The wood is valued for use in making canoes because of its durability (Nigeria: 15, 16; Liberia: 9), and in Gabon it is used for the ribs of boats (23). The wood is reported to be a good firewood (20) and to produce good charcoal (21).

The bark is reddish-brown, shaggy and peels off in large flakes. When slashed it hisses and exudes a clear sticky amber-brown sap with a sweet sickly scent resembling that of sugarcane (9, 19–21). In the coastal region of Ivory Coast sap is used in draught as a febrifuge and for abdominal pains, and diluted in hip-baths to promote muscle tone in newly-delivered women (5). The bark itself is often sold in markets in sheets or rolls as a bitter for adding to palm-wine or gin to add flavour and potency (1, 3, 9, 11, 13, 15, 16, 25). The latter action may perhaps be due to the inhibiting of souring bacteria (24). In Liberia (21) and Sierra Leone (19) its addition to palm-wine is recognised as aphrodisiac, and in parts of the former country only men take the treated wine. In Sierra Leone a bark-decoction is a stomachic (19), and in Nigeria a fever prophylactic (11, 17). In Gabon bark is taken as an emetic and is used as a fish-poison (22, 23) while in Congo (Brazzaville) a decoction, with other drugs, is added to a bath in the treatment of vaginal infections and of children with fever (4). In serious cases of ovarian and vaginal trouble an infusion may also be drunk.

Tests of the bark have not shown the presence of alkaloids, flavones, nor steroids, but only *tannin* in appreciable amount with a trace of *saponins* (5). Tests for action on avian malaria have given negative results (7) and the roots have shown no insecticidal capacity (12).

The leaves, which may attain a length of 15 cm by 7 cm wide, are said to be used in Sierra Leone by nursing mothers to 'clean up' the baby; hence the Mende name meaning 'excrement tree' (19).

The fruit is a drupe 2.7–3.5 cm long by 2.5–3.0 cm wide, known to the Liberians as a 'cherry'. The exocarp is said to be sweet and edible (21, 23), and is sometimes eaten by Gabonese children (23). The seeds are also edible (23). The fruit is much relished by monkeys, elephants and other forest animals who disperse the seeds (3, 9–11, 21). The endocarp has a number of cavities filled with a sweet resinous substance which is much sought after by bees (9). The seeds are oil-bearing; but the oil is without recorded usage.

References:

1. Adams, RFG, 1947: 23–24. 2. Anon. s.d., a. 3. Aubréville, 1959: 1: 368. 4. Bouquet, 1969: 136. 5. Bouquet & Debray, 1974: 95. 6. Chevalier 33166, K. 7. Claude & al., 1947: 145–74. 8. Cooper 68, K. 9. Cooper & Record, 1931: 49, with timber characters. 10. Cuatrecacas, 1961: 172–4. 11. Dalziel, 1937: 134. 12. Heal & al. 1950: 121. 13. Herbalists, Joru village, 1973. 14. Irvine, 1961: 205. 15. King-Church 14 E.P., K. 16. Mackay & Wardrop 1939: 65–74. 17. Moloney 30, K. 18. Sankey 8, K. 19. Savill & Fox, 1967: 136–7. 20. Taylor, 1960: 179. 21. Voorhoeve, 1965: 116. 22. Walker, 1953,b: 277. 23. Walker & Sillans, 1961: 203. 24. Herington, 1979: 34. 25. Iwu, 1984: 144.

HYDROCHARITACEAE

Hydrilla verticillata Presl

An aquatic herb of tanks, ponds and streams, native of the Old World tropics, and now dispersed to many temperate countries of Europe and Asia, and into S and Central America. It is recorded present in Ivory Coast in the Region.

The plant is a pest of water-storage systems in India hindering anti-malarial measures and fish-culture. It is, however, palatable to some fish. It is suitable for indoor and outdoor aquaria as a good oxygenator. Where growth is too vigorous and ponds have to be cleaned, it is deemed to be a good green manure.

Reference:

Burkill, IH, 1935: 1210. Sastri, 1959: 145.

Ottelia ulvifolia (Planch) Walp.

FWTA, ed. 2, 3: 7. UPWTA, ed. 1, 464.
West African: **SENEGAL** FULA-TUKULOR kumubu (LeClerq fide JMD) MANDING-MAN-DINKA kunto (JB) **SIERRA LEONE** KISSI komachumja (?) (FCD) KONO yi-sasa (FCD) MENDE nɛŋgbɛ (NWT; FCD) **LIBERIA** MANO yi ba bo (Har.)

An aquatic plant with submerged leaves and yellow or white flowers above water-level, in moving water; throughout the Region and widespread in tropical Africa.

The plant is not particularly troublesome but is potentially so in dams and slow-moving water (8). It is planted in wells in Tanganyika ostensibly to prevent rapid evaporation of the water (1).

The leaves are succulent and are eaten in Liberia (4). In Gabon they have medicinal use (7). In the Central African Republic the plant is equated with *Nymphaea lotus* (Nymphaeaceae), being given the same vernacular names, and use for cardiac trouble (2, 6) and pounded to a mush for poulticing fractures and as an embrocation on fever-patients and sufferers of goitre (6). Calcined and lixiviated it is a source of vegetable salt.

The fruit is edible and is eaten (Leclerq fide 3, 5).

References:

1. Allnutt 11, K. 2. Bouquet, 1969: 137. 3. Dalziel, 1937: 464. 4. Harley s.n., 3/5/32, K. 5. Raynal-Roques, 1978: 335. 6. Vergiat, 1970, a: 88. 7. Walker & Sillans, 1961: 204. 8. Wild, 1961: 34.

Vallisneria aethiopica Fenzl

FWTA, ed. 2, 3: 7.

An aquatic herb of fresh or brackish water, recorded only in Senegal, Ghana and S Nigeria in the Region, and extending into Ethiopia and Zimbabwe.

The plant is often dense-growing with a fibrous root-system. It is recorded as being troublesome in slow-flowing waterways in Zimbabwe (1).

It is very closely related to *V. spiralis* Linn. of European distribution, which is popularly grown in aquaria. *V. aethiopica* may have the same utility.

Reference:

1. Wild, 1961: 34.

HYDROPHYLLACEAE

Hydrolea floribunda Kotschy & Peyr

FWTA, ed. 2, 2: 317.

West African: SENEGAL MANDING-BAMBARA kononia (JB)

An erect herb, simple or branched, stem spongy, holding much water, to 90 cm high, of swampy sites in the savanna zone from Senegal to W Cameroons, and extending to Sudan and Uganda.

Hydrolea palustris (Aubl.) Raeusch.

FWTA, ed. 2, 2: 316–7, as *H. glabra* Schum. & Thomm.

West African: SIERRA LEONE LIMBA butiɔgɔ (NWT) MANDING-MANDINKA ningya (NWT) MENDE kabawe (NWT) SUSU fundɛŋwuri, santuxemi (NWT) TEMNE ka-fut or 'the female kafut'; cf. Bacopa decumbens (Ferwald) F. N. Williams (Scrophulariaceae), the 'male kafut' (Jordan) ka-fut-ka-bɛra (FCD) NIGERIA IJO-IZON (Kolokuma) ofóróróbenibóu (KW&T) YORUBA oníyèníyè from òye: intelligence (Verger)

A glabrous herb with spongy stem and root, ascending or weakly erect to 40–60 cm high, of wet places from Senegal to S Nigeria, and into E Cameroun, the Congo basin and Madagascar.

The plant invades rice-padis in E Nigeria and becomes a weed (2). Yoruba doctors use the plant in a treatment to develop intelligence, hence òyé: Yoruba for 'intelligence', is compounded in the Yoruba name for the plant (see above) (3). In Sierra Leone Mendes grind up the leaves to rub on babies suffering headache (1). Ijo of SE Nigeria make unspecified medicinal use of the plant (4).

References:

1. Deighton 2447, K. 2. Tuley 71, K. 3. Verger, 1972: 39. 4. Williamson, K & Timitimi, 1983.

HYPOXIDACEAE

Curculigo pilosa (Schum. & Thonn.) Engl.

FWTA, ed. 2, 3: 174. UPWTA, ed. 1, 513.

English: African crocus (Morton).

French: manioc de lièvre (Berhaut).

West African: SENEGAL BASARI anan (JB) FULA-PULAAR (Senegal) batdi (JB) meresembu (JB) MANDING-MANDINKA toyhõ (JB) toyo (JB) MANINKA (Senegal) ologuélé (JB) tihongo (JB) togo (JB) SERER fuloh a kay (JB) fuloh diad (JB) fuloh no diad (JB) fuloh non hon (JB) lukit mõn (JB) lukit mõn (JB) WOLOF puloh i dar (JB) puloh i dar (JB) puloh i diombor (JB) THE GAMBIA MANDING-MANDINKA sulu tambo = spear of the hyena (Hayes) GUINEA BASARI a-nán (FG&G) MANDING-MANINKA kokun (Brossart; A.Chev.) ologuelé (CHOP) GHANA ADANGME bebengaa (Bunting; FRI) ADANGME-KROBO bebena = francolin's grass (Glover; JMD) bebengaa = francolin's grass (GBunting; FRI) NIGERIA EDO oriẹma (JMD) FULA-FULFULDE (Nigeria) biriji jire = ground-squirrel's ground-nut (JMD) nofru wamnde (pl. noppi bamɗi) = ear of the donkey (JMD) HAUSA dóóyàr kùréégéé = yam of the ground-squirrel (JMD; ZOG) mùrúchín mákwárwáá = francolin's 'muruchi' (the radicle of sprouted Borassus seeds) (JMD; ZOG) TIV kpeve (JMD)

A herbaceous plant with stout, erect rhizomes bearing a cluster of grass-like leaves to 60 cm long and flower shoots to 20 cm at the end of the dry season; of seasonally marshy savanna; widely dispersed from Senegal to W Cameroons, and over much of tropical Africa and Madagascar.

The foliage is said to be eaten by herbivorous animals in Sudan (1), perhaps in search of the seed capsules which are edible (2). The peduncle of the flower on fertilisation retracts the capsule down to the ground at the base of the plant.

The root-stock is held to be non-poisonous (2, 6). It has medicinal usages: in N Nigeria as a purgative (2, 6), and in Congo (Brazzaville) as a remedy for hernia (5). In Central African Republic the root reduced to a pulp is applied topically to swellings held to be of fetish origin by porcupines (3) or by aardvark (4).

References:

1. Andrews A.765, K. 2. Dalziel, 1937: 513. 3. Fay 4572, K. 4. Fay 4573, K. 5. Geerinck, 1971, 2–4. 6. Morton, 1961.

Hypoxis angustifolia Lam.

FWTA, ed. 2, 3: 172.

English: slender star lily (Morton).

An herbaceous plant with a small black bulb to $2\frac{1}{2}$ cm diameter and grass-like leaves, dwarf or to 25 cm long; of upland savanna; fairly common across the Region from Guinea to W Cameroons, and extending to NE and E Africa, and Madagascar.

The flower has a green outer perianth and a yellow inner one. The plant is worthy of cultivation as an ornamental.

The bulb is deemed to be edible (3). The white flesh is eaten by children of Masai and Kipsigi in Kenya (2) who also use the bulbs as playthings (1). The bulb is also eaten by francolin (4).

References:

1. Glover & al. 617, K. 2. Glover & al. 1225, K. 3. Graham Q.544, K. 4. Magogo & Glover 498, K.

Hypoxis recurva Nel

FWTA, ed. 2, 3: 172. UPWTA, ed. 1, 514.

English: star lily (Morton).
West African: **WEST CAMEROONS** KPE fawa songo (Nel fide JMD)

An herbaceous plant with subterranean rhizome producing several erect rhizomes bearing succulent white roots and old leaf fibres; inflorescences appearing before the leaves during the dry season; in montane grassland of S Nigeria and W Cameroons, and in E Cameroun.

The flowers are yellow and crocus-like, to 2 cm across, attractive, making the plant worthy of cultivation (2). They are much sought after by bees (1).

References:

1. Hepper 2111, K. 2. Morton, 1961.

Hypoxis urceolata Nel

FWTA, ed. 2, 3: 172.

English: silky star lily (Morton).

An herbaceous plant with a long rhizome bearing fibrous leaf-bases and inflorescences to 25 cm high appearing before the leaves which reach up to 40 cm long; of upland savanna in N Nigeria and in NE and E Africa.

The root, which is up to 4 cm diameter, is deemed eatable in Sudan (1).

The flowers are large, 2½ cm across, and showy, yellow, making the plant attractive and worthy of cultivation (2).

References:

1. Andrews A.1909, K. 2. Morton, 1961.

Molineria capitulata (Lour.) Herb.

FWTA, ed. 2, 3: 172, as *M. capitata* (Lour.) Merr.

A tufted herb with plicate leaves and dense clusters of yellow star-like flowers; of the tropical Himalayas from Nepal to S China and Australasia, and dispersed by man to many tropical countries including Sierra Leone in the Region.

The foliage is attractive, for which it is frequently cultivated in gardens. It is reported commonly grown in Freetown, Sierra Leone. The leaves yield a fibre which is used by hill tribes in the Philippines to make false hair (1, 2). Fish-nets are made of it in Borneo.

The fruit is edible (2).

References as *Curculigo capitulata* Ktze:

1. Burkill, IH, 1935: 703–4. 2. Sastri 1950: 400.

ICACINACEAE

Alsodeiopsis poggei Engl. var. robynsii Boutique

A shrub 2–5 m tall on wet sites of the closed-forest or in swamp-forest; native of Cameroun to Zaïre, and now recorded at Tarkwa, Ghana (1).

The wood of var. *poggei* is used in Zaïre to make bows (2), and its roots are reputed in Cameroun (3), Gabon (4) and Zaïre (2) to be aphrodisiac.

References:

1. Andoh 5576, K. 2. Boutique, 1960: 271–2. 3. Villiers, J-F, 1973, a: 28–30. 4. Villiers, J-F, 1973, b: 28–30.

Alsodeiopsis rowlandii Engl.

FWTA, ed. 2, 1: 638.

A shrub to 3 m high; on stream-banks of the closed-forest in S Nigeria, and through E Cameroun, Gabon to Zaïre.

The roots are reputed in E Cameroun (2), Gabon (3) and Zaïre (1) to be aphrodisiac.

References:

1. Boutique, 1960: 272. 2. Villiers, J-F, 1973, a: 27–28. 3. Villiers, J-F, 1973, b: 27–28.

Alsodeiopsis staudtii Engl.

FWTA, ed. 2, 1: 638.
West African: GHANA AKAN-WASA bonsa dua = *Bonsa River tree* (auctt.)

A shrub or small tree 2–5 m high; of forest undergrowth from Ivory Coast to S Nigeria, and in E Cameroun to Zaïre.

The roots are recognised as aphrodisiac in Gabon (5) and E Cameroun (4).

ICACINACEAE

The fruit is edible (1-3).

References:

1. Burtt Davy & Hoyle, 1937: 58. 2. Irvine, 1961: 463. 3. Vigne 168, K. 4. Villiers, J-F, 1973, a: 24, 26. 5. Villiers, J-F, 1973, b: 24, 26.

Desmostachys vogelii (Miers) Stapf

FWTA, ed. 2, 1: 639. UPWTA, ed. 1, 291.
West African: **LIBERIA** KRU-BASA gboe-kpar = *dog bone* (C; C&R) ju-eh-ye-ne-chu = *elephant's tusk tree* (C; C&R) **IVORY COAST** ? (Soumié) djo pu (Aub.) ABE diombo (Aub.)

A shrub, often scandent, or small tree to 9–10 m tall; of the forest; of limited distribution in Sierra Leone to Ivory Coast.

The wood is yellow and very hard. It is fine-textured, takes a glossy polish and is probably durable (2). It is used in Sierra Leone for implement-handles, house-poles and hunting-traps. Its hardness is reflected in the Basa names given above (1).

References:

1. Cooper & Record, 1931: 81, with timber characters. 2. Dalziel, 1937: 291.

Icacina mannii Oliv.

FWTA, ed. 2, 1: 641.
West African: **SIERRA LEONE** LIMBA buntumatuge (NWT) LOKO kεmoi (NWT) MENDE banje (NWT) **IVORY COAST** KYAMA akin (B&D) **GHANA** AKAN-TWI mutuo (BD&H; FRI) **NIGERIA** IGBO ututo ogiri (BNO)

A scandent shrub to 6 m high with a large tuber; of evergreen forest, swamp-forest, galleried forest and in clearings; from Sierra Leone to S Nigeria, and on to Zaïre.

It is recognised as a medicinal plant in Gabon, but no manner of usage is reported (3). In Ivory Coast the leaf-sap with guinea-grains (*Aframomum melegueta* K Schum., Zingiberaceae) added constitute an embrocation applied in massage to areas of costal pain (2). In Congo (Brazzaville) the leaf-sap is added to that of *Tetrorchidium sp.* (?*T. didymostemon* (Baill.) Pax & K Hoffm.) for use on sores (1). A leaf-decoction is also taken by draught as an expectorant and emetic to relieve bronchial affections and cough, though the root may sometimes be used (1).

In Ivory Coast the plant (part not stated) is reputed to be a purgative and a powerful diuretic; it is given by mouth for oedema and female sterility (2). In Congo (Brazzaville) the tuber is hashed up and macerated for 2–3 days in water which is drunk for gastro-intestinal troubles and for dysentery. If the patient does not respond, a leaf must be placed beneath the bed, and if it becomes rolled up or perforated the patient is thus exposed as a sorcerer who, if a cure is desired, must confess; then a draught of the tuber-macerate will effect a cure (1).

A fanciful attribute of the leaf is believed in Congo (Brazzaville): if a leaf is held in the hand during sexual intercourse, ejaculation will be delayed, and will only happen when the leaf is placed on the head! (1).

References:

1. Bouquet, 1969: 138. 2. Bouquet & Debray, 1974: 96. 3. Walker & Sillans, 1961: 206.

Icacina oliviformis (Poiret) J Raynal

FWTA, ed. 2, 1: 639, as *I. senegalensis* A Juss. UPWTA, ed. 1, 291, as *I. senegalensis* A Juss.
English: false yam.

West African: SENEGAL BADYARA mana sè (JB) BALANTA foe (JB) foya (K&A; JB) mtazi
(JB) songol (JB) tiafi (K&A) BANYUN ba dingalí (JB) dia vogõ (JB) dia wogõ (K&A) BASARI a-
nagan (K&A; JB) a-narham (K&A; JB) BEDIK gi-ndél (FG&G) ma-rél (auctt.) CRIOULO
mânganasa (JB) DIOLA bu bambulaf (JB) bu tima (JB) fu timay (JB) furabá (JB) kurabã
(auctt.) u rabã (JB) DIOLA (Fogny) furabân (K&A) DIOLA (Tentouck) butima (K&A) futima
(K&A) DIOLA-FLUP butéé ma (K&A) butima (K&A) hu timo (JB) u timo (JB) FULA-PULAAR
(Senegal) mânganatié (JB) mânkanase (K&A) mankanaso (JB) silla (K&A; JB) KONYAGI
dagan (JB) a-ndagana (JB) ndarham (K&A) MANDING-BAMBARA bankanazé (A.Chev.) MAN-
DINKA bâkanazé (K&A) mâkanas (K&A) mâkanaso (K&A) ˙SOCE˙ bâkanazé (K&A) mâkanas
(K&A) makanaso (K&A) mânganaso (JB) MANDYAK be nasia (JB) mânkanas (JB) MANKANYA
be nasin (JB) SERER 'ba (JB) iba (JB; K&A) WOLOF bâkanas (JB; K&A) bânkanas (JB)
ndiangam (JB) mâkanas (K&A) **THE GAMBIA** DIOLA kuraban (Hallam) FULA-PULAAR (The
Gambia) bankanase (DAP) bankanesi (DAP) manankaso (DAP) sila (DRR) MANDING-MAN-
DINKA bankanase (DAP) bankanesi (DAP) manankaso (auctt.) manankoso (def.) (DRR; ST)
mankanaso (JMD; Hallam) wamgaso (def.) (ST) wangaso (JMD) WOLOF bananafi (Hallam)
bankanafi (JMD) bankanase (DAP) bankanesi (DAP) manankasa (DRR) manankaso (DAP)
mbahanasa (ST) **GUINEA-BISSAU** BALANTA fóè (auctt.) sóngol (auctt.) BIAFADA manasse
(JDES) BIDYOGO em-hándú (JDES; RdoF) CRIOULO manganace (JDES) manganaz (auctt.)
FULA-PULAAR (Guinea-Bissau) mancanadje (pl.) (JDES) manganadje (RdoF) silá (D'O) MAND-
ING-MANDINKA manacossô (JDES; RdoF) mancanassô (JDES; RdoF) MANDYAK unasse
(JDES; RdoF) MANKANYA unasse (JDES; RdoF) PEPEL unássem (auctt.) **GUINEA** BASARI a-
nàxàn (FG&G) KONYAGI a-ndàxán (FG&G) u-ryakán (FG&G) **GHANA** AKAN-ASANTE abubu
ntɔpe = breaks hoe (BD&H; FRI) BRONG lalakwei (auctt.) larekwe (FRI) DAGBANI takwara
(BD&H; FRI) tankoru (FRI) GBE-VHE miña (Veldkamp) **NIGERIA** IGBO ututo ogiri (BNO)

An erect woody herb or scandent shrub to 1 m tall of a number of stems
arising from a large subterranean tuber with long roots; of roadsides and waste
places in the savanna scrub; from Senegal to S Nigeria, and in the Central
African Republic and Sudan.

The plant is a weed of cultivation, and is common around villages and on old
farmland, especially in yam-fields. It is vigorous and rapidly invades newly
cleared and cultivated land (2, 9). Eradication is fraught with difficulty because
of its large, or very large, tuber, which may reach 30–50 cm diameter (9) or 25
kg in weight (5), set 30 cm or so deep in the soil, and its long creeping roots (24).
The Asante name, abubu ntɔpe: breaks hoes, reflects this situation (7).

The leaves are toxic to sheep and cause death. It is recorded that pasturalists
bringing their sheep into the Casamance area (Tambacoumba), where the plant
is common, used to muzzle them to prevent them grazing it (6, 16). The plant is
one of the grand panaceas in the Casamance pharmacopoeia, the leaf, always
prescribed alone, being used for numerous diverse treatments. Leafy twigs in
decoction are used for internal haemorrhages and in baths and washes, for
cough and all chest affections, and for feverish states. For the last-named, the
patient should sleep the night on a bed of newly-cut leaves (18, 19, 21). A
decoction of twigs and roots is given to adults in Senegal for general debility of
an undiagnosed origin (20), and a leaf-decoction by draught for snake-bite (17).
Tenda in SE Senegal apply heated leaves topically to points of pain, particularly
in elephantiasis, as an analgesic (8). Leaf-sap is used in The Gambia for eye
infections (10).

The leaf is used in The Gambia as a container for carrying cashew nuts (10).
Tenda use them to blacken pottery (8).

The leaf enters into Tenda ritual in a burial ceremony: a head-pad is placed
under the head of the dead person (8).

The enormous tuber is fleshy and on the whole is poisonous. With suitable
treatment it has in the past served as a famine-food (1, 4, 5, 9, 10, 12, 13, 21). It
needs to be cut up and placed in running water for several days to remove
bitterness and to facilitate maceration. When dried it is pounded and the fibres
strained out, then boiled down to a paste. The product is mainly starch and is
said to be palatable (7, 13–15). If consumed without detoxification colic and
dysentery will result (10), and if eaten in quantity death will follow within a few
days. As a common weed of many savanna areas, it could be exploited where
crop-failure is likely (15). Selection for ameliorated strains should be considered.

The root is put to various medicinal uses. In Casamance a decoction is given by draught for dermatitis, headache, chest-complaints, kidney-troubles, etc. It is considered tonic and is given for senescence, debility, internal pain, etc., and to children with rickets, as also for use against dental caries (17, 20, 21). A root-macerate is given by draught or in baths for general fatigue. Tenda administer the tip ends of roots as an abortifacient (8), while Asante in western Ghana take it for impotence (14).

The poisonous principle of the tuber is a bitter; alkaloid is also present (3), and a gum-resin at 0.9–2.8% concentration (15).

The fruit when ripe is red with a thin sweet gelatinous pulp which is edible (7–10, 15, 19, 22). In Senegal the fruit is eaten by chimpanzees (23).

The kernel is eaten in time of dearth after pounding and drying to yield a flour (15, 17). Normally it is deemed poisonous and is soaked for a week with daily changes of water, dried in the sun for 2 days and then ground to flour. Tenda prepare the latter with millet or beans (*enap*) to produce a palatable food said to be 'good enough for guests' (8, 22).

The flowers are sweet-scented and are visited by bees and flies (11). There is no report on the quality of bee honey from this source.

References:

1. Abbiw GC. 47242, K. 2. Bally B.140, K. 3. Bouquet, 1972: 29, as *Icacina sp.* 4. Burtt Davy & Hoyle, 1937: 57, as *I. senegalensis* A Juss. 5. Busson, 1965: 359, as *I. senegalensis* A Juss. 6. Chevalier, 1937: 171, as *I. senegalensis* A Juss. 7. Dalziel, 1937: 291, as *I. senegalensis* A Juss. 8. Ferry & al., 1974: sp. no. 140, *I. senegalensis.* 9. Fox 95, K. 10. Hallam, 1979: 114–5, as *I. senegalensis.* 11. Hepper 2339, K. 12. Irvine, 1952, a: 29, as *I. senegalensis.* 13. Irvine, 1952, b: as *Icacina sp.* 14. Irvine, 1961: 464, as *I. senegalensis* A Juss. 15. Kay, 1973: 67, as *I. senegalensis* A Juss. 16. Kerharo, 1967: as *I. senegalensis* A Juss. 17. Kerharo & Adam, 1962: as *I. senegalensis* A Juss. 18. Kerharo & Adam, 1964, a: 421, as *I. senegalensis* A Juss. 19. Kerharo & Adam, 1964, b: 551–2, as *I. senegalensis* A Juss. 20. Kerharo & Adam, 1964, c: 310, as *I. senegalensis* A Juss. 21. Kerharo & Adam, 1974: 486–7, as *I. senegalensis* A Juss. 22. Tattersall, 1978: as *I. senegalensis* A Juss. 21. Tutin 48, K. 24. Williams 152, K.

Icacina triacantha Oliv.

FWTA, ed. 2, 1: 641. UPWTA, ed. 1, 291.
West African: **NIGERIA** IGBO ji-muo (Iwu) ututo ògìrì (BNO) IGBO ('Ibugo') ọgwá (NWT) IGBO (Umuahia) unumbe (AJC) IGBO (Umudike) unumbia (Ariwaodo) YORUBA gbégbé (auctt.)

A shrub to 2 m with scandent growth above, and with a very large tuber; of forest and jungle vegetation in S Nigeria.

The plant is reported to become a weed of rice-padis in Bendel State (4).

The leaf is said to be used as a wrapper for castor-oil seeds (7), but the purpose of wrapping them is not disclosed.

The thick yam-like root attains a large size. Yoruba say that it is edible alone, or dried and pounded to a white powder called *gbẹ-wutu* (Yoruba-Ijebu dialect) which is used in soup, or added to a food known as *igbālò* made from the roasted seeds of *Citrullus lanatus* (Cucurbitaceae) (q.v.) (3). Igbo treat it as a famine-food eating the flour after prolonged maceration and repeated washings (5). The tuber is inflammable and when burning gives out so fierce a heat as to be unapproachable (Migoed fide 3).

Though the root is thus reported edible, a strong presence of *alkaloid* has been reported in Nigerian material (1), also *benzophenones* (5). The leaf contains a trace only (1).

The fruit is a drupe about 2½ cm long with a soft sweet outer pulp which is edible (2, 3, 5, 6). The kernel appears to be not eaten (3).

Igbo consider the plant (part not stated) to be aphrodisiac, and they use it on soft tumours (5).

References:

1. Adegoke & al., 1968: 13–33. 2. Carpenter AJC.437 (UIH 2434), UCI. 3. Dalziel, 1937: 291. 4. Gill & Ene, 1978: 182–3. 5. Iwu, 1986: 141. 6. Okafor FHI.34969, K. 7. Thomas NWT.1999 (Nig. ser.), K.

Iodes africana Welw.

FWTA, ed. 2, 1: 643.
West African: NIGERIA YORUBA àṣẹ = *power* (Verger)

A liane, said to be weak in S Nigeria (4), but able in E Cameroun to Zaïre to reach the top storey of the evergreen forest (1, 3, 7, 8); on dry or riverine soils liable to inundation.

The leaf has a strong poisonous and sharp smell, especially when old. In Congo (Brazzaville) leaf-sap is instilled into the nose as a decongestant of the respiratory passages and sinuses in cases of colds, bronchitis, sinusitis and headaches (1). A decoction of the whole plant is taken for stomach-troubles, diarrhoea and blennorrhoea; this is prepared as a vapour-bath for trypanosomiasis, and is also sometimes instilled into the eye to kill filaria in the mucosae of the eyelid (1).

Traces of *saponin* are reported in the leaf, bark and root; *tannin* is present in bark and root with a little in the leaf, and *steroids* and *terpines* are in the bark and root (2).

Yoruba invoke the plant in an Odu incantation against sleeplessness (7, 8). In Congo (Brazzaville) a man wishing to seduce a woman rubs his body with some palm-wine in which leaves have been macerated (1).

References:

1. Bouquet, 1969: 138–9. 2. Bouquet, 1972: 29. 3. Breteler & al. 2434, K. 4. Keay, 1958: 643. 5. Verger, 1967: no. 52. 6. Verger, 1986. 7. Villiers, J-F, 1973, a: 6–8. 8. Villiers, J-F, 1973, b: 6–8.

Iodes liberica Stapf

FWTA, ed. 2, 1: 643–4.

A climbing shrub or liane to 20 m high in forested areas from Guinea to Ivory Coast, and also in Zaïre.

The plant has a place in ju-ju in Liberia: it is buried in rice-fields to make them yield more plentifully (1).

Reference:

1. Cooper 204, K.

Lasianthera africana P Beauv.

FWTA, ed. 2, 1: 638. UPWTA, ed. 1, 291.
West African: NIGERIA IGBO kpurugiza (Chesters fide JMD) kpuruziza (Chesters) uyoro (AJC fide JMD) IGBO (Umuahia) nka-nka (AJC) WEST CAMEROONS KPE belele (Waldau fide JMD) MUNI itebele (Waldau fide JMD)

A shrub to 4 m high; of the understorey of secondary jungle and thickets in S Nigeria, W Cameroons and Fernando Po, and extending to Zaïre.

The wood is made into axe-handles in Cameroun (7) and Gabon (5, 8). It is very fibrous and decayed wood can be used as a sponge (4, 6).

In W Cameroons the Baakpe use a leaf-infusion for stomach-ache (Santesson fide 4, 6) while in Congo (Brazzaville) a leaf-decoction is taken to relieve stomach-complaints with or without diarrhoea (1). A decoction is considered vermifugal. In E Cameroun (7) and in Gabon (5, 8) a decoction is used cold as a wash against headache. Unspecified medicinal use in Gabon is also noted (9). In Congo (Brazzaville) leaf-sap is considered anti-psoric; also the leaf crushed in warm water, which raises an abundant froth, is used to bathe feverish infants (1). Pulped leaf or pulped bark is made up into a dressing to bind over fractures (1).

Tests for *alkaloids* have given mild response for the leaf, moderate for the bark and strong for the root, and *tannin* is present in all these parts (2).

Notwithstanding the presence of alkaloid and tannin, Igbo of S Nigeria put the leaf in soup (3).

The plant enters into certain fetish rites in Gabon (9).

References:

1. Bouquet, 1969: 139. 2. Bouquet, 1972: 29. 3. Carpenter AJC.353 (UIH 2443), UCI. 4. Dalziel, 1937: 291. 5. Hallé 1625, K. 6. Irvine, 1961: 464–5. 7. Villiers, J-F, 1973, a: 14–17. 8. Villiers, J-F, 1973, b: 14–17. 9. Walker & Sillans, 1961: 207.

Lavigeria macrocarpa (Oliv.) Pierre

FWTA, ed. 2, 1: 641.

A large forest liane, stems to 4 cm diameter; in S Nigeria, W Cameroons and Fernando Po, and extending to Zaïre.

The fruit is a large drupe 5–11 cm long by 4–4½ cm across covered with a fleshy pericarp which is edible. The kernel is starchy. Both parts are held to be aphrodisiac (Cameroun: 2; Gabon: 3; Zaïre: 1).

References:

1. Boutique, 1960: 268–70. 2. Villiers, J-F, 1973, a: 65–66. 3. Villiers, J-F, 1973, b: 65–66.

Leptaulus daphnoides Benth.

FWTA, ed. 2, 1: 637. UPWTA, ed. 1, 291–2.

West African: **SENEGAL** DIOLA-FLUP o diaba (JB) **SIERRA LEONE** MENDE bongawi (*def.* -i) (FCD; S&F) bongowi (L-P) doborumbe (Aylmer) faffeo (Aylmer) TEMNE ɛ-limri-ɛ-ro-kant (NWT) *ka*-propri (auctt.) **LIBERIA** ? (Nimba Mt.) yallah (Mus. Stockholm) **IVORY COAST** AKYE parandedi (A.Chev.) ANYI eboro-dumuen = *bush-lime* (A.Chev.) KRU-GUERE (Wobe) pain'tu (A.Chev.)

A shrub or small tree to 20 m tall; bole straight, sometimes fluted, by 90 cm girth; in high forest and common in secondary jungle; distributed throughout from Senegal to W Cameroons, and to Sudan, Uganda, Zaïre and Cabinda.

The wood is hard, close-grained and cream-coloured (2, 4, 5). The stems are used for hut-poles in Liberia (1).

The bark is reported to have unspecified medicinal use in Ghana (5).

The leaf is used in Gabon (7, 8) and E Cameroun (6) as an emetic.

The fruit, a pointed ellipsoid drupe about 1.25 cm long, contains a kernel which is a favourite food of mona monkeys (*Cercopithecus mona*) in W Cameroons (3).

References:

1. Cooper & Record, 1931: 81. 2. Dalziel, 1937: 292. 3. Gartlan 3, K. 4. Keay & al., 1964: 203. 5. Irvine, 1961: 465. 6. Villiers, J-F, 1973, a: 59. 7. Villiers, J-F, 1973, b: 59. 8. Walker & Sillans, 1961: 207.

Leptaulus zenkeri Engl.

FWTA, ed. 2, 1: 637.

A shrub or small tree to 5 m high; of the evergreen forest in S Nigeria, and extending to Zaïre.
The leaf has unspecified medicinal use in Gabon (1).

Reference:

1. Walker & Sillans, 1961: 207.

Polcephalium capitatum (Baill.) Keay

FWTA, ed. 2, 1: 642. UPWTA, ed. 1, 290, as *Chlamydocarya capitata* Baill.
West African: SIERRA LEONE KRIO dɔg-bɔbi (FCD) LIMBA buhago (NWT) MENDE ɲgeta-nyina (FCD) kpewuli (NWT) pɛmbue (NWT) LIBERIA KRU-BASA te-ohn-way-doo (C; C&R) IVORY COAST DAN gnontenkannè (A&AA)

A small liane, recorded from Guinea to Ivory Coast.
Liberian midwives prepare a decoction from the leaf for putting into baths and to make a draught for weakling babies to make them strong (1, 2).

References:

1. Cooper 363, K. 2. Dalziel, 1937: 290–1, as *Chlamydocarya capitata* Baill.

Pyrenacantha acuminata Engl.

FWTA, ed. 2, 1: 642.
West African: LIBERIA MANO gele wê bele (Har.)

A slender liane of the closed-forest, flooded or not liable to flooding; from Sierra Leone to Ivory Coast in the Region, and in E Cameroun, Gabon and Zaïre.

Pyrenacantha staudtii (Engl.) Engl.

FWTA, ed. 2, 1: 642. UPWTA, ed. 1, 292.
West African: NIGERIA EDO ohogha (Kennedy fide JMD) IGBO (Umueleke) nhia (Ariw-aodo) YORUBA ahara (Thompson)

An arborescent liane to 6 m long by 5–10 cm diameter; of secondary jungle in S Nigeria and W Cameroons, and across central Africa to Uganda and Angola.
Palm-wine in which the plant (? part) has been boiled is drunk in Congo (Brazzaville) in treatment for blennorrhoea (1). The plant (?leaf) is used as an analgesic for intestinal pain and for hernia, the medication being partly by mouth, and partly by topical dressing at the point of pain (1).

Reference:

1. Bouquet, 1969: 139.

Pyrenacantha vogeliana Baill.

FWTA, ed. 2, 1: 642.
West African: SIERRA LEONE KISSI kole-pundo = *kola rope* (FCD) LOKO tɛnde (NWT)

A slender woody liane; of marshy sites in swamp-forest, and on river-banks; widely dispersed from Sierra Leone rather disparately to S Nigeria, and on to Zaïre and Tanganyika.

No usage is recorded, but the Kissi name meaning 'kola rope' (1) is suggestive.

Reference:

1. Deighton 2411, K.

Rhaphiostylis beninensis (Hook.f.) Planch.

FWTA, ed. 2, 1: 638. UPWTA, ed. 1, 292.
West African: SENEGAL DIOLA bu kita (JB; K&A) FULA-PULAAR (Senegal) buru tâgol (K&A; JB) GUINEA FULA-PULAAR (Guinea) buru tiangol (O.Caille) SIERRA LEONE MENDE foklobe (FCD) fokobi (L-P) pɛwɛla (FCD) MENDE (Kpa) pɛgɛla (FCD) TEMNE am-purepure (FCD) ra-thɔnk (FCD) LIBERIA MANO pro pro lah (JMD) IVORY COAST AKYE kpè-kpè (A&AA) GAGU logbapawkpawkla (B&D) KRU-GUERE (Chiehn) bogdrobo (B&D) KWENI malé-niru (B&D) ˙NÉKÉDIÉ˙ kwékora (B&D) GHANA AKAN-TWI ɔkwakora-gyahene = *old man's shin-bone* (BD&H; FRI) NIGERIA IGBO (Obompa) kp'ɔlɔkɔtɔ (NWT) IGBO (Uburubu) kpɔlɔ-kɔtɔ (NWT fide JMD) ucici n'efifiè = *ucici of the afternoon* (NWT fide JMD) IGBO (Umuahia) oke ikpokrikpo *from* oke: *strong* (AJC) YORUBA idia pata (Millen) itá para (EWF; JMD)

A scandent or twining shrub, occasionally arborescent to 10 m; of the closed-forest throughout the Region from Senegal to S Nigeria, and extending across central Africa to Uganda, Zambia and Angola.

Goats eat the plant, evoking in Igbo the epithet *ike* meaning 'strong' (3). The idea of imparting strength is also manifest in the root and bark which are sold in Lagos medicine market for the purpose of preparing an infusion taken by pregnant Yoruba women to strengthen the baby, such preparation also being given as a tonic to infants up to the age of 2–3 years (4). A decoction of the leaf (and stem) is drunk, a glassful at a time, by Yoruba to cure *aiperi* (convulsions) and a sickness called *afun* where the skin becomes whitish and transparent in children up to 1 year old (6). The leaf is boiled in Sierra Leone and the liquid used as a mouthwash (4, 5), while in Igboland such a preparation is used to wash sores (4, 10).

The bark-macerate is considered by medicine-men in Casamance to be the laxative of choice for babies up to 3 months old with chronic constipation (7, 8). The root, stem and leaf are decocted in Ghana and Nigeria and the liquor is drunk to kill and expell roundworm (*Ascaris*) (6).

In Ivory Coast the leaves are put into wet dressings for rheumatism and haemorroids, and decocted the draught is taken for bronchial troubles and used in eye-instillations for ophthalmias (2).

Alkaloid is reported to be strongly present in the bark (2) and the root (1).

The leaf when touched by fire crackles violently, thus conjuring up magical use in Ivory Coast to chase away spirits to quieten fits of madness in possessed persons (2).

The seed is edible (4, 6). They are eaten in Lagos (9).

References:

1. Adegoke & al., 1968: 13–33. 2. Bouquet & Debray, 1974: 96. 3. Carpenter AJC.356 (UIH.2447), UCI. 4. Dalziel, 1937: 292. 5. Deighton 857, K. 6. Irvine, 1961: 466. 7. Kerharo & Adam, 1963, b. 8. Kerharo & Adam, 1974: 487. 9. Millen 22, K. 10. Thomas NWT.2273 (Nig. Ser.), K.

Rhaphiostylis preussii Engl.

FWTA, ed. 2, 1: 639.
West African: SIERRA LEONE LOKO bialui (NWT) MENDE banja (NWT) ngwaho (NWT) tɛle (NWT) tɛrai (NWT) TEMNE fɛŋke (NWT) ɛ-kbɛkbe (NWT)

A scandent shrub or liane of the forest from Sierra Leone to W Cameroons and into E Cameroun and Gabon.

This plant is similar to *R. beninensis*. Though a number of vernaculars is recorded from Sierra Leone, no usage is known – perhaps a case of confusion.

Stachyanthus occidentalis (Keay & Miège) Boutique

FWTA, ed. 2, 1: 643, as *Neostachyanthus occidentalis* Keay & Miège.

A slender liane of the forest zone in Ivory Coast to S Nigeria.

The lianous stems are very thin and are used by Asante as string in hut-building (1–3).

References:

1. Burtt Davy & Hoyle, 1937: 58, as *Desmostachys tenuifolius* Oliv. 2. Irvine, 1961: 466, as *Pyrenacantha sp. B.* 3. Vigne 1620, K.

IRIDACEAE

Aristea alata Bak.

FWTA, ed. 2, 3: 139.

A perennial herbaceous plant with blue inflorescences to 60 cm high; of rocky hill situations on the Jos and Mambila Plateaux of N Nigeria and upland areas of W Cameroons; widely distributed in NE, E, central and S central Africa.

Masai children in Kenya use the leaf-sap to dye their faces yellow (2). In Uganda the plant with sap from bracken (*Pteridium aquilinum* (Linn.) Kuhn, Pteridophyta) is used to produce a tattoo dye (3); the colour is not stated.

In Kenya the plant is said to be poisonous (1).

References:

1. Edwards 1812, K. 2. Glover & al. 1090, K. 3. Greenway, 1941: sp. no. 37.

Crocosmia aurea (Hook.) Planch.

FWTA, ed. 2, 3: 138.

English: montbretia.

A perennial grass-like herb, erect to 1 m high from a corm 3 cm diameter; flowers orange; native of tropical southern Africa, and extensively naturalised in E Africa, and into the Cameroon Mountain area of the Region.

No usage is recorded in W Africa. In E Africa leaf-sap and a decoction of the corm is drunk for malaria, and a root-decoction is drunk or the plant-ash with castor oil is embrocated into scarifications for arthritic rheumatism (2).

The flowers yield a yellow dye which can be used as a substitute for saffron (1, 3).

References:

1. Greenway, 1941: sp. no. 91. 2. Haerdi, 1964: 201. 3. Uphof, 1968: 159.

Crocus sativus Linn.

English: crocus; saffron; saffron crocus.
French: crocus; safran.
Portuguese: croco; açafrão.

IRIDACEAE

West African: MALI ARABIC (Mali) zafran (HM)

An herbaceous perennial plant with lilac/purple flowers to 10 cm long pushed up from a perennial woody corm, preceeding the foliage; a cultivar known only in cultivation arising from the Mediterranean–Near East region. It is of ancient cultivation and was certainly known in the Levant at the time of King Solomon. It is presently recorded on the northern limit of the Region in the Hoggar area (2).

The plant is the source of saffron which is obtained from the dried stigmas and anthers used for colouring and flavouring food. Cultivation is carried out in several Old World countries. Besides use in food, fabrics are dyed with it in E Africa (1).

References:

1. Greenway 1941: sp. no. 92. 2. Maire, H, 1933: 231.

Gladiolus Linn.

FWTA, ed. 2, 3: 141, incl. *Acidanthera* Hochst., 3: 139. UPWTA, ed. 1, 487–8, incl. *Acidanthera* Hochst., 487.

English: gladiolus; corn flag; sword lily.

French: glaïeul; glaïeul d'Afrique.

Portuguese: espadana.

A cormiferous herbacous genus well represented in W Africa, but few have uses. The genus consists of some 250–300 species of the Mediterranean and tropical and southern African regions. The gladiolus of temperate and subtemperate horticulture is the product of much hybridising, especially of species from S Africa. Several of the W African species are beautiful plants (1) and offer scope for breeding and selection work for the improvement of local gardening. Cultivation is easy from cormlets and seed.

Reference:

1. Morton, 1961.

Gladiolus aequinoctialis Herb.

FWTA, ed. 2, 3: 139, as *Acidanthera aequinoctialis* (Herb.) Bak. and *A. divina* Vaupel. UPWTA, ed. 1, 487, as *A. aequinoctialis* Hochst.

English: wild acidanthera (Morton, as *A. aequinoctialis* Bak.).

An herbaceous plant with a corm producing foliage and inflorescences to 30 cm high; of rocky crevices in montane situations from Guinea to W Cameroons.

It is a handsome lily-like plant with white flowers, vinous-red or purple blotched in the throat with bracts often similarly coloured (1, 2). Propagation is by corms or cormlets.

References as *Acidanthera aequinoctialis* Baker:

1. Dalziel, 1937: 487. 2. Morton, 1961.

Gladiolus atropurpureus Bak.

FWTA, ed. 2, 3: 144, as *G. unguiculatus* Bak.

English: dingy gladiolus (Morton).

West African: SENEGAL MANDING-BAMBARA taku diurhõ (JB) NIGERIA BIROM dyék (LB) HAUSA rumáánáá (LB)

424

A slender herb to about 30 cm tall with small corms; flowers dull purple fading to white; of moist places in the savanna across the whole Region, and generally widespread in tropical Africa. The corms are starchy and are eaten in the Region in soup. They are often an article of market trade (3). The roots of an unspecified species which are eaten by the Tenda of SE Senegal are perhaps of this plant (2). In the Central African Republic a root-decoction is drunk for testicular hernia (1).

References:

1. Fay 5417, K. 2. Ferry & al., 1974: sp. no. 141. 3. Irvine, 1952, a: 28.

Gladiolus daleni van Geel

FWTA, ed. 2, 3: 141, as *G. psittacinus* Hook., in part.
English: yellow gladiolus (Morton).
West African: GUINEA-BISSAU FULA-PULAAR (Guinea-Bissau) djabreguele (EPdS; JDES) SIERRA LEONE ? (Freetown) banda-karafia (Burbridge) GHANA AKAN-ASANTE wisapokoru (TFC) NIGERIA BASSA (Kwomu) igara (Lamb) BIROM dyék (LB) FULA-FULFULDE (Nigeria) ndayewu (*pl.* ndaweji) (JMD) GWARI kokowi yako *a large corm used in medicine* (JMD) HAUSA rumáánáá (auctt.) rumáánan dooki *a large corm used in medicine* (JMD) IDOMA ukpẹ́ndú *for the edible corm; of general application to other G.spp. also* (Armstrong fide RB) IGALA okperdo (Lamb; JMD) KAMBARI-KIMBA kàrụ́mpàsàngízò = *spiders' onion* (RB) SALKA àarụ̀mbàsà kwíisɔ̀ = *spiders' onion* (RB) TIV neman, waiyol (JMD) YORUBA agunmu *a mixture of foodstuffs made of corm, prepared as medicine* (auctt.) bààká *applied to any corm* (JMD; Verger)

An erect robust herb with sword-like leaves and inflorescences to over 1 m high from a woody corm; of the rocky savanna in montane locations from Guinea and Mali to S Nigeria, and widespread elsewhere in the savanna of tropical and southern Africa.

This is the largest of the W African *Gladiolus* species. It has large, showy, pure white or yellow flowers variously spotted or heavily mottled dark orange. It merits cultivation as an ornamental (3, 10, 11).

The corm has edible and medicinal usages, for which it is often cultivated (9, 10). The corm is starchy. It serves as a supplementary food pounded and cooked up with guinea-corn flour to make a sweet pap or beverage (7-9), known as *kunun zaƙi* in Hausa (4). This is also an invalid diet in cases of mild fever (2). A decoction of the corm is a cure for dysentery, dysmenorrhoea and diarrhoea, and is given as an enema, or part of a general treatment, for rheumatism and lumbago (2). Also the corm with ginger is given as a potent enema for constipation and dysentery. The same mixture is applied to snake-bite, and together with bitter kola (?*Garcinia kola* Heckel, Guttiferae) and onion is drunk with lime-juice or warm water as a vermifuge (2). The corms are sold in Lagos market as *bàká* (Yoruba) as a purgative to clear the system of gonorrhoea and other conditions; compounded with onions and food this medicine is known as *agunumu* (Yoruba) (4). For persons in a faint some finely powdered corm is put in the nostrils, and if sneezing results, there is hope of recovery (4).

In Kordofan the dried corm is rasped and taken with fat against constipation (3). Its use in Lesotho is particularly prescribed for diarrhoea when there is bleeding (5). In Tanganyika scrapings of the corm are applied as a wound-dressing, and sap expressed from the powdered corm is dripped into the ear for inflammation of the middle ear (6). A decoction of the corm is a Lesotho medicine for colds and dysentery, or the powdered corm is taken for the latter, and smoke from a burning corm is inhaled for the former (12).

In veterinary medicine a large corm, probably of this species, known as *rumáánan dooki* (Hausa) and *kokowi yoko* (Gwari) is used as a horse-medicine administered by rectal injection for mucal diarrhoea (4).

A strong presence of *alkaloid* in this and other *Gladiolus spp.* is reported (1).

IRIDACEAE

References:

1. Adegoke & al., 1968: 13–33, as *Gladiolus spp.* 2. Ainslie, 1937: sp. no. 166. 3. Baumer, 1975: 100, as *G. psittacinus* Hook.f. 4. Dalziel, 1937: 487, as *Gladiolus spp.* 5. Guillarmod, 1971: 430, as *G. psittacinus.* 6. Haerdi, 1964: 202, as *G. psittacina* Hook. 7. Irvine, 1952, a: 27, as *G. quartinianus*; 6: 28, as *Gladiolus spp.* 8. Irvine, 1952, b, as *Gladiolus* corms. 9. Lamb 88, K, and in litt., 12/3/1915, K. 10. Morton, 1961, as *Gladiolus* and *G. primulinus.* 11. Walker & Sillans, 1961: 207, as *G. quartinianus* A. Rich. 12. Watt & Breyer-Brandwijk, 1962: 505, as *G. psittacinus* Hook.

Gladiolus gregarius Welw.

FWTA, ed. 2, 3: 144, incl. *G. klattianus* Hutch.

West African: SENEGAL DIOLA be sémin (JB) bésémin (JB; K&A) ediobufay = *poison antidote* (JB) nâdal (JB) DIOLA (Fogny) nâdal = *purgative* (K&A) DIOLA-FLUP édoluay = *poison antidote* (K&A) WOLOF nâdal (K&A) niâdal (JB) GUINEA MANDING-MANINKA baga diura (Brossart ex A.Chev.) GHANA ADANGME-KROBO samana tʃupa (FRI) yaane (FRI) AKAN-TWI botofufuo (FRI) GBE-VHE dzogbekoẽ *from* dzogbe: *grassland*; koẽ, *soap; because it lathers in water* (JMD) NIGERIA BIROM dyék (LB) HAUSA rumáánáá (auctt.) IDOMA ukpẹ́ndá *for the edible corm; of general application to other G.spp. also* (Armstrong fide RB) YORUBA baka (JRA)

A slender perennial herb, with leaves and erect inflorescence to over 1 m tall from a woody corm; of moist savanna; dispersed across the Region from Senegal to S Nigeria, and extending to Zimbabwe and Angola.

The corms are smaller than those of *G. daleni*, but they are starchy and are eaten as a supplementary food (1, 2). They are sometimes sold in markets (2).

The corm has a strong reputation in the Casamance as a medication, its Diola-Flup name (see list above) meaning 'poison antidote' (3, 4). It is taken dried and powdered to produce emesis (3).

References:

1. Elliott 193, K. 2. Irvine, 1952, a: 28, as *G. klattianus*. 3. Kerharo & Adam, 1963, a: as *G. klattianus* Hutch. 4. Kerharo & Adam, 1974: 488, as *G. klattianus* Hutch.

Gladiolus melleri Bak.

FWTA, ed. 2, 3: 144.

An herbaceous plant with near leafless stems of up to 90 cm tall; of rocky upland savanna in N Nigeria.

It is an attractive plant with pale pink flowers. It is worthy of cultivation as an ornamental (1).

Reference:

1. Morton, 1961.

Gladiolus oligophlebius Bak.

FWTA, ed. 2, 3: 144.

A delicate herbaceous plant with fine grass-like leaves and slender inflorescences to 60 cm tall; of rocky crevices and thin soil situations in the uplands of Mali and N Nigeria.

The flowers are pink with yellow inside, attractive and worthy of cultivation (1).

Reference:

1. Morton, 1961.

Gladiolus sp. indet.

West African: GUINEA BASARI a-tekarabɔ FG&G a-tekéréɓ (FG&G)

Lapeirousia erythrantha (Kl.) Back.

FWTA, ed. 2, 3: 141, as *L. rhodesiana* NE Br.
West African: NIGERIA HAUSA luwon (RES)

An herbaceous plant to 15 cm high from a small woody rootstock; flowers purple; of shallow soils on rock pans in montane locations of N Nigeria, and also in Tanganyika and Zimbabwe.

No usage is reported for this species, but related species have medicinal use in southern Africa (1).

Reference:

1. Watt & Breyer-Brandwijk, 1962: 510, cf. *Lapeirousia coerulea* Schinz and *L. grandiflora* Bak.

Moraea schimperi (Hochst.) Pichi-Serm.

FWTA, ed. 2, 3: 138.

An herbaceous plant 20–40 cm high from a corm about 2 cm diameter; flowers purple-blue; of montane locations in N Nigeria and W Cameroons, and widespread in other upland regions of Africa from Ethiopia, E and S central Africa to Angola.

Some *Moraea* species, principally *M. polyanthos* Linn.f. and *M. polystachya* Ker., are highly toxic. The two named cause an affliction called 'tulp poisoning' in sheep, goats, horses and mules in S Africa, symptomised by weakness, stupor, lacrymation, suffusion of the eyelids and sinking of the eyeballs, laboured breathing and accelerated pulse (2). The status of *M. schimperi* in the Njombe District of Tanganyika has been reported on (1), but whether or not it is poisonous is not known. In the Region this plant should, however, be regarded as being toxic until definitely proved otherwise.

References:

1. Gillett 17781, K. 2. Watt & Breyer-Brandwijk, 1962: 505, as *Morea*.

Neomarica caerulea (Lodd.) Sprague

FWTA, ed. 2, 3: 138.

An herb to 60 cm tall with leaves over 1 m long by 4 cm wide, and flower-spikes to 6 cm tall with light blue/lilac flowers 10 cm across; native of Brazil, and introduced and naturalised in the Region.

The plant is attractive and is grown as an ornamental.

IRIDACEAE

Neomarica gracilis (Herb.) Sprague

FWTA, ed. 2, 3: 138.

An erect or reclining herb to 60 cm high, leaves to 80 cm long by 2½ cm wide; flowers: sepals white with green/yellow/brown markings and petals yellow shading to blue, and with a faint smell of violets; native of the American tropics, Mexico to Brazil, and now naturalised in W Africa.

It is cultivated as an ornamental in Sierra Leone. The mature plant is floriferous, flowering continuously with 1-day flowers opening at dawn and fading before dusk (1).

Reference:

1. Deighton 4697, 5328, K.

Trimezia martinicensis (Jacq.) Herb.

FWTA, ed. 2, 3: 138.
English: dragon's blood (Trinidad, Wong).

An herbaceous plant with flowering stems to 80 cm tall bearing yellow flowers about 2 cm across with purple-brown spots at the base; native of W Indies and S America, and now introduced to many tropical countries and into temperate countries under greenhouse cultivation. It is naturalised in W Africa.

No usage is recorded in the Region. In Trinidad an infusion of the corm is drunk for flu, oliguria and amenorrhoea (1). A resin, a fatty oil, a volatile oil and astringent substances have been reported in the corm (1).

Reference:

1. Wong, 1976: 111.

Zygotritonia bongensis (Pax) Mildbr.

FWTA, ed. 2, 3: 144, as *Z. crocea* Stapf. UPWTA, ed. 1, 488, as *Z. crocea* Stapf.
West African: GUINEA MANDING-MANINKA nkoko diuru (Brossart fide JMD) NIGERIA FULA-FULFULDE (Nigeria) ndayewu, ndayewu huɗowu (JMD)

An erect herb with leaves to 2 cm wide, and leaves and inflorescence to 60 cm high; flowers yellow-green or brown reddish; on shallow soil over rock on the savanna in Guinea, Mali and N and S Nigeria, and in NE Africa.

The fruits and corms are eaten in Guinea in time of famine (1–3).
as *Z. crocea* Stapf.:

1. Dalziel, 1937: 488. 2. Irvine, 1952, a: 30. 3. Irvine, 1952 b: 16.

Zygotritonia praecox Stapf

FWTA, ed. 2, 3: 144. UPWTA, ed. 1, 488.
West African: SENEGAL BASARI otikan (JB) GUINEA BASARI ɔ-tíkàn (FG&G) MANDING-MANINKA nkoko diuru (Brossart fide JMD) NIGERIA FULA-FULFULDE (Nigeria) ndayewu, ndayewu huɗowu (JMD)

A slender herb with narrow leaves and inflorescence bearing pink or white flowers, to 15 cm high from a corm about 2½ cm diameter; of alluvial soil in wet sites; in Senegal, The Gambia, Guinea, Mali and N Nigeria.

Tenda of SE Senegal pound the corm to produce a sort of cake for eating in time of famine (1).

Reference:

1. Ferry & al., 1974: sp. no. 142.

IXONANTHACEAE

Desbordesia glaucescens (Engl.) van Tiegh.

FWTA, ed. 2, 1: 693–4. UPWTA, ed. 1, 311.

West African: NIGERIA BOKYI kawo (Catterall fide JMD) IJO-IZON bou ogbóín *perhaps correctly* Irvingia gabonensis (KW) ogbóín (KW) WEST CAMEROONS DUALA bwiba njoc (JMD)

A large tree attaining 45 m high, with a long straight bole, to 3.75 m diameter, high thin buttresses, of the rain-forest area of S Nigeria and W Cameroons, and extending to Gabon, Congo (Brazzaville) and Cabinda.

The wood is pinkish-white to yellowish, and very hard. No usage is recorded in the Region, but in Gabon it comes in for construction-work (5, 6).

A bark-decoction is administered for stomach-ache in Congo, where this is held also to be aphrodisiac. The bark is made into an ointment with palm-oil for use in chicken-pox, and for application to the temples in headache (1).

Tannin has been reported in the bark and roots, and some *saponin* in the latter (2).

The fruits are winged with one or two seeds containing an oily edible kernel which is used in Gabon after crushing in a mortar to make a sort of *dika bread* (3, 4, 6).

References:

1. Bouquet, 1969: 140. 2. Bouquet, 1972: 29. 3. Busson, 1965: 331. 4. Dalziel, 1937: 311. 5. Keay & al., 1964: 247–8. 6. Walker & Sillans, 1961: 207, as *D. oblonga* A. Chev.

Irvingia gabonensis (Aubry-Lecomte) Baill.

FWTA, ed. 2, 1: 693. UPWTA, ed. 1, 312–3.

English: native mango; wild mango; bush mango; African mango; dika nut tree; dika bread tree.

West African: SENEGAL CRIOULO mãngo bravo (JB) DIOLA bah é nab, ku èl é kobol (JB) DIOLA-FLUP bu diay kola, hu lukutu, sosuso (JB) MANDING-MANINKA ko sosuké (JB) GUINEA-BISSAU CRIOULO mango bravo (JDES) NALU *n*corobaque (JDES) uncorobaque (JDES) GUINEA LOMA g'ba (FB) SIERRA LEONE KONO *g*bele (FCD; S&F) mbei (S&F) *k*pele (FCD; S&F) KRIO tolah (Pyne) LOKO kɛɛga (FCD; S&F) MENDE bɔbɔ (*def.* -i) (FCD; S&F) SUSU tolah (Pyne) TEMNE *an*-gbere (FCD; S&F) LIBERIA DAN kpeh (GK) KRU-GUERE (Krahn) *g*belle-tu (GK) IVORY COAST ABE boboru (auctt.) pobolu (Aub.; FB) AKYE be *small-leaved form, fruit edible* (auctt.) losioko *large-leaved form, fruit inedible* (Aub.) DAN gô *large-leaved form, fruit inedible* (Aub.) kpé *small-leaved form, fruit edible* (Aub.) FULA-FULFULDE (Ivory Coast) asro (Aub.) KRU-BETE sakosu (Aub.) GUERE kplétu *small-leaved form, fruit edible* (Aub.) zran mana *large-leaved form, fruit inedible* (Aub.) GUERE (Wobe) poletu (A.Chev.; FB) KWENI kakuru (Aub.) kalo (Aub.) KYAMA akuhia (A.Chev.) brêtié (Aub.) NZEMA wanini (A.Chev.) GHANA AKAN-ASANTE abisebuo (CJT; E&A) TWI abesebuo (FD; E&A) bisebuo (CJT; E&A) WASA bisebuo (CJT; E&A) ANYI-AOWIN ehyiriwa (FRI) SEHWI abesẽbuo (FRI) ehyiriwa (FRI) NZEMA abesẽbuo (FRI) bisebuo (CJT) DAHOMEY YORUBA-NAGO oro (Aub.) NIGERIA ? ujio (LC) ABUA ọmeh (DRR; JMD) ANAANG ùyó (KW) ùyó ùmánì 'opobo *the papaya-like fruit whose fibres stick in the teeth, and which remains green when ripe* (KW) BOKYI bojep (auctt.) dzhep (Osang fide RB) EDO ogwi, ogwe *the tree* (auctt.) ọhẹrẹ *the kernel* (JMD) ọkẹrẹ *the kernel* (JMD) okeri (LC) okheri *the nut* (DRR) EFIK ùyó (auctt.) EJAGHAM osing (auctt.) EJAGHAM (Keaka) nsing (DRR) oshing (DRR) EJAGHAM-ETUNG nsing (DRR) EKPEYE igidi (DRR; JMD) ENGENNI ogweyi (DRR; JMD) HAUSA góóròn bírìì = *monkey's kola* (ZOG) góóròn ruwa = *water kola* (ZOG) hakokari (ZOG) IBIBIO ùyó (BAC) IDOMA ohupi (Odoh) IGALA okru, oro, ọ̀rọ̀ aikpélẹ̀ (Odoh; RB) IGBO abwọno *the seeds* (JMD; KW) àgbòlò *the seeds* (BNO; KW) àgbǫ̀nǫ̀ the seeds (JMD; KW) àgbǫ̀nǫ̀ the seed (LC) obono the seeds (KO&S; KW) ǫ̀gbǫ̀nǫ̀, ǫ̀gbǫ̀lǫ̀ the seeds (KW) ugilĩ the tree (JMD; KW) ugirĩ the tree (auctt.) ugiri àgbǫ̀lǫ̀ the tree (BNO; KW) ujirĩ the tree (JMD; KW) ukwukwa (Iwu) IGBO (Amakalu) ǫkpǫpa (DRR; JMD) IGBO (Arochukwu) ujirĩ (DRR) IGBO (Awka) àgbònò the fruit (DRR) ugirĩ the tree(DRR) IGBO (Nkalagu) ǫkpopa (DRR; JMD) IGBO (Okigwe) ugirĩ (DRR) IGBO (Okpanam) àgbǫ̀nà (NWT) IGBO (Owerri) opopa (auctt.) IGBO (Uburubu) àgbǫ̀nǫ̀ the seed (NWT) IGBO (Umuakpo) ugirĩ (DRR) IGBO-IKWERE (Isiokpo) igiri-ǫhĩa = bush igiri (DRR; JMD) UKWUANI ujirĩ (auctt.) IJO-IZON (Kolokuma) bou ogbóin = bush mango (KW&T) ìzǫ̀n ógbóin = the people's (Izon) mango (KW&T) ogboin (auctt.) KAKIBA (Kalabari) ndisok (LC) LOKE keyamma (Berry & Guy) NKEM iriring (DRR) NUPE pekpeara (auctt.) OKOBO ogirĩ (DRR; JMD) OLULUMO (Okuni) kokunor (DRR) PERO dúupè said to have fruit like a mango - probably this sp. but perhaps Pentadesma butyracea (Guttiferae) (ZF) YALA oropa (auctt.) YALA (Ikom) ò̀rǫ̂kpaá (Ogo fide RB) YALA (Ogoja) ǫ̀lǫ̂kpá (Onoh fide RB) YORUBA àà̀pǫ̀n the fruit (JMD; Verger) òro the tree (auctt.) WEST CAMEROONS BAFOK bope (JMD) ·BUIVA· bakwiri (DRR) DUALA bwiba (JMD) bwĩbà ba mbalὲ (JMD; Ittmann) bwĩbà ba njǫ̀ù (Ittmann) KPE bwiwa (JMD) KUNDU wewe (JMD) LONG bopek (JMD) NYANGA nsinga (DRR)

A large tree reaching 35 m height in the western part of the Region, but less in the east, high buttresses, sometimes to 6 m, straight bole to 1 m diameter, less in the East, slightly fluted, carrying a dense compact crown, of the evergreen dense rain-forest from Senegal to W Cameroons, and widely dispersed from Sudan to Uganda and Angola (15, 16, 20–22).

Forms are recognised by Africans. In Liberia there are two: one small-leaved with inedible seeds, and the other larger-leafed and edible seeds (22). In Nigeria var. gabonensis has sweet edible fruit-pulp, and var. excelsa (Mildbr.) Okafar has bitter inedible pulp but is slimy and is added to soup for this quality (17). In Lower Dahomey a variety with a thick edible pulp is cultivated (12, 14). In Ivory Coast large-leaved and small-leaved forms are given separate vernacular names, the former being considered inedible, which differences on taxonomic standards, however, appear untenable (3). Nevertheless, the distinction between edibility and inedibility is of basic pragmatic importance, and this and other distinctions recognised by countrymen in the field merit serious study to gain a proper understanding. In Igalaland (Nigeria), as no doubt elsewhere, it is considered one of the most important trees of the bush, and individual Igala lay claim to the produce of a tree by farming around it or by clearing the land beneath (26).

Sap-wood is light brown, and heart-wood a slightly darker or greenish-brown (11, 15, 16, 20, 21). The wood is very hard, and not easy to cut limiting its usefulness where only simple implements are available. It is very heavy and durable, but its weight is said to preclude it from all but the most rugged construction-work, e.g., for railway-ties, etc. (8, 22). It is immune to termite attack and is used for house-building. Canoes can be made from the trunk, and pestles for yam-mortars. There is a fine moderately close grain and a good polished finish can be achieved. It is suitable for boards, planking, ships' decking, paving blocks, and the like (11-13).

Tests for paper manufacture have shown cellulose content 48.8%, fibre length 1.5 mm, and the resultant dark brown paper to be inferior, rather weak and soft, and not bleachable (8, 10).

The bark-slash is brown with lighter brown to orange-yellow stripes or spots. A small amount of clear sap is expissated (20). This is said to be sweet (21) though the bark itself is bitter and has the usual usages of bitter barks: in Sierra Leone it is ground up with water for rubbing on to the body for pains (12); in Ghana it is used in an enema for an unspecified purpose (15); in Gabon it is added to palm-wine to increase potency (12), and scrapings are taken in a baked banana or prepared in enemas to relieve diarrhoea and dysentery (24); in Congo (Brazzaville) the bark has use in mouth-washes for toothache, pulped on sores and wounds, internally as a purgative for gastro-intestinal and liver conditions, for sterility, hernias and urethral discharge, and is considered by some to be a powerful aphrodisiac and to be beneficial in cases of senility (4).

Tannin has been reported present in both the bark and the roots (5), also a strong presence of *alkaloid* in the bark, though none in the roots (1). Root and wood extracts have proved ineffective in avian malaria (9), though Igbo use a leaf-decoction as a febrifuge (28). A wax has been extracted from the plant which has been found useful as an adjunct in making medicinal tablets (27).

As already indicated above, the fruit is variable, with special forms. It is the most important part of the tree. It is a drupe, resembling a mango, with a fibrous pulp surrounding the hard-shelled nut. The pulp of some trees is edible with a turpentine flavour, and of others inedible, bitter and acrid. The edible ones are a source of vitamins (18). In the forest animals readily search out the fruit for the pulp, but the seed is protected by its hard endocarp (21). The pulp is used at Lagos to prepare a black dye for cloth (21).

The kernel is an important source of vegetable oil. In season the fallen fruits are collected in the forest and stacked till the pulp has rotted away. The nuts are opened and the cotyledons removed and dried. The cotyledons are a common item of market produce and are used in soups and as a food flavouring. They are said to have a pleasant taste with a lingering slight bitterness. They are rich in oil, but there is a wide variation in quantity and composition; even so they are considered a suitable source of industrial and edible oils. Total fat content has been recorded as 54–68% and a series of five analyses from Nigeria (in brackets) and central Africa gave the following assays: *lauric acid* 0–39 (39) %; *myristic* 31–69 (51) %; *oleic* 0–22 (10) %; *stearic* 0–4 (0) % and *palmitic* 0–65 (0) % (2, 7). The principal domestic use is for the preparation of *odika*, or *dika bread*, also known as *Gabon chocolate*. For this the cotyledons are ground and heated in a pot, lined with banana leaves, to melt the fat, and then left to cool. The resultant grey-brown greasy mass is dika bread. It has a slightly bitter and astringent taste with a more or less aromatic odour. Pepper and other spices may be added, and it may perhaps be subjected to woodsmoke. The end product may be made up into cylindrical packets wrapped in a basket-like or leaf-wrapping. It can be kept for a long time without going off and it is used as a food-seasoner (3, 6, 12, 14, 15).

An alternative method of preparation, more akin to the making of vegetable butters, is to take the fresh or stored cotyledons and pound them into a paste. This can be done in quantities according to the immediate requirement. A third preparation, known in Gabon as *ovéke*, is to soak the kernels for 15–20 days till soft and then to knead them by hand into a cheese-like paste (3, 7, 12, 14, 15, 23, 24). A fourth practice is known in Sierra Leone, in which the cotyledons are dried and ground to a brown 'flour' in which form it can be stored for use as an additive to food as and when required (19). The crude *dika* paste yields on heating or boiling 70–80% of a pale yellow or nearly white solid fat, *dika butter*, which has qualities comparable with cacao-butter, and is, in fact, a possible adulterant or substitute for the latter in chocolate manufacture. Freed from its slight odour it can also be regarded as suitable for margarine manufacture. It is also suitable for soap-making (12).

Following processes in which the fat is removed the residual cake, rich in protein, is a cattle feed-stuff similar to copra cake. The following composition is recorded, crude protein 31%, fat 10%, carbohydrate 39%, fibre 3% and ash (minerals) 6% (25).

In Nigeria and Cameroon the split shells of the fruit are used in divination – if one falls convex side up and the other in reverse, the omens are good (12).

References:

1. Adegoke & al., 1968: 13–33. 2. Adriaens, 1931: 234–8. 3. Aubréville, 1959: 2: 122. 4. Bouquet, 1969: 140. 5. Bouquet, 1972: 30. 6. Bouquet & Debray, 1974: 97. 7. Busson, 1965: 33; seed analyses, 333–5. 8. Chalk & al., 1933: 91. 9. Claude & al., 1947: 145–74. 10. Coomber, 1952–3: 13–27. 11. Cooper & Record, 1931: 88. 12. Dalziel, 1937: 312–3. 13. Deighton 2606, K. 14. Irvine, 1948: 231, 265. 15. Irvine, 1961: 506–7. 16. Keay & al., 1964: 246. 17. Okafor, 1975: 211–21. 18. Okiy, 1960: 121. 19. Pyne, 102, K. 20. Savill & Fox, 1967: 141. 21. Taylor, CJ, 1960: 328. 22. Voorhoeve, 1965: 357. 23. Walker, 1930: 209–18. 24. Walker & Sillans, 1961: 207. 25. Watt & Breyer-Brandwijk, 1962: 941. 26. Blench, 1981–82. 27. Iwu, 1984: citing Udeala & al., Pharm. Pharmacol. 32 (1980). 28. Iwu, 1986: 141.

Irvingia grandifolia (Engl.) Engl.

FWTA, ed. 2, 1: 693. UPWTA, ed. 1, 314, as *Klainedoxa grandifolia* Engl.
French: manguier sauvage; dika or udika (from a Gabonese name).
West African: NIGERIA EDO agbara (Kennedy; JMD) akhuekhue (Kennedy; JMD) IGBO
(Awka) akpųlų (Kennedy; JMD) IJO-IZON ogbo-ako (Kennedy; JMD) YORUBA apepere (Ken-
nedy; JMD) epologun (Kennedy; JMD) YORUBA (Egba) ifainaki (JMD) YORUBA (Ijebu)
karakoro (JMD) odudu (JMD) YORUBA (Ondo) alukan raba (JMD) odudu (JMD) WEST
CAMEROONS DUALA ngondo (JMD)

A large tree to 45 m high, buttressed to 3 m height, bole straight and fluted, to
4.75 m diameter, deciduous with the leaves turning bright red before being shed,
an unusual character found in only a few trees of the rain-forest, occuring in S
Nigeria and W Cameroons, and extending to Zaïre and Angola.

The wood is white (6) and very hard (5). In W Cameroons it is said to have no
value as timber (1).

The bark has a number of medicinal uses in Congo (Brazzaville): a decoction
is used for bathing children with fever and as an eyewash for ophthalmias; a
decoction is taken internally for stomach and kidney-complaints and to treat
menstrual and vaginal affections; and an ointment is made of the crushed bark
with palm-oil for topical application in muscular pain, arthritis, rheumatism,
sprains, fractures, oedemas, etc. (2). Some *saponin* and *tannin* has been found in
the bark and roots (3).

Extracts from the stems have been examined for activity against avian
malaria with negative result (4).

References:

1. Aninze, FHI.24720, K. 2. Bouquet, 1969: 140. 3. Bouquet, 1972: 30. 4. Claude & al., 1947: 145–
74, as *Klainedoxa grandiflora*. 5. Keay & al., 1964: 247. 6. Louis 227, K.

Irvingia smithii Hook.f.

FWTA, ed. 2, 1: 693. UPWTA, ed. 1, 313.
West African: NIGERIA GWARI shihi (Edgar; JMD) HAUSA góóròn rúwá = *waterside
kola* (JMD; ZOG) góròn bíríì = *monkey's kola* (JMD; ZOG) KAMBARI-KIMBA màshókò màmìnì
(RB) SALKA mèdèlè màmìnì (RB) NUPE suchi (Yates; JMD)

A tree to about 28 m high with trunk to nearly 1 m diameter, of river-banks,
especially in the savanna in Nigeria and extended across Africa to Sudan,
Congo (Brazzaville), Zaïre and Angola.

Sap-wood is yellowish-white and heart-wood reddish-brown, very hard and
durable (1, 2, 7).

The bark is used in Congo (Brazzaville) in decoction for dysentery (3).
Examination of the roots for action against avian malaria has given negative
result (5). *Saponin* and *tannin* are reported present in the bark and roots (4).

The fruit is bright red and elongated up to 5 cm long by 0.8 cm wide. The pulp
is eaten in Sudan (8) though another report (2) states the fruit to be of no value.
The kernel is eaten by Salka in NW Nigeria (12), and in Ubangui (6) and in
Zaïre (9). It is rich in fatty matter. Zaïrean material has been reported that the
cotyledon represents 15% of the seed, and is itself 75% fatty matter. The
vegetable butter constitutes *dika* of good quality and is of the same characteristics as that from *I. gabonensis* (1).

The plant is said to be a source of an essential oil used locally in perfumery
(10, 11), but no further detail is given.

References:

1. Adriaens, 1931: 234–8. 2. Broun & Massey, 1929: 227–8. 3. Bouquet, 1969: 141. 4. Bouquet, 1972: 30. 5. Claude & al., 1947. 6. Dalziel, 1937: 313. 7. Keay & al., 1964: 246–7. 8. Myers, 9183, K. 9. Shanz & al., 644, K. 10. Uphof, 1968: 285. 11. Usher, 1974: 322. 12. Blench, 1985.

Irvingia sp. indet.

West African: GHANA AKAN-FANTE ɔkur (FRI)

An unidentified tree species of *Irvingia* is recorded from the closed-forest in Ghana to yield a copious quantity of white latex which turns red on exposure, and which has been an adulterant of superior rubbers. The timber provides firewood, and the bark enters into local medicine.

References:

1. Burtt Davy & Hoyle, 1937: sp. no. 134. 2. Irvine, 1930: 242. 3. Irvine, 1961: 507.

Klainedoxa gabonensis Pierre

FWTA, ed. 2, 1: 693. UPWTA, ed. 1, 314.

Trade: eveuss, kroma (Liberia, Voorhoeve, Kunkel).
West African: SENEGAL DIOLA ku el é kobol (JB) GUINEA-BISSAU BIAFADA bissám-bana (JDES) NALU nbambete (JDES) umbambete (JDES) SUSU cossòssúquè (JDES) SIERRA LEONE KISSI silɔ-lando (FCD; S&F) MENDE g-bewɔ (def. -ei) (auctt.) g-biwɔ (FCD) kpenye (def. -i) (Fox) SUSU kofuri (FCD) kopuri (FCD) TEMNE ka-kupər (FCD; S&F) LIBERIA DAN goh (GK) gooh (ACV) KRU-BASA goe (auctt.) GUERE (Krahn) dalu (GK) IVORY COAST ABE akuabo (FB) akwabo (A.Chev.) aquabo (Aub.; RS) AKYE akpabeu (A&AA) akwabo (A.Chev.) ANYI kroma (auctt.) DAN bloseri (Aub.; RS) 'KRU' tobo (Aub.; RS) KRU-GUERE bolobaille (Aub.) GUERE (Wobe) blotué (auctt.) KYAMA adionkué (Aub.; RS) adiumkué (A.Chev.) NZEMA kroma (auctt.) 'SOUBRE' botu (Aub.; RS) GHANA VULGAR kroma (FD) AKAN-ASANTE bisiabo (auctt.) kokodebu (Soward; FRI) koroma (auctt.) kroma (E&A) kroma (E&A) FANTE koko-debu (BD&H; FRI) TWI kroma (auctt.) WASA bisiabo (auctt.) koroma (auctt.) ANYI kroma (BD&H; FRI) NZEMA kroma (FRI; CJT) NIGERIA BOKYI kechi-kelim (Catterall; JMD) kelim-kelim (KO&S) EDO ighozo (Lowe) oguegodin (auctt.) MBEMBE otukpo (DRR) YORUBA epologun *for trees without spines* (auctt.) odudu (auctt.) YORUBA (Egba) ifainaki (JMD) YORUBA (Ijebu) karaboro (Sankey; JMD) odudu (Sankey; JMD) YORUBA (Ondo) alukan raba (Sankey; JMD) odudu (Sankey; JMD) WEST CAMEROONS DUALA ngondo (Hédin fide JMD)

A large tree reaching 40 m height, trunk to over 1.20 m diameter, buttressed, bole straight, cylindrical to 25 m long, with open spreading evergreen crown, one of the largest trees of the humid rain-forest; from Senegal to W Cameroons, and extending to Uganda and Tanganyika. Several species and a number of varieties have been recognised in what is a variable aggregate of which *K. gabonensis* var. *oblongifolia* Engl. is presently held to be the regional representative. A revision of the genus is necessary.

The sap-wood is thin, light brown, and though hard is liable to borer attack (14, 15). The heart-wood is attractively coloured reddish to golden brown with wide dark veining, and often with zigzag markings. It is extremely hard and one of the heaviest of West African timbers. It is durable in contact with the ground and poles are often used in hut-construction. The wood, however, is too hard for cutting with domestic implements, so that its local usages are very limited. Where forest is being cleared for agriculture the tree is usually left standing and becomes a conspicuous feature of the landscape. Its presence in secondary vegetation is a fair indication of previous occupation. With specialist equipment the timber can be worked and use as railway-sleepers and for heavy construction projects is suggested. It can be made into attractive furniture. In primitive dwellings in Sierra Leone, the buttresses are sometimes used to make doors (14).

The wood is deemed a good firewood. The branches and the stems of young trees are flexible and are used to make spring traps (3, 8–11, 14–16).

The bark-slash exudes a watery clear or honey-coloured sap with a musky smell (12, 14, 16). The cut surface, at first white, becomes purplish on exposure (15). The bark displays analgesic properties. It is ground to a fine powder in Sierra Leone, then mixed with clay and water for rubbing onto rheumaticy joints (14). In Gabon the bark is used for rheumatic pains said to precede yaws (18), and in Ubangi a patient suffering from lumbago has the affected area exposed to bark smoke (Vergiat fide 13, 19). In Congo (Brazzaville) the pulped-up bark is prepared as an ointment with palm-oil for application to areas of rheumatism, and in that country too a bark-decoction is put in baths and lotions for buccal infections, small-pox and chickenpox, and a preparation is taken by mouth for venereal disease, sterility and impotence (4).

The roots have been examined for activity against avian malaria, with a negative result (7).

Tannin is reported present in both the bark and the root (5).

At the end of the rainy season (about October) the tree puts on a spectacular flush of brilliant red new leaf. Flowering takes place at the same time and the crown is covered with a purplish hue (16). The young leaves are characterised by the presence of long linear stipules which are soon shed, though on some trees they are lacking, a character recognised by the Yoruba who have a separate name for the stipule-less form (see vernacular name list above). The young stipuled leaves are eaten with palm kernels by the Akye of Ivory Coast as an aphrodisiac (2). In Gabon they are also held to be aphrodisiac (17, 18). A trace of *alkaloid* is reported present in the leaves (1). In Congo they are eaten with vegetables, oil, salt, fish or meat as a stomachic, and are used as an analgesic for topical or internal application in various cases of pain (4).

The fruit, 5–8 cm long, is usually 5-angled, each angle with a hard woody nut, and the whole in a fibrous covering containing some gummy substance (15). Elephants in Liberia are said to eat the fruit (12) presumably for the nutritious kernels of the seeds. The kernels are much relished for human use. They are eaten fresh or roasted, or crushed to a paste, and enter into cooking in numerous ways. The kernel prepared as a vegetable butter is a substitute for shea (karite) butter (*Vitellaria paradoxa* Gaertn.f, Sapotaceae). The seed contains about 65% oil consisting, on a dry weight basis, of *myristic acid* 42%, *oleic acid* 22%, *lauric acid* 16%, *palmitic acid* 11%, and others (6).

The kernel of 'var. *microphylla*', is used in Gabon in washes and frictions on furuncles (18).

References:

1. Adegoke & al., 1968: 13–33. 2. Adjanohoun & Aké Assi, 1972: 153. 3. Aubréville, 1959: 2: 121. 4. Bouquet, 1969: 141. 5. Bouquet, 1972: 30. 6. Busson, 1965: 333–5, with kernel analysis. 7. Claude & al., 1947. 8. Cooper & Record, 1931: 88, with timber characters. 9. Dalziel, 1937: 314. 10. Irvine, 1961: 507–8. 11. Keay & al., 1964: 245. 12. Kunkel, 1965: 132. 13. Portères, s.d. 14. Savill & Fox, 1967: 142–3. 15. Taylor, CJ, 1960: 329. 16. Voorhoeve, 1965: 354. 17. Walker, 1953,b: 278. 18. Walker & Sillans, 1961: 209. 19. Vergiat, 1970,c: 326–7.

Phyllocosmus africanus (Hook.f.) Klotz.

FWTA, ed. 2, 1: 355, as *Ochthocosmus africanus* Hook.f. UPWTA, ed. 1, 38, as *O. africanus* Hook.f.

West African: **GUINEA-BISSAU** BALANTA *n*boi (JDES; EPdS) **SIERRA LEONE** KISSI tundui-balo (FCD; S&F) KONO tɔwanɛ (S&F) tuafa (FCD; S&F) KORANKO buwulekoloma (S&F) LIMBA sasakate (NWT) LOKO fɛgurugooŋgo (FCD; S&F) mahoŋgi (NWT) umowainana (NWT) MENDE tɔwanyɛi = *pumpkin of the fish* (S&F) twanyɛ (auctt.) TEMNE *ka*-thɔŋai (FCD; S&F) **IVORY COAST** ABE abrahassa (A.Chev.; Aub.) AKYE bokubi (A.Chev.) **GHANA** VULGAR akokorabeditoa (DF) **NIGERIA** IGBO araba-uji (KO&S)

A shrub or medium-sized tree of the lower forest canopy to 20 m high by 2 m girth, occurring from Guinea-Bissau to W Cameroons, and on to Zaïre.

In areas under managed forestry the tree is considered a weed species retarding the growth of more desirable species (7). The poles are commonly used for hut-posts (4), an use which may account for one of the Mende names, though there is ambiguity of orthography and hence of the derivation. The stems are pliable and are used in Sierra Leone to make traps (7). The sap-wood is creamy-white (5) and the heart-wood very hard, close-grained and red (2, 4, 6) though in Zaïre it is said to be yellowish-brown, and good for carpentry (9). It is a popular firewood in Sierra Leone and converts into good charcoal (7) favoured by blacksmiths in Ivory Coast (2).

In Zaïre the leaves are attacked by an unspecified insect which is said to be eatable (9).

Some *tannins, steroids* and *terpenes* have been recorded in the leaves, bark and roots (3).

The plant, presumably the seeds, is recognised in Congo (Brazzaville) as oil-bearing (8). The fruit in Sudan is deemed edible (1).

References:

1. Andrews, A.1541, K. 2. Aubréville, 1959: 1: 364. 3. Bouquet, 1972: 31. 4. Dalziel, 1937: 38. 5. Irvine, 1961: 206. 6. Keay & al., 1960: 239. 7. Savill & Fox, 1967: 144. 8. Sillans, 1953,a: 92. 9. Wilczek & Boutique, 1958: 36.

REFERENCES

ADAM, J.G., 1954: Note sur les graminées fourragéres de la presqu'île du Cap-Vert (Sénégal), in *Revue d'Élevage et de Médecin Vétérinaire des Pays tropicaux*, Vigot Frères, Editeurs, Paris.
— 1960, a: Quelques plantes adventrices des rizières de Richard-Toll, *Bull. Inst. Franç. Afr. Noire*, A. 22: 361–84.
— 1965: La végétation du delta du Sénégal en Mauritanie. (Le cordon littoral et l'île de Thiong.), *loc. cit.* 27: 121–38.
— 1966, a: Les pâturages naturels et postculturaux du Sénégal, *Bull. Inst. fond. Afr. Noire*, A. 28: 450–537.
— 1966, b: Composition chemique de quelques herbes mauritaniennes pour dromadaires, *J. Agr. trop. Bot. appl.* 13: 337–42.
— 1968: Flore et végétation de la lisière de la forêt dense en Guinée, *Bull. Inst. fond. Afr. Noire*, A. 30: 920–52.
— 1971: Quelques utilisations de plantes par les Manon du Libéria (Monts Nimba), *J. Agr. trop. Bot. appl.* 18: 372–8.
— 1972: Un Panicum américain colonise l'Ouest Africain. Panicum laxum Sw., *loc. cit.* 19: 59–62 (with chemical analyses).
ADAM, J.G., N. ECHARD & M. LESCOT, 1972: Plantes médicinales Hausa de l'Ader (République du Niger), *Agr. trop. Bot. appl.*
ADAMS, R.F.G., 1943: Efik vocabulary of living things, I. *Nigerian Field* 11.
— 1947: *op. cit.*, II, *loc. cit.* 12: 23–34.
ADANDÉ, A., 1953: Le Maïs et ses usages dans le Bas-Dahomey, *Bull. Franç, Afr. Noire*, 15: 220–82.
ADEGOKE, E.A., A. AKISANYA & S.H.Z. NAQVI, 1968: Studies of Nigerian medicinal plants, I: A preliminary survey of plant alkaloids, *J.W. Afr. Sci. Ass.* 13: 13–33.
ADEWUNMI, C.O. & E.A. SOFOWORA, 1980: Preliminary screening of some plant extracts for molluscicidal activity, *Pl. Med.* 39: 57–65.
ADJANOHOUN, E. & L. AKÉ ASSI, 1972[?]: *Plantes pharmaceutiques de Côte d'Ivoire*, Abidjan, Ivory Coast [mimeo].
ADRIAENS, L., 1931: Quelques oléagineux du Congo-belge, *Ann. Soc. Scient. Brux.* 51: 228–46.
— 1946: *Recherches sur la composition chimique des Flacourtiacées a huile chaulmoogrique du Congo Belge*, Brussels.
AHMED, EL H.M., A.K. BASHIR & Y.M. EL KHEIR, 1984: Investigations of Molluscicidal Activity of certain Sudanese Plants used in folk-medicine, IV, *Pl. Med.* 50: 74–77.
AINSLIE, J.R., 1937: A list of plants used in native medicine in Nigeria, *Imp. Forest, Inst.*, Oxford, Inst. Paper No. 7 [mimeo].
AKINNIYI, J.A. & M.U.S. SULTANBAWA, 1983: A glossary of Kanuri names of plants, with botanical names, distribution and uses, *Ann. Borno*, 1.
AKIWUMI, F.A., 1986: *pers. comm.*, 22/5/86.
AKUBUE, P.I. & G.C. MITTAL, 1982: Clinical evaluation of a traditional herbal practice in Nigeria: a preliminary report, *J. Ethno-Pharmacol.* 6: 355–9.
ALLISON, P., 1969: Collecting Nigerian Antiquities, *Nigerian Field* 34: 99–114.
AMIN, M.A., A.A. DAFFALA & O.A.EL MONEIM, 1972: Preliminary report on the mollusci-cidal properties of habat el-mollok, Jatropha sp., *Trans. R. Soc. Trop. Med. Hyg.* 66: 805.
AMPOFO, OKU, 1983: *First Aid in Plant Medicine*, Ghana Rural Reconstruction Movement, Mampong-Akwapim.
ANDERSON, D.E., 1961: Taxonomy and Distribution of the Genus Phalaris, *Iowa St. Coll. J. Sci.* 36: 1–96.
ANON., 1913: *E. Afr. Standard*, 8/11/1913.
ANON., 1924, b: Report on Brachiaria brizantha and Panicum trichocladum hay from Uganda, *Ann. Rep., Dept. Agr., Uganda.*
ANON., 1930: Feeding value of Para rubber seed-meal, *Bull. Imp. Inst.* 28: 459–60.
ANON., 1936: Composition of some African Foods and Feeding-stuffs mainly of vegetable origin, *Imp. Bur. Anim. Nutrit.*, Tech. Comm. 6.
ANON., 1939, a: Cassava, *Bull. Imp. Inst.* 37: 205.
ANON., 1940, a: Adlay, *loc. cit.*, 38: 182–4.
ANON., 1940, b: Sugar-cane Wax, *loc. cit.* 40: 11–12.
ANON., 1943, c: Sugar-cane Wax, *loc. cit.* 41: 86–90.
ANON., 1944, a: Euphorbia tirucalli resin from South Africa, *loc. cit.*, 42: 1–13.
ANON., 1971, a: L'opération 'Pan de Mil' prend l'Ampleur au Tchad. (La Mission de M. Henry Smeets, O.N.U.), *Trop. Sci.* 13: 215–7.

REFERENCES

ANON., 1973: A report on rice straw, *Daily Times (Lagos)*, 28/9/1973.

ANON., s.d., a: Use guide for Ghanian timbers, *Inf. Bull. No. 1, Forest Products Research Institute*, Kumasi, Ghana.

ANON., s.d., e: MS re Digitaria exilis, Herb. K.

ANON., s.d., g: Echinochloa, ad not., Herb. K.

APPA RAO, S., M.H. MENGESHA & D. SHARMA, 1985: Collection and Evaluation of Pearl Millet, *Pennisetum americanum* germplasm from Ghana, *Econ. Bot.* 39: 25–38.

APPIA, B., 1940: Superstitions Guinéennes et Sénégalaises, *Bull. Inst, Franç. Afr. Noire* 2: 358–95.

ARANA, E. & M. SWODESH, 1967: *Diccionario Analytico del Mampruli*, Museo de las Culturas, Mexico.

ARENAS, P. & R.M. AZORERO, 1977: Plants of common use in Paraguayan Folk Medicine for regulating Fertility, *Econ. Bot.* 31: 298–301.

ARORA, R.K., 1977: Job's tears (*Coix-lacryma-jobi*) – a minor food and fodder crop of Northeastern India, *Econ. Bot.* 31: 358–66.

AUBRÉVILLE, A., 1950: *Flore Forestière Soudano-guinéenne, A.O.F.-Cameroun – A.E.F.*, Société d'Editions Geographiques, Maritimes et Coloniales, Paris.

— 1959: *La Flore forestière de la Côte d'Ivoire*, Ed. 2, Vols. 3, Centre Technique Forestier tropicale, Nogent-sur-Marne.

BABAWA TUBRA (MALLAM), 1985: Ebina (Bàná) plant names and plant medicinal uses in Dumne Village, Adamawa, Nigeria, fide R. BLENCH, *pers. comm.*

BAILEY, L.H., 1901: *Cyclopedia of American Horticulture*, Macmillan, London.

BALLY, P.R.O., 1937: Native Medicinal and Poisonous Plants of East Africa, *Bull. Misc. Inf.* 1937: 10–26.

BAMPS, P., 1970: *Guttiferae*, in R. BOUTIQUE: *Flore du Congo, du Rwanda et du Burundi*, Brussels.

BARDIN, A., 1937, b: Le Gorli et les plantes antilepreuses en Côte d'Ivoire, *Ann. agric. Afr. occid. fr.* 1: 356–65.

BARTH, H., 1857–58: *Travels and Discoveries in North and Central Africa, being a journal of an expedition undertaken under the auspices of H.B.M.'s Government in the Years 1849–1855*, 5 vols., Ed. 2, Longman, Brown, Green, Longmans & Roberts, London.

BATTANDIER, J.A. & L. TRABUT, 1911: *Bull. Soc. bot. Franç.*, 58: 623–9, 669–77.

BAUMER, M.C., 1975: Catalogue des Plantes utiles du Kordofan (République du Soudan), particulièrement du point de vue pastoral, *J. Agr. trop. Bot. appl.* 22: 81–119.

BAYNARD, Lieut., 1947: Aspects principaux et consistance des Dunes (Mauritanie), *Bull. Inst. Franç. Afr. Noire*, 9: 1–17.

BEILLE, H., 1927: *Phyllanthus*, in M.H. LECOMTE, *Flore générale de l'Indo-Chine*, pp. 571–608, Masson & Cie, Paris.

BENNETT, H., 1950: Alchornea cordifolia leaves and bark from Nigeria, *Colon. Pl. Anim. Prod.* 1: 132–4.

BERHAUT, J., 1967: *Flore du Sénégal*, Ed. 2, Clairafrique, Dakar.

— 1975, a: *Flore illustrée du Sénégal, III, Connaracées à Euphorbiacées*, Dakar.

— 1975, b: *Flore illustrée du Sénégal, Dicotylédones*, Vol. IV – Ficoidées à Legumineuses, Dakar.

BERRY, E. & J. GUY, s.d. 1973[?]: *Reading and writing Yakurr*, S.I.L., Joss.

BERTHO, J. 1951, a: Quatre dialectes mandé du Nord-Dahomey et de la Nigeria anglaise, *Bull. Inst. Franç. Afr. Noire*, 13: 1265–71.

— 1951, b: La Place des dialectes Géré et Wobê par rapport aux autres dialectes de la Côte d'Ivoire, *op. cit.* 13: 1272–80.

— 1953: La Place des dialectes Dogon (dogõ) de la falaise de Bandiagara parmi les autres groupes linguistiques de la zone soudanaise, *op. cit.* 15: 405–41.

BÉZANGER-BEAUQUESNE, L., 1955: Contribution des Plantes à la défense de leurs semblages, *Bull. Soc. Bot. France* 102: 548–75.

BHANDARI, M.M., 1974: Famine Foods in the Rajasthan Desert, *Econ. Bot.* 28: 73–81.

BHIDE, N.K. & R.A. AIMEN, 1959: Pharmacology of a tranquilizing principle in *Paspalum scrobiculatum* grains, *Nature* 183: 1735–6.

BLASS, R. [Ed.], 1975: *Sisaala-English, English-Sisaala Dictionary*, Inst. Linguistics, Univ. Ghana.

BLENCH, R., 1981–1986: *pers comm.*

— 1985: *Field notes*, ined.

— 1992: *pers. comm.*, 10/8/1992.

BLENCH, R., K. Williamson & B. CONNELL, in press: The Diffusion of Maize in Nigeria: a historical and linguistic investigation, *Sprache und Geschichte in Afrika* (1994).

BOBOH, 1974: *Comm.*, 24/4/1974.

BOR, N.L., 1953: *Manual of Indian Forest Botany*, Oxford Univ. Press.

— 1960: *Grasses of Burma, Ceylon, India and Pakistan (excluding Bambuseae)*, Pergamon Press.

438

REFERENCES

BOUQUET, A., 1969: Féticheurs et Médecines traditionnelles du Congo (Brazzaville), *Mém. O.R.S.T.O.M.* 36.
— 1972: Plantes médicinales du Congo-Brazzaville: Uvariopsis, Pauridiantha, Diospyros, etc., *Trav. Doc. O.R.S.T.O.M.* 13.
BOUQUET, A. & M. DEBRAY, 1974: Plantes médicinales de la Côte d'Ivoire, *Trav. Doc. O.R.S.T.O.M.*, 32.
BOUQUIAUX, L, 1971–72: Les noms de plantes chez les Birom, *Afrika und Ubersee*, 55.
BOURKE, D. O'D, 1963: The West African Millet crop and its Improvement, *Sols Afr.* 8: 121–32.
BOURONVILLE, D. de, 1967: Contribution à l'étude du chimpanzé en République de Guinée, *Bull. Inst. Franç. Afr. Noire*, 29. A: 1188–1269.
BOURY, N'D.J., 1962: Végétaux utilisés dans la Médecine africaine, dans la région de Richard-Toll (Sénégal), in J.G.ADAM, Les Plantes utiles en Afrique occidentale, *Notes Afr.* 93: 14–16.
BOUTIQUE, R., 1960: Icacinaceae, in *Flore du Congo, du Rwanda et du Burundi*, 9: 237–78.
— 1972: Gentianaceae, in *Flore d'Afrique Centrale (Zaïre – Rwanda – Burundi)*.
BOWDEN, B.N., 1971: Studies on *Andropogon gayanus* Kunth. VI; The leaf nectaries of *Andropogon gayanus* var. *bisquamulatus* (Hochst.) Hack. (Gramineae), *J. Linn, Soc. (Bot.)* 64: 77–80.
BOYD, C.E. & P.S. MCGINTY, 1981: Percentage Digestible Dry Matter and Crude Protein in Dried Aquatic Weeds, *Econ. Bot.* 35: 296–9.
BRANDIS, D., 1911: *Indian Trees*, 3rd imp., London: Constable.
BRASSEUR, G., 1952: Un type d'habitat au Bas-Dahomey, *Bull. Inst. Franç. Afr. Noire* 14: 669–76.
BRAY, G.T. [G.T.B.], 1947: Oil of Tetracarpidium conophorum, *Bull. Imp. Inst.* 45: 131–3.
BROTHERTON, J.G.H., 1969: The Nomadic Fulani, *Nigerian Field* 34: 126–36.
BROUN, A.F. & R.E. MASSEY, 1929: *Flora of Sudan*, Sudan Govt. Office, London.
BROWN, E. & W.S.A. MATTHEWS, 1951: Notes on the aromatic grasses of commercial importance, *Col. Pl. Anim. Prod.* 2: 174–87.
BRUNKEN, J.N., 1977: A systematic study of Pennisetum, Sect. Pennisetum (Gramineae), *Am. J. Bot.* 64: 161–76.
BURKILL, I.H., 1935: *A dictionary of the economic products of the Malay Peninsula*, Crown Agents for the Colonies, London.
BURTT DAVY J. & A.C. HOYLE [Ed.], 1937: *Check-lists of the Forest Trees and Shrubs of the British Empire: No. 3, Draft of the first description Check-list of the Gold Coast*, Imp. For. Inst., Oxford.
BUSNEL, R.G., 1959: Étude d'un appeau acoustique pour la pêche, utilisé au Sénégal et au Niger, *Bull. Inst. Franç. Afr. Noire A.* 21: 346–60.
BUSSON, F., 1965: *Plantes alimentaires de l'Ouest africain*, Leconte, Marseille.
BUT, P.P-H, CHIA, L-C, FUNG, H-L, & HU S-Y, 1985: *Hong Kong Bamboos*, Hong Kong.
CALAME-GRIAULE, G., 1968: *Dictionnaire Dogon*, C. Klinksiek, Paris.
CAMUS, A. & P. VIGUIER, 1937: Riz flottants du Soudan, *Rev. Bot. appl. Agr. trop.* 187: 201–3.
CHADHA, Y.R., 1962: Sources of Starch in Commonwealth Countries, IV: Maize, *Trop. Sci.* 4: 13–37.
— [Ed.], 1972: *The Wealth of India; Raw materials*, 9(Rh–So), C.S.I.R., India, New Delhi.
— [Ed.], 1976, a: *op. cit.* 10 (Sp–W).
— [Ed.], 1985, *op. cit.* 1(A), revised edition.
CHALK, L., J. BURTT DAVY, H.E. DESCH & A.C. HOYLE, 1933: *Twenty West African Timber Trees*, Clarendon Press, Oxford.
CHAMPAULT, A., 1970: Étude caryosystématique et écologique de quelques Euphorbiacées herbacées et arbustives africaines, *Bull. Soc. Bot. France* 117: 137–68.
CHAPMAN, J.D., 1992: *pers. comm.*, 7/1/92.
CHEN, S-L, & CHIA, L-C, 1988: *Chinese bamboos*, Dioscorides Press, Portland, U.S.A.
CHEVALIER, A., 1920: *Exploration botanique de l'Afrique occidentale française, I: Énumération des Plantes récoltées*, Paul Lechevallier, Paris.
— 1935: Les Iles du Cap Vert. Flore de l'Archipel, *Rev. Bot. appl. Agr. trop.* 15: 733–1090.
— 1937, b: Une equête sur les Plantes médicinales de l'Afrique occidentale, *Rev. Bot. appl. Agr. trop.* 187: 165–75.
— 1950, a: Sur l'origine des *Digitaria* cultivés, *Rev. int. Bot. appl.* 30: 329–30.
CHIPP, T.F., 1922, a: *The Forest Officers' Handbook of the Gold Crest, Ashanti and the Northern Territories*, London.
CHITTENDEN, A.E. & H.E. COOMBER, 1948: Castor Stems from Ceylon, *Bull. Imp. Inst.* 46: 223–7.
CHITTENDEN, A.E. & D. MORTON, 1951: Sorghum stalks from the Gold Coast as a paper-making material and as a source of furfural, *Col. Pl. Anim. Prod.* 2: 299–305.

REFERENCES

CHITTENDEN, A.E. & L.J. FLAWS, 1964: The use of Rice Hulls as aggregate in light-weight concrete *Trop. Sci.* 6: 187–99.

CHITTENDEN, A.E., C.G. JARMAN, D. MORTON & G.B. PICKERING, 1954: Three timbers from Kenya, *Neoboutonia macrocalyx* Pax., *Macaranga kilimandscharica* Pax. and *Croton macrostachys* Hochst. as paper-making materials, *Col. Pl. Anim. Prod.* 4: 46–52.

CHUMBOW, B.S., 1982: *pers. comm.*, ex K. Williamson.

CLAUDE F. et al., 1947: Survey of Plants for anti-malarial activity, *Lloydia* 10: 145–74.

CLAYTON, W.D., 1965: Studies in Gramineae, VI, *Kew Bull.* 19: 287–96.

— 1969: A revision of the genus Hyparrhenia, *Kew Bull., Addit. Ser.* II.

— 1972: *Gramineae*, in F.N. HEPPER: *Flora of West Tropical Africa*, 3: 349–512.

— 1974, a: *Cynodon*, in W.D. CLAYTON, S.M. PHILLIPS & S.A. RENVOIZE, *The Flora of Tropical East Africa*, Gramineae, pt. 2, pp. 316–21.

— 1975: The *Paspalum scrobiculatum* complex in Tropical Africa, *Kew Bull.* 30: 101–5.

CLAYTON, W.D. & S.A. RENVOIZE, 1982: *Paniceae*, in R.M. POLHILL, *Flora of Tropical East Africa*, Gramineae, 3: 451–859.

— 1986: Genera Graminarum. Grasses of the World, *Kew Bull., Addit. Ser.*, XIII.

COLE, N.H.A., 1968, a: *The Vegetation of Sierra Leone*, Njala University College Press, Sierra Leone.

— 1977: *pers. comm.*, 13/9/1977.

CONNELL, B.A., 1991–92: *pers. comm.*

COOK, T.N., s.d. (ined.): Notes for a Benue-Congo comparative word-list.

COOMBER, H.E., 1952–53: Pulping studies with colonial tropical hardwoods as paper-making materials, *Col. Pl. Anim. Prod.* 3: 13–27.

COOPER, G.P. & S.J. RECORD, 1931: The Evergreen Forests of Liberia, *Yale Univ., Sch. For. Bull.* 31.

CRAWFORD, M., S.W. HANSON & M.E.S. KOKER, 1975: The structure of cymbopogone, a novel triterpenoid from lemon-grass, *Tetrahedron Lett.* 35: 3099–3102.

CROIZART, L., 1938: Euphorbia (Diacanthium) deightonii. A new succulent from West Africa, with brief notes on some allied species, *Bull. Misc. Inf.* 1938: 53–59.

CRUZ, A.A. de la, 1978: The Production of Pulp from Marsh Grass, *Econ. Bot.* 32: 46–50.

CUATRECACAS, J., 1961: A taxonomic revision of the Humiriaceae, *Contr. U.S. Nat. Herb.* 35: 25–214.

CYFFER, N. & J. HUTCHINSON, s.d., (ined.), *Kanuri Dictionary*, (msc.), ex R. Blench.

DALE, I.R. & P.J. GREENWAY, 1961: *Kenya Trees and Shrubs*, Buchanan's Kenya Estates, Ltd & Hatchards, London.

DALZIEL, J.M., 1937: *The Useful Plants of West Tropical Africa*, Crown Agents for the Colonies, London.

DANIELL, W.F., 1852: On the Zea Mays and other Cerealia of Western Africa, *Pharm. J.* 11: 395–401.

DAVEY, J.T., M. DESCAMPS & R. DEMANGE, 1959: Notes on the Acrididae of the French Sudan with special reference to the central Niger Delta, I. *Bull. Inst. Franç. Afr. Noire*, A. 21: 60–112.

DEBRAY, M., H. JACQUEMIN & R. RAZAFINDRAMBAO, 1971: Contribution à l'inventaire des plantes médicinales de Madagascar, *Trav. Doc. O.R.S.T.O.M.* 8.

DECARY, H., 1946: Plantes et animaux utiles de Madagascar, *An. Mus. Colon. Marseille*, ser. 6, 4: 6–234.

DEIGHTON, F.C., 1957: *Vernacular botanical vocabulary for Sierra Leone*, Crown Agents for Overseas Governments and Administrations, London.

DESHAPRABHU, S.B. [Ed.], 1966: *Wealth of India. Raw Materials.* 7 (N–Pe), C.S.I.R., India, New Delhi.

DE WET, J.M.J., 1978: Systematics and Evolution of Sorghum Sect. Sorghum (Gramineae), *Am. J. Bot.* 65: 477–84.

DIARRA, N'G., 1977: Quelques plants vendues sur les marchés de Bamako, *J. Agr. trad. Bot. appl.* 24: 41–49.

DOGGETT, H., 1970: *Sorghum*, Longmans, London.

D'OREY, J. & M.C. LIBERATO, 1970: Adimento a flora da Guiné Portuguesa, *Bol. Soc. Brot.*, sêr. 2, 44: 307–42.

— 1971: *Flora da Guiné Portuguesa-Papilionaceae*, Lisbon.

DOUGALL, H.W. & A.V. BOGDAN, 1960: The Chemical Composition of the Grasses of Kenya, II, *E. Afr. Agr. J.* 25: 241–4.

DUCKWORTH, E.H., 1947: Jatropha curcas, *Nigerian Field* 12: 58.

DUCROZ, J.M. & M.C. CHARLES, 1978: *Lexique Soney-Français*, Paris.

DUFRENÉ, H., 1887: *La Flore Sanskrite*, Maisonneuve Frères & L. Leclerc.

EGGELING, W.J. & I.R. DALE, 1952: *The Indigenous Trees of the Uganda Protectorate*, Govt. Printer, Uganda.

REFERENCES

EGUNJOBI, J.K., 1969: Some Common Weeds of Western Nigeria, *Bull. Res. Div., Min. Agr., Nat. Resources, W. State, Nigeria.*

EL-HAMIDI, 1970: Drug-plants of the Sudan Republic in native medicine, *Pl. Med.* 18: 278–80.

ELUGBE, B.O. 1982 (ined.): *Lists of plant names from Edoid languages.*

ENTI, A.A., 1981–82, 1991: *pers. comm.*

ESPIRITO SANTO, J. Do., 1963: Nomes vernáculos de algumes plantas da Guiné Portuguesa, *Estudos e Documentos,* No. 104, Junta de Investigacões do Ultramar, Lisbon.

ETKIN, N.L., 1981: A Hausa Herbal Pharmacopoeia: Biomedical Evaluation of commonly used Plant Medicines, *J. Ethno-Pharmacol.* 4: 75–98.

FARNSWORTH, N.R., R.N. BLOMSTER, W.M. MESSMER, J.C. KING, G.J. PERSINOS & J.D. WILKES, 1969: A phytochemical and biological review of the genus Croton, *Lloydia* 32: 1–28.

FEIJÃO, R. d'O., 1960–63: *Elucidário Fitológico. Plantas vulgares de Portugal Continental, insular e ultramarino, (classificacao, nomes vernaculos e aplicacoes),* 3 vols., Instituto Botanico de Lisboa.

FERNANDEZ, J.W., 1972: *Tabernanthe iboga: narcotic ecstasis and the work of ancestors,* in P.T. FURST, *Flesh of the Gods,* Praeger, New York.

FERRY, M.P., M. GESSAIN & R. GESSAIN, 1974: Ethnobotanique Tenda, *Docums Centre Rech. anthrop., Mus. Homme,* No. 1.

FOGGIE, A., 1957: Forest problems in the Closed Forest zone of Ghana, *J.W. Afr. Sci. Ass.* 3: 131–47.

FORMAN, L.L., 1971: Into the Bush, *Herb. Newsletter, Kew,* No. 691.

FOSTER, W.H. & E.J. MUNDAY, 1961: Forage Species of Northern Nigeria, *Trop. Agr.* 38: 311–8. [Reprinted in *Samuru Res. Bull.* 14, 1961].

FRAJZYNGIER, Z., 1985: *A Pero-English and English-Pero Vocabulary,* Marburger Studien zur Afrika und Asien Kunde, Verlag von Dietrich Reimer, Berlin.

FREEMAN, W.G. & R.O. WILLIAMS, 1928: *The Useful and Ornamental Plants of Trinidad and Tobago,* Ed. 2, Govt. Printer, Trinidad, Port-of-Spain.

FRISON [Ed.], 1942: De la présence de corpuscules siliceux dans le bois tropicaux en général et en particulier dans le bois du *Parinari glabra* Oliv. et du *Dialium klainei* Pierre. Utilisation de ces bois en construction maritime, *Bull. Agric. Congo Belge* 33: 91–105.

FRITZ, F. & G. GAZET DU CHATELIER, 1967: Sur le *Tragia benthami* Baker, Euphorbiacée africaine. Etude botanique et pharmacodynamique, *J. Agr. trop. Bot. appl.* 14: 339–58.

FROELICH, J.C., 1954: *La Tribu Konkomba du Nord Togo,* Inst. Franç. Afr. Noire, Dakar.

FURLONG, J.R., 1942: Nigerian Cassava Starch, *Bull. Imp. Inst.* 40: 257–68.

— 1944: *Vossia cuspidata* grass from Nigeria for paper-making, *Bull. Imp. Inst.* 42: 149–53.

FURNISS, G., 1990: *pers. comm.,* 29/1/90.

GALINAT, W.C., 1977: *The Origin of Corn,* in G.F. SPRAGUE [Ed.], Corn and Corn Improvement, *Am. Soc. Agron.* 18.

GAMBLE, J.S., 1902: *A Manual of Indian Timbers,* Ed. 2, London, Sampson, Low, Manston & Co.

GARDNER, I.D., 1975 [mimeo]: *Odual-English word-list,* Institute of Linguistics, Joss.

GARNIER, P., 1976 [mimeo]: *Noms de Plantes en: Langue Mandingue, Langue Baoulé. Essai de Classification logique des Noms Populaires de Plantes,* Marseille.

GAST, M., 1972: Céréales et pseudo-céréales de cuillette du Sahara Central (Ahaggar), *J. Agr. trop. Bot. appl.* 19: 51–59, with chemical analyses.

GBILE, Z.O., 1980: *Vernacular names of Nigerian Plants (Hausa),* For. Res. Inst. Nigeria. Ibadan.

GEERINCK, D., 1971: *Hypoxidaceae,* in *Flore du Congo, du Rwanda et du Burundi; Spermatophyta,* Brussels.

GETAHUN, A., 1975 (ined.): *Some common medicinal and poisonous plants used in Ethiopian folk-medicine,* MS in Herb. K.

GILL, L.S. & J.C. ENE, 1978: Weed Flora of Rice Fields of Bendel State, Nigeria, *Nigerian Field* 43: 182–3.

GIRAULT, R.P.L., 1958: Un remède à serpents efficace, *Not. Afr.* 77.

— 1960: Processions rogatoires au Pays Dagara, *Not. Afr.* 88: 119–20.

GLEDHILL, D., 1966: An African Relict, or an Introduction from America?, *J.W. Afr. Sci. Ass.* 11: 20–27.

GOLDING, F.D. & A.M. GWYNN, 1939: Notes on the Vegetation of the Nigerian shore of Lake Chad, *Bull. Misc. Inf.* 1939: 631–43.

GOMES E SOUSA, A. de F., 1930: Subsidios para o conhecimento da Flora da Guiné Portugesa, *Mem. Soc. Brot.* 1.

GOODING, E.G.B., A.R. LOVELESS & G.R. PROCTOR, 1965: *Flora of Barbados,* H.M. Stationery Office.

GRALL, Lieut., 1945: Le Secteur nord du Cercle de Gouré, *Bull. Inst. Franç. Afr. Noire,* 7: 1–46.

441

REFERENCES

GREENWAY, P.J., 1941: Dyeing and Tanning plants in East Africa, *Bull. Imp. Inst.* 39: 222–45.
GRÉGOIRE, H.C., 1975: Etude de la lingue Gouro, *Doc. Linguistiques*, 31/32, Univ. d'Abidjan, Ivory Coast.
GRIST, D.H., 1975: *Rice*, Ed. 5, Longmans, London.
GROUNDS, R., 1989: *Ornamental Grasses*, Christopher Helm (Publishers), London.
GUILLARMOD, A.J., 1971: *Flora of Lesotho (Basutoland)*, J. Cramer.
GUPTA, R.K. & S.K. SHARMA, 1971: Grasses of the Rangelands in arid Rajasthan, *J. Agr. trop. Bot. appl.* 18: 50–99.
HAERDI, F., 1964: Die Eingeborenen-Heilpflanzen des Ulanga – Distriktes Tanganyikas (Ostafrika) in F. HAERDI, J. KERHARO & J.G. ADAM, *Afrikanisches Heilpflanzen/Plantes médicinales africaines*, Basel.
HALL, J.B., P. PIERCE & G. LAWSON, 1971: *Common Plants of the Volta Lake*, Univ. Legon, Ghana.
HALLAM, G., 1979: *Medicinal Uses of Flowering Plants in The Gambia*, Dept. For., Yundum, The Gambia.
HAMPSON, B.H., 1958 (ined.): *Collection of cultivated Digitaria species*, ad not. Herb. K.
HARGREAVES, B.J., 1978: Kill and Curing: *Soc. Malawi J.* 31: 21–30.
HARRISON, S.G., 1950: Manna and its Sources, *Kew Bull.* 5: 407–17.
HARTWELL, J.L., 1969, b: Plants used against cancer, A survey, *Lloydia* 32: 153–205.
HEAL, R.E., E.F. ROGERS, R.T. WALLACE & O. STARNES, 1950: A Survey of plants for insecticidal activity, *Lloydia* 13: 89–162.
HEMMERLY, T.E., 1983: Traditional method of making Sorghum molasses, *Econ. Bot.* 37: 406–9.
HENRY, T.A., 1939: *The Plant Alkaloids*, Ed. 3, Churchill, London.
HEPPER, F.N., 1965: The vegetation and flora of the Vogel Peak Massif, Northern Nigeria, *Bull. Inst. Franç. Afr. Noire*, A. 27: 413–513.
— 1984: A visit to a lake stilt village in Ghana, *Nigerian Field* 49: 45–51.
Herbalists, Joru Village, Sierra Leone: Meeting of Herbalists, Nov., 1973.
Herbaria, BM: British Museum (Natural History), London.
— E: Royal Botanic Garden, Edinburgh.
— FBC: Fourah Bay College, Sierra Leone.
— IFE: University of Ife, Nigeria.
— IFAN: University of Dakar, Senegal.
— K: Royal Botanic Gardens, Kew.
— P: Museum d'Histoire naturelle, Paris.
— SL: Njala University College, Sierra Leone.
— UCI: Ibadan University, Nigeria.
HERRINGTON, G.N., 1979: Palm-wine Tapping in Nigeria, *Nigerian Field* 44: 29–35.
HEYNE, K., 1927: *De nuttige planten van Nederlandsch-Indie*, Ed. 2, Buitenzorg.
HILU, K.W. & J.M.J. de WET, 1976: Domestication of Eleusine coracana, *Econ. Bot.* 30: 199–208.
HOLLAND, J.H., 1908–22: The useful Plants of Nigeria, *Bull. Misc. Inf. Addit. Ser.* IX.
HOWES, F.N., 1945: *Plants and Bee-keeping*, Faber & Faber, London.
— 1946: Fence and Barrier Plants in warm climates, *Kew Bull.* 1: 51–87.
— 1949, a: *Vegetable Gums and Resins*, Chronica Botanica, Waltham, Mass., USA.
HUBBARD, C.E., 1954: *Grasses: A guide to their structure, identity, uses and distribution in the British Isles*, Penguin Books.
HUNTING TECHNICAL SERVICES LTD, 1964: *Report on the Survey of Geology, Geo-Morphology and Soils. Vegetation and Present Land-use. Land and water use survey in Kordofan Province of the Republic of the Sudan*, Doc. DOX-SUD-A26 (May 1964) UNSF-FAO.
HUNTING TECHNICAL SURVEYS, LTD, 1968: *Land and Water Resources Survey of the Jebel Marra area, Republic of the Sudan: Reconnaissance vegetation Survey*, LA: SF/SUD/17, FAO, Rome.
HUTCHINSON, J. & J.M. DALZIEL, 1937: Tropical African Plants, XV, *Bull. Misc. Inf.* 1937: 54–63.
IRVINE, F.R., 1930: *Plants of the Gold Coast*, Oxford Univ, Press.
— 1948: The indigenous food-plants of West African peoples, *New York Bot. Gard. J.* 49: 225–36, 254–67.
— 1952, a: Supplementary and Emergency Food Plants of West Africa, *Econ. Bot. (New York Bot. Gard.)* 6: 23–40.
— 1952, b: Food Plants of West Africa, *Lejeunia*, 16.
— 1956: Cultivated and Semi-cultivated Leafy Vegetables of West Africa, *Materiae veg.* 2.
— 1961: *Woody Plants of Ghana*, Oxford Univ. Press, London.
— s.d., a: *Reliquae*, Library, Herb. E.

442

REFERENCES

— s.d., b [mimeo]: Vocabularies of Plant Names in Nigerian Languages with botanical equivalents (2 volumes).

ISAWUMI, M.A., 1978, a: Nigerian Chewing Sticks, I, *Nigerian Field* 43: 50–58.

— 1978, b: *op. cit.*, II, *loc. cit.* 43: 111–21.

ITHMANN, J, 1976: *Worterbuch der Duala Sprache.*

IWU, M.M., 1980: Anti-diabetic properties of Bridelia ferruginea leaves, *Pl. Med.* 39: 247.

— 1984: *The state of studies in Igbo Pharmacon and Therapy*, Uwa ndi Igbo, no. 1, Univ. Nigeria, Nsukka.

— 1986: *Empirical Investigations of Dietary Plants used in Igbo Ethno?- medicine*, pp. 131-56, in N.L. ETKIN, *Plants in Indigenous Medicine and Diet. Behavioral approaches*, Redgrave Publishing Co., New York.

IWU, M.M. & B.N. Anyanwu, 1982: Phytotherapeutic profile of Nigerian herbs, I: Anti-inflammatory and anti-arthritic agents, *J. Ethno-Pharmac.* 6: 263–74.

JACKSON, G., 1973 (ined.): *MS re Fulani in N. Nigeria*, Herb. UCI.

JACQUES-FÉLIX, H., 1964: Contribution de René Caillié à l'ethno-botanique africaine au cours de ses voyages en Mauritanie et à Tombouctou, *J. Agr. trop. Bot. appl.* 10: 551–602.

JAEGER, P., 1964: *Icones Plantarum Africanarum*, fasc. VI, nos. 121–144, I.F.A.N., Dakar.

— 1965: Espèces végétales de l'étage altitudinal des Monts Loma (Sierra Leone), *Bull. Inst. Franç. Afr. Noire*, A. 27: 34–120.

JAIN, S.K. & N. DAM, 1979: Some Ethnobotanical Notes from Northeastern India, *Econ. Bot.* 33: 52–56.

JANARDHANAN, K.K., M.L. GUPTA & AKHTAR HUSSEIN, 1982: A new commercial strain of Ergot adapted from a wild grass, *Pl. Med.* 44: 166–7.

JENNINGS, D.L., 1957: Further studies in breeding cassava for virus resistance, *E. Afr. Agric. J.* 22: 213–9.

JOHNSON, R.M. & W.D. RAYMOND, 1964, a: The chemical composition of some tropical food-plants: I, Finger and Bulrush Millet, *Trop. Sci.* 6: 6–11.

— 1965: *op. cit.*: IV, Manioc, *loc. cit.* 7: 109–15.

JORU, 1973 (ined.): *Notes of a herbalists meeting with the author at Joru, SE Sierra Leone.*

KAY, D.E., 1973: Root Crops, *TPI Crop and Product Digest*, No. 2.

KEAY, R.W.J., 1954: *Flora of West Tropical Africa*, Ed. 2, vol. 1, pt. 1, Crown Agents for Overseas Governments and Administrations, London.

— 1958, b: Icacinaceae, in *Flora of West Tropical Africa*, Ed. 2, 1: 636–44.

KEAY, R.W.J., C.F.A. ONOCHIE & D.P. STANFIELD, 1960: *Nigerian Trees*, Vol. 1, Govt. Printer, Lagos.

— 1964: *op. cit*, Vol. 2, Ibadan, Nigeria.

KENNEDY, J.D., 1936: *Forest flora of Southern Nigeria*, Govt. Printer, Lagos.

KENNEDY-O'BYRNE, J., 1957: Notes on African Grasses: XXIX. A new species of Eleusine from Tropical and South Africa, *Kew Bull.* 12: 65–72.

KERHARO, J., 1967: A propos de la pharmacopée sénégalaise: aperçu historique concernant les recherches sur la flore et des plantes médicinales du Sénégal, *Bull. Inst. fond. Afr. Noire* A. 29: 1391–434.

— 1973: Pharmacognosie de quelques graminées sénégalaises, *Bull. Soc. Méd. Afr. Noire, Lang. Franç.* 18: 1–13.

KERHARO, J. & J.G. ADAM, 1962: Premier inventaire des plantes médicinales et toxiques de la Casamance (Sénégal), *Ann. Pharm. Franç.* 20: 726–44, 823–41.

— 1963, a: Deuxième inventaire des plantes médicinales et toxiques de le Casamance (Sénégal), *loc. cit.* 21: 773–92.

— 1963, b: *op. cit.*, *loc. cit.* 21: 853–70.

— 1964, a: Note sur quelques plantes médicinales des Bassari et des Tandanké du Sénégal oriental, *Bull. Inst. Franç. Afr. Noire*, A. 26: 403–37.

— 1964, b: Plantes médicinales et toxiques des Peul et des Toucouleur du Sénégal, *J. Agr. trop. Bot. appl.* 11: 384–444, 543–99.

— 1964, c: Les plantes médicinales, toxiques et magiques des Niominka et des Socé des Iles du Saloum (Sénégal), in F. HAERDI, J. KERHARO & J.G. ADAM, *Afrikanisches Heilpflanzen/Plantes médicinales africaines*, Basel.

— 1974: *La Pharmacopme Sénégalaise traditionnelle. Plantes médicinales et toxiques*, Vigot Frères, Paris.

KERHARO, J. & A. BOUQUET, 1950: *Plantes médicinales et toxiques de la Côte d'Ivoire-?- Haute-Volta*, Vigot Frères, Paris.

KERHARO, J. & J.P. PACCIONI, 1974: Considérations d'Actualité sur un vieux remède Sénégalais tombé dans l'oubli: le remède de Joal ('Garab u Djoula' ou 'Garab Diafan'), *J. Agr. trop. Bot. appl.* 21: 345–50.

KINGHORN, A.D. & F.J. EVANS, 1974: Occurence of Ingenol in Elaeophorbia species, *Pl. Med.* 26: 150–4.

REFERENCES

KINTZ, D. & B. TOUTAIN, 1981: *Lexique commente Peul-Latin des Flores de Haute?-Volta.* Etude Botanique No. 10; Institut d'Elevage et de Médecin vétérinaire des Pays Tropicaux.

KRAFT, C.H. [Ed.], 1981: *Chadic Word-lists,* 3 vols.; in Marburger: *Studien zur Afrika- und Asienkunde,* Dietrich Reimer, Berlin.

KRISHNAMURTHI, A. [Ed.], 1969: *Wealth of India. Raw Materials,* vol. 8 [Ph–Re], C.S.I.R., Calcutta.

KROPP, M.E., 1967: *Word-lists of Lefana, Akpafu and Avatime,* Inst. African Studies, Univ. Ghana, Legon.

KROPP-DAKUBU, M.E., 1973: *Ga-English Dictionary,* Inst. African Studies, Univ. Ghana, Legon.

KUBITZIKI, K., 1969: Monographie der Hernandiaceen, *Bot. Jahrb.* 89.

KULLENBERG, B., 1955: Quelques observations sur les Apides en Côte-d'Ivoire faites en août, 1954, *Bull. Inst. Franç. Afr. Noire,* A. 17: 1125–31.

KUNKEL, G., 1965: *The Trees of Liberia,* German Forestry Mission to Liberia; Report No. 3, Munich.

LANCASTER, P.A. & J.E. BROOKS, 1983: Cassava Leaves as Human Food, *Econ. Bot.* 37: 331–48.

LAPERRINE, General, 1919: Notice sur la cendre d'Aferegak, *Bull. Soc. Hist. Nat. Afr. Nord.* 10: 31.

LAUFER, B., 1919: Sino-Iranica. Chinese Contributions to the History of Civilisation in Ancient Iran, *Fld. Mus., Nat. Hist.* Publ. 201, *Anthropological Ser.* vol. 15.

LAUTER, W.M., L.E. FOX & W. ARIAIL, 1952: Investigation of the toxic principles of *Hippomane mancinella* L.: I, A historical review, *J. Am. Pharm. Ass.* 41: 199–201.

LEANDRI, J., 1958: Le problème du *Casearia bridelioides* Mildbr. ex Hutch. et Dalz., *Bull. Soc. bot. France* 105: 512–7.

LE BRIS, P. & A. PROST, 1981: *Dictionnaire Bobo-Français,* Lacito 44, Paris.

LEMORDANT, D., 1970: Contribution à l'Ethnobotanique Éthiopienne, I, *J. Agr. trop. Bot. appl.* 17: 1–35.

— 1971: *op. cit.,* II, *loc. cit.* 17: 142–79.

LEONARD, J., 1955: Notulae Systematicae, XIX: Observations sur divers *Bridelia africains* (Euphorbiaceae), *Bull. Jard. bot. Nation. Belg.* 25: 359–74.

— 1960: *op. cit.,* XXIX: Révision des *Cleistanthus* d'Afrique continental (Euphorbiacées), *loc. cit.* 30: 421–61.

— 1961: *op. cit.,* XXXII: Observations sur des espèces africaines de *Clutia, Ricinodendron* et *Sapium* (Euphorbiacées), *loc. cit.* 31: 391–406.

— 1962, b: *Flore du Congo et du Rwanda-Burundi, Spermatophytes* 8(1), 71, Euphorbiaceae (part), I.N.E.A.C., Brussels.

LERICHE, A., 1952: De l'enseignement arabe féminin en Mauritanie, *Bull. Inst. Franç. Afr. Noire* 14: 975–83.

LEROUX, H., 1948: Animisme et Islam dans la Subdivision de Maradi, *Bull. Inst. Franç. Afr. Noire* 10: 595–697.

LETOUZEY, R. & F. WHITE, 1970, a: *Ebenacées,* in A. AUBRÉVILLE & J.F. LEROY, *Flore du Cameroun,* 11, Mus. Nation. Hist. Nat., Paris.

— 1970, b: *Ebenacées,* in A. AUBRÉVILLE & J.F. LEROY, *Flore du Gabon,* 18, Mus. Nation. Hist. Nat., Paris.

LHOTE, H., 1952: Les boîtes moulées en peau du Soudan, dites 'Bata', *Bull. Inst. Franç. Afr. Noire* 14: 919–55.

LOCK, J.M., s.d. (ined.): *Notes on the Genus Cenchrus* (Gramineae) in the Arid Zone, MS. E.C.O.S./S.E.P.A.S.A.L. Herb. K.

LOWE, J., 1976 (ined.): List of chew-stick plants for the Ibadan area, Nigeria by B.I. ASUQUO – *pers. comm.,* 29/9/1976.

— 1984: Gnetum in West Africa, *Nigerian Field* 49: 99–104.

— 1989: *pers. comm.,* 8/7/1989.

LUCAS, E.B., 1967: The properties of some savanna Timber Trees, *Fed. Dept. For. Res., Ibadan,* For. Prod. Res. Repts, No. F.P.R.L./11.

LUKAS, J., 1939: *Die Sprache der Buduma in zentralen Sudan.*

MACFOY, C.A., & A.M. SAMA, 1983: Medicinal Plants in Pujehum District of Sierra Leone, *Journal of Ethnopharmacology,* 8: 215–23.McINTOSH, M., 1978–79: *pers. comm.*

MACKAY, J.H. & T.N. WARDROP, 1939: The forest Trees of Southern Nigeria, II: Common waterside trees, *Nigerian Field* 8: 65–74.

MACMUNN, G., 1933: *The Underworld of India,* London, Jarrolds.

McVAUGH, R., 1944: The Genus Cnidoscolus: generic limits and intro-generic groups, *Bull. Torrey Botan. Club* 71: 457–74.

MAIRE, H., 1933: *Études sur la Flore et la Végétation du Sahara central, Mém. Soc. Hist. Nat. Afr. Nord* 65.

MAIRE, R., 1925: Sur le *Chrozophora brocchiana* Schweinf., *Bull. Soc. Hist. Nat. Afr. Nord* 16: 42.

MALCOLM, S.A. & E.A. SOFOWORA, 1969: Antimicrobial activity of selected Nigerian folk remedies and their constituent plants, *Lloydia* 32: 512–7.

MALIKI, A.B., 1981: Ngaynaaka – Herding according to the Wodaabe, *Discussion Paper No. 2, Republic of Niger, Min. Rural Development, Niger Range and Livestock Project.* [mimeo].

MANESSY, G., 1975: *Langues et Civilisations a Tradition Orale, No. 15: Les Langues Oti-Volta. Classification généalogiques d'un groupe de langues Voltaiques,* Société d'Études Linguistiques et Anthropologiques de France Paris.

MANGELSDORF, P.C., 1974: *Corn. Its Origin, Evolution and Improvement,* Harvard Univ. Press.

MANJUNATH, B.L. [Ed.], 1948: *The Wealth of India. Raw Materials,* Vol. 1 (A–B), C.S.I.R., New Delhi.

MARCHAL, E., 1959: Variation de la population anophélienne d'une mare à salinité variable de la region de Konakri (Guinée française), *Bull. Inst. Franç. Afr. Noire,* A. 21: 180–203.

MARCHESSE, L., 1979: *Atlas Linguistique Kru. Essai de Typologie.* Institut de Linguistique Appliquée, vol. 73, Univ. d'Abidjan, Ivory Coast.

MASSAQUOI, J. (herbalist, Magbena, Sierra Leone), Nov. 1973: herbal prescriptions, *verb. comm.*

MATSUSHITA, S., 1974: *A comparative vocabulary of Gwandara dialects,* ILCAA, Tokyo.

MAUNY, R., 1953: Notes historiques autour des principales plantes cultivées d'Afrique occidentale, *Bull. Inst. Franç. Afr. Noire,* 15: 684–730.

MEEK, C.K., 1931: *Tribal Studies in Northern Nigeria,* 2 vols., Kegan Paul, Trench Trubner.

MENINGER, E.A., 1948: Oncoba – The 'Chic' of Araby, *J. New York Bot. Gard.* 49: 249–51.

MIÉGE, J., 1963: Les blés de Mauritanie, *J. Agr. trop. Bot. appl.* 10: 335–43.

MILNE-REDHEAD, E., 1949: Euphorbia geniculata Orteg. (Euphorbiaceae), *Kew Bull.* 1948: 457–8.

MODEBE, A.N.A., 1963, a: Preliminary studies on the effect of limited and unlimited systems of feeding and of tropical-type grass/legume clippings fed in dry-lot, on the growth rate, economy of gain, and carcass quality of growing/fattening pigs, *J.W. Afr. Sci. Ass.* 7: 116–26.

— 1963, b: Preliminary trial on the value of dried cassava (Manihot utilissima Pohl) for pig feeding, *J.W. Afr. Aci. Ass.* 7: 127–33.

MONOD, Th., 1950: *Vocabulaire Botanique Teda,* in R. MAIRE & Th. MONOD, Études sur la flore et la végétation du Tibesti, *Mém. Inst. Franç. Afr. Noire,* 8.

MONTEIL, V., 1953: *Institut des Hautes Études Marocaines. Notes & Documents, VI. Contribution à l'étude de la Flore du Sahara occidental, II,* Larose, Paris.

MOOMO, D.O., 1982: ad not. in K. Williamson, 1973–82: *pers. comm.*

MORETTI, C. & P. GRENAND, 1982: Les Nivrées ou plantes ichtyotoxiques de la Guyane Française, *J. Ethno-Pharmacol.* 6: 139–60.

MORTIMER, W.G., 1901: *Peru. History of Coca. 'The Divine Plant of the Incas'.* Vail & Co., New York.

MORTIMORE, M., 1987 (ined.): *Plant foods used in famine in Northern Kano State, Nigeria, 1972–74, pers. comm.* 15/1/1987.

MORTON, J.K., 1961: *West African Lilies and Orchids,* in H.J. SAVORY, *West African Nature Handbooks,* Longmans, Green and Co. Ltd., London.

MOTTE, E., 1980: A propos des thérapeutes Pygmées Aka de la Région de la Lobaye (Centrafrique), *J. Agr. trad. Bot. appl.* 27: 113–32.

MURRAY, H.J.R., 1952: *A history of Board Games other than Chess,* Clarendon Press, Oxford.

MURRAY, K.C., 1972: Pottery of the Ibo of Ohuhu-Ngwa, *Nigerian Field* 37: 148–75.

NAEGELÉ, A., 1958, a: Contributions à l'étude de la flore et des groupements végétaux de la Mauritanie, I: Note sur quelques plantes recoltées à Chinguetti (Adrar Tmar), *Bull. Inst. Franç. Afr. Noire,* A. 20: 293–305.

— 1958, b: *op. cit.,* II: Plantes recueillÉs par Mlle Odette de Puigaudeau en 1950, *loc. cit.* A. 20: 876–908.

N'DIAYE, S., 1964: Notes sur les Engins de Pécher chez les Sérèr, *Notes Afr.* 104: 116–20.

NEWMAN, P., 1974: *The Kanakuru language,* West African Language Mon. 9. Inst. Modern English Language Studies, Leeds & West African Linguistic Society.

NICKLIN, K., 1975: Agiloh: the Giant Mbube Xylophone, *Nigerian Field* 40: 148–58.

NICOLAS, F.J., 1952: Mythes et êtres mythiques des l'Éla de la Haute-Volta, *Bull. Inst. Franç. Afr. Noire,* 14: 1353–84.

— 1953: Onomastique personnelle des L'éla de la Haute-Volta, *Bull. Inst. Franç. Afr. Noire* 15: 818–47.

NICOLE, J., 1979: *Phonologie et Morphophonologie de Nawdm,* Dept. de Linguistique, Univ. de Benin (Togo).

REFERENCES

NICOLS, R.F.W., 1947: Breeding Cassava for virus resistance, *E. Afr. Agr. J.* 12: 184–94.
NORMAND, D., 1937: Le Bois de Landa, Erythroxylum du Cameroun, *Rev. Bot. appl. Agr. trop.* 196: 883–9.
ODEI, M.A., 1973: Observations on some weeds of malacological importance in the Volta Lake, *Bull. Inst. fond. Afr. Noire*, A. 35: 57–66.
ODOH, G.O.A., 1978[?], a: Some trees bearing Edible Fruits and their Igala Names, *Elaeis*, 1(1).
— 1978[?], b: Some Farm Grasses in Igala/Idomaland, *op. cit.*.
OGBE, F.M.D. & J.T. WILLIAMS, 1978: Evolution in Indigenous West African Rice, *Econ. Bot.* 32: 59–64.
OGO, O.E., s.d.: *Word-list – Yala (Ikom)*, ms. fide R. Blench, *pers. comm.*
OGOR, E. & D.W. HEDRICK, 1963: Management of Natural Pasturage, *J.W. Afr. Sci. Ass.* 7: 145–53.
OKAFOR, J.C., 1975: Varietal delimitation in *Irvingia gabonensis* (Irvingiaceae), *Bull. Jard. Bot. Nat. Belg.* 45: 211–21.
OKE, O.L., 1966, a: Chemical Studies on some Nigerian Foodstuffs – Kpokpogari (processed cassava), *Trop. Sci.* 8: 23–27.
OKIGBO, B.N., 1964: Studies of seed germination in star grasses: I, The effect of Nitrate and alternating Temperature, *J.W. Afr. Sci. Ass.* 8: 141–58.
— 1980: *1980 Ahiajoku Lecture: Plants and Food in Igbo Culture*, Ministry of Information, Culture, Youth and Sports, Govt. Printer, Owerri.
OKIY, G.E.O., 1960: Indigenous Nigerian Food Plants, *J.W. Afr. Sci. Ass.* 6: 117–21.
OLIVER, B., 1960: *Medicinal Plants in Nigeria*, Nigerian College of Arts, Science and Technology.
OLIVER-BEVER, B., 1982: Medicinal Plants in Tropical West Africa, I: Plants acting on the cardiovascular system, *J. Ethno-Pharmacol.* 5: 1–71.
— 1983: *op. cit.*, II: Plants acting on the nervous system, *loc. cit.* 7: 1–93.
ONÒH, C.I., s.d.: *Word-list of Yala (Ogoja)*, ms. fide R. Blench, *pers. comm.*
OSANG, T., s.d.: *Word-list – Bokyi (Iruan)*, ms. fide R. Blench, *pers. comm.*
PAGEAUD, P., 1972: Lexique Bete *Afrique et Language*, no. 7 Paris.
PASCH, H., 1980: *Linguistische Aspekte der Verbreitung Lateinamerikischer Nutzpflanzen in Afrika*, Univ. Cologne, thesis for Master's degree.
PELLEGRIN, F., 1959: Guttifères d'Afrique Équatoriale, *Bull. Soc. Bot. France* 106: 216–30.
PERCIVAL, D.A., 1968: *The Common Trees and Shrubs of The Gambia*, Govt. Printer, The Gambia.
PICHL, W.J., 1963: *Sherbro-English – English-Sherbro Vocabulary*, Fourah Bay College, Freetown.
PICHON, M., 1953: Monographie des Landolphiées. (Classification des Apocynacées, XXXV), *Mém. IFAN*, 35.
PLANCHE, O., 1948: Le Psorospermum guineense Hochr., ou 'Karidiakouma' de la Guinée française, *Ann. pharm. Franç.* 6: 546–65.
POBÉGUIN, H., 1912: *Plantes médicinales de la Guinée*, Paris.
PORTÈRES, R., 1935: Plantes toxiques utilisées par les peuplades Dan et Guéré de la Côte d'Ivoire, *Bull. Comité d'Études Historiques et scientifiques de l'AOF* 18.
— 1946: L'aire culturale du *Digitaria iburua*, *Agron. trop. Nogent* 1: 589–92.
— 1950, a: Naturalisation du Chrysopogon aciculus Trinius à la Côte d'Ivoire, *Bull. Soc. Bot. France*, 97: 101–2.
— 1950, b: Vieilles agricultures de l'Afrique intertropicale, *Agron. trop.* 5: 489–507.
— 1951, a: *Comptes rendus. Première conférence internationale des Africanistes de l'Ouest*, I.F.A.N., Dakar.
— 1951, c: *Eleusine coracana* Gaertner. Céréale des humanités pauvres des pays tropiceaux, *Bull. Inst. Franç. Afr. Noire* 13: 1–78.
— 1955: Les céréales mineures du genre *Digitaria* en Afrique et en Europe, *J. Agr. trop. Bot. appl.*, 2: 349–86, 477–510, 620–75.
— 1956: a: Taxonomie agrobotanique des riz cultivés. *O. sativa* et *O. glaberrima*, *J. Agr. trop. Bot. appl.* 3: 341–84, 541–80, 627–700, 821–56.
— 1956, b: Un Riz précoce estimé en petite culture dans L'Ouest-Africain, le Toulou-oule ou Konko (*O. sativa*), *J. Agr. trop. Bot. appl.* 3: 50–59.
— 1958-59: Les appelations des Céréales en Afrique, *J. Agr. trop. Bot. appl.* 5 & 6.
— 1962: Berceaux agricoles primaires sur le Continent Africain, *J. Afr. Hist.* 3. 2: 195–210.
— 1966: Les noms des riz en Guinée, MS. in Lab. Ethnobot. Mus. Nat. Hist. Nat., Paris – reprint from *J. Agr. trop. Bot. appl.*
— 1974: Un curieux élément culturel Arabico-Islamique et Neo-Africain: les baguetes végétales machées servant de frottes-dents, *J. Agr. trop. Bot. appl.* 21: 1–36, 111–49.
— s.d.: *Reliquiae*, Lab. d'Ethnobotanique, Paris.

REFERENCES

POUQUET, J., 1956: Le Plateau du Labé (Guinée française, A.O.F.) Remarques sur le charactére dramatique des phénomènes d' érosion des sols et sur les remédes proposés, *Bull Inst. Franç. Afr. Noire* A. 18: 1–24.

PROST, A., 1971, 1972[?] [mimeo]: Elements de Sembla, *Afrique et Language*, Doc. 5.

— 1980: *La Langue des Kronroumba ou Akurumfe*, A Schendl, Vienna.

PURSEGLOVE, J.W., 1972: *Tropical Crops. Monocotyledons*, 2 vols., Longman Group, London.

QUISUMBING, E., 1951: Medicinal Plants of the Philippines, *Dept. Agric. Nat. Resources, Tech. Bull.*, 16, Manila.

RAO, K.V., 1974: Toxic principles of Hippomane mancinella, *Pl. Med.* 25: 166–71.

RAYMOND, W.D., 1961: Castor Beans as Food and Fodder, *Trop. Sci.* 3: 19–27.

RAYMOND, W.D., J.A. SQUIRES & J.B. WARD, 1954: The Milling of Sorghum in Nigeria, *Colon. Pl. Anim. Prod.* 4: 152–8.

RENVOIZE, S.A., 1974: *Chloris*, in W.D. CLAYTON, S.M. PHILLIPS & S.A. RENVOIZE, *Flora of Tropical East Africa, Gramineae*, pt. 2, pp. 337–47.

RESPLANDA, A., 1955: Détection d'un alcaloïde dans le liquide séminal d'une Guttiferacée Africaine: Allanblackia parviflora A Chev., *J. Agr. trop. Bot. appl.* 2: 543–6.

RAYNAL, A., 1967: Sur un Sebaea africain saprophyte (Gentianaceae), *Adansonia* 2, 7: 207–19.

RAYNAL-ROQUES, A., 1978: Les Plantes aquatiques alimentaires, *Adansonia* 2, 18: 327–43.

RIDLEY, H.N., 1906: Malay Drugs, *Agr. Bull. Straits & Fed. Malay States* 5: 193–206, 245–54, 269–82.

RIVIER, L. [Ed.], 1981: Papers presented at the Symposium on Erythroxylon – Historical and Scientific Aspects: Quito, Equador, Dec., 1979, *J. Ethno?- Pharmacol.* 3.

ROBERTY, G., 1953: Notes de botanique Ouest-africaine, VI: Plantes banales dans le Sahel de Nioro, *Bull. Inst. Franç. Afr. Noire* 15: 442–52.

— 1955: Notes sur la flore de l'Ouest-Africain, VI, *Bull. Inst. Franç. Afr. Noire*, A. 17: 12–79.

— 1958: Végégation de la guelta de Soungout (Mauritanie méridionale) en mais, 1955, *Bull. Inst. Franç. Afr. Noire* A. 20: 869–75.

— 1961: Les Andropogonées ouest-africaines, *Bull. Inst. Franç. Afr. Noire* A. 23: 638–702.

ROBIN, J., 1947: Description de la Province de Dosso, *Bull. Inst. Franç. Afr. Noire* 9: 56–98.

ROBINSON, C.H., 1913: *Dictionary of the Hausa Language, Vol. 1: Hausa-English Vol. 2: English-Hausa*, Cambridge Univ. Press.

ROBINSON, J.B.D. & K.J. TREHARNE, 1985: Exploited Plants. Maize, *Biologist* 32: 199–207.

ROBYNS, W., 1948: *Flore du Congo Belge et du Ruanda-Urundi*, I, 4: Gnetaceae.

ROIA, F.C. & R.A. SMITH, 1977: The Antibacterial Screening of some common Ornamental Plants, *Econ. Bot.* 31: 28–37.

ROSEVEAR, D.R., 1961 (ined.): *Gambia Trees and Shrubs*. [Notes to accompany the author's 'Forestry conditions in The Gambia', *Emp. For. J.* 16, 1937. MS in Herb. K].

ROULON, P., 1980: Le Bois de Feu chez les Gbaya-Kara – 'Bodoe: essai de métholodogie et d'analyse ethnolinguistique, *J. Agr. trad. Bot. appl.* 27: 221–46.

RUPPER, Fr. G., O.S.B., 1989, a: pers. comm., 16/6/1989.

— 1989, b: *Sunnhemp 'Marejea'*, *Crotala ochroleuca*, , Benedictine Publications Ndanda Peramiho, Tanzania.

RYTZ, O., [Ed.], s.d.: *Gonja-English Dictionary and Spelling Book*, Inst. African Studies, Univ. Ghana, Legon.

SAMIA AL AZHARIA JAHN, 1977: pers. comm., 8/12/1977.

SAMPSON, H.C., 1936: Cultivated Crop Plants of the British Empire and the Anglo? Egyptian Sudan (Tropical and Sub-tropical), *Kew Bull., Addit. Ser.* XII.

SANDBERG, F. & A. CRONLUND, 1982: An ethnopharmacological inventory of medicinal and toxic plants from Equatorial Africa, *J. Ethno-Pharmacol.* 5: 187–204.

SASTRI, B.N. [Ed.], 1950: *Wealth of India. Raw Materials*, 2(C), C.S.I.R., New Delhi.

— 1952: *op. cit.*, 3(D–E).

— 1959: *op. cit.*, 5(H–K).

— 1962: *op. cit.*, 6(L–M).

SAVILL, P.S. & J.E.D. FOX, 1967 (ined.): *Trees of Sierra Leone*, For. Dept., Sierra Leone, Freetown [mimeo].

SCARPA, A. & A. GUERCI, 1982: Various uses of the Castor Oil plant (Ricinus communis L.). A review, *J. Ethno-Pharmacol.*, 5: 117–37.

SCHNELL, R., 1950, b: *Manuels Ouest-africains, I. La forêt dense. Introduction à l'étude botanique de la région forestière d'Afrique occidentale*, Paris.

— 1953, a: *Icones Plantarum Africanarum*, Fasc. I, nos. 1–24, I.F.A.N., Dakar.

— 1953, b: *op. cit.*, Fasc. II, nos. 25–48.

— 1960: *op. cit.*, Fasc. V, nos. 97–120.

SCHUNCK DE GOLDFIEM, J., 1945: Flore utilitaire de la Thyrénéide, *Bull. Soc. bot. France* 92: 152–3.

SHARLAND, R.E., 1978: pers. comm., 27/9/1978, 26/10/1978.

REFERENCES

SHIMIZU, J., s.d. 1985: *Field Notes re Jukun of Wukari*, ms. ined. fide R. Blench, *pers. comm.*

SIDIBE, M., 1939: Famille, vie sociale et vie religieuse chez les Birifer et les Oulé (region de Diébougou, Côte d'Ivoire), *Bull. Inst. Franç. Afr. Noire* 1: 697–742.

SIKES, S.K., 1972: *Lake Chad*, Eyre Methuen, London.

SILLANS, R., 1953, b: Plantes Alimentaires spontanées d'Afrique Centrale, *Bull. Inst. Études Centafr.* 5: 77–99.

SINGHA, S.C., 1965: *Medicinal Plants of Nigeria*, Nigerian National Press, Apapa, Nigeria.

SLEUMER, H., 1973: Révision du genre Homalium Jacq. (Flacourtiacées) en Afrique (y compris Madagascar et les Mascareignes), *Bull. Jard. bot. Nation. Belg.* 43: 239–328.

SNOWDEN, J.D., 1936: *The Cultivated Races of Sorghum*, London.

SODAH AYERNOR, G.K., & J.S. MATTHEWS, 1971: The sap of the palm, Elaeis guineensis Jacq., as raw material for alcoholic fermentation in Ghana, *Trop. Sci.*, 13: 71–83.

SOENARKO, S., 1977: The Genus Cymbopogon Sprengel (Gramineae), *Reinwardtia* 9: 225–375.

SOUSA, E.P.de, 1956: Contribuições para o conhecimento da Flora da Guiné portuguesa, VII, *Anais Jta Invest. Ultramar* 9: 5–38.

SPEKE, J.H., 1863: *Journal of the Discovery of the Source of the Nile*, Blackwood, Edinburgh & London.

SPICKETT, R.G.W., J.A. SQUIRES & J.B. WARD, 1955: Gari from Nigeria, *Colon. Pl. Anim. Prod.* 5: 230–5.

STANER, P., 1932: Une plante toxique pour le bétail, *Revue Zool. Bot. Afr.* 23.

STAPF, O. & C.E. HUBBARD, 1934: *Pennisetum*, in D. PRAIN, *Flora of Tropical Africa*, 9. London.

SURUGUE, B., 1979: Études Gulmance, *SELAF*, 75, 76.

SUZUKI, S., 1978: *Index to Japanese Bambusaceae*, Gakken Co., Tokyo.

TATTERSALL, S.L., 1978: *The Lesser-known Food Plants of The Gambia – a study*, Dept. Agr., The Gambia, [mimeo].

TAYLOR, C.J., 1960: *Synecology and Silviculture in Ghana*, Thos. Nelson & Sons.

TAYLOR-SMITH, R., 1966, a: Investigations on plants of West Africa, III: Phytochemical studies of some plants of Sierra Leone, *Bull. Inst. Franç. Afr. Noire* A. 28: 538–41.

TAYLOR-SMITH, R. & D.E.B. CHAYTOR, 1966: Investigations on West African plants, III: Studies Three West African medicinal plants, *Bull. Inst. fond. Afr. Noire* A. 28: 895–8.

TCHAMDA, N., 1976 [mimeo]: *Nufi Dictionnaire*, fide R. Blench, *pers. comm.*

TERRY, P.J., s.d.: Tropical Pesticides Research Institute, Miscellaneous Report No. 924, Arusha, Tanzania.

TERSIS, N., 1967: *Essai pour une phonologie du Gurma parlé à Kpana (Nord Togo)*, SELAF, 4, Paris.

THOMAS, N.W., s.d.: *Reliqueae, Field Notes*, Herb. K.

TIEMO, G.O.E., 1968: Egbema-Igo Customs and Traditions, *Occ. Publ. No. 17, Inst. Afr. Studies*, Univ. Ibadan.

TROCHAIN, J., 1940: La végétation du Sénégal, *Mém. Inst. Franç. Afr. Noire* 2.

TULLEY, P., 1966: The Obudu Plateau. Utilisation of high altitude tropical grassland, *Bull. Inst. fond. Afr. Noire*, A. 28: 899–911.

TYLER, V.E., 1966: The physiological properties and chemical constituents of some habit-forming plants, *Lloydia* 29: 275–92.

UGBE, P., 1983: *Word-list of plant names in Bete*, fide K. Williamson, *pers. comm.*

UNWIN, A.H., 1920: *West African Forests and Forestry*, Unwin, London.

UPHOF, J.C.Th., 1968: *Dictionary of Economic Plants*, Cramer.

USHER, G., 1974: *A Dictionary of Plants used by Man*, Constable, London.

VANDEN BERGHEN, C., 1979: La végétation des sables maritimes de la Basse Casamance méridionale (Sénégal), *Bull. Jard. bot. Nation. Belg.* 49: 185–238.

VERGER, P., 1965: Culte de Iyami Osoronga among the Yoruba, *J. Africanistes* 35: 141–243.

— 1967: *Awon ewe osanyin (Yoruba medicinal leaves)*, Univ. Ife, Nigeria.

— 1972: Automatisme verbal et communication du savoir chez les Yoruba, *Rev. franç. anthropologia* 12: 5–46.

— 1986: *pers. comm.*, 9/12/86.

VERGIAT, A.M., 1970: Plantes magiques et médicinales des Féticheurs de l'Oubangui (Région de Bangui), *J. Agr. trop. Bot. appl.* 17: (a) 60–91, (b) 171–99, (c) 295–339.

VIEILLARD, G., 1940: Le Chant de l'Eau et du Palmier doum. Poème bucolique du marais nigérien par Tiello Hamgourdo, Tionudio et Sigua Tiosoho, *Bull. IFAN*, 2: 299–315.

VILLIERS, J.F., 1973, a: *Icacinacées, Olacacées, Pentadiplandracées, Opiliacées, Octoknemacées*, in A. AUBRÉVILLE & J.F. LEROY, *Flore du Cameroun*, 15.

— 1973, b: *ibid.*, in A. AUBRÉVILLE & J.F. LEROY, *Flore du Gabon*, 20.

VISSER, L.E., 1975: Plantes médicinales de la Côte d'Ivoire, *Meded. Landb. Hoogesch., Wageningen* 75-15.

VOORHOEVE, A.G., 1965: *Liberian high forest trees*, Wageningen.

WALKER, A.R., 1930: Plantes oléifères du Gabon, *Rev. Bot. appl. Agr. trop.* 10: 209–18.

— 1953, a: Usages pharmaceutiques des plantes spontanées du Gabon, II, *Bull. Inst. Études Centrafr.* n.s., 5: 19–40.

— 1953, b: *op. cit.*, III, *loc. cit.* n.s., 6: 275–329.

WALKER, A.R. & R. SILLANS, 1961: *Les Plantes utiles du Gabon*, Paris.

WANG, D-J. & S-J. SHEN, 1987: *Bamboos of China*, Christopher Helm, London.

WANG, S-C. & J.B. HUFFMAN, 1981: Botanochemicals: Supplements to Petrochemicals, *Econ. Bot.* 35: 369–82.

WATT, J.M., 1967: African Plants potentially useful in mental health, *Lloydia* 30: 1–22.

WATT, J.M. & M.G. BREYER-BRANDWIJK, 1962: *The Medicinal and Poisonous Plants of Southern and Eastern Africa*, Ed. 2., Livingstone, Edinburgh & London.

WEBSTER, G.L., 1957: A monographic study of the West Indian species of Phyllanthus, II, *J. Arnold Arbor.* 38: 51–79.

WEISS, E.A., 1973: Some indigenous trees and shrubs used by local fishermen on the East African coast, *Econ. Bot.* 27: 174–92.

— 1979: Some indigenous plants used domestically by East African coastal fishermen, *Econ. Bot.* 33: 35–51.

WET, J.M.de, K.E. PRASADA RAO, M.H. MENGESHA & D.E. BRINK, 1983, a: Diversity in Kodo millet, Paspalum scrobiculatum, *Econ. Bot.* 37: 159–63.

WHITE, F., 1957: Notes on Ebenaceae, III, *Bull. Jard. bot. Nation. Belg.* 27: 515–31.

WIGBOLDUS, J.S., 1986: Trade and Agriculture in Coastal Benin c.1470–1660; an examination of Manning's early-growth thesis, *Afd. Agrar, Geschied. Bijd.* 28: 299–383.

WILCZEK, R. & R. BOUTIQUE, 1958: Linaceae – Ochthocosmus Benth., in R. WILCZEK, *Flore du Congo Belge et du Ruanda-Urundi, Spermatophytes*, vol. 7.

WILD, H., 1961: Harmful Aquatic Plants in Africa and Madagascar, *Kirkia* 2: 1–66.

WILLAMAN, J.J. & H-L. LI, 1970: Alkaloid-bearing plants and their contained alkaloids, *Lloydia* 33(3A).

WILLIAMS, J.T. & R.M. FARIAS, 1972: Utilisation and Taxonomy of the Desert Grass, Panicum turgidum, *Econ. Bot.* 26: 13–20.

WILLIAMS, R.O., 1949: *The useful and ornamental plants in Zanzibar and Pemba*, Zanzibar.

WILLIAMSON, J., 1955, 1956[?]: *Useful Plants of Nyasaland*, Govt. Printer, Zomba.

WILLIAMSON, K., 1970: Some food-plant names in the Niger Delta, *Int. J. Amer. Linguistics*, 36: 156–67.

— 1973–92: *pers. comm.*

— s.d., a (ined.): *Field-notes re Niger Delta collections*, MS in Herb. UCI.

WILLIAMSON, K. & A.O. TIMITIMI [Ed.], 1983: *Short Izon-English Dictionary*, Delta Ser. No. 3, Univ. Port Harcourt Press.

WILSON, R.T. & W.G. MARIAM, 1979: Medicine and Magic in Central Tigre: A contribution to the Ethnobotany of the Ethiopian Plateau. *Econ. Bot.* 33: 29–34.

WOLFF, H., 1969: *A comparative vocabulary of Abuan dialects*, NUP, Evanston.

WOLVERTON, B.C., 1982: Hybrid Waste-water Treatment System using anaerobic micro-organisms and reed (*Phragmites communis*) *Econ. Bot.* 36: 373–80.

WONG, W., 1976: Some folk medicinal plants from Trinidad, *Econ. Bot.* 30: 103–42.

ZEVEN, A.C., 1974: Indigenous bread wheat varieties from Northern Nigeria, *Acta Bot. Neerl.* 23: 137–44.

— 1980: The Spread of Bread Wheat over the Old World since Neolithicum as indicated by its genotype for hybrid necrosis, *J. Agr. trad. Bot. appl.* 27: 19–53.

Authorities are given in this list either by the full name, or names in cases of joint publications, or by cypher which may be identified as indicated, and then taken up in the general biographic references.

A&AA – Adjanohoun & Aké Assi, 1972.
AAE – Enti, pers. comm.
A&S – Akinniyi & Sultanbawa, 1983.
A. Chev. – A. Chevalier; ex auctt.
ABM – Maliki, 1981.
AE&L – Adam, Echard & Lescot, 1972.
AEK – Kitson; Herb. K; ex Irvine.
AGV – Voorhoeve, 1965; Herb. K.
AHU – Unwin, 1920; Herb. K; ex Dalziel.
AJC – Carpenter; Herb. K, UCI.
AJML – Leeuwenberg; Herb. K.
AN – Naegelé, 1958.
APDJ – Jones; Herb. K.
AS – Sébire; Herb. K; ex auctt.
AS Thomas: Herb. K; ex Irvine.
AST – Thornewill; Herb. K.
AT Johnstone: Herb. K; ex Dalziel.
Abraham: ex Pasch, 1980.
Abrahams: ex Irvine.
Adames: Herb. K.
Adams: Herb. K.
Adams: ex Pasch, 1980.
Adandé: 1953.
Adelodun: Herb. K.
Akubue & Mittal: 1982.
Anape: ex K. Williamson.
Anderson: Herb. K.
Andoh: Herb. K; ex Irvine.
Ankrah: Herb. K.
Appa Rao & al.: 1985.
Arana & Swodesh: 1967.
Ariwaodo: Herb. K.
Armstrong: ex Blench; ex Pasch, 1980.
Aub. – Aubréville. 1950, 1959: ex auctt.
auctt. – citation of three or more authorities.
Aylmer: Herb. K.; ex Dalziel.
B&D – Bouquet & Debray, 1974.
B&T – Battandier & Trabut, 1911; ex Dalziel.
Bégué: ex Kerharo & Bouquet, 1950.
BAC – Connell, 1991–92.
BD&H – Burtt Davy & Hoyle, 1937; ex Irvine.
BM – Moiser; Herb. K.
BNO – Okigbo, 1980.
Banfield: ex Dalziel; ex Pasch, 1980.
Bardin: 1937.
Bargery: ex Dalziel.
Barker: Herb. K.
Barth: 1857–58; ex Dalziel.
Bartha: Herb. K.
Bartlett: Herb. UCI.
Baumer: 1975.
Bayard.
Becker-Donner: ex Pasch, 1980.
Benton: ex Dalziel.
Berry & Guy: s.d.
Bertho: 1951.
Blaikie: Herb. K.
Blass: 1975.
Bojang: Herb. K.
Bouronville: 1967.
Brent: Herb. K; ex Dalziel, ex Irvine.
Brew: ex Pasch, 1980.
Brosnahan: ex Pasch, 1980.
Brossart: ex Chevalier; ex Dalziel.
Brotherton: 1969.

Bunting: ex Dalziel; ex Irvine.
Burbridge: Herb. K.
Burton: Herb. K.
Busnel: 1959.
C – Cooper: Herb. K; ex Cooper & Record, 1931.
C&H – Cyffer & Hutchinson. s.d.
C&R – Cooper & Record, 1931.
CG – Calme-Griaule, 1968.
CHOP – Pobéguin; ex auctt.
CJT – C.J. Taylor, 1960.
CV – Vigne: ex Dalziel; ex Irvine.
Caille: ex Chevalier; ex Dalziel.
Castelhain: ex Pasch, 1980.
Catterall: ex Dalziel.
Chapman: Herb. K.
Cheron: ex Pasch, 1980.
Chesters: ex Dalziel.
Chumbow: ex K. Williamson.
Clarke: Herb. K.
Clusters: Herb. K.
Cole: 1968.
Cook: s.d.
Correira: ex Pasch, 1980.
Coull: Herb. K; ex Dalziel; ex Irvine.
Curasson: ex Kerharo & Bouquet, 1950.
D&C – Ducroz & Charles, 1978.
D'O – D'Orey & Liberato, 1970.
DA – Departments of Agriculture (various territories).
DAP – Percival: Herb. K.
DF – Departments of Forests (various territories).
DRR – Rosevear, 1961.
DS – Small: Herb. K.
Dalby: ex Pasch, 1980.
Davey: Herb. K.
David: Herb. K.
Dawe: Herb. K.
Dawodu: Herb. K; ex Dalziel.
De la Croix: ex Kerharo.
Delafosse: ex Pasch, 1980.
Dennett: Herb. K; ex Dalziel.
Derive: ex Pasch, 1980.
Diarra: 1977.
Diop. Fal: ex Pasch, 1980.
Dodd: Herb. K.
Dolphyne: ex Pasch, 1980.
Donneux: ex Pasch, 1980.
Dubois: ex Kerharo & Bouquet; ex Pasch, 1980.
Dumas & Renoux: ex Dalziel.
Dumas: ex Dalziel.
Dumestre: ex Pasch, 1980.
Dunlap: Herb. K; ex Dalziel.
Dunnett: Herb. K; ex Dalziel; ex Pichon, 1953.
Duveyrier: ex Dalziel.
E – Engler.
E&A – Enti & Abbiw, pers. comm.
EPdS – De Sousa; Herb. K.
EWF – Foster: Herb. K; ex Dalziel.
Eady: Herb. K.
Easmon: ex Dalziel; ex Irvine.
Edgar: ex Dalziel.
Eggeling: Herb. K.
Egunjobi: 1969.
Ejiofor: Herb. K.
Elugbe: 1969.
Enti: Herb. K.
Etkin: 1981.
FB – Busson, 1965.
FCD – Deighton, 1957.

451

LIST OF AUTHORITIES CITED FOR VERNACULAR NAMES

FD – Forest Departments (various territories).
FG&G – Ferry, Gessain & Gessain, 1974.
FNH – Hepper: Herb. K.
FRI – Irvine, 1930, 1961, Herb. K, E.
FWHM – Migoed: ex Dalziel.
Fahaule & Russell: Herb. K.
Farmar: Herb. K; ex auctt.
Farquhar: Herb. K; ex Dalziel.
Fisher: Herb. K.
Foggie: ex Irvine.
Foureau: ex Chevalier; ex Dalziel.
Fox: Herb. K.
Frajzyngier: 1985.
Freeman: Herb. K.
Frith: Herb. K.
Froelich: 1954.
Funke: ex Pasch, 1980.
GFSE – Scott-Elliot: Herb. K; ex Dalziel.
GK – Kunkel, 1965.
GM – Manessy, 1975.
GR – Roberty.
Gaisser: ex Dalziel.
Gardner: 1975.
GeS – Gomes e Sousa, 1930.
Geissler: Herb. K.
Gerhard: ex Pasch, 1980.
Gillett: Herb. K.
Girault: 1958.
Giwa: Herb. UCI.
Glanville: Herb. K; ex auctt.
Glover: Herb. K.
Goldie: ex Pasch, 1980.
Golding: Herb. K.
Golding & Gwynn: 1939.
Graham: Herb. K; ex Dalziel.
Grall: 1945.
Gray: Herb. K.
Green: ex Pasch, 1980.
Greene: Herb. K.
Grégoire: 1975.
Grove: Herb. K.
Gwynn: in Golding & Gwynn, 1939.
H&D – Hutchinson & Dalziel.
H-Hansen – Huisberg-Hansen: ex auctt.
Hédin: ex Dalziel.
Höftmann: ex Pasch, 1980.
HM – H Maire, 1933.
Hair: ex Pasch, 1980.
Hallam: 1979.
Hambler: Herb. K.
Hampson: Herb. K.
Har. – the Harleys: Herb. K.
Hardcastle: Herb. K.
Harris: ex Pasch, 1980.
Haswell: Herb. K, FBC.
Hayes: ex Dalziel.
Heine: ex Pasch, 1980.
Henry: 1936; ex Dalziel.
Hepburn: Herb. K; ex Dalziel.
Herault: ex Pasch, 1980.
Herington: 1979.
Heudelot: Herb. K; ex Dalziel.
Heydorn: ex Pasch, 1980.
Hoffmann: ex K. Williamson.
Hopkinson: ex Dalziel.
Hovis: ex Pasch, 1980.
Howes: Herb. K; ex auctt.
Huntting: Herb. K; ex Dalziel.
IFE – Herb., Dept Pharmaceutics, Ife University.
Ilonah: pers. comm.
Innes: ex Pasch, 1980.
Isawumi: 1978.

Ithmann: 1976.
Ivanoff: ex Kerharo & Bouquet.
Iwu: 1980.
J – G. Jackson, 1973.
J&D – Jackson & David; ex Jackson.
JB – Berhaut: ex Kerharo & Adam.
JDES – Espirito Santo, 1963.
JGA – J.G Adam; ex Kerharo & Adam.
JLT – Trochain, 1940.
JMD – Dalziel; ex auctt.
JRA – Ainslie, 1937; ex auctt.
Jackson: 1973.
Jeffreys: ex Pasch, 1980.
Jenkins: Herb. K.
Johnston: Herb. K; ex Dalziel; ex Irvine.
Johnstone: Herb. K; ex Dalziel.
Jordan: Herb. K, FBC.
Joru: 1973.
Jumelle: ex Dalziel.
K – Kerharo.K&A – Kerharo & Adam; ex auctt.
K&B – Kerharo & Bouquet.
K&T – Kintz & Toutain, 1981.
KD – Kropp, Kropp-Dabuku, 1967, 1973.
KM – Miscewski: Herb. K.
KO&S – Keay, Onochie & Stanfield, 1960.
KW – K. Williamson: Herb. K, UCI; ex auctt.
KW&T – K. Williamson & Timitimi, 1983.
Kaliai: ex Pasch, 1980.
Kastenholz: ex Pasch, 1980.
Kennedy: Herb. K; ex Dalziel.
Kersting: Herb. K; ex Dalziel; ex Pichon.
Kesby: Herb. K.
Knops: ex Pasch, 1980.
Koelle: ex Pasch, 1980.
Konneh: ex Pasch, 1980.
Kpolovie: ex K. Williamson.
Kraft: 1981.
Krass: ex Pasch, 1980.
Krause: ex Irvine.
Kropp: 1967.
Kumah: ex Pasch, 1980.
L-P – Lane-Poole: Herb. K.
Lässig: ex Pasch, 1980.
LAKC – King-Church: Herb. K.
LB – Bouquiaux, 1971–72.
LC – Chalk et al., 1933.
Labouret: ex Pasch, 1980.
Ladefoged: ex Pasch, 1980.
Laffitte: ex Kerharo & Bouquet.
Lamb: Herb. K.
Lancaster: Herb. K.
Langdale-Brown: Herb. K.
Le Bris & Prost: 1981.
Le Riche: Herb. K.
Le Clerq: ex Dalziel.
Lean: Herb. K.
Lely: Herb. K.
Leroux: Herb. K.
Letham: ex Dalziel; ex Pasch, 1980.
Ll-W – Lloyd-Williams: Herb. K.
Lock: s.d.
Lowe: pers. comm.
Lukas: 1939.
Lynn: ex Dalziel; ex Irvine.
M&H – Melville & Hooker: Herb. K.
MM – M. McIntosh: pers. comm.
Macdonald: Herb. K.
Macfoy & Sama: 1983.
Macgregor: Herb. K.
McLeod: Herb. K; ex Dalziel; ex Pichon.
Macluskie: Herb. K.
Maitland: Herb. K; ex Dalziel; ex Lowe.
Marchal: 1959.

452

Marchesse: 1979.
Martineau: Herb. K.
Martinson: ex Irvine.
Matsushita: 1974.
Matthes: Herb. K.
Maule: Herb. K.
Mauny: 1953.
Mayo: Herb. K.
Meek: 1931; ex Dalziel.
Mellin: ex auctt.
Melzian: ex Pasch, 1980.
Metzer: Herb. K.
Mfonyam: ex K. Williamson.
Migoed: Herb. K.
Mildbraed: ex Dalziel.
Millen : Herb. K; ex Dalziel.
Millson: Herb. K; ex Dalziel.
Mohrang: ex Pasch, 1980.
Monteil: ex Pasch, 1980.
Moomo: ex K. Williamson.
Mortimore: pers. comm.
Mus. Stockholm – Museum Stockholm; Herb. K.
Musa Daggash: Herb. K.
N'Diaye: 1964.
N'Diaye-Correard: ex Pasch, 1980.
NWT – N.W. Thomas; Herb. K; ex auctt.
Nel: Herb. K.
Newman: 1974.
Newton: ex Dalziel.
Nicklin: 1975.
Nicolas: 1953.
Nicole: 1979.
Noel: ex Pasch, 1980.
OA – Ampofo, 1983, pers. comm.
Oblete: ex Blench.
Oche: Herb. K.
Odoh: 1978.
Ogo: ex Blench.
Oke: 1986.
Okusi: Herb. UCI; ex Carpenter.
Oldeman: Herb. K.
Onochie: Herb. K.
Onoh: ex Blench.
Osang: ex Blench.
PG – Garnier, 1976.
Pageaud: 1972.
Palmer: Herb. K.
Pasch: 1980.
Patterson: Herb K.
Pax: ex Dalziel.
Pedder: Herb. K.
Pirie: ex Dalziel.
Philcox: Herb. K.
Phillips: Herb. K; ex Dalziel.
Pichl: 1963.
Port Develop, Synd.: Herb. K.
Prost: Herb. K. IFAN; ex auctt.
Pyne: Herb. K.
RB – R. Blench.
RES – R.E. Sharland; Herb. K; pers. comm.
RFGA – Adams, 1943; Herb. K.
RGA – R.G. Armstrong; ex Blench, 1985.
RM – R. Maire.
RP – R. Portères.
RS – Schnell.
RdoF – Feijão.
Reece: Herb K.
Richards: Herb. K, BM.
Robin: 1947.
Rogeon: ex Dalziel.
Ross: Herb. K; ex Dalziel.

Roth-Lely: ex Pasch, 1980.
Rowland: Herb. K.
Rytz: s.d.
S&F – Savill & Fox.
SKS – S.K. Samai; Herb. K, FBC.
SOA – Alasoadura: Herb. K, UCI.
ST – Tattersall, 1978.
Sampson: Herb. K.
Sankey: Herb. K, UCI; ex Dalziel.
Saunders: Herb. K; ex Irvine.
Scholz: Herb. K.
Schuh: 1978, 1981.
Schule: Herb. K.
Seidel: ex Pasch, 1980.
Shimizu: s.d.; ex Blench.
Sidibe: Herb. K.
Sikes: 1972.
Singha: 1965.
Sodah Ayenor & Matthews: 1971.
Soward: Herb. K; ex auctt.
Stocks: Herb. K.
Surugue: 1979.
Swarbrick: ex Lowe.
Symington: Herb. K.
TFC – Chipp, 1922; Herb. K.
Taiwo: Herb. K.
Taylor – C.J. Taylor, 1960.
Taylor: ex Dalziel, several persons.
Tchamda: ex Blench.
Terpstra: ex Pasch, 1980.
Terry: s.d.
Tersis: ex Blench; ex Pasch, 1980.
Thatcher: Herb. K.
Thillard: ex Dalziel.
Thoiré: ex Pasch, 1980.
Thomann: ex Pasch, 1980.
Thompson: Herb. K.
Thorold: Herb. K.
Tiemo: ex K. Williamson.
Trenga: ex Pasch, 1980.
Trifkovic: ex Pasch, 1980.
Tuley: Herb. K.
Turnbull: Herb. K.
Twi Dict. – Twi Dictionary; ex Irvine.
Udofia: ex Lowe.
Ugbe: ex K. Williamson.
Veldkamp: Herb. K.
Verger: 1965–86; pers. comm.
Vermeer: Herb. UCI.
Vieillard: 1940.
Virgo: Herb. K.
Visser: 1975.
Volkens: ex Dalziel; ex Irvine.
Volkens & Gaisser: ex Dalziel.
Von Doering: Herb. K; ex Dalziel.
Von Duisberg: ex Dalziel.
Vuillet: ex Dalziel.
WG&A – unidentified.
Waldau: ex Dalziel.
Wallace: Herb. K.
Ward: Herb. K.
Westermann: ex auctt.
White: 1957.
Wigboldus: 1986.
Williams: Herb. K.
Williams & Farias: 1972.
Wolff: 1969.
Yates: Herb. K; ex Dalziel.
ZF – Frajzngier, 1985.
ZOG – Gbile, 1980.

453

457

458

INDEX OF PLANT AND PRODUCT NAMES

In English (Eng.), French (Fr.), Portuguese (Port.) and some other languages, and trade names.

A

Abrasin (Fr.) – *Vernicia cordata*, **Euphorbiaceae**
Abricotier d'Afrique (Fr.) – *Mammea africana*, **Guttiferae**
Abricotier d'Amérique (Fr.) – *Mammea americana*, **Guttiferae**
Açafrão (Port.) – *Crocus sativus*, **Iridaceae**
Acha grass (Eng.) – *Digitaria exilis*, **Gramineae**
Achimenes (Eng.) – *Achimenes spp.*, **Gesneriaceae**
Adjansi (trade) – *Margaritaria discoidea*, **Euphorbiaceae**
Adlai (Eng.) – *Coix lacryma-jobi*, **Gramineae**
African apple (Eng.) – *Mammea africana*, **Guttiferae**
African apricot (Eng.) – *Mammea africana*, **Guttiferae**
African crocus (Eng.) – *Curculigo pilosa*, **Hypoxidaceae**
African ebony (Eng.) – *Diospyros kamerunensis*, *D. sanza-minika*, **Ebenaceae**
African love-grass (Eng.) – *Eragrostis curvula*, **Gramineae**
African mammee-apple (Eng.) – *Mammea africana*, **Guttiferae**
African mango (Eng.) – *Irvingia gabonensis*, **Ixonanthaceae**
African millet (Eng.) – *Eleusine coracana*, **Gramineae**
African nut tree (Eng.) – *Ricinodendron heudelotii*, **Euphorbiaceae**
African oak (Eng.) – *Oldfieldia africana*, **Euphorbiaceae**
African rice (Eng.) – *Oryza glaberrima*, **Gramineae**
African violet (Eng.) – *Saintpaulia ionantha*, **Gesneriaceae**
African walnut (Eng.) – *Plukenetia conophora*, **Euphorbiaceae**
African wood (Eng.) – *Ricinodendron heudelotii*, **Euphorbiaceae**
African wood-oil nut tree (Eng.) – *Ricinodendron heudelotii*, **Euphorbiaceae**
'Ajandhaman' pennisetum (Eng.) – *Cenchrus setigerus*, **Gramineae**
Akossika à grandes feuilles (Fr.) – *Scottellia chevalieri*, **Flacourtiaceae**
Akossika à petites feuilles (Fr.) – *Scottellia coriacea*, **Flacourtiaceae**
Alexandrian laurel (Eng.) – *Calophyllum inophyllum*, **Guttiferae**
Annual beard grass (Eng.) – *Polypogon monspeliensis*, **Gramineae**
Annual poinsettia (Eng.) – *Euphorbia heterophylla*, **Euphorbiaceae**
Antelope grass (Eng.) – *Echinochloa pyramidalis*, **Gramineae**

Apricot of San Domingo (Eng.) – *Mammea americana*, **Guttiferae**
Arbre à beurre (Fr.) – *Pentadesma butyracea*, **Guttiferae**
Arbre à savon du Gabon (Fr.) – *Tetrochidium didymostemon*, **Euphorbiaceae**
Arbre à tabatières (Fr.) – *Oncoba spinosa*, **Flacourtiaceae**
Arbre au corail (Fr.) – *Jatropha multifida*, **Euphorbiaceae**
Arbre de Saint Sébastien (Fr.) – *Euphorbia tirucalli*, **Euphorbiaceae**
Arruz (Port.) – *Oryza sativa*, **Gramineae**
Asas (trade) – *Bridelia micrantha*, **Euphorbiaceae**
Asiatic or Asian rice (Eng.) – *Oryza sativa*, **Gramineae**
Assas (trade) – *Bridelia micrantha*, **Euphorbiaceae**
Atala (Eng.) – *Sacoglottis gabonensis*, **Humiriaceae**
Australian couch (Eng.) – *Cynodon dactylon*, **Gramineae**
Australian (or Queensland) asthma herb (Eng.) – *Euphorbia hirta*, **Euphorbiaceae**
Awran (trade) – *Diospyros kamerunensis*, **Ebenaceae**
Awusa (trade) – *Plukenetia conophora*, **Euphorbiaceae**

B

Bahama grass (Eng.) – *Cynodon dactylon*, **Gramineae**
Balan des savanes (Fr.) – *Securinega virosa*, **Euphorbiaceae**
Balsam spurge (Eng.) – *Euphorbia balsamifera*, **Euphorbiaceae**
Bamboo (Eng.) – *Bambusa vulgaris*, *Oxytenanthera abyssinica*, **Gramineae**
Bambou (Fr.) – *Bambusa vulgaris*, *Oxytenanthera abyssinica*, **Gramineae**
Bambu-vulgar (Port.) – *Bambusa vulgaris*, **Gramineae**
Barbados nut (Eng.) – *Jatropha curcas*, **Euphorbiaceae**
Barbon (Fr.) – *Hyparrhenia hirta*, **Gramineae**
Barley (Eng.) – *Hordeum vulgare*, **Gramineae**
Barnyard millet (Eng.) – *Echinochloa crus-galli*, **Gramineae**
Barra grass (Eng.) – *Pennisetum pedicellatum*, *P. polystachion*, **Gramineae**
Bastard mahogany (Eng.) – *Mammea africana*, **Guttiferae**
Bastard millet (Eng.) – *Paspalum scrobiculatum*, **Gramineae**
Beignefala (Fr.) – *Cymbopogon giganteus*, **Gramineae**

467

E

Ébano (Port.) – *Diospyros crassiflora,*
Ebenaceae
Ebénier de l'Ouest Africain (Fr.) – *Diospyros
mespiliformis,* **Ebenaceae**
Ebénier véritable du Gabon (Fr.) – *Diospyros
crassiflora,* **Ebenaceae**
Ebony (Eng.) – *Diospyros mespiliformis, D.
sanza-minika,* **Ebenaceae**
Efwatakala grass (Eng.) – *Melinis minutiflora,*
Gramineae
Egyptian corn (Eng.) – *Zea mays,* **Gramineae**
Eho (trade) – *Ricinodendron heudelotii,*
Euphorbiaceae
Elephant grass (Eng.) – *Pennisetum purpureum,*
Gramineae
Eleusine (Fr.) – *Eleusine coracana, E. indica,*
Gramineae
Epi collant (Fr.) – *Setaria verticillata,*
Gramineae
Épine du Christ (Fr.) – *Euphorbia milii,*
Euphorbiaceae
Erimado (trade) – *Ricinodendron heudelotii,*
Euphorbiaceae
Erapo grass (Eng.) – *Loudetia phragmitoides,*
Gramineae
Erua-dos-rosários (Port.) – *Coix lacryma-jobi,*
Gramineae
Erva Príncipe (Port.) – *Cymbopogon citratus,*
Gramineae
Erva-da-Guiné (Port.) – *Panicum maximum,*
Gramineae
Escalracho (Port.) – *Panicum repens,*
Gramineae
Espadana (Port.) – *Gladiolus,* **Iridaceae**
Essang (trade) – *Ricinodendron heudelotii,*
Euphorbiaceae
Essessang (trade) – *Ricinodendron heudelotii,*
Euphorbiaceae
Étoile de Noel (Fr.) – *Euphorbia pulcherrima,*
Euphorbiaceae
Euphorbe bicolore (Fr.) – *Euphorbia
heterophylla,* **Euphorbiaceae**
Euphorbe cactiforme (Fr.) – *Euphorbia
sudanica,* **Euphorbiaceae**
Euphorbe candélabre (Fr.) – *Euphorbia
balsamifera,* **Euphorbiaceae**
Euphorbe de Cayor (Fr.) – *Euphorbia
balsamifera,* **Euphorbiaceae**
Euphorbe du Soudan (Fr.) – *Euphorbia
sudanica,* **Euphorbiaceae**
Euphorbe écarlate (Fr.) – *Euphorbia milii,*
Euphorbiaceae
Euphorbe lactée (Fr.) – *Euphorbia lactea,*
Euphorbiaceae
Eveuss (trade) – *Klainedoxa gabonensis,*
Ixonanthaceae
Evila (Fr.) – *Diospyros crassiflora,* **Ebenaceae**

F

False chew-stick tree (Eng.) – *Garcinia
gnetoides, G. smeathmannii,* **Guttiferae**
False kola (Eng.) – *Garcinia kola,* **Guttiferae**

False yam (Eng.) – *Icacina oliviformis,*
Icacinaceae
Fausse citronelle (Fr.) – *Cymbopogon citratus,*
Gramineae
Fausse-cane à sucre (Fr.) – *Pennisetum
purpureum,* **Gramineae**
Faux ebène (Fr.) – *Diospyros gabunensis,*
Ebenaceae
Faux ebénier (Fr.) – *Diospyros mannii,*
Ebenaceae
Faux (fausse) colatier (Fr.) – *Garcinia kola,*
Guttiferae
Faux palétuvier (Fr.) – *Uapaca guineensis,*
Euphorbiaceae
Fel-da-terra (Port.) – *Centaurium pulchellum,*
Gentianaceae
Feve d'enfer (Fr.) – *Jatropha curcas,*
Euphorbiaceae
Fig nut (Eng.) – *Jatropha curcas,*
Euphorbiaceae
Fiji couch (Eng.) – *Cynodon dactylon,*
Gramineae
Finger comb grass (Eng.) – *Dactyloctenium
aegyptium,* **Gramineae**
Finger grass (Eng.) – *Digitaria debilis, D.
horizontalis,* **Gramineae**
Finger millet (Eng.) – *Eleusine coracana,*
Gramineae
Fish-pole bamboo (Eng.) – *Phyllostachys
aervea,* **Gramineae**
Flint bark (Eng.) – *Diospyros sanza-minika, D.
canaliculata,* **Ebenaceae**
Flint-bark tree (Eng.) – *Diospyros gabunensis,*
Ebenaceae
Fogno (Fr.) – *Digitaria exilis,* **Gramineae**
Fonio (Fr.) – *Digitaria exilis, D. longiflora,*
Gramineae
Fonio à grosses graines (Fr.) – *Brachiaria
deflexa,* **Gramineae**
Fonio noire (Fr.) – *Digitaria iburua,* **Gramineae**
Fonio sauvage (Fr.) – *Digitaria longiflora,
Panicum laetum,* **Gramineae**
French physic nut (Eng.) – *Jatropha multifida,*
Euphorbiaceae
Fried egg tree (Eng.) – *Oncoba spinosa,*
Flacourtiaceae
Fundi (Fr.) – *Digitaria exilis,* **Gramineae**

G

Gaboon ebony (Eng.) – *Diospyros viridicans,*
Ebenaceae
Gazon bleu à Dakar (Fr.) – *Cynodon dactylon,*
Gramineae
Gazon des Bermudes (Fr.) – *Cynodon dactylon,*
Gramineae
Giant St. John's wort (Eng.) – *Hypericum
revolutum,* **Guttiferae**
(Giant) star grass (Eng.) – *Cynodon
plectostachys,* **Gramineae**

470

K

Kafir-corn (Eng.) – *Sorghum bicolor*,
Gramineae
Kaki de brousse (Fr.) – *Diospyros
mespiliformis*, **Ebenaceae**
Kamala (Eng.) – *Mallotus philippinensis*,
Euphorbiaceae
Kamerun grass (Eng.) – *Sorghum
arundinaceum*, **Gramineae**
Karamani wax (trade) – *Symphonia globulifera*,
Guttiferae
Khus-khus (Eng.) – *Vetiveria zizanioides*,
Gramineae
Kikuyu grass (Eng.) – *Pennisetum clandestinum*,
Gramineae
Koda (millet) (Eng.) – *Paspalum scrobiculatum*,
Gramineae
' Kodra (millet) (Eng.) – *Paspalum
scrobiculatum*, **Gramineae**
Korakan (Eng.) – *Eleusine coracana*,
Gramineae
Kram-kram (Fr.) – *Cenchrus biflorus*,
Gramineae
Kroma (trade) – *Klainedoxa gabonensis*,
Ixonanthaceae
Kurkan (Eng.) – *Eleusine coracana*, **Gramineae**
Kuskus grass (Eng.) – *Echinochloa obtusiflora*,
Gramineae

L

Lacewood (trade) – *Allanblackia floribunda*,
Guttiferae
Lágrimas de Job (Port.) – *Coix lacryma-jobi*,
Gramineae
Lake Chisi grass (Eng.) – *Leptochloa fusca*,
Gramineae
Lalang (Eng.) – *Imperata cylindrica*, **Gramineae**
Lami (Fr.) – *Pentadesma butyracea*, **Guttiferae**
Landa (trade) – *Erythroxylum mannii*,
Erythroxylaceae
Large-seeded millet grass (Eng.) – *Brachiaria
brizantha*, **Gramineae**
Larmes de Job (Fr.) – *Coix lacryma-jobi*,
Gramineae
Larmier de Job (Fr.) – *Coix lacryma-jobi*,
Gramineae
Lawn-grass (Eng.) – *Axonopus compressus*,
Gramineae
Lehmann's love grass (Eng.) – *Eragrostis
lehmanniana*, **Gramineae**
Lemon grass (Eng.) – *Cymbopogon citratus*,
Gramineae
Liane brûlante (Fr.) – *Tragia*, **Euphorbiaceae**
Louisiana grass (Eng.) – *Axonopus compressus*,
Gramineae
Loureiro-de-Alexandria (Port.) – *Calophyllum
inophyllum*, **Guttiferae**
Love grass (Eng.) – *Chrysopogon aciculatus*,
Gramineae

M

Mabola persimmon (Eng.) – *Diospyros discolor*,
Ebenaceae
Mabolo (Eng.) – *Diospyros discolor*, **Ebenaceae**
Madagascar plum (Eng.) – *Flacourtia indica*,
Flacourtiaceae
Magbel's millet (Eng.) – *Panicum afzelii*,
Gramineae
Mahogany (Eng.) – *Sacoglottis gabonensis*,
Humiriaceae
Maïs (Fr.) – *Zea mays*, **Gramineae**
Maize (Eng.) – *Zea mays*, **Gramineae**
Male kola (Eng.) – *Garcinia kola*, **Guttiferae**
Malnommée (Fr.) – *Euphorbia hirta*,
Euphorbiaceae
Mammee apple (Eng.) – *Mammea africana, M.
americana*, **Guttiferae**
Mammy apple (Eng.) – *Mammea americana*,
Guttiferae
Mammy supporter (Eng.) – *Mammea africana*,
Guttiferae
Manchineel (Eng.) – *Hippomane mancinella*,
Euphorbiaceae
Mancinillier (Fr.) – *Hippomane mancinella*,
Euphorbiaceae
Mandinka rice (Eng.) – *Oryza barthii*,
Gramineae
Mandioc (Fr.) – *Manihot esculenta*,
Euphorbiaceae
Mandioca (Port.) – *Manihot esculenta*,
Euphorbiaceae
Mangostão (Port.) – *Garcinia mangostana*,
Guttiferae
Mangosteen (Eng.) – *Garcinia mangostana*,
Guttiferae
Manguier sauvage (Fr.) – *Irvingia gabonensis*,
Ixonanthaceae
Mani wax (trade) – *Symphonia globulifera*,
Guttiferae
Manicoba (Port.) – *Manihot glaziovii*,
Euphorbiaceae
Maniçoba (Fr.) – *Manihot glaziovii*,
Euphorbiaceae
Manioc (Fr.) – *Manihot esculenta*,
Euphorbiaceae
Manioc de Céara (Fr.) – *Manihot glaziovii*,
Euphorbiaceae
Manioc de lièvre (Fr.) – *Curculigo pilosa*,
Hypoxidaceae
Manne noire (Fr.) – *Digitaria iburua*,
Gramineae
Marua (Eng.) – *Eleusine coracana*, **Gramineae**
Matting grass (Eng.) – *Pennisetum
pedicellatum, P. polystachion*, **Gramineae**
Mauritius grass (Eng.) – *Brachiaria mutica*,
Gramineae
Mealie (Eng.) – *Zea mays*, **Gramineae**
Médicinier (Fr.) – *Jatropha gossypiifolia, J.
multifida*, **Euphorbiaceae**
Médicinier batard (Fr.) – *Jatropha
gossypiifolia*, **Euphorbiaceae**
Médicinier béni (Fr.) – *Jatropha curcas*,
Euphorbiaceae
Médicinier d'Espagne (Fr.) – *Jatropha
multifida*, **Euphorbiaceae**

Médicinier rouge (Fr.) – *Jatropha gossypiifolia*, **Euphorbiaceae**
Médicinier sauvage (Fr.) – *Jatropha gossypiifolia*, **Euphorbiaceae**
'Medlar' (Eng.) – *Uapaca esculenta*, **Euphorbiaceae**
Mevini (Fr.) – *Diospyros crassiflora*, **Ebenaceae**
Miglio zaburro (Port.) – *Sorghum bicolor*, **Gramineae**
Mil (Fr.) – *Pennisetum glaucum*, *Sorghum bicolor*, **Gramineae**
Mil à chandelle (Fr.) – *Pennisetum glaucum*, **Gramineae**
Mil chandelle (Fr.) – *Pennisetum glaucum*, **Gramineae**
Mil sauvage (Fr.) – *Sorghum arundinaceum*, **Gramineae**
Milhã-verde (Port.) – *Setaria viridis*, **Gramineae**
Milhã-verticilada (Port.) – *Setaria verticillata*, **Gramineae**
Milho (Port.) – *Zea mays*, **Gramineae**
Milho grande (Port.) – *Zea mays*, **Gramineae**
Milho grosso (Port.) – *Zea mays*, **Gramineae**
Milho-maeês (Port.) – *Zea mays*, **Gramineae**
Milk bush (Eng.) – *Euphorbia tirucalli*, **Euphorbiaceae**
Millepertuis velu (Fr.) – *Psorospermum corymbiferum*, **Guttiferae**
Millet de Yokohama (Fr.) – *Eleusine coracana*, **Gramineae**
Millet digitaire (Fr.) – *Digitaria exilis*, **Gramineae**
Mission grass (Eng.) – *Pennisetum polystachion*, **Gramineae**
Molasses grass (Eng.) – *Melinis minutiflora*, **Gramineae**
Monkey guava (Eng.) – *Diospyros mespiliformis*, **Ebenaceae**
Monkey's dinner bell (Eng.) – *Hura crepitans*, **Euphorbiaceae**
Montbretia (Eng.) – *Crocosmia aurea*, **Iridaceae**

Niger plum (Eng.) – *Flacourtia flavescens*, **Flacourtiaceae**
Noix chandelle (Fr.) – *Vernicia cordata*, **Euphorbiaceae**
Noix de Bancoil (Fr.) – *Vernicia cordata*, **Euphorbiaceae**
Ntala (Eng.) – *Sacoglottis gabonensis*, **Humiriaceae**
'Number one' (Eng.) – *Mareya micrantha*, **Euphorbiaceae**

O
Oatgrass (Eng.) – *Monocymbium ceresiiforme*, **Gramineae**
Obota (Fr.) – *Mammea africana*, **Guttiferae**
Odoko (trade) – *Scottellia coriacea*, **Flacourtiaceae**
Ohoué à grandes feuilles (Fr.) – *Grossera vignei*, **Euphorbiaceae**
Okwen (trade) – *Ricinodendron heudelotii*, **Euphorbiaceae**
Old citronella grass (Eng.) – *Cymbopogon winterianus*, **Gramineae**
Old lands grass (Eng.) – *Chloris virgata*, **Gramineae**
Oncoba tabatières (Fr.) – *Oncoba spinosa*, **Flacourtiaceae**
Orge (Fr.) – *Hordeum vulgare*, **Gramineae**
Orogbo kola nut (Eng.) – *Garcinia kola*, **Guttiferae**
Orroz (Port.) – *Oryza*, **Gramineae**
Osol (trade) – *Symphonia globulifera*, **Guttiferae**
Otaheite gooseberry (Eng.) – *Phyllanthus acidus*, **Euphorbiaceae**
Otutu (trade) – *Diospyros kamerunensis*, **Ebenaceae**
Ozouga (Eng.) – *Sacoglottis gabonensis*, **Humiriaceae**

N
Nadi blue grass (Eng.) – *Dichanthium caricosum*, **Gramineae**
Napier grass (Eng.) – *Pennisetum purpureum*, **Gramineae**
Napier's fodder (Eng.) – *Pennisetum purpureum*, **Gramineae**
Natal grass (Eng.) – *Melinis repens*, **Gramineae**
Native mango (Eng.) – *Irvingia gabonensis*, **Ixonanthaceae**
'Nettle' (Eng.) – *Tragia*, **Euphorbiaceae**
New citronella grass (Eng.) – *Cymbopogon nardus*, **Gramineae**
N'gart (trade) – *Plukenetia conophora*, **Euphorbiaceae**
Ngavi à gros fruits (Fr.) – *Diospyros mannii*, **Ebenaceae**
Ngavi à petites feuilles (Fr.) – *Diospyros heudelotii*, **Ebenaceae**
Ngavi du fourré littoral (Fr.) – *Diospyros ferrea*, **Ebenaceae**

P
Paddy (Eng.) – *Oryza sativa*, **Gramineae**
Padi (Eng.) – *Oryza sativa*, **Gramineae**
Palétuvier d'eau douce (Fr.) – *Uapaca guineensis*, **Euphorbiaceae**
Palha carga (Port.) – *Imperata cylindrica*, **Gramineae**
Palma Christi (Eng.) – *Ricinus communis*, **Euphorbiaceae**
Para grass (Eng.) – *Brachiaria mutica*, **Gramineae**
Para rubber (Eng.) – *Hevea brasiliensis*, **Euphorbiaceae**
Pearl millet (Eng.) – *Pennisetum glaucum*, **Gramineae**
Pé-de-galinha (Port.) – *Brachiaria ramosa*, *Dactyloctenium aegyptium*, **Gramineae**
Pé-de-galo (Port.) – *Eleusine indica*, **Gramineae**
Pepper stick (Eng.) – *Drypetes aubrevillei*, **Euphorbiaceae**
Pepperbark (Eng.) – *Drypetes aubrevillei*, **Euphorbiaceae**

Sierra Leone guinea-corn (Eng.) – *Sorghum bicolor*, **Gramineae**

Silky star lily (Eng.) – *Hypoxis urceolata*, **Hypoxidaceae**

Silt grass (Eng.) – *Paspalum vaginatum*, **Gramineae**

Slender star lily (Eng.) – *Hypoxis angustifolia*, **Hypoxidaceae**

Snake grass (Eng.) – *Eragrostis cilianensis*, **Gramineae**

Snow bush (Eng.) – *Breynia disticha*, **Euphorbiaceae**

Snuff-box tree (Eng.) – *Oncoba spinosa*, **Flacourtiaceae**

Sorgho (Fr.) – *Sorghum bicolor*, **Gramineae**

Sorgo (Port.) – *Sorghum bicolor*, **Gramineae**

Sorgo d'Alep (Port.) – *Sorghum X drummondii*, **Gramineae**

'Sour grapes' (Eng.) – *Phyllanthus reticulatus*, **Euphorbiaceae**

Sour grass (Eng.) – *Paspalum conjugatum*, **Gramineae**

South African 'Kyasuwa' (Eng.) – *Cenchrus ciliaris*, **Gramineae**

South African pennisetum (Eng.) – *Cenchrus setigerus*, **Gramineae**

Southern burgrass (Eng.) – *Cenchrus echinatus*, **Gramineae**

Spanish physic nut (Eng.) – *Jatropha multifida*, **Euphorbiaceae**

Spear grass (Eng.) – *Heteropogon contortus*, *Imperata cylindrica*, **Gramineae**

Spiked millet (Eng.) – *Pennisetum glaucum*, **Gramineae**

Star lily (Eng.) – *Hypoxis recurva*, **Hypoxidaceae**

Stink grass (Eng.) – *Eragrostis cilianensis*, *Melinis minutiflora*, **Gramineae**

Strongly scented love-grass (Eng.) – *Eragrostis cilianensis*, **Gramineae**

Sudan grass (Eng.) – *Sorghum X drummondii*, **Gramineae**

Sugar plum (Eng.) – *Uapaca esculenta*, *U. guineensis*, **Euphorbiaceae**

Sugar-cane (Eng.) – *Saccharum officinarum*, **Gramineae**

Surelle (Fr.) – *Phyllanthus acidus*, **Euphorbiaceae**

Surette de la Martinique (Fr.) – *Phyllanthus acidus*, **Euphorbiaceae**

Swamp ebony (Eng.) – *Diospyros mespiliformis*, **Ebenaceae**

Sweet grass (Eng.) – *Chloris virgata*, **Gramineae**

Sweet sorghum (Eng.) – *Sorghum bicolor*, **Gramineae**

Sword lily (Eng.) – *Gladiolus*, **Iridaceae**

T

Tabas grass (Eng.) – *Lasiurus hirsutus*, **Gramineae**

Tala (Eng.) – *Sacoglottis gabonensis*, **Humiriaceae**

Tallow (Eng.) – *Pentadesma butyracea*, **Guttiferae**

Tallow tree (Eng.) – *Allanblackia floribunda*, *Pentadesma butyracea*, **Guttiferae**

Tamba millet (Eng.) – *Eleusine coracana*, **Gramineae**

Tambuki grass (Eng.) – *Andropogon gayanus*, **Gramineae**

Tanner grass (Eng.) – *Brachiaria arrecta*, **Gramineae**

Tapioca (Eng.) – *Manihot esculenta*, **Euphorbiaceae**

Tègne (Fr.) – *Sorghum bicolor*, **Gramineae**

Tequie manincoba (Eng.) – *Manihot dichotoma*, **Euphorbiaceae**

Thatch grass (Eng.) – *Anadelphia leptocoma*, **Gramineae**

Thatchgrass (Eng.) – *Anadelphia afzeliana*, **Gramineae**

Tigne (Fr.) – *Sorghum bicolor*, **Gramineae**

Torpedo grass (Eng.) – *Panicum repens*, **Gramineae**

Tree cassava (Eng.) – *Manihot glaziovii*, **Euphorbiaceae**

Tricholène rose (Fr.) – *Melinis repens*, **Gramineae**

Trigo (Port.) – *Triticum aestivum*, **Gramineae**

Tsauri grass (Eng.) – *Cymbopogon giganteus*, **Gramineae**

Turkish corn (Eng.) – *Zea mays*, **Gramineae**

U

Udika (Fr.) – *Irvingia gabonensis*, **Ixonanthaceae**

Utrasum bean tree (Eng.) – *Elaeocarpus angustifolius*, **Elaeocarpaceae**

Uva grass (Eng.) – *Gynerium sagittatum*, **Gramineae**

V

Vetiver (Eng.) – *Vetiveria nigritana*, *V. zizanioides*, **Gramineae**

Vétiver (Fr.) – *Vetiveria nigritana*, **Gramineae**

W

Walking-stick ebony (Eng.) – *Diospyros monbuttensis*, **Ebenaceae**

Water grass (Eng.) – *Brachiaria mutica*, **Gramineae**

Weeping love-grass (Eng.) – *Eragrostis curvula*, **Gramineae**

West African bamboo (Eng.) – *Oxytenanthera abyssinica*, **Gramineae**

West African ebony (Eng.) – *Diospyros mespiliformis*, **Ebenaceae**

Wheat (Eng.) – *Triticum aestivum*, **Gramineae**

White acha (Eng.) – *Digitaria exilis*, **Gramineae**

White buffel (Eng.) – *Cenchrus pennisetiformis*, **Gramineae**

White kotowuli (Eng.) – *Drypetes afzelii*, **Euphorbiaceae**

PLANT SPECIES BY USAGES

Plant species are listed here by recorded usage. Many usages are well-established and are indisputable. Others are not so well known and readers are enjoined to make their own subjective judgements on the evidence offered and their own additional experience, particularly in instances where risk of human damage may result. There is much yet to be learnt regarding the harvesting of, say, famine-foods, and their preparation for consumption, and in the phytochemical/pharmacological field on edaphic and phenological factors and on methods by which medicinal materials are collected and prepared, as, indeed, on safe and certain identification of the plants themselves.

A: Food;   1, general
             2, special diets
             3, sauces, condiments, spices, flavourings
             4, sweets, sweetmeats
             5, masticatory
B: Drink;   1, water/sap
             2, sweet, milk substitutes
             3, infusions
             4, water-purifiers
             5, alcoholic, stimulant
C: Medicines;   1, generally healing
             2, skin, mucosae
             3, cutaneous, subcutaneous parasitic infection
             4, skeletal structure
             5, paralysis, epilepsy, convulsions, spasm
             6, insanity
             7, brain, nervous system
             8, heart
             9, arteries, veins
             10, blood disorders
             11, pain-killers
             12, sedatives, etc.
             13, arthritis, rheumatism, etc.
             14, eye treatments
             15, ear treatments
             16, oral treatments
             17, naso-pharyngeal affections
             18, pulmonary troubles
             19, stomach troubles
             20, 'intestines'
             21, emetics
             22, antemetics
             23, laxatives, etc.
             24, diarrhoea, dysentery
             25, cholera
             26, vermifuges
             27, liver, etc.
             28, kidneys, diuretics
             29, anus, haemorrhoids
             30, genital stimulants/depressants
             31, menstrual cycle
             32, pregnancy, antiabortifacients
             33, lactation stimulants (incl. veterinary)
             34, abortifacients, ecbolics
             35, venereal diseases
             36, febrifuges
             37, small-pox, chicken-pox, measles, etc.
             38, leprosy
             39, yaws
             40, dropsy, swellings, oedema, gout
             41, tumours, cancers
             42, malnutritrion, debility
             43, food poisoning

44, antidotes (venomous stings, bites, etc.)
45, homeopathic
46, diabetes

D: Phytochemistry;   1, alkali salts (excl. common salt)
2, soap and substitutes
3, salt and substitutes
4, mineral salts
5, fatty acids, etc.
6, aromatic substances
7, starch, sugar
8, mucilage
9, alkaloids
10, glycosides, saponins, steroids
11, tannins, astringents
12, flavones
13, resins
14, hydrogen cyanide
15, fish-poisons
16, insecticides, arachnicides
17, rodenticides, mammal and bird poisons
18, reptile-repellents
19, molluscicides
20, arrow-poisons
21, ordeal-poisons
22, depilatory
23, antibiotic, bacteristatic, fungistatic
24, urticant
25, miscellaneously poisonous or repellent

E: Agri-horticulture;   1, ornamental, cultivated or partially tended
2, fodder
3, veterinary medicine
4, bee/honey plants, insect plants
5, land conservation
6, composting, manuring
7, indicators (soil, water)
8, indicators (weather, season, time)
9, shade-trees
10, hedges, markers
11, fence-posts, poles, sticks
12, weeds, parasites
13, biotically active

F: Products;   1, building materials
2, carpentry and related applications
3, farming, forestry, hunting and fishing apparatus
4, fuel and lighting
5, household, domestic and personal items
6, pastimes—carving, musical instruments, games, toys, etc.
7, containers, food-wrappers
8, abrasives, cleaners, etc.
9, chew-sticks, etc.
10, fibre
11, floss, stuffing and caulking
12, withies and twigs
13, pulp and paper
14, pottery
15, beehives
16, dyes, stains, inks, tattoos and mordants
17, tobacco, snuff
18, exudations—gums, resins, etc.
19, manna and other exudations

G: Social;   1 religion, superstition, magic
2 ceremonial
3 sayings, aphorisms

478

A: Food; 1, cooked or uncooked; staple, supplementary or famine foods (including those eaten only in limited quantity or requiring special treatment before eating to reduce or eliminate toxic substances) EBENACEAE: *Diospyros barteri* (fruit-pulp), *D. dendo* (fruit), *D. discolor* (fruit), *D. elliotii* (fruit-pulp), *D. ferrea* (fruit-pulp), *D. heterotricha* (fruit), *D. heudelotii* (fruit-pulp), *D. kamerunensis* (fruit), *D. mespiliformis* (leaf, fruit), *D. soubreana* (fruit-pulp), *D. thomasii* (fruit-pulp), *D. viridicans* (fruit). ELAEOCARPACEAE: *Elaeocarpus angustifolius* (fruit-pulp). ERYTHROXYLACEAE: *Erythroxylum emarginatum* (berry). EUPHORBIACEAE: *Acalypha ciliata* (leaf), *Alchornea cordifolia* (fruit), *A. floribunda* (leaf), *Aleurites moluccana* (kernel cake), *Antidesma venosum* (fruit), *Bridelia atroviridis* (seed), *B. micrantha* (fruit), *B. ndellensis* (fruit), *B. scleroneura* (fruit), *Cleistanthus* spp. (fruit), *Cnidoscolus acontifolius*, *Cyrtogonone argentea* (seed), *Drypetes floribunda* (fruit pulp), *D. gilgiana* (fruit pulp), *D. ivorensis* (fruit-pulp), *Erythrococca chevalieri*, *E. welwitschiana*, *Euphorbia balsamifera* (young shoots), *E. hirta* (young shoots), *Hevea brasiliensis* (seed), *Hymenocardia acida* (young leaf, fruit), *Jatropha curcas* (young leaf), *J. multifida* (root), *Maesobotrya barteri* (fruit), *M. dusenii* (fruit), *M. staudtii* (fruit), *Manihot esculenta* (leaf, tuber), *M. glaziovii* (leaf, root), *Manniophyton fulvum* (seed-kernel), *Margaritaria discoidea* (fruit), *Micrococca mercurialis* (plant), *Phyllanthus muellerianus* (leaf, fruit), *P. reticulatus* (fruit), *Plukenetia conophora* (leaf, young shoot, kernel, seed-oil, seed-cake), *Ricinodendron heudelotii* (seed-kernel), *Ricinus communis* (seed), *Sapium ellipticum* (fruit), *Securinega virosa* (fruit), *Uapaca esculenta* (fruit), *U. guineensis* (fruit), *U. paludosa* (fruit), *U. staudtii* (fruit), *U. togoensis* (fruit). FLACOURTIACEAE: *Caloncoba brevipes* (mucilage in fruit, seed), *C. echinata* (fruit-pulp), *C. gilgiana* (fruit-pulp), *C. glauca* (fruit-pulp), *C. welwitschii* (fruit-pulp), *Flacourtia flavescens* (fruit), *F. indica* (leaf, fruit), *F. vogelii* (fruit), *Lindackeria dentata* (fruit aril), *Oncoba spinosa* (fruit-pulp). GNETACEAE: *Gnetum africanum* (leaf, fruit-pulp, seed), *G. buchholzianum* (leaf). GOODENIACEAE: *Scaevola plumeri* (fruit-pulp). GRAMINEAE: *Anthephora nigritana* (grain), *A. pubescens* (grain), *Aristida* (pith), *A. stipoides* (node centre), *Bambusa vulgaris* (young new shoot), *Brachiaria comata* (grain), *B. deflexa* (grain), *B. jubata* (grain), *B. lata* (grain), *B. ramosa* (grain), *B. serrifolia* (grain), *B. stigmatisata* (grain), *B. villosa* (grain), *Cenchrus biflorus* (grain), *C. ciliaris*, *C. prieurii*, *C. setigerus* (grain), *Chloris lamproparia* (grain), *Coix lacryma-jobi* (grain), *Cynodon dactylon* (root), *Dactyloctenium aegyptium* (grain), *Digitaria barbinodis*, *D. ciliaris* (grain), *D. debilis* (grain), *D. delicatula* (?), *D. diagonalis* (leaf), *D. exilis* (grain), *D. iburua* (grain), *D. leptorhachis* (grain), *D. longiflora* (?grain), *D. nuda* (grain), *Diheteropogon hagerupii* (?grain), *Echinochloa colona* (grain, young shoots), *E. crus-galli* (young shoot, grain), *E. obtusiflora* (grain), *E. pyramidalis* (grain), *E. stagnina* (grain, sap), *Eleusine coracana* (vegetative parts, seed), *E. indica* (seed), *Enteropogon prieurii* (grain), *Eragrostis cilianensis* (grain), *E. ciliaris* (grain), *E. minor* (grain), *E. pilosa* (grain), *E. tenella*, *E. tremula* (grain), *E. turgida* (grain), *Eriochloa fatmensis* (grain), *Hemarthria altissima* (rhizome), *Hordeum vulgare* (grain), *Hyparrhenia nyassae* (grain), *Imperata cylindrica* (rhizome), *Ischaemum afrum* (grain), *I. rugosum* (grain), *Lasiurus hirsutus* (grain), *Leptothrium senegalense* (grain), *Oryza barthii* (grain), *O. glaberrima* (grain), *O. longistaminata* (grain), *O. punctata* (grain), *O. sativa* (grain), *Oxytenanthera abyssinica* (young leaf, seed), *Panicum fluviicola* (grain), *P. laetum* (grain), *P. maximum* (grain), *P. pansum* (grain), *P. subalbidum* (grain), *P. turgidum* (grain), *P. walense* (grain), *Paspalidium geminatum* (? grain), *Paspalum scrobiculatum* (cultivar grain), *Pennisetum divisum* (grain), *P. glaucum* (grain), *P. purpureum* (young leaf/shoots, culm ash), *P. unisetum* (grain), *Phragmites australis* (root), *Phyllostachys aervea* (young shoot), *Rottboellia cochinchinensis* (grain), *Saccharum spontaneum* (young shoot, grain), *Sacciolepis africana* (grain), *Setaria palmifolia* (grain), *S. pumila* (grain), *S. sphacelata* (grain), *S. verticillata* (grain), *Sorghum arundinaceum* (grain), *S. bicolor* (grain), *S. bicolor* (grain), *S. bicolor* (grain), *S. bicolor* (grain), *S. bicolor* (grain), *S. bicolor* (grain), *S. halepense* (grain), *Sporobolus africanus* (?grain), *S. festivus* (grain), *S. panicoides* (grain), *S. pyramidalis* (grain), *S. spicatus* (grain), *S. virginicus* (grain), *Stipagrostis plumosa* (grain), *S. pungens* (grain), *Themeda triandra* (seed), *T. villosa* (young shoots), *Triticum aestivum* (grain), *Urochloa mosambicensis* (grain), *U. trichopus* (grain), *Zea mays* (grain, oil). GUTTIFERAE: *Allanblackia floribunda* (seed), *Calophyllum inophyllum* (immature fruit), *Garcinia afzelii* (fruit-pulp), *G. epunctata* (fruit-pulp), *G. kola* (fruit-pulp), *G. livingstonei* (fruit-pulp), *G. mangostana* (fruit), *G. punctata* (kernel), *G. smeathmannii* (fruit-pulp), *Harungana madagascariensis* (fruit), *Hypericum peplidifolium* (fruit), *Mammea africana* (fruit, seed), *M. americana* (fruit), *Pentadesma butyracea* (immature fruit, immature seed, seed-fat), *Psorospermum febrifugum* (fruit). HUMIRIACEAE: *Sacoglottis gabonensis* (fruit, seed). HYDROCHARITACEAE: *Ottelia ulvifolia* (leaf, fruit). HYPOXIDACEAE: *Curculigo pilosa* (seed-capsule), *Hypoxis angustifolia* (bulb), *H. urceolata* (root), *Molineria capitulata* (fruit). ICACINACEAE: *Alsodeiopsis staudtii* (fruit), *Icacina oliviformis* (tuber, fruit-pulp, kernel), *I. triacantha* (tuber, fruit-pulp), *Lasianthera africana* (leaf), *Lavigeria macrocarpa* (fruit), *Leptaulus daphnoides* (kernel (monkeys)), *Rhaphiostylis beninensis* (seed). IRIDACEAE: *Gladiolus atro-purpureus* (corm), *G. daleni* (corm), *G. gregarius* (corm), *Zygotritonia bongensis* (corm, fruit), *Z. praecox* (corm). IXONANTHACEAE: *Desbordesia glaucescens* (seed-kernel), *Irvingia gabonensis* (fruit-pulp, seed), *I. smithii* (fruit-pulp, kernel), *Klainedoxa gabonensis* (seed-kernel), *Phyllocosmus africanus* (fruit).

A: Food; 2, special diets
EUPHORBIACEAE: *Manihot esculenta* (tuber), *Phyllanthus muellerianus* (leaf), *P. reticulatus* (leaf).

A: Food; 3, sauces, condiments, spices, flavourings
EUPHORBIACEAE: *Acalypha segetalis* (leaf), *Antidesma laciniatum* (leaf), *Croton zambesicus* (fruit), *Drypetes aubrevillei* (bark), *D. paxii* (bark), *D. pellegrinii* (bark), *Erythrococca africana* (leaf), *Euphorbia pulcherrima* (leaf), *Maesobotrya barteri* (fruit), *M. dusenii* (fruit), *M. staudtii* (fruit), *Phyllanthus acidus* (fruit). FLACOURTIACEAE: *Flacourtia indica* (leaf, bark, fruit). GRAMINEAE: *Cymbopogon citratus, C. giganteus, Echinochloa pyramidalis* (grain), *E. stagnina* (sap). IRIDACEAE: *Crocus sativus* (stigma, stamen).

A: Food; 4, sweets, sweetmeats, suckers
EUPHORBIACEAE: *Hura crepitans* (inflorescence), *Ricinodendron heudelotii* (kernel). GRAMINEAE: *Echinochloa stagnina* (sap), *Saccharum officinarum* (sap), *Sorghum bicolor* (sweet culms), *S. bicolor* (culm), *S. bicolor* (culm).

A: Food; 5, masticatory
ERYTHROXYLACEAE: *Erythroxylum coca* (leaf). GUTTIFERAE: *Garcinia kola* (seed).

B: Drink; 1, water/sap
EUPHORBIACEAE: *Phyllanthus muellerianus* (stem). GRAMINEAE: *Saccharum officinarum* (sap).

B: Drink; 2, sweet, milk substitutes
FLACOURTIACEAE: *Caloncoba echinata* (fruit-pulp). GRAMINEAE: *Cenchrus biflorus* (grain), *Echinochloa stagnina* (sap). IRIDACEAE: *Gladiolus daleni* (corm).

B: Drink; 3, infusions, tisanes, etc., including substitutes, adulterants
GENTIANACEAE: *Canscora decussata* (leaf). GRAMINEAE: *Cymbopogon citratus* (leaf), *C. nardus* (leaf), *Elionurus ciliaris* (root), *Trachypogon spicatus, Vetiveria fulvibarbis* (root).

B: Drink; 4, water purifiers
GRAMINEAE: *Vetiveria nigritana* (root), *V. zizanioides* (root).

B: Drink; 5, alcholic, stimulant, including additives in fermentation
EBENACEAE: *Diospyros mespiliformis* (fruit). EUPHORBIACEAE: *Alchornea floribunda, Bridelia ferruginea* (bark), *Phyllanthus muellerianus* (bark). FLACOURTIACEAE: *Flacourtia indica* (fruit). GRAMINEAE: *Coix lacryma-jobi* (grain), *Digitaria exilis* (grain), *Echinochloa colona* (grain), *E. stagnina* (sap), *Eleusine coracana* (seed), *E. indica* (seed), *Imperata cylindrica* (rhizome), *Oryza sativa* (grain), *Oxytenanthera abyssinica* (seed), *Panicum anabaptistum, P. heterostachyum, Pennisetum glaucum* (grain), *P. unisetum* (grain), *Saccharum officinarum* (sap), *Sorghum bicolor* (grain), *S. bicolor* (grain), *S. bicolor* (grain), *Zea mays* (grain). GUTTIFERAE: *Garcinia kola* (bark), *G. livingstonei* (fruit), *Harungana madagascariensis* (fruit), *Mammea americana* (fruit). HUMIRIACEAE: *Sacoglottis gabonensis* (bark). IXONANTHACEAE: *Irvingia gabonensis* (bark).

C: Medicines; 1, generally healing (cicitrisant, haemostatic, antiseptic, stimulant, tonic, drawing, etc.
EBENACEAE: *Diospyros abyssinica* (bark, root), *D. crassiflora* (bark), *D. gabunensis* (inner bark, leaf), *D. hoyleana* (leaf), *D. mespiliformis* (bark), *D. preussii* (bark), *D. soubreana* (leaf). ELAEOCARPACEAE: *Elaeocarpus angustifolius* (fruit-pulp). ELATINACEAE: *Bergia suffruticosa* (root). EUPHORBIACEAE: *Acalypha ciliata* (leaf), *A. ornata* (leaf), *A. racemosa, A. villicaulis* (leaf), *Alchornea cordifolia* (leaf), *A. floribunda* (leaf-sap, root-sap), *Anthostema aubryanum* (latex), *Antidesma membranaceum* (bark), *A. venosum* (bark), *Bridelia ferruginea* (bark, leaf, root), *B. micrantha* (wood, bark), *B. scleroneura* (root), *Chrozophora brocchiana* (plant ash), *Croton eluteria* (bark), *C. lobatus, C. longiracemosus* (leaf), *C. nigritanus, C. zambesicus* (leaf, shoot, root), *Discoglypremna caloneura* (bark), *Drypetes aubrevillei* (bark, fruit), *D. gossweileri* (root), *Erythrococca chevalieri, E. welwitschiana, Euphorbia macrophylla,*

E. prostrata, E. thymifolia (leaf, seed), *Grossera vignei* (leaf), *Hymenocardia acida* (leaf, bark), *Jatropha chevalieri* (latex), *J. curcas* (leaf, sap, root-bark), *J. multifida* (root, root-bark), *Maesobotrya barteri* (leaf, sap), *Mallotus oppositifolius* (leaf), *M. subulatus* (leaf), *Manihot esculenta* (leaf, sap, root), *Manniophyton fulvum* (leaf, sap), *Maprounea membranacea* (leaf), *Margaritaria discoidea* (leaf, bark), *Oldfieldia africana* (bark), *Phyllanthus amarus, P. muellerianus* (leaf, root, root-bark), *P. reticulatus* (leaf), *P. urinaria, Ricinodendron heudelotii* (stem-bark), *Ricinus communis* (twig-sap, bark, fruit), *Securinega virosa* (charcoal), *Tragia preussii* (leaf), *Uapaca guineensis* (leaf). FLACOURTIACEAE: *Caloncoba echinata* (leafy twig), *C. welwitschii* (leaf), *Dovyalis zenkeri* (root), *Oncoba spinosa* (leaf). FLAGELLARIACEAE: *Flagellaria guineensis*. GENTIANACEAE: *Canscora decussata, C. diffusa*. GNETACEAE: *Gnetum africanum* (leaf), *G. buchholzianum* (leaf). GOODENIACEAE: *Scaevola plumeri* (leaf). GRAMINEAE: *Arthraxon lancifolius, Arundinella nepalensis, Coix lacryma-jobi* (fruit), *Cymbopogon giganteus, C. nardus, C. schoenanthus, Cynodon aethiopicus, C. dactylon, Dactyloctenium aegyptium, Danthoniopsis chevalieri, Eleusine indica, Eragrostis ciliaris* (plant ash, flower/seed ash), *E. gangetica, Hyparrhenia glabriuscula, Imperata cylindrica* (floss), *Olyra latifolia* (leafy culm, root, seed), *Oryza sativa* (rice flour/water), *Oxytenanthera abyssinica* (hairs from culm sheath, rhizome pith), *Panicum anabaptistum, P. antidotale, P. maximum* (sap), *P. turgidum* (culm), *Pennisetum clandestinum, P. polystachion* (sap), *P. purpureum, Phacelurus gabonensis* (root), *Phragmites karka* (root), *Pogonarthria squarrosa, Setaria megaphylla* (plant, leaf, ash), *Sporobolus africanus, S. pyramidalis, Streptogyna crinita, Urelytrum agropyroides* (root-ash), *Vetiveria nigritana* (root), *Zea mays* (flour). GUTTIFERAE: *Garcinia smeathmannii* (latex), *Harungana madagascariensis* (sap), *Hypericum lalandii, Psorospermum febrifugum* (flower-stalk), *Vismia guineensis* (young leaf, bark), *V. rubescens* (gum). HYDROCHARITACEAE: *Ottelia ulvifolia* (leaf). ICACINACEAE: *Icacina mannii, I. oliviformis, Leptaulus daphnoides* (bark), *L. zenkeri* (leaf), *Rhaphiostylis beninensis*. IRIDACEAE: *Gladiolus daleni* (corm). IXONANTHACEAE: *Irvingia gabonensis* (bark), *I. sp. indet.* (bark).

C: Medicines; 2, skin, mucosae (dermal eruptions, inflamations, ulcers, pruritus and skin diseases generally)

EBENACEAE: *Diospyros mespiliformis* (root-bark). ELATINACEAE: *Bergia suffruticosa*. ERICACEAE: *Agauria salicifolia* (whole plant). EUPHORBIACEAE: *Alchornea cordifolia* (fruit), *A. floribunda* (leaf-sap, root-sap), *Croton lobatus, C. longiracemosus* (leaf), *Discoglypremna caloneura* (wood-ash), *Drypetes aubrevillei* (bark), *D. ivorensis* (bark, fruit), *Elaephorbia drupifera* (latex), *E. grandifolia* (latex), *Euphorbia balsamifera, E. deightonii* (latex), *E. forskaolii* (latex), *E. glomerifera, E. hirta, E. lateriflora, E. prostrata, E. serpens, E. thymifolia, E. tirucalli, Excoecaria guineense* (leaf), *Grossera macrantha* (bark), *Hippomane mancinella* (sap), *Jatropha gossypiifolia* (leaf, leaf-sap), *Mallotus oppositifolius* (leaf), *M. philippinensis* (hairs from fruit surface), *Manihot esculenta* (leaf, root), *Pedilanthus tithymaloides* (sap), *Phyllanthus amarus, P. muellerianus* (leaf), *P. reticulatus* (leaf), *Ricinus communis* (leaf), *Tetrochidium didymostemon* (bark), *Tragia benthamii* (leaf), *Uapaca paludosa* (leaf, bark). FLACOURTIACEAE: *Caloncoba echinata* (bark, root, seed), *C. welwitschii* (leaf, bark), *Oncoba spinosa* (seed-oil). GUTTIFERAE: *Calophyllum inophyllum* (seed-oil), *Garcinia kola* (bark), *G. smeathmannii* (bark, latex), *Harungana madagascariensis* (leaf, sap), *Mammea africana* (bark, fruit), *Psorospermum alternifolium, P. febrifugum* (bark), *P. lanatum* (bark), *P. senegalense, P. tenuifolium* (bark), *Vismia guineensis* (gum). ICACINACEAE: *Lasianthera africana* (leaf). IXONANTHACEAE: *Klainedoxa gabonensis* (seed-kernel).

C: Medicines; 3, cutaneous/subcutaneous parasites (Guinea-worm, filaria, jiggers, scabies, etc.)

EUPHORBIACEAE: *Acalypha lanceolata* (leaf), *A. ornata* (leaf), *Alchornea cordifolia* (leaf), *A. hirtella* (bark), *Antidesma venosum* (leafy twig, root), *Bridelia ferruginea* (root-bark), *B. micrantha* (bark), *Croton aubrevillei, C. lobatus* (leaf), *C. macrostachyus* (leaf), *Elaephorbia drupifera* (leaf, latex), *E. grandifolia* (bark), *Erythrococca anomala* (leafy twig), *E. chevalieri* (leaf-sap), *E. welwitschiana* (leaf-sap), *Euphorbia balsamifera* (latex), *E. forskaolii* (latex), *E. hirta, E. lateriflora* (latex), *E. thymifolia* (leaf, oil), *Excoecaria grahamii* (whole plant, leaf, latex), *Hymenocardia acida* (bark), *Jatropha curcas* (leaf, seed-oil), *Mallotus oppositifolius* (leaf), *Manihot esculenta* (leaf, sap, root), *M. glaziovii* (stem, root), *Manniophyton fulvum* (sap), *Mareya micrantha* (bark), *Margaritaria discoidea* (leaf, twig), *Phyllanthus maderaspatensis, Plagiostyles africana* (leaf), *Ricinodendron heudelotii* (leaf, seed-oil), *Ricinus communis* (leaf, seed-oil), *Sapium ellipticum* (leaf, bark), *Securinega virosa* (leaf-sap), *Uapaca guineensis* (bark), *U. heudelotii* (leaf, bark, sap). FLACOURTIACEAE: *Caloncoba brevipes* (seed, seed-oil), *C. echinata* (seed, seed-oil), *C. glauca* (seed-oil), *C. welwitschii* (bark). FLAGELLARIACEAE: *Flagellaria guineensis* (whole plant, berry). GRAMINEAE: *Axonopus compressus, Cymbopogon schoenanthus* (flower), *Eleusine indica, Isachne buettneri, Olyra latifolia* (leaf, root), *Oplismenus burmannii* (leaf), *Oryza sativa* (rice flour), *Pennisetum purpureum* (ash), *Streptogyna crinita* (leaf), *Zea mays* (seed). GUTTIFERAE: *Harungana madagascariensis* (sap, ash), *Mammea africana* (bark, sap, root, fruit), *M. americana* (resin), *Pentadesma butyracea* (bark, seed-fat), *Psorospermum corymbiferum* (leaf/twig-oil, root), *P. febrifugum* (stem-bark, root, root-bark), *P. senegalense* (leaf, leaf-oil, stem-bark, root-bark), *Symphonia globulifera* (resin), *Vismia guineensis*(bark, gum). ICACINACEAE: *Icacina oliviformis*(root), *Iodes africana, Lasianthera africana* (leaf).

C: Medicines; 4, skeletal structure, bones, limbs, deformity, rickets
EBENACEAE: *Diospyros bipindensis*(root), *D. mannii*(bark). EUPHORBIACEAE: *Chrozophora senegalensis, Euphorbia forskaolii, E. prostrata, E. thymifolia* (leaf), *E. tirucalli, Hymenocardia acida* (bark). FLACOURTIACEAE: *Caloncoba welwitschii* (leaf). GRAMINEAE: *Paspalum conjugatum* (leaf), *P. paniculatum* (leaf), *Phacelurus gabonensis* (root). ICACINACEAE: *Lasianthera africana* (leaf, bark). IXONANTHACEAE: *Irvingia grandifolia* (bark).

C: Medicines; 5, paralysis, epilepsy, convulsions, spasm
EBENACEAE: *Diospyros sanza-minika*. EUPHORBIACEAE: *Alchornea cordifolia* (leaf), *Bridelia ferruginea* (bark), *B. scleroneura* (root), *Croton lobatus* (leaf), *C. zambesicus* (leaf), *Discoglypremna caloneura*(bark), *Hymenocardia acida*(bark), *Mallotus oppositifolius*(leaf), *M. subulatus, Phyllanthus fraternus, P. muellerianus* (leaf), *P. odontadenius* (leaf), *P. reticulatus* (leaf, root), *Ricinus communis*(leaf), *Tetrochidium didymostemon*(leaf). FLACOURTIACEAE: *Caloncoba welwitschii* (root). GENTIANACEAE: *Canscora decussata, C. diffusa, Hoppea dichotoma*(root). GRAMINEAE: *Cymbopogon densiflorus*(plant-sap), *C. giganteus, C. nardus, Eleusine indica, Setaria barbata, S. megaphylla* (leaf). GUTTIFERAE: *Mammea africana* (bark), *Psorospermum febrifugum* (bark). IRIDACEAE: *Gladiolus daleni* (corm).

C: Medicines; 6, insanity
ELAEOCARPACEAE: *Elaeocarpus angustifolius*(fruit-pulp). EUPHORBIACEAE: *Acalypha ornata* (leaf), *Manniophyton fulvum* (leaf), *Plagiostyles africana, Securinega virosa* (leaf). FLACOURTIACEAE: *Lindackeria dentata*. GRAMINEAE: *Cymbopogon schoenanthus* (smoke), *Panicum antidotale, Setaria megaphylla* (leaf), *Vetiveria nigritana* (root). ICACINACEAE: *Rhaphiostylis beninensis* (leaf).

C: Medicines; 7, brain/nervous diseases, encephalitis, meningitis
EUPHORBIACEAE: *Jatropha podagrica, Sapium ellipticum* (root). GUTTIFERAE: *Garcinia kola* (fruit-pulp). HERNANDIACEAE: *Illigera pentaphylla* (sap). HYDROPHYLLACEAE: *Hydrolea palustris*.

C: Medicines; 8, heart (tonal, disease, tremor, palpitations, etc.)
ELAEOCARPACEAE: *Elaeocarpus angustifolius* (fruit-pulp). EUPHORBIACEAE: *Alchornea cordifolia* (leaf), *Maesobotrya barteri* (leaf), *Manniophyton fulvum* (leaf). GRAMINEAE: *Cynodon dactylon, Eleusine indica, Paspalum conjugatum* (leaf), *P. paniculatum* (leaf). GUTTIFERAE: *Harungana madagascariensis* (leaf), *Symphonia globulifera* (bark). HYDROCHARITACEAE: *Ottelia ulvifolia* (leaf).

C: Medicines; 9, arteries, veins
EUPHORBIACEAE: *Plagiostyles africana, Pycnocoma angustifolia* (sap), *Ricinus communis* (leaf). GRAMINEAE: *Cymbopogon densiflorus*. GUTTIFERAE: *Harungana madagascariensis* (sap). ICACINACEAE: *Icacina oliviformis* (leaf).

C: Medicines; 10, blood disorders, anaemia
EUPHORBIACEAE: *Euphorbia lateriflora, Jatropha gossypiifolia*(leaf), *Manniophyton fulvum* (sap). GRAMINEAE: *Coelachyrum brevifolium*.

C: Medicines; 11, pain-killers (revulsive, anodynal, analgesic, anaesthetic)
EBENACEAE: *Diospyros bipindensis*(root), *D. canaliculata*(bark), *D. hoyleana*(leaf, leaf-sap), *D. iturensis*(root-bark), *D. mespiliformis*(sap), *D. sanza-minika* (leaf). ELAEOCARPACEAE: *Elaeocarpus angustifolius* (fruit-pulp). ERICACEAE: *Agauria salicifolia* (leaf, sap). ERYTHROXYLACEAE: *Erythroxylum coca* (leaf), *E. mannii* (bark). EUPHORBIACEAE: *Acalypha brachystachya* (leaf-sap), *A. indica* (leaf, sap), *A. ornata* (leaf, plant ash), *A. wilkesiana* (leaf), *Alchornea cordifolia* (leaf), *A. hirtella* (sap, root), *Antidesma venosum* (leaf, root), *Bridelia ferruginea* (bark), *B. grandis* (bark, leaf), *B. micrantha* (bark, leaf), *B. scleroneura* (leaf, bark?, wood, root), *Croton aubrevillei, C. lobatus* (leaf), *C. macrostachyus, C. zambesicus* (leaf), *Dalechampia ipomoeofolia, Dichostemma glaucescens*(leaf), *Drypetes aubrevillei*(bark, fruit), *D. capillipes* (bark, leaf), *D. gossweileri* (bark), *D. klainei* (bark), *D. roxburghii, Elaephorbia drupifera* (latex), *E. grandifolia* (latex), *Erythrococca anomala* (leaf, fruit-pulp), *E. chevalieri* (root), *E. welwitschiana*(root), *Euphorbia balsamifera* (latex), *E. forskaolii, E. hirta, E. poissonii,*

E. scordifolia, E. thymifolia (leaf), *Hymenocardia acida* (leaf, bark, charcoal, root), *Jatropha gossypiifolia* (pith), *Macaranga spinosa, Maesobotrya barteri* (bark), *Mallotus oppositifolius* (leaf), *M. subulatus* (seed), *Manihot esculenta* (leaf, root), *Manniophyton fulvum* (leaf, stem, root), *Maprounea africana* (leaf), *Mareya micrantha, Margaritaria discoidea* (bark, root), *Micrococca mercurialis* (sap), *Phyllanthus amarus, P. capillaris* (root), *P. maderaspatensis, P. muellerianus* (sap), *P. niruroides* (sap), *P. pentandrus, P. reticulatus* (root), *Plagiostyles africana* (bark), *Plukenetia conophora* (leaf), *Ricinodendron heudelotii* (leaf, bark), *Ricinus communis* (leaf, root, seed-oil), *Sapium ellipticum* (leaf), *Securinega virosa* (root), *Tetrochidium didymostemon* (leaf, bark), *Tetrorchidium oppositifolium* (bark), *Thecacoris stenopetala, Tragia preussii, T. spathulata* (sap), *Uapaca guineensis* (leaf, root), *U. heudelotii* (leaf, bark), *U. paludosa* (root, bark). FLACOURTIACEAE: *Caloncoba brevipes* (bark), *C. glauca* (leaf-sap), *C. welwitschii* (leaf, leaf-sap, bark), *Camptostylus mannii* (root), *Flacourtia indica* (bark), *Homalium smythei* (bark-ash), *Lindackeria dentata* (root), *Oncoba spinosa* (wood, fruit-pulp), *Scottellia coriacea.* FLAGELLARIACEAE: *Flagellaria guineensis* (ash of whole plant, leaf). GENTIANACEAE: *Canscora decussata, Exacum oldenlandioides* (plant). GNETACEAE: *Gnetum africanum* (stem), *G. buchholzianum* (stem). GRAMINEAE: *Aristida sieberana* (root), *Cymbopogon citratus* (leaf, rhizome), *C. giganteus, C. schoenanthus* (leaf), *Cynodon aethiopicus, Eleusine indica, Elionurus muticus* (root), *Gramineae spp. indet.* (sp. no. 4), *Olyra latifolia* (root), *Oplismenus burmannii* (leaf), *Panicum maximum, Paspalum conjugatum* (leaf), *Pennisetum glaucum* (bran), *P. hordeoides, P. polystachion, P. purpureum* (seed), *Phacelurus gabonensis* (root), *Puelia olyriformis* (root), *Setaria megaphylla* (leaf, sap, ash), *Streptogyna crinita.* GUTTIFERAE: *Allanblackia floribunda* (bark), *Garcinia afzelii* (bark), *G. elliotii* (seed), *G. epunctata* (bark), *G. kola* (bark, seed), *G. punctata* (bark), *Harungana madagascariensis* (leaf), *Mammea africana* (bark), *Psorospermum corymbiferum* (leaf), *P. lanatum, P. senegalense* (bark), *Symphonia globulifera* (bark), *Vismia guineensis* (leaf). HUMIRIACEAE: *Sacoglottis gabonensis* (sap). HYDROPHYLLACEAE: *Hydrolea palustris* (leaf). ICACINACEAE: *Icacina mannii* (leaf-sap), *I. oliviformis* (leaf), *Iodes africana* (leaf-sap), *Lasianthera africana* (leaf), *Pyrenacantha staudtii.* IXONANTHACEAE: *Desbordesia glaucescens* (bark), *Irvingia gabonensis* (bark, bark-sap), *I. grandifolia* (bark), *Klainedoxa gabonensis* (leaf, bark).

C: Medicines; 12, sedatives, tranquillisers, hallucinogens, narcotics

EUPHORBIACEAE: *Alchornea cordifolia* (leaf), *Bridelia atroviridis* (leaf), *Euphorbia convolvuloides, E. hirta, Phyllanthus maderaspatensis, Plagiostyles africana, Securinega virosa* (leaf, root), *Tetrochidium didymostemon* (bark). FLACOURTIACEAE: *Camptostylus mannii* (root), *Flacourtia indica* (leaf). GRAMINEAE: *Imperata cylindrica* (floss), *Paspalum scrobiculatum* (seed), *Setaria megaphylla* (leaf).

C: Medicines; 13, arthritis, rheumatism, lumbago, muscular and body-pains, stiffness (see also C: 11)

ERICACEAE: *Agauria salicifolia* (leaf, sap). EUPHORBIACEAE: *Chrozophora senegalensis, Euphorbia lactea.* GRAMINEAE: *Aristida sieberana* (cortex), *Chloris virgata, Chrysopogon aciculatus, Cymbopogon densiflorus* (leaf), *Heteropogon contortus, Paspalum vaginatum, Phragmites karka, Streptogyna crinita, Zea mays* (style/stigma). ICACINACEAE: *Rhaphiostylis beninensis* (leaf). IRIDACEAE: *Crocosmia aurea* (root, plant-ash), *Gladiolus daleni* (corm). IXONANTHACEAE: *Irvingia grandifolia* (bark), *Klainedoxa gabonensis* (bark, bark-smoke).

C: Medicines; 14, eye treatments (see also C: 11)

EBENACEAE: *Diospyros crassiflora* (leaf sap). EUPHORBIACEAE: *Alchornea cordifolia* (leaf), *Bridelia micrantha, Chrozophora senegalensis* (fruit-juice), *Erythrococca anomala* (leaf-sap), *Euphorbia convolvuloides, E. hirta, E. thymifolia, Hymenocardia acida* (leaf), *H. heudelotii* (bark-sap), *Mallotus oppositifolius* (leaf), *Manihot esculenta* (leaf, sap, latex), *Margaritaria discoidea* (leaf), *Micrococca mercurialis* (sap), *Phyllanthus acidus* (fruit-juice), *P. fraternus, P. muellerianus* (leaf, sap), *P. reticulatus* (stem-sap), *Plagiostyles africana* (bark, cambium), *Ricinodendron heudelotii* (bark-sap), *Ricinus communis* (leaf), *Sapium ellipticum* (leaf, root-sap), *Securinega virosa* (root), *Tetrochidium didymostemon* (sap), *Tetrorchidium oppositifolium* (bark). FLACOURTIACEAE: *Camptostylus mannii* (bark). GOODENIACEAE: *Scaevola plumeri* (leaf). GRAMINEAE: *Aristida sieberana* (root), *Coix lacryma-jobi* (culm-sap), *Eleusine indica, Eriochloa fatmensis* (culm), *Oplismenus burmannii* (leaf), *Paspalum conjugatum* (leaf), *P. scrobiculatum* (sap), *P. vaginatum, Pennisetum polystachion* (sap), *P. purpureum.* GUTTIFERAE: *Calophyllum inophyllum* (leaf), *Endodesmia calophylloides* (leaf), *Garcinia smeathmannii* (latex), *Hypericum peplidifolium* (leaf). HERNANDIACEAE: *Illigera pentaphylla* (leaf-sap). ICACINACEAE: *Icacina oliviformis* (leaf-sap), *Rhaphiostylis beninensis* (leaf). IXONANTHACEAE: *Irvingia grandifolia* (bark).

C: Medicines; 15, ear treatments (*see also* C: 11)
EBENACEAE: *Diospyros mespiliformis* (leaf). EUPHORBIACEAE: *Croton macrostachyus* (seed), *Hymenocardia acida* (leaf, root), *Jatropha curcas* (sap), *Manniophyton fulvum* (leaf, sap), *Margaritaria discoidea* (leaf), *Micrococca mercurialis* (sap), *Phyllanthus amarus, P. niruroides* (sap), *P. pentandrus* (whole plant, plant-ash). GRAMINEAE: *Eleusine indica, Olyra latifolia* (root), *Pennisetum polystachion* (inflorescence), *P. purpureum.* IRIDACEAE: *Gladiolus daleni* (corm).

C: Medicines; 16, oral (*see also* C: 11 for toothache treatment)
EUPHORBIACEAE: *Alchornea cordifolia* (leaf, twig, root-pith), *Bridelia ferruginea* (bark), *B. micrantha* (bark, leaf), *Hymenocardia acida* (root), *Jatropha curcas* (leaf), *Macaranga spinosa* (bark), *Phyllanthus muellerianus* (leaf), *P. niruroides* (leaf-sap), *P. reticulatus* (leaf), *Ricinus communis* (root), *Tetrochidium didymostemon* (bark), *Uapaca guineensis* (bark), *U. heudelotii* (bark), *U. paludosa* (stem-bark, root-bark). FLAGELLARIACEAE: *Flagellaria guineensis* (leaf). GRAMINEAE: *Cymbopogon giganteus, Digitaria perrottetii* (root), *Olyra latifolia* (leaf, culm), *Pennisetum purpureum, Streptogyna crinita.* GUTTIFERAE: *Allanblackia floribunda* (bark), *Garcinia kola* (root). ICACINACEAE: *Icacina oliviformis* (root), *Rhaphiostylis beninensis* (leaf). IXONANTHACEAE: *Irvingia gabonensis* (bark), *Klainedoxa gabonensis* (bark).

C: Medicines; 17, naso-pharyngeal (catarrh, stuffiness, sneezing, sore-throat, cough, phlegm, etc.)
ELAEOCARPACEAE: *Elaeocarpus angustifolius* (fruit-pulp). EUPHORBIACEAE: *Acalypha ciliata, A. indica, A. villicaulis* (root), *A. wilkesiana* (leaf), *Alchornea cordifolia* (leaf, pith), *Bridelia atroviridis* (bark), *Croton pseudopulchellus* (root), *C. zambesicus, Drypetes chevalieri* (leaf), *Elaephorbia grandifolia* (leaf), *Euphorbia thymifolia, Hymenocardia acida* (leaf, bark), *Jatropha gossypiifolia* (pith), *Phyllanthus amarus, P. muellerianus* (bark), *P. niruroides* (sap), *Uapaca guineensis* (root), *U. paludosa* (root). FLACOURTIACEAE: *Caloncoba welwitschii, Camptostylus mannii* (root), *Flacourtia indica* (leaf, bark). GRAMINEAE: *Chloris virgata, Coix lacryma-jobi* (fruit), *Cymbopogon citratus* (leaf), *C. giganteus, Echinochloa pyramidalis* (root), *Eragrostis cilianensis* (root), *Imperata cylindrica* (whole plant), *Olyra latifolia* (leaf, culm), *Panicum antidotale, Paspalum conjugatum* (leaf), *Pennisetum unisetum* (root), *Perotis patens.* GUTTIFERAE: *Vismia guineensis* (leaf). ICACINACEAE: *Iodes africana* (leaf-sap). IRIDACEAE: *Gladiolus daleni* (corm), *Trimezia martinicensis* (corm).

C: Medicines; 18, pulmonary (chest, lungs, pneumonia, bronchitis, pleurisy, tuberculosis, asthma, etc.)
EBENACEAE: *Diospyros bipindensis* (root), *D. iturensis* (root-bark), *D. mannii* (bark), *D. mespiliformis* (leaf, bark). ERYTHROXYLACEAE: *Erythroxylum mannii* (bark). EUPHORBIACEAE: *Acalypha hispida* (bark, flower, root), *Alchornea cordifolia* (leaf, pith), *Antidesma venosum* (bark), *Bridelia atroviridis* (bark), *B. micrantha* (tree, bark), *Croton pseudopulchellus* (leaf, root), *Discoglypremna caloneura* (leaf), *Drypetes aubrevillei* (bark), *D. chevalieri, Elaephorbia grandifolia* (leaf), *Erythrococca chevalieri* (leaf-sap), *E. welwitschiana* (leaf-sap), *Euphorbia convolvuloides, E. hirta, E. thymifolia, Hymenocardia acida* (leaf), *Jatropha curcas* (leaf), *Macaranga barteri, M. beillei* (leaf), *M. heterophylla* (bark), *M. hurifolia* (bark, twig), *M. spinosa, M. spp. indet.* (young leaf), *Maesobotrya barteri, Mallotus oppositifolius* (leaf-sap, root), *Manniophyton fulvum* (shoot, bark, sap, root, nut), *Phyllanthus fraternus, P. muellerianus* (leaf), *P. niruroides* (sap), *Plagiostyles africana* (bark), *Ricinus communis* (leaf), *Sapium ellipticum* (root), *Uapaca guineensis* (unripe fruit), *U. paludosa* (root). FLACOURTIACEAE: *Flacourtia indica* (leaf, bark). GRAMINEAE: *Coix lacryma-jobi* (fruit), *Cymbopogon citratus* (leaf, root), *C. densiflorus* (plant-sap, flower-head), *C. giganteus, Cynodon dactylon, Dactyloctenium aegyptium, Eleusine indica, Imperata cylindrica* (whole plant), *Leersia hexandra, Paspalum conjugatum* (leaf), *Pennisetum glaucum* (grain), *Streptogyna crinita* (leaf, rhizome), *Vetiveria zizanioides* (root), *Zea mays* (seed). GUTTIFERAE: *Allanblackia floribunda* (leaf, bark), *Garcinia afzelii* (bark), *G. epunctata* (bark), *G. kola* (leaf, bark, root-bark, seed), *G. punctata* (bark), *Harungana madagascariensis* (leaf, leaf-bud, twig, bark, sap), *Psorospermum corymbiferum* (root), *P. senegalense* (leaf), *Symphonia globulifera* (bark). HERNANDIACEAE: *Illigera pentaphylla* (leaf-sap). ICACINACEAE: *Icacina mannii* (leaf, tuber), *I. oliviformis* (leaf, root), *Rhaphiostylis beninensis* (leaf).

C: Medicines; 19, stomachic (*see also* C: 11 for stomachache and other pains)
EBENACEAE: *Diospyros monbuttensis* (twig, bark). ERICACEAE: *Agauria salicifolia* (bark). EUPHORBIACEAE: *Alchornea cordifolia* (leaf), *Antidesma laciniatum* (bark), *A. venosum* (leaf), *Bridelia scleroneura* (root), *Chrozophora senegalensis, Croton aubrevillei, Cyrtogonone argentea* (bark), *Dichostemma glaucescens* (leaf), *Hymenocardia acida* (leaf), *Jatropha gossypiifolia* (leaf), *Macaranga spinosa, Maesobotrya barteri* (bark), *Manniophyton fulvum* (twig), *Maprounea africana* (leaf), *Margaritaria discoidea* (bark), *Securinega virosa* (fruit), *Uapaca*

guineensis (root-bark). FLACOURTIACEAE: *Dovyalis zenkeri, Oncoba spinosa* (wood, fruit-pulp). GENTIANACEAE: *Canscora decussata.* GRAMINEAE: *Andropogon gayanus, A. tectorum* (root), *Cymbopogon citratus* (leaf), *C. giganteus, C. nardus, C. schoenanthus, Cynodon aethiopicus, Digitaria exilis* (grain), *Eragrostis ciliaris, Imperata cylindrica* (rhizome), *Paspalum conjugatum* (leaf), *Phacelurus gabonensis* (root), *Puelia olyriformis* (root), *Stenotaphrum secundatum, Urelytrum agropyroides* (root). GUTTIFERAE: *Harungana madagascariensis* (leaf), *Hypericum peplidifolium* (leaf), *H. revolutum* (leaf), *Psorospermum senegalense* (leaf), *Symphonia globulifera* (bark). HUMIRIACEAE: *Sacoglottis gabonensis* (bark). ICACINACEAE: *Iodes africana, Lasianthera africana* (leaf), *Pyrenacantha staudtii.* IXONANTHACEAE: *Irvingia grandifolia* (bark), *Klainedoxa gabonensis* (leaf).

C: Medicines; 20, 'intestines' (unspecified)

EUPHORBIACEAE: *Drypetes chevalieri, Hymenocardia acida* (root), *Macaranga hurifolia* (leafy twig), *Margaritaria discoidea* (root), *Phyllanthus fraternus, P. muellerianus* (root). GRAMINEAE: *Elionurus muticus* (root). GUTTIFERAE: *Garcinia kola* (leaf, bark). ICACINACEAE: *Icacina oliviformis* (leaf).

C: Medicines; 21, emetic

EUPHORBIACEAE: *Acalypha ciliata, A. indica, Anthostema senegalense* (latex), *Argomuellera macrophylla* (leaf-sap), *Dichostemma glaucescens* (bark), *Discoglypremna caloneura* (bark), *Euphorbia balsamifera* (root-bark), *E. cervicornu* (latex), *Excoecaria guineense* (bark), *Hymenocardia acida* (stem-bark, root-bark), *Jatropha curcas* (fruit), *J. gossypiifolia* (seed-oil), *Maprounea africana* (leaf, bark, root), *Pedilanthus tithymaloides* (root), *Phyllanthus muellerianus* (root-charcoal), *Pycnocoma cornuta* (bark), *Ricinus communis* (seed-oil), *Uapaca heudelotii* (sap). GOODENIACEAE: *Scaevola plumeri.* GUTTIFERAE: *Harungana madagascariensis* (bark, sap, fruit). HUMIRIACEAE: *Sacoglottis gabonensis* (bark). ICACINACEAE: *Icacina mannii* (leaf), *Leptaulus daphnoides* (leaf). IRIDACEAE: *Gladiolus gregarius* (corm).

C: Medicines; 22, antemetic

EUPHORBIACEAE: *Phyllanthus capillaris* (whole plant), *Uapaca guineensis* (bark). GNETACEAE: *Gnetum africanum* (leaf). GRAMINEAE: *Pennisetum polystachion* (root), *Phragmites karka* (rhizome, root).

C: Medicines; 23, laxatives, purgatives, drastics, enemas

EBENACEAE: *Diospyros canaliculata* (root), *D. heudelotii* (root), *D. mespiliformis* (leaf). EUPHORBIACEAE: *Acalypha indica, A. villicaulis* (root), *Alchornea cordifolia* (leaf, leafy twig, fruit), *Anthostema aubryanum* (latex, seed), *A. senegalense* (latex, young leaf), *Antidesma venosum* (leaf), *Argomuellera macrophylla* (leaf-sap), *Bridelia atroviridis* (bark, leaf), *B. ferruginea* (bark, leaf), *B. grandis* (bark), *B. micrantha* (leaf, bark, root), *B. scleroneura* (bark?), *Claoxylon hexandrum, Croton aubrevillei, C. lobatus* (leaf), *C. macrostachyus* (bark, seed-oil, root), *C. pyrifolius* (young leaf), *C. sylvaticus* (seed, seed-oil), *C. tiglium* (seed, seed-oil), *C. zambesicus* (root), *Cyrtogonone argentea* (bark), *Elaephorbia drupifera* (latex), *E. grandifolia* (latex), *Erythrococca africana* (leaf), *E. anomala* (leaf), *Euphorbia cervicornu* (latex), *E. convolvuloides* (whole plant, latex), *E. forskaolii, E. glomerifera, E. hirta, E. kamerunica* (latex), *E. lateriflora* (latex), *E. poissonii* (latex), *E. schimperiana* (leaf, root), *E. tirucalli, Excoecaria grahamii, E. guineense* (stem-bark, root-bark), *Hura crepitans* (seed-oil), *Jatropha curcas* (leaf, seed, seed-oil), *J. gossypiifolia* (leaf, seed, seed-oil), *J. multifida* (leaf, sap, seed), *J. podagrica, Macaranga barteri, M. heterophylla* (root), *M. hurifolia* (bark), *Maesobotrya barteri* (bark), *Manihot esculenta* (seed-oil), *Maprounea africana* (twig, stem-bark, root-bark), *M. membranacea* (bark), *Mareya micrantha, Margaritaria discoidea* (bark, root), *Neoboutonia mannii* (root-bark), *Phyllanthus acidus* (root, seed), *P. fraternus, P. maderaspatensis* (root,seed), *P. muellerianus* (leaf, flower), *P. niruroides, P. pentandrus* (leaf-sap), *P. reticulatus* (root), *Pycnocoma cornuta* (leaf, stem, bark, root), *Ricinodendron heudelotii* (root-bark), *Ricinus communis* (leaf, root-bark, seed-oil), *Sapium ellipticum* (bark, root), *Securinega virosa* (leaf), *Tetrochidium didymostemon* (bark), *Uapaca guineensis* (bark). FLACOURTIACEAE: *Caloncoba echinata* (leafy twig), *Casearia barteri* (leaf), *Flacourtia flavescens.* GENTIANACEAE: *Canscora decussata, C. diffusa, Exacum oldenlandioides* (plant). GOODENIACEAE: *Scaevola plumeri.* GRAMINEAE: *Cynodon dactylon, Melinis minutiflora, Pennisetum purpureum, Saccharum officinarum* (sap). GUTTIFERAE: *Allanblackia floribunda* (seed-kernel fat), *Garcinia kola* (leaf), *G. smeathmannii* (bark), *Harungana madagascariensis* (bark, sap, fruit), *Pentadesma butyracea* (leaf, bark), *Psorospermum febrifugum.* HYPOXIDACEAE: *Curculigo pilosa* (root-stock). ICACINACEAE: *Icacina mannii, Rhaphiostylis beninensis* (bark). IRIDACEAE: *Gladiolus daleni* (corm). IXONANTHACEAE: *Irvingia gabonensis* (bark).

PLANT SPECIES BY USAGES

C: Medicines; 24, diarrhoea, dysentery
EBENACEAE: *Diospyros thomasii* (inner bark). EUPHORBIACEAE: *Alchornea cordifolia* (leaf, root-bark, root), *Antidesma venosum* (leaf), *Bridelia atroviridis* (bark), *B. ferruginea* (bark), *B. grandis* (bark), *B. scleroneura* (bark?, root), *Chrozophora senegalensis* (whole plant, root), *Croton zambesicus* (leaf), *Crotonogyne preusii, Discoglypremna caloneura, Drypetes chevalieri* (sap), *Euphorbia convolvuloides* (leaf), *E. hirta, E. polycnemoides, E. scordifolia, Hymenocardia acida* (leaf, bark), *Jatropha gossypiifolia* (leaf), *Macaranga heudelotii, M. spinosa, Maesobotrya barteri* (bark), *Mallotus oppositifolius* (all parts), *M. subulatus* (leaf, root, fruit), *Manniophyton fulvum* (sap), *Phyllanthus fraternus, P. muellerianus* (leaf, root), *P. pentandrus* (young shoot), *Ricinodendron heudelotii* (leaf, root-bark, seed), *Securinega virosa* (root), *Uapaca paludosa* (stem-bark, root-bark). FLACOURTIACEAE: *Flacourtia flavescens* (stem), *F. indica* (sap), *Oncoba spinosa* (root). GOODENIACEAE: *Scaevola plumeri* (pith). GRAMINEAE: *Dactyloctenium aegyptium, Desmostachya bipinnata* (culm), *Eleusine coracana, E. indica, Imperata cylindrica* (rhizome), *Oryza glaberrima* (root), *O. sativa* (rice), *Paspalum conjugatum* (root), *Pseudechinolaena polystachya, Vetiveria nigritana* (root), *Zea mays* (whole plant (veterinary)). GUTTIFERAE: *Allanblackia floribunda* (bark), *Garcinia kola* (seed), *G. mannii* (root-bark), *Harungana madagascariensis* (leaf, sap), *Psorospermum febrifugum* (bark). ICACINACEAE: *Icacina mannii* (tuber), *Iodes africana, Lasianthera africana* (leaf). IRIDACEAE: *Gladiolus daleni* (corm). IXONANTHACEAE: *Irvingia gabonensis* (bark), *I. smithii* (bark).

C: Medicines; 25, cholera
EUPHORBIACEAE: *Pycnocoma macrophylla* (all parts). GRAMINEAE: *Pennisetum glaucum* (bran).

C: Medicines; 26, vermifuges, parasite expellants
EBENACEAE: *Diospyros mespiliformis* (leaf, bark). EUPHORBIACEAE: *Acalypha indica, Anthostema senegalense* (latex), *Antidesma venosum* (root), *Bridelia micrantha* (root-sap), *Chrozophora senegalensis, Croton macrostachyus* (leaf, seed), *C. zambesicus* (leaf), *Drypetes gossweileri* (bark), *D. klainei* (bark), *Erythrococca anomala, Euphorbia balsamifera* (root, flower), *E. forskaolii, E. hirta, E. prostrata, E. thymifolia, E. tirucalli, Excoecaria grahamii, Jatropha curcas* (root, fruit, seed), *J. multifida* (root), *Macaranga barteri* (leaf, bark), *Mallotus oppositifolius* (leaf), *M. philippinensis* (hairs from fruit surface), *Maprounea africana* (stem-bark, root-bark), *M. membranacea* (bark), *Mareya micrantha, Margaritaria discoidea* (bark), *Phyllanthus amarus* (ripe fruit), *P. niruroides* (leaf-sap), *P. reticulatus* (root), *Plagiostyles africana* (leaf, bark), *Ricinus communis* (leaf, seed-oil), *Sapium ellipticum* (leaf, bark), *Securinega virosa* (root), *Tetrochidium didymostemon* (leaf). FLACOURTIACEAE: *Caloncoba welwitschii* (leaf, bark), *Flacourtia indica* (leaf-sap, root). GRAMINEAE: *Aristida stipoides* (gall), *Axonopus compressus, Coix lacryma-jobi* (root), *Cymbopogon nardus, Digitaria argyrotricha* (leaf), *Imperata cylindrica* (rhizome), *Olyra latifolia, Pennisetum glaucum* (grain), *Streptogyna crinita* (leaf). GUTTIFERAE: *Harungana madagascariensis* (sap, resin), *Pentadesma butyracea* (root). ICACINACEAE: *Lasianthera africana* (leaf), *Rhaphiostylis beninensis* (leaf, stem, root). IRIDACEAE: *Gladiolus daleni* (corm).

C: Medicines; 27, liver, gall bladder, spleen (biliousness, jaundice, 'yellow fever')
ELAEOCARPACEAE: *Elaeocarpus angustifolius* (fruit-pulp). EUPHORBIACEAE: *Dichostemma glaucescens* (leaf), *Drypetes roxburghii* (leaf, fruit), *Euphorbia lateriflora, Jatropha podagrica, Macaranga spinosa, Securinega virosa* (root). FLACOURTIACEAE: *Flacourtia flavescens*. GRAMINEAE: *Cymbopogon giganteus, C. schoenanthus, Pogonarthria squarrosa* (leaf). GUTTIFERAE: *Harungana madagascariensis* (bark). IXONANTHACEAE: *Irvingia gabonensis* (bark).

C: Medicines; 28, kidneys, micturition, diuresis
EBENACEAE: *Diospyros heudelotii*. EUPHORBIACEAE: *Acalypha hispida* (leaf, flower), *Anthostema aubryanum* (latex), *Bridelia atroviridis* (bark), *B. ferruginea* (leaf, bark, root-bark), *B. grandis* (bark), *B. scleroneura* (bark?), *Croton sp.* (root), *Euphorbia hirta, Jatropha curcas* (leaf), *J. multifida* (young leaf), *J. podagrica, Macaranga hurifolia* (root), *Mallotus oppositifolius* (leaf), *Maprounea africana* (leaf, bark, root), *Margaritaria discoidea* (bark), *Phyllanthus fraternus* (leaf), *P. niruroides, P. pentandrus* (leaf, shoot, root), *P. reticulatus* (leaf, bark), *Securinega virosa* (root). GERANIACEAE: *Geranium ocellatum*. GOODENIACEAE: *Scaevola plumeri* (leaf). GRAMINEAE: *Aristida stipoides* (gall), *Coix lacryma-jobi* (fruit), *Cymbopogon giganteus, C. nardus, Cynodon dactylon, Desmostachya bipinnata* (culm), *Digitaria exilis* (grain),

486

Eleusine indica, Heteropogon contortus, Imperata cylindrica (rhizome), *Oryza sativa* (rice water), *Oxytenanthera abyssinica* (leaf), *Pennisetum pedicellatum* (whole plant), *P. purpureum* (all parts), *Phragmites australis* (root), *Puelia olyriformis* (root), *Stenotaphrum dimidiatum, S. secundatum, Streptogyna crinita, Zea mays* (whole plant, or parts). GUTTIFERAE: *Garcinia kola* (bark), *Harungana madagascariensis* (leaf), *Psorospermum* (leaf), *P. senegalense* (leaf), *Symphonia globulifera* (resin). ICACINACEAE: *Icacina mannii, I. oliviformis* (root). IRIDA-CEAE: *Trimezia martinicensis* (corm). IXONANTHACEAE: *Irvingia grandifolia* (bark).

C: Medicines; 29, anus, haemorrhoids
EUPHORBIACEAE: *Acalypha ornata* (leaf, root), *Alchornea cordifolia* (leaf, bark), *Jatropha gossypiifolia* (leaf), *Manniophyton fulvum* (sap), *Phyllanthus amarus* (leaf), *Ricinus communis* (leaf), *Uapaca guineensis* (bark), *U. paludosa* (bark). GENTIANACEAE: *Hoppea dichotoma* (whole plant). GUTTIFERAE: *Calophyllum inophyllum* (leaf), *Harungana madagascariensis* (sap). ICACINACEAE: *Rhaphiostylis beninensis* (leaf).

C: Medicines; 30, genital stimulants/depressants
EUPHORBIACEAE: *Alchornea floribunda, Antidesma membranaceum* (bark), *A. vogelianum* (root), *Argomuellera macrophylla* (leaf), *Bridelia ferruginea* (leaf, bark), *B. grandis* (bark), *B. scleroneura* (bark?), *Croton aubrevillei, Drypetes gossweileri* (bark), *Erythrococca chevalieri, E. welwitschiana, Euphorbia balsamifera, E. hirta, E. sp. indet., E. tirucalli, Macaranga spinosa, Maesobotrya barteri* (bark), *Mallotus oppositifolius* (root), *Margaritaria discoidea* (root), *Phyllanthus beillei* (root), *P. muellerianus* (leaf), *P. pentandrus, P. reticulatus* (root), *Plukenetia conophora* (seed), *Ricinodendron heudelotii* (root), *Sapium aubrevillei* (root), *Securinega virosa* (leaf), *Uapaca guineensis* (root). FLACOURTIACEAE: *Caloncoba glauca, Oncoba spinosa.* FLAGELLARIACEAE: *Flagellaria guineensis* (leaf). GOODENIACEAE: *Scaevola plumeri* (pith). GRAMINEAE: *Cenchrus biflorus* (root), *Cymbopogon schoenanthus* (rhizome), *Imperata cylindrica* (young shoot), *Oplismenus burmannii* (leaf), *Pennisetum polystachion, Puelia olyriformis* (root), *Zea mays* (cob). GUTTIFERAE: *Garcinia afzelii* (root-bark, seed), *G. epunctata* (root), *G. kola* (bark, root, seed), *G. mannii* (root), *Harungana madagascariensis* (bark), *Pentadesma butyracea* (bark). HUMIRIACEAE: *Sacoglottis gabonensis* (bark). ICACINA-CEAE: *Alsodeiopsis poggei* (root), *A. rowlandii* (root), *A. staudtii* (root), *Icacina mannii, I. oliviformis* (root), *Lavigeria macrocarpa* (fruit pericarp, kernel). IXONANTHACEAE: *Desbordesia glaucescens* (bark), *Irvingia gabonensis* (bark), *Klainedoxa gabonensis* (young leaf, bark).

C: Medicines; 31, menstrual cycle
EBENACEAE: *Diospyros crassiflora* (bark). EUPHORBIACEAE: *Alchornea cordifolia* (leaf, root), *Anthostema senegalense* (latex), *Bridelia micrantha, Croton lobatus* (leaf), *C. macrostachyus, C. zambesicus* (shoot, root), *Hymenocardia acida* (bark, root), *Jatropha gossypiifolia* (bark), *Macaranga heterophylla* (root), *Mallotus oppositifolius, Manniophyton fulvum* (sap), *Maprounea membranacea* (bark), *Margaritaria discoidea* (root), *Phyllanthus amarus* (leaf), *P. niruroides, P. urinaria, Ricinodendron heudelotii* (leaf, bark), *Ricinus communis* (leaf), *Securinega virosa* (root). FLACOURTIACEAE: *Caloncoba echinata* (root). GERANIACEAE: *Monsonia senegalensis.* GOODENIACEAE: *Scaevola plumeri* (leaf). GRAMINEAE: *Coix lacryma-jobi* (root), *Cymbopogon nardus, Desmostachya bipinnata* (culm), *Eleusine indica* (root), *Panicum repens* (rhizome), *Setaria palmifolia* (leaf), *Streptogyna crinita, Themeda triandra* (root). GUT-TIFERAE: *Harungana madagascariensis* (leaf, sap), *Mammea africana* (bark). IRIDACEAE: *Gladiolus daleni* (corm), *Trimezia martinicensis* (corm). IXONANTHACEAE: *Irvingia grandifolia* (bark).

C: Medicines; 32, conception, pregnancy promotion, antiabortifacient
EBENACEAE: *Diospyros hoyleana* (leaf). EUPHORBIACEAE: *Alchornea cordifolia* (root-pith), *Antidesma laciniatum* (leaf), *A. membranaceum* (bark), *Bridelia micrantha* (bark), *Dichostemma glaucescens* (bark), *Hymenocardia acida* (leaf-sap, root), *Jatropha curcas* (leaf), *Macaranga monandra* (bark), *Maprounea africana* (stem-bark, root-bark), *Ricinodendron heudelotii* (bark), *Securinega virosa* (root, fruit), *Uapaca guineensis* (bark, root), *U. heudelotii* (bark), *U. paludosa* (stem-bark, root-bark). GRAMINEAE: *Olyra latifolia* (root), *Pseudechinolaena polystachya, Streptogyna crinita.* GUTTIFERAE: *Garcinia kola* (bark), *Harungana madagascariensis* (sap), *Hypericum roeperanum* (root).

C: Medicines; 33, lactation stimulants (incl. veterinary)

EUPHORBIACEAE: *Bridelia grandis* (bark), *Dichostemma glaucescens* (bark), *Euphorbia balsamifera, E. convolvuloides* (latex), *E. forskaolii, E. hirta, E. scordifolia, E. sp. indet., E. thymifolia, Hymenocardia acida* (root), *Manihot esculenta* (root), *Phyllanthus fraternus* (root), *Plagiostyles africana* (wood), *Ricinus communis* (leaf). GRAMINEAE: *Aristida adscensionis, Dinebra retroflexa, Eragrostis japonica, E. tremula, Imperata cylindrica* (rhizome), *Sporobolus helvolus.* GUTTIFERAE: *Harungana madagascariensis* (sap).

C: Medicines; 34, abortifacients, ecbolics, parturition stimulants

EBENACEAE: *Diospyros mespiliformis* (root, root-bark). EUPHORBIACEAE: *Alchornea cordifolia* (leaf), *Croton macrostachyus* (seed), *C. tiglium* (root), *Discoglypremna caloneura, Euphorbia balsamifera* (leaf), *E. thymifolia, Maesobotrya barteri* (bark, sap), *Mareya micrantha* (leaf), *Neoboutonia mannii* (root-bark), *Phyllanthus amarus, P. fraternus, P. niruroides* (sap), *P. urinaria, Ricinus communis* (twig), *Tragia benthamii* (leaf), *Uapaca guineensis* (root). GNETACEAE: *Gnetum africanum* (stem), *G. buchholzianum* (stem). GRAMINEAE: *Cymbopogon giganteus, C. schoenanthus* (inflorescence), *Digitaria horizontalis* (whole plant), *Paspalum scrobiculatum* (root, rhizome), *Setaria megaphylla* (root), *Vetiveria zizanioides* (root). GUTTIFERAE: *Garcinia kola* (bark), *Harungana madagascariensis* (sap), *Mammea africana* (bark), *Symphonia globulifera* (bark). ICACINACEAE: *Icacina oliviformis* (root).

C: Medicines; 35, venereal diseases, treatment, prophylaxis (*see also* C: 28 for diuretics)

ERICACEAE: *Agauria salicifolia* (whole plant). EUPHORBIACEAE: *Alchornea cordifolia* (leaf), *Antidesma venosum* (bark, root), *Bridelia atroviridis* (bark), *B. ferruginea* (leaf, bark), *B. grandis* (bark), *B. scleroneura* (bark?), *Chrozophora senegalensis, Croton macrostachyus* (fruit, root), *C. pseudopulchellus* (leaf), *Drypetes gossweileri* (bark), *Erythrococca chevalieri, E. welwitschiana, Euphorbia balsamifera, E. convolvuloides* (whole plant), *E. hirta, E. lateriflora, E. prostrata, E. thymifolia, E. tirucalli, Hippomane mancinella* (sap), *Hymenocardia acida* (bark), *Jatropha chevalieri* (latex), *J. curcas* (sap, root, seed), *J. multifida* (root), *Macaranga barteri* (leaf), *M. heterophylla* (young leaf), *Mallotus oppositifolius, Manniophyton fulvum* (stem), *Maprounea africana* (stem-bark, root, root-bark), *M. membranacea* (leaf, bark, root), *Mareya micrantha, Margaritaria discoidea* (leaf, root), *Oldfieldia africana* (bark), *Phyllanthus amarus, P. fraternus, P. muellerianus* (root), *P. pentandrus* (whole plant), *P. urinaria, Ricinodendron heudelotii* (bark, seed), *Sebastiana chamelaea* (sap), *Securinega virosa* (leaf, root), *Tetrochidium didymostemon* (leaf), *Tragia benthamii* (leaf). FLACOURTIACEAE: *Caloncoba glauca* (seed-oil), *Dovyalis zenkeri* (leaf), *Oncoba spinosa* (leaf, root). FLAGELLARIACEAE: *Flagellaria guineensis* (leaf, berry). GOODENIACEAE: *Scaevola plumeri.* GRAMINEAE: *Coix lacrymajobi* (root), *Cymbopogon giganteus, Eleusine indica* (root), *Imperata cylindrica* (rhizome), *Pennisetum glaucum* (grain), *P. purpureum, Setaria megaphylla* (leaf, root), *Zea mays* (culm). GUTTIFERAE: *Calophyllum inophyllum, Garcinia kola* (bark, gum, seed), *Mammea africana* (bark), *Psorospermum febrifugum* (bark), *Symphonia globulifera* (bark, resin). HUMIRIACEAE: *Sacoglottis gabonensis* (bark). ICACINACEAE: *Iodes africana, Pyrenacantha staudtii.* IXONANTHACEAE: *Irvingia grandifolia* (bark), *Klainedoxa gabonensis* (bark).

C: Medicines; 36, febrifuges, sudorifics, temperature control, rigor control, etc.

EBENACEAE: *Diospyros mespiliformis* (leaf, root), *D. monbuttensis* (twig, bark). ERYTHROXYLACEAE: *Erythroxylum mannii* (leafy twig). EUPHORBIACEAE: *Alchornea cordifolia* (leaf), *Bridelia atroviridis* (bark, leaf), *B. ferruginea* (leaf, bark), *B. grandis* (bark), *B. scleroneura* (bark?), *Croton macrostachyus* (leafy shoot), *C. pyrifolius* (Leaf), *C. zambesicus* (bark, leaf, shoot, root), *Drypetes aubrevillei* (bark, fruit), *D. gossweileri* (bark), *Erythrococca anomala* (leaf), *Euphorbia balsamifera* (leaf), *E. heterophylla* (root, bark), *E. hirta, E. thymifolia, Hymenocardia acida* (leaf, bark, root), *Jatropha curcas* (leaf, sap), *J. multifida* (leaf, fruit), *J. podagrica, Macaranga spinosa, Maesobotrya barteri, Manihot esculenta* (leaf), *Mareya micrantha* (leaf), *Margaritaria discoidea* (bark, sap), *Micrococca mercurialis, Mildbraedia paniculata* (leaf), *Phyllanthus amarus, P. fraternus, P. muellerianus* (leaf,root), *P. pentandrus* (leaf, root), *P. reticulatus* (leaf), *P. urinaria, Ricinodendron heudelotii* (bark), *Ricinus communis* (leaf), *Sapium ellipticum* (root), *Securinega virosa* (leaf), *Tetrochidium didymostemon* (bark), *Tetrorchidium oppositifolium* (bark). FLACOURTIACEAE: *Camptostylus mannii* (bark), *Flacourtia indica* (sap, root), *Oncoba spinosa* (seed-oil). GENTIANACEAE: *Centaurium pulchellum* ((?)). GOODENIACEAE: *Scaevola plumeri* (leaf). GRAMINEAE: *Cymbopogon citratus* (leaf, root), *C. giganteus, C. nardus, C. schoenanthus* (inflorescence), *Eleusine indica, Paspalum conjugatum* (leaf), *Phragmites australis* (root), *P. karka* (rhizome, root), *Setaria longiseta, S. megaphylla, Sporobolus helvolus, Stenotaphrum dimidiatum, Vetiveria zizanioides* (root). GUTTIFERAE: *Harungana madagascariensis* (leaf, sap, bark, root), *Psorospermum* (leaf), *P. alternifolium, P.*

PLANT SPECIES BY USAGES
febrifugum (bark), *P. senegalense* (bark), *Vismia guineensis* (leaf, bark). HUMIRIACEAE: *Sacoglottis gabonensis* (bark, sap). HYDROCHARITACEAE: *Ottelia ulvifolia* (leaf). ICACI-NACEAE: *Icacina oliviformis* (leaf), *Lasianthera africana* (leaf). IRIDACEAE: *Crocosmia aurea* (leaf-sap, corm). IXONANTHACEAE: *Irvingia grandifolia* (bark).

C: Medicines; 37, chicken-pox, measles, etc.
EBENACEAE: *Diospyros monbuttensis* (leaf). EUPHORBIACEAE: *Croton macrostachyus* (leafy shoot), *Euphorbia convolvuloides, E. polycnemoides, E. unispina* (latex), *Hymenocardia acida* (leaf), *Maesobotrya barteri* (bark), *Mallotus oppositifolius, Manihot esculenta* (leaf), *Maprounea membranacea* (bark). FLACOURTIACEAE: *Caloncoba echinata* (leafy twig). GRAMINEAE: *Panicum antidotale.* IXONANTHACEAE: *Desbordesia glaucescens* (bark), *Klainedoxa gabonensis* (bark).

C: Medicines; 38, leprosy treatment
EBENACEAE: *Diospyros canaliculata* (bark), *D. mespiliformis* (leaf, bark), *D. monbuttensis* (twig, bark), *D. viridicans* (leafy twig). EUPHORBIACEAE: *Acalypha hispida* (leaf), *A. ornata* (root), *Alchornea cordifolia* (root), *Anthostema senegalense* (latex), *Euphorbia balsamifera, E. deightonii* (latex), *E. kamerunica* (latex), *E. paganorum* (plant-ash), *E. sudanica* (plant-ash), *E. unispina* (latex), *Excoecaria grahamii, Jatropha chevalieri* (latex), *J. curcas* (seed), *J. gossypiifo-lia* (seed-oil), *Maesobotrya floribunda* (bark), *Mallotus oppositifolius, Manniophyton fulvum* (leaf-sap), *Maprounea africana* (bark, root), *M. membranacea* (bark), *Mareya micrantha, Uapaca guineensis* (bark). FLACOURTIACEAE: *Caloncoba echinata, C. glauca* (seed-oil), *C. welwitschii* (seed-oil), *Lindackeria dentata* (seed-oil). GRAMINEAE: *Cymbopogon nardus* (oil), *Pennisetum glaucum* (grain). GUTTIFERAE: *Calophyllum inophyllum* (seed-oil), *Harungana madagascariensis* (sap).

C: Medicines; 39, yaws
EBENACEAE: *Diospyros crassiflora* (bark). EUPHORBIACEAE: *Alchornea cordifolia* (leaf), *Euphorbia balsamifera* (latex), *Manihot esculenta* (leaf). FLACOURTIACEAE: *Lindackeria dentata* (seed-oil). IXONANTHACEAE: *Klainedoxa gabonensis* (bark).

C: Medicines; 40, dropsy, swellings, oedemas, gout
EBENACEAE: *Diospyros monbuttensis* (twig, bark). EUPHORBIACEAE: *Bridelia grandis* (bark), *B. micrantha, Croton sylvaticus* (wood shavings), *Euphorbia tirucalli, Hippomane mancinella* (sap), *Maprounea africana* (stem-bark, root-bark), *M. membranacea* (bark), *Phyllan-thus amarus, P. fraternus, P. muellerianus* (root-bark), *P. pentandrus, Ricinodendron heudelotii* (leaf, bark), *Sapium ellipticum* (leaf), *Tetrochidium didymostemon* (bark), *Tragia benthamii* (leaf). FLACOURTIACEAE: *Caloncoba welwitschii* (leaf), *Dovyalis zenkeri, Homalium letestui* (bark), *Lindackeria dentata* (root-ash), *Scottellia coriacea.* GRAMINEAE: *Cynodon dactylon, Eleusine indica* (sap), *Gynerium sagittatum* (root), *Leptaspis zeylanica* (leaf-sap), *Olyra latifolia* (root), *Oxytenanthera abyssinica* (leaf), *Rottboellia cochinchinensis, Zea mays* (tassel). GUTTI-FERAE: *Allanblackia floribunda* (fruit). HYDROCHARITACEAE: *Ottelia ulvifolia.* HYPOX-IDACEAE: *Curculigo pilosa* (root-stock). ICACINACEAE: *Icacina mannii, I. oliviformis* (leaf), *Pyrenacantha staudtii.* IRIDACEAE: *Gladiolus atropurpureus* (root). IXONANTHACEAE: *Irvingia grandifolia* (bark).

C: Medicines; 41, tumours, cancers
EUPHORBIACEAE: *Croton macrostachyus, Jatropha curcas* (leaf, seed-oil), *J. gossypiifolia* (whole plant), *J. multifida* (latex), *Manihot esculenta* (leaf), *Margaritaria discoidea* (root), *Phyllanthus muellerianus* (root), *P. reticulatus* (root). GRAMINEAE: *Coix lacryma-jobi* (fruit), *Cymbopogon schoenanthus, Imperata cylindrica* (leaf, stem), *Oryza sativa, Saccharum offici-narum, Vetiveria zizanioides* (root). GUTTIFERAE: *Garcinia kola* (bark).

C: Medicines; 42, malnutritrion, debility
EUPHORBIACEAE: *Hymenocardia acida* (bark), *Macaranga barteri, Mallotus oppositifolius* (leaf, root), *Margaritaria discoidea* (leaf), *Phyllanthus muellerianus* (leaf, root), *P. reticulatus, Plukenetia conophora* (kernel), *Ricinodendron heudelotii* (bark), *Securinega virosa* (leaf, root), *Uapaca guineensis* (bark), *U. heudelotii* (leaf, bark), *U. paludosa* (bark), *U. togoensis* (leaf). FLACOURTIACEAE: *Oncoba spinosa* (root). GRAMINEAE: *Cymbopogon densiflorus, Digitaria leptorhachis, Paspalum conjugatum* (leaf). GUTTIFERAE: *Psorospermum corymbiferum* (leaf). ICACINACEAE: *Icacina oliviformis* (leaf, root), *Polcephalium capitatum* (leaf).

489

PLANT SPECIES BY USAGES

C: Medicines; 43, food poisoning
EBENACEAE: *Diospyros heudelotii.* EUPHORBIACEAE: *Bridelia ferruginea* (bark), *Discoglypremna caloneura* (bark), *Elaephorbia grandifolia* (latex), *Tetrochidium didymostemon* (bark), *Uapaca heudelotii* (sap), *U. paludosa* (stem-bark, root-bark). GRAMINEAE: *Sporobolus pyramidalis.* GUTTIFERAE: *Harungana madagascariensis* (leaf).

C: Medicines; 44, poison antidotes (venomous stings, bites, etc.)
EBENACEAE: *Diospyros mespiliformis* (leaf). ERICACEAE: *Agauria salicifolia* (leaf). EUPHORBIACEAE: *Alchornea cordifolia* (root), *A. floribunda* (leaf), *Bridelia ferruginea* (bark), *Croton lobatus* (leaf), *C. tiglium* (leaf), *Euphorbia balsamifera* (latex), *E. convolvuloides, E. hirta, E. polycnemoides, E. schimperiana, E. thymifolia* (leaf, sap), *Jatropha curcas* (sap), *Macaranga heterophylla, M. spp. indet.* (bark), *Mareya micrantha* (root), *Pedilanthus tithymaloides, Phyllanthus amarus* (leaf), *P. fraternus, P. maderaspatensis, Plagiostyles africana* (bark), *Ricinodendron heudelotii* (bark). GENTIANACEAE: *Hoppea dichotoma* (whole plant). GNETACEAE: *Gnetum africanum* (leaf). GRAMINEAE: *Chrysopogon aciculatus, Cymbopogon schoenanthus, Elionurus elegans, Olyra latifolia* (seed), *Oplismenus burmannii* (leaf), *Pennisetum glaucum* (grain), *Saccharum officinarum* (sap), *Sporobolus africanus, S. pyramidalis.* GUTTIFERAE: *Garcinia kola* (seed), *G. punctata* (bark), *G. smeathmannii* (bark), *Psorospermum febrifugum* (bark). ICACINACEAE: *Icacina oliviformis* (leaf). IRIDACEAE: *Gladiolus daleni* (corm), *G. gregarius* (corm).

C: Medicines; 45, homeopathic, Theory of Signatures
EBENACEAE: *Diospyros monbuttensis* (spine). GUTTIFERAE: *Allanblackia floribunda* (fruit).

C: Medicines; 46, diabetes
GRAMINEAE: *Eleusine coracana, Oxytenanthera abyssinica* (leaf), *Panicum turgidum* (grain), *Phragmites karka* (rhizome, root).

D: Phytochemistry; 1, alkali salts—excl. common salt. (*see also* D: 2)
EUPHORBIACEAE: *Euphorbia convolvuloides.*

D: Phytochemistry; 2, soap, soap-substitutes
EUPHORBIACEAE: *Phyllanthus amarus* (leaf), *Ricinodendron heudelotii* (wood-ash), *Tetrochidium didymostemon* (bark). FLACOURTIACEAE: *Caloncoba echinata* (oil), *Scottellia coriacea* (ash). GRAMINEAE: *Sacciolepis africana* (spikelet), *Zea mays* (oil). GUTTIFERAE: *Mammea africana* (bark), *Pentadesma butyracea* (seed-fat). IXONANTHACEAE: *Irvingia gabonensis* (seed-kernel).

D: Phytochemistry; 3, salt, salt-substitutes
EUPHORBIACEAE: *Croton macrostachyus* (leaf), *Jatropha curcas* (plant-ash), *Macaranga heterophylla* (wood-ash), *Manihot esculenta* (leaf, twig), *Ricinodendron heudelotii* (wood-ash). FLACOURTIACEAE: *Lindackeria dentata.* GRAMINEAE: *Chloris gayana, C. lamproparia* (?), *C. robusta, Gramineae spp. indet.* (sp. nos. 1 & 2), *Pennisetum polystachion, Saccharum spontaneum, Setaria megaphylla, Sorghum X drummondii, Sporobolus ioclados, S. pyramidalis, Zea mays* (culm). GUTTIFERAE: *Mammea africana* (bark). HYDROCHARITACEAE: *Ottelia ulvifolia.*

D: Phytochemistry; 4, mineral salts
EUPHORBIACEAE: *Margaritaria discoidea* (wood-ash). GRAMINEAE: *Echinochloa colona* (ash), *E. pyramidalis, E. stagnina, Panicum turgidum* (root), *Setaria verticillata.*

D: Phytochemistry; 5, fatty acids, oils, waxes
ERICACEAE: *Agauria salicifolia* (bark, leaf). EUPHORBIACEAE: *Acalypha indica, Aleurites moluccana* (seed), *Croton eluteria* (bark), *C. macrostachyus* (seed), *Euphorbia tirucalli* (latex), *Hevea brasiliensis* (seed), *Hura crepitans* (seed), *Jatropha curcas* (ash, bark, seed, seed-oil), *J.*

multifida (seed), *Mallotus philippinensis* (kernel), *Manihot glaziovii* (seed-oil), *Manniophyton fulvum*(seed), *Phyllanthus maderaspatensis*(seed), *Plukenetia conophora*(kernel), *Ricinodendron heudelotii* (kernel), *Ricinus communis* (seed-oil). FLACOURTIACEAE: *Caloncoba echinata* (seed, oil), *C. glauca* (seed), *C. welwitschii* (seed), *Lindackeria dentata* (seed), *Oncoba spinosa* (seed-oil). GRAMINEAE: *Cenchrus biflorus* (grain), *Coix lacryma-jobi* (grain), *Cymbopogon giganteus*, *Imperata cylindrica*(rhizome), *Oryza sativa*. GUTTIFERAE: *Allanblackia floribunda* (seed), *Garcinia mangostana* (kernel), *Mammea africana* (seed), *Mesua ferrea*, *Pentadesma butyracea* (seed), *Psorospermum corymbiferum* (leaf, twig). HUMIRIACEAE: *Sacoglottis gabonensis* (seed). IXONANTHACEAE: *Irvingia gabonensis* (plant, kernel), *I. smithii* (seed-kernel), *Klainedoxa gabonensis* (seed-kernel), *Phyllocosmus africanus* (seed).

D: Phytochemistry; 6, aromatic substances (scent, cosmetics, coumarin, musk, incense, etc.)
EUPHORBIACEAE: *Croton zambesicus* (fruit, seed), *Euphorbia thymifolia*. GRAMINEAE: *Alloteropsis cimicina*, *A. paniculata*, *Andropogon schirensis*(root), *Bothriochloa bladhii*(foliage), *B. insculpta* (foliage), *Ctenium elegans* (leaf), *C. newtonii* (leaf), *Cymbopogon citratus*, *C. densiflorus*, *C. giganteus*(rhizome, panicle), *C. nardus*, *C. schoenanthus*, *C. winterianus*, *Elionurus ciliaris* (leaf, root), *E. muticus* (root), *Eragrostis scotelliana*, *E. tremula* (root), *Melinis minutiflora*, *Phacelurus gabonensis* (root), *Sporobolus africanus*, *Vetiveria zizanioides* (root). GUTTIFERAE: *Hypericum revolutum* (leaf). IRIDACEAE: *Trimezia martinicensis* (corm).

D: Phytochemistry; 7, starch, sugar
GRAMINEAE: *Andropogon gayanus* (leafsheath, pseudo-petiole), *Imperata cylindrica* (rhizome), *Oryza sativa* (grain).

D: Phytochemistry; 9, alkaloids
EBENACEAE: *Diospyros canaliculata* (bark), *D. crassiflora* (root), *D. discolor* (leaf, stem), *D. hoyleana* (leaf). ERICACEAE: *Agauria salicifolia* (bark, leaf). ERYTHROXYLACEAE: *Erythroxylum coca* (leaf), *E. mannii*. EUPHORBIACEAE: *Acalypha ciliata* (leaf), *A. indica*, *Alchornea cordifolia* (stem, root), *A. floribunda*, *A. hirtella* (stem-bark, root-bark), *Antidesma venosum* (leaf), *Croton lobatus*, *C. zambesicus* (leaf, stem), *Discoglypremna caloneura* (stem-bark, root-bark), *Drypetes gossweileri* (bark, root), *Erythrococca anomala* (leaf, twig, root-bark, seed), *E. chevalieri* (leaf, root), *Euphorbia hirta*, *Hymenocardia acida* (leaf, stem-bark, root-bark), *Jatropha curcas* (leaf), *Macaranga barteri* (leaf), *Mallotus subulatus* (root), *Mareya micrantha*, *Margaritaria discoidea*, *Phyllanthus fraternus*, *P. urinaria*, *Plukenetia conophora* (bark), *Ricinodendron heudelotii* (leaf,stem), *Securinega virosa* (leaf, bark), *Tetrochidium didymostemon* (leaf, bark), *Tragia* (leaf, root). FLACOURTIACEAE: *Homalium africanum* (bark, root), *Lindackeria dentata* (leaf, stem, bark, root). GRAMINEAE: *Acroceras zizanioides* (leaf), *Anadelphia afzeliana*(root), *Chloris pilosa*, *Cymbopogon citratus*(leaf, rhizome), *Eleusine indica*, *Imperata cylindrica*, *Oxytenanthera abyssinica* (leaf, culm), *Pennisetum purpureum*(leaf), *Phragmites australis*, *Rottboellia cochinchinensis*, *Saccharum officinarum*, *Thelepogon elegans*. GUTTIFERAE: *Allanblackia floribunda* (fruit), *Garcinia kola* (bark, seed), *Harungana madagascariensis* (bark), *Psorospermum corymbiferum* (root). HERNANDIACEAE: *Gyrocarpus americanus* (leaf, stem, bark, fruit). ICACINACEAE: *Icacina oliviformis* (root), *I. triacantha* (leaf, tuber), *Lasianthera africana* (leaf, bark, root), *Rhaphiostylis beninensis* (bark, root). IRIDACEAE: *Gladiolus daleni* (corm). IXONANTHACEAE: *Irvingia gabonensis* (bark), *Klainedoxa gabonensis* (leaf).

D: Phytochemistry; 10, glycosides, saponins, steroids
EBENACEAE: *Diospyros canaliculata* (bark), *D. hoyleana* (leaf, bark, root), *D. iturensis* (leaf, bark, root), *D. mespiliformis* (wood). EUPHORBIACEAE: *Alchornea cordifolia* (leaf, bark), *Antidesma venosum* (leaf), *Bridelia atroviridis* (bark), *B. ferruginea*, *B. grandis* (bark), *B. micrantha* (bark), *B. scleroneura*, *Croton lobatus*, *Drypetes gossweileri* (root), *Euphorbia forskaolii*, *E. hirta*, *Jatropha multifida*(leaf), *Keayodendron bridelioides*, *Macaranga monandra*(leaf, bark, root), *M. spinosa*, *Mallotus subulatus* (root), *Maprounea africana* (leaf, bark, root), *M. membranacea* (leaf, bark, root), *Phyllanthus fraternus*, *Uapaca heudelotii* (leaf). FLACOURTIACEAE: *Caloncoba glauca* (leaf, bark, root), *C. welwitschii* (leaf, bark, root), *Casearia barteri*, *Oncoba spinosa*(leaf, bark, root). GRAMINEAE: *Andropogon perligulatus*, *Chasmopodium afzelii*, *Dactyloctenium aegyptium*, *Echinochloa pyramidalis*, *Eleusine indica* (root), *Paspalum conjugatum*, *Phragmites australis*, *Sorghum X drummondii*, *S. bicolor* (young plants), *S. halepense*. GUTTIFERAE: *Allanblackia floribunda* (bark, root), *Garcinia ovalifolia*

(bark, root), *Harungana madagascariensis* (leaf), *Mammea africana* (bark, root), *Pentadesma butyracea* (bark, root), *Symphonia globulifera* (bark). HUMIRIACEAE: *Sacoglottis gabonensis* (bark). ICACINACEAE: *Iodes africana* (leaf, bark, root). IXONANTHACEAE: *Desbordesia glaucescens* (root), *Irvingia grandifolia* (bark, root), *I. smithii* (bark, root), *Phyllocosmus africanus* (leaf, bark, root).

D: Phytochemistry; 11, tannins, astringents, bitters

EBENACEAE: *Diospyros canaliculata* (bark), *D. dendo* (bark), *D. hoyleana* (leaf, bark, root), *D. iturensis* (leaf, bark, root), *D. mespiliformis.* ERICACEAE: *Agauria salicifolia* (bark, leaf). EUPHORBIACEAE: *Alchornea cordifolia* (leaf, bark), *Anthostema aubryanum* (leaf), *Bridelia atroviridis* (bark), *B. ferruginea, B. grandis* (bark), *B. micrantha* (bark), *B. scleroneura, Crotonogyne chevalieri* (root-bark), *Euphorbia convolvuloides* (latex), *E. hirta, Hura crepitans* (seed), *Hymenocardia acida* (leaf, bark, root), *Jatropha gossypiifolia* (leaf), *Macaranga monandra* (leaf, bark, root), *M. schweinfurthii, M. spinosa, Mallotus subulatus* (bark, root), *Maprounea africana* (leaf, bark, root), *M. membranacea* (leaf, bark, root), *Margaritaria discoidea* (bark), *Phyllanthus muellerianus, P. urinaria, Pycnocoma macrophylla* (fruit), *Securinega virosa* (bark). FLACOURTIACEAE: *Caloncoba glauca* (bark, root), *C. welwitschii* (root), *Camptostylus mannii* (root), *Lindackeria dentata* (leaf, bark, root), *Oncoba spinosa* (leaf, bark, root). GERANIACEAE: *Geranium ocellatum.* GOODENIACEAE: *Scaevola plumeri* (bark). GUTTIFERAE: *Allanblackia floribunda* (bark, root), *Garcinia afzelii* (stem-bark, root-bark), *G. epunctata* (bark, root), *G. kola* (bark), *G. mangostana* (fruit-shell), *Harungana madagascariensis* (leaf), *Mammea africana* (bark, root), *Pentadesma butyracea* (bark, root), *Psorospermum febrifugum* (root-bark), *P. senegalense, Symphonia globulifera* (bark). HUMIRIACEAE: *Sacoglottis gabonensis* (bark). ICACINACEAE: *Iodes africana* (leaf, bark, root), *Lasianthera africana* (leaf, bark, root). IRIDACEAE: *Trimezia martinicensis* (corm). IXONANTHACEAE: *Desbordesia glaucescens* (bark, root), *Irvingia gabonensis* (bark, root), *I. grandifolia* (bark, root), *I. smithii* (bark, root), *Klainedoxa gabonensis* (bark, root), *Phyllocosmus africanus* (leaf, bark, root).

D: Phytochemistry; 12, flavones

EUPHORBIACEAE: *Macaranga schweinfurthii, Phyllanthus maderaspatensis.* FLACOURTIACEAE: *Camptostylus mannii* (root). GRAMINEAE: *Andropogon perligulatus, A. schirensis, Echinochloa pyramidalis.* GUTTIFERAE: *Mammea africana* (bark, root), *Pentadesma butyracea* (bark, root), *Psorospermum febrifugum* (root-bark).

D: Phytochemistry; 13, resin. (*see also* F: 18)

EUPHORBIACEAE: *Acalypha indica, Croton eluteria* (bark), *C. macrostachyus* (seed), *Euphorbia heterophylla, E. tirucalli.* GRAMINEAE: *Saccharum officinarum.* ICACINACEAE: *Icacina oliviformis* (root).

D: Phytochemistry; 14, hydrogen cyanide

EBENACEAE: *Diospyros discolor* (stem-bark, root-bark, leaf). EUPHORBIACEAE: *Manihot esculenta* (tuber), *M. glaziovii* (leaf). FLACOURTIACEAE: *Caloncoba glauca* (leaf, bark, root, seed-oil), *C. welwitschii* (leaf, bark, root), *Lindackeria dentata* (leaf, stem-bark, root-bark). GRAMINEAE: *Cymbopogon citratus* (leaf, root), *Cynodon dactylon, Dactyloctenium aegyptium, Eleusine coracana* (vegetative parts), *E. indica, Leersia hexandra* (leaf, stem, root), *Melinis repens, Panicum repens* (root, seed), *Pennisetum purpureum, Saccharum officinarum, Setaria megaphylla, S. palmifolia* (root), *Sorghum X drummondii, S. bicolor* (young plants), *S. halepense, Stipagrostis uniplumis, Themeda triandra.*

D: Phytochemistry; 15, fish-poisons

EBENACEAE: *Diospyros canaliculata* (bark, leaf, fruit, sap, seed), *D. piscatoria* (fruit, bark). EUPHORBIACEAE: *Anthostema aubryanum* (bark), *A. senegalense* (leafy twig, inflorescence), *Antidesma venosum* (leafy twig), *Bridelia ferruginea* (root-sap), *Croton macrostachyus* (seed), *C. tiglium* (seed-oil), *Drypetes gossweileri* (bark, fruit), *D. klainei* (bark, fruit), *Elaephorbia drupifera* (latex, fruit), *E. grandifolia* (latex), *Euphorbia poissonii, E. pulcherrima, E. tirucalli, Hura crepitans* (latex), *Jatropha curcas* (bark), *Oldfieldia africana* (bark, seed), *Phyllanthus fraternus* (whole plant), *P. urinaria, Plagiostyles africana* (leaf), *Pycnocoma cornuta* (bark), *P. macrophylla* (bark), *Securinega virosa* (bark), *Spondianthus preussi* (bark-sap). GUTTIFERAE: *Mammea africana* (bark), *Pentadesma butyracea* (bark). HUMIRIACEAE: *Sacoglottis gabonensis* (bark). ICACINACEAE: *Iodes africana* (leaf).

D: Phytochemistry; 16, insecticides, arachnicides
EBENACEAE: *Diospyros canaliculata* (fruit-pulp). ERICACEAE: *Agauria salicifolia* (leaf). EUPHORBIACEAE: *Anthostema senegalense, Euphorbia lateriflora* (latex), *E. sp.* indet., *E. tirucalli, Hymenocardia acida* (root), *Jatropha curcas* (leaf, seed-oil), *Manihot esculenta* (leaf,-stem), *Maprounea membranacea* (stem), *Oldfieldia africana* (leaf, bark), *Phyllanthus maderaspatensis, Ricinus communis* (the plant, seed-oil), *Tetrochidium didymostemon* (bark). FLACOURTIACEAE: *Caloncoba echinata* (seed), *C. welwitschii, Lindackeria dentata* (leaf). GRAMINEAE: *Cenchrus biflorus* (bur-hooks), *Cymbopogon citratus, Cynodon dactylon, Melinis minutiflora, Oxytenanthera abyssinica* (leaf), *Vetiveria nigritana, V. zizanioides* (root). GUTTIFERAE: *Garcinia kola* (leaf), *Mammea americana* (seed), *Psorospermum lanatum* (bark), *P. senegalense, Vismia guineensis* (gum).

D: Phytochemistry; 17, rodenticides, mammal and bird poisons
ERICACEAE: *Agauria salicifolia* (leaf, root). EUPHORBIACEAE: *Drypetes ivorensis* (bark), *Euphorbia poissonii, Jatropha curcas* (seed), *Spondianthus preussi* (bark). FLACOURTIACEAE: *Caloncoba glauca* (seed). GUTTIFERAE: *Calophyllum inophyllum* (mature fruit), *Psorospermum spp.* indet..

D: Phytochemistry; 18, reptile-repellents
EUPHORBIACEAE: *Drypetes gossweileri* (bark), *Euphorbia lateriflora, E. polycnemoides, Jatropha gossypiifolia.*

D: Phytochemistry; 19, molluscicide
ELATINACEAE: *Bergia suffruticosa.* EUPHORBIACEAE: *Acalypha ornata* (leaf, root), *Cyrtogonone argentea* (root), *Euphorbia forskaolii, Jatropha curcas* (seed), *Manihot glaziovii, Phyllanthus fraternus* (root).

D: Phytochemistry; 20, arrow-poisons (incl. adhesives)
EBENACEAE: *Diospyros canaliculata* (bark, fruit pulp). EUPHORBIACEAE: *Croton lobatus, C. tiglium* (leaf, bark), *Elaephorbia grandifolia* (latex), *Euphorbia balsamifera* (latex), *E. deightonii* (latex), *E. kamerunica* (latex), *E. lateriflora* (stem), *E. paganorum* (latex), *E. poissonii* (stem), *E. unispina* (latex), *Excoecaria grahamii* (root), *Hippomane mancinella* (sap), *Hura crepitans* (latex), *Jatropha curcas* (seed), *Sapium ellipticum* (sap), *Uapaca guineensis* (bark, flower). GERANIACEAE: *Monsonia senegalensis* (root). GRAMINEAE: *Sorghum bicolor* (young plants). GUTTIFERAE: *Harungana madagascariensis.*

D: Phytochemistry; 21, ordeal-poisons
EUPHORBIACEAE: *Euphorbia balsamifera* (latex), *E. kamerunica* (latex), *Mareya micrantha.* GUTTIFERAE: *Vismia guineensis* (bark).

D: Phytochemistry; 23, antibiotic, bacteristatic, fungistatic
EBENACEAE: *Diospyros canaliculata* (bark), *D. mespiliformis* (leaf, root-bark, fruit), *D. preussii* (bark), *D. tricolor* (bark). EUPHORBIACEAE: *Acalypha wilkesiana* (leaf), *Bridelia ferruginea* (stem, root), *Euphorbia hirta, E. pulcherrima* (leaf, stem, flower), *Jatropha podagrica, Ricinus communis.* GRAMINEAE: *Trachypogon spicatus, Vetiveria zizanioides, Zea mays* (flour). GUTTIFERAE: *Garcinia kola* (stem, root, seed), *G. livingstonei* (leaf, flower), *Harungana madagascariensis.*

PLANT SPECIES BY USAGES

D: Phytochemistry; 24, urticant
EBENACEAE: *Diospyros canaliculata* (bark, fruit pulp), *D. mespiliformis* (sawdust).
EUPHORBIACEAE: *Dalechampia ipomoefolia* (flower, fruit), *D. scandens* (leaf, stem, fruit),
Euphorbia deightonii (latex), *E. poissonii* (latex), *E. pulcherrima*, *E. tirucalli* (latex), *E. unispina*
(latex), *Excoecaria grahamii* (root-latex), *Plagiostyles africana* (bark), *Sapium ellipticum* (sap),
Tragia , *T. benthamii, T. preussii, T. spathulata*. GRAMINEAE: *Bambusa vulgaris* (leaf-
sheaths), *Eragrostis* spp. indet., *Hackelochloa granularis* (leaf-hairs), *Panicum anabaptistum,
Saccharum spontaneum* (leaf), *Setaria poiretiana* (leaf).

D: Phytochemistry; 25, miscellaneously poisonous or repellent. (*see also* D: 15 to 21)
EUPHORBIACEAE: *Crotonogyne strigosa, Elaephorbia drupifera* (latex), *E. grandifolia,
Euphorbia baga, E. balsamifera* (whole plant), *E. calyptrata, E. deightonii* (latex), *E. heterophylla, E. poissonii* (plant-stem), *E. pulcherrima* (leaf, bark, root), *E. schimperiana, E. sudanica*
(latex), *Excoecaria grahamii, E. guineense* (sap), *Hippomane mancinella, Jatropha chevalieri, J.
gossypiifolia* (sap), *Mallotus oppositifolius* (seed), *Manihot esculenta* (old tuber), *Manniophyton
fulvum* (root-bark), *Maprounea africana, M. membranacea* (root), *Mareya micrantha* (leaf,
fruit), *Phyllanthus reticulatus* (root, fruit), *Pycnocoma cornuta, P. macrophylla, Ricinus communis* (seed-oil), *Sapium ellipticum* (bark), *Sebastiana chamelaea, Securinega virosa* (bark), *Spondianthus preussi* (all parts). FLACOURTIACEAE: *Caloncoba welwitschii* (seed). GRAMI-
NEAE: *Aristida sieberana, Chasmopodium afzelii, Digitaria diagonalis, Heteropogon contortus,
Loudetia arundinacea* (seed-hairs), *Panicum maximum, Paspalum scrobiculatum, Pennisetum
glaucum* (?), *Phalaris minor* (when fresh), *Phragmites australis, Setaria megaphylla* (plant, seed),
S. sphacelata (grain), *S. verticillata* (mature plant), *Sorghum arundinaceum, S. bicolor* (young
plants), *S. halepense, Stipagrostis uniplumis, Themeda triandra* (seed). GUTTIFERAE: *Calo-
phyllum inophyllum* (fruit), *Hypericum roeperanum* (root). ICACINACEAE: *Icacina oliviformis*.
IRIDACEAE: *Aristea alata, Moraea schimperi* (?).

E: Agri-horticulture; 1, ornamental, cultivated or partially tended
ELAEOCARPACEAE: *Elaeocarpus angustifolius*. ERYTHROXYLACEAE: *Erythroxylum
emarginatum*. EUPHORBIACEAE: *Acalypha racemosa, A. wilkesiana, Breynia disticha,
Codiaeum variegatum* (foliage), *Croton zambesicus* (tree), *Drypetes floribunda, Duvigneaudia
inopinata, Euphorbia cervicornu, E. deightonii, E. heterophylla, E. leucocephala, E. milii, E.
poissonii, E. pulcherrima, Jatropha gossypiifolia, J. multifida* (plant, flower), *J. podagrica* (plant,
flower), *Manihot glaziovii* (plant), *Plukenetia conophora, Ricinodendron heudelotii, Ricinus
communis, Securinega virosa, Suregardia ivorense*. FLACOURTIACEAE: *Caloncoba gilgiana,
Dissomeria crenata, Homalium letestui, Oncoba brachyanthera, O. spinosa*. GESNERIACEAE:
Achimenes spp., *Saintpaulia ionantha*. GRAMINEAE: *Anadelphia afzeliana, Andropogon gayanus, Axonopus affinis, A. compressus, Bambusa vulgaris, Chrysopogon aciculatus, Cymbopogon
citratus, C. densiflorus, C. nardus, Cynodon dactylon, Dactyloctenium aegyptium, Dichanthium
annulatum, Digitaria ciliaris, D. debilis, D. exilis, D. fuscescens, D. longiflora, Echinochloa
colona, Eleusine coracana, Eragrostis tenella, Gynerium sagittatum, Hackelochloa granularis,
Hyparrhenia rufa, Leersia hexandra, Loudetia phragmitoides, Melinis repens, Oryza sativa,
Paspalum conjugatum, P. notatum, P. vaginatum, Pennisetum clandestinum, P. polystachion, P.
purpureum, P. setaceum, P. thunburgii, P. villosum, Perotis patens, Phacelurus gabonensis,
Phragmites australis, Phyllostachys aervea, Polypogon monspeliensis, Polytrias diversiflora,
Saccharum spontaneum, Sorghum arundinaceum, S. bicolor, Sporobolus pyramidalis, S. spicatus,
S. tenuissimus, Stenotaphrum dimidiatum, S. secundatum, Themeda villosa, Urochloa mosambi-
censis, Vetiveria nigritana, V. zizanioides*. GUTTIFERAE: *Garcinia kola, G. mangostana,
Mammea africana, M. americana, Mesua ferrea, Pentadesma butyracea*. HYDROCHARITA-
CEAE: *Ottelia ulvifolia, Vallisneria aethiopica* (?). HYPOXIDACEAE: *Hypoxis recurva, H.
urceolata, Molineria capitulata* (foliage). IRIDACEAE: *Crocus sativus, Gladiolus, G. aequinoc-
tialis, G. daleni* (corm), *G. melleri, G. oligophlebius, Neomarica caerulea, N. gracilis*. IXONAN-
THACEAE: *Irvingia gabonensis*.

E: Agri-horticulture; 2, fodder (grazing, browsing, or eaten for lack of better)
EBENACEAE: *Diospyros mespiliformis* (leaf), *D. suaveolens* (fruit-pulp), *D. viridicans* (fruit),
D. zenkeri (fruit). ELATINACEAE: *Bergia suffruticosa*. EUPHORBIACEAE: *Acalypha
ciliata, A. crenata, A. indica, A. ornata, Bridelia micrantha* (young foliage), *Chrozophora
senegalensis, Croton lobatus, C. macrostachyus* (older foliage), *Euphorbia balsamifera, E.
convolvuloides, E. depauperata, E. forskaolii, E. schimperiana, E. scordifolia, Hymenocardia
acida* (foliage), *Manihot esculenta* (foliage, tuber), *Manniophyton fulvum* (foliage), *Margaritaria*

discoidea (young leaf, flower, fruit), *Oldfieldia africana* (seed), *Phyllanthus fraternus, P. mader-aspatensis* (when young), *P. pentandrus, P. reticulatus* (foliage), *Ricinus communis* (treated seed-cake), *Securinega virosa* (foliage, fruit). FRANKENIACEAE: *Frankenia pulverulenta.* GERA-NIACEAE: *Monsonia senegalensis.* GRAMINEAE: *Acrachne racemosa, Acritochaete volken-sis, Acroceras amplectens, A. gabunense, A. zizanioides, Aeluropus lagopoides, Alloteropsis cimicina, A. paniculata, A. semialata, Anadelphia afzeliana, A. leptocoma, A. trepidaria* (?), *Andropogon africanus, A. amethystinus, A. auriculatus, A. canaliculatus, A. chinensis, A. fastigiatus, A. gayanus, A. lima, A. macrophyllus, A. pinguipes, A. pseudapricus, A. schirensis, A. tectorum, Anthephora nigritana, A. pubescens, Aristida , A. adscensionis, A. funiculata, A. hordacea, A. kerstingii, A. mutabilis, A. sieberana, A. stipoides, Arthraxon lancifolius, A. micans, Arundinella nepalensis, Axonopus affinis, A. compressus, A. flexuosus, Bambusa vulgaris, Bewsia biflora, Bothriochloa bladhii, B. insculpta, Brachiaria arrecta, B. brizantha, B. comata, B. deflexa, B. humidicola, B. jubata, B. lata, B. mutica, B. orthostachys, B. plantaginea, B. ramosa, B. reptans, B. ruziziensis, B. serrata, B. serrifolia, B. stigmatisata, B. villosa, B. xantholeuca, Brachypodium flexum* (potential), *Bromus leptoclados, Cenchrus biflorus, C. ciliaris, C. penniseti-formis, C. prieurii, C. setigerus, Centotheca lappacea, Centropodia forskalii, Chasmopodium afzelii, Chloris barbata, C. gayana, C. lamproparia, C. pilosa, C. pycnothrix, C. virgata, Chrysopogon aciculatus, C. plumulosus, Coelachyrum brevifolium, Coelorhachis afraurita, Coix lacryma-jobi, Crypsis vaginiflora, Ctenium elegans, Cymbopogon citratus* (leaf residue), *C. commutatus, C. giganteus* (when young), *C. nardus* (young leaf), *C. schoenanthus, Cynodon aethiopicus, C. dactylon, C. nlemfuensis, C. plectostachys, Dactyloctenium aegyptium, Dantho-niopsis chevalieri, Desmostachya bipinnata, Dichanthium annulatum, D. caricosum, D. foveola-tum, Digitaria abyssinica, D. acuminatissima, D. argillacea, D. ciliaris* (grain), *D. debilis, D. delicatula, D. diagonalis, D. exilis* (haulm, straw), *D. gayana, D. horizontalis, D. leptorhachis, D. longiflora, D. milanjiana, D. nuda, D. perrottetii, D. ternata, D. velutina, Diheteropogon amplectens, D. hagerupii* (?), *Dinebra retroflexa, Echinochloa callopus, E. colona, E. crus-galli, E. crus-pavonis, E. pyramidalis, E. rotundiflora, E. sp. indet., E. stagnina, Eleusine coracana* (vegetative parts), *E. indica, Elionurus ciliaris, E. elegans, E. hirtifolius, E. muticus, E. platypus, E. royleanus, Elytrophorus spicatus, Enneapogon desvauxii, E. persicus, Enteropogon macrosta-chyus, E. prieurii, E. rupestris, Eragrostis aegyptiaca, E. aspera, E. atrovirens, E. barrelieri, E. barteri, E. camerunensis, E. cenolepsis* (?), *E. cilianensis, E. ciliaris, E. curvula, E. cylindriflora, E. gangetica, E. japonica, E. lehmanniana, E. lingulata, E. macilenta, E. minor, E. mokensis, E. paniciformis, E. pilosa, E. squamata, E. superba, E. tenella, E. tenuifolia, E. tremula, E. turgida, E. viscosa, E. welwitschii, Eriochloa fatmensis, Euclasta condylotricha, Festuca* (potential), *Hackelochloa granularis, Helictotrichon elongatum, Hemarthria altissima, Heteropogon contor-tus, H. melanocarpus, Hyparrhenia bracteata, H. collina, H. cymbaria, H. diplandra, H. figariana, H. filipendula, H. hirta, H. newtonii, H. nyassae, H. poecilotrichia, H. quarrei, H. rudis, H. rufa, H. subplumosa, H. umbrosa, H. violascens, H. welwitschii, Hyperthelia dissoluta, Ichnanthus pallens, Imperata cylindrica* (young foliage), *Ischaemum afrum, I. indicum, I. rugosum, I. timorense, Koeleria capensis, Lasiurus hirsutus, Leersia drepanothrix, L. hexandra, Leptochloa caerulescens, L. fusca, L. panicea, L. uniflora, Leptothrium senegalense, Loudetia annua, L. arundinacea, L. coarctata* (? possible), *L. flavida, L. hordeiformis, L. kagerensis, L. simplex, L. togoensis, Melinis macrochaeta, M. minutiflora, Michrochloa indica, Microglossa kunthii, Monocymbium ceresiiforme, Oplismenus burmannii, O. hirtellus, Oropetium capense, Oryza barthii, O. eichingeri, O. longistaminata, O. sativa* (foliage, grain, straw), *Oxytenanthera abyssinica* (leaf), *Panicum anabaptistum, P. antidotale, P. brazzavillense* ((?)), *P. brevifolium, P. calvum, P. coloratum, P. fluviicola, P. griffonii, P. hochstetteri, P. laetum, P. laxum, P. maximum, P. pansum, P. porphyrrhizos, P. repens, P. subalbidum, P. turgidum, P. walense, Paspalidium geminatum, Paspalum conjugatum, P. dilatatum, P. lamprocaryon, P. notatum, P. scrobiculatum, P. urvillei, P. vaginatum, Pennisetum clandestinum, P. divisum, P. glaucum, P. hordeoides, P. macrourum, P. pedicellatum, P. polystachion, P. purpureum, P. ramosum, P. setaceum, P. thunburgii, P. trachyphyllum, P. unisetum, P. villosum, Perotis indica, P. patens, Phalaris minor* (when dry), *Phragmites australis, P. karka* (young shoot), *Poa , Pogonarthria squarrosa, Polypogon monspeliensis, Polytrias diversiflora, Pseudechinolaena polystachya, Rhy-tachne gracilis, R. rottboellioides, R. triaristata, Rottboellia cochinchinensis, Saccharum sponta-neum, Sacciolepis africana, S. cymbiandra, S. indica, S. myosuroides, S. typhura, Schizachyrium brevifolium, S. exile, S. maclaudii, S. platyphyllum, S. ruderale, S. rupestre, S. sanguineum, Schmidtia pappophoroides, Schoenefeldia gracilis, Sehima ischaemoides, Setaria barbata, S. incrassata, S. longiseta, S. megaphylla, S. poiretiana, S. pumila, S. sphacelata, S. verticillata, Sorghastrum bipennatum, S. stipoides* (flower-head), *Sorghum X drummondii, S. arundinaceum, S. bicolor* (old plants, grain), *S. bicolor* (culm, grain), *S. halepense, S. purpureo-sericeum, Sporobolus africanus, S. cordofanus, S. festivus, S. helvolus, S. ioclados, S. jacquemontii, S. microprotus, S. montanus, S. myrianthus, S. nervosus, S. panicoides, S. paniculatus, S. pellucidus, S. pyramidalis, S. spicatus, S. stapfianus, S. stolzii, S. tourneuxii, S. virginicus, Stenotaphrum dimidiatum, S. secundatum, Stipagrostis hirtigluma, S. plumosa, S. pungens* (young culm, grain),

S. uniplumis, Tetrapogon cenchriformis, Thelepogon elegans, Themeda triandra, T. villosa (young shoots), *Trachypogon spicatus, Tragus berteronianus, T. racemosus, Tricholaena monachne, Trichoneura mollis, Tripogon minimus, Triraphis pumilio, Urelytrum giganteum, U. muricatum* (exceptionally), *Urochloa mosambicensis, U. trichopus, Vetiveria fulvibarbis, V. nigritana, V. zizanioides, Vossia cuspidata, Vulpia bromoides.* GUTTIFERAE: *Allanblackia floribunda* (fruit), *Hypericum peplidifolium, H. revolutum* (foliage). HYDROCHARITACEAE: *Hydrilla verticillata* (foliage (fish fodder)). HYPOXIDACEAE: *Curculigo pilosa* (?foliage, seed-capsule), *Hypoxis angustifolia* (bulb). IXONANTHACEAE: *Irvingia gabonensis* (seed-cake).

E: Agri-horticulture; 3, veterinary medicine
EBENACEAE: *Diospyros mespiliformis* (bark). EUPHORBIACEAE: *Chrozophora brocchiana, Euphorbia balsamifera, E. forskaolii, Mallotus philippinensis* (hairs from fruit surface), *Ricinus communis* (root-bark). GRAMINEAE: *Cymbopogon giganteus* (root), *C. sp., Digitaria argyrotricha, Heteranthoecia guineensis, Paspalum conjugatum* (leaf), *P. paniculatum, Perotis patens, Sporobolus helvolus, Vetiveria nigritana* (root), *Zea mays* (whole plant). GUTTIFERAE: *Harungana madagascariensis* (bark), *Mammea africana* (bark). HYPOXIDACEAE: *Molineria capitulata* (fibre). IRIDACEAE: *Gladiolus daleni* (corm).

E: Agri-horticulture; 4, bee/honey plants, insect plants
EBENACEAE: *Diospyros mespiliformis.* ELATINACEAE: *Bergia suffruticosa.* EUPHORBIACEAE: *Croton macrostachyus* (flower), *C. pseudopulchellus* (flower), *Drypetes afzelii* (flower), *Euphorbia paganorum, Manihot esculenta, M. glaziovii, Margaritaria discoidea, Securinega virosa.* FLACOURTIACEAE: *Phyllobotryum soyauxianum.* GRAMINEAE: *Andropogon tectorum, Heteropogon contortus, Zea mays* (pollen). GUTTIFERAE: *Garcinia livingstonei* (flower). HYPOXIDACEAE: *Hypoxis recurva.* ICACINACEAE: *Icacina oliviformis.*

E: Agri-horticulture; 5, land conservation, cover-plants, sand-binders, erosion prevention, pioneer species, fire-resistant, etc.
EUPHORBIACEAE: *Croton lobatus, C. pseudopulchellus, Euphorbia balsamifera, E. glaucophylla, Macaranga barteri, Phyllanthus amarus, Ricinus communis* (the plant), *Uapaca guineensis, U. heudelotii.* GOODENIACEAE: *Scaevola plumeri.* GRAMINEAE: *Aeluropus lagopoides, Alloteropsis cimicina, Anthephora ampullacea, Aristida adscensionis, A. funiculata, A. sieberana, Axonopus compressus, Bambusa vulgaris, Brachiaria jubata, Cenchrus ciliaris, C. setigerus, Centropodia forskalii, Chloris gayana, Chrysopogon aciculatus, Cymbopogon citratus, Cynodon dactylon, Dactyloctenium aegyptium, Dichanthium foveolatum, Digitaria abyssinica, D. argyrotricha, D. debilis, Echinochloa crus-galli, E. pyramidalis, E. stagnina, Eleusine indica, Elionurus elegans* (straw), *Eragrostis cilianensis, E. ciliaris, E. curvula, E. domingensis, Heteropogon contortus, Imperata cylindrica, Lasiurus hirsutus, Leptochloa caerulescens, Loudetia hordeiformis, Melinis minutiflora, M. repens, Panicum antidotale, P. repens, P. turgidum, Paratheria glaberrima, Paspalum conjugatum, P. notatum, P. scrobiculatum, P. vaginatum, P. virgatum, Pennisetum clandestinum, P. prostrata, P. purpureum, Phragmites australis, Pseudechinolaena polystachya, Saccharum spontaneum, Sacciolepis africana, Schizachyrium pulchellum, Sehima ischaemoides, Sporobolus microprotus, S. robustus, S. spicatus, S. virginicus, Stenotaphrum dimidiatum, S. secundatum, Tricholaena monachne, Vetiveria nigritana, V. zizanioides, Vossia cuspidata.* GUTTIFERAE: *Harungana madagascariensis.* HYPOXIDACEAE: *Molineria capitulata* (fibre).

E: Agri-horticulture; 6, composting, green-manuring, fertilizer, etc.
EUPHORBIACEAE: *Ricinodendron heudelotii* (seed-cake), *Ricinus communis* (seed-cake). GRAMINEAE: *Echinochloa crus-galli, Paspalum paniculatum, Setaria sphacelata, Sorghum bicolor* (ash), *Vetiveria zizanioides.* HYDROCHARITACEAE: *Hydrilla verticillata.*

E: Agri-horticulture; 7, indicators (soil, water)
EUPHORBIACEAE: *Pycnocoma macrophylla.* GRAMINEAE: *Cenchrus biflorus, Chloris gayana, Ctenium newtonii, Digitaria milanjiana, Echinochloa stagnina, Hackelochloa granularis, Heteropogon contortus, Imperata cylindrica, Leersia hexandra, Leptochloa fusca, Melinis repens, Paspalum vaginatum, Pennisetum polystachion, Phragmites karka, Polypogon monspeliensis, Schizachyrium scintallans, Sporobolus ioclados, S. robustus, S. spicatus, Themeda triandra, Vetiveria nigritana.*

E: Agri-horticulture; 8, indicators (weather, season, time)
EUPHORBIACEAE: *Alchornea cordifolia, Claoxylon hexandrum* (fruit). GRAMINEAE: *Setaria barbata.*

E: Agri-horticulture; 9, shade-trees, avenue-trees, nursery-plants
EBENACEAE: *Diospyros discolor, D. mespiliformis.* ELAEOCARPACEAE: *Elaeocarpus angustifolius.* EUPHORBIACEAE: *Bridelia micrantha, Croton macrostachyus* (tree), *C. sylvaticus, Euphorbia sudanica, Hura crepitans, Manihot esculenta, Margaritaria discoidea.* GUTTIFERAE: *Calophyllum inophyllum, Mammea africana.*

E: Agri-horticulture; 10, living hedges, markers
ERYTHROXYLACEAE: *Erythroxylum coca.* EUPHORBIACEAE: *Acalypha wilkesiana, Breynia disticha, Elaephorbia grandifolia, Euphorbia balsamifera, E. deightonii, E. desmondii, E. kamerunica, E. lactea, E. laro, E. lateriflora, E. tirucalli, Jatropha curcas, J. gossypiifolia, J. multifida, Macaranga , Manihot esculenta, M. glaziovii, Pedilanthus tithymaloides, Securinega virosa.* FLACOURTIACEAE: *Flacourtia flavescens, F. indica, F. vogelii.* GRAMINEAE: *Andropogon gayanus, Hyparrhenia diplandra, H. rufa, Panicum phragmitoides, Sorghum arundinaceum, Vetiveria nigritana, V. zizanioides.* GUTTIFERAE: *Calophyllum inophyllum.* HYPOXIDACEAE: *Molineria capitulata* (leaf).

E: Agri-horticulture; 11, fence-posts, yam-poles, bean-sticks etc.
EUPHORBIACEAE: *Anthostema senegalense* (poles), *Hura crepitans, Jatropha curcas, Macaranga barteri, M. hurifolia, Mallotus oppositifolius, Mareya micrantha* (stem), *Securinega virosa, Tetrochidium didymostemon.* FLACOURTIACEAE: *Caloncoba brevipes, Homalium africanum, H. letestui, H. longistylum, H. smythei.* GRAMINEAE: *Andropogon tectorum.* GUTTIFERAE: *Allanblackia floribunda, Garcinia afzelii, G. smeathmannii, Harungana madagascariensis, Pentadesma butyracea* (poles), *Vismia guineensis.* HUMIRIACEAE: *Sacoglottis gabonensis.* IXONANTHACEAE: *Phyllocosmus africanus.*

E: Agri-horticulture; 12, weeds, parasites
EUPHORBIACEAE: *Acalypha ciliata, A. crenata, A. segetalis, Caperonia fistulosa, C. serrata, Chrozophora plicata, Euphorbia heterophylla, E. hirta, Hymenocardia lyrata, Macaranga barteri, Mallotus oppositifolius, Phyllanthus amarus, P. maderaspatensis, P. muellerianus, P. reticulatus, P. sublanatus, Tetrochidium didymostemon, Uapaca guineensis.* FLAGELLARIACEAE: *Flagellaria guineensis.* GENTIANACEAE: *Sebaea oligantha.* GRAMINEAE: *Acroceras amplectens, Alloteropsis paniculata, Andropogon gayanus, Anthephora nigritana, Arthraxon micans, Brachiaria comata, B. deflexa, B. lata, B. ramosa, B. reptans, B. villosa, B. xantholeuca, Cenchrus biflorus, C. pennisetiformis, Chloris pycnothrix, C. virgata, Cynodon dactylon, Dactyloctenium aegyptium, Dichanthium foveolatum, Digitaria abyssinica, D. acuminatissima, D. argillacea, D. argyrotricha, D. gayana, D. leptorhachis, D. longiflora, D. milanjiana, D. nuda, D. patagiata, D. ternata, D. velutina, Dinebra retroflexa, Echinochloa colona, E. crus-galli, E. crus-pavonis, E. obtusiflora, E. pyramidalis, E. stagnina, Eleusine indica, Elytrophorus spicatus, Eragrostis arenicola, E. aspera, E. barrelieri, E. cilianensis, E. ciliaris, E. cylindriflora, E. gangetica, E. japonica, E. macilenta, E. mokensis, E. paniciformis, E. pilosa, E. superba, E. tenella, E. tenuifolia, E. tremula, E. welwitschii, Euclasta condylotricha, Hackelochloa granularis, Imperata cylindrica, Ischaemum rugosum, Leersia hexandra, Melinis repens, Oplismenus burmannii, Oryza barthii, O. longistaminata, Panicum brevifolium, P. calvum, P. laxum, P. pansum, P. parvifolium, P. repens, P. subalbidum, P. walense, Parahyparrhenia annua, Paspalidium geminatum, Paspalum dilatatum, P. scrobiculatum, P. vaginatum, Pennisetum hordeoides, P. ramosum, P. thunburgii, Phragmites australis, Poa , Rhytachne rottboellioides, Rottboellia cochinchinensis, R. purpurascens, Sacciolepis africana, S. indica, S. micrococca, Schizachyrium brevifolium, S. exile,*

PLANT SPECIES BY USAGES

Setaria pumila, *S. verticillata*, *S. viridis*, *Sorghum halepense*, *Sporobolus jacquemontii*, *S. microprotus*, *S. molleri*, *S. paniculatus*, *S. pyramidalis*, *Streptogyna crinita*, *Trachypogon spicatus*, *Tragus racemosus*, *Urochloa trichopus*, *Vossia cuspidata*, *Vulpia bromoides*. HYDRO-CHARITACEAE: *Hydrilla verticillata*, *Ottelia ulvifolia*, *Vallisneria aethiopica*. HYDRO-PHYLLACEAE: *Hydrolea palustris*. ICACINACEAE: *Icacina oliviformis*, *I. triacantha*. IXONANTHACEAE: *Phyllocosmus africanus*.

E: Agri-horticulture; 13, biotically active
EUPHORBIACEAE: *Croton hirtus*, *C. tiglium*, *Euphorbia sp. indet..* GRAMINEAE: *Cenchrus biflorus*, *Digitaria abyssinica* (?), *Imperata cylindrica*, *Paspalum conjugatum*.

F: Products; 1, building materials
EBENACEAE: *Diospyros canaliculata*, *D. cooperi*(bole), *D. gabunensis*, *D. monbuttensis*(pole), *D. sanza-minika* (timber). ERYTHROXYLACEAE: *Erythroxylum emarginatum*, *E. mannii* (timber). EUPHORBIACEAE: *Anthostema senegalense* (timber), *Antidesma venosum* (pole), *Bridelia atroviridis*, *B. ferruginea* (timber), *B. grandis* (timber), *B. micrantha*, *B. ndellensis*, *Croton longiracemosus* (timber), *C. macrostachyus*, *C. pseudopulchellus* (pole), *C. pyrifolius* (timber), *C. zambesicus*(pole), *Drypetes afzelii*(wood), *D. aubrevillei*, *D. aylmeri*, *D. floribunda*, *D. molunduana*, *Euphorbia tirucalli*, *Hevea brasiliensis*, *Hymenocardia acida*, *Klaineanthus gaboniae*(timber), *Macaranga barteri*, *Maesobotrya barteri*, *Margaritaria discoidea*, *Neobouto-nia mannii*, *Oldfieldia africana*, *Sapium ellipticum*, *Tetrochidium didymostemon*, *Uapaca heudelo-tii*, *U. staudtii*. EUPHROBIACEAE: *Protomegabaria stapfiana*. FLACOURTIACEAE: *Caloncoba gilgiana*, *Homalium letestui*, *H. smythei*, *Scottellia chevalieri*. FLAGELLARIACEAE: *Flagellaria guineensis* (stem). GRAMINEAE: *Anadelphia afzeliana* (haulm), *A. leptocoma*, *A. trepidaria* (?), *Andropogon canaliculatus*, *A. chinensis*, *A. fastigiatus*, *A. gayanus*, *A. pseudapricus*, *A. schirensis*, *A. tectorum*, *A. tenuiberbis*, *Aristida adscensionis*(culm), *A. funiculata*(straw), *A. mutabilis*, *A. sieberana* (culm), *A. stipoides*, *Arundinella nepalensis*, *Bambusa vulgaris*, *Bothriochloa insculpta*, *Chasmopodium afzelii* (cane), *Chloris gayana* (culms), *Coelorhachis afraurita*, *Ctenium elegans*, *C. newtonii*, *Cymbopogon commutatus*, *C. giganteus* (culm), *C. nardus* (culm), *C. schoenanthus*, *Danthoniopsis chevalieri*, *Desmostachya bipinnata* (culm), *Digitaria diagonalis*, *D. exilis*(straw), *Diheteropogon amplectens*(culm), *Echinochloa pyramidalis* (culm), *E. stagnina*, *Elymandra androphyla*, *Enteropogon macrostachyus*, *E. rupestris*, *Eragrostis atrovirens* (culm), *E. cilianensis* (culm), *E. tremula* (culm), *Eriochloa fatmensis* (straw), *Gynerium sagittatum* (leaf, culm), *Heteropogon contortus*, *Hyparrhenia collina*, *H. cyanescens*, *H. cymbaria*, *H. figariana*, *H. filipendula*, *H. finitima*, *H. involucrata*, *H. newtonii*, *H. nyassae*, *H. quarrei*, *H. rufa*, *H. smithiana*, *H. subplumosa*, *H. umbrosa*, *H. violascens*, *H. welwitschii*, *Hyperthelia dissoluta*, *Imperata cylindrica*, *Lasiurus hirsutus*, *Loudetia arundinacea*, *L. kagerensis*, *L. simplex*, *Loudetiopsis* (?), *Monocymbium ceresiiforme*, *Oryza longistaminata* (straw), *O. sativa* (husk, straw), *Oxytenanthera abyssinica* (culm), *Panicum anabaptistum* (culm), *P. maximum*(straw), *P. phragmitoides*(culm), *P. turgidum*(culm), *Paspalum scrobiculatum*, *Pennisetum glaucum* (culm), *P. pedicellatum*, *P. polystachion* (culm), *P. purpureum*, *Phacelurus gabonensis* (haulm), *Phragmites australis*, *P. karka* (culm), *Schizachyrium exile*, *S. sanguineum* (culm), *Schoenefeldia gracilis*, *Setaria incrassata* (culm), *S. megaphylla* (culm), *S. pumila*, *S. sphacelata*, *Sorghum bicolor*(culm), *Sporobolus africanus*, *S. festivus*, *S. helvolus*, *S. pyramidalis*, *Themeda triandra*, *Trachypogon spicatus*, *Triticum aestivum* (straw), *Vetiveria nigritana* (culm), *Vossia cuspidata*. GUTTIFERAE: *Allanblackia floribunda*, *Calophyllum inophyllum*, *Garcinia afzelii* (wood), *Harungana madagascariensis*, *Hypericum revolutum*, *Mammea africana*, *Mesua ferrea*, *Pentadesma butyracea*, *Psorospermum alternifolium*, *Symphonia globulifera*. HUMIRIACEAE: *Sacoglottis gabonensis*. ICACINACEAE: *Desmostachys vogelii*, *Leptaulus daphnoides* (pole), *Pyrenacantha vogeliana* (?liane), *Stachyanthus occidentalis* (liane). IXONANTHACEAE: *Desbordesia glaucescens*, *Irvingia gabonensis*, *I. smithii*, *Klainedoxa gabonensis*.

F: Products; 2, carpentry and related applications
EBENACEAE: *Diospyros crassiflora* (timber), *D. dendo*(timber), *D. gracilens*, *D. kamerunensis* (timber), *D. mannii*(heart-wood), *D. mespiliformis*(timber), *D. sanza-minika* (wood). ELAEO-CARPACEAE: *Elaeocarpus angustifolius* (wood). ERYTHROXYLACEAE: *Erythroxylum mannii*. EUPHORBIACEAE: *Alchornea cordifolia*, *Anthostema senegalense* (timber), *Bridelia grandis* (timber), *B. micrantha*, *Drypetes roxburghii*, *Hippomane mancinella*, *Hura crepitans*, *Keayodendron bridelioides*(wood), *Macaranga hurifolia*(wood), *Margaritaria discoidea*, *Necepsia afzelii*, *Oldfieldia africana*, *Plukenetia conophora* (seed-cake), *Ricinodendron heudelotii*, *Securinega virosa*, *Spondianthus preussi*, *Uapaca esculenta*, *U. guineensis*, *U. heudelotii*, *U.

498

staudtii. EUPHROBIACEAE: *Protomegabaria stapfiana.* FLACOURTIACEAE: *Caloncoba gilgiana, C. glauca, Homalium africanum, H. letestui, Oncoba spinosa, Scottellia chevalieri, S. coriacea.* GRAMINEAE: *Bambusa vulgaris, Oxytenanthera abyssinica* (culm), *Streptogyna crinita* (inflorescence). GUTTIFERAE: *Allanblackia floribunda, Calophyllum inophyllum, Garcinia afzelii* (wood), *G. kola* (wood), *G. mangostana, Harungana madagascariensis, Mammea africana, M. americana, Pentadesma butyracea, Symphonia globulifera, Vismia guineensis.* HOPLESTIGMATACEAE: *Hoplestigma kleineanum.* HUMIRIACEAE: *Sacoglottis gabonensis.* IXONANTHACEAE: *Irvingia gabonensis, Klainedoxa gabonensis, Phyllocosmus africanus.*

F: Products; 3, farming, forestry, hunting and fishing apparatus
EBENACEAE: *Diospyros canaliculata, D. crassiflora* (young trees), *D. elliotii* (wood), *D. ferrea* (wood), *D. heudelotii* (stems), *D. iturensis* (stem), *D. kamerunensis, D. monbuttensis* (wood, branch), *D. sanza-minika* (wood), *D. thomasii* (wood, smaller branches). ELATINACEAE: *Bergia suffruticosa.* EUPHORBIACEAE: *Acalypha ornata* (stems), *Anthostema senegalense* (latex), *Antidesma venosum* (wood, fruit), *Bridelia grandis* (timber), *B. micrantha, Cleistanthus* spp. (fruit), *Croton macrostachyus, Discoglypremna caloneura* (seed), *Drypetes floribunda* (pole), *Euphorbia balsamifera* (latex), *E. lateriflora* (latex), *E. poissonii, Euphorbiacea indeterminata, Mallotus oppositifolius, Manniophyton fulvum* (stem), *Margaritaria discoidea, Phyllanthus reticulatus, Ricinodendron heudelotii* (wood), *Securinega virosa, Spondianthus preussi, Uapaca esculenta, U. guineensis.* FLACOURTIACEAE: *Caloncoba glauca.* FLAGELLARIACEAE: *Flagellaria guineensis* (stem). GNETACEAE: *Gnetum africanum* (stem). GRAMINEAE: *Aristida adscensionis, Bambusa vulgaris, Chasmopodium afzelii* (cane), *Cymbopogon giganteus* (culm), *C. nardus* (culm), *Echinochloa pyramidalis* (culm), *E. stagnina, Eragrostis tremula* (leaf), *Gramineae* spp. *indet.* (sp. no. 3), *Gynerium sagittatum* (culm), *Hyparrhenia barteri, H. hirta, Loudetia phragmitoides* (culm), *Monocymbium ceresiiforme, Olyra latifolia* (culm, culm internode), *Oxytenanthera abyssinica* (culm), *Pennisetum purpureum, P. setaceum, Phacelurus gabonensis* (cane), *Phragmites australis, P. karka* (culm), *Phyllostachys aervea* (culm), *Saccharum spontaneum* (cane), *Schoenefeldia gracilis, Setaria pumila, S. sphacelata, Sorghum bicolor* (culm), *Sporobolus festivus, S.* sp. *indet., Stipagrostis pungens, Tricholaena monachne, Urelytrum giganteum* (culm), *Vetiveria nigritana* (culm, root). GUTTIFERAE: *Calophyllum inophyllum* (wood), *Garcinia ovalifolia* (wood), *Harungana madagascariensis, Mammea africana, Psorospermum* spp. *indet., Symphonia globulifera.* ICACINACEAE: *Alsodeiopsis poggei* (wood), *Desmostachys vogelii, Lasianthera africana* (wood). IXONANTHACEAE: *Irvingia gabonensis, Klainedoxa gabonensis, Phyllocosmus africanus* (stem).

F: Products; 4, fuel, illuminant, tinder, smoke-maker, etc.
EBENACEAE: *Diospyros dendo* (wood), *D. discolor* (wood), *D. elliotii* (wood), *D. ferrea* (wood), *D. sanza-minika* (wood). ERYTHROXYLACEAE: *Erythroxylum mannii.* EUPHORBIACEAE: *Aleurites moluccana, Anthostema aubryanum, Antidesma laciniatum, A. venosum, A. vogelianum* (stem), *Bridelia atroviridis, B. ferruginea, B. grandis* (wood), *B. micrantha, B. scleroneura, Croton macrostachyus, C. tiglium* (oil), *Dichostemma glaucescens, Drypetes aylmeri, D. floribunda, Euphorbia poissonii, E. tirucalli, Hevea brasiliensis, Hymenocardia acida, H. lyrata, Jatropha gossypiifolia* (seed-oil), *J. multifida* (seed-oil), *Macaranga barteri* (wood), *M. hurifolia* (wood), *Maesobotrya barteri* (wood), *Margaritaria discoidea, Phyllanthus muellerianus, P. reticulatus, Ricinus communis* (seed-oil), *Securinega virosa, Uapaca esculenta, U. guineensis, U. heudelotii, U. paludosa, U. togoensis, Vernicia cordata* (seed-oil). EUPHROBIACEAE: *Protomegabaria stapfiana.* FLACOURTIACEAE: *Caloncoba echinata* (oil), *C. gilgiana, Homalium smythei.* GRAMINEAE: *Bambusa vulgaris, Oryza sativa* (husk), *Panicum turgidum* (culm), *Pennisetum glaucum* (culm), *Sorghum arundinaceum* (culm), *S. bicolor* (culm). GUTTIFERAE: *Allanblackia floribunda* (twig), *Garcinia kola* (twig), *Harungana madagascariensis, Mesua ferrea* (seed-oil), *Pentadesma butyracea, Symphonia globulifera* (resin). HUMIRIACEAE: *Sacoglottis gabonensis.* IXONANTHACEAE: *Irvingia* sp. *indet., Klainedoxa gabonensis, Phyllocosmus africanus.*

F: Products; 5, household, domestic and personal items
EBENACEAE: *Diospyros abyssinica, D. discolor* (wood), *D. gabunensis, D. iturensis, D. mannii* (heart-wood), *D. tricolor.* ELAEOCARPACEAE: *Elaeocarpus angustifolius* (stone). EUPHORBIACEAE: *Alchornea laxiflora* (stem, branchlets), *Antidesma oblongum* (sap), *Bridelia atroviridis* (wood), *Croton tiglium* (oil), *Cyrtogonone argentea* (wood), *Drypetes chevalieri, D. floribunda, D. ivorensis, D. roxburghii, Hymenocardia acida, Jatropha curcas* (sap), *Mallotus subulatus* (seed), *Maprounea membranacea* (wood), *Margaritaria discoidea, Necepsia afzelii,*

Phyllanthus muellerianus (fruit-pulp), *Plagiostyles africana, Plukenetia conophora* (seed-oil), *Ricinodendron heudelotii, Sapium ellipticum* (wood), *Securinega virosa, Uapaca heudelotii, U. staudtii.* FLACOURTIACEAE: *Caloncoba echinata* (oil), *Homalium smythei, Oncoba brachyanthera* (fruit-shell), *O. spinosa* (fruit-shell), *Ophiobotrys zenkeri, Scottellia coriacea.* FLAGELLARIACEAE: *Flagellaria guineensis* (stem). GNETACEAE: *Gnetum africanum* (stem). GRAMINEAE: *Andropogon canaliculatus, A. tectorum, Aristida adscensionis* (culm), *A. funiculata* (straw), *A. mutabilis, A. sieberana* (culm), *A. stipoides, Bambusa vulgaris, Coelorhachis afraurita, Coix lacryma-jobi* (fruit), *Cymbopogon commutatus, C. giganteus* (culm), *C. nardus* (culm), *C. schoenanthus, Dactyloctenium aegyptium, Digitaria gayana, D. longiflora, Echinochloa pyramidalis, Elionurus royleanus* (straw), *Enteropogon rupestris, Eragrostis cilianensis* (straw), *E. ciliaris* (culm), *E. spp. indet., E. tremula* (culm), *Heteropogon contortus, Hyparrhenia barteri, H. involucrata* (culm), *H. rufa, H. violascens, Hyperthelia dissoluta, Imperata cylindrica, Koeleria capensis, Loudetia arundinacea* (culm), *L. simplex, L. togoensis, Olyra latifolia* (culm internode), *Oryza sativa* (grain flour), *Oxytenanthera abyssinica* (culm), *Panicum anabaptistum* (culm), *P. maximum* (culm), *P. phragmitoides* (culm), *P. porphyrrhizos, P. subalbidum* (culm, leaf), *P. turgidum* (culm, root), *Pennisetum glaucum* (culm), *P. pedicellatum* (culm), *P. polystachion* (culm), *P. ramosum, P. unisetum* (culm), *Phacelurus gabonensis* (cane), *Phragmites australis, P. karka* (culm), *Phyllostachys aervea* (culm), *Rhytachne rottboellioides, R. triaristata, Rottboellia cochinchinensis, Saccharum spontaneum* (cane), *Schizachyrium brevifolium, S. exile, S. sanguineum* (culm), *Schoenefeldia gracilis, Setaria incrassata* (culm), *S. sphacelata, S. verticillata, Sorghastrum stipoides* (culms), *Sorghum bicolor* (culm, foliage), *S. purpureo-sericeum, Sporobolus africanus, S. ioclados, S. myrianthus* (culms), *S. natalensis, S. pyramidalis, Stipagrostis pungens, Urelytrum digitatum* (culm), *U. giganteum* (culm), *Vetiveria nigritana* (culm, root), *V. zizanioides* (root), *Vossia cuspidata, Zea mays* (cob axis). GUTTIFERAE: *Garcinia livingstonei* (stick), *G. mangostana* (wood), *Pentadesma butyracea, Symphonia globulifera.* HUMIRIACEAE: *Sacoglottis gabonensis* (leaf). IXONANTHACEAE: *Irvingia gabonensis, I. smithii* (seed-oil).

F: Products; 6, pastimes (carving, musical instruments, games, toys, etc.)
EBENACEAE: *Diospyros iturensis, D. kamerunensis, D. mannii* (heart-wood), *D. mespiliformis, D. thomasii* (sapling). EUPHORBIACEAE: *Claoxylon hexandrum* (seed), *Discoglypremna caloneura* (wood), *Euphorbia spp. indet., 'cactiform'* (sap), *E. tirucalli, Pycnocoma macrophylla* (wood), *Ricinodendron heudelotii, Sapium ellipticum.* FLACOURTIACEAE: *Oncoba brachyanthera* (fruit-shell), *O. spinosa* (fruit-shell). GRAMINEAE: *Andropogon gayanus* (culm), *Aristida stipoides* (galled branch), *Cymbopogon giganteus* (culm), *Loudetia phragmitoides* (culm), *Olyra latifolia* (culm), *Oxytenanthera abyssinica* (culm), *Panicum subalbidum* (culm), *Paspalum conjugatum* (stolon), *Pennisetum unisetum* (culm), *Phragmites australis, P. karka* (culm), *Setaria verticillata.* GUTTIFERAE: *Harungana madagascariensis.* HYPOXIDACEAE: *Hypoxis angustifolia* (bulb).

F: Products; 7, containers, food-wrappers
EUPHORBIACEAE: *Alchornea laxiflora* (leaf), *Macaranga barteri* (leaf), *Neoboutonia mannii* (wood), *Uapaca guineensis* (leaf), *U. togoensis* (leaf). GRAMINEAE: *Eleusine indica, Hyperthelia dissoluta, Oxytenanthera abyssinica* (culm), *Phacelurus gabonensis* (cane), *Rhytachne rottboellioides, Sporobolus festivus, Zea mays* (sheath). ICACINACEAE: *Icacina oliviformis* (leaf), *I. triacantha* (leaf).

F: Products; 8, abrasives, polishers, cleaners, sponges
GRAMINEAE: *Leptaspis zeylanica* (leaf), *Vetiveria nigritana* (root). ICACINACEAE: *Lasianthera africana* (wood).

F: Products; 9, chew-sticks, tooth-cleaners
EBENACEAE: *Diospyros barteri, D. elliotii* (twigs), *D. ferrea* (twigs), *D. heudelotii* (twigs), *D. mespiliformis, D. tricolor.* EUPHORBIACEAE: *Alchornea floribunda* (root), *Antidesma venosum* (twig), *Bridelia ferruginea* (wood, root), *Cleidion gabonicum* (twig), *Drypetes floribunda, Jatropha curcas* (twig), *Mallotus oppositifolius* (twig), *Phyllanthus muellerianus* (twig), *P.*

reticulatus (twig), *Securinega virosa* (twig), *Tetrochidium didymostemon* (twig). FLACOUR-TIACEAE: *Casearia barteri* (twig). GRAMINEAE: *Cymbopogon citratus* (rhizome), *Saccharum officinarum.* GUTTIFERAE: *Allanblackia floribunda* (twig), *Garcinia afzelii* (twig, root), *G. epunctata* (wood, twig), *G. kola* (wood, root), *G. mangostana* (wood), *G. smeathmannii, Vismia guineensis* (twig).

F: Products; 10, fibre for ties, cordage and bark-cloth
EUPHORBIACEAE: *Caperonia fistulosa* (stem), *Cleistanthus spp.* (bark, root), *Manniophyton fulvum* (bark), *Margaritaria discoidea* (bark), *Ricinus communis* (bark). GRAMINEAE: *Andropogon pinguipes* (culm), *Eleusine indica, Eragrostis tremula, Hemarthria altissima, Lasiurus hirsutus* (root), *Phragmites karka, Setaria incrassata* (culm), *Sporobolus pyramidalis, Stipagrostis pungens* (leaf), *Urelytrum muricatum* (haulm), *Zea mays* (sheath). GUTTIFERAE: *Harungana madagascariensis* (bark).

F: Products; 11, floss, stuffing and caulking
EUPHORBIACEAE: *Bridelia micrantha* (bark), *Cleistanthus spp.* (leaf, bark, root), *Ricinodendron heudelotii* (saw-dust). GRAMINEAE: *Imperata cylindrica* (floss), *Sporobolus pyramidalis, Zea mays* (sheath). IXONANTHACEAE: *Klainedoxa gabonensis.*

F: Products; 12, withies (basketry), twigs (brooms), matting, etc.
EUPHORBIACEAE: *Cleistanthus spp.* (leaf, bark, root), *Phyllanthus reticulatus* (stem).

F: Products; 13, pulp and paper
EUPHORBIACEAE: *Croton macrostachyus, Hevea brasiliensis, Manihot esculenta, Ricinodendron heudelotii* (wood), *Ricinus communis* (stem). GRAMINEAE: *Bambusa vulgaris, Chasmopodium afzelii* (cane), *Ctenium elegans, Desmostachya bipinnata, Hyparrhenia filipendula, H. hirta, H. rufa, H. subplumosa, Hyperthelia dissoluta, Imperata cylindrica* (foliage), *Pennisetum purpureum* (culm), *Phragmites australis, P. karka, Saccharum spontaneum* (cane), *Schizachyrium sanguineum* (culm), *Sorghum bicolor* (culm), *Sporobolus pyramidalis, Themeda triandra, T. villosa, Trachypogon spicatus, Vetiveria zizanioides, Vossia cuspidata.* IXONANTHACEAE: *Irvingia gabonensis.*

F: Products; 14, pottery
EUPHORBIACEAE: *Alchornea cordifolia* (fruit), *Bridelia ferruginea* (wood, bark).

F: Products; 15, beehives
GRAMINEAE: *Andropogon gayanus* (culm), *Chasmopodium afzelii* (cane), *Ctenium elegans, Phacelurus gabonensis* (cane), *Sorghum bicolor* (culm).

F: Products; 16, dyes, stains, inks, tattoos and mordants
EBENACEAE: *Diospyros cooperi* (leaf), *D. dendo* (bark). EUPHORBIACEAE: *Alchornea cordifolia* (various parts), *Bridelia ferruginea* (leaf), *B. micrantha* (young leaf, twig, bark), *B. scleroneura* (bark), *Chrozophora plicata* (fruit), *C. senegalensis, Euphorbia hirta, Excoecaria grahamii* (latex), *Hymenocardia acida* (bark), *Jatropha curcas* (sap), *Mallotus philippinensis* (hairs from fruit surface), *Manihot esculenta, Phyllanthus fraternus* (leaf, stem), *P. maderaspatensis* (seed), *P. muellerianus, P. reticulatus* (bark, root), *Securinega virosa* (leaf, bark, fruit), *Uapaca guineensis* (gum), *U. heudelotii* (sap). GRAMINEAE: *Oryza sativa* (grain), *Pennisetum glaucum* (inflorescence), *Sorghum arundinaceum* (leaf-sheath, leaf-blade), *S. bicolor* (leaf-sheath, stem), *S. bicolor* (leaf-sheath). GUTTIFERAE: *Garcinia mangostana* (fruit-shell), *Harungana madagascariensis* (gum/sap), *Vismia rubescens* (gum). ICACINACEAE: *Icacina oliviformis* (leaf). IRIDACEAE: *Aristea alata, Crocosmia aurea* (flower), *Crocus sativus* (stigma, stamen).

F: Products; 17, tobacco, snuff
EUPHORBIACEAE: *Hamilcoa zenkeri* (latex). GRAMINEAE: *Zea mays* (cob).

PLANT SPECIES BY USAGES

F: Products; 18, exudations (gum, resin, etc.)
EBENACEAE: *Diospyros mespiliformis*(bark). EUPHORBIACEAE: *Anthostema aubryanum*, *A. senegalense*(all parts), *Bridelia ferruginea*, *B. micrantha* (bark), *Croton macrostachyus*(bark, young branch), *Drypetes afzelii*, *Duvigneaudia inopinata* (latex), *Elaephorbia drupifera* (latex), *E. grandifolia*, *Euphorbia balsamifera* (latex), *E. cervicornu* (latex), *E. convolvuloides* (latex), *E. deightonii* (latex), *E. depauperata* (latex), *E. desmondii* (latex), *E. heterophylla* (latex), *E. hirta* (latex), *E. kamerunica* (latex), *E. lateriflora* (latex), *E. macrophylla*, *E. paganorum* (latex), *E. poissonii* (latex), *E. prostrata* (latex), *E. pulcherrima*, *E. schimperiana*, *E. sudanica* (latex), *E. unispina* (latex), *Excoecaria grahamii* (latex), *Hevea brasiliensis* (latex), *Hura crepitans* (latex), *Jatropha chevalieri*(latex), *J. multifida* (latex), *Macaranga heterophylla* (gum), *M. schweinfurthii* (mucilage, jelly, resin), *Manihot dichotoma* (rubber), *M. glaziovii* (rubber, wax), *Manniophyton fulvum* (red sap), *Oldfieldia africana* (watery sap), *Pedilanthus tithymaloides* (milky sap), *Plagiostyles africana* (latex), *Sapium aubrevillei* (latex), *S. ellipticum* (gummy latex), *Securinega virosa* (gum), *Tetrochidium didymostemon* (sap), *Uapaca guineensis*(red sap/gum), *U. heudelotii* (sap). FLACOURTIACEAE: *Casearia barteri*(gum), *C. calodendron* (gum), *Homalium letestui* (sap), *H. stipulaceum* (gum), *Scottellia coriacea* (clear sap/gum). GUTTIFERAE: *Allanblackia floribunda* (resin), *Garcinia epunctata* (resin), *G. gnetoides*(resin), *G. kola* (gum), *G. livingstonei* (gum), *G. mannii* (sap), *G. smeathmannii* (latex), *Harungana madagascariensis* (gum/sap), *Mammea africana* (sap), *M. americana* (resin), *Pentadesma butyracea* (latex), *Symphonia globulifera* (resin), *Vismia guineensis* (gum), *V. rubescens* (gum). HUMIRIACEAE: *Sacoglottis gabonensis* (sap). IXONANTHACEAE: *Irvingia gabonensis* (sap), *I. sp. indet.* (latex), *Klainedoxa gabonensis* (sap).

F: Products; 19, manna and other exudations
GRAMINEAE: *Oxytenanthera abyssinica* (flower-head), *Phragmites australis*.

G: Social; 1 religion, superstition, magic
EBENACEAE: *Diospyros mespiliformis* (plant, bark, root). ELAEOCARPACEAE: *Elaeocarpus angustifolius*(stone). ELATINACEAE: *Bergia capensis*. ERYTHROXYLACEAE: *Erythroxylum mannii* (bark). EUPHORBIACEAE: *Acalypha ciliata*, *A. ornata* (root), *Alchornea cordifolia*, *A. floribunda*, *A. laxiflora*, *Antidesma membranaceum*(fruit), *Bridelia grandis*(bark), *B. micrantha*, *Croton lobatus*, *C. zambesicus*(tree), *Elaephorbia drupifera* (plant), *E. grandifolia* (plant), *Erythrococca anomala*, *Euphorbia balsamifera*, *E. convolvuloides*, *E. kamerunica*, *E. lateriflora*, *E. paganorum*, *E. prostrata*, *E. tirucalli*, *E. unispina*, *Excoecaria grahamii*, *Hymenocardia acida*, *Jatropha gossypiifolia*, *J. multifida*, *Macaranga barteri* (leaf), *Manihot esculenta*, *Oldfieldia africana*, *Phyllanthus amarus*, *P. muellerianus*, *P. niruroides*, *P. odontadenius*, *P. rotundifolius*, *Plagiostyles africana*, *Plukenetia conophora*, *Ricinodendron heudelotii*, *Ricinus communis*, *Spondianthus preussi* (fruit), *Tragia* , *T. benthamii*. FLACOURTIACEAE: *Caloncoba glauca* (bark, root), *Homalium letestui*, *Oncoba spinosa*. FLAGELLARIACEAE: *Flagellaria guineensis* (stem). GRAMINEAE: *Andropogon gayanus*, *Aristida adscensionis*, *A. sp. indet.*, *Cenchrus biflorus*(root, grain), *Chloris pilosa*, *Chrysopogon aciculatus*, *Coix lacryma-jobi* (fruit), *Cymbopogon citratus*, *C. densiflorus*, *C. giganteus*, *Cyrtococcum chaetophoron*, *Digitaria exilis*, *Eleusine indica*, *Eragrostis ciliaris*, *E. spp. indet.*, *E. tenella*, *Hyparrhenia subplumosa*, *Imperata cylindrica*, *Ischaemum rugosum*, *Melinis repens*, *Monocymbium ceresiiforme*, *Oplismenus burmannii*(leaf), *Panicum turgidum*, *Pennisetum glaucum*, *P. polystachion*, *P. purpureum*, *Phacelurus gabonensis* (root), *Pseudechinolaena polystachya*, *Setaria verticillata*, *Sporobolus pyramidalis*, *Streptogyna crinita* (inflorescence), *Vetiveria nigritana* (root), *V. zizanioides*(root), *Zea mays*(cob). GUTTIFERAE: *Harungana madagascariensis*, *Mammea africana*, *Pentadesma butyracea*, *Psorospermum senegalense*, *Vismia guineensis*. HYPOXIDACEAE: *Curculigo pilosa* (root-stock). ICACINACEAE: *Icacina mannii*, *Iodes africana*, *I. liberica*, *Rhaphiostylis beninensis* (leaf). IXONANTHACEAE: *Irvingia gabonensis*.

G: Social; 2 ceremonial
EBENACEAE: *Diospyros chevalieri* (leaf). EUPHORBIACEAE: *Elaephorbia drupifera* (latex). GRAMINEAE: *Andropogon tectorum*, *Coix lacryma-jobi* (fruit), *Cymbopogon schoenanthus*, *Oryza sativa* (grain), *Pennisetum glaucum*, *P. polystachion* (inflorescence), *P. purpureum*, *Sorghum bicolor* (flour, beer), *Sporobolus pyramidalis*, *Vetiveria nigritana* (culm). ICACINACEAE: *Icacina oliviformis* (leaf), *Lasianthera africana*.

G: Social; 3 sayings, aphorisms
GRAMINEAE: *Eleusine indica*, *Imperata cylindrica*, *Zea mays* (cob).

502

507

INDEX

batánka	**LIMBA** EUPHORBIACEAE *Manihot esculenta*
batarpolì	**SERER** EUPHORBIACEAE *Ricinus communis*
batdi	**FULA - PULAAR** HYPOXIDACEAE *Curculigo pilosa*
bategné ni	**KRU - BETE** EUPHORBIACEAE *Jatropha curcas*
batere	**MANDING - MANDINKA** EUPHORBIACEAE *Manihot esculenta*
g-bati	**MENDE** GRAMINEAE *Acroceras zizanioides*
bati	**SERER** GRAMINEAE *Anadelphia afzeliana*
ɓátlâ (pl. -átiín)	**NGIZIM** GRAMINEAE *Pennisetum glaucum*
batu mendé	**BALANTA** EUPHORBIACEAE *Hymenocardia acida*
b'atue	**MENDE** GUTTIFERAE *Psorospermum alternifolium*
batué	**KRU - GUERE** EUPHORBIACEAE *Macaranga heterophylla*
batukari	**FULA - PULAAR** EUPHORBIACEAE *Euphorbia balsamifera*
bau	**HAUSA** GRAMINEAE *Oryza longistaminata*
baud̃ (pl.)	**FULA - FULFULDE** GRAMINEAE *Pennisetum glaucum*
ba-udiga	**NANKANNI** EUPHORBIACEAE *Bridelia scleroneura*
baugassongau	**TAMACHEK** GRAMINEAE *Paspalidum geminatum*
baukarara	**MOORE** GRAMINEAE *Echinochloa pyramidalis*
baul	**NON** EUPHORBIACEAE *Chrozophora senegalensis*
ɓauren kiyeshi	**HAUSA** EUPHORBIACEAE *Chrozophora senegalensis*
baushin kurege	**HAUSA** EUPHORBIACEAE *Manihot esculenta*
bausiri dama	**DOGON** EUPHORBIACEAE *Antidesma venosum*
mbawi	**AKAN - ASANTE** EUPHORBIACEAE *Cleidion gabonicum*
mbawol	**FULA - FULFULDE** EUPHORBIACEAE *Manihot esculenta*
bawpoh	**KRU - BASA** EUPHORBIACEAE *Oldfieldia africana*
bawu	**HAUSA** GRAMINEAE *Oryza longistaminata*
baya	**HAUSA** GRAMINEAE *Echinochloa colona* **KANURI** *Aristida mutabilis, A. sieberana*
báyáá	**HAUSA** GRAMINEAE *Panicum laetum*
báyaá sa're	**HAUSA** GRAMINEAE *Panicum laetum*
bayan maraya	**HAUSA** GRAMINEAE *Anadelphia afzeliana, Andropogon fastigiatus*
mbayeeri	**FULA - FULFULDE** GRAMINEAE *Sorghum bicolor*
mbayeeri mbod̃eeri	**FULA - FULFULDE** GRAMINEAE *Sorghum bicolor*
bayori	**GURMA** GRAMINEAE *Setaria pumila*
bayuri	**GURMA** GRAMINEAE *Zea mays*
ba-zaɓa, dan-zaɓi	**HAUSA** EUPHORBIACEAE *Manihot esculenta*
bàzàyyánáó	**HAUSA** GRAMINEAE *Aristida adscensionis*
bazinga fufué	**KYAMA** EUPHORBIACEAE *Macaranga spinosa*
bazure	**MOORE** GRAMINEAE *Setaria sphacelata*
bazuré	**MOORE** GRAMINEAE *Setaria sphacelata*
be	**AKYE** IXONANTHACEAE *Irvingia gabonensis*
mbe(-i)	**LOKO** GUTTIFERAE *Harungana madagascariensis*
be	**MENDE** GRAMINEAE *Sacciolepis africana*
bē	**DAN** EUPHORBIACEAE *Manihot esculenta*
be gof	**MANKANYA** EUPHORBIACEAE *Phyllanthus reticulatus*
be lasi	**MANDYAK** EUPHORBIACEAE *Erythrococca africana*
be lora	**MANKANYA** EUPHORBIACEAE *Alchornea cordifolia*
be mével	**MANDYAK** FLACOURTIACEAE *Oncoba spinosa*
be mièl	**MANDYAK** EUPHORBIACEAE *Macaranga heterophylla*
be ñakat, bi ñañak	**MANDYAK** GUTTIFERAE *Harungana madagascariensis*
be nakenten, be vana	**MANKANYA** GUTTIFERAE *Harungana madagascariensis*
be nasia	**MANDYAK** ICACINACEAE *Icacina oliviformis*
be nasin	**MANKANYA** ICACINACEAE *Icacina oliviformis*
be pal	**MANKANYA** EUPHORBIACEAE *Uapaca guineensis*
be sak	**MANKANYA** EUPHORBIACEAE *Bridelia micrantha/stenocarpa*
be sémin	**DIOLA** IRIDACEAE *Gladiolus gregarius*
be tafé	**MANKANYA** EUPHORBIACEAE *Anthostema senegalensis*
be tange	**MENDE** EUPHORBIACEAE *Manihot esculenta*
be tasi	**MANKANYA** EUPHORBIACEAE *Anthostema senegalensis*
be tira	**MANDYAK** EUPHORBIACEAE *Alchornea hirtella*
bealibe ti	**KPE** GRAMINEAE *Cymbopogon citratus* **WOVEA** *C. citratus*
mbɛbambe	**MENDE** GUTTIFERAE *Vismia guineensis*
bébé	**AKYE** EUPHORBIACEAE *Manihot esculenta* **KYAMA** *M. esculenta*
ɔ-bébéĩ (pl. sì-)	**AKPAFU** GRAMINEAE *Oxytenanthera abyssinica*
bébela siraba	**MANDING - BAMBARA** GRAMINEAE *Aristida stipoides*
bebena	**ADANGME - KROBO** HYPOXIDACEAE *Curculigo pilosa*
bebengaa	**ADANGME** HYPOXIDACEAE *Curculigo pilosa* **ADANGME - KROBO** *C. pilosa*
bechna	**ARABIC** GRAMINEAE *Pennisetum glaucum, Sorghum bicolor*
bédé	**AKAN - ASANTE** EUPHORBIACEAE *Manihot esculenta*
mbédé	**KYAMA** EUPHORBIACEAE *Manihot esculenta*
mbèdè	**AKYE** EUPHORBIACEAE *Manihot esculenta*
bɛ̃de	**MANDING - DYULA** EUPHORBIACEAE *Manihot esculenta*
mbɛdɛ	**ABE** EUPHORBIACEAE *Manihot esculenta*
bɛdɛ	**ALADYÃ** EUPHORBIACEAE *Manihot esculenta* 'EOTILE' *M. esculenta* **NZEMA** *M. esculenta*
bédé bro	**KYAMA** EUPHORBIACEAE *Manihot esculenta*
bede kofi	**GBE - VHE** EUPHORBIACEAE *Manihot esculenta*
bédé mango	**KYAMA** EUPHORBIACEAE *Manihot esculenta*
bédé nana	**KYAMA** EUPHORBIACEAE *Manihot esculenta*
bédé putu	**KYAMA** EUPHORBIACEAE *Manihot esculenta*
gbedi	**GBE - VHE** EUPHORBIACEAE *Manihot esculenta*
bedibɛsa	**AKAN - TWI** EUPHORBIACEAE *Drypetes floribunda*
bedie kofie	**AKAN - ASANTE** EUPHORBIACEAE *Manihot esculenta*
bedrine	**ARABIC** EUPHORBIACEAE *Phyllanthus rotundifolius*

bobo, bobo-yamba	KONO GRAMINEAE *Setaria megaphylla*
bobo, mbobo	LOKO GRAMINEAE *Setaria megaphylla*
am-bobo	TEMNE GRAMINEAE *Andropogon tectorum*
bɔbɔ	LOKO GRAMINEAE *Andropogon tectorum*
bɔbɔ (def. -i)	MENDE IXONANTHACEAE *Irvingia gabonensis*
bobogia	KRIO GRAMINEAE *Andropogon tectorum*
boboi (?)	MENDE EUPHORBIACEAE *Drypetes afzelii*
boboni-vɔlɔ	MENDE GRAMINEAE *Coix lacryma-jobi*
boborasien	MANDING - DYULA GRAMINEAE *Cymbopogon giganteus*
boboru	ABE IXONANTHACEAE *Irvingia gabonensis*
a-bɔbɔruni	TEMNE GRAMINEAE *Chasmopodium afzelii caudatum*
bobuat	BOKYI GUTTIFERAE *Mammea africana*
bɔbwɛ	MENDE EUPHORBIACEAE *Bridelia atroviridis*
bodé	WOLOF GRAMINEAE *Imperata cylindrica*
boďehi	FULA - FULFULDE EUPHORBIACEAE *Hymenocardia acida*
bô-diara	MANDING - BAMBARA EUPHORBIACEAE *Phyllanthus pentandrus*
bodiè	BAULE EUPHORBIACEAE *Mareya micrantha*
bodooko	FULA - FULFULDE GRAMINEAE *Hyperthelia dissoluta*
bodri	KRU - BETE EBENACEAE *Diospyros canaliculata*
bodrui	KRU - BETE EBENACEAE *Diospyros canaliculata*
boduchié	KWENI EUPHORBIACEAE *Phyllanthus muellerianus*
bõe	KRU - BASA EUPHORBIACEAE *Manihot esculenta*
gbɔɛ	KONO EUPHORBIACEAE *Ricinodendron heudelotii* GRAMINEAE *Coix lacryma-jobi*
bofan	BOKYI FLACOURTIACEAE *Ophiobotrys zenkeri*
bofinhè	BIAFADA GRAMINEAE *Digitaria exilis*
bofua	AKAN - ASANTE GUTTIFERAE *Harungana madagascariensis* TWI *H. madagascariensis*
bofuakechi	BOKYI GUTTIFERAE *Pentadesma butyracea*
bogbo	KRU - GUERE EUPHORBIACEAE *Mareya micrantha*
gbɔgbɔ(-i)	MENDE EUPHORBIACEAE *Protomegabaria stapfiana*
bo-gboho	KISSI EUPHORBIACEAE *Ricinodendron heudelotii*
bogboi	KONO EUPHORBIACEAE *Protomegabaria stapfiana*
bogdrobo	KRU - GUERE ICACINACEAE *Rhaphiostylis beninensis*
boylẽ	GBE - VHE GRAMINEAE *Saccharum officinarum*
boylẽ biri	GBE - VHE GRAMINEAE *Saccharum officinarum*
boylẽ fe	GBE - VHE GRAMINEAE *Saccharum officinarum*
boylẽ yibɔ	GBE - VHE GRAMINEAE *Saccharum officinarum*
boylẽ-biri	GBE - VHE GRAMINEAE *Saccharum officinarum*
boylẽ-fe	GBE - VHE GRAMINEAE *Saccharum officinarum*
boylẽ-yibɔ	GBE - VHE GRAMINEAE *Saccharum officinarum*
bogo dolori	FULA - PULAAR GRAMINEAE *Pennisetum hordeoides, P. polystachion*
bógòbodje	FULA - PULAAR GRAMINEAE *Cynodon dactylon*
bogodollo	FULA - FULFULDE GRAMINEAE *Pennisetum pedicellatum*
bogodollo (pl. -ooji)	FULA - FULFULDE GRAMINEAE *Pennisetum polystachion*
bogodollooji	FULA - FULFULDE GRAMINEAE *Pennisetum pedicellatum*
ɓogol boje	FULA - FULFULDE GRAMINEAE *Cynodon dactylon*
bogonliŋxumbe	SUSU EBENACEAE *Diospyros thomasii*
gbogontiŋyi	SUSU EBENACEAE *Diospyros heudelotii*
boguntinyi	SUSU EBENACEAE *Diospyros thomasii*
bogyamono	AKAN - ASANTE GRAMINEAE *Oplismenus burmannii, O. hirtellus*
bohdô	MANDING - BAMBARA GRAMINEAE *Andropogon pinguipes*
boho	MANDING - MANDINKA GRAMINEAE *Oxytenanthera abyssinica*
bɔho	KISSI GRAMINEAE *Oxytenanthera abyssinica*
bɔhɔnaga	LOKO EUPHORBIACEAE *Croton dispar*
bɔhɔri	SUSU GRAMINEAE *Coix lacryma-jobi*
bohwe	AKAN - ASANTE GUTTIFERAE *Allanblackia floribunda* TWI *A. floribunda*
bɔhwe	AKAN - ASANTE GUTTIFERAE *Garcinia smeathmannii* TWI *G. smeathmannii*
nboi	BALANTA IXONANTHACEAE *Phyllocosmus africanus*
bõi	MANO EUPHORBIACEAE *Manihot esculenta*
bɔi	GA EUPHORBIACEAE *Manihot esculenta*
boidien	AKAN - ASANTE EUPHORBIACEAE *Phyllanthus muellerianus*
bojep	BOKYI IXONANTHACEAE *Irvingia gabonensis*
bojunde	KPE EUPHORBIACEAE *Spondianthus preussii*
boka	KUNDU EUPHORBIACEAE *Macaranga schweinfurthii*
bõki	GBE - VHE EUPHORBIACEAE *Manihot esculenta*
bokpone	KISSI EUPHORBIACEAE *Phyllanthus capillaris*
bɔku (def. -i)	MENDE FLACOURTIACEAE *Lindackeria dentata*
boku-balondo	MBONGE EUPHORBIACEAE *Ricinus communis*
bokubi	AKYE IXONANTHACEAE *Phyllocosmus africanus*
bokuri	BAFOK EUPHORBIACEAE *Ricinus communis* LONG *R. communis*
bol	SERER EUPHORBIACEAE *Chrozophora senegalensis*
bola	MENDE GRAMINEAE *Imperata cylindrica*
bolapan	NDUT EUPHORBIACEAE *Securinega virosa*
gbɔlei	MENDE EUPHORBIACEAE *Ricinodendron heudelotii*
gbolesɛhrɛ-na	YALUNKA GRAMINEAE *Anadelphia afzeliana*
bolidé yassua	BAULE EUPHORBIACEAE *Erythrococca anomala*
bɔlɔ	IJO - KAKIBA GUTTIFERAE *Mammea africana*
bɔlɔ, borlor	IJO - NEMBE GUTTIFERAE *Mammea africana*
bɔ́lɔ́	IJO - IZON GUTTIFERAE *Mammea africana*
gbɔlɔ	MENDE EUPHORBIACEAE *Ricinodendron heudelotii*
gbɔlɔ̃	KISSI GRAMINEAE *Coix lacryma-jobi*
bòlɔ ánala	DOGON GRAMINEAE *Paspalum scrobiculatum*

borikbori	TEMNE EUPHORBIACEAE *Euphorbia glaucophylla*
borikio	ABE EUPHORBIACEAE *Uapaca esculenta, U. guineensis*
mborno	MANDING - MANDINKA EUPHORBIACEAE *Chrozophora senegalensis*
borobali	LOMA EUPHORBIACEAE *Jatropha curcas*
boroboro	LIMBA GRAMINEAE *Panicum subalbidum*
boroï	TOMA EUPHORBIACEAE *Ricinodendron heudelotii*
borọlọ, owosokotọ	ISEKIRI GUTTIFERAE *Mammea africana*
bòromọ́	YORUBA GRAMINEAE *Sorghum bicolor*
borooje	FULA - FULFULDE EUPHORBIACEAE *Euphorbia unispina*
boror	SERER GRAMINEAE *Coix lacryma-jobi*
bororamga	LOKO EUPHORBIACEAE *Manihot glaziovii*
g-borotu	KRU - GUERE EUPHORBIACEAE *Elaeophorbia grandifolia*
gbörr	SENUFO EUPHORBIACEAE *Manihot esculenta*
bosambi	KUNDU EUPHORBIACEAE *Uapaca staudtii*
bòsambi	DUALA EUPHORBIACEAE *Uapaca guineensis, U. staudtii*
mbosi	ADYUKRU EUPHORBIACEAE *Manihot esculenta*
bosisang	LUNDU EUPHORBIACEAE *Ricinodendron heudelotii*
bɔtɛmbã	KISSI EUPHORBIACEAE *Jatropha curcas, J. gossypiifolia, J. multifida*
boto	'KRU' EUPHORBIACEAE *Elaeophorbia grandifolia*
botofufuo	AKAN - TWI IRIDACEAE *Gladiolus gregarius*
botong	SISAALA FLACOURTIACEAE *Flacourtia flavescens*
botu	'SOUBRE' IXONANTHACEAE *Klainedoxa gabonensis*
botué	KRU - GUERE EUPHORBIACEAE *Discoglypremna caloneura*
bòtújẹ̀	YORUBA EUPHORBIACEAE *Jatropha multifida*
bòtújẹ̀	YORUBA EUPHORBIACEAE *Jatropha curcas*
bòtújẹ̀ pupa	YORUBA EUPHORBIACEAE *Jatropha curcas, J. gossypiifolia*
bòtújẹ̀-pupa	YORUBA EUPHORBIACEAE *Jatropha multifida*
bòtújẹ̀-úbọ̀	YORUBA EUPHORBIACEAE *Jatropha curcas*
bou obírima	IJO - IZON EUPHORBIACEAE *Euphorbia hirta*
bou ogbóin	IJO - IZON IXONANTHACEAE *Desbordesia glaucescens, Irvingia gabonensis*
bou pulóú	IJO - IZON GUTTIFERAE *Harungana madagascariensis*
boua	DAN FLACOURTIACEAE *Lindackeria dentata*
boué	KRU - GREBO EUPHORBIACEAE *Uapaca esculenta*
mbovi	MENDE GRAMINEAE *Setaria megaphylla*
gbɔvui-hini (*def.* -hinei)	MENDE EUPHORBIACEAE *Discoglypremna caloneura*
gbɔ-wai	KONO FLACOURTIACEAE *Dovyalis zenkeri*
gbɔ-wai-kɔnɛ	KONO FLACOURTIACEAE *Dovyalis zenkeri*
bowal	FULA - PULAAR GRAMINEAE *Hyparrhenia* gen.
gbɔ-wali	KONO FLACOURTIACEAE *Dovyalis zenkeri*
m-bowi	MENDE GRAMINEAE *Acroceras zizanioides*
mbowi	MENDE GRAMINEAE gen., *Acroceras amplectens, Brachiaria villosa, Setaria megaphylla*
gbɔwi, kɔndoi, kɔndui	MENDE GUTTIFERAE *Garcinia smeathmannii*
mbowi-hei	MENDE GRAMINEAE *Pennisetum purpureum, Sporobolus pyramidalis*
mbowo	MENDE GRAMINEAE *Setaria megaphylla*
mbowo-la	MENDE GRAMINEAE *Setaria megaphylla*
mboworo	LOKO GRAMINEAE *Setaria megaphylla*
boxefɔrotai	SUSU EUPHORBIACEAE *Euphorbia forskaolii, E. prostrata, E. thymifolia*
bɔxefɔrotai	SUSU EUPHORBIACEAE *Euphorbia thymifolia*
boxifɔrotai	SUSU EUPHORBIACEAE *Euphorbia hirta* GRAMINEAE *Heteranthoecia guineensis*
bóyè	GUANG - GONJA EUPHORBIACEAE *Uapaca paludosa*
gbɔyɛ	KONO GRAMINEAE *Coix lacryma-jobi*
mbɔyɛ (*def.* -i)	MENDE EUPHORBIACEAE *Pycnocoma angustifolia*
brakassa	KYAMA EUPHORBIACEAE *Margaritaria discoidea*
brasi	BALANTA EUPHORBIACEAE *Anthostema senegalensis*
mbraua	ABE EUPHORBIACEAE *Protomegabaria stapfiana*
brêtié	KYAMA IXONANTHACEAE *Irvingia gabonensis*
bridié	DAN EBENACEAE *Diospyros viridicans*
brigué	KRU - GUERE EUPHORBIACEAE *Mallotus oppositifolius*
brindé	BAULE EUPHORBIACEAE *Hymenocardia acida*
bro-kpar	KRU - BASA FLACOURTIACEAE *Homalium letestui, H. smythei*
brɔmabene	NZEMA GUTTIFERAE *Garcinia gnetoides*
bromabine	AHANTA GUTTIFERAE *Pentadesma butyracea* NZEMA *P. butyracea*
brozi	BALANTA EUPHORBIACEAE *Anthostema senegalensis*
bru	'NÉKÉDIÉ' EUPHORBIACEAE *Alchornea cordifolia*
brue	KRU - GUERE EUPHORBIACEAE *Uapaca guineensis*
bruété	BAULE EUPHORBIACEAE *Croton lobatus*
bruǰe	BALANTA EUPHORBIACEAE *Bridelia micrantha/stenocarpa*
brumbarongo	MANDING - MANDINKA EUPHORBIACEAE *Securinega virosa*
brusus	CRIOULO EUPHORBIACEAE *Alchornea cordifolia*
bu	KRU - GUERE EUPHORBIACEAE *Manihot esculenta* MANO GRAMINEAE *Oryza sativa*
bu areiba	ARABIC GRAMINEAE *Tetrapogon cenchriformis*
bu bambulaf	DIOLA ICACINACEAE *Icacina oliviformis*
bu begel	DIOLA EUPHORBIACEAE *Uapaca guineensis, U. togoensis*
bu bon a tsimen	DIOLA EUPHORBIACEAE *Chrozophora brocciana*
bu bon kashla	DIOLA EUPHORBIACEAE *Jatropha curcas*
bu bun a vana	DIOLA - FLUP EUPHORBIACEAE *Antidesma membranaceum*
bu diay kola, hu lukutu, sosuso	DIOLA - FLUP IXONANTHACEAE *Irvingia gabonensis*
butue	KRU - GUERE FLACOURTIACEAE *Caloncoba brevipes*
butugama	LIMBA EUPHORBIACEAE *Necepsia afzelii*
butumpu	LIMBA EUPHORBIACEAE *Macaranga heudelotii*

549

INDEX

INDEX

kālofi — MANKANYA EUPHORBIACEAE *Manihot esculenta*
kalogozhi (*pl.* a-) — MALA EUPHORBIACEAE *Manihot esculenta*
kaloiéga — MOORE GRAMINEAE *Andropogon pseudapricus*
kalo-na — YALUNKA GRAMINEAE *Rottboellia cochinchinensis*
kalum — WOLOF EBENACEAE *Diospyros mespiliformis*
kaluwana — DAGBANI GRAMINEAE *Zea mays* MAMPRULI *Z. mays*
kaluwandi — MOBA GRAMINEAE *Zea mays*
kàlùwúyà — GUANG GUTTIFERAE *Psorospermum corymbiferum*
kám áàne — NANKANNI GRAMINEAE *Zea mays*
kam bon yulè — KONKOMBA EUPHORBIACEAE *Manihot esculenta*
kama — AKAN - ASANTE GRAMINEAE *Eleusine indica* BAULE *Paspalum conjugatum*
    MANDING - MANDINKA *Zea mays* MBOI *Sorghum bicolor*
kamaande — MOORE GRAMINEAE *Zea mays*
kamame mano — MANDING - MANDINKA GRAMINEAE *Oryza barthii*
kaman — DAGAARI GRAMINEAE *Zea mays*
kamana — GRUSI GRAMINEAE *Zea mays*
kamanaari — FULA - FULFULDE GRAMINEAE *Zea mays*
kamandi — LIMBA FLACOURTIACEAE *Homalium africanum*
kamaré — SONINKE - SARAKOLE GRAMINEAE *Vetiveria nigritana*
kambarahi — GWARI GRAMINEAE *Heteropogon contortus*
kambiri — BOKYI EBENACEAE *Diospyros zenkeri*
kambo-hando — BASSARI EUPHORBIACEAE *Manihot esculenta*
kambonjule — DAGBANI EUPHORBIACEAE *Manihot esculenta*
kambu — HAUSA GRAMINEAE *Vetiveria nigritana*
kambura fage — HAUSA GRAMINEAE *Eragrostis arenicola*
k'ameh, el k'ameh — ARABIC - SHUWA GRAMINEAE *Triticum aestivum*
kaméré — SONINKE - SARAKOLE GRAMINEAE *Elionurus elegans*
kamétêô — MANDING - MANDINKA GRAMINEAE *Leersia hexandra*
kamfalwa — HAUSA GRAMINEAE *Digitaria gayana*
nkamfo-barima — AKAN - TWI EUPHORBIACEAE *Euphorbia lateriflora*
kamfua — TEM EUPHORBIACEAE *Margaritaria discoidea*
kamimbi — MANDING - BAMBARA GRAMINEAE *Brachiaria stigmatisata*
kã-moɛ — BULOM GRAMINEAE *Zea mays*
kampana — BISA GRAMINEAE *Zea mays*
kamsuvan doki — HAUSA GRAMINEAE *Chasmopodium afzelii caudatum*
kámsuwa — HAUSA GRAMINEAE *Pennisetum pedicellatum, P. polystachion*
kamwii — SUSU EUPHORBIACEAE *Manniophyton fulvum*
ɛ-kan — TEMNE EUPHORBIACEAE *Erythrococca africana*
kán — JUKUN GRAMINEAE *Gramineae indet.*
kán másáráá — HAUSA GRAMINEAE *Zea mays*
ƙan suwaa — HAUSA GRAMINEAE *Pennisetum pedicellatum, P. polystachion*
nkanaa — AKAN - ASANTE EUPHORBIACEAE *Securinega virosa*
kana-lɛ — BULOM GRAMINEAE *Bambusa vulgaris, Oxytenanthera abyssinica*
kanarin doki — HAUSA GRAMINEAE *Brachiaria ramosa*
kánàssàarâ — KAMBARI - AUNA GRAMINEAE *Oryza gen.*
kanda, kandei — MENDE EUPHORBIACEAE *Manihot esculenta*
ɛ-kanda — TEMNE EUPHORBIACEAE *Manihot esculenta*
kaŋ-dɛ — BULOM GRAMINEAE *Zea mays*
kandié kwékwé — BAULE EUPHORBIACEAE *Hymenocardia acida*
kàndíírìn béégúwáá — HAUSA EUPHORBIACEAE *Manihot esculenta*
kandiolé — SENUFO - TAGWANA EUPHORBIACEAE *Hymenocardia acida*
kandioli — SENUFO - TAGWANA EUPHORBIACEAE *Hymenocardia acida*
kanɛ — AKAN - FANTE EUPHORBIACEAE *Elaeophorbia drupifera*
kaneadua — AKAN - FANTE EUPHORBIACEAE *Jatropha curcas*
kãŋ'fo — GA EUPHORBIACEAE *Euphorbia lateriflora*
kàngálè sawadə — KANURI GRAMINEAE *Sorghum bicolor*
kanggeli — YEDINA GRAMINEAE *Sorghum bicolor*
kanggeli aali — YEDINA GRAMINEAE *Sorghum bicolor*
kania — HAUSA EBENACEAE *Diospyros mespiliformis* TAMACHEK *D. mespiliformis*
kanido — ARABIC - SHUWA GRAMINEAE *Gramineae indet.* KANURI *G. indet.*
kankan — AKAN - FANTE EUPHORBIACEAE *Elaeophorbia drupifera*
kánkán dìká — YORUBA FLACOURTIACEAE *Caloncoba glauca, Flacourtia flavescens ,*
    Oncoba spinosa
kankutuma ndimo — MANDING - MANINKA EUPHORBIACEAE *Hymenocardia acida*
kanlé — DAN EUPHORBIACEAE *Maesobotrya barteri*
kanran — YORUBA EBENACEAE *Diospyros crassiflora, D. mespiliformis*
kansina — TEM GRAMINEAE *Sorghum bicolor*
kansuwa — HAUSA GRAMINEAE *Pennisetum hordeoides*
kanthinkɔ-lɛ — BULOM EUPHORBIACEAE *Manniophyton fulvum*
kantinyi — SUSU EUPHORBIACEAE *Drypetes afzelii*
kantu — KRU - GUERE EUPHORBIACEAE *Uapaca guineensis*
kántù — PERO GRAMINEAE *Saccharum officinarum*
kánya — HAUSA EBENACEAE *Diospyros mespiliformis*
kanyaà — HAUSA EBENACEAE *Diospyros mespiliformis*
kanyan — HAUSA EBENACEAE *Diospyros mespiliformis*
kanyané maka — SONINKE - SARAKOLE GRAMINEAE *Hackelochloa granularis*
kanyang yallo — MANDING - MANDINKA GRAMINEAE *Cymbopogon citratus*
kanyia — HAUSA EBENACEAE *Diospyros mespiliformis*
kapai — NUPE GRAMINEAE *Pennisetum glaucum*
kapé — FULA - PULAAR EUPHORBIACEAE *Manihot esculenta*
kapia — MENDE GRAMINEAE *Paspalum scrobiculatum*
kapie — MENDE GRAMINEAE *Paspalum conjugatum, P. lamprocaryon*
kapika — MENDE GRAMINEAE *Paspalum scrobiculatum*

562

567

kuntan	AKAN - ASANTE EUPHORBIACEAE *Uapaca guineensis, U. heudelotii* TWI *U. guineensis, U. heudelotii, U. paludosa* WASA *U. guineensis, U. heudelotii, U. paludosa*
kuntan ɛsirem	AKAN - ASANTE EUPHORBIACEAE *Uapaca togoensis*
kuntan-akoa	AKAN - ASANTE EUPHORBIACEAE *Uapaca heudelotii*
kuntan-nua	AKAN - ASANTE EUPHORBIACEAE *Uapaca paludosa*
kunto	MANDING - MANDINKA HYDROCHARITACEAE *Ottelia ulvifolia*
kunu	HAUSA GRAMINEAE *Digitaria exilis, Pennisetum glaucum*
kununding mano	MANDING - MANDINKA GRAMINEAE *Panicum subalbidum*
kùnuu	HAUSA GRAMINEAE *Pennisetum glaucum*
kuono	SAMO GRAMINEAE *Rottboellia cochinchinensis*
ka-kupər	TEMNE IXONANTHACEAE *Klainedoxa gabonensis*
kupika	TEMNE GRAMINEAE *Eragrostis gangetica*
kupintaŋ	LIMBA EUPHORBIACEAE *Manihot esculenta*
kupuruku	YORUBA GRAMINEAE *Leptochloa caerulescens*
ƙur ƙura	HAUSA GRAMINEAE *Sorghum bicolor*
kurabã	DIOLA ICACINACEAE *Icacina oliviformis*
kuraban	DIOLA ICACINACEAE *Icacina oliviformis*
kuradagi	SUSU GRAMINEAE *Sporobolus pyramidalis*
ƙurar shanu	HAUSA EUPHORBIACEAE *Euphorbia forskaolii*
kurdiãngdiãng	MANDING - MANDINKA EUPHORBIACEAE *Croton lobatus*
kúrè	DERA GRAMINEAE *Sorghum bicolor*
kuremoke	SUSU EBENACEAE *Diospyros cooperi*
kureteburaxe, kurutuburaxe	SUSU GUTTIFERAE *Garcinia smeathmannii*
kurkutumâdi	MANDING - 'SOCE' GUTTIFERAE *Psorospermum senegalensis*
kurni	HAUSA EUPHORBIACEAE *Bridelia ferruginea*
kurozi	BALANTA EUPHORBIACEAE *Anthostema senegalensis*
kurtu	HAUSA GRAMINEAE *Dactyloctenium aegyptium*
kurudɛrafumfuri	SUSU GRAMINEAE *Panicum brevifolium*
kurukondo	MANDING - MANDINKA EUPHORBIACEAE *Hymenocardia acida*
kurukuru	SENUFO GRAMINEAE *Cymbopogon giganteus*
kurunkondo	MANDING - MANDINKA EUPHORBIACEAE *Hymenocardia acida*
kurutinjengo	MANDING - MANDINKA EUPHORBIACEAE *Croton zambesicus*
kuse togun	BUSA GRAMINEAE *Euclasta condylotricha*
kuseta-ka-gbil	TEMNE GRAMINEAE *Panicum brevifolium*
kusibiri	VULGAR EBENACEAE *Diospyros sanza-minika* AKAN *D. mespiliformis*
okusibiri	AKAN - ASANTE EBENACEAE *Diospyros sanza-minika*
kusibiru	AKYE EBENACEAE *Diospyros sanza-minika*
okusie-dua	AKAN - ASANTE GUTTIFERAE *Allanblackia floribunda*
kusigbɔro	LIMBA EUPHORBIACEAE *Jatropha curcas, J. gossypiifolia, J. multifida*
kus-kus	KRIO GRAMINEAE *Pennisetum glaucum, Sorghum bicolor*
kusomi	LIMBA EUPHORBIACEAE *Tragia gen.*
kussein	? GRAMINEAE *Brachiaria arrecta, B. mutica, Echinochloa pyramidalis, Paspalum scrobiculatum*
kusua	AKAN - ASANTE GUTTIFERAE *Harungana madagascariensis*
kutanki	LIMBA GRAMINEAE *Zea mays*
kuté	BOLE GRAMINEAE *Sorghum bicolor*
kútè	NGAMO GRAMINEAE *Sorghum bicolor*
kutɛgea	LIMBA EUPHORBIACEAE *Macaranga heudelotii*
kutoso	LIMBA GUTTIFERAE *Harungana madagascariensis*
kùttùng	SURA GRAMINEAE *Eleusine corocana*
kutugblɛtʃo	ADANGME - KROBO EUPHORBIACEAE *Jatropha curcas* GA *J. curcas*
kutukku	HAUSA GRAMINEAE *Dactyloctenium aegyptium*
kutuku	HAUSA GRAMINEAE *Dactyloctenium aegyptium*
kutumbo	MANDING - MANDINKA EUPHORBIACEAE *Euphorbia macrophylla*
kuturta	HAUSA GRAMINEAE *Pennisetum glaucum, Sorghum bicolor*
kututu	HAUSA GRAMINEAE *Zea mays*
kùtúwè	GUANG GRAMINEAE *Sorghum bicolor*
kùtúwé pèpèr	GUANG GRAMINEAE *Sorghum bicolor*
kuuku	FULA - PULAAR EBENACEAE *Diospyros mespiliformis* TUKULOR *D. mespiliformis*
kuùrar shaánuu	HAUSA EUPHORBIACEAE *Euphorbia prostrata*
kuwie	VAI EUPHORBIACEAE *Bridelia micrantha/stenocarpa*
kuya	MANDING - BAMBARA GRAMINEAE *Pennisetum glaucum*
kuyi, kuyi beyigi	NUPE GRAMINEAE *Sorghum bicolor*
kuyu (*pl.* ayu)	GUANG GRAMINEAE *Gramineae gen.*
kwaadidi-tʃo	GA EUPHORBIACEAE *Jatropha curcas*
ƙwad-da-gayya	HAUSA GRAMINEAE *Zea mays*
kwadiditʃo	GA EUPHORBIACEAE *Jatropha curcas*
kwadrega	AKAN - FANTE EUPHORBIACEAE *Maesobotrya barteri* WASA *M. barteri* NZEMA *M. barteri*
kwakadili	MOORE FLACOURTIACEAE *Flacourtia flavescens*
kwakka	HAUSA EUPHORBIACEAE *Euphorbia balsamifera*
kwaku nsanson	AKAN - ASANTE EUPHORBIACEAE *Tragia spathulata*
kwama	NZANGI GRAMINEAE *Sorghum bicolor*
kwamossè, kwamurè, pamussè, yoè	KRU - GUERE GRAMINEAE *Streptogyna crinita*
ƙwanbarbaɗa	HAUSA GRAMINEAE *Sorghum bicolor*
kwang ufa	LIBO GRAMINEAE *Zea mays*
kwanla (*pl.* kwama)	LONGUDA GRAMINEAE *Sorghum bicolor*
ƙwar biyu	HAUSA GRAMINEAE *Sorghum bicolor*
kwar-kar	HAUSA EUPHORBIACEAE *Euphorbia balsamifera*
kwarkwaroo	HAUSA GRAMINEAE *Pennisetum unisetum*

ladikoro	YORUBA GRAMINEAE *Brachiaria arrecta, B. mutica*
láfa	SUSU EBENACEAE *Diospyros heudelotii*
a-láfasyèn	BASARI GRAMINEAE *Loudetiopsis tristachyoides*
lafiya	HAUSA GRAMINEAE *Sorghum bicolor*
lagani	MANDING - BAMBARA EUPHORBIACEAE *Jatropha curcas*
laguiri	MANDING - MANINKA GUTTIFERAE *Harungana madagascariensis*
lah	WOLOF EUPHORBIACEAE *Alchornea cordifolia*
lahédìi	FULA - PULAAR EUPHORBIACEAE *Alchornea cordifolia*
lahéhì	FULA - PULAAR EUPHORBIACEAE *Alchornea cordifolia*
láhwe	CHOMO GRAMINEAE *Pennisetum glaucum*
lakaawal (*pl.* lakaaje)	FULA - FULFULDE GRAMINEAE *Sorghum bicolor*
lakatblel	FULA - FULFULDE GRAMINEAE *Sorghum bicolor*
lakawal	FULA - FULFULDE GRAMINEAE *Sorghum bicolor*
lake	BURA GRAMINEAE *Saccharum officinarum*
lakemera	ARABIC GRAMINEAE *Eragrostis pilosa*
laki davangel	FULA - PULAAR GRAMINEAE *Setaria pumila*
laki davangel, lakiérande, mbihuri	FULA - PULAAR GRAMINEAE *Setaria sphacelata*
lakio wanduho	FULA - PULAAR GRAMINEAE *Schoenefeldia gracilis*
an-lal	TEMNE GRAMINEAE *Pennisetum purpureum*
lalakwei	AKAN - BRONG ICACINACEAE *Icacina oliviformis*
lalame	HAUSA GRAMINEAE *Pennisetum glaucum*
lale shamuwa	HAUSA GRAMINEAE *Aristida adscensionis, Leptochloa caerulescens*
lalemo	HAUSA GRAMINEAE *Hyparrhenia subplumosa*
lallaki	HAUSA GRAMINEAE *Oryza gen.*
lállàki	HAUSA GRAMINEAE *Oryza longistaminata*
lállàkíi	HAUSA GRAMINEAE *Oryza barthii*
lallen biri	HAUSA GRAMINEAE *Sporobolus paniculatus*
lállèn birii	HAUSA GRAMINEAE *Sporobolus festivus*
lállèn shamuwa	HAUSA GRAMINEAE *Sporobolus festivus*
lalume	HAUSA GRAMINEAE *Pennisetum glaucum*
lamapèti	SERER EUPHORBIACEAE *Ricinus communis*
lamarudu	FULA - FULFULDE GRAMINEAE *Saccharum officinarum*
lamba	KISSI EUPHORBIACEAE *Mareya micrantha*
lambami	HAUSA GRAMINEAE *Vossia cuspidata*
lambani	HAUSA GRAMINEAE *Gramineae gen.*
lambusan	NANKANNI GRAMINEAE *Brachiaria lata*
làmɔ̀tó	BATA - BACAMA GRAMINEAE *Pennisetum glaucum*
lami	SUSU GUTTIFERAE *Pentadesma butyracea*
lammulammugel	FULA - FULFULDE GRAMINEAE *Pogonarthria squarrosa*
lammulammuki	FULA - FULFULDE EUPHORBIACEAE *Bridelia ferruginea*
lampèti	SERER EUPHORBIACEAE *Ricinus communis*
làmụ́tí	BATA - BACAMA GRAMINEAE *Pennisetum glaucum*
lamyondon	MOORE GRAMINEAE *Schoenefeldia gracilis*
lamzudu	MOORE GRAMINEAE *Ctenium newtonii*
ɛ-lan-ɛ-ra-wuto	TEMNE FLACOURTIACEAE *Homalium smythei*
lansango	MANDING - MANDINKA GRAMINEAE *Gramineae gen.*
ka-lant	TEMNE GRAMINEAE *Paspalum conjugatum*
lanyere	FULA - FULFULDE GRAMINEAE *Andropogon gayanus*
a-lǎp ɔ-xèdi	BASARI GRAMINEAE *Dactyloctenium aegyptium*
làpá làpá	YORUBA EUPHORBIACEAE *Jatropha curcas*
làpálàpá	YORUBA EUPHORBIACEAE *Jatropha multifida*
làpálàpá adẹ́tẹ̀	YORUBA EUPHORBIACEAE *Ricinus communis*
làpálàpá pupa	YORUBA EUPHORBIACEAE *Jatropha gossypiifolia*
lapet	SERER EUPHORBIACEAE *Ricinus communis*
lapoka	LOKO EUPHORBIACEAE *Phyllanthus muellerianus*
laque dawa	FULA - PULAAR GRAMINEAE *Loudetia coarctata*
lárà	YORUBA EUPHORBIACEAE *Ricinus communis*
larba	MANDING - KHASONKE GRAMINEAE *Brachiaria stigmatisata*
larekwe	AKAN - BRONG ICACINACEAE *Icacina oliviformis*
laruha	MANDING - KHASONKE GRAMINEAE *Paspalum scrobiculatum*
las a koy	SERER GRAMINEAE *Enteropogon prieurii*
las a tat	SERER GRAMINEAE *Loudetia hordeiformis*
las o fam	SERER GRAMINEAE *Enteropogon prieurii*
las pis	SERER GRAMINEAE *Enteropogon prieurii*
ta-lasoi	TEMNE GRAMINEAE *Sorghum bicolor*
lassa	BISA GRAMINEAE *Hyparrhenia subplumosa*
a-lath	TEMNE GRAMINEAE *Imperata cylindrica*
lawrehe	FULA - FULFULDE GRAMINEAE *Andropogon schirensis*
lawula	SENUFO - NIAGHAFOLO GRAMINEAE *Rottboellia cochinchinensis*
lē	MANO GUTTIFERAE *Pentadesma butyracea*
lébèl	FULA - PULAAR EUPHORBIACEAE *Phyllanthus fraternus, P. pentandrus*
lebooji	FULA - FULFULDE GRAMINEAE *Oxytenanthera abyssinica*
leébèn raaƙúmii	HAUSA GRAMINEAE *Sorghum bicolor*
leébèn raaƙúmìi	HAUSA GRAMINEAE *Sorghum bicolor*
lɛfɛbuiyɛ	KORANKO GRAMINEAE *Paspalum scrobiculatum*
lefora	MANDING - DYULA EUPHORBIACEAE *Excoecaria grahamii*
lefyɔg	NGYEMBOON GRAMINEAE *Bambusa vulgaris*
legabɔl	KONKOMBA EBENACEAE *Diospyros mespiliformis*
legamma, shegar	KANURI GRAMINEAE *Triticum aestivum*
leggal roogo	FULA - FULFULDE EUPHORBIACEAE *Manihot esculenta*
legsaïbö	ARABIC GRAMINEAE *Centropodia forskalii*
légun oko	YORUBA GUTTIFERAE *Psorospermum corymbiferum, P. febrifugum*
legun soko	YORUBA GUTTIFERAE *Harungana madagascariensis*

pamplo	ADANGME - KROBO GRAMINEAE *Oxytenanthera abyssinica* GBE - VHE *O. abyssinica*
pampló	GA GRAMINEAE *Oxytenanthera abyssinica*
pamploo	ADANGME GRAMINEAE *Oxytenanthera abyssinica* GA *O. abyssinica*
ɔ-pampotoporopoo	AKAN - TWI EUPHORBIACEAE *Bridelia atroviridis, B. micrantha/stenocarpa*
pampotoprofo	AKAN - TWI EUPHORBIACEAE *Bridelia micrantha/stenocarpa*
pámpró	AVATIME GRAMINEAE *Oxytenanthera abyssinica*
pampru	GBE - VHE GRAMINEAE *Oxytenanthera abyssinica*
pamu	'KRU' GRAMINEAE *Zea mays*
kpaŋ	VAI GRAMINEAE *Saccharum officinarum*
kpaŋgba	KONO GRAMINEAE *Axonopus compressus, A. flexuosus*
kpaŋgba-tava	KONO GRAMINEAE *Axonopus compressus, A. flexuosus*
pankasaxi	SUSU GRAMINEAE *Hyparrhenia subplumosa*
pankoko	ANYI EUPHORBIACEAE *Bridelia grandis*
panu	HWANA GRAMINEAE *Zea mays*
panya	BOKYI EUPHORBIACEAE *Manihot esculenta*
ú-pànyá (*pl. i-*)	BETTE - BENDI EUPHORBIACEAE *Manihot esculenta*
pànyá	ICHEVE EUPHORBIACEAE *Manihot esculenta*
panyirakul-la	YALUNKA GRAMINEAE *Pennisetum polystachion*
panyuma	GURMA EUPHORBIACEAE *Manihot esculenta*
kpaola(-i)	MENDE EUPHORBIACEAE *Oldfieldia africana*
papa-yɛ	AKAN - ASANTE FLACOURTIACEAE *Dissomeria crenata*
papira	SURA GRAMINEAE *Sorghum bicolor*
pãplo	ADANGME - KROBO GRAMINEAE *Oxytenanthera abyssinica* GA *O. abyssinica*
papodé	BADYARA EUPHORBIACEAE *Anthostema senegalensis*
papu	KONO EUPHORBIACEAE *Macaranga barteri, M. heterophylla, M. heudelotii, M. hurifolia*
para	BAMILEKE ERICACEAE *Philippia mannii*
paradio	KULANGO EUPHORBIACEAE *Hymenocardia acida*
an-paraŋ	TEMNE EUPHORBIACEAE *Elaeophorbia grandifolia*
parandedi	AKYE ICACINACEAE *Leptaulus daphnoides*
parkatari	FULA - PULAAR GRAMINEAE *Paspalum scrobiculatum*
paroko	YORUBA EBENACEAE *Diospyros ferrea*
pasẽẽ	ANYI - AOWIN GUTTIFERAE *Mammea africana* SEHWI *M. africana*
paseng	ANYI - SEHWI GUTTIFERAE *Mammea africana*
ɛ-pasɔr	TEMNE GRAMINEAE *Oryza glaberrima , O. sativa*
pasi	YORUBA GRAMINEAE *Schizachyrium sanguineum*
pasin	ANYI - SEHWI GUTTIFERAE *Mammea africana*
patroa	AKAN - ASANTE EUPHORBIACEAE *Maesobotrya barteri*
pattoyi	FULA - FULFULDE EUPHORBIACEAE *Hymenocardia acida*
pau bicho	CRIOULO EUPHORBIACEAE *Oldfieldia africana*
pau branco	CRIOULO EUPHORBIACEAE *Tetrorchidium didymostemon*
pau raio	CRIOULO EUPHORBIACEAE *Jatropha gossypiifolia*
pau-lai	KRU - BASA EUPHORBIACEAE *Oldfieldia africana* MENDE *O. africana*
pawdo	FULA - FULFULDE EUPHORBIACEAE *Excoecaria grahamii*
mpawi	AKAN - ASANTE EUPHORBIACEAE *Cleidion gabonicum*
mpawuo	AKAN - ASANTE EUPHORBIACEAE *Cleidion gabonicum*
a-pè	KONYAGI EUPHORBIACEAE *Anthostema senegalensis*
pebife	MANDYAK GRAMINEAE *Digitaria longiflora*
pedu kúú	DOGON GRAMINEAE *Enteropogon prieurii*
mpeduro	AKAN - ASANTE EUPHORBIACEAE *Grossera vignei*
pɛgɛla	MENDE ICACINACEAE *Rhaphiostylis beninensis*
mpégui baté	AKYE EUPHORBIACEAE *Phyllanthus amarus*
pegya	AKAN - WASA GUTTIFERAE *Mammea africana* NZEMA *Pentadesma butyracea*
peingyɔ	KISSI EUPHORBIACEAE *Argomuellera macrophylla*
pei-poto	BULOM GRAMINEAE *Cymbopogon citratus*
peja	AKAN - FANTE GUTTIFERAE *Pentadesma butyracea*
pejui	MENDE GRAMINEAE *Digitaria horizontalis*
pekpeara	NUPE IXONANTHACEAE *Irvingia gabonensis*
pekpéla	BAATONUN EUPHORBIACEAE *Bridelia ferruginea*
pekun odo	YORUBA GRAMINEAE *Setaria sphacelata*
pekunodo	YORUBA GRAMINEAE *Setaria sphacelata*
k-pel	TEMNE GUTTIFERAE *Harungana madagascariensis*
a-pɔla (*pl.*)	TEMNE GRAMINEAE *Oryza sativa*
a-pɔla-*pa*-krifi	TEMNE GRAMINEAE *Oryza barthii*
a-pɔla-*pa*-ro-bat	TEMNE GRAMINEAE *Oryza sativa*
a-pɔla-*pa*-ro-gban	TEMNE GRAMINEAE *Oryza sativa*
kpele	KONO IXONANTHACEAE *Irvingia gabonensis*
pɔlè	BULOM GRAMINEAE *Oryza sativa*
pɛlɛ	BULOM GRAMINEAE *Oryza glaberrima , O. sativa*
kpele-nyɔ	MENDE GRAMINEAE *Pennisetum glaucum*
péleti	FULA - PULAAR EUPHORBIACEAE *Hymenocardia acida*
péléti	FULA - PULAAR EUPHORBIACEAE *Hymenocardia acida*
pɛlis	LIMBA GRAMINEAE *Sporobolus pyramidalis*
peliteró	FULA - PULAAR EUPHORBIACEAE *Hymenocardia acida*
pelitoro	FULA - PULAAR EUPHORBIACEAE *Hymenocardia acida*
pélitoro	FULA - PULAAR EUPHORBIACEAE *Hymenocardia acida*
pelitoro pété	FULA - PULAAR EUPHORBIACEAE *Hymenocardia lyrata*
pel-lɛ	BULOM GUTTIFERAE *Harungana madagascariensis*
pellen	WOLOF GRAMINEAE *Rottboellia cochinchinensis , Schoenefeldia gracilis*
pelôfé	MANDING - MANDINKA EUPHORBIACEAE *Manihot esculenta*

609

612

618

INDEX

INDEX